Physics
Matters

Physics Matters

An Introduction to Conceptual Physics

James Trefil

Robert M. Hazen

George Mason University

WILEY

JOHN WILEY & SONS, INC.

ACQUISITION EDITOR: Stuart Johnson
SENIOR DEVELOPMENT EDITOR: Nancy Perry
OUTSIDE DEVELOPMENT EDITOR: David Chelton
SENIOR MARKETING MANAGER: Bob Smith
DIRECTOR OF PRODUCTION: Pam Kennedy
PRODUCTION EDITOR: Barbara Russiello
SENIOR DESIGNER/COVER DESIGNER: Harold Nolan
TEXT DESIGN: Circa 86, Inc.
ILLUSTRATION EDITOR: Anna Melhorn
ELECTRONIC ILLUSTRATIONS: Matrix Art Services
SENIOR PHOTO EDITOR: Sara Wight
PHOTO RESEARCHER: Elyse Rieder
SUPPLEMENTS EDITOR: Geraldine Osnato
SENIOR NEW MEDIA EDITOR: Martin Batey
COVER PHOTO: © Lawrence Englesberg/Image State

This book was set in 10/12 Times Ten Roman by Matrix Publishing and printed and bound by PrintPlus. The cover was printed by PrintPlus.

Library of Congress Cataloging-in-Publication Data

Trefil, James S., 1938–
    Physics matters : an introduction to conceptual physics / James Trefil, Robert M. Hazen.
        p. cm.
    Includes index.
    ISBN 978-0-471-15058-9
    1. Physics. I. Hazen, Robert M., 1948– II. Title.

QC23.2.T74 2004
530—dc22

                                                                2003065703

ISBN 978-0-471-15058-9

10 9 8 7 6

Dedicated to our children

Anne Trefil

Benjamin Hazen

Carl Ehrhardt

Dominique Waples-Trefil

Elizabeth Hazen

Flora Waples-Trefil

James K. Trefil

Stefan Trefil

Tomas Waples-Trefil

Contents in Brief

Contents

29 | Cosmology   626

Preface

 ## TO THE INSTRUCTOR

We all know how important all of the sciences, particularly physics, have become to modern society. All too often, however, physics is taught as if it were an isolated field of study, with no connections to other sciences, to history, or to society at large. What we have tried to do in this book is to present the traditional principles of physics in the rigor appropriate to a liberal arts course, while at the same time emphasizing how those principles are connected to the rest of human experience. We feel strongly that when students see physics as part of life, as something relevant to what they see in the news of the day, the learning process will become more meaningful.

Organization of the Book

Physics Matters follows the traditional historical approach to the presentation of physics, beginning with mechanics, thermodynamics, and electromagnetism, and progressing from them to the physics of the twentieth and twenty-first centuries. This division reflects both the history of the science and a difference in focus and subject matter. Although we do not make an explicit distinction, the division between classical and modern physics is pretty clear in the text.

Making the Connection It has been our experience that each of these areas—classical and modern physics—offers opportunities and difficulty as far as teaching is concerned. Classical physics is close to everyday experience, close to things our students see every day. We have stressed this aspect of the science in many ways, hoping to draw students into the subject. On the other hand, if presented unimaginatively, classical physics can also be a little dull. To see the connection between a block sliding down an inclined plane and the latest mission of space exploration, or to understand how simple experiments of rubbing glass jars to produce electric charge led to generators and power lines are the sorts of goals appropriate to classical physics.

To stress this point, we open each chapter with a section titled **Physics Around Us . . .**, which recounts common experiences and relates them to the principles of physics that will be studied in that chapter. It is our hope that these opening sections will serve to convince students that there is something worthwhile to be gained from studying the chapter.

Modern physics, particularly relativity and quantum mechanics, has an innate ability to fire the imagination, to engage the intellect. On the other hand, it is also farther from everyday experience than its classical counterpart. In this section of the book, we emphasize continually the connection of seemingly obscure laws to everyday life. A student who looks at a computer screen needs to know that she is dealing with the results of quantum mechanics, a student getting an MRI that he is benefiting from studies of the atomic nucleus. As with classical physics, *connectedness is the key*.

In some cases, we have looked at this connection explicitly. In the chapter dealing with cosmology, for example, we have gone into some detail about how the laws of physics apply to the large-scale structure and history of the universe. In other cases, we have pointed out the interconnectedness of physics with other sciences in each chapter in sections titled **Connections.** These special sections flow integrally from the text and are not presented as isolated boxes. In this way, students will not get the message that applications of physics to other sciences is somehow superfluous or added on. Instead, they will view them as central components of the idea we are trying to get across: Physics is at the basis of a logical, interconnected web of ideas and concepts that helps us understand the world around us.

The Issues of the Day It is hard to imagine being a citizen in the twenty-first century without having some understanding of science, at least as it applies to the political issues of the day. Because we believe that the connection between science and important issues of the day should be stressed in any university course, at the end of every chapter in this book you will find a section titled **Thinking More About. . . .** In this section, we talk about how the topics covered in that chapter affect some important political or social issue, and invite the student to apply what he or she has just learned to an area outside of science. Science and technology do not exist by themselves in some isolated world. The student needs to learn to think critically about them, to make judgments about the myriad of science-related issues that will cross his or her horizon every day. This need was constantly in our minds as we wrote this book, and the section at the end of each chapter is only one manifestation of that concern.

A Spirit of Investigation and Discovery In each chapter you will also find integrated sections titled **Physics in the Making,** which tell the fascinating story of our unfolding understanding and illustrate how the scientific method has been applied to bring us to where we are today. Often, these stories contain a good deal of human interest as well as scientific principles, and we make use of the stories to portray physics (and science) as a human endeavor, one that requires as much creativity and imagination as writing a novel or a symphony.

The process of learning about physics, like the process of learning about almost anything, is not passive. It is necessary to get into the subject, to think about what you are reading and learning. To help the student with this process, you will find short sections labeled **Develop Your Intuition** embedded throughout the text. When students get to one of these, encourage them to pause for a moment and try to think about how to answer the question being posed, given what they have just learned. When they have done this, they compare their answer to the one given in the text immediately following the question. In this way, students are encouraged to see for themselves how even the most complex of phenomena can be understood in terms of processes that are at bottom simple and understandable.

The same philosophy can be seen in our use of numerical problems at the end of the chapters. In many chapters you will find worked examples, fully explained, at the beginning of the Problem section. Again, the purpose of these **Problem-solving Examples** is to illustrate that the kind of reasoning used in mathematics can be expressed in simple language. Once this point is made, of course, the student is then given ample opportunity to apply that sort of reasoning in different situations.

   All of these special features, described more fully below, are designed with one goal in mind—to show our students how physics is an integral part of their lives and of their futures.

Special Features of *Physics Matters*

In an effort to aid student learning and make the world of physics both accessible and exciting, we have incorporated several distinctive features throughout the book. These features have two basic goals:

1. to emphasize the interconnectedness of the great ideas of physics and so help students connect key concepts into a coherent whole, and
2. to inspire students to explore the principles behind the many practical applications that fuel our world and our society.

Key Ideas Each chapter begins with a statement of a key idea in physics, so that students immediately grasp the chief concept of that chapter. These statements are not intended to be recited or memorized, but rather to provide a framework for placing the concepts students encounter in the chapter.

Physics Around Us . . . Each chapter begins with a Physics Around Us section, in which we tie the chapter's main theme to a common experience, such as riding in a train, riding a roller coaster, relaxing at the beach, going to the doctor, or using the Internet. In this way we emphasize that the great principles of physics are constantly part of our lives.

Develop Your Intuition Students testing *Physics Matters* before publication commented on the helpfulness of having immediate applications of the concepts presented, pointing in particular to the Develop Your Intuition sections. In these sections, which appear throughout the book, we ask students to pause and think about the implications of what they have learned. We do this by asking them to consider questions related to their daily life, such as, "Why when you're in an elevator do you usually feel heavier when the elevator starts going up and lighter when it starts down?" or "Why is it that when you look at your face in the bathroom mirror, the image is reversed?"

---

4 Isaac Newton and the Laws of Motion

KEY IDEA

Newton's laws of motion describe the behavior of objects on Earth and in space.

PHYSICS AROUND US . . . Dealing with Momentum

You get in your car, put your books on the seat next to you, and drive away. A squirrel runs across the road and you jam on your brakes to avoid hitting it. As you do so, the pile of books slides onto the floor of the car.

Later that day, in the cafeteria, you watch a fellow student hurrying across the floor with her tray. Stopping suddenly, she reaches out to grab her soft-drink cup to keep it from falling over and some of the drink sloshes over on her hand.

That night, you turn on your TV to watch a hockey game. The puck comes free and two players skate for it at top speed. They collide just as they reach the puck and bounce off each other, taking themselves out of the action and leaving the puck for another player to pick up.

All of these occurrences (and countless more) illustrate a quantity called momentum, whose properties follow from Newton's laws of motion. And believe it or not, in addition to being involved in everyday events, momentum also governs such large-scale processes as the formation of planets and stars. As a result, momentum is one of the most important physical attributes of an object in motion.

Develop Your Intuition: Flashbulbs and Baseball

   You'll often see hundreds of flashbulbs go off at a stadium when a famous baseball player comes up to bat during a night game. Based on what you know about the inverse square relationship, do these flashes help the would-be photographers?

   The intensity of light drops off as the inverse square of the distance, so flashbulbs are ineffective at distances greater than a few dozen feet. All the popping flashbulbs make a great sight, but they don't help photographs taken over the great distances of a stadium.

Connection
Buckyballs—A Technology of the Future?

An extraordinary discovery, announced in 1990, reveals the close relati[on] among pure scientific research, applied research, and technology. For c[enturies] the element carbon was known in only two basic forms—graphite, the so[ft] mineral used in pencils, and diamond, the hard transparent gemstone. H[owever,] in 1985 chemists at Sussex University in England and Rice University i[n Texas] found evidence for a totally new form of carbon—one in which 60 carbo[n atoms] bond together in a ball-shaped molecule (Figure 1-5). The distinctive lin[k among] the atoms, much like the geodesic domes of architect-inventor Buck[minster] Fuller, led scientists to dub the new material buckminsterfullerene, or bu[ckyballs] for short. Buckyballs, although completely unexpected, at first excited l[ittle at]tention outside a small research community becau[se the discovery] had no[...]

Connection   Special sections highlight applications of physics concepts to other areas of science, including biology, earth science, astronomy, materials science, and critical modern technologies. These recurring discussions flow naturally from the text and are identified by a specific icon of a bridge. Topics include the clotting of blood, applications of ultrasound, television and computer screens, and inertial navigation systems.

Special visual features   One of our goals has been to support the verbal exposition with visually exciting and pedagogically meaningful illustrations. We know that many students can be drawn into the text if inspired by compelling figures and photographs. Student reviewers have praised the way the numerous illustrations and photos supported their understanding of the text. Three kinds of illustrations in particular help students grasp physics concepts and serve as handy reviews of basic concepts:

- **Looking at . . .** figures familiarize students with the range of magnitudes common for various physical quantities, such as length, energy, temperature, and voltage.
- **Physics and Modern Technology** figures illustrate the many systems and devices generated by application of individual physics concepts, such as the atomic structure of a diamond or the orbits of satellites.
- **Physics and Daily Life** figures show how physical concepts and relations appear in the context of everyday, familiar activities, from playing volleyball to holding a coffee cup to using the telephone.

Physics in the Making   These integrated essays illustrate the scientific method by describing the way in which specific new understanding was gained, with particular emphasis on pivotal experimental and observational advances. Students have commented that inclusion of such background information helps them retain information. Examples include the development of conservation laws, Galileo's experiments with inertia, and the advance of the Copernican model of the solar system.

Ongoing Process of Science   Science is a never-ending process of asking questions and seeking answers. In these features, integrated with the text as are the other recurring features, we examine some of the most exciting questions currently being addressed by physicists.

Equations and Worked Examples   Perhaps the most difficult aspect of physics for the new student is the use of mathematics. In this book, we try to demystify the presence of mathematics in the sciences, while at the same time retaining the basic laws of physics in their familiar mathematical forms.

The most important concept to grasp, as we emphasize in Chapter 2, is the fact that an equation is simply a translation into the language of mathematics of an idea that can also be expressed in an ordinary English sentence. To get this idea across, each equation in the text is given in three separate forms. It is first stated as a simple sentence which contains the main idea of the relationship, then as an equation in which words are substituted for symbols, and only then in its full mathematical form. We do this to emphasize the fact that, as mysterious as mathematics may be to some students, the ideas contained in the equations are really simple.

Physics in the Making
Measuring Time Without a Watch

In our age of stop watches, digital timers, and atomic clocks, with Olympic races routinely measured to a hundredth or even a thousandth of a second, it's hard to imagine measuring time without an accurate instrument. But when Galileo set out to study accelerated motion in the 1600s, measuring time was a formidable technological challenge. Think about how you might determine small time intervals if all you had was a clock that ticked off seconds but could record no shorter times.

Galileo wanted to document as accurately as possible the way falling objects accelerate, but these measurements required knowledge of both distance and time. Since objects that fall straight down moved much too fast for him to mea-

Ongoing Process of Science
The Value of G

It might surprise you to learn that the measurement of G is still of great interest to scientists around the world. It turns out that most fundamental physical constants, such as the mass of an electron or the electrical force between two charged particles, are known with great precision and accuracy, to many decimal places. But G is still only known, at best, to the fourth decimal place.

In November 1998 a group of 45 physicists met in London to celebrate the 200th anniversary of Cavendish's original experiment and to compare notes on several new determinations of the constant. All of these workers performed meticulous experiments: they enclosed their apparatus in a vacuum, they eliminated all stray magnetic and electrical fields, they checked and rechecked every aspect of their work.

Acceleration

Acceleration measures the rate of change of velocity.

1. In words:

   Acceleration *is the change in velocity divided by the time it takes for that change to occur.*

2. In an equation with words:

$$\text{Acceleration} = \frac{\text{Final velocity} - \text{Initial velocity}}{\text{Time}}$$

3. In an equation with symbols:

$$a = \frac{\Delta v}{t}$$

where Δv indicates the change in velocity. Like velocity, acceleration requires information about the direction and is therefore a vector.

Key Terms   Key terms appear in **bold** type within each chapter. They are highlighted in the chapter **Summary** and are also listed at the end of the chapter, along with their definitions and the number of the page on which they are first explained. Many other terms are important, although more specialized; we have highlighted these terms in italics within each chapter. Both sets of terms are defined in the **Glossary** at the end of the text.

Looking Deeper   These sections are intended to be assigned by professors who prefer a somewhat higher-level discussion of certain topics, such as collisions in two dimensions, diet and calories, and efficiency. They are set off in screened blue boxes for easy identification and flexibility in assignment.

Thinking More About   Each chapter ends with a section that addresses a social or philosophical issue tied to physics, such as our place in the ordered universe, what constitutes proof that something is real, and the effects of electric and magnetic fields on wildlife.

THINKING MORE ABOUT

Momentum: Why Isaac Newton Would Wear His Seat Belt

One of the authors (JT) knows a highway patrolman who makes a point about auto safety by saying, "I never pulled a dead man from behind a seat belt." Although some people do indeed die in car crashes while wearing seat belts, there is no question that seat belts greatly improve your chances of walking away from a crash. Given what you know about momentum and impulse, why should this be so?

Look at it this way: when you are sitting in a

If you're not wearing a seat belt, you keep moving forward when the car stops, and your motion stops when your head hits the steering wheel or the windshield. Because these surfaces are hard, the time of the impact is short—a fraction of a second. Consequently, the force needed to stop you must be large. In addition, this force is applied to a small area of your skull, greatly increasing the pressure. Such a large focused force can be deadly.

If you're wearing a flexible seat belt, however, a smaller force is applied over a longer time interval to produce the same change in momentum. What's more, the broad, flat seat belt distributes that force over a much larger area of your body.

Questions   We feature four levels of end-of-chapter questions, giving instructors the flexibility to choose the types of questions appropriate for their classes.

- **Review** questions test important factual information covered in the text and are provided to emphasize key points.
- **Questions** are also based on chapter material but involve more exploration and analysis of physics concepts. Some questions include diagrams to help students visualize geometric relationships, whereas other questions build on the same or similar situations to test students' understanding in greater depth.
- **Problems** are quantitative questions that require students to use mathematical operations, typically those introduced in worked examples. **Problem-solving Examples,** placed directly before the start of the Problem set in some chapters, help prepare students to work the Problems.
- **Investigations** require additional research outside the classroom.

WWW Resources   At the end of each chapter, you will find a list of relevant websites. In addition, the www icon appears in the margin throughout the book to highlight links that amplify the text. These links have been well researched and provide additional content on many fascinating topics. You can also visit the *Physics Matters* home page at **www.wiley.com/college/trefil** to explore materials developed to support and enhance the study of physics.

● SUPPLEMENTS

The *Lab Manual* by Robert Ehrlich and Anna Wyczalkowski of George Mason University contains enjoyable and interesting experiments that will reinforce the concepts learned during lecture. There are minimal equipment requirements for the labs so they can be implemented with ease at both small and large departments.

The *Activity Book* by Michael Tammaro of the University of Rhode Island is an excellent resource for self-study and homework assignments. Activities build skills in making hypotheses, taking measurements, and plotting data. Follow-up questions emphasize conceptual understanding.

The *Instructor's Resource and Solution Manual* by Michael Tammaro of the University of Rhode Island is an essential item that contains the solutions and answers to every end-of-chapter problem and discussion question. It also contains lecture outlines, lecture tips, additional discussion questions (with answers), and demonstration ideas. Both new and experienced instructors are sure to find this manual useful.

The *Test Bank* contains over 1500 test items. Instructors will find this a time-saver in exam preparation. Many of the questions have been specifically written to promote conceptual understanding rather than rote memorization.

The *Instructor Resource CD-ROM* includes an *Image Bank* with all of the line art from the text as well as a *Computerized Test Bank*. Use the image bank to enhance classroom presentations, create custom handouts, and add images to course websites. The computerized test bank by Brownstone allows instructors a wide range of functionality in an easy-to-use format.

The *Wiley Physics Demonstration Videos* by David Maiullo of Rutgers University consist of over 100 classic physics demonstrations that will engage and instruct your students. Filmed, edited, and produced by a professional film crew, the demonstrations include

Demonstrating rocket motion with a fire extinguisher cart.

- 55-Gallon Drum: Illustrates atmospheric pressure through the dramatic collapsing of a large metal drum.
- Bed of Nails: Shows how having more nails on a bed of nails actually makes it safer.
- Breaking Glass with Sound: Demonstrates how sound resonance can be used to break a glass.
- Penny and Feather: Illustrates that objects will fall at the same rate in the absence of atmosphere.
- Standing Waves on a String: Shows how vibrating a large rope with the right frequencies can produce standing waves on the rope.
- Light the Match (uses mirrors and a heat source): Shows that the infrared spectra may act in the same way as the visible light spectra.
- Magnet over Superconductor (Meissner Effect): Illustrates the strange properties of superconductors by floating a magnet over one.
- Polaroids with a Light Source: Shows how light can be polarized and filtered through the use of two Polaroids.

Each demonstration is labeled according to the Physics Instructional Resource Association's demonstration classification system. This system identifies the area, topic, and concept presented in each demonstration. Go to www.pira.nu for more information about the Physics Instructional Resources Association and to download a spreadsheet of the demonstration classification system.

The *Physics Matters* website offers a wealth of material for both students and instructors. You can view demonstration videos, get suggestions for group projects and quantitative skill building exercises, or take practice quizzes. To access the *Physics Matters* website go to **www.wiley.com/college/trefil**. Select *Physics Matters* from the list of titles and enter either the instructor or student website.

John Wiley & Sons may provide complimentary supplements to qualified adopters. Please contact your sales representative for more information. If you receive supplements you do not need, please return them to your sales representative or send them to:

Attn: Returns Department
Heller Park Center
360 Mill Road
Edison, NJ 08817

ACKNOWLEDGMENTS

Student Involvement

We are grateful to Professor Michael Tammaro, University of Rhode Island, and the students in his Ideas of Physics class, who used some of this material while it was in manuscript form and gave us invaluable feedback that helped shape the final book: Sarah Allard, Brett Vincent Beaubien, Erin Bonsall, Carlton Bradshaw, Joe Cinaglia, Nicholas DePetrillo, Elizabeth Feuer, Abby Heredia, Letitia Tyler Kane, Zachary Karp, Katie Lidestri, Holly Osberg, Adi Pekmezovic, Steven Rangel, Eric Simpson, Andrew Stover, and Trevor Utley.

Faculty Input

We are also grateful to Kurt Gareiss, who contributed many of the end-of-chapter questions. Besides class-testing the manuscript, Professor Michael Tammaro, University of Rhode Island, provided invaluable editorial suggestions, offered many creative end-of-chapter questions, and prepared the Activity Book and the comprehensive Instructor's Resource and Solutions Manual. We are also grateful to Prof. Dan MacIsaac, SUNY–Buffalo State College, for identifying many of the WWW Resources entries listed at the end of each chapter.

We gratefully acknowledge the significant contributions of many reviewers who commented in detail on the various drafts of this text. Their insights and teaching experience have been invaluable to us:

Elizabeth C. Behrman, *Wichita State University*
Edward R. Borchardt, *Minnesota State University, Mankato*
Arthur J. Braundmeier, *Southern Illinois University, Edwardsville*
William D. Bruton, Stephen F. *Austin State University*
Tom Carter, *College of DuPage*
Marco Ciocca, *Eastern Kentucky University*
Doug Davis, *Eastern Illinois University*
Doyle V. Davis, *New Hampshire Community Technical College*
John Farrell, *Winona State University*
Lyle Ford, *University of Wisconsin, Eau Claire*
Kyle Forinash, *Indiana University Southeast*
Mike Franklin, *Northwestern Michigan College*
Nevin D. Gibson, *Denison University*
Yuan K. Ha, *Temple University*
Robert B. Hallock, *University of Massachusetts, Amherst*
Joseph H. Hamilton, *Vanderbilt University*
Adam Johnston, *Weber State University*
Hsu-Feng Lai, *Santa Monica College*
David Lamp, *Texas Tech University*
Teresa L. Larkin, *American University*
Mary Lu Larsen, *Towson University*
Andrew J. Leonardi, *University of South Carolina, Spartanburg*
Ilan Levine, *Indiana University, South Bend*
Steve Lindaas, *Minnesota State University, Moorhead*
Michael C. LoPresto, *Henry Ford Community College*
John A. McClelland, *University of Richmond*

Allen Miller, *Syracuse University*
Peter Morse, *Santa Monica College*
Russell Palma, *Sam Houston State University*
Russ Poch, *Howard Community College*
Brian Raue, *Florida International University*
Russell A. Roy, *Santa Fe Community College*
Toni D. Sauncy, *Angelo State University*
Donald Schnitzler, *Linfield College*
Michael I. Sobel, *Brooklyn College*
Chris Sorensen, *Kansas State University*
Jimmy H. Stanton, *Cameron University*
Francis M. Tam, *Frostburg State University*
Michael J. Tammaro, *University of Rhode Island*
Frederick J. Thomas, *Sinclair Community College*
Lawrence B. Weinstein, *Old Dominion University*
Jeff Williams, *Bridgewater State College*

Iris Knell at George Mason University assisted in manuscript preparation and offered valuable suggestions for editorial improvements, while providing an atmosphere of cheerful efficiency.

Publisher Support

We are deeply indebted to the dedicated staff of John Wiley & Sons, who first proposed this textbook and who have contributed substantially during every aspect of its development and production. *Physics Matters* would not have been possible without their extraordinary efforts. Stuart Johnson oversaw the project from its inception with vision and a commitment to excellence. Development editor Nancy Perry managed the complex coordination of text, images, special features, reviews, and more with unflagging enthusiasm and consummate professionalism. Supplements Editor Geraldine Osnato oversaw with creativity and dedication the development of the print and electronic supplements and worked with Senior New Media Editor Martin Batey to develop the *Physics Matters* website and demonstration videos. We also thank Marketing Manager Bob Smith for his continuing efforts to ensure that this book reaches a wide audience.

David Chelton played a central role in creative aspects of writing, editing, revising and illustrating the book. His major contributions are reflected on every page and are most gratefully acknowledged.

We also thank Wiley's exceptional production team. The project was skillfully managed by Production Editor Barbara Russiello, who dealt with the countless technical details associated with an introductory physics book. Many thanks to Gloria Hamilton for her careful proofreading and requests for the clearest language. We are also grateful to Illustration Editor Anna Melhorn and to Matrix Art Services, Photo Editors Hilary Newman and Sara Wight for their tireless efforts in securing the many distinctive images that amplify our text, as well as to Senior Designer Harry Nolan and design studio Circa 86 who created the book's distinctive and handsome layout.

Science shapes our lives, touches things we do every day, gives us the necessities and conveniences that make our lives what they are. It is also a force that drives change in our world, that makes your life different from your parents' lives, and that will make your children's lives different from yours.

But there is another reason to study science. When you enjoy a sunset or watch the wind blow ripples across a lake, your appreciation and understanding of the experience can be deepened and enriched by understanding something about the processes that produce what you are seeing. At a deeper level, knowing something about how the universe works brings with it a sense of philosophical satisfaction, a sense of grounding, that is hard to obtain any other way.

Why Study Physics?

Physics is the most basic of the sciences. Open any textbook on modern science—chemistry, astronomy, and even biology—and you will find the concepts you'll encounter in the pages that follow. The sorts of things that you'll study in this course, such as energy, atoms, and electricity, form the solid core on which scientists have built our modern picture of the world.

And what a picture it is! We live in a world that is both complex and simple, easy and difficult to understand. Historically, the science of physics has been concerned with finding simplicity in nature, even when that simplicity is hidden under seeming complexity.

This historic quest is intimately tied up with the development of a new way for human beings to learn about nature. We call this technique the *scientific method,* and it is so important that it is dealt with in the first chapter of this book. It represents nothing less than a new way of understanding and controlling the world around us. All of the benefits of science and technology we mentioned above have flowed from our ability to apply this method to different areas of experience, from the motion of electric charges to the strange behavior of atoms. Each new advance, each increase in our understanding, has come as the result of someone applying the scientific method in a real world situation.

It would be wrong, though, to think that the great discoveries of physics affect only the esoteric world of the professional scientist. In fact, the principles of physics are all around you, affecting everything you do every day of your life. It would also be wrong to think that what you will be studying in this book applies only in a narrow way to the science of physics. In fact, the principles you will be learning are used by all scientists, no matter what their field of expertise. For science in general, and physics in particular, has never been only a collection of unconnected facts. Science—and physics—are instead, a logical, interconnected web of ideas and concepts.

Finally, it is important to realize that the scientific process never really ends—that we never know everything. All the answers are not in, all knowledge is not to be found in books like this one. In fact, the same process that began with ancient astronomers tracking the paths of the planets through the heavens goes on today, with physicists probing ever deeper into the fundamental structure of matter, astronomers studying ever more distant galaxies, and biologists uncovering more and more of the complexity of the things we call "living systems."

All of this magnificent saga, from Greek philosophers thinking about the nature of the universe two thousand years ago to the scientists taking data from complex modern instruments as you read this, is the story of physics. When you have finished this course, you will know something about how we got to where we are and will have a picture of where we may be going in the future.

J. T. & R. M. H.

1 Science: A Way of Knowing

KEY IDEA

Science is a way of answering questions about the physical universe.

PHYSICS AROUND US . . . Making Choices

Our lives are filled with choices. What should I eat? Is it safe to cross the street? Should I bother to recycle an aluminum can or just throw it in the trash? Every day we make dozens of decisions; each choice is based, in part, on the knowledge that actions in a physical world have predictable consequences.

When you drive from home to school, for example, you have to decide which of several possible routes to take. Your choice might depend on many factors: the time of day, road construction and repair schedules, traffic reports, and perhaps even the weather. Some routes might be shorter in distance but rely on roads with lots of long traffic lights. Other routes might require driving a longer distance but at higher speeds. Over time you test many different routes, observing the time and convenience of each.

In the end, you develop an excellent sense of the alternatives and choose your route accordingly.

This simple example illustrates one way we learn about the universe. First, we look at the world to see what is there and to learn how it works. Then we generalize, making rules that seem to fit what we see. And finally, we apply these general rules to new situations we've never encountered before, and we fully expect the rules to work.

There doesn't seem to be anything earth-shattering about discovering the best driving route to school. However, the same analytical procedure of observation and testing can be applied in a more formal and quantitative way when we want to understand the workings of a distant star or a living cell. In these cases, the enterprise is called science.

THE SCIENTIFIC METHOD

Science is a discipline that asks and answers questions about the working of the physical world. It is not primarily a set of facts or a catalog of answers, but rather a way of conducting an ongoing investigation of our physical surroundings. The people who conduct these investigations about our world are called **scientists.** Like any human activity, science is enormously varied and rich in subtleties. Nevertheless, a few basic steps taken together can be said to comprise the **scientific method.**

What we discuss below presents several elements that characterize science as a way of knowing. Most of the time, scientists more or less follow this scheme, but you shouldn't think of it as a rigid cookbook procedure. Many scientists have made fundamental contributions while deviating from the outline shown in Figure 1-1.

Observation

If our goal is to learn about the world, the first thing we have to do is look around us and see what's there. This statement may seem obvious to us; yet throughout much of history, learned men and women have rejected the idea that you can understand the world simply by observing it.

The Greek philosopher Plato, living during the Golden Age of Athens, argued that we cannot deduce the true nature of the universe by trusting our senses. The senses lie, he said. Only the use of reason and the insights of the human mind can lead us to true understanding. In his famous book *The Republic,* Plato compared human beings to people living in a cave, watching shadows on a wall but unable to see the objects causing the shadows. In just the same way, he argued, observing the physical world can never put us in contact with reality but will doom us to a lifetime of wrestling with shadows. Only with the eye of the mind can we break free from illusion and arrive at the truth.

During the Middle Ages in Europe, a similar frame of mind existed. However, at that time, a devout trust in wisdom passed down from classical scholars and theologians replaced human reason as the ultimate tool in the search for truth. A story (probably apocryphal) recounts a debate in an Oxford college on the question, "How many teeth does a horse have?" One learned scholar quoted the Greek scientist Aristotle on the subject; another quoted the theologian St. Augustine to put forward a different answer. Finally, a young monk at the back

FIGURE 1-1. The scientific method can be represented as a cycle of collecting observations (data), identifying patterns and regularities in the data (synthesis), forming hypotheses, and making predictions, which lead to more observations.

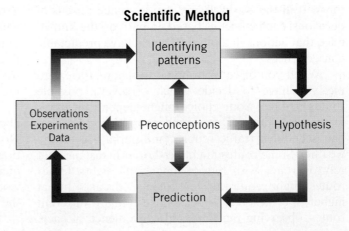

of the hall got up and noted that since there was a horse outside, they could settle the question by looking in its mouth. At this point, the manuscript states, the assembled scholars "fell upon him, smote him hip and thigh, and cast him from the company of educated men."

As both of these examples illustrate, we can develop strategies for learning about the physical world using our reasoning powers alone or relying on accepted authority, without actually making observations. However, such approaches are not what we call the scientific method, nor do they produce the kinds of advanced technologies and knowledge we associate with modern societies. These other attempts to understand the physical world were, however, perfectly serious and were pursued by people every bit as intelligent as we are. In the next chapter, we will see how human beings gradually came to understand that observation complements pure reasoning and thus has an important role to play in learning about the universe.

In the remainder of this book, we differentiate between **observations,** in which we observe nature without manipulating it, and **experiments,** in which we manipulate some aspect of nature and observe the outcome. An astronomer, for example, might observe distant stars without changing them, while a physicist might experiment by heating materials and measuring changes in their properties.

Plato compared observing nature to watching shadows on a wall; the underlying reality is inaccessible.

Identifying Patterns and Regularities

When we observe a particular phenomenon over and over again, we begin to get a sense of how nature behaves. We start to recognize patterns in nature. Eventually, we generalize our experience into a synthesis that summarizes what we have learned about the way the world works. We may, for example, notice that whenever we drop a book, it falls. With this statement, we're incorporating the results of many observations.

Scientists often summarize the results of their observations in mathematical form, particularly if they have been making quantitative measurements. In the case of a falling book, for example, they might measure the time it takes a book to fall a certain distance, rather than just noticing that it falls. Their next step would probably be to collect their data in the form of a table (see Table 1-1). These data could also be presented in the form of a graph, in which distance is plotted against time (see Figure 1-2).

After preparing tables and graphs of their data, our scientists might notice that the longer the time something falls, the greater the distance it travels. Furthermore, the distance isn't simply proportional to the time of fall. That is, if a book falls for twice as long, it does not travel twice as far. Rather,

TABLE 1-1 Measurements of a Falling Object	
Time of fall (seconds)	**Distance of fall (meters)**
0	0
1	4.9
2	19.6
3	44.1
4	78.4

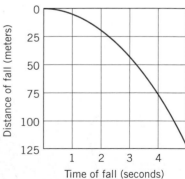

FIGURE 1-2. Measurements of a falling object can be presented visually in the form of a graph. Time of fall (on the horizontal axis) is plotted versus distance of fall (on the vertical axis).

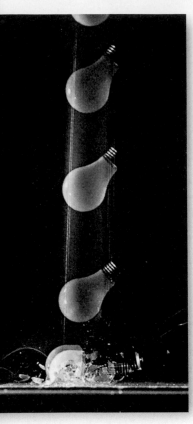

A time-lapse photograph of a falling lightbulb.

if one book falls for twice as long as another book, it travels four times as far. If it falls three times longer, it travels nine times as far, and so on. This statement can be summarized in three ways (a format that we'll use throughout this book):

1. In words:

The distance traveled by a falling object is proportional to the square of the time of the object's travel. Thus, during each tick of the clock, the distance the object falls is greater than during the previous tick.

2. In an equation with words:

$$\text{Distance} = \text{constant} \times \text{time} \times \text{time}$$
$$= \text{constant} \times (\text{time})^2$$

3. In an equation with symbols:

$$d = k \times t^2$$

The symbol k is a *constant* that defines the quantitative mathematical relationship between distance and time squared. The value of this constant has to be determined from measurements. (We'll return to the subject of constants in Chapter 2.)

Mathematics is a concise language that allows scientists to communicate their results and make very precise predictions (see Chapter 2). However, anything that can be said in an equation can also be said (although in a less concise way) in a plain English sentence. When you encounter equations in your science courses, you should always ask, "What English sentence does this equation represent?" This routine will keep the mathematics from obscuring the simple ideas that lie behind most equations.

Not every scientific idea can be or has to be stated this precisely, however. A scientist studying the formation of tornadoes, for example, might notice that certain combinations of atmospheric conditions—low pressure, high humidity, and strong vertical temperature differences, for example—always precede the formation of a tornado. The scientist might conclude that a number of criteria favor tornado formation. This conclusion can be tested, and so it is a part of scientific inquiry.

Hypothesis and Theory

Once we have summarized experimental and observational results, we can form a **hypothesis**—a tentative, educated guess—about how the world works for the behavior under study. In the case of our everyday experience with many different kinds of falling objects, we can formulate a hypothesis very easily. We can say, "When I drop a solid object, it falls." In other cases, the formation of the hypothesis may be more complicated, and the hypothesis may be stated in the form of mathematical equations. When confronted with a new phenomenon, scientists often weigh several different hypotheses at once, much as a detective in a murder mystery may consider several different suspects. It's also quite common to start with a hypothesis and then look for observations that support or disprove it. However, you must be careful that you don't let a favorite hypothesis prejudice your observations.

The word **theory** refers to a description of the world that covers a relatively large number of phenomena and has met and explained many observational and experimental tests. After observing hundreds of dropped objects, for example, we might state a theory such as "In the absence of wind resistance, all objects fall a distance proportional to the square of the time of the fall." Just as a detective announces a solution at the conclusion of a murder mystery, so do scientists reach a logical conclusion based on their observations of nature.

One word of caution: scientists don't rely on a rigorous definition when they use the words "theory" and "law." Their use often follows historical precedent, and many bodies of knowledge that are called "theories" are among the best-verified aspects of our knowledge of the world. Two examples are the theory of relativity (see Chapter 28) and the theory of evolution in biology.

Prediction and Testing

In science, every hypothesis must be tested. We test hypotheses by using them to make **predictions** about how a particular system will behave; then we observe nature to see if the system behaves as predicted. For example, if we hypothesize that all objects fall when they are dropped, then we can test this idea by dropping all sorts of objects—a lightbulb, a book, a glass of water. Each drop constitutes a test of our prediction. The more tests that give the same result, the more confidence we have that the hypothesis is correct.

So long as we restrict our tests to solids or liquids on the Earth's surface, the hypothesis is consistently confirmed. Test a helium-filled balloon, however, and we discover a clear exception to the rule. The balloon "falls" up. The original hypothesis, which worked so well for most objects, fails for certain gases. More tests would show that's not the only limitation. If you were an astronaut in the space shuttle, every time you held something out and let it go, it would neither fall nor rise. It would float in space. Evidently, our hypothesis is invalid in orbit, as well.

This example illustrates an important aspect about testing hypotheses. Tests do not necessarily prove or disprove a hypothesis; instead, they often serve to

In orbit, an object does not fall if you drop it, but continues floating. This is an example for which the simple hypothesis "When I drop a solid object, it falls" doesn't work.

define the range of situations under which the hypothesis is valid. For example, we may observe that nature behaves in a certain way only at high temperatures or only at low ones, or only at low velocities or only at high ones. Such limitations indicate that the original hypothesis doesn't cover enough ground and has to be replaced by something more general. In the example of falling objects, we will see that the hypothesis "Objects fall when dropped" has to be replaced by a more sophisticated and general set of hypotheses called Newton's laws of motion and the law of universal gravitation. These laws describe and predict the motion of dropped objects both on the Earth and in space and are, therefore, a more successful set of statements than the original hypothesis. (We discuss Newton's laws in more detail in Chapters 4 and 5.)

Testing and retesting of hypotheses under many different circumstances lies at the heart of science. Any scientific hypothesis must be subject to modification or even rejection based on new observations and experiments. The famous astronomer and popular science writer Carl Sagan once said that "the essence of science is that it is self-correcting." No scientific idea, no matter how cherished, is immune from the power of new facts. For example, the idea that the Sun moves around the Earth was considered an established fact for at least 2000 years. But later, more accurate measurements of planetary motions could not be explained by this idea, and eventually it was replaced by today's understanding that the Earth moves around the Sun.

When a hypothesis has been tested extensively and seems to apply everywhere in the universe—when we have had enough experience with it to have a lot of confidence that it is true—we generally elevate the hypothesis to a new status. We call it a **law of nature.** We will encounter many such physical laws in this book, all of them backed by countless observations and measurements. It is important, however, to remember where these laws come from. They are not written on tablets of stone, nor are they simply good ideas that someone once had. They arise from repeated and rigorous observation and testing—observations and testing that you could duplicate yourself. They represent our best understanding of how nature works.

Remember, scientists never stop questioning the validity of their hypotheses, even after we call them laws. Scientists constantly think up new, more rigorous experiments to test the limits of theories and laws. In fact, one of the central tenets of science is:

Every law of nature is subject to change based on new observations.

The Scientific Method in Operation

Together, the elements of observation, hypothesis formation, prediction, and testing make up the scientific method. In an idealized sense, you can think of the method as working as shown in Figure 1-1. In this never-ending cycle, observations lead to hypotheses, which lead to more observations.

If observations are consistent with a hypothesis, then more exacting tests may be devised. If the hypothesis fails, then the new observations may be used to modify it, after which the revised hypothesis is tested again. Scientists continue this process until they reach the limits of existing equipment, in which case they often try to develop better instruments to do even more rigorous tests. If it appears that there's just no point to going further—when decades of experiments support a given hypothesis—then scientists may eventually call the hypothesis a law of nature. But even the most thoroughly tested law of nature is subject to change if new observations warrant.

Several important points should be made about the scientific method.

1. While scientists attempt to be objective, they often observe nature with preconceptions about what they are going to find. Most experiments and observations are designed and undertaken with a specific hypothesis in mind, and most researchers have a strong hunch about whether that hypothesis is right or wrong. Given human nature and the difficulty of many state-of-the-art experiments, the history of science has many examples of discoveries that were later shown to be false, from N-rays to polywater to cold fusion. In most of these cases, researchers thought they saw the evidence that they wanted to see, but the results could not be reproduced by other people. Perhaps the most important point about the scientific method is that scientists have to believe the results of their experiments and observations, whether the results fit preconceived notions or not. Science doesn't ask that we enter the cycle of Figure 1-1 with no preconceptions or hunches, but it demands that we be ready to change those ideas if the evidence forces us to do so.

2. There is no one correct place to enter the cycle. Scientists often start their work by making extensive observations, but they can also start with a hypothesis and test it. Wherever they enter the cycle, the scientific process takes them all the way around.

3. Observations and experiments must be reported in such a way that anyone with the proper equipment can verify the results. In other words, scientific results must be *reproducible*.

4. There is no end to the cycle. Science does not provide final answers, nor is it always a search for ultimate truth. Science is a way of producing successively more detailed and exact descriptions of the physical world—descriptions that allow us to predict the behavior of that world with higher and higher levels of confidence.

5. Finally, the orderly cycle shown in Figure 1-1 provides a useful idealized framework to help us think about science, but science is not a rigid cookbook-style set of steps to follow. Science is often an intensely creative activity, undertaken by a wide variety of human beings. Scientific discovery often involves occasional bursts of intuition, sudden leaps of understanding, a joyful breaking of the rules, and all the other sorts of spontaneity we associate with human activities.

Physics in the Making

Dmitri Mendeleev and the Periodic Table

The discoveries of previously unrecognized patterns in nature provide scientists with some of their most exhilarating moments. Dmitri Mendeleev (1834–1907), a popular chemistry professor at the Technological Institute of St. Petersburg in Russia, experienced such a breakthrough in 1869 as he was tabulating data for a new chemistry textbook.

The mid-nineteenth century was a time of great excitement in physics and chemistry. Almost every year saw the discovery of new physical phenomena and one or two new chemical elements, while new apparatus and processes were greatly expanding the repertoire of laboratory and industrial scientists. In such a stimulating field, it wasn't easy to keep up to date with all the developments and summarize them in a textbook. In an effort to consolidate the current state of knowledge about the most basic chemical building blocks, Mendeleev listed various physical and chemical properties of the 63 known chemical elements (substances that could not be divided by chemical means). He arranged his list in order of increasing atomic weight and then noted the distinctive characteristics of each element.

Examining his list, Mendeleev recognized an extraordinary pattern: elements with similar physical and chemical properties appeared at regular, or *periodic,* intervals. One group of elements, including lithium, sodium, potassium, and rubidium (he called them Group I elements), were soft silvery metals that formed compounds with chlorine in a one-to-one ratio. Immediately following the Group I elements in the list were beryllium, magnesium, calcium, and barium—Group II elements that form compounds with chlorine in a one-to-two ratio—and so on.

As other similar patterns emerged from his list, Mendeleev realized that the elements could be arranged in the form of a table (Figure 1-3). Not only did this periodic table highlight previously unrecognized relationships among the elements, but it also revealed obvious gaps—places where as yet undiscovered elements must lie.

The power of Mendeleev's periodic table of the elements was demonstrated when he used it to predict the properties of several elements that were not known at the time. Within 15 years, chemists isolated the elements gallium and germanium, whose properties were strikingly close to Mendeleev's predictions. In subsequent years, Mendeleev's table provided physicists with a foundation for understanding the structure of the atom and quantum mechanics (see Chapter 22).

ПЕРИОДИЧЕСКАЯ СИСТЕМА ЭЛЕМЕНТОВ

ГРУППЫ ЭЛЕМЕНТОВ

		I	II	III	IV	V	VI	VII	VIII			0
1	I	H 1 1,008										He 2 4,003
2	II	Li 3 6,940	Be 4 9,02	5 B 10,82	6 C 12,010	7 N 14,008	8 O 16,000	9 F 19,00				Ne 10 20,183
3	III	Na 11 22,997	Mg 12 24,32	13 Al 26,97	14 Si 28,06	15 P 30,98	16 S 32,06	17 Cl 35,457				Ar 18 39,944
4	IV	K 19 39,096	Ca 20 40,08	Sc 21 45,10	Ti 22 47,90	V 23 50,95	Cr 24 52,01	Mn 25 54,93	Fe 26 55,85	Co 27 58,94	Ni 28 58,69	
	V	29 Cu 63,57	30 Zn 65,38	31 Ga 69,72	32 Ge 72,60	33 As 74,91	34 Se 78,96	35 Br 79,916				Kr 36 83,7
5	VI	Rb 37 85,48	Sr 38 87,63	Y 39 88,92	Zr 40 91,22	Nb 41 92,91	Mo 42 95,95	Ma 43 —	Ru 44 101,7	Rh 45 102,91	Pd 46 106,7	
	VII	47 Ag 107,88	48 Cd 112,41	49 In 114,76	50 Sn 118,70	51 Sb 121,76	52 Te 127,61	53 J 126,92				Xe 54 131,3
6	VIII	Cs 55 132,91	Ba 56 137,36	La 57 * 138,92	Hf 72 178,6	Ta 73 180,88	W 74 183,92	Re 75 186,31	Os 76 190,2	Ir 77 193,1	Pt 78 195,23	
	IX	79 Au 197,2	80 Hg 200,61	81 Tl 204,39	82 Pb 207,21	83 Bi 209,00	84 Po 210	85 —				Rn 86 222
7	X	87 —	Ra 88 226,05	Ac 89 227	Th 90 232,12	Pa 91 231	U 92 238,07					

* ЛАНТАНИДЫ 58–71

Ce 58 140,13	Pr 59 140,92	Nd 60 144,27	61 —	Sm 62 150,43	Eu 63 152,0	Gd 64 156,9
Tb 65 [illegible]	Dy 66 162,46	Ho 67 164,94	Er 68 167,2	Tu 69 169,4	Yb 70 173,04	Cp 71 174,99

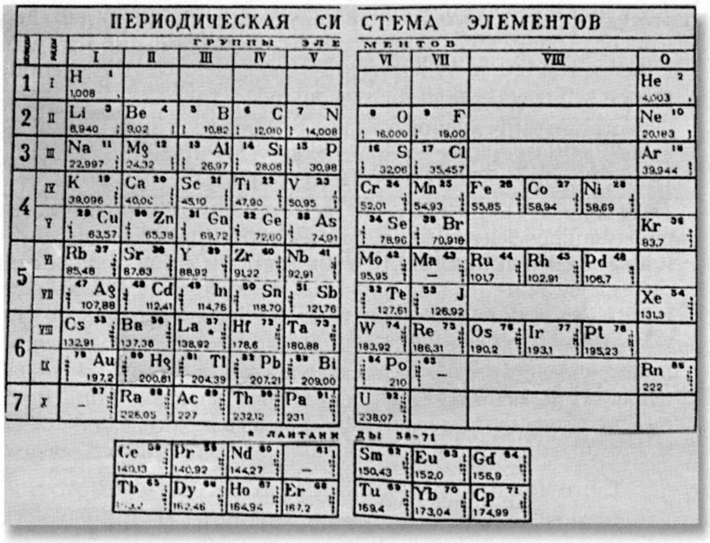

FIGURE 1-3. The first published version of Dmitri Mendeleev's periodic table of the elements revealed regular patterns in the chemical behavior of known elements, as well as obvious gaps where as yet undiscovered elements must lie. Elements in the same column (which Mendeleev called Groups I, II, etc.) have similar chemical properties.

The discovery of the periodic table ranks as one of the great achievements of science. It was so important, in fact, that Mendeleev's students carried a copy of it behind his coffin in his funeral procession. ●

 # OTHER WAYS OF KNOWING

The central idea of science revolves around the notion that by observing and measuring, we can discover laws that describe how nature works. *Every* idea in science is subject to testing. If an idea cannot be tested, it may not be wrong, but it simply isn't part of science.

For example, a scientist can hypothesize that a particular painting was executed in the seventeenth century. She could use various chemical tests to analyze the composition of the paint, document the age of the canvas, X-ray the structure of the painting, and so on. Her statement about the age of the painting may turn out to be wrong (the painting may, for example, turn out to be a modern forgery). The key here is that the statement can be tested and is, therefore, an acceptable scientific hypothesis.

The methods of science can determine the age of a painting, but they cannot tell us if the painting is beautiful.

But some questions cannot be answered by the methods of science. No physical or chemical test can tell us whether the painting is beautiful or important or valuable, nor can any test be devised that can tell us how we are to respond to it. These questions are simply outside the realm of science. In fact, the methods of science are not the only way to answer many questions that matter in our lives. While science provides us with a way of tackling questions about the physical world, such as how it works and how we can shape it to our needs, many questions—you might argue the most important questions—lie beyond the scope of science and the scientific method. Some of these questions are deeply philosophical: What is the meaning of life? Why does the world hold so much suffering? Is there a God? Other important questions are more personal: What career should I choose? Whom should I marry? Should I have children? These questions cannot be answered by the cycle of observation, hypothesis, and testing. For answers, we turn instead to religion, philosophy, and the arts.

A symphony, a poem, and a painting are not, in the end, objects to be studied scientifically. These art forms address different human needs and they use different methods than science. The same can be said about religious faith. Strictly speaking, no conflict should exist between science and religion because they deal with different aspects of life. Conflicts arise only when zealots on either side try to push their methods into areas where they aren't applicable.

Pseudoscience

Many kinds of inquiry—extrasensory perception (ESP), unidentified flying objects (UFOs), astrology, crystal power, reincarnation, and the myriad claims of psychic phenomena you see advertised in magazines and on TV—fail the elementary test that defines science. None of these subjects, collectively labeled **pseudoscience,** is subject to reproducible testing in the sense we are using that term. No test that you can devise will convince those who believe in these notions that their ideas are wrong. Yet, as we have seen, the central property of scientific ideas is that they can be independently tested and that they may, at least in principle, be wrong. Untestable pseudoscientific ideas thus lie outside the domain of science.

The rejection of these subjects may at first glance make scientists seem close-minded and narrow, but nothing could be further from the truth. Scientists thrive on discovering the strange and remarkable in nature. That's how they make their reputations and increase human understanding. So if the physical remains of a UFO were to be discovered and made available for study or if reproducible experiments pointed to as yet unknown abilities of the human brain, scientists would jump at the chance to study them.

LOOKING DEEPER

Astrology

Astrology is a very old system of beliefs that most modern scientists would call a pseudoscience. The central belief of astrology is that the positions of objects in the sky at a given time (at the moment of a person's birth, for example) determine a person's future. Astrology was part of a complex set of omen systems developed by the Babylonians and was practiced by many famous astronomers well into modern times.

If you were in a spaceship above the Earth's atmosphere, you could see the Sun and the stars at the same time. As the Earth traveled around the Sun, you would see the backdrop of stars change. The band of background stars through which the Sun appears to move is called the zodiac (Figure 1-4). The stars of the zodiac are customarily divided into twelve constellations, called signs

or houses. At any time, the Sun, the Moon, and the planets all appear in one or another of these constellations, and a diagram showing these positions is called a horoscope. The constellation in which the Sun appeared at the time of your birth is your Sun sign, or simply your sign.

Astrologers have a complex (and far from unified) system in which each combination of heavenly bodies and signs is believed to signify particular things. The Sun, for example, is thought to indicate the outgoing, expressive aspects of one's character, the Moon represents the inner-directed ones, and so on.

Scientists reject astrology for two reasons. First, there is no known way that planets and stars could exert a significant influence on a child at birth. It is true, as we learn in Chapter 5, that stars and planets exert a tiny gravitational force on the infant, but the gravitational force exerted by the delivering physician (who is much smaller but much closer) is far greater than that exerted by any celestial object.

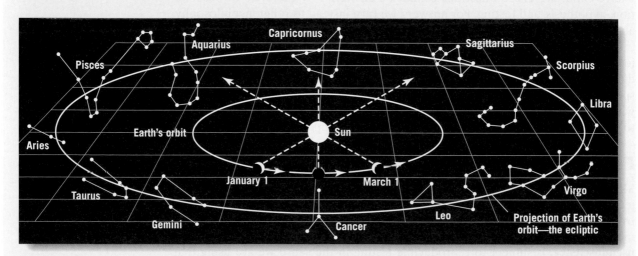

FIGURE 1-4. The stars of the zodiac form a band in the same plane as the Earth's equator.

More important, scientists reject astrology because it just doesn't work. Over the millennia, no evidence at all shows that positions of the stars can predict the future.

You can test the ideas of astrology for yourself, if you like. Try this: have a member of the class take the horoscopes from yesterday's newspaper and type them on a sheet of paper without indicating which horoscope goes with which sign. Then ask members of your class to indicate the horoscope that best matches the day they actually had. Have them write their birthday (or sign) on the paper as well.

If people just picked horoscopes at random, you could expect about 1 person in 12 to pick the horoscope corresponding to his or her sign. Are the results of your survey any better than that? What does this tell you about the predictive power of astrology?

THE STRUCTURE OF SCIENCE

Scientists investigate all sorts of natural objects and phenomena: the tiniest elementary particles, microscopic living cells, the human body, forests, the Earth, stars, and the entire cosmos. Throughout this vast sweep, the same scientific method is applied. Men and women have been carrying out this task for hundreds of years, and by now we have a pretty good idea about the way that many parts of our universe work. In the process, scientists have also developed a social structure that provides unity to the pursuit of scientific knowledge, as well as the recognition of important disciplinary differences within the larger scientific framework.

The Specialization of Science

Science is a human endeavor, and humans invariably form themselves into groups with shared interests. When modern science first started in the seventeenth century, it was possible for one person to know almost all there was to know about the physical world and the three kingdoms—animals, vegetables, and minerals. In the seventeenth century, Isaac Newton could do pioneering research in astronomy, the physics of moving objects, the behavior of light, and mathematics. Thus, for a time prior to the mid-nineteenth century, scholars who studied the workings of the physical universe formed a more or less cohesive group, who called themselves natural philosophers. However, as human understanding expanded and knowledge of nature became more detailed and technical, science began to fragment into increasingly specialized disciplines and subdisciplines.

Today, our knowledge and understanding of the world is so much more sophisticated and complex that no one person could possibly be at the frontier in such a wide variety of fields. Scientists today must choose a field—biology, chemistry, physics, and so on—and study one small part of the subject in great depth. Within each of these broad disciplines there are hundreds of different subspecialties. In physics, for example, a student may elect to study the behavior of light, the properties of materials, the nucleus of the atom, elementary particles, or the origin of the universe. The amount of information and expertise required to get to the forefront in any of these fields is so large that most students have to ignore almost everything else to learn their specialty.

Science is further divided because scientists within each subspecialty approach problems in different ways. Some scientists are *field researchers,* who go into natural settings to observe nature at work. Other scientists are *experimentalists,* who manipulate nature with controlled experiments. Still other scientists, called *theorists,* spend their time imagining how the universe might work in areas where we still don't have detailed explanations. These different kinds of scientists need to work together to make progress.

The fragmentation of science into disciplines was formalized by a peculiar aspect of the European university system. In Europe, each academic department can have only one professor. All other teachers, no matter how famous and distinguished, have less prestigious titles. As the number of outstanding scientists grew in the nineteenth century, universities were forced to create new departments to attract new professors. Several German universities, for example, supported separate departments of theoretical and experimental physics. At one time, Cambridge University in England had seven different departments of chemistry!

In North America, each academic department generally has many professors. Nevertheless, American science faculties are often divided into several departments, including physics, chemistry, astronomy, geology and biology—the branches of science.

The Branches of Science

Several branches of science are distinguished by the scope and content of the questions they address.

Physics is the search for laws that describe the most fundamental aspects of nature: matter, energy, forces, motion, heat, light, and other phenomena. All natural

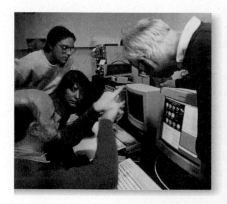

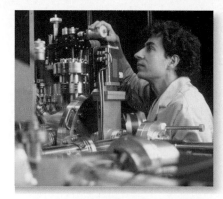

Scientists come from all kinds of backgrounds but share a curiosity about the world and a desire to learn about it.

systems, including planets, stars, cells, and people, display these basic properties, so physics is the starting point for almost any study of how nature works.

Chemistry is the study of atoms in combination. Chemicals form every material object of our world, while chemical reactions initiate vital changes in our environment and our bodies. Chemistry is thus an immensely practical (and profitable) science. Most of the largest industries in the world today are chemical industries, including petroleum refining, agrochemicals, plastics, mining, and pharmaceuticals.

Astronomy is the study of stars, planets, and other objects in space. We are in an era of unprecedented astronomical discovery, thanks to human and robotic space exploration.

Geology is the study of the history, evolution, and present state of our home, planet Earth. Many geology departments also emphasize the study of other planets as a way to understand the unique character of our own world.

Biology is the study of living systems. Biologists document life at many scales, from individual microscopic molecules and cells to expansive ecosystems.

The Scope of Physics

Physics is the study of nature at its most basic level. Everything from the particles that make up atoms to the stars and galaxies we see in deep space fall within its scope. One of the most remarkable aspects of physics is that its laws apply to this wide variety of objects and systems.

Among the most basic questions to ask about any physical object are how does it move and what makes it move that way. Every object moves to some extent, whether it be the space shuttle zipping along at 17,000 miles per hour or the atoms in rocks at the summit of Mount Everest vibrating back and forth in place. Important information about an object can be found by describing its motion and trying to determine why it moves as it does. This study of motion and the causes of motion is called mechanics and is discussed in Chapters 3–8 of this book.

Energy, which enables an object to move, provides a unifying concept in mechanics. But the transfer of energy from one object to another has its own rules and limitations. The study of these rules is called thermodynamics and is presented in Chapters 9–13.

Mechanics deals with most objects as simple particles or rigid bodies in motion. It turns out that waves offer a model of motion that is just as valid as the concept of a particle moving along a line. In fact, we will see that it is not always possible to tell the difference between particle motion and wave motion. We discuss waves in Chapter 14 and present some of their most common applications in Chapter 15 on sound.

The next part of our survey of physics examines electricity and magnetism. In combination, these forces help hold atoms and molecules together, so that when you sit in a chair you remain you and don't become part of the chair. Electricity and magnetism are manifested in many familiar appliances, but they are also remarkably intertwined in the phenomenon of light. We discover the amazing world of electromagnetism in Part V of the book (Chapters 16–20), including brief presentations of electrical circuits and optics.

Once we've learned the concepts and vocabulary from all these areas of physics, we can put them together to study the structure of matter itself. What makes a material a gas, a liquid or a solid? What happens when iron rusts or wood burns? These are questions that chemists can answer, but the underlying principles of atoms in combination are part of physics. These concepts have led to today's information age of computers, the Internet, and the devices that enable them to operate; we talk about these topics in Chapters 21–25.

Delving still deeper within the atom, we enter the realm of the smallest known entities—the subatomic particles. We touch on the mind-expanding topics of nuclear physics and particle physics in Chapters 26 and 27.

Among the most astonishing discoveries of physics is the realization that the laws of nature need to be modified when applied to objects that are very small, such as atoms, or very large, such as stars, or that are moving very fast. Some of these modifications are discussed in Chapter 21 on quantum mechanics, while others are presented in Chapter 28 on special and general relativity. In the book's final chapter, we examine connections between subatomic particles and the largest known physical systems, galaxies, and the universe itself.

By the time you have finished this journey, you will have touched on many of the great truths about the physical universe that scientists have deduced over the centuries. You will discover how the different parts of our universe operate and how all the parts fit together, and you will know that there are still great unanswered questions that drive scientists today. You will understand some of the great scientific and technological challenges that face our society and, more important, you will know enough about how the world works to deal with many of the new problems that will arise in the future.

SCIENCE, TECHNOLOGY, AND SOCIETY

We can study the physical universe in many ways and for many reasons. Many scientists are driven by curiosity and the pure joy of discovering how the world works—they are interested in knowledge for its own sake. These scientists are engaged in **basic research.** They might study the behavior of distant stars, matter at extremely cold temperatures, or subatomic particles. Although discoveries made by basic researchers may have profound effects on society (for example, see the discussion in Chapter 17 of the discovery of the electric generator), that is not the primary goal of these scientists.

Other scientists approach their work with specific goals in mind. They wish to develop **technology,** in which they apply the results of science to specific commercial or industrial goals. These scientists are said to be doing **applied research,** and their ideas are often translated into practical systems by large-scale **research and development (R&D)** projects. For example, physicists have determined that some materials emit electrically charged particles (electrons) when you shine a light on them. This is an example of basic research. Later, physicists and engineers applied this concept and developed devices such as electric-eye door openers, solar panels for generating electricity, CD players, and optical bar code scanners, among many other products. Each of these products came about after years of research and development, but they all use the same basic idea of physics. For another example, see Physics and Modern Technology on page 16.

Government laboratories, colleges and universities, and private industries all support both basic and applied research; however, most large-scale R&D (as well as most applied research) is done in government laboratories and private industry (Table 1-2).

TABLE 1-2	*Some Important Research Laboratories in the United States*	
Facility	**Type**	**Location**
Argonne National Laboratory	Govt/Univ	near Chicago, IL
Bell Laboratories	Industrial	Middletown, NJ
Brookhaven National Laboratory	Government	Upton, NY
DuPont Central Research & Development	Industrial	Wilmington, DE
Fermi National Accelerator Laboratory	Govt/Univ	Batavia, IL
IBM Watson Research Center	Industrial	Yorktown Hts, NY
Keck Observatory	University	Kamuela, HI
Los Alamos National Laboratory	Government	Los Alamos, NM
National Institutes of Health	Government	Bethesda, MD
Oak Ridge National Laboratory	Government	Oak Ridge, TN
Stanford Linear Accelerator Center	Govt/Univ	Menlo Park, CA
Texas Center for Superconductivity	University	Houston, TX
United States Geological Survey	Government	Reston, VA
Woods Hole Oceanographic Institution	University	Woods Hole, MA

Connection

Buckyballs—A Technology of the Future?

An extraordinary discovery, announced in 1990, reveals the close relationships among pure scientific research, applied research, and technology. For centuries the element carbon was known in only two basic forms—graphite, the soft black mineral used in pencils, and diamond, the hard transparent gemstone. However, in 1985 chemists at Sussex University in England and Rice University in Texas found evidence for a totally new form of carbon—one in which 60 carbon atoms bond together in a ball-shaped molecule (Figure 1-5). The distinctive linkage of the atoms, much like the geodesic domes of architect-inventor Buckminster Fuller, led scientists to dub the new material buckminsterfullerene, or buckyballs for short. Buckyballs, although completely unexpected, at first excited little attention outside a small research community because the material had no known uses and it could only be produced in minute quantities. Nevertheless, the lure of the new form of an important chemical element kept several research groups busy studying the stuff. With no obvious practical applications, these early buckyball studies were examples of basic research.

A major advance came in May 1990, when a small team of German chemists discovered a way to produce and isolate large quantities of buckyball crystals in a simple and inexpensive device. With the possibility of commercial-scale production, an explosion of applied buckyball research followed. Thousands of scientists, including teams at most of the major industrial and government

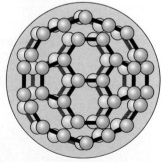

FIGURE 1-5. Buckyballs are soccer-ball-like molecules of 60 carbon atoms (red spheres) that form crystals in which these round molecules stack together like oranges at the grocery store.

Physics and Modern Technology—Diamonds

You probably know that diamonds make beautiful and expensive jewelry, and you may know that a diamond is one of the hardest substances known. But physicists have shown that diamond has other remarkable properties, including low-friction surfaces, transparency to infrared radiation, and high thermal conductivity. These properties make industrial diamonds useful in many areas of modern technology.

Miniature high-pressure cells use diamond anvils to squeeze samples of rock or metal without breaking, forming new kinds of materials.

Diamond saws enable fast, focused street repair without noisy and destructive jackhammers.

Grinding wheels with diamonds embedded in the edges are used to cut stone and metal materials.

Gem-quality diamonds are beautiful forever.

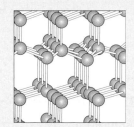

The diamond structure consists of carbon atoms held together in a strong network of chemical bonds.

One-hour eyeglasses are made possible by diamond grinders and polishers linked to a computer.

The space shuttle has hinges lined with thin diamond coatings for strength and low friction. Future spacecraft may have diamond windows as well.

laboratories, jumped on the buckyball bandwagon. Hundreds of scientific articles documented an astonishing range of chemical and physical properties for the new carbon.

Among the extraordinary findings, scientists found that buckyballs and closely related materials may contribute to a new generation of versatile electronic materials, powerful lightweight magnets, atom-sized ball bearings, and super-strong building materials. With such extraordinary prospects on the horizon, buckyball investigations may soon become the domain of engineers developing new technologies—new kinds of batteries for automobiles, carbon-based girders for skyscrapers, unparalleled lubricants, and other products as yet undreamed of.

Buckyball products may soon appear at your hardware store when engineers take the results of applied scientific research and use them to design large-scale production facilities. When the discovery is big enough, the transition from small, basic research to new technologies may be rapid indeed! ●

Funding for Science

An overwhelming proportion of funding for American scientific research comes from various agencies of the federal government—your tax dollars at work (see Table 1-3 and Figure 1-6). In 2002, the United States government's total research and development budget was about $118 billion. The *National Science Foundation,* with an annual budget of almost $5 billion, supports research and education in all areas of science. Other agencies, including the Department of Energy, the Department of Defense, the Environmental Protection Agency, and the National Aeronautics and Space Administration, fund research and science education in their own particular areas of interest, while Congress may appropriate additional money for special projects.

An individual scientist seeking funding for research usually submits a grant proposal to the appropriate federal agency. Such a proposal includes an outline

TABLE 1-3 Your Tax Dollars: Federal Science Funding, 2002	
Agency or Department	**Funding (in millions of dollars)**
Department of Agriculture	1,182
Department of Defense	50,134
Department of Energy	21,209
United States Environmental Protection Agency	592
United States Geological Survey	950
National Aeronautics and Space Administration	14,902
National Institutes of Health	23,333
National Institutes of Standards and Technology	493
National Oceanographic and Atmospheric Administration	836
National Science Foundation	4,789

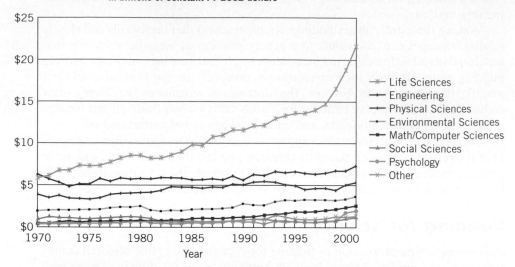

Trends in Federal Research by Discipline, FY 1970–2001
in billions of constant FY 2002 dollars

- Life Sciences
- Engineering
- Physical Sciences
- Environmental Sciences
- Math/Computer Sciences
- Social Sciences
- Psychology
- Other

FIGURE 1-6. A graph showing the spending by the federal government on scientific research (in constant dollars), 1970–2001. (*Source:* National Science Foundation data and graph © 2002 AAAS)

of the planned research, together with a statement of why the work is important. The agency evaluating the proposals asks panels of independent scientists to rank the proposals in order of importance and funds as many as it can. Depending on the field, a proposal has anywhere from about a 10% to 40% chance of being successful. This money from federal grants buys experimental equipment and computer time, pays the salaries of researchers, and supports advanced graduate students. Without this support, science in the United States would all but come to a halt. The funding of science by the federal government is one place where the opinions and ideas of citizens, through their elected representatives, have a direct effect on the development of science.

As you might expect, scientists and politicians engage in many debates about how this research money should be spent. One constant point of contention concerns the question of basic versus applied research. How much money should we put into applied research, which can be expected to show a quick payoff, as opposed to the basic sciences, which may not have a payoff for years (if at all)?

THINKING MORE ABOUT

Science: Research Priorities

Sometimes questions of research funding get caught up in questions of public policy. For example, there have been 800,000 cases of AIDS diagnosed in the United States since 1980, primarily among male homosexuals and intravenous drug users. Over that same period, more than 5,000,000 Americans have died of cancer. Yet by 1990, the budget for AIDS research at the National Institutes of Health exceeded that for cancer.

Critics of this policy argue that research money should be spent on those diseases that affect the greatest number of people and that a vocal minority has distorted the federal policy.

Supporters argue that AIDS, an incurable and invariably fatal disease, represents a *potential* threat to many more people than cancer. They point to the tens of millions of heterosexual men and women who have died of the disease in Africa as a portent of what could happen if a cure or vaccine for the disease is not found soon.

As so often happens, there is no scientific solution to this problem. What do you think the proper course for the government ought to be? Should we spend more to combat a disease that is already killing many people or one that is relatively minor now but could potentially be even more deadly? What nonscientific arguments should be brought to bear in making decisions such as these?

Summary

Science, a way of learning about our physical universe, is undertaken by women and men called **scientists.** The **scientific method** relies on making reproducible **observations** and **experiments,** which may suggest general trends and **hypotheses,** or **theories.** Hypotheses, in turn, lead to **predictions** that can be tested with more observations and experiments. Successful hypotheses may, after extensive testing, be elevated to the status of **laws of nature,** but are always subject to further testing. Science and the scientific method differ from other ways of knowing, including religion, philosophy, and the arts, and differ from **pseudosciences.**

Science is organized around a hierarchy of fundamental principles. **Physics,** the most fundamental science, focuses on overarching concepts about forces, motion, matter, and energy—phenomena that apply to all scientific disciplines. Other scientific disciplines—**chemistry, astronomy, geology,** and **biology**—address specific aspects of the natural world. This body of scientific knowledge forms a seamless web, in which every detail fits into a larger, integrated picture of our universe.

Scientists engage in **basic research,** whose goal is solely the acquisition of knowledge, and in **applied research** and **research and development (R&D)**, which are aimed at specific problems. This process develops **technology.**

Key Terms

applied research The type of research performed by scientists with specific and practical goals in mind. This research is often translated into practical systems by large-scale research and development projects. (p. 14)

basic research The type of research performed by scientists who are interested simply in finding out how the world works, in knowledge for its own sake. (p. 14)

experiment The manipulation of some aspect of nature to observe the outcome. (p. 3)

hypothesis A tentative, educated guess, after summarizing experimental and observational results, about how the world works for the behavior under study. (p. 4)

law of nature An overarching statement of how the universe works, following repeated and rigorous observation and testing of a hypothesis or group of related hypotheses. (p. 6)

observation The act of noting nature without manipulating it. (p. 2)

physics The search for laws that describe the most fundamental aspects of nature: matter, energy, forces, motion, heat, light and other phenomena. (p. 12)

prediction The behavior of a system that will confirm or deny a hypothesis. (p. 5)

pseudoscience The types of inquiry, such as extrasensory perception (ESP), unidentified flying objects (UFOs), astrology, crystal power, reincarnation, and the myriad claims of psychic phenomena, that fail the elementary test that defines science. (p. 9)

research and development (R&D) The process of bringing new discoveries to practical use, often in industrial or governmental laboratories. (p. 14)

science A method for answering questions about the working of the physical world. (p. 2)

scientific method A cycle of collecting observations (data), identifying patterns and regularities in the data (synthesis), forming hypotheses, and making predictions, which lead to more observations. (p. 2)

scientist A person who studies questions about our world for a living. (p. 11)

technology The application of science to specific commercial or industrial goals. (p. 14)

theory A description of the world that covers a relatively large number of phenomena and has met and explained many observational and experimental tests. (p. 4)

Review

1. Describe the steps in the scientific method.
2. What is the difference between an experiment and an observation?
3. Why do scientists use equations?
4. How does a hypothesis differ from a guess?
5. What is pseudoscience? Give an example.
6. What distinguishes a theory from a natural law?
7. How does a scientist choose between competing hypotheses?
8. Must scientists always conduct their research without preconceptions?
9. What does it mean that scientific experiments must be reproducible?
10. What are some ways of knowing that are not science?
11. How is Mendeleev's development of the periodic table of the elements an example of the scientific method at work?
12. Who pays for most scientific research in the United States?
13. What are the five major branches of science?
14. What kind of experiment might a chemist perform?
15. What kind of observation might an astronomer make?
16. What is the difference between basic and applied research?
17. How do research efforts of theorists, experimentalists and field scientists differ?
18. What is the National Science Foundation?

Questions

1. Which of the following statements can be tested scientifically? Explain your reasoning.

   a. Most of the energy coming from the Sun is in the form of visible light.
   b. Unicorns exist.
   c. Shelley wrote beautiful poetry.
   d. The Earth was created over 4 billion years ago.
   e. Diamond is harder than steel.
   f. Diamond is more beautiful than ruby.
   g. A virus causes the flu.
   h. Chocolate ice cream tastes better than strawberry ice cream.

2. Ten balls have ten different weights, but they all have the same size and surface texture. Each ball is dropped from the same elevation and allowed to fall to the ground. You measure the time of fall for each ball and notice that the heavier the ball, the less time it takes to fall. Can you conclude from the results of this experiment that heavier things fall faster than lighter things? Explain your reasoning.

3. The claim is sometimes made that the cycle of the scientific method produces closer and closer approximations to reality. Is this a scientific statement? Why or why not?

4. Scientists are currently investigating whether certain microscopic organisms can clean up toxic wastes. Suppose that one scientist proposed the following experiment: "In a plastic bucket, mix toxic waste and some microscopic organisms. If the toxicity of the waste is reduced, then the microscopic organisms tested are effective at cleaning up toxic waste." Is this a good scientific experiment? Why or why not?

5. Categorize the following examples as basic research or applied research.

   a. The discovery of a new galaxy.
   b. The development of a better method to fabricate rubber tires.
   c. The breeding of a new variety of disease-resistant chicken.
   d. A study of the diet of parrots in a tropical rain forest.
   e. The identification of a new chemical element.
   f. The improvement of a method to extract the element gold from stream gravels.

6. A recent television commercial claimed that an antacid consumed "47 times its own weight in excess stomach acid." How would you test this statement in the laboratory? As a consumer, what additional questions should you ask before deciding to buy this product? Are all of these questions subject to the scientific method?

7. State whether each of the following situations is an example of an observation or an experiment and explain the reason(s) for your choice.

   a. Recording the position of the setting sun on the horizon during the year.
   b. Recording the changes in atmospheric pressure while a cold front moves through your hometown.
   c. Measuring the amount of snowfall or rain on the roof of your physical science building.
   d. Recording the boiling point for a beaker of water while you add different amounts of salt to the water.
   e. Recording the times of free fall for different objects that have been thrown off the roof of your science building.
   f. Measuring the difference between the temperature inside and outside your dormitory room window during the winter and relating this difference to the presence and absence of fog on the window.

8. One form of pseudoscience goes by the name of Ancient Astronauts. Part of this argument is that ancient monuments such as the pyramids could not have been built by Egyptian engineers but required the help of extraterrestrials. How would you go about investigating such a claim? (*Hint:* You might start by finding out what ancient Egyptian engineers knew how to do.)

Problems

1. Susan has kept careful records of driving speed versus fuel efficiency. She has noted that in traveling 10 miles per hour (mph) she averages 22 miles per gallon (mpg) of gasoline. Similarly, she gets 26 mpg at 20 mph, 29 mpg at 30 mph, 31 mpg at 40 mph, 32 mpg at 50 mph, 28 mpg at 60 mph, and 24 mpg at 65 mph. Describe and illustrate some of the ways you might present these data. What additional data would you like to obtain to improve your description?

2. Measure the height and weight of 10 friends and present these data both in a table and graphically. What trends do you observe? Why might physicians find such a table useful?

3. Follow these instructions for the next six experiments or observations. First, describe and identify in words the pattern that you observe. Then, using a graph, plot the data that will best illustrate this pattern.

a. Your friend is at a stop sign on the way to the pizza shop. You record the speedometer readings every 2 seconds as she travels to the next stop sign. The following table gives the times and speedometer readings for her car.

Time (seconds)	Speed (miles per hour)
0	0
2	10
4	20
6	30
8	40
10	50
12	50
14	50
16	50
18	35
20	20

b. The following table gives the measured times for a simple pendulum (a washer at the end of a string) to complete one full oscillation (one period), depending on the length of the string. (*Hint:* Try graphing the string length against the square of the period and see if you get a pattern.)

String length (centimeters)	Period (seconds)
5	0.45
10	0.63
15	0.78
20	0.90
25	1.00
50	1.42

c. An exercise therapist has recorded the maximum heart rate for a sample of typical human beings. A table of the maximum heart rate and age of this sample of human beings is given in the following table.

Heart rate (beats per minute)	Age (years)
200	20
195	25
190	30
180	40
170	50
155	65
140	80

d. While waiting for the gas station attendant to fill up your car's 10-gallon tank, you record the time it takes for the pump to reach every 2 gallons. A table of your findings is given next.

Volume (gallons)	Time (seconds)
0	0.0
2	2.5
4	5.0
6	7.5
8	10.0
10	12.5

e. Every day, Cowboy Joe used to set his clock to noontime by noting the time when the shadow of the corral gate was the shortest during the day. Cowboy Joe, taking a keen interest in geometry, used this information to calculate the altitude of the Sun at noon (i.e., the angle above the horizon that the Sun is at noon). Joe also noticed that the length of this noontime shadow and the altitude of the Sun changed throughout the year. The table here gives the noontime altitude of the Sun and the length of the corral gate's shadow during the year.

Date	Altitude of Sun (degrees)	Shadow of corral gate (feet)
January 22	19	29
February 20	28	19
March 21	39	12
April 21	51	8
May 20	58	6
June 21	62	5
July 21	59	6
August 21	50	8
September 21	39	12
October 24	27	19
November 22	19	29
December 21	16	35

f. The estimated world population is given in the following table.

Decade (AD)	Population (in billions)
1650	0.5
1850	1.0
1920	2.0
1990	5.5

4. Make the following predictions based on the data presented in the six tables of Problem 3 and the trends that you observed.

    a. What will the speed of your friend's car be at time 22 s? 24 s? (See Problem 3a.)

    b. What will the period of the simple pendulum be for a string of length 75 cm? (See Problem 3b.)
    What is the string length if the period of the pendulum is 2 s? (See Problem 3b.)

    c. What is the maximum heart rate for a 90-year-old woman? A 15-year-old boy? (See Problem 3c.)

    d. How long will it take you to pump 15 gallons of gas using the pump in Problem 3d?

    e. How long will the shadow of the corral gate be on May 1? (See Problem 3e.)
    What will the altitude of the Sun be on August 1? (See Problem 3e.)

    f. What will be the world population in 2010 A.D.? (See Problem 3f.)

Investigations

1. Find a science story in a newspaper or magazine. Did it originate at a scientific meeting? Which one?

2. Which is the closest major government research laboratory to your school? Which is the closest industrial laboratory? What kind of research do they perform?

3. How did your representatives in Congress vote on funding for the space station? Why did they vote that way?

4. How are animals used in scientific experimentation? What limits should scientists accept in research using animals?

5. Malaria, the deadliest infectious disease in the world, kills more than 2 million people (mostly children in poor countries) every year. The annual malaria research budget in the United States is less than $1 million—a minuscule fraction of the spending on cancer, heart disease, and AIDS. Should the United States devote more research funds to this disease, which is uncommon in North America? Why or why not?

6. Describe a program of scientific research carried out by a member of your school's faculty. How is the scientific method employed in this research?

7. Identify a current piece of legislation relating to science or technology (perhaps an environmental or energy bill). How did your representatives in Congress vote on this issue?

 WWW Resources

See the *Physics Matters* home page at **www.wiley.com/college/trefil** for valuable web links.

1. www.uwgb.edu/dutchs/CosmosNotes/cosmos3.htm A discussion of the third episode of the PBS video series *Cosmos: A Personal Journey* by astronomer Carl Sagan, discussing the pseudoscience of astrology and some of the history linking astrology to modern science.

2. www.periodic.lanl.gov/ The Periodic Table of the Elements. A sophisticated online periodic table of the elements created by the Los Alamos National Laboratory.

3. mathforum.org/alejandre/workshops/buckyball.html A webpage dedicated to buckminsterfullerenes (buckyballs), their discovery and uses (including animations).

4 www.project2061.org/tools/sfaaol/chap1.htm *Chapter 1: The Nature of Science* from the Project 2061 website at the American Association for the Advancement of Science (AAAS).

5. merlot.org/ An expert-reviewed website dedicated to reviewing web resources. Click on *Science and Technology* and then *Physics* to view thousands of annotated websites discussing physics.

6. www.sciencenews.org/ The online version of a weekly newsmagazine of science for all the latest happenings.

2 | The Language of Science

PHYSICS AROUND US . . . Math in the Kitchen

Have you ever baked a loaf of bread from scratch? There's something special about the taste and smell of bread fresh from the oven. It does take some effort, however, and you have to follow the directions carefully.

In one popular recipe, you start by combining yeast and a tablespoon of sugar (at 75°F) with $\frac{1}{2}$ cup of water (at 85°F)—a mixture that activates the yeast. After 10 minutes, you add other ingredients: 1 beaten egg, 8 cups of flour, and so forth, to make the dough. After kneading and shaping the dough, you place it into a preheated oven (400°F) for 15 minutes, reduce the temperature to 375°F, and then bake for an additional 25 minutes.

Even the simplest recipe for bread involves dozens of numbers—times, temperatures, weights, and volumes—all of which are essential to the task. Without these exact quantitative instructions, baking bread would be a messy, frustrating business.

In science, as in cooking and countless other aspects of our lives, mathematics is the essential language for communicating ideas with accuracy and precision.

QUANTIFYING NATURE

Take a stroll outside and look carefully at a favorite tree. Think about how you might describe the tree in as much detail as possible so that a distant friend could envision exactly what you see and distinguish that tree from all others.

A cursory description would note the rough brown bark, branching limbs, and canopy of green leaves, but that description would do little to distinguish your tree from most others. You might use adjectives such as "lofty," "graceful," or "stately" to convey an overall impression of the tree. Better yet, you could identify the exact kind of tree and specify its stage of growth—a sugar maple at the peak of autumn color, for example. However, even then your friend would have relatively little to go on.

There are many ways to describe a tree, from focusing on its shape to a discussion of the complex chemistry that goes on in its leaves.

Giving exact dimensions of the tree—its height, the distance spanned by its branches, or the diameter of the trunk—would enhance your description. You could document the shape and size of leaves, the texture of the bark, the angles and spacing of the branching limbs, and the tree's approximate age. You could even estimate the number of board feet of lumber the tree could produce.

To provide more detail, you might examine the tree for moss and insects living on the trunk and for evidence of disease on the leaves. The more detailed your description, the more varied the vocabulary you would need to command and the more precise your measurements of the tree's many parts would have to be. Photographs and other illustrations might be included to supplement your written report. Ultimately, you might even probe the tree at the microscopic level, examining the cells and molecules that give the tree its unique characteristics.

For each different kind of description of the tree, there is an appropriate language. For some uses, words might be sufficient. For example, if you were doing a census of a particular group of trees, a simple "oak tree" might be enough to get the job done. For other uses, such as including an image of the tree in a decorating scheme, you might want a picture or a geometrical shape. For still others, such as a quantitative description of the tree's energy balance or its economic value as lumber, you would need to use numbers. All of these are useful descriptions of the tree, but each is appropriate for answering a different question about the tree.

Scientists constantly grapple with the challenge of describing our world. Their solution to the problem invariably involves developing a complex vocabulary, coupled with appropriate mathematical expressions. In the words of Galileo Galilei, "The book [of science] is written in the mathematical language . . . without whose help it is humanly impossible to comprehend a single word of it, and without which one wanders in vain through a dark labyrinth."

Language and Physics

What makes a language useful? First and foremost, a language must be able to communicate a wide range of expression without ambiguity or confusion. In most day-to-day activities, two or three thousand words suffice for basic communication. However, as soon as you deal with a complex system, such as an automobile, the vocabulary increases dramatically. Think about the last time you had to have your car repaired, for example. The repair shop first had to know which of the hundreds of makes, models, years, and engine types you own. The mechanic then had to identify which of the thousand or so automobile parts was defective. Just to describe the problem, the mechanic has to master thousands of words—the specialized vocabulary of automobiles.

A catalog of parts, alone, however, is insufficient to describe your automobile and how it works. Other statements are needed to describe the car's operation. An engine must idle at a prescribed speed, for example, and the tires must be inflated to a safe pressure. All of these conditions and hundreds more are measured by various gauges and sensors, which are critical to the operation of your car. Numbers, not words, best describe these quantities. Indeed, almost everything to do with the mechanics of driving—speed, acceleration, distance, time—is expressed by numbers.

To be an automobile mechanic, you have to learn a specialized vocabulary that you don't need just to run a car.

The same situation applies to many other things we do in everyday life. To prepare your meals you must know the complex vocabulary of food, including numerous varieties of fruits and vegetables, dozens of cuts of meat and types of seafood, shelves of herbs and spices, and so on. But any cook needs numbers—quantifiable information—as well, to communicate the details of a recipe: how much, how hot, and how long? Similarly, virtually all sports have evolved specialized vocabularies, and they often employ sophisticated mathematical scales to measure performance: earned run average (baseball), third down efficiency (football), serving percentage (tennis), and a host of other parameters that enliven sports reporting.

Communication in science poses special challenges because, like your automobile, natural systems are complex in design and they operate according to strict quantitative guidelines. And, like cooking and sports, science involves complex procedures that must be documented with precision so that others can try the activity for themselves.

Learning the Language of Science

Memorizing complex vocabulary is an integral part of learning *to do* science. Doctors and medical researchers, for example, must be able to refer to thousands of different bones, muscles, nerves, and other anatomical features. Chemists must have command of the names of more than a hundred elements and countless chemical compounds. And physicists must master the intricate vocabulary of mechanics, electromagnetism, thermodynamics, and particle physics. Without this detailed vocabulary, communication between specialists would be all but impossible.

Specialized vocabulary is primarily for the experts. You don't have to learn all the mechanic's jargon to know if your car is running properly; however, if you decide to become a mechanic yourself, you'll need a lot of specialized training, which includes the vocabulary. Similarly, you don't have to be a master chef to enjoy good food or be a star athlete to appreciate sports. The same is true of science—you can appreciate science without having to become a scientist and mastering its specialized vocabulary.

DESCRIBING THE PHYSICAL WORLD

The challenge of describing the vast and complex universe may be divided roughly into two tasks. First, scientists must describe all kinds of physical objects, from atoms to stars. Then they must document how these objects interact and change over time. Both of these jobs rely, in large measure, on mathematics.

Describing an Object: What Is It?

We can't understand how the universe works unless we know its components. For hundreds of years, astronomers plotted the position of every visible star, while geographers mapped the features of our globe. Naturalists traveled to the ends of the Earth collecting every possible rock, shell, flower, and other curio for their museum collections. In our own century, discoveries of vast numbers of galaxies, disease-causing viruses, and a complex zoo of subatomic particles have transformed our understanding of the universe.

Describing new objects requires the ability to identify enough features that distinguish one object from all others. To a certain extent, these descriptions rely on words, which is why the vocabulary of science has become so complex. For example, you might describe a rock as rose-pink, fine-grained, silica-rich, and intrusive. However, eventually such a description has to incorporate numbers for added precision. What is the average size of the grains? How much silica is contained in the rock? What are the light-absorbing properties that give the rock a pink color?

Scalars and Vectors

All of the descriptions we've discussed so far can be expressed as a single number—you buy one gallon of paint, or you drive 10 miles to work. Any quantity that can be expressed as a single number is called a **scalar.** Scalars are crucial to the description of the physical world. As a consumer you are surrounded by scalars: the wattage of a light bulb, the octane rating of gasoline, the efficiency of appliances, and the voltage of your car's battery. You pay for coffee by the pound, fabric by the yard, milk by the gallon, and electricity by the kilowatt-hour. In science we measure the size of microbes, the mass of stars, the density of crystals, and the temperature of our bodies. We will even find that the colors of light may be represented as a scalar quantity (see Chapter 19).

However, sometimes you can't give a description in terms of a single number. If you were giving a friend instructions to your favorite restaurant, for example, you might say something like, "Go north 3 miles on Main Street." Here you have to give a scalar (3 miles) and some additional information (in this case, the direction north). Similarly, physicists describe the velocity of an object in terms of both its speed *and* direction. When a mathematical quantity, such as velocity, requires two numbers in its definition—both a magnitude and a direction—it is called a **vector.**

Vector Addition

Physicists often have to calculate the sum of two or more vectors to analyze real-world problems. The easiest situation to analyze occurs when two vectors lie along the same line. For example, a rower trying to move against a current will find his actual velocity (speed and direction) determined by the sum of two vectors. One of these vectors is the velocity the rower would achieve if he were rowing on still water; the other is the velocity of the current. The sum of the two vectors in this case is just the difference between the two velocities. If, for example, the rower could achieve 10 km/h in still water and the current is against him at 2 km/h, then his net velocity is $(10 - 2) = 8$ km/h.

Vectors and the Crash of the *Stardust*

On August 2, 1947, a British South American Airways flight, the *Stardust,* from Buenos Aires, Argentina, to Santiago, Chile, crashed into the high Andes Mountains, killing six passengers and three crew. They had encountered heavy cloud cover, but the experienced crew had made the routine flight many times before—fly at 18,000 feet for 3 hours, ascend to 26,000 feet to avoid the treacherous spine of the Andes, and then descend to Santiago near the Pacific Coast. But on this flight they flew straight into the mountains. What happened?

Fifty years after the mysterious accident, a team of scientists visited the crash site and discovered its cause: the ill-fated passengers and crew were victims of un-usually high winds. On that fateful day the *Stardust,* a Lancaster Mark III aircraft, flew west at 300 miles per hour—a speed that made it easy to calculate a flight path. But the crew didn't realize that they were flying straight into an unusually strong jet stream—a 100-mile-per-hour wind blowing to the east.

A plane's net speed (its ground speed) is the sum of two separate speeds. The first is the speed of the plane in still air, while the second is the wind speed. If the wind blows in the same direction as the plane's flight, it adds to the plane's speed in still air. But if the wind is opposite to the direction of the plane, it lessens the plane's speed. The *Stardust's* actual ground speed was the combination of two vectors: 300 miles per hour west *plus* 100 miles per hour east. The vector sum is only 200 miles per hour west—much slower than the crew thought. Thus the plane had not traveled as far west as the crew thought it had, and they descended too soon.

Today, meteorologists constantly monitor the shifting jet streams of Earth's upper atmosphere, and vector addition of wind velocity and plane velocity is a critical part of every flight plan.

Develop Your Intuition: Going with the Flow

What if the rower on page 26 is moving with the current instead of against it? In this case, the rower's velocity is in the same direction as the current. The current increases the rower's speed, so he moves at a speed of $(10 + 2) = 12$ km/h. When we add two vectors along the same line, the result is the sum of their magnitudes if they are in the same direction and the difference between their magnitudes if they are in the opposite direction.

A more complicated situation arises when the two vectors do not lie along the same line. In this case, the easiest way to add vectors is to use a graph (Figure 2-1). Each vector can be represented as an arrow on an *x-y* plot. For example, a plane flying west 300 miles per hour would appear as an arrow pointing to the left with its tail at the origin (0,0) and a length of 300. If there is a crosswind of 300 miles per hour, we would find the actual velocity of the airplane by connecting the tail of the vector representing the crosswind to the head of the vector representing the airplane's velocity in still air. The sum of the two vectors is a new arrow that extends from the origin point to the head of the second vector, as shown in Figure 2-1.

Vector addition is a lot like giving directions to a friend. "Go three blocks north on Main Street, turn right on Maple, and it's the fourth house on the left." Do you recognize the three vectors described by these instructions?

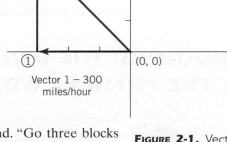

FIGURE 2-1. Vector graph of the velocity of a plane; addition of two vectors.

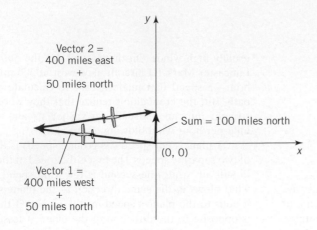

FIGURE 2-2. Vector addition for the path of a plane in a cross wind.

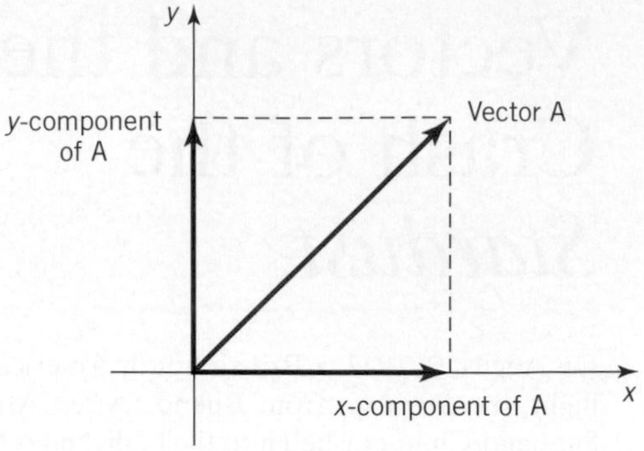

FIGURE 2-3. Any vector can be decomposed into two perpendicular vectors.

Adding Vectors

EXAMPLE 2-1

A plane flies due west at 400 miles per hour for 1 hour, and then flies back due east at the same speed for another hour. Meanwhile, a steady 50-mile-per-hour wind blows from the south. Where does the plane end up relative to its starting position?

SOLUTION: Intuitively, we know that a wind blowing from the south will push the plane toward the north. Since the plane's flight plan is straight east and west, with no intentional veering off to the north or south, we expect the plane will wind up somewhere north of its intended flight path.

To work out the solution exactly, we consider this as a problem of vector addition involving three vectors. During the plane's 2-hour flight, the west leg of 400 miles exactly cancels the east leg of 400 miles. But the steady 50-mile-per-hour southerly wind acts for 2 hours, shifting the plane's position 100 miles to the north. Therefore, as shown in Figure 2-2, the plane winds up 100 miles due north of its starting position. ●

Vector Decomposition Just as vectors can be added together, so too can one vector be broken down into two or more component vectors. In many physical situations, it's useful to decompose a vector into two parts at right angles to each other. Any vector on an *x-y* graph, for example, can be decomposed into one vector parallel to the *x* axis plus a second vector parallel to the *y* axis (Figure 2-3). This property of vectors will become especially useful when we analyze the momentum of a system (Chapter 6).

EQUATIONS: THE DYNAMICS OF THE PHYSICAL WORLD

If scientists just described objects in the universe, science would seem pretty boring. What makes science fascinating and useful is that systems change. Science is a search to understand and predict these changes—the dynamics of our physical world.

Change can be described in words, in tables of numbers, or visually through the use of graphs. Of special importance are **equations,** which define a precise mathematical relationship among two or more measurements. Let's look at an example to see how the same physical behavior can be described in these different ways.

As you will see in Chapter 11, a bar of iron expands when heated. A researcher might carefully measure the length of a 1-meter iron bar at a series of temperatures and prepare a table like Table 2-1. We see a systematic trend in these data; both temperature and length increase. These data can be described in several ways.

1. We can describe what happens to the iron bar in words:

When we heat a 1-meter iron bar, it gets longer.

2. We can express this idea as an equation with words:

The length of the bar equals the original 1-meter length plus a constant times the change in temperature.

3. We can express this idea as an equation in symbols and numbers (approximately):

$$L = 1 + (0.00006 \times \Delta T)$$

where L is the length and ΔT is the change in temperature. (Note that Δ, the capital Greek letter delta, is often used in physics to denote a change in some quantity. It is not used by itself, but always with the symbol for that changing quantity. So ΔT denotes a change in temperature, ΔL is a change in length, and so on. Note also that the number 0.00006 comes from dividing the increase in length given in Table 2-1 by the corresponding change in temperature; this is the change in length per 1-degree change in temperature.)

4. Finally, the data might be displayed in graphical form, as a plot of temperature versus length (Figure 2-4).

Similar relationships are found in all scientific literature. Researchers graphically document changes in the volume of a gas with pressure, changes in the distance objects fall with time, changes in the growth rate of bacteria with concentration of nutrients, and countless other trends.

The example of the expanding iron bar illustrates why scientists use the language of mathematics. Table 2-1 certainly represents the data, though in a modern experiment that list of numbers could easily run to thousands or even millions

Blacksmith with a hot iron bar.

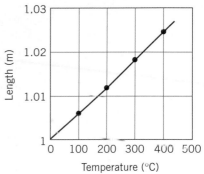

FIGURE 2-4. A graph of temperature versus the length of an iron bar illustrates how two scalar properties are related.

TABLE 2-1 Thermal Expansion of an Iron Bar	
Temperature (°C)	**Length (meters)**
0	1.0000
100	1.0060
200	1.0125
300	1.0183
400	1.0240

of entries. All of that information can be packaged into a one-line equation. Thus, the use of mathematics allows us to express the results of experiments in a highly compressed and convenient form.

As we shall see later, equations have the added advantage of providing us with the best way to make predictions about the behavior of our surroundings. In addition, they transcend national barriers in that they have exactly the same meaning all over the world.

Develop Your Intuition: Fuel Efficiency

How would you describe the gas efficiency of your automobile? A colloquial answer might be, "I get pretty good mileage, especially on interstate highways." Most people would accept that answer, but it wouldn't be very useful in trying to compare two different cars.

To give a more accurate answer, you could keep exact records of your car's mileage and the amount of gas purchased each time you fill up the tank. By dividing the total miles driven by the number of gallons purchased, you could calculate:

$$\text{Miles per gallon} = \frac{\text{Total miles driven}}{\text{Gallons of gas purchased}}$$

Then, you could reply with a scalar quantity, "I get about 30 miles per gallon."

Your answer could be even more precise if you record additional notes. What was the brand and grade of fuel? Was the driving between each fill-up in the city or on high-speed roads? Did you use the air conditioner? Did you recently have an oil change? Were the tires properly inflated? What were the weather and road conditions? With sufficiently detailed records you might be able to say, "My car, when properly serviced and fueled with regular unleaded gas, averages 33.7 miles per gallon when traveling 55 miles per hour on level, dry interstate highways, and approximately 25.5 miles per gallon in city traffic. The use of the air conditioner reduces these values by about 2.5 miles per gallon."

Automobile manufacturers, who must document the fuel efficiency of their vehicles, carry this process a step further by running carefully controlled mileage experiments on dozens of cars in special laboratories. There, engineers develop graphs and equations that relate fuel consumption to numerous other variables. Many of these tests are now mandated by the Environmental Protection Agency to provide consumers with an accurate measure of each brand's fuel efficiency.

Modeling the World

Scientists have devised many ways to describe the natural world. As shown in the previous example of the expanding iron bar, the behavior of a physical system may be documented in words, tables of numbers, or graphs. But no description is more compact and efficient than an equation. A brief survey will help you to visualize the everyday reality underlying four common types of equations used in this book: direct, inverse, power law, and inverse square. These equations may be used to describe all manner of natural phenomena.

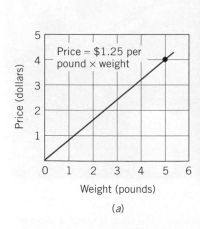

(a)

(b)

Figure 2-5. (a) A graph of a direct relationship between price of fruit and its weight. (b) The price per pound can be treated as a constant of proportionality between weight and cost.

1. **Direct Relationships** The simplest equations consist of a *direct relationship* between two variables, *A* and *B*, in the form:

$$A = k \times B$$

where *k* is called a "constant of proportionality." You use a direct relationship every time you buy gas by the gallon or food by the pound:

$$\text{Cost} = \text{Price per pound} \times \text{Weight}$$

In this case, two variables, the weight and the cost, are related by a constant of proportionality called the price per pound.

   In a direct relationship the two variables change together: if weight doubles, so must the cost; if weight triples, so does the cost. We say that the cost is *proportional* to the weight. The graph of such a relationship is a straight line (Figure 2-5). In subsequent chapters we will find many direct relationships between pairs of variables, including:

Acceleration is proportional to force (Chapter 4).

Electric power is proportional to electric current (Chapter 18).

Wave frequency is proportional to wave velocity (Chapter 14).

2. **Inverse Relationships** In many everyday situations, one variable increases as another decreases, a situation called an *inverse relationship*:

$$A = \frac{k}{B}$$

where *k* is a constant. For example, consider an assembly line at which workers who work at different speeds produce automobile parts. The shorter the time, *t*, it takes for a worker to produce one part, the greater the number, *N*,

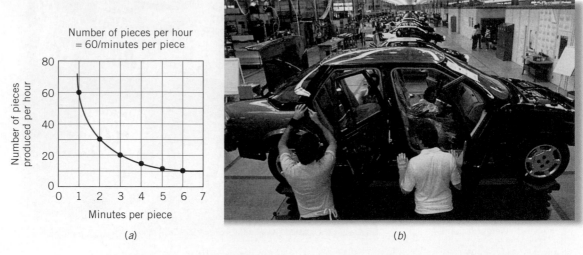

FIGURE 2-6. (a) A graph of an inverse relationship between the time it takes a worker to assemble an item and the worker's hourly output. The shorter the assembly time per item, the greater the hourly output. (b) An auto assembly line.

of cars produced per hour (Figure 2-6). We say that the number of cars produced in a given time period is *inversely proportional* to the production time:

$$\text{Output of cars} = \frac{\text{Constant}}{\text{Production time per car}}$$

or

$$N = \frac{k}{t}$$

Think about the behavior of the two variables, production time and output in this case. If you make an automobile part in half the time of a fellow worker, you will produce twice as many parts in any given time period. If you produce a part in a third of the time, you'll produce three times as many, and so forth. Inverse relationships thus lead to the distinctive kind of curving graph illustrated in Figure 2-6. We will encounter many examples of such inverse relationships, such as:

For a given force, acceleration is inversely proportional to the mass being accelerated (Chapter 4).

The wavelength of light is inversely proportional to the frequency of light (Chapter 19).

3. **Power Law Relationships** You may recall from a geometry class the equation that defines the area of a square, *A,* in terms of the edge length, *L:*

$$A = L \times L$$

which can be rewritten as:

$$A = L^2$$

Area is said to be equal to length *squared* (Figure 2-7a and b). Squared relationships are common in our daily lives. For example, in Chapter 3 we see that a dropped object falls a distance that is proportional to the time of fall squared.

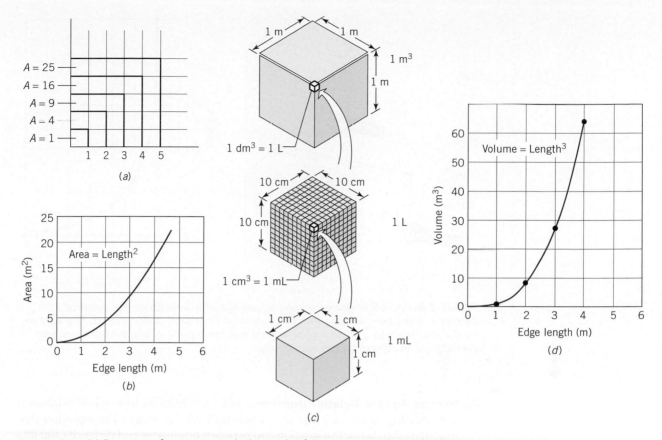

FIGURE 2-7. (a) Diagram of a square grid; (b) graph of the squared relationship between the area and edge length of a square. (c) Diagram of stacked cubes; (d) graph of the cubed relationship of volume. An object 10 cm on a side holds $10 \times 10 \times 10 = 1000 \ cm^3$ or 1 liter, while an object 1 meter on a side holds 1000 liters.

A similar kind of relationship is found between the *volume* of a cube, *V*, and its edge length, *L*:

$$V = L \times L \times L$$

or, in mathematical notation:

$$V = L^3$$

Note that volume, which is the amount of space an object occupies, is measured in units of distance *cubed*, such as cubic meters (Figure 2-7c and *d*). Many systems of measurement adopt special volume units, such as liters or gallons.

Squared-type and cubed-type equations are special cases of a more general class, called "power law equations." These common equations have the form:

$$A = k \times B^n$$

where *n* is any number and *k* is a constant. A general feature of these relationships is that one variable changes much more quickly than another. Double the edge of a cube, for example, and the volume increases eightfold; triple the edge length and the volume becomes 27 times larger. The result is a steeply rising graph, as shown in the examples in Figure 2-7.

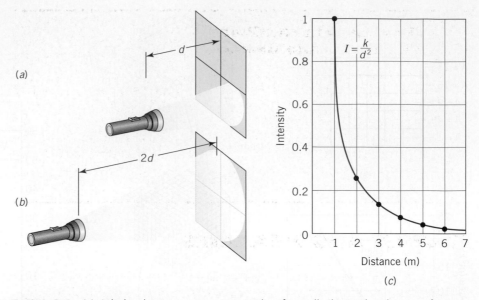

(a)

(b)

$$I = \frac{k}{d^2}$$

(c)

FIGURE 2-8. (a) A light shines on one square tile of a wall. (b) As the distance from light to wall doubles, an area four times as large (four tiles) is illuminated. The light intensity decreases to $\frac{1}{4}$, in an inverse square relationship. (c) A graph of distance versus intensity for an inverse square relationship.

4. **Inverse Square Relationship** Have you ever noticed how much brighter a car's headlights are up close than when they are far away? The equation that describes this distinctive relationship between brightness and distance (and lots of other natural phenomena, as well) is called the *inverse square relationship*.

Imagine shining a flashlight on a wall with a regular array of square tiles (Figure 2-8a). First, you hold the flashlight close to the wall so that all the light falls onto just one square. If you measured that brightness with a photographic light meter, it might register 100 on the meter's scale.

If you move the flashlight twice that distance from the wall (Figure 2-8b), then the light illuminates an area twice as high and twice as wide. The light is spread out over an area four times as large as before. Consequently, the light meter reads a brightness of about 25—only $\frac{1}{4}$ as strong as before. This inverse square relationship between light intensity, *I*, and distance, *d,* is

$$I = \frac{k}{d^2}$$

where *k* is a constant. The general inverse square relationship is

$$A = \frac{k}{B^2}$$

A graph of an inverse square relationship (Figure 2-8c) reveals the sharp fall-off of one variable as the other changes gradually. We find in later chapters that inverse square relationships are common in nature; for example, inverse square relationships describe the change of magnitude of everyday electrostatic and gravitational forces as a function of the distance between two objects (see Chapters 5 and 16).

Develop Your Intuition:
Flashbulbs and Baseball

You'll often see hundreds of flashbulbs go off at a stadium when a famous baseball player comes up to bat during a night game. Based on what you know about the inverse square relationship, do these flashes help the would-be photographers?

The intensity of light drops off as the inverse square of the distance, so flashbulbs are ineffective at distances greater than a few dozen feet. All the popping flashbulbs make a great sight, but they don't help photographs taken over the great distances of a stadium.

UNITS AND MEASURES

As soon as we begin using numbers to describe any physical system, we have to deal with the issue of units. Walk into any hardware store in the United States and you will notice immediately that the things for sale are measured in many different ways. You can buy paint by the gallon, insulation in terms of how many BTU will leak through it, and grass seed by the pound. In some cases, the units are strange indeed—nails, for example, are measured in an archaic unit called the penny (abbreviated "d" for denarius, a small Roman coin). A 16d nail is a fairly substantial thing, perfect for holding the framework of a house together, while a 6d nail might find use tacking up a wall shelf.

No matter what the material, there is a unit to measure how much is being sold. In the same way, in all areas of science, systems of units have been developed to measure how much of a given quantity there is. We will encounter many of these units in this book—the newton as a measure of force, for example, and the degree as a measure of temperature. Every quantity used in the sciences has an appropriate unit or combination of units associated with it.

We customarily use certain kinds of units together, in what is called a **system of units.** In a given system, units are assigned to fundamental quantities such as mass (or weight), length, time, and temperature. Someone using that system uses only those units and ignores the units associated with other systems.

The International System

In the United States, two systems of units are in common use. The one encountered most often in daily life is the *English system*. This traditional system of units has roots that go back to the Middle Ages. The basic unit of length is the *foot* (which was actually defined in terms of the average length of men's shoes outside a certain church on a certain day), and the basic unit of weight is the *pound*.

Throughout this book, and throughout most of the world outside of the United States, the metric system or, more correctly, the **International System** (abbreviated **SI** for the French Système International) is preferred. In this system, the unit of length is the *meter* and the unit of mass is the *kilogram*. In both the SI and the English system, the basic unit of time is the *second*.

In all probability, the unit from the metric system with which you are most familiar is the *liter,* a measure of volume. A liter is the volume enclosed by a cube 10 centimeters on a side or 1000 cubes 1 centimeter on a side. Soft drinks and other liquids are routinely sold in 1- and 2-liter bottles in the United States. The cubic meter—the volume contained in a cube 1 meter on a side—is also often used as a volume measure in the metric system. The measure of volume in the English system is the cubic foot, but liquids are commonly measured in gallons (3.79 liters), quarts ($\frac{1}{4}$ gallon), and pints ($\frac{1}{8}$ gallon).

Within the SI, units are based on multiples of 10. Thus, the centimeter is one-hundredth the length of a meter, the millimeter is one-thousandth of a meter, and so on. In the same way, a kilometer is 1000 meters, a kilogram is 1000 grams, and so on. This systematic organization differs from the English system, in which 12 inches equals 1 foot, 3 feet makes 1 yard, and 1760 yards makes 1 mile. A list of metric prefixes follows.

METRIC PREFIXES

If the prefix is:	Multiply the basic unit by:	Example with abbreviation
giga-	billion (thousand million)	gigameter, Gm
mega-	million	megagram, Mg
kilo-	thousand	kilometer, km
hecto-	hundred	hectogram, hg
deka-	ten	dekameter, dam

If the prefix is:	Divide the basic unit by:	Example with abbreviation
deci-	ten	decigram, dg
centi-	hundred	centimeter, cm
milli-	thousand	milligram, mg
micro-	million	micrometer, μm
nano-	billion	nanogram, ng

Physics in the Making
A Brief History of Units

Ever since humans started engaging in commerce, there has been a need for agreements on weights and measures. Merchants needed to be assured that they were buying and selling the same quantity of goods, that the buyer was getting what he paid for and the seller was receiving full value for her wares. This meant that someone (often a government) had to set up and maintain a system of standard weights and lengths.

The oldest weight standard we know about is the Babylonian "mina," which weighed between 1 and 2 pounds. Archaeologists have found standard stone weights carved in the shapes of ducks (5 mina) and swans (10 mina). In medieval Europe, almost every town maintained its own system of weights and measures, and the only institutions pushing for universal standards were the great trade fairs. The keeper of the fair in Champagne, France, for example, kept an iron bar against which all bolts of cloth sold at the fair had to be measured. The Magna Carta, signed by King John of England in 1215 and generally reckoned to be one of the key documents in the history of democracy, required that "There shall be standard measures of wine, ale, and corn throughout the kingdom." The English system of units eventually evolved from the welter of medieval systems.

The metric system, on the other hand, was a product of the French Revolution at the end of the eighteenth century. In 1799, the French Academy recommended that the length standard be the meter, then defined to be 1/10,000,000 of the distance between the equator and the North Pole at the longitude of Paris and that the gram be defined to be the mass of a cubic centimeter of water at 4°C. In the Connection section in this chapter, we discuss the modern definitions of these quantities. ●

King John grants the Magna Carta (Great Charter) to his barons in England. The charter required the establishment of common standards for weights and lengths throughout the kingdom.

Conversion Factors

All systems of units help us describe physical objects and events. Confusion may arise, however, when switching back and forth between two different systems. **Conversion factors,** which are used to shift from one system of units to another, are thus vital in both science and commerce. If you have ever visited a foreign country, you have had direct experience with this process; you had to use conversion factors all the time when converting dollars to some other currency.

Hundreds of conversion factors apply to the physical world. One person may give temperature in degrees Fahrenheit, another in degrees Celsius. Distance may be recorded in centimeters or in inches. Some important conversion factors are tabulated in Appendix A.

Driving in North America

Suppose you are driving in Canada. The odometer on your car reads 20,580 miles. You see a sign that reads, "Toronto 87 kilometers." What will your odometer read when you get to that city?

EXAMPLE
2-2

REASONING: Since the odometer reads in miles, the first thing to do is convert 87 kilometers to miles by using the conversion tables in Appendix A. We then add that mileage to the current reading to get our answer.

SOLUTION: From Appendix A, the conversion factor from kilometers to miles is 0.6214. When you see the sign, then, the distance to Toronto is:

$$87 \text{ kilometers} \times 0.6214 \text{ mile/kilometer} = 54 \text{ miles}$$

When you have traveled this far the odometer will read:

$$20,580 + 54 = 20,634 \text{ miles} \bullet$$

LOOKING DEEPER

Powers of Ten

Very large or very small numbers may be written conveniently in a compact way—a way that doesn't involve writing down a lot of zeroes. The system called "powers of ten" notation (also called "exponential notation") accomplishes this goal. The basic rules for the notation are:

1. Every number is written as a number between 1 and 10 followed by 10 raised to a power, or an exponent.
2. If the power of 10 is positive, it means "move the decimal point this many places to the right."
3. If the power of 10 is negative, it means "move the decimal point this many places to the left."

Following these rules, 3.56×10^3 is equivalent to 3560, and 7.87×10^{-4} equals 0.000787.

Multiplying or dividing numbers in powers of ten notation requires special care. If you are multiplying two numbers, such as 2.5×10^3 and 4.3×10^5, you multiply 2.5 and 4.3, but you add the two exponents:

$$(2.5 \times 10^3) \times (4.3 \times 10^5) = (2.5 \times 4.3) \times 10^{3+5}$$
$$= 10.75 \times 10^8$$
$$= 1.075 \times 10^9$$

When dividing two numbers, such as 4.3×10^5 divided by 2.5×10^3, you divide 4.3 by 2.5, but you subtract the denominator exponent from the numerator exponent:

$$(4.3 \times 10^5)/(2.5 \times 10^3) = (4.3/2.5) \times 10^{5-3}$$
$$= 1.72 \times 10^2$$
$$= 172$$

For more examples of powers of ten notation, see Looking at Length (p. 38).

Looking at Length

You are about 100 million times larger than a virus, but the Earth is about 10 million times larger than you. That sounds pretty big, but the Sun's diameter is about 100 times larger than that of the Earth; the Earth is only the size of a sunspot. And the Sun is only a tiny dot in the Milky Way galaxy, which contains a hundred billion stars just like it.

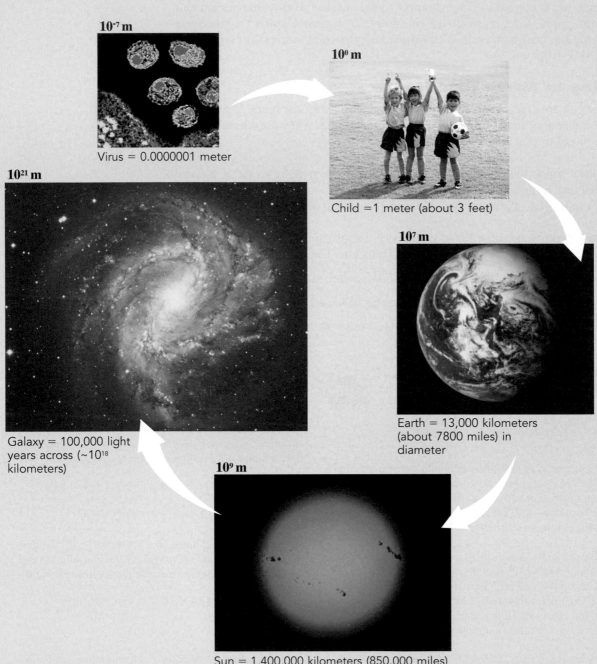

10^{-7} m

Virus = 0.0000001 meter

10^{0} m

Child =1 meter (about 3 feet)

10^{21} m

Galaxy = 100,000 light years across (~10^{18} kilometers)

10^{7} m

Earth = 13,000 kilometers (about 7800 miles) in diameter

10^{9} m

Sun = 1,400,000 kilometers (850,000 miles) in diameter

Connection

Maintaining Standards

Systems of units are one place where governments become intimately involved with science, since the maintenance of standards has traditionally been the task of governments. When you buy a pound of meat in a supermarket, for example, you know that you are getting full weight for your money because the scale is certified by a state agency, which relies, ultimately, on international standards of weight maintained by a treaty among all nations.

Originally, the standards were kept in sealed vaults at the International Bureau of Weights and Measures near Paris, with secondary copies kept at places such as the National Institutes of Standards and Technology (formerly National Bureau of Standards) in the United States. For instance, the meter was defined as the distance between two marks on a particular bar of metal, and the kilogram was defined as the mass of a particular block of iridium-platinum alloy. The second was defined as a certain fraction of the length of the year.

Today, however, only the kilogram is still defined in this way. Since 1967, the second has been defined as the time it takes for 9,192,631,770 crests of a light wave (see Chapter 19) of a certain type of light emitted by a cesium atom to pass by a given point. In 1960, the meter was defined as the length of 1,650,763.73 wavelengths of the radiation from a krypton atom, and in 1983 it was redefined to be the distance light travels in 1/299,792,458 seconds. In both these cases, the old standards have been replaced by numbers relating to atoms—standards that any reasonably equipped laboratory can maintain for itself. Atomic standards have the additional advantage of being truly universal—every cesium atom in the universe is equivalent to any other. Only mass is still defined in the old way, in relation to a specific block of material kept in a vault, and scientists are working hard to replace that standard by one based on the mass of individual atoms. ●

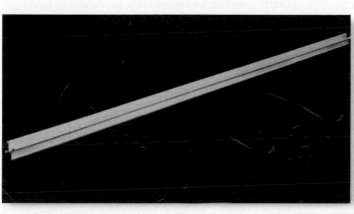

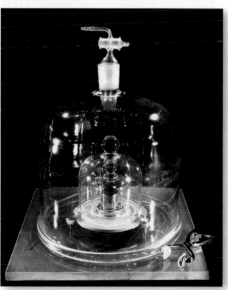

(a) The old standard meter. (b) The standard kilogram. The meter, which used to be defined in terms of the distance between marks on a bar like this, is now defined in terms of measurements on atoms.

Units You Use in Your Life

The metric and English systems each give a comprehensive set of units that could, in principle, be used to measure everything we encounter in our lives. In point of fact, for various historical and technical reasons, we often use units that don't fit easily into either system all the time. How many of the following units do you recognize?

acre—used to measure land area in the United States (43,560 square feet, or $\frac{1}{640}$th of a square mile)

barrel—international unit for oil production (42 gallons; although many different specialized definitions of barrel exist for other commodities, including wine, spirits, and cranberries)

bushel—used to measure production of grains in the United States (1.24 cubic feet)

caliber—used to measure diameter of bullets and gun barrels (0.01 inches)

carat—used to measure size of gemstones (0.2 grams)

fathom—used to measure depth of navigable water (6 feet)

knot—used to measure speed of ships (1.85 kilometers per hour)

ounce—used to measure the weight of produce ($\frac{1}{16}$ pound)

Troy ounce—used to measure precious metals ($\frac{1}{12}$ pound)

THINKING MORE ABOUT

Units: Conversion to Metric Units

Why does the United States still use English units long after most of the rest of the world has converted to SI units? It may have to do with nonscientific factors such as the geographical isolation of the country, the size of our economy (the world's largest), and, perhaps most important, the expense of making the conversion. (For example, think of the cost to change all the road signs from miles to kilometers on the entire interstate highway system.)

To understand the debate over conversion, you have to realize one important point about units. There is no such thing as a "right" or "scientific" system of units. Units can only be convenient or inconvenient. Thus, U.S. manufacturers who sell significant quantities of goods in foreign markets long ago converted to metric standards to make those sales easier. Builders, on the other hand, whose market is largely restricted to the United States, have not.

By the same token, very few scientists actually use the SI exclusively in their work. Almost every discipline, including physics, chemistry, geology, biology, and astronomy, has its own preferred non-SI units for some measurements. Astronomers, for example, often measure distance in light years or in parsecs; geologists usually measure pressure in kilobars; and many physicists prefer to record energy in electron volts. In the United States, engineers use English units almost exclusively—indeed, when the federal government was considering a tax on energy use in 1993, it was referred to as a BTU tax (the BTU, or British thermal unit, is the unit for energy in the English system). Medical professionals use the cgs system, in which the unit of length is the centimeter, the unit of mass is the gram, and the unit of time is the second. Next time you have blood drawn, take a look at the syringe. It will be calibrated in cubic centimeters (cc).

Sometimes the use of different systems of units on the same scientific or engineering project can lead to trouble. A notable example occurred in fall 1999 when a spacecraft costing $125 million crashed into the planet Mars instead of orbiting it as planned. It turned out that the company that built the rocket for slowing down the spacecraft as it got close to Mars reported its thrust as pounds per square foot, but the flight engineers assumed the number was in metric units. The difference was enough to throw the spacecraft off course by a few hundred miles, leading to the crash.

Given the wide range of units actually in use, how much emphasis should the U.S. government give to metric conversion? How much should the government be willing to spend on the conversion process—how many new signs as opposed to how many repaired potholes on the roads?

Summary

Language allows people to communicate information and ideas. In their efforts to describe the physical world with accuracy and efficiency, scientists have created many new words to distinguish the many different kinds of objects in the universe.

These descriptive terms are amplified by mathematics, which allows scientists to quantify their observations. **Scalars** are numbers that indicate a quantity—mass, length, temperature, and time are familiar scalar quantities. Other quantities, such as velocity and change in position, must be described with a **vector,** which combines information on both magnitude (a scalar) and direction.

Many scientists work to find mathematical relationships between two or more properties—temperature and the length of a metal bar, for example. Such relationships may be presented in the form of an **equation** or a graph.

Scientific measurements rely on a **system of units,** particularly the metric system or **International System (SI). Conversion factors** are used to change from one unit to another.

Key Terms

conversion factor Established mathematical quantity used to shift from one system of units to another. (p. 37)

equation The definition of a precise mathematical relationship among two or more measurements. (p. 29)

International System or SI (Système International) An internally consistent system of units within the metric system; also known as the metric system. (p. 35)

scalar Any quantity that can be expressed as a single number and without a direction. (p. 26)

system of units Units assigned to fundamental quantities such as mass (or weight), length, time, and temperature. (p. 35)

vector A quantity that requires two numbers in its definition—a magnitude and a direction. (p. 26)

Review

1. Why do physicists and other scientists require a specialized vocabulary?

2. Why is scientific vocabulary still growing?

3. What is a scalar quantity? Give an everyday example.

4. What is a vector quantity? Give an everyday example.

5. What is the role of equations in science?

6. What is a direct relationship? Give an example.

7. What is an inverse relationship? Give an example.

8. What is a power law relationship? Give an example.

9. What is an inverse square relationship? Give an example.

10. Describe the ways a scientist might present quantitative data.

11. Why do we need standards of units and measures?

12. What is a system of units? What system do most scientists use?

13. Discuss the relative advantages of the English system and SI of units and measurements.

14. What are the units of length, mass, and volume in the English system of units?

15. What are the units of length, mass, and volume in the metric system of units?

16. What is a conversion factor?

17. Why do scientists often use powers of ten notation?

18. Identify three units of measurement that you might encounter at a grocery store.

19. Which branch of the U.S. government regulates standard weights and measures?

Questions

1. Categorize the following terms as either a scalar or vector quantity. Explain your reasoning for each choice.

  a. a 100-yard dash
  b. 20,000 leagues under the sea
  c. a $75.00 dress
  d. 4 days
  e. 100 degrees Celsius
  f. 1996 A.D.
  g. 1 mile northeast
  h. the lower 40 acres
  i. 12:45 P.M.
  j. a cubic yard of cement
  k. a force of 20 newtons
  l. 40 m/s west
  m. 20 kg
  n. 30 revolutions per minute

2. What is the principal difference between a vector and a scalar quantity, and why is that important in science?

The following graphs are for Questions 3–5.

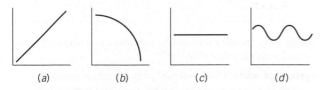

(a)   (b)   (c)   (d)

3. You have entered a pie-eating contest to eat as many pieces of pie in the shortest amount of time until you can't eat another bite. Your physics instructor watches you during the contest and makes a graph of the number of pieces of pie eaten per minute (vertical axis) versus the time elapsed (horizontal axis). Which of the graphs most likely represents your instructor's data? Express this trend in words.

4. To maximize your chances of winning the lottery, you have decided to buy multiple lottery tickets. Which graph best represents the chance to win the lottery (vertical axis) versus the number of tickets bought (horizontal axis)? Explain this trend in words.

5. The water in the oceans rises and falls with the rotation of the Earth. These cycles are called tides. Which graph best represents the height of the ocean (vertical axis) versus the time of day (horizontal axis)? Explain the trend in words.

6. As discussed in the chapter, scientists at places such as the National Institutes of Standards and Technology near Washington, D.C., take great pains to protect the standard kilogram, encasing it in a vault filled with nitrogen. Why do you suppose this has to be done, while no one seems to want to do the same for the meter and the second?

7. A woman can paddle her kayak at a speed of 3 kilometers per hour through still water. She is paddling upstream in a river that has a flow speed of 2 kilometers per hour. Draw an arrow that represents the velocity of the kayak through the water and another arrow that represents the velocity of the water. Draw a third arrow that represents the actual velocity of the kayak.

8. Suppose a jet airplane travels at 500 miles per hour in still air. On a recent flight the plane was pointed east and was encountering a 100-mile-per-hour side wind, blowing toward the north. Draw an arrow that represents the velocity of the plane without the wind. Draw another arrow that represents the velocity of the wind. Draw a third arrow that represents the velocity of the plane in the presence of the side wind.

9. Decompose the following vectors into a horizontal component and a vertical component. (Draw a horizontal vector and a vertical vector whose sum equals the vector shown.)

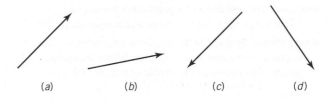

(a)   (b)   (c)   (d)

10. As a person grows taller, he or she usually gets heavier, too. Is the relationship between the weight of a person and a person's height a *direct relationship*? Why or why not? Give examples to support your conclusion.

11. Consider the relationship between a person's body fat percentage and their top running speed. Is this more likely to be a *direct relationship* or an *inverse relationship*? Explain.

12. You have been hired to paint circles of various diameters as a decorative feature on a new building. Consider the relationship between the amount of paint needed to paint a circle and the diameter of that circle. Is this a *direct, inverse,* or *power law relationship*?

13. Assuming the price of musical CDs has remained fairly constant over the past two years, characterize the relationship between the amount of money a person has spent on CDs and the number of CDs they've bought in the last two years. Is it a *direct, inverse,* or *power law* relationship?

14. Would you rather be given 10 kilo-cents or 1,000,000 nano-dollars? Explain.

15. About how many seconds are there in a year: 30 mega-seconds, 30 kilo-seconds, or 30 milli-seconds?

Problem-Solving Examples

Vectors on the Water

EXAMPLE 2-3

An inexperienced canoeist sets out straight across a 1-mile-wide river, paddling at 5 miles per hour. The average current of the river is 6 miles per hour. Where does she land on the opposite side of the river?

SOLUTION: This problem involves two vectors. The first vector is 1 mile long in a direction perpendicular to the flow of the river (Figure 2-9). The second vector, in a direction parallel to the river's flow, is due to the 6-mile-per-hour current acting on the canoe while it's in the water. In order to calculate the length of this second vector, we have to determine how long the canoe is in the water:

$$\text{Time} = \frac{\text{Distance}}{\text{Velocity}}$$

$$= \frac{1 \text{ mile}}{5 \text{ miles/hour}}$$

$$= 0.2 \text{ hour}$$

So the length of the second vector is:

$$\text{Distance} = \text{Velocity} \times \text{Time}$$
$$= 6 \text{ miles/hour} \times 0.2 \text{ hour}$$
$$- 1.2 \text{ miles}$$

The canoeist winds up more than a mile downstream from where she left. ●

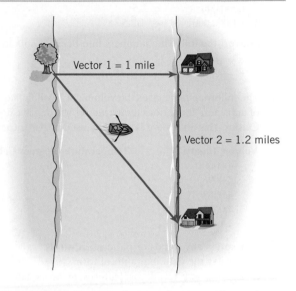

FIGURE 2-9. A person in a rowboat crossing a river is also carried downstream by the current.

Running the Dash

EXAMPLE 2-4

American athletes used to run an event called the 100-yard dash. If an athlete could run the 100-yard dash in 10 seconds, what time would you expect her to have in the 100-meter dash?

REASONING AND SOLUTION: The first step is to use the conversion factor in Appendix A to convert 100 meters to a distance in feet. From Appendix A, the conversion factor for meters to feet is 3.281. Consequently, 100 meters is:

$$100 \text{ meters} \times 3.281 \text{ feet/meter} = 328.1 \text{ feet}$$

We then have to divide the number of feet by 3 to convert to yards:

$$\frac{328.1 \text{ feet}}{3 \text{ feet/yard}} = 109.4 \text{ yards}$$

If we assume that the runner travels at the same speed in the two races, the 100-meter race (equal to a 109.4-yard race) will take longer than the 100-yard race. The time of the 100-meter race is proportional to the distances of the two races multiplied by 10 seconds:

$$\text{Time} = 10 \text{ seconds} \times \frac{109.4 \text{ yards}}{100 \text{ yards}}$$

$$= 10.94 \text{ seconds}$$

(The women's world record for the 100-meter dash is 10.49 seconds, set in 1988 by American Florence Griffith Joyner. At the same speed, would she have run 100 yards in more or less than 10.0 seconds?) ●

Problems

1. In Canada a speed limit sign says 70 kilometers per hour. What is the legal speed in miles per hour?

2. How many liters are in a half-gallon container of milk?

3. A runner consistently completes a 1-mile race in 4 minutes. What is his expected time in a 1500-meter race?

4. Write the following numbers in powers of ten notation:
   a. 1,000,000
   b. 1/1,000,000
   c. 2.5
   d. 1/2.5

5. Convert the following numbers into decimal notation.
   a. 7×10^4
   b. 7×10^{-4}
   c. 6.41×10^6
   d. 6.41×10^{-6}

6. The following table gives the volume of 1 gallon (4 quarts) of antifreeze for various temperatures starting at 6 degrees Celsius. This antifreeze is typically used in American cars.

Volume (quarts)	Temperature (degrees Celsius)
4.0000	6
4.0096	12
4.0288	24
4.0672	48
4.1440	96

   a. Express any trends or patterns in words.
   b. Display the data in graphical form.
   c. Express any trends or patterns in an equation with words.
   d. Express any trends or patterns in an equation with symbols.

7. Veronica was doing a laboratory assignment in which she was investigating the time of descent, final speed, acceleration, and incline angle of a marble rolling down an inclined track and how each was related to the other. Her data are given next.

Incline Angle (degrees)	Time of Descent (s)	Final Speed (m/s)	Acceleration [(m/s)/s]
5	3.75	3.2	0.85
10	2.65	4.5	1.70
15	2.17	5.5	2.54
20	1.89	6.3	3.40
30	1.56	7.6	5.08

   A. Consider only the incline angle and the acceleration.
      a. Express any trends or patterns in words.
      b. Display the data in graphical form.
      c. Express any trends or patterns in an equation with words.

   B. Consider only the incline angle and the time of descent.
      a. Express any trends or patterns in words.
      b. Display the data in graphical form.

   c. Express any trends or patterns in an equation with words.

   C. Consider only the incline angle and the final speed.
      a. Express any trends or patterns in words.
      b. Display the data in graphical form.
      c. Express any trends or patterns in an equation with words.

8. An industrious student decided that she wanted to prove certain laws about gases and the relationships among pressure, volume, and temperature. In Sarah's science laboratory, she collected the following data.

Temperature (kelvins)	Volume (liters)	Pressure (atmospheres)
100	1000	1.0
100	500	2.0
100	250	4.0
100	125	8.0
200	2000	1.0
200	1000	2.0
200	500	4.0
300	750	4.0
600	1500	4.0

   a. Show by using a graph, an equation, or a written statement that the volume is directly proportional to the temperature if the pressure is held constant.
   b. Use these data to show Boyle's law, which states that at a constant temperature the pressure and volume vary inversely.

9. The brightness of a lightbulb can be measured by a light meter in a unit named lumens. Jeremy decided to investigate how the brightness of a certain lightbulb changes with the distance from the lightbulb. Jeremy recorded the following data.

Distance from bulb (feet)	Brightness (lumens)
1	1600
2	400
3	178
4	100
5	64
10	16
20	4

   a. Express any trends or patterns in words.
   b. Display the data in graphical form.
   c. Express any trends or patterns in an equation with words.
   d. Express any trends or patterns in an equation with symbols.

10. Ron used 6980 kW-h (kilowatt-hours) of electricity last year and Jennifer used 5235 kW-h in only 10 months.

   a. Who used, on the average, the most electricity per month?

   b. What, on the average, were Ron and Jennifer's daily uses of electricity? (Assume a 30-day month.)

   c. If Ron paid 8 cents per kW-h and Jennifer paid 10 cents per kW-h, who paid more for electricity last year?

11. P.J. maintains the following exercise schedule. Every Sunday she runs for 30 minutes using 12 calories per minute. On Monday, Wednesday, and Friday, P.J. takes brisk walks for 1 hour each day (4 calories per minute), and every Tuesday and Thursday P.J. plays volleyball for 1.5 hours (5.5 calories per minute). On Saturday she does not exercise.

   a. Calculate the total amount of calories P.J. expends during the week.

   b. On average, how many calories per day (include Saturday) does P.J. expend?

12. The G-7 countries (the United States, Canada, Japan, Germany, the United Kingdom, France, and Italy) are the leading industrialized countries in the world. Their populations, total energy use, total wilderness areas, and total solid waste disposed for 1991 are given in the following table.

Country	Population (millions)	Annual energy use (10^9 BTU)	Wilderness area (square miles)	Annual solid waste (10^6 tons)
United States	249.2	76,355	379,698	230.1
Canada	26.5	10,309	2,473,308	18.1
Japan	123.5	15,707	9,276	53.2
Germany	77.5	13,881	19,128	21.0
United Kingdom	57.2	8,575	17,912	22.0
France	56.1	8,355	18,454	30.2
Italy	57.1	6,579	5,021	19.1

   a. Which country uses the most energy per person? The least energy per person?

   b. Which country has the most wilderness area per person? The least wilderness area per person?

   c. Which country disposes of the most solid waste per person? The least solid waste per person?

   d. If you combined the four European countries into one country, would your answers to a, b, and c change?

13. Express the following quantities in powers of ten notation.

   a. 150 gigadollars

   b. 43 hectofeet

   c. 23 micrometers

   d. 92 nanoseconds

   e. 74 milligrams

   f. 617 kilobucks

   g. 43 microbreweries

14. Multiply the following.

   a. $(4.3 \times 10^6) \times (7.4 \times 10^{-7})$

   b. $(1.2 \times 10^{-8}) \times (3.4 \times 10^{-5})$

   c. $(5.5 \times 10^3) \times (6.7 \times 10^7)$

   d. $(6.6 \times 10^2) \times 120$

   e. $(2.3 \times 10^{12}) \times (4.9 \times 10^8)$

15. Divide the following.

   a. $\dfrac{3.3 \times 10^{12}}{3.0 \times 10^{-4}}$

   b. $\dfrac{7.6 \times 10^{-6}}{8.2 \times 10^8}$

   c. $\dfrac{1.5 \times 10^2}{5.0 \times 10^7}$

   d. $\dfrac{2.2 \times 10^{11}}{4.5 \times 10^8}$

16. Convert the given quantities to the units shown in parentheses.

   a. 40 acres (square miles)

   b. 23,000 bushels (cubic yards)

   c. 50 barrels (liters)

   d. 125 bushels (cubic meters)

   e. 50 caliber (millimeters)

   f. 50,000 carats (grams)

   g. 20 fathoms (meters)

   h. 600 knots (kilometers per second)

   i. 540 knots (meters per second)

Investigations

1. Identify 20 specialized terms that relate to your favorite sport. What statistics are commonly recorded, and how are they calculated?

2. Investigate the history of temperature scales. Why are two different scales, Celsius and Fahrenheit, still in use?

3. Are scientists the only people who have devised specialized vocabulary? What other fields have their own jargon?

4. Describe your favorite tree so that another student can identify it.

5. Read a history of the French Revolution. Why did this political movement lead to a new system of weights and measures?

6. Investigate the history of systems of units used in a well-established profession, such as surveying, agriculture, cooking, or maritime commerce. Which of these specialized units are still used today?

7. The computer age has led to a wide variety of new units of measurement (e.g., "gigabyte" or "megaflop"). Identify some of these units and investigate their history. When were they introduced, by whom, and why?

8. Both common temperature scales, the Fahrenheit and the Celsius, use the freezing and boiling points of water for their reference points. Since the choice of units is largely a matter of convenience, what sort of temperature scale do you suppose a beer manufacturer, who works with alcohol, might use? A jewelry maker working in gold? A beekeeper who has to prepare honey for bottling? (*Hint:* Honey flows easiest when it is warm but starts to change chemically at temperatures around 160°F.)

 ## WWW Resources

See the *Physics Matters* home page at **www.wiley.com/college/trefil** for valuable web links.

1. **powersof10.com** This interactive web site is based on a famous film by Charles and Ray Eames. You can guide yourself through the metric system of prefixes.

2. **helios.physics.uoguelph.ca/tutorials/vectors/vectors.html** The vectors tutorial at the Department of Physics, University of Guelph.

3. **www.pa.uky.edu/~phy211/VecArith/** Another vector tutorial—this one is an interactive graphical Java applet.

4. **www.nrlm.go.jp/keiryou-e.html** The official scientific standards website of Japan, with a nice set of descriptions of how the basic standards are established.

5. **www.velocity.net/~trebor/prelude.html** A published essay, *A Prelude to the Study of Physics,* discussing the role of models and problem-solving in physics.

3 | Motions in the Universe

KEY IDEA

The regular motions in the universe can be discovered by observation and described mathematically.

PHYSICS AROUND US . . . Calculated Moves

How many times have you crossed a street today? Once? A dozen times? More? It's such an ordinary thing to do that you probably don't even keep track. Yet a simple act like crossing a street contains, within itself, a very important lesson about the way the universe works.

Think about what happens when you see a car approaching. You watch it for a while, estimate its speed, make an unconscious calculation about how long it will take before the car gets to your corner, and only then do you make a decision about whether to start across the street. You couldn't carry out this ordinary process if you didn't have an understanding of how objects moved—an understanding born of long experience with cars and their behavior. Early in your life you observed cars in motion, came to some conclusions about their properties, and have used (and tested) that knowledge ever since. We cannot tell whether the car will turn left or right at the intersection or how much gas is in the tank, but we don't need to know that for estimating how quickly the car is approaching. Our observation and experience enable us to determine what we need to know for deciding when to cross the street.

In the same way, physicists observe the world and summarize their conclusions, often using a series of mathematical laws to do so. This process forms the core of what we call science.

47

● PREDICTABILITY

Among the most predictable objects in the universe are the lights we see in the sky at night—the stars and planets. People who live in today's large metropolitan areas no longer pay much attention to the richness of the night sky's shifting patterns. But think about the last time you were out in the country on a clear moonless night, far from the lights of town. There, the stars seem very close, very real. Now try to imagine what it was like before the development of artificial lighting in the nineteenth century. Human beings often experienced jet-black skies that were filled with brilliant pinpoint stars.

The stars in the night sky, showing constellations. Most of these stars are not visible in the glare of city lights.

If you observe the sky closely, you notice that it changes; it's never quite the same from one night to the next. Our ancestors also observed regularities in the arrangement and movement of stars and planets, and they wove these patterns into their religion and mythology. They knew, based on their observations, that when the Sun rose in a certain place, it was time to plant crops because spring was on its way. They came to know that there were certain times of the month when a full Moon would illuminate the ground, allowing them to continue harvesting and hunting after sunset. To these people, knowing the behavior of the sky was not an intellectual game or an educational frill. It was an essential part of their lives. No wonder, then, that astronomy, the study of the heavens, was one of the first sciences to develop.

By relying on their observations and records of the regular motion of the stars and planets, ancient observers of the sky were perhaps the first humans to accept the most basic tenet of science:

The universe is predictable and quantifiable.

Without the predictability of physical events, as we saw in Chapter 1, and our ability to quantify what we observe, as discussed in Chapter 2, the scientific method could not proceed.

Stonehenge and the Cycle of Seasons

No better symbol exists of humankind's discovery of the predictability of nature than Stonehenge, the great prehistoric stone monument on Salisbury Plain in southern England. The structure consists of a large circular bank of earth, surrounding a ring of single upright stones, which, in turn, encircle a horseshoe-shaped structure of five giant stone archways. Each arch is constructed from three massive blocks—two vertical supports several meters tall capped by a great stone lintel. The open end of the horseshoe aligns with an avenue that leads northeast to another large stone, called the "heel stone" (Figure 3-1).

Stonehenge is testimony to the predictability of the Earth's seasons over the centuries.

Stonehenge was built in spurts over a long period of time, starting about 2800 B.C. Despite various legends assigning it to the Druids, Julius Caesar, or the magician Merlin (who was supposed to have levitated the stones from Ireland), archaeologists have shown that it was built by several groups of people, none of whom had a written language and some of whom even lacked metal tools. Stonehenge, like many similar structures scattered around the world, was built to mark the passing of time—a calendar based on the movement of objects in the sky. At Stonehenge, the seasons were marked by the alignment of the stones with astronomical events. On midsummer's morning, for example, someone standing in the center of the monument sees the Sun rising directly over the heel stone.

Building a structure such as Stonehenge required the accumulation of a great deal of knowledge about the sky—knowledge that could only have been gained through many years of observation. Without a written language, people needed to pass complex information about the movements of the Sun, the Moon, and the planets from one generation to the next. How else could they have aligned their stones so perfectly that modern-day Druids in England can still greet the midsummer sunrise over the heel stone?

But as impressive as Stonehenge the monument might be, Stonehenge the symbol of universal regularity and predictability is even more impressive. If the universe were not regular and predictable—if repeated observation could not show us patterns that occur over and over again—the very concept of a monument such as Stonehenge would be impossible. And yet, it continues to stand after 4000 years, a testament to human ingenuity and to the possibility of predicting the behavior of the universe in which we live.

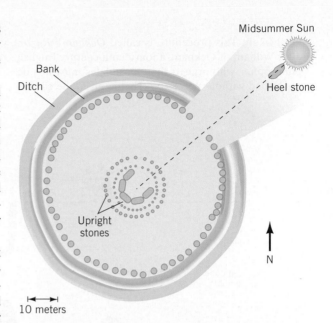

FIGURE 3-1. Stonehenge is built so that someone standing at the center will see the Sun rise over the heel stone on midsummer's morning.

LOOKING DEEPER

Stonehenge and Ancient Astronauts

Who built Stonehenge? Confronted by such an awesome stone monument, with its precise orientation and epic proportions, some writers evoke outside intervention by extraterrestrial visitors. Many ancient monuments, including the pyramids of Egypt, the Mayan temples of Central America, and the giant statues of Easter Island, have been ascribed to these mysterious aliens. One point often made is that the stones in the monument are simply too large to have been moved by people who didn't even have the wheel.

The largest stone at Stonehenge, about 10 meters (more than 30 feet) in length, weighs about 50 metric tons (50,000 kilograms or about 100,000 pounds) and had to be moved overland some 30 kilometers (20 miles) from quarries to the north. Could primitive people equipped only with wood and ropes have moved this massive block?

While Stonehenge was being built, the climate was cooler than it is now and it snowed frequently in southern England, so the stones could have been hauled on sleds. A single person can easily haul 100 kilograms on a sled (think of pulling a couple of your friends). How many people would it take to haul a 50,000-kilogram stone? To estimate the answer, we divide the total weight of the largest stone by the weight an individual can move:

$$\frac{50,000 \text{ kg}}{100 \text{ kg pulled by each person}} = 500 \text{ people}$$

Organizing 500 people for the job would have been a major social achievement in ancient times, but it was certainly physically possible (Figure 3-2).

Scientists cannot absolutely disprove the possibility that Stonehenge was constructed by some strange, forgotten technology. But why invoke such alien intervention when the concerted actions of a dedicated, hard-working human society would have sufficed? All of us are fascinated and awed by the mysterious and unknown, and an ancient structure such as Stonehenge, standing stark and bold on the Salisbury plain, certainly evokes these feelings.

When confronted with phenomena in a physical world, we should accept the simplest explanation as the most likely. This procedure is called *Ockham's Razor,* after William of Ockham, a fourteenth-century English philosopher who argued that "postulates must not be multiplied without necessity." That is, given the choice,

the simplest solution to a problem is most likely to be right. Scientists thus reject the notion of ancient astronauts building Stonehenge, and they relegate such speculation to the realm of pseudoscience.

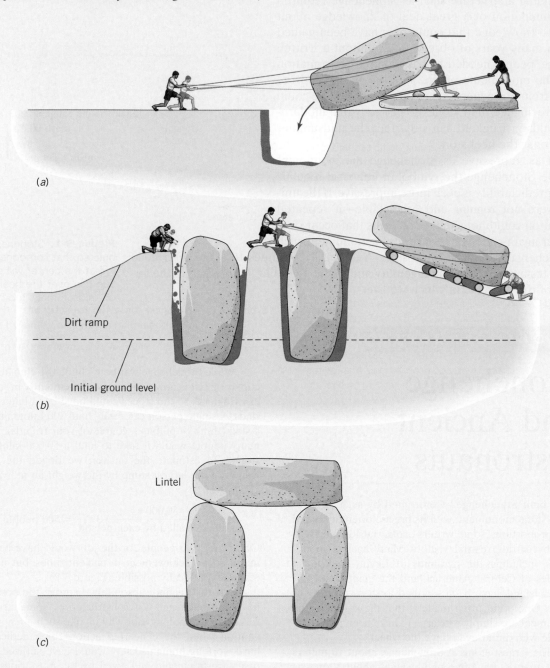

FIGURE 3-2. Perhaps the most puzzling aspect of the construction of Stonehenge is the raising of the giant lintel stones. As shown in this reconstruction, three steps in the process were probably (*a*) dig a pit for each of the upright stones; (*b*) pile dirt into a long sloping ramp up to the level of the two uprights so that the lintel stones could be rolled into place; and (*c*) cart away the dirt, thus leaving the stone archway.

THE BIRTH OF MODERN ASTRONOMY

When you look up at the night sky, you see a dazzling array of objects. Thousands of visible stars fill the heavens and appear to move each night in stately, circular arcs centered on Polaris, the North Star. The relative positions of these stars never seem to change, and closely spaced groups of stars, called *constellations,* have been given names such as the Big Dipper and Leo the Lion. Moving across this fixed starry background are the Earth's Moon, with its regular succession of phases, and half a dozen planets that wander through the sky. You might also see swiftly streaking meteors or long-tailed comets, transient objects that grace the night sky from time to time.

The motion of planets can be especially complex. From night to night most planets, most of the time, appear to move gradually from east to west against the backdrop of the stars. But occasionally a planet seems to reverse its course, seemingly traveling backward with respect to the stars in retrograde motion for a few weeks (Figure 3-3). How can that be?

The Historical Background— Ptolemy and Copernicus

Claudius Ptolemy, an Egyptian-born Greek astronomer and geographer who lived in Alexandria in the second century AD, proposed the first plausible explanation for such complex celestial motions. Working with the accumulated observations of earlier Babylonian and Greek astronomers, he put together a singularly successful theory about how the

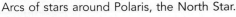

Arcs of stars around Polaris, the North Star.

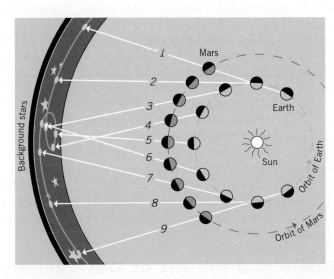

FIGURE 3-3. Today we know that the retrograde motion of Mars can be explained by the fact that as the Earth passes Mars in orbit, the position of Mars against the background of stars seems to reverse itself temporarily. This explanation can only work if the Earth moves, so Greek astronomers could not have evoked it.

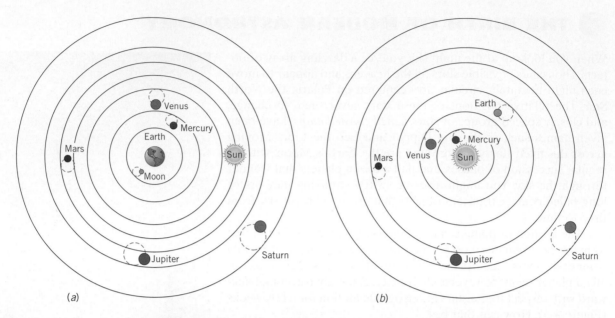

FIGURE 3-4. The Ptolemaic (*a*) and Copernican (*b*) systems. Both systems used circular orbits. The fundamental difference is that Copernicus placed the Sun at the center.

heavens had to be arranged to produce the display we see every night (Figure 3-4*a*). Earth sits unmoving at the center of Ptolemy's universe. Around it, on a series of concentric rotating spheres, move the stars and planets. The model was carefully crafted to take account of observations. The planets, for example, were attached to small spheres rolling inside of the larger spheres so that their uneven retrograde motion across the sky could be understood. This system remained the best explanation of the universe for almost 1500 years. It successfully predicted planetary motions, eclipses, and a host of other heavenly phenomena and was one of the longest-lived scientific theories ever devised.

During the first decades of the sixteenth century, however, a Polish cleric by the name of Nicolas Copernicus (1473–1543) proposed a competing hypothesis that was to herald the end of Ptolemy's crystal spheres. His ideas were published in 1543 under the title *On the Revolutions of the Spheres*. Copernicus retained the notion of a spherical universe with circular orbits, and even kept the idea of spheres rolling within spheres, but he asked a simple and extraordinary question. Is it possible to construct a model of the heavens whose predictions are as accurate as Ptolemy's, but in which the Sun, rather than the Earth, is at the center? We do not know how Copernicus, a busy man of affairs in medieval Poland, conceived this question, nor do we know why he devoted his spare time for most of his adult life to answering it. We do know, however, that in 1543, for the first time in over a millennium, a serious alternative to the Ptolemaic system was presented (see Figure 3-4*b*).

A portrait of Nicolas Copernicus.

Tycho Brahe and the Art of Observation

With the publication of the Copernican theory, two competing models of the universe confronted astronomers. The Ptolemaic and Copernican systems differed in a fundamental way that had far-reaching implications about the place of humanity in the universe. They both described possible universes, but in one the Earth, and by implication humankind, was no longer at the center. The astronomers' task was to decide which model best describes our universe.

To resolve this question, astronomers had to compare the predictions of the two competing hypotheses. When astronomers tried to make comparisons, however, a fundamental problem became apparent. Although the models of Ptolemy and Copernicus made different predictions about the position of a planet at midnight or the time of moonrise, the differences were too small to be measured with equipment that was available at the time. The telescope had not yet been invented and astronomers had to record planetary positions by depending entirely on naked-eye measurements with awkward instruments. Until the accuracy of measurement was improved, the question of whether or not the Earth was at the center of the universe could not be decided.

Some scientists thrive on experimental challenges and they revel in devising new tricks for making measurements better than anyone else before. The Danish nobleman Tycho Brahe (1546–1601) was such a scientist. Abducted in infancy by his uncle, Tycho was raised in comfort and given the best possible education. His scientific reputation was firmly established at the age of 25, when he observed and described a new star in the sky (in fact, a type of exploding star called a supernova). By the age of 30, Brahe received from the Danish king the island of Hveen off the coast of Denmark and funds to build an observatory there.

Brahe built his career on designing and using vastly improved observational instruments. He determined each star or planet position with a *quadrant,* a large sloping device something like a gunsight. With this sort of instrument, you can record the position of a star or planet by measuring two angles—for example, the angle up from the horizon and the angle around from due north. Brahe constructed his sighting device of carefully selected materials, and he learned to correct his measurements for the inevitable contraction of brass and iron components that occurred during the cold Danish nights. Over a period of 25 years, he accumulated precise data on the positions of the planets with these instruments, compiling the most accurate record of planetary positions of his day.

A portrait of Tycho Brahe in his observatory on the island of Hveen.

This instrument, called a quadrant, measures the angular position of stars and planets to the Earth.

Kepler's Laws

After Tycho Brahe died in 1601, his data passed into the hands of his assistant, Johannes Kepler (1571–1630), a German mathematician who had joined Tycho 2 years before. Kepler was skilled in mathematics, and he was able to analyze Tycho Brahe's decades of planetary data in new ways. In the end, Kepler found that the data could be summarized in three basic mathematical statements about the solar system, known as **Kepler's laws of planetary motion.** The most important of these (shown in Figure 3-5) states that all planets, including the Earth, orbit the Sun in elliptical paths, not in perfect circles as had been previously assumed. In this picture, the spheres within spheres are gone. Not only do Kepler's laws give a more accurate description of what is observed in the sky, but they present a simpler picture of the solar system as well. And as we have seen in our discussion of Stonehenge, simple explanations are often a sign of deeper understanding of how nature works.

An *ellipse* is defined as a curve drawn so that the sum of the distances from any point on the curve to two fixed points is always the same. Imagine tacking

Johannes Kepler (1571–1630).

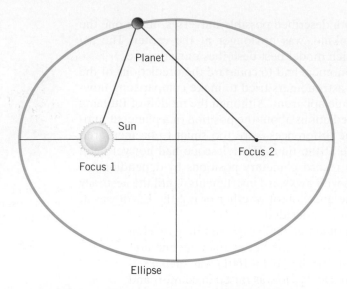

FIGURE 3-5. Kepler's first law, shown schematically. An ellipse is a geometrical figure in which the sum of the distances to two fixed points (each of which is called a focus) is always the same. For the planets, the Sun is at one focus of the ellipse.

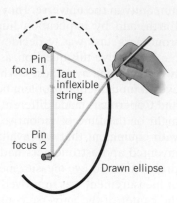

FIGURE 3-6. You can draw an ellipse by bracing a pencil against a string that is tacked down at each end by a pin, and drawing a curve while keeping the string taut. The two pins are at the foci of the ellipse. Note that as the two pins are brought closer together, the ellipse more nearly approximates a circle.

down a length of loose string at two points, then drawing a curve by bracing a pencil in the now-taut string, as shown in Figure 3-6. Each of the two fixed points is called a *focus*. What Kepler found was that the orbits of all planets known at the time, from Mercury to Saturn, have one focus at the Sun. The statement that the planets have elliptical orbits with one focus at the Sun is known as *Kepler's first law of planetary motion*.

Kepler's second law describes the speed at which the planets move in their elliptical orbits. If you remember the playground game in which you ran toward a post, then grabbed it on the fly, and swung around, you have a pretty good notion of the way the planets move. They speed up as they get closer to the Sun and then slow down in the farther parts of their orbits. Kepler's second law is usually stated in terms of equal areas, a concept illustrated in Figure 3-7. Imagine that a line drawn from the Sun to an orbiting planet sweeps out an area in a fixed period of time. *Kepler's second law* says that for a given time interval, this swept-out area is the same, no matter where the planet is in its orbit. A glance at Figure 3-7 should convince you that this means that planets move fastest when they are nearest the Sun and slowest when farther away.

Finally, Kepler turned his attention to the *period* of a planet's revolution—the time it takes for one complete orbit, or its "year." Planets farther from the Sun have a longer year than those that are closer in for two reasons: (1) they have farther to go as they make their circuit, and (2) the outer planets travel more slowly than the inner ones. The net effect is a systematic relationship between the planet's period and its orbital distance from the Sun. *Kepler's third law* expressed this relationship between a planet's distance from the Sun and its period as a simple equation that allows us to predict the behavior of orbiting objects.

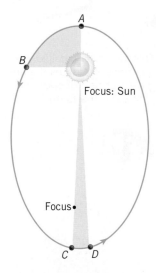

FIGURE 3-7. Kepler's second law of equal areas states that planets move fastest when they are closest to the Sun and slowest when at the farthest point in their orbits. Thus, a planet in an elliptical orbit sweeps out equal areas in equal times.

1. In words:

The farther a planet is from the Sun, the longer its year.

2. In an equation with words:

The square of the period of a planet's orbit is proportional to the cube of its average distance from the Sun.

3. In an equation with symbols:

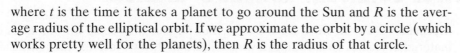

$$\frac{t^2}{R^3} = \text{constant}$$

or

$$t^2 = \text{constant} \times R^3$$

where t is the time it takes a planet to go around the Sun and R is the average radius of the elliptical orbit. If we approximate the orbit by a circle (which works pretty well for the planets), then R is the radius of that circle.

If t is measured in Earth years (so for Earth, $t = 1$) and R is measured in terms of the distance between the Sun and the Earth (so for Earth, $R = 1$), then the constant in Kepler's law is also equal to 1.

Jupiter's Year

Jupiter is about 5.2 times farther from the Sun than the Earth is. How long is Jupiter's year?

EXAMPLE
3-1

REASONING AND SOLUTION: Kepler's third law relates a planet's year, t, and its distance from the Sun, R. In this case, $R = 5.2$ and we want to find t by using the equation

$$\frac{t^2}{R^3} = 1$$

Substituting the known value of R:

$$\frac{t^2}{5.2^3} = \frac{t^2}{141} = 1$$

so

$$t = \sqrt{141} = 11.9 \text{ years}$$

So Jupiter orbits the Sun once every 11.9 Earth years. ●

Decades of careful observations by Brahe and mathematical analysis by Kepler firmly established that the Earth is not at the center of the universe, that planetary orbits are not circular, and that neither Ptolemy nor Copernicus were correct in their models of the universe (although the Copernican model was much closer to the modern view than Ptolemy's). This research also illustrates a recurrent point about scientific progress. The ability to answer scientific questions often depends on the quality of instruments scientists have at their disposal. The person dealing with the grubby details of how a telescope measures the position of a star may not appear to be doing something glamorous, but such people often provide insights into the most fundamental scientific questions. In Tycho's case, his meticulous attention to experimental detail provided an important step in answering the age-old question, "What is the location of Earth in the universe?"

At the end of this historical episode, astronomers had Kepler's laws to describe *how* the planets in the solar system moved; however, they had no idea of *why* planets behaved the way they did. In essence, they had completed the first two steps of the scientific method—observation and pattern identification—that we describe in Chapter 1. But the next step, leading to a fundamental theoretical understanding of the sky, had yet to be taken. Kepler's laws, as important as they are, give no insight into the basic mechanisms that make the solar system operate. The answer to that question was to come from an unexpected source.

THE BIRTH OF MECHANICS AND EXPERIMENTAL SCIENCE: GALILEO GALILEI

Mechanics is an old word for the branch of physics that deals with motions of material objects. A rock rolling down a hill, a ball thrown into the air, and a sailboat skimming over the waves are all fit subjects for mechanics. For sixteenth-century military leaders concerned with the behavior of cannonballs and other projectiles, mechanics was a science of practical interest. Since ancient times, philosophers had speculated on why objects move the way they do, but it wasn't until about 1600 that our modern understanding of the subject began to emerge.

The Italian physicist and philosopher Galileo Galilei (1564–1642) was in many ways a forerunner of the twentieth-century scientist. A professor of mathematics at the University of Padua, he quickly became an advisor to the powerful court of the Medici at Florence as well as a consultant at the Arsenal of Venice, the most advanced naval construction center in the world at that time. He invented many practical devices, such as the first thermometer, the pendulum clock, and the proportional compass that draftsmen still use today. Galileo gained fame as the first person to observe the heavens with a telescope, which he built after hearing of the instrument from others. He was the first to see many astronomical phenomena, including the moons of Jupiter (now called the "Galilean moons"), craters and other surface features of Earth's Moon, and sunspots.

From the scientist's point of view, Galileo's greatest achievement was his work on experimental technique. You can see why by considering his research on the behavior of objects thrown or dropped on the surface of the Earth—work that we discuss in more detail next. Greek philosophers had taught the reasonable idea that heavier objects must fall faster than light ones, because the

What Galileo saw through his telescopes and what we can see today. (*a*) **Telescopes:** Left, telescopes built and used by Galileo; right, the Kech Observatory Mauna Kea, Hawaii, has twin 394-inch telescopes.

(*a*)

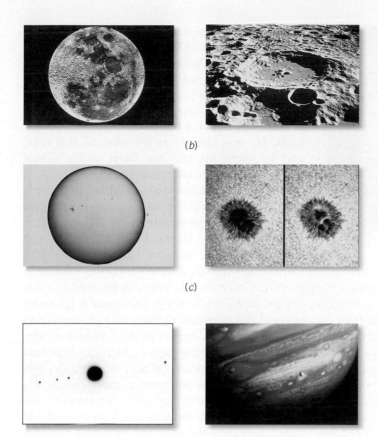

(b)

(c)

(d)

What Galileo saw through his telescopes and what we can see today. (b) **Craters on the Moon:** Left, some cratering is visible from Earh; right, close-up views taken from orbit around the Moon. (c) **Sunspots:** Left, sunspots are visible in reflected images of the Sun; right, close-up view of an Earth-sized sunspot, taken from a satellite orbiting the Sun. (d) **Moons of Jupiter:** Left, four major moons are visible in a common telescope; right, Io and Europa viewed in front of Jupiter, as seen by the Voyager spacecraft.

heavier ones want to get to the center of the Earth (and of the universe) more than lighter ones. In a series of classic experiments, Galileo showed that this idea, as reasonable as it may seem, was not correct. In the process, he demonstrated that at the surface of the Earth, all objects fall at the same rate.

Physics in the Making

The Heresy Trial of Galileo

Galileo is famous for the wrong reason. Despite the fact that he was a founder of modern experimental science and was the first to make a systematic survey of the sky with a telescope, he is remembered primarily because of his trial in 1633 on suspicion of heresy.

Galileo published a summary of his telescopic observations in a book called *The Starry Messenger.* This book was written in Italian, the language of common people, rather than Latin, the language of scholars. Thus Copernican ideas, including the disturbing concept that the Earth is not the center of the universe, became available to the educated public. Some readers complained that these ideas violated Church doctrine and, in 1616, Galileo was called before the College of Cardinals. What happened at this meeting is not clear. The Church later claimed that Galileo had been warned not to discuss Copernican ideas unless he treated them, as Copernicus had, as a hypothesis. Galileo, on the other hand, claimed he had not been given any such warning.

A painting of the trial of Galileo.

In any case, the situation remained in this unsettled state until 1632, when Galileo published a book called *A Dialogue Concerning Two World Systems,* which was a long defense of the Copernican system. This publication led to the famous trial, in which Galileo excused himself of charges of heresy by denying that he held the views in his book. He was already an old man by this time, and he spent his last few years under virtual house arrest in his villa near Florence.

The legend of the trial of Galileo, in which a rigid hierarchy crushes an earnest seeker after truth (as in Bertoldt Brecht's play *Galileo*), bears little resemblance to the historical events. The Catholic Church had not banned Copernican ideas; indeed, seminars on the Copernican system were given at the Vatican in the years before Galileo. Furthermore, Galileo's arguments in favor of the system were not very convincing. For example, much of the *Dialogues* is taken up by a completely incorrect discussion of the tides and Galileo had no convincing answers for why things fall if the Earth is not the center of the universe or why we don't notice any effects of the Earth's motion if it moves. His confrontational tactic of putting the Pope's favorite arguments into the mouth of a foolish character in the book brought a predictable reaction that earlier, more reasonable approaches had not. As often happens, under close inspection the simple myth associated with an historical event dissolves into something much more complex.

A footnote: In 1992, the Vatican reopened the case and, in effect, issued a retroactive not guilty verdict in the case of Galileo. The grounds for the reversal were that the original judges had not separated questions of faith from questions of scientific fact. ●

DESCRIBING MOTION

To lay the groundwork for understanding Galileo's study of moving objects, we have to begin with precise definitions of three familiar terms: speed, velocity, and acceleration. These terms are of basic importance throughout all areas of physics and we will use them often. We also note that these and many other terms common in everyday language are used in physics with precise definitions. Usually the definitions are not very different from what you would expect from their common use, but sometimes there are important differences. You need to be aware of these differences when you talk about the concepts of physics and try to explain how they apply in the world around us.

Speed

Speed is one of the many everyday words that has a precise definition in physics.

1. In words:

> **Speed** *is the distance an object travels divided by the time that it takes to travel that distance.*

2. In an equation with words:

$$\text{Speed} = \frac{\text{Distance traveled}}{\text{Time of travel}}$$

3. In an equation with symbols:

$$s = \frac{d}{t}$$

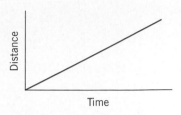

FIGURE 3-8. Graph of the distance traveled by an object moving at a constant speed.

where *s* is speed, *d* is distance traveled, and *t* is time of travel. This *direct relationship* between speed and distance is illustrated in graphical form in Figure 3-8. If you know the distance traveled and the time elapsed during the travel, you can calculate the speed.

From time to time we will need to use two variations of this equation. First, if you know the average speed, *s*, of an object and the time of travel, *t*, you can calculate how far the object has traveled, *d*:

$$\text{Distance traveled} = \text{Average speed} \times \text{Time of travel}$$

or $$d = s \times t$$

Second, if you know the total distance traveled and the average speed of travel, you can calculate how long the journey takes:

$$\text{Time of travel} = \frac{\text{Distance traveled}}{\text{Average speed}}$$

or $$t = \frac{d}{s}$$

Driving Your Car

If the speedometer on your car reads a constant 50 kilometers (31 miles) per hour, how far will you go in 15 minutes?

EXAMPLE
3-2

REASONING AND SOLUTION: We are given a speed and a time and we want to find a distance. We can apply the equation that relates time, speed, and distance in the form:

$$\text{Distance traveled} = \text{Average speed} \times \text{Time of travel}$$

However, this question also involves changing units. First, we must know the travel time in hours:

$$\frac{15 \text{ minutes}}{60 \text{ minutes/hour}} = \frac{1}{4} \text{ hour}$$

Then, using the relationship between distance and time given, we find:

$$\text{Distance} = 50 \text{ kilometers/hour} \times \frac{1}{4} \text{ hour}$$

$$= 12.5 \text{ kilometers} \quad (7.7 \text{ miles})$$

Your car will travel 12.5 kilometers (or 7.7 miles) in 15 minutes. ●

A word about units: You may have noticed that in Example 3-2 we put $\frac{1}{4}$ hour into the equation for the time instead of 15 minutes. The reason we did this was that we needed to be consistent with the units in which an automobile speedometer measures speed. Since the speedometer dial reads in kilometers (or miles) per hour, we also put the time in hours to make the equation balance. A useful way to deal with situations such as this is to imagine the units are quantities that can be canceled in fractions, just like numbers. In this case, we have:

$$\text{Distance} = \text{kilometers/hour} \times \text{hour}$$
$$= \text{kilometers}$$

If, however, we put the time in minutes, we'd have:

$$\text{Distance} = \text{kilometers/hour} \times \text{minutes}$$

and there would be no cancellation.

Whenever you do a problem like this, it's a good idea to check to make sure the units come out correctly. This important process is known as *dimensional analysis.*

Velocity

Velocity has the same numerical value as speed, but it is a vector quantity that also includes information about the direction of travel (see Chapter 2). The speed of a car might be 40 kilometers per hour, for example, while the velocity is 40 kilometers per hour due west. Velocity and speed are measured in units of distance per time, such as meters per second, feet per second, or miles per hour.

Acceleration

Acceleration measures the rate of change of velocity.

1. In words:

> **Acceleration** *is the change in velocity divided by the time it takes for that change to occur.*

2. In an equation with words:

$$\text{Acceleration} = \frac{\text{Final velocity} - \text{Initial velocity}}{\text{Time}}$$

3. In an equation with symbols:

$$a = \frac{\Delta v}{t}$$

where Δv indicates the change in velocity. Like velocity, acceleration requires information about the direction and is therefore a vector.

When velocity changes, it may be by a certain number of feet per second or meters per second in each second. Consequently, the units of acceleration are meters per second per second, usually described as meters per second squared (and abbreviated m/s^2), where the first meters per second refers to the velocity, and the last per second refers to the time it takes for the velocity to change.

To understand the difference between acceleration and velocity, think about the last time you were behind the wheel of a car, driving along a straight highway and glancing at your speedometer. If the needle is unmoving (at 50 kilometers per hour, for example), you are moving at a constant speed. Suppose, however, that the needle isn't stationary on the speedometer scale (perhaps because you have your foot on the gas pedal or on the brake). Your speed is changing and, by our definition, you are accelerating. The higher the acceleration, the faster the needle moves. If the needle doesn't move, however, this doesn't mean you and the car aren't moving. As we have seen, an unmoving needle simply means that you are traveling at a constant speed without acceleration. Motion at a constant speed in a single direction is called **uniform motion.**

Deceleration

If you're driving a car and step on the brakes, the car slows down, or *decelerates*. This process involves a change in velocity, so it is actually acceleration. We use *deceleration* in everyday speech because we normally associate the term *acceleration* with speeding up. If you look at the definition of acceleration, however, you will notice that if the final velocity is less than the initial velocity (which is what happens when you step on the brakes), the acceleration is a negative number. So deceleration is simply a negative acceleration.

Average and Instantaneous Velocity

While it is fairly straightforward to measure speeds and velocities for objects that aren't accelerating, measuring becomes more complicated when objects accelerate. If your speedometer needle climbs steadily from 30 to 40 kilometers per hour, the speed of your car is constantly changing. At no point during this time interval is speed (or velocity) a constant. So how do we deal with objects whose speed is changing? How do we answer the question "How fast is it moving right now?"

To tackle this problem we need to make a distinction between average velocity and instantaneous velocity. The *average velocity* is simply the total distance traveled divided by the total time it takes to travel that distance. If the distance is a meter and it takes a second to travel that distance, then the average velocity is one meter per second.

Instantaneous velocity, on the other hand, is the velocity at a specific time. We can determine the instantaneous velocity of an accelerating object by thinking of the process this way: Suppose you had marked out a short distance on the pavement—a few feet, for example, or even a fraction of an inch. As the accelerating object goes by, you can time how long it takes to cross that small distance. If the time interval is short enough, the velocity won't change very much as the object crosses the small distance. You can think of the instantaneous velocity as the average velocity measured over a very small time interval. (This concept is explored numerically in Example 3-3 on page 73.)

For the record, what is actually measured in your car's speedometer is the time it takes a particular gear in the transmission to make one revolution. When the car moves a known distance forward, this gear turns once. The number of turns it makes is eventually translated into the number displayed on your speedometer.

Physics in the Making
Measuring Time Without a Watch

In our age of stop watches, digital timers, and atomic clocks, with Olympic races routinely measured to a hundredth or even a thousandth of a second, it's hard to imagine measuring time without an accurate instrument. But when Galileo set out to study accelerated motion in the 1600s, measuring time was a formidable technological challenge. Think about how you might determine small time intervals if all you had was a clock that ticked off seconds but could record no shorter times.

Galileo wanted to document as accurately as possible the way falling objects accelerate, but these measurements required knowledge of both distance and time. Since objects that fall straight down moved much too fast for him to measure, he devised an experiment in which balls rolled down a gently inclined plane. However, even these balls moved too fast to time with available clocks (see Figure 3-9).

Galileo tried a variety of different methods to measure these small increments of time. He experimented with his own heartbeat, but his pulse proved too irregular. He tested rapidly swinging pendulums, but they were difficult to start and stop precisely. He had more success measuring the weight of water that accumulated when a steady flow was started and stopped to coincide with the period of an object's fall, but irregularities in the flow and uncertainties in starting and stopping the water limited the accuracy of that method.

Galileo's most ingenious solution for measuring time intervals relied on his musical training. Galileo stretched lute strings across the rolling ball's path, so that there was a discernible twang when the ball passed over the string. He then adjusted the distance between the strings until he heard the notes coming at precisely equal intervals. A musician with a good ear can tell if notes in a series are off by as little as $\frac{1}{64}$ second. Thus, even though he did not have clocks capable of measuring time intervals better than a second or so, with this scheme he could be certain that several very short time intervals were the same.

When Galileo got the time intervals between twangs just right, he measured the distances between lute strings. He found that if a ball had traveled 1 inch in the first twang, then it traveled 4 inches by the end of two twangs, 9 inches by

Galileo's apparatus: inclined plane

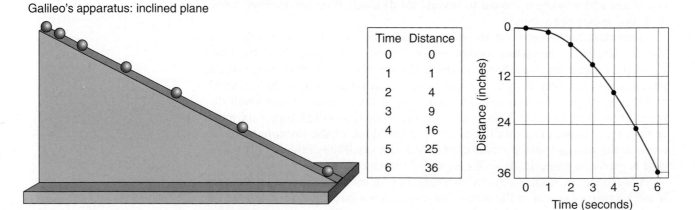

Time	Distance
0	0
1	1
2	4
3	9
4	16
5	25
6	36

FIGURE 3-9. Galileo's falling-ball apparatus, with a table of measurements and a graph of distance versus time.

the end of three twangs, 16 inches by the end of four twangs, and so on. In other words, he found that the distance traveled by an accelerating object depends on the *square* of the time, not just on the time itself.

Modern versions of the rolling ball experiment, relying on laser beams and electronic timers, have greatly improved the accuracy of Galileo's experiment, but they produce exactly the same result. ●

RELATIONSHIPS AMONG DISTANCE, VELOCITY, AND ACCELERATION

The Velocity of an Accelerating Object

Galileo's experiments revealed a simple relationship between an object's acceleration and the distance it travels.

1. In words:

When an object is accelerated in a uniform way from a standing start, the distance it covers in a given time depends on the square of the time.

2. In an equation with words:

$$\text{Distance traveled} = \frac{1}{2} \times \text{Acceleration} \times \text{Time}^2$$

3. In an equation with symbols:

$$d = \frac{1}{2} at^2$$

This relationship between the distance traveled and time can also be represented in a graph, as shown in Figure 3-10. This is the same kind of graph that we show in Figure 2-8 for a squared relationship.

Suppose that at the instant the object starts accelerating (time equals zero), the velocity is also zero, but the object begins to accelerate at a uniform rate. In this case, the instantaneous velocity of the falling body is given by:

1. In words:

The instantaneous velocity of a uniformly accelerating object that started at rest equals the acceleration multiplied by the total time of acceleration.

2. In an equation with words:

$$\text{Velocity} = \text{Acceleration} \times \text{Time}$$

3. In an equation with symbols:

$$v = a \times t$$

The *direct relationship* between the velocity and time of travel for a given distance is illustrated in Figure 3-11.

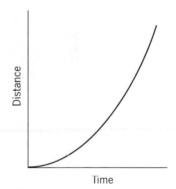

FIGURE 3-10. Graph of the distance traveled by a uniformly accelerating object.

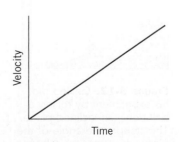

FIGURE 3-11. Graph of the velocity of a uniformly accelerating object.

The relationships among distance, time, velocity, and acceleration enable you to calculate a lot of information about an object's motion from some simple measurements you can make with a ruler and a stop watch. The thing to remember is that during motion with uniform velocity, the distance covered is the average velocity multiplied by the elapsed time; however, during accelerated motion, the distance covered is the acceleration multiplied by one-half the square of the elapsed time. Example 3-3 at the end of the chapter shows how to use these equations to analyze the motion of a sprinter.

Connection

The Evolution of Speed

A hallmark of advancing technology is an increase in speed. Year by year, cars and planes are faster, computers are faster, medical procedures are faster—even food service is faster (most of the time).

Thousands of years ago humans learned to increase their natural speed by domesticating and riding horses. The top human speed in modern times is about 22 miles per hour in a sprint or 16 miles per hour in a 1-mile run. But racehorses can run over 35 miles per hour in a $1\frac{1}{2}$-mile run. People on horseback could hunt animals such as buffalo for food and escape other predators such as wolves. Greater speed meant improvements in life.

By the nineteenth century, machines surpassed horses in speed. Around 1830, Peter Cooper's steam-engine locomotive, the *Tom Thumb,* pulled ahead of a horse-drawn carriage pulling the same load in a race, but broke down before the end and the horse won. Journalists of the time speculated about the dangers to the human body of such high speeds as 60 miles per hour, but by 1860, railroads had expanded across the country. Their ability to carry huge amounts of freight changed the world's economy. Today, of course, freight trains are still a commonplace, but land speed belongs to the passenger trains. High-speed bullet trains in France and Japan routinely cruise at 175 miles per hour between cities.

Automobiles powered by steam engines were developed as early as the eighteenth century, but were not practical. True land vehicles that did not run on rails had to wait until the development of the internal combustion engine, patented in 1879. Within a few decades, some commercial automobiles could exceed 100 miles per hour on the open road.

However, the true kings of speed are airplanes. The Wright brothers' first airplane flight in 1903 lasted 12 seconds and covered 120 feet. Five years later they were flying 10 miles at a stretch at average speeds of 40 miles per hour. Today, planes can fly more than three times faster than the speed of sound. The SR-71 Blackbird, the fastest vehicle ever built other than spacecraft, has achieved speeds of almost 2200 miles per hour and is one of the great achievements of modern technology. (See Looking at Speed on page 65.) ●

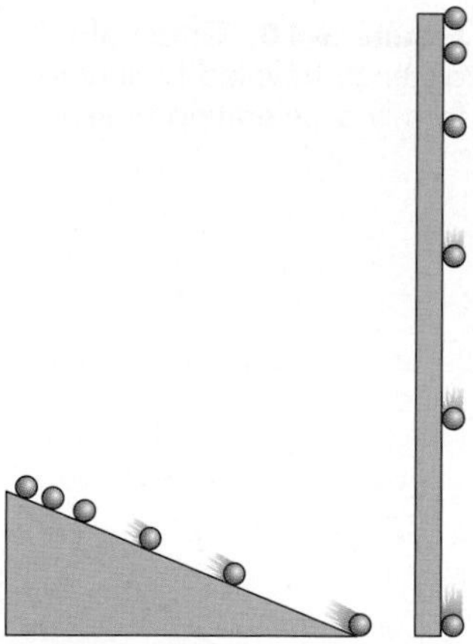

FIGURE 3-12. Galileo did his experiment by rolling a ball down an inclined plane. The steeper the angle of the plane, the faster the fall. A plane at a right angle to the ground corresponds to a freely falling object.

Acceleration due to Gravity

In Galileo's experiments, he produced greater accelerations by increasing the angle of the incline down which the balls rolled (Figure 3-12). Eventually, at very steep angles, the motion of the balls became too fast to measure the time intervals. In our everyday lives, most falling objects fall straight down, equivalent to a plane tilted at an angle of 90 degrees.

Looking at Speed

You can probably walk much faster than a tortoise, but you wouldn't want to race against a cheetah. Not unless you drove in a car, which could easily pass any animal, or in a jet plane, which is faster than anything other than spacecraft. And speaking of spacecraft, some probes have been launched to take close-up looks at comets, which can move far faster than anything on Earth.

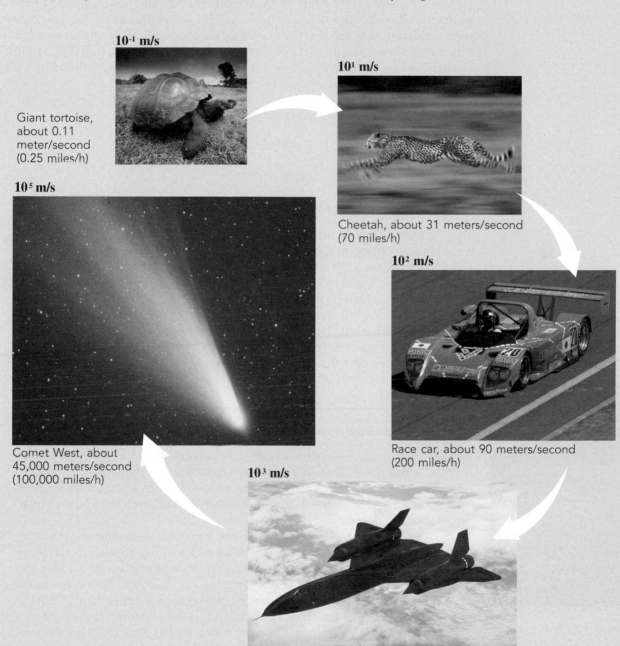

10^{-1} m/s

Giant tortoise, about 0.11 meter/second (0.25 miles/h)

10^1 m/s

Cheetah, about 31 meters/second (70 miles/h)

10^5 m/s

Comet West, about 45,000 meters/second (100,000 miles/h)

10^2 m/s

Race car, about 90 meters/second (200 miles/h)

10^3 m/s

SR-71 Blackbird jet, about 1100 meters/second (2500 miles/h)

You can verify that there is an acceleration involved by dropping an object and watching it fall. Notice that at the instant you release the object it barely moves. Can you see that it's moving faster at the end of its fall than at the beginning?

The acceleration of a freely falling body is so important that physicists give it a special name, **acceleration due to gravity,** and assign it a special letter, **g.** Galileo's experiments led him to the hypothesis that any object dropped near the Earth's surface, no matter how heavy or light, falls with exactly the same constant acceleration. In the absence of complications such as air resistance or wind, which may slow down or alter the direction of a falling object, it makes no difference how massive an object is—all objects experience exactly the same acceleration. (In Chapter 4 you will learn why this is so.)

The numerical value of g can be determined by measuring the actual motion of objects. The modern value of g is given by:

$$g = 32 \text{ feet/s}^2 = 9.8 \text{ m/s}^2$$

The velocity of a falling object is given by:

$$\text{Velocity of a falling object} = g \times \text{Time}$$

These equations tell us that in the first second, a falling object accelerates from a stationary position to a velocity of 9.8 m/s (about 22 miles per hour) straight down. After two seconds, the velocity doubles to 19.6 m/s; after three seconds, it triples to 29.4 m/s, and so on.

Ironically, Galileo probably never performed the one experiment for which he is most famous—dropping two different weights from the leaning Tower of Pisa to see which would land first. Had he done so, in fact, the effects of friction between the air and the falling bodies probably would have slowed the lighter one slightly more than the heavier object, so that the two would have been perceived to fall at slightly different rates.

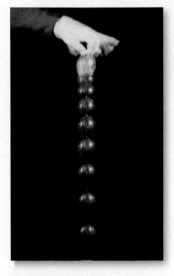

FIGURE 3-13. A multiple-exposure photograph captures the accelerated motion of a falling apple. In each successive time interval, the apple falls farther.

Develop Your Intuition: Freely Falling Objects

According to Galileo's results, every object at the surface of the Earth, once released, should accelerate downward at the same rate (Figure 3-13). But we know that if a leaf and an acorn fall from a branch at the same time, the acorn reaches the ground well before the leaf, even if they have equal mass. Was Galileo wrong?

To understand this problem, think for a minute about *how* the leaf and the acorn fall. The acorn plummets straight down to the ground, accelerating as it goes. The leaf, on the other hand, flutters down slowly. It appears that the air has a much greater effect on the leaf than on the acorn. The acorn, being compact, has a much smaller resistance relative to its weight exerted on it by the air than does the leaf. In fact, if we repeated this experiment on the Moon (where there is no air) or in a tube from which air has been removed, the leaf would not flutter and the two objects would fall at exactly the same rate. (This experiment has actually been done, both on the Moon and in a vacuum tube, with a feather and a rock. Both reached the ground at the same time.)

MOVEMENT IN TWO DIMENSIONS: THE LAW OF COMPOUND MOTION

Until now, we have talked only about motion along a straight line—called motion in one dimension. However, if you throw a baseball or ride in a car on a highway, you experience motion in more than one dimension. The baseball, for example, follows an arching path as it travels forward. The car goes around turns as well as moving ahead. In Galileo's time, analyzing the motion of projectiles such as baseballs was particularly important because the cannon had just been introduced into warfare, and cannonballs, like baseballs, move in two-dimensional arcs.

The central question in analyzing two-dimensional motion is this: how does the motion in one direction affect motion in the other dimension? For example, when a baseball is thrown, does its speed in the horizontal direction affect how high it goes? Galileo proposed what we now call **the law of compound motion.** It states that:

Motion in one dimension has no effect on motion in another dimension.

Projectile Motion

Let's look at an example of projectile motion to see how this law works. Suppose you stand on a high cliff and throw a rock outward, as shown in the graph of Figure 3-14. The law of compound motion tells us that we should think of the rock's path as being made up of two separate parts or components. In the horizontal direction, the rock moves at a constant velocity imparted to it by your hand. In the vertical direction, the rock accelerates downward like any other falling object.

Suppose, for example, that the rock is moving 12 meters per second in the horizontal direction. At the end of 1 second, the rock is 12 meters from the cliff. At the same time, however, the rock is falling with an acceleration of 9.8 meters per second per second, so at the end of 1 second it is 4.9 meters down (remember, distance $= \frac{1}{2}at^2$, where in this case $a = g = 9.8$ m/s^2). The rock's position, then, is 12 meters out and 4.9 meters down, as shown in Figure 3-14. A second later, it is 24 meters out and 19.6 meters down, and so on. The path, illustrated in Figure 3-15, is called a *parabola.*

The law of compound motion also applies to projectiles thrown up from the ground, such as a baseball hit by a bat or a football thrown or kicked by a player. In this case, the vertical motion of the ball is an ever-slowing motion upward, followed by a fall like that in Figure 3-14 (if the effects of wind are negligible). The horizontal motion is, as in Figure 3-14, movement at a constant velocity. The combination is an arc, also in the shape of a parabola, that begins at the ground, goes up to the peak in a path that is the mirror image of that in Figure 3-14, and then goes back to the ground exactly as in Figure 3-14.

Physics in the Making

The Range of a Cannonball

The distance a projectile travels before it comes back to the ground is called its range. In the fifteenth century, when cannons were first being introduced into the military affairs of Europe, finding the range of a cannonball was of more than

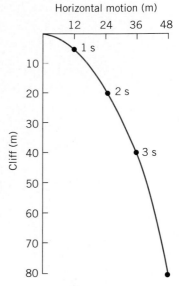

FIGURE 3-14. The trajectory of a rock thrown off a cliff. The motion in the horizontal direction is independent of the motion in the vertical direction.

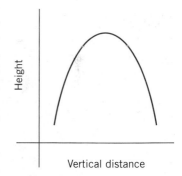

FIGURE 3-15. An object thrown upward at the Earth's surface follows an arching path, called a parabola.

academic interest. The Duke of Milan, having acquired some cannons, wanted to know how to get the maximum range from his new purchases. He called in his chief engineer, a man named Tartaglia (The Stutterer), and told him to find out how to do it. What followed was one of the earliest episodes of the new experimental method in science.

The prevailing thought about projectiles at the time, following the teachings of Aristotle, was that motion could be divided into two classes: natural motion, in which an object followed its natural inclination to fall toward the center of the Earth, and violent motion, which was imposed by an outside agency such as gunpowder. Scholars believed that a cannonball would move off in a straight line until the violent motion was expended and then fall straight down. They, like Aristotle, arrived at this conclusion by simply thinking about the situation.

Tartaglia, on the other hand, took a different approach. He went out to a field outside Milan and shot off cannonballs, noting the distance the ball went for different elevations of the barrel and different gunpowder charges. He was the first person to realize that the maximum range occurs when the projectile leaves the ground at an angle of 45 degrees. (This is true in the absence of air resistance. The maximum range with air resistance requires a slightly smaller launch angle above the ground; for example, 43 degrees for a baseball, 38 degrees for a golf ball.) More important, however, Tartaglia showed that observation and experiment were an essential piece of the scientific method. ●

Motion in a Circle

The motion of an object moving in an arc or circle presents a more difficult challenge to analyze than a batted baseball or a rock thrown from a cliff. The basic problem is that when we look at the regular motion of an object moving in a circle at constant speed, our first impression may be that it is not accelerating. *But it is!*

If you have difficulty envisioning this acceleration, you are in good company. Many famous scientists, including Galileo, tried to work out the properties of uniform circular motion and failed. But recall from the definition of velocity that velocity involves *both speed and direction.* The ordinary kind of accelerated motion we discuss earlier, such as stepping on your car's gas pedal or brake, involves a change of speed without a change of direction. We have no difficulty recognizing this motion as acceleration. Uniform circular motion, on the other hand, involves a change of direction without a change of speed. This, too, changes the velocity, and therefore requires a compensating acceleration.

Consider a simple example illustrating this point. Imagine holding one end of a string that has a ball attached to the other end and whirling the ball around your head (Figure 3-16). You have to pull on the string all the time—you can feel the tension in your hand. If you let go or if the string breaks, the ball doesn't keep moving in a circle; it flies off in a straight line tangent to the circle. Just as your car won't pick up speed unless you keep your foot on the gas pedal, the ball won't move in a circle unless you keep pulling the string. In both cases, your actions produce an acceleration.

In fact, the acceleration of an object moving in a circle of radius *r* with a velocity whose magnitude (speed) is *v* turns out to be:

1. In words:

The acceleration of an object moving in a circle is equal to the square of its speed divided by the radius of the circle.

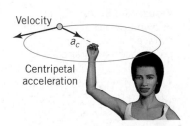

Velocity

a_c

Centripetal acceleration

FIGURE 3-16. The acceleration of an object moving in a circle. The object moves at the same speed, and only the direction changes.

2. In an equation with words:

$$\text{Acceleration (centripetal)} = \frac{\text{Velocity}^2}{\text{Radius}}$$

3. In an equation with symbols:

$$a_c = \frac{v^2}{r}$$

Another interesting fact about this acceleration is that it is directed inward toward the center of the circle, perpendicular to the velocity. (To grasp this point, remember that when you whirl the ball around your head you are actually pulling inward on the string.) For this reason, it is called the *centripetal acceleration* (which means center-seeking acceleration). We encounter centripetal acceleration again in Chapter 5 when we discuss the orbits of satellites.

With Galileo's work, then, physicists began to isolate and observe the motion of material objects in nature and to summarize their results in mathematical relationships. However, why bodies behave this way remained undiscovered. The man who made this discovery was the great English mathematician and physicist Isaac Newton, who was born in 1642, the year of Galileo's death. We discuss Newton's extraordinary contributions in the next two chapters.

THINKING MORE ABOUT

Our Place in the Ordered Universe

Galileo's work and writings had great significance in human society beyond their central importance in science. To understand their importance, we need to recognize the role of Ptolemy's Earth-centered solar system and Aristotle's principles of physics within the context of the general culture and religious beliefs of the time.

The Catholic Church held the idea that humans were the supreme achievement of God's creation. In the eyes of religious leaders, each person's spiritual salvation was the primary concern of our life on Earth, far outweighing issues such as the best model of the solar system. Ptolemy's model placed our planet at the center of the universe, which fit with the Church's sense of the importance of humans in God's overall plan. The Copernican model made Earth just one of several planets orbiting the Sun, which undermined the idea of humans' being central to the rest of the universe. Church leaders feared that such a model might confuse people and lead to doubt about the importance of individual salvation. This fear was not confined to the Catholic Church; both Martin Luther and John Calvin, who spearheaded the Protestant Reformation, condemned the Copernican model.

Church views about how the world worked were based on the explanations of Aristotle, which by the time of Galileo had been unchallenged for some 2000 years. Thus, when Galileo's experiments led to conclusions that differed from those of Aristotle, they also seemed to challenge the authority of the Church. Galileo reported that moons orbited the planet Jupiter, but the Church believed that everything in the heavens revolved around the Earth. Galileo noted dark spots on the Sun but, according to Church doctrine, God created the Sun as a perfect source of light, without blemish. Galileo showed that objects of different mass fell at the same rate, but religious philosophers, following Aristotle's teaching, said that heavier objects fell faster so as to reach the center of the Earth, and thus the center of the universe, more quickly.

Galileo did recant his discoveries, thus avoiding the fate of Giordano Bruno, who did not renounce his belief in the Copernican model and was burned at the stake in 1600. But the Church was fighting a losing battle. By the nineteenth century, not only was the Copernican model of the solar system considered as established fact, but it was beginning to be recognized that the Sun is simply a star, not essentially different from other stars in the sky except for its proximity.

Galileo's troubles arose because his research challenged the prevailing mores of his day. Under what circumstances should society at large restrict the research activities of scientists? What research topics are considered immoral or illegal today? Who should decide these limits?

Summary

Since before recorded history, astronomers have observed regularities in the heavens and have built monuments such as Stonehenge to help establish order in their lives. Models, such as the Earth-centered system of Ptolemy and the Sun-centered system of Copernicus, attempted to explain these regular motions of stars and planets. Astronomers such as Tycho Brahe made ever more precise measurements of star and planet positions. These data led mathematician Johannes Kepler to propose his **laws of planetary motion,** which, among other things, state that planets orbit the Sun in elliptical orbits, not circular orbits as had been previously assumed.

Meanwhile, Galileo Galilei and other scientists investigated the science of **mechanics,** which is the study of how objects move near the Earth's surface. These workers recognized two fundamentally different kinds of motion: **uniform motion,** which means constant **speed** and direction **(velocity),** and **acceleration,** which entails a change in either speed or direction of travel. Galileo devised experiments to study falling objects, and he discovered that all things fall with the constant rate of acceleration of 9.8 meters per second per second, which is called the **acceleration due to gravity (g).** He also discovered the **law of compound motion,** which states that the motion in one dimension has no effect on motion in another dimension.

Key Terms

acceleration The change in velocity divided by the time it takes for that change to occur. Acceleration can involve changes of speed, changes in direction, or both. (p. 60)

acceleration due to gravity (g) The velocity change of a freely falling body at the Earth's surface. (p. 66)

Kepler's laws of planetary motion Three basic mathematical statements about the solar system: *Kepler's first law of planetary motion* states that the planets have elliptical orbits with one focus at the Sun; *Kepler's second law* says that for a given time interval, the swept-out area is the same, no matter where the planet is in its orbit; *Kepler's third law* expresses the relationship between a planet's distance from the Sun and its period as a simple equation that allows scientists to predict the behavior of orbiting objects. (p. 53)

law of compound motion Galileo's proposition that motion in one dimension has no effect on motion in another dimension. (p. 67)

mechanics The branch of physics that deals with motions of material objects. (p. 56)

speed The distance an object travels divided by the time that it takes to travel that distance. (p. 58)

uniform motion Motion at a constant speed in a single direction. (p. 61)

velocity A vector quantity that has the same numerical value as speed but also includes information about the direction of travel. (p. 60)

Key Equations

$$\text{Speed} = \frac{\text{Distance}}{\text{Time}}$$

$$\text{Acceleration} = \frac{\text{Final velocity} - \text{Initial velocity}}{\text{Time}}$$

For an object starting at rest and experiencing constant acceleration:

$$\text{Distance} = \tfrac{1}{2} \times \text{Acceleration} \times \text{Time}^2$$

$$\text{Velocity of falling object} = g \times \text{Time}$$

$$\text{Acceleration due to gravity} = g = 9.8 \text{ meters/second}^2$$

For an object in circular motion:

$$\text{Acceleration} = \frac{\text{Velocity}^2}{\text{Radius}}$$

Review

1. How did Stonehenge allow ancient people to make predictions?

2. Why do scientists argue that ancient astronauts did not build Stonehenge?

3. What are the characteristic movements of some of the objects you see in the night sky?

4. Describe the main features of the Ptolemaic and Copernican systems of the universe. In what ways are they similar?

5. What did Tycho Brahe try to do to resolve the question of the structure of the universe?

6. What was Kepler's role in interpreting Tycho Brahe's data?

7. What is Kepler's first law of planetary motion? What assumption of the Copernican system did this law refute?

8. What is Kepler's second law of planetary motion? According to this law, at what point in its orbit does a planet move fastest?

9. What is Kepler's third law of planetary motion? Given this law, what are the relative lengths of the year for Earth, Venus (next closest planet to the Sun), and Mars (next farthest away planet from the Sun)?

10. How are the works of Tycho Brahe and Kepler an example of the scientific method?

11. What is mechanics? Provide an example of an event that might be studied by this discipline.

12. What new observations did Galileo make with his telescope? Why were these discoveries controversial?

13. Define speed. How can you calculate speed from a known distance traveled and time of travel?

14. What is the difference between speed and velocity?

15. Define acceleration. How can you tell if your car is accelerating by looking at the speedometer?

16. What is instantaneous velocity?

17. What two quantities did Galileo have to measure in his rolling ball experiment? How might you improve on his experiment using modern technology?

18. How did Galileo slow down the rate at which objects fall in the laboratory?

19. How did Galileo measure time?

20. Why is the measurement of time important in mechanics?

21. How are Galileo's experiments in mechanics examples of the scientific method?

22. What is *g*?

Questions

1. The asteroid belt is a large collection of rocks and boulders that lies about three times as far from the Sun as the Earth does. How long is the orbital period, or year, for one of these rocks?

2. Which of the following objects are in uniform motion and which are in accelerated motion? Explain each response.

   a. A car heading north at 35 mph
   b. A car going around a curve at 50 mph
   c. A dolphin leaping out of the water
   d. An airplane cruising at 30,000 feet at 500 mph
   e. A book resting on your desk
   f. The Moon

3. People who put oversized wheels on their cars often find that their actual speed is significantly greater than the speedometer indicates. Can you explain why this is so?

4. A planet orbits the Sun in an elliptical orbit (see figure). The distance along the orbit from A to B is the same as the distance from C to D. Compare the time it takes the planet to move from A to B to the time it takes the planet to move from C to D. Explain your reasoning.

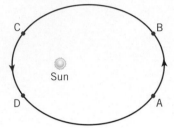

5. Suppose two different moons, X and Y, follow the same elliptical orbit around a planet. Which moon is moving faster according to the figure? Will the faster moon ever catch up to the slower one? Explain.

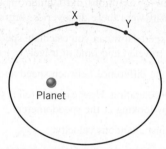

6. By extending the logic used to define instantaneous velocity, define instantaneous acceleration.

7. Consider a comet with a very elongated, elliptical orbit around the Sun. Using Kepler's three laws of motion, describe the speed of the comet as it orbits the Sun.

8. Astronomers investigating other solar systems have found many systems in which the planets have very elongated, elliptical orbits, rather than almost circular orbits as in our own system.

   a. Does the existence of highly elliptical orbits for planets violate Kepler's laws? Why or why not?

   b. What effect do you think a highly elliptical orbit might have on the chances of the planet developing life? (*Hint:* What would happen to Earth's oceans if the planet spent time far away from the Sun?)

9. As you drive north on the highway at 65 miles per hour, the cars in the opposing lane are traveling south at 65 miles per hour. Do the cars in the opposing lane have the same speed as you do? Do they have the same velocity? Explain.

10. Unfortunately, your car has developed an oil leak. One drop of oil falls from your engine every 3 seconds, leaving a trail of oil drops on the road. In the figure are four patterns of oil drops you've left over the same 200-meter stretch of road. For which one(s) is your car accelerating?

For which one(s) is your car moving at a constant speed? For which one is your average speed the greatest?

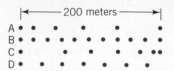

11. Unfortunately, your car has developed an oil leak. One drop of oil falls from your engine every 3 seconds, leaving a trail of oil drops on the road. In the figure are two patterns of oil drops you've left over the same 200-meter stretch of road. In which case do you achieve the highest instantaneous speed? In which case do you have the highest average speed? In which case do you achieve the greatest instantaneous acceleration (acceleration at a specific point)?

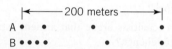

12. Is it possible for your speed to be zero when your acceleration is not zero? Explain.

13. A bowling ball and a volleyball are dropped at the same time from the top of a tall building. Neglecting air drag, which one will hit the ground first? Would your answer change if we did not neglect air resistance? How?

14. A cannon fires a shot from a high cliff as shown in the figure. Where in the cannonball's trajectory is its acceleration the greatest, A, B, C, or D? Where is its speed the greatest? Where is its speed 0? Is its acceleration ever 0? If so, where?

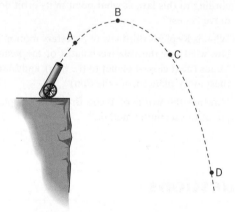

15. You roll a ball off a horizontal tabletop. In which case will the ball take longer to hit the floor: if it is moving fast or if it is moving slowly? Explain.

16. Two balls roll off a horizontal tabletop. One is moving fast and one is moving slowly. Which one hits the ground with a higher speed? Explain.

17. A race car driver is driving on a circular track. If he doubles his speed, how much greater will his centripetal acceleration be? What if he triples his speed?

18. A car is rounding a corner on an cold winter day. If the road suddenly turns to ice (at the X in the figure), where will the car run off the road?

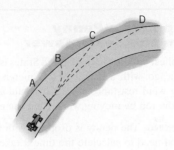

Problem-Solving Examples

EXAMPLE
3-3

Out of the Blocks

A sprinter accelerates from the starting blocks to a speed of 11 meters per second in 1.5 seconds. Answer the following questions about the sprinter's speed, acceleration, time, and distance run. In each case, answer the question by substituting into the appropriate motion equation.

1. What is his acceleration?

SOLUTION: Acceleration is defined as the change in velocity divided by the time interval of that change.

$$\text{Acceleration} = \frac{\text{Final velocity} - \text{Initial velocity}}{\text{Time}}$$

In this case, the sprinter starts from rest at the beginning of the race, so his initial velocity is 0.

$$\text{Acceleration} = \frac{11 \text{ meters/second}}{1.5 \text{ seconds}}$$
$$= 7.3 \text{ m/s}^2$$

2. How far does the sprinter travel during this acceleration?

SOLUTION: We saw from Galileo's experiments that distance traveled is proportional to the square of the time interval.

$$\text{Distance} = \tfrac{1}{2} \times \text{Acceleration} \times \text{Time}^2$$
$$= \tfrac{1}{2} \times (7.3 \text{ m/s}^2) \times (1.5 \text{ s})^2$$
$$= 8.2 \text{ meters}$$

3. How fast is the sprinter going when he's halfway through the period of acceleration?

SOLUTION: This question asks for the runner's instantaneous velocity. We can find it from the values of the acceleration (7.3 m/s²) and the time interval (0.75 s).

$$\text{Instantaneous velocity} = \text{Acceleration} \times \text{Time}$$

At $\tfrac{3}{4}$ second, his instantaneous velocity is

$$\text{Velocity} = 7.3 \text{ m/s}^2 \times 0.75 \text{ s}$$
$$= 5.5 \text{ m/s}$$

4. How far has the sprinter traveled in the first 0.75 second?

A sprinter gets a fast start by pushing off from angled blocks.

SOLUTION: This question is the same as part 2 but over a different time interval.

$$\text{Distance} = \tfrac{1}{2} \times \text{Acceleration} \times \text{Time}^2$$
$$= \tfrac{1}{2} \times (7.3 \text{ m/s}^2) \times (0.75 \text{ s})^2$$
$$= \tfrac{1}{2} \times (7.3 \text{ m/s}^2) \times 0.56 \text{ s}^2$$
$$= 2.05 \text{ m}$$

Notice that halfway through the 1.5-second period of acceleration the sprinter has not covered half of the 8.2 meters (see part 2). This feature is common to accelerated motion. The sprinter moves much faster and farther during the second half of the period of acceleration and therefore covers more ground.

5. Assuming the sprinter covers the remaining 91.8 meters at a constant speed of 11 m/s, what will be his time for the event?

SOLUTION: We have already calculated that the time to cover the first 8.2 meters is 1.5 seconds. The time required to cover the remaining 91.8 meters at a constant velocity of 11 meters per second is

$$\text{Time} = \frac{\text{Distance}}{\text{Velocity}}$$
$$= \frac{91.8 \text{ m}}{11 \text{ m/s}}$$
$$= 8.35 \text{ s}$$

Thus,   Total time = 1.5 + 8.35 = 9.85 seconds

For reference, the world record for the 100-meter dash, set by Tim Montgomery of the United States in 2002, is 9.78 seconds. ●

Dropping a Penny from the Sears Tower

EXAMPLE 3-4

The tallest building in the United States is the Sears Tower in Chicago, with a height of 443 meters (1454 feet). Ignoring wind resistance, how fast would a penny dropped from the top be moving when it hit the ground?

REASONING: The penny is dropped with 0 initial velocity. We first need to calculate the time it takes to fall 443 meters. From this time we can calculate the velocity at impact.

SOLUTION:

Step 1. Time of fall. The distance traveled by an accelerating object is:

$$\text{Distance} = \tfrac{1}{2} \times \text{Acceleration} \times \text{Time}^2$$
$$= \tfrac{1}{2} \times 9.8 \text{ m/s}^2 \times t^2$$
$$= 4.9 \text{ m/s}^2 \times t^2$$

The given distance of the fall equals 443 meters, so rearranging gives:

$$t^2 = \frac{443 \text{ m}}{4.9 \text{ m/s}^2}$$
$$= 90.41 \text{ s}^2$$

Taking the square root of both sides gives the time of fall:

$$t = 9.5 \text{ s}$$

Step 2. Velocity at impact. The velocity of an accelerating object is:

$$\text{Velocity} = \text{Acceleration} \times \text{Time}$$
$$= 9.8 \text{ m/s}^2 \times 9.5 \text{ s}$$
$$= 93.1 \text{ m/s}$$

This velocity is about 200 miles per hour—a high speed indeed. A penny traveling at such a velocity could easily kill a person, so *don't* try this experiment!

In fact, most objects dropped in air do not accelerate indefinitely. Because of air resistance, an object accelerates only until it reaches its *terminal velocity;* and it continues falling at a constant velocity after that point. The terminal velocity for a penny is somewhat less than 200 miles an hour, still fast enough to cause serious injury. We return to the topic of terminal velocity in Chapter 4, after we have studied more about force and motion. ●

Throwing a Ball Straight Up

EXAMPLE 3-5

Suppose you throw a ball straight up into the air with an initial speed of 25 meters per second (about 55 miles per hour).

1. How high will it go?
2. How long will it take to return to the ground?

REASONING AND SOLUTION:

1. The motion of the ball is the result of two effects: the velocity, pointing up, and the acceleration, pointing down. When the ball is thrown straight up, it decelerates because of the effects of gravity. It moves more and more slowly as it climbs, until finally it stops and starts to fall back down. We can use the equation for the velocity of an accelerating body to tell us how long it takes for the velocity to be reduced to 0. Then we can use the equation for the distance traveled by a decelerating body to tell us how far it has traveled (and hence how high it will go).

   The velocity of the ball as it moves up is:

   Final velocity = Initial velocity + Acceleration × Time,

   At the top of the throw, $v = 0$, so

   $$0 = (25 \text{ m/s}) - (9.8 \text{ m/s}^2) \times t$$

   and the time it takes for the ball to stop moving up is

   $$t = \frac{25 \text{ m/s}}{9.8 \text{ m/s}^2} = 2.55 \text{ s}$$

Then, to calculate the distance traveled by the ball in 2.55 seconds, apply the distance equation, making sure to include the effects of both the velocity pointing up and the acceleration pointing down:

$$d = [v_0 \times t] - \left[\tfrac{1}{2} g \times t^2\right]$$
$$= \left[(25 \text{ m/s}) \times 2.55 \text{ s}\right] - \left[\tfrac{1}{2}(9.8 \text{ m/s}^2) \times (2.55 \text{ s})^2\right]$$
$$= 63.75 \text{ m} - 31.86 \text{ m}$$
$$= 31.89 \text{ m}$$

2. There are two ways to determine how long it takes for the ball to return to the ground. One simple way is to note that it takes the ball just as long to fall as it did to climb up, so that the total time of flight is 2×2.55 seconds, or 5.10 seconds.

   The other way is to note that the problem of tracing the ball's path once it starts down is exactly the same as dropping a ball from rest from a height of 31.89 meters. If we work out how long it takes for the ball to fall and add it to the 2.55 seconds it took to get to the top of the throw, we'll have the total time the ball was in the air.

   The time it takes a ball to fall 31.89 meters is

   $$d = \tfrac{1}{2} g \times t^2$$
   $$31.89 \text{ m} = \tfrac{1}{2} \times 9.8 \text{ m/s}^2 \times t^2$$
   $$t^2 = \frac{31.89 \times 2}{9.8} = 6.51 \text{ s}^2$$
   $$t = 2.55 \text{ s}$$

Thus, the total time the ball is in the air is $2.55 + 2.55 = 5.10$ seconds.

Note in this example how we handled the calculation of a distance due to an upward velocity and a downward acceleration. Both factors affect how high the ball goes; the distance traveled is the sum of the distances due to each factor separately. ●

EXAMPLE 3-6

The Human Cannonball

One of the attractions you may see at the circus is the human cannonball (Figure 3-17). This person is lowered into the barrel of a giant cannon, only to be shot out, travel through the air, and finally land in a safety net to the sound of applause from the audience. Suppose he emerges from the cannon's mouth with a vertical velocity of 20 m/s and a horizontal velocity of 8 m/s. How far away would you have to place the net to make sure he landed safely?

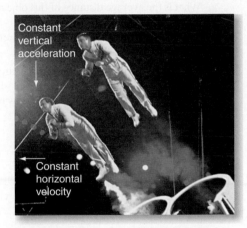

Constant vertical acceleration

Constant horizontal velocity

FIGURE 3-17. The motion of a human cannonball illustrates the law of compound motion. The accelerated up-and-down motion in the vertical direction is completely independent of the uniform motion in the horizontal direction.

REASONING: The way to approach this problem is to remember that the horizontal and vertical motions are independent of each other. In the vertical direction we have a problem just like Example 3-5, in which an object is thrown upward but there is a constant downward acceleration equal to g. In this direction, the object slows down as it moves up, then stops and falls back down. The time for the up-and-down trip is twice the time it takes to get from the ground to the top of the arc. Then we can use the total time of up-and-down travel, multiplied by 8 m/s, to tell us the distance the object travels horizontally, and hence where to place the net.

SOLUTION:
Step 1. How long will it take to get to the top of the arc?

The velocity in the vertical direction is:

Final velocity = Initial velocity + Acceleration × Time

where initial velocity is 20 m/s and final velocity at the top of the arc is 0. We get

$$0 = 20 \text{ m/s} + (-9.8 \text{ m/s}^2 \times t)$$

where all values are in SI units. Note that the acceleration due to gravity has a negative sign because the acceleration is downward. Rearranging this equation, we can solve for time, t:

$$9.8 \text{ m/s}^2 \times t = 20 \text{ m/s}$$
$$t = \frac{20 \text{ m/s}}{9.8 \text{ m/s}^2}$$
$$= 2.04 \text{ s}$$

In other words, the human cannonball takes about 2 seconds to get to the top of the arc and two more seconds to come down, for a total flight time of about 4 seconds.

Step 2. Where should you place the net?

In the horizontal direction, the human cannonball is travelling at a steady 8 m/s. In four seconds the horizontal distance traveled is:

$$\text{Distance} = \text{Velocity} \times \text{Time}$$
$$= 8 \text{ m/s} \times 4.08 \text{ s}$$
$$= 32.6 \text{ m}$$

The net, then, should be placed 32.6 meters (about 106 feet) from the mouth of the cannon.

Note that in the absence of air resistance, the vertical motion is the same for the ball in Example 3-5 and the human in Example 3-6. This equivalence is part of the great strength and beauty of physics: the same principles apply to what may at first seem to be very different situations. ●

Problems

1. If a race car completes a 3-mile oval track in 58 seconds, what is its average speed? Did the car accelerate during the 58 seconds?

2. If your car goes from 0 to 60 miles per hour in 6 seconds, what is your average acceleration?

3. The hare and the tortoise are at the starting line together. When the gun goes off, the hare moves off at a constant speed of 10 meters per second. (Ignore the acceleration required to get the animal to this speed.) The tortoise starts more slowly, but accelerates at the rate of 2 meters per sec-

ond per second. Make a table showing the positions of the two racers after 1 second, 2 seconds, 3 seconds, and so forth. How long will it be before the tortoise passes the hare?

4. Someone in a car going past you at the speed of 20 meters per second drops a small rock from a height of 2 meters. How far from the point of the drop will the rock hit the ground? (*Hint:* Find how long it will take the rock to fall and then apply the law of compound motion.)

5. The Statue of Liberty weighs nearly 205 metric tons. If a person can pull an average of 100 kg, how many people would it take to move the Statue of Liberty?

6. The weight of the space shuttle is about 4.5 million pounds. How many people would it take to move it? (See problem 5.)

7. The eccentricity of an ellipse is a measure of how elongated, or oval, it is. It is defined for a planet's orbit as the distance between the two foci divided by twice the average distance to the sun, which resides at one of the foci (see Figure 3-18). A perfect circle has an eccentricity of zero since the two foci are in the same position.

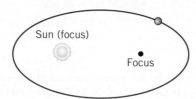

FIGURE 3-18. The eccentricity of a planetary orbit is defined as the ratio of the distance between the foci and twice the average distance to the Sun.

a. Calculate the eccentricities for the following solar system objects. All data are in terms of the average distance of the Earth from the Sun, called the astronomical unit (AU).

Object	$f_1 f_2$ **(AU)**	Average distance
Earth	0.017	1.0
Mars	0.14	1.52
Pluto	9.8	39.5
Halley's comet	17.4	17.9

b. Which object has the most nearly circular orbit? Which object has the most elliptical orbit?

8. For the planets and comet in the list in problem 7, calculate the orbital periods using Kepler's third law of planetary motion.

9. Consider the orbit of a typical comet around the sun given in Figure 3-19, which is marked at five different positions, A, B, C, D, and E. Using Kepler's second law of planetary

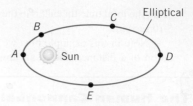

FIGURE 3-19. The elliptical orbit of a hypothetical comet around the Sun is shown with five positions along the orbit.

motion, rank those positions in order of their relative speeds, with the position for the fastest speed first.

10. Imagine that a new asteroid is discovered in the solar system with a circular orbit and an orbital period of 8 years.

a. What is the average distance of this object from the Sun in Earth units?

b. Between which planets would this new asteroid be located?

11. The four Galilean moons of Jupiter are Io, Europa, Ganymede, and Callisto. Their average distances from Jupiter and orbital periods are listed below in terms of Io's values.

Moon	Relative average distance	Relative orbital period
Io	1.00	1.00
Europa	1.59	2.00
Ganymede	2.54	4.05
Callisto	4.46	9.42

a. Plot the square of the relative orbital period versus the cube of the relative average distance for each moon. In words, state the pattern you find in your graph.

b. From this information, do you agree or disagree that Kepler's third law (as applied to the moons of Jupiter) holds for Jupiter's four Galilean moons? Explain.

12. An average person can walk 1 kilometer in 10 minutes.

a. What is the speed in miles per hour? In kilometers per hour?

b. How long would it take an average person to walk 3.5 miles? To walk 10 km?

c. How far can an average person walk in 45 minutes? In 1.5 hours?

13. The North American continental plate is moving away from the European continental plate at a constant speed of 4.2 cm per year.

a. If the average distance between the two plates is 7000 km and the two plates maintained their constant speeds, how long ago were the two continental plates together?

b. In 1 million years (10^6 years), how large will the separation be between the two plates?

14. A typical motorist in the United States travels 25,000 miles in his or her car every year. If you assume that the average speed of the car while traveling is 45 miles per hour (remember that the car is not always moving), calculate the total number of hours an average motorist spends in his or her car. How many hours per day is this? Do you think that the average speed is a reasonable estimate? Explain.

15. The typical airborne speed of an intercontinental B747 jet is 530 miles per hour, while the airborne speed of the supersonic Concorde is 1500 miles per hour. If each airliner were to circumnavigate the Earth (25,000 miles), what would be the difference in air time spent by the two aircraft?

16. It takes light (speed = 3.0×10^8 m/s) 8.33 minutes to travel from the Sun to the Earth and 1.3 seconds from the Moon to the Earth. What is the Sun's average distance from the Earth? The Moon's?

17. While traveling out in the country at 50 miles per hour, your car's engine (and brakes) stops working and you coast to a stop in 25 seconds. What was your average acceleration during the time after the motor shut off?

18. Starting from rest, a train reaches a final, constant speed in 35 seconds while accelerating at a constant rate of 3 km/hour/s.

    a. What is the final speed of the train?
    b. What is the total distance traveled by the train during this period of constant acceleration? (Be careful here with your units.)

19. A rock falls to the bottom of a tall canyon, falling freely with no air resistance, for 4.5 seconds. Make a table of the distance traveled by the rock and its velocity after 1.0, 2.0, 3.0, 3.5, and 4.5 seconds.

20. In a safety test, a car traveling at 65 miles per hour crashes directly into a wall, coming to a complete stop. The time of contact for the crash was 0.25 s. What is the deceleration of the car in terms of the acceleration of gravity (i.e., the number of gs)?

21. A baseball is hit off the edge of a cliff horizontally at a speed of 30 m/s. It takes the ball 3 seconds to reach the ground, with no air resistance.

    a. How far from the cliff wall does the ball land?
    b. How high is the cliff wall?

22. Sammy Sosa pops up a baseball directly over the batter's box. It takes the ball 5.0 seconds to reach the waiting glove of the catcher.

    a. What is the instantaneous speed of the ball at the top of the ball's path?
    b. What is the instantaneous speed of the ball immediately after it was in contact with the bat?
    c. How far above the ground did the ball travel? (Assume that the ball was caught at the same height that it was hit.)

23. Two balls are released simultaneously from the same height, 10 meters above the ground. The first ball is released at rest and the second ball is released with a horizontal velocity of 15 m/s. Which ball reaches the ground first? Why?

24. A girl grabs a bucket of water and swings it around her in a horizontal circle, at a constant speed of 2 m/s at an arm's length of 0.7 meters. What is the centripetal acceleration of the bucket of water?

25. The space shuttle orbits the Earth in a near-circular orbit at a constant speed approximately 100 miles above the Earth's surface. If we assume that the centripetal acceleration is equal to the acceleration due to gravity at sea level (9.8 m/s^2) and the orbital radius is equal to the radius of the Earth (6380 km):

    a. What is the average speed of the space shuttle?
    b. How long does the space shuttle take to make one orbit around the Earth?

26. The Moon moves around the Earth in a near-circular orbit of radius 3.84×10^8 m in 27.3 days. What is the centripetal acceleration of the Moon in m/s^2?

27. Some people who study the history of life on Earth have suggested that every 26 million years a hitherto unknown companion star to the sun comes near the solar system, sending a storm of comets into the inner solar system. (In this scheme, the dinosaurs were wiped out when one of these comets hit the Earth 65 million years ago). From Kepler's laws, what would the axis of the elliptical orbit of this companion, named Nemesis, have to be for its period to be 26 million years? Compare that axis to the distance to the nearest star.

28. Scientists who study the Earth have found that Europe and North America are separating from each other at the rate of about 5 centimeters per year. Assuming this rate has been constant throughout history, estimate the age of the Atlantic Ocean. (*Hint:* How wide is the ocean now?) Compare this number to the age of the Earth.

Investigations

1. Read the Bertold Brecht play *Galileo*, which dramatizes Galileo Galilei's heresy trial. Discuss the dilemma faced by scientists whose discoveries offend conventional ideas. What areas of scientific research does today's society find offensive or immoral? Why?

2. What other kinds of models of the universe did old civilizations develop? Look up those of the Mayans, the Chinese, and the Indians of the American Southwest, and describe some of their models. What features do these models have in common?

3. Find out how Galileo came to the idea of the pendulum clock. What did he actually observe that led him to this development?

4. Drop a wadded-up sheet of paper and a flat one side by side. Which reaches the ground first? Why? What do you think would happen if this experiment were done in a vacuum?

5. When you are in a car traveling at a constant speed, throw a ball up and describe its motion as you see it. Is there a difference when the car is being accelerated?

6. Drop a helium-filled balloon. Does it fall with acceleration *g*? Why? What do you think would happen if you dropped the balloon in a vacuum?

7. Investigate different technologies that scientists use to measure time. What is the shortest time interval that can be measured and how is such a measurement accomplished? What sorts of experiments require this kind of measurement?

 # WWW Resources

See the *Physics Matters* home page at **www.wiley.com/college/trefil** for valuable web links.

1. **www.mcm.acu.edu/academic/galileo/ars/arshtml/mathofmotion1.html** A website based on a video series, *The Art of Renaissance Science,* which includes a discussions of Galileo's contributions to the mathematics of motion, to science, and to art via the development of painting perspective.

2. **galileo.imss.firenze.it/museo/b/egalilg.html** The Galileo room of the History of Science Museum in Florence, Italy. The museum contains originals and models of apparatus described in this chapter, and Galileo's preserved right middle finger.

3. **observe.arc.nasa.gov/nasa/education/reference/orbits/orbits.html** A partially animated NASA tutorial on satellite motion and Kepler's laws.

4. **liftoff.msfc.nasa.gov/toc.asp?s=Satellites** Contains tutorials on types of satellites (including a section on geosynchronous satellites) and an extensive section on tracking current Earth-orbiting spacecraft live via the web.

5. **www.fourmilab.ch/solar/solar.html** An online solar system orrery showing positions of the solar planets and a few comets.

6. **jersey.uoregon.edu/vlab/Cannon/index.html** The cannon Java applet simulates projectile motion, allowing control of launch conditions.

7. **www.mcm.acu.edu/academic/galileo/ars/arshtml/mathofmotion1.html** A website based on a video series, *The Art of Renaissance Science,* which includes a discussion of Galileo's contributions to the mathematics of motion, to science and to art via the development of painting perspective.

4 Isaac Newton and the Laws of Motion

KEY IDEA

Newton's laws of motion describe the behavior of objects on Earth and in space.

PHYSICS AROUND US . . . Getting Around

Travelling fast is so much a part of our modern society that we barely give it a thought. You board a train, sit back in a comfortable seat, and begin to read. You barely notice the feeling of being pushed back into your seat as the train accelerates and leaves the station.

An hour into your journey, you walk back to the snack bar for a doughnut and a cup of coffee. The train sways slightly as you return to your seat, but you're moving so steadily that you don't spill a drop, you make it back to your seat with no mishaps. You hardly no-

tice the motion of the train again until you feel a slight forward pull as you slow down and stop at a station.

The high-speed bullet train above can reach a speed of 175 miles per hour in a straight line. However, the laws that govern its motion are the same no matter what speed it attains.

Every moment of our lives we are subjected to forces and motions. They're so much a part of our daily routine that we scarcely notice the deep and satisfying order that underlies all of these events—an order codified in Newton's laws of motion.

CLASSICAL MECHANICS

Kepler's laws of planetary motion and Galileo's work on falling bodies (see Chapter 3) each describe the motion of objects in a particular situation, but neither is a general description of motion. Kepler's laws, for example, don't tell you how to calculate the trajectory of a space probe. Similarly, Galileo's law of compound motion isn't much help if you want to know how quickly your car can accelerate from 0 to 60 miles per hour. To answer these sorts of questions, we need a theory that tells us how any object is affected by any force. Traditionally, the field of study that deals with these sorts of questions is known as *classical mechanics.*

The man who took the scientific process from the work of Galileo and Kepler to a full-fledged science of mechanics was an Englishman named Isaac Newton (1642–1727). Arguably the greatest scientist who ever lived, Newton made significant contributions to many areas of science and mathematics (see Physics in the Making). However, his most important contributions for our present discussion came in two closely related areas. First, he developed a science of mechanics based on what we now call **Newton's laws of motion.** These three simple laws describe how any object in the universe behaves when acted on by any force. Second, he described the effects of one of the fundamental forces that govern the universe—the force of gravity. Using Newton's discoveries, subsequent generations of scientists were able to develop a comprehensive view of motion in the universe. That sweeping view incorporated everything—from the flow of blood in an artery to the rotation of a distant galaxy—into a single coherent framework. Because of Newton's role in these developments, the area of physics you are about to study is often referred to as *Newtonian mechanics* and the philosophical ideas that grew from it as the *Newtonian worldview.*

Isaac Newton (1642–1727).

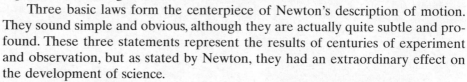

Three basic laws form the centerpiece of Newton's description of motion. They sound simple and obvious, although they are actually quite subtle and profound. These three statements represent the results of centuries of experiment and observation, but as stated by Newton, they had an extraordinary effect on the development of science.

Physics in the Making

Newton's Miraculous Year

Sixteen-sixty-five was not a good year for Cambridge University. Bubonic plague (the Black Death) was making one of its periodic appearances in England, and people fleeing the cities were spreading the disease everywhere. In some towns the situation got so bad that convicts who had been sentenced to be hanged were given the chance to escape their fate by burning the bodies of plague victims. Each morning they wheeled their carts through the streets, calling, "Bring out your dead," and then took the bodies outside the city for cremation.

In situations like these, even university administrations take action. They closed Cambridge University, sending the students back to their homes until the plague abated. For most students and teachers it was probably just a vacation, but for one young man named Isaac Newton the break provided a chance to think on his own, without the distractions of the university. By the time the university opened again 18 months later he had developed (1) a new branch of mathematics known as calculus, (2) the basis for the modern theory of color, (3) the statements of the laws of motion, (4) the law of universal gravitation (which, as we'll see, connected the sciences of physics and astronomy), and, in his spare time, polished off a few mathematical problems that had eluded solution up to that time. In his words, "In those days I was in the prime of my age for invention, and minded Mathematics and Philosophy more than at any time since."

Newton's work formed the basis for most of the scientific and technological discoveries that led to the Industrial Revolution. Even today his ideas are still essential for our basic understanding of science. Little wonder that historians refer to 1665 as Newton's "annus mirabilis," or miraculous year. ●

● THE FIRST LAW

Newton's first law of motion links the key concepts of motion and force.

A moving object will continue moving in a straight line at a constant speed, and a stationary object will remain at rest, unless acted on by an unbalanced force.

Read Newton's first law again, and think about what it means. An object that is moving will continue to move, and an object that is just sitting on a table will continue to sit on the table, *unless* you exert a force on it. It seems self-evident to us that if you leave something alone, it won't change its state of motion. A bowling ball keeps rolling down the lane until it hits the pins. Your car won't change direction unless you turn the steering wheel. A skater glides across the ice until she does something to change direction.

Baseball hit for a home run.

However, virtually all scientists from the ancient Greeks to Copernicus would have argued that Newton's first law of motion is wrong. They were not able to take the step that Galileo and Newton took by separating out the effects of friction on an object's motion. These scientists believed that since the circle is the most perfect geometrical shape, heavenly bodies move in circles unless something interferes. This is why, in their theories, the heavenly spheres kept turning forever.

Newton, basing his arguments on observations and the work of his predecessors, especially Galileo, turned this notion around. A moving object left to itself travels in a straight line. If you want to get it to move in a circle, you have to apply a force. You know this is true from your own experience. If you swing something around your head, such as a ball tied to a string, it moves in a circle only so long as you hold on to it. Let go, and off it flies in a straight line.

Newton's first law tells us that when we see a change in an object's motion, then something must have acted to produce that change. As we saw in Chapter 3, *acceleration* is defined as any change in speed, any change in direction, or any combination of the two.

The first law states that any acceleration is a consequence of the action of a force. In this way, the first law of motion establishes the intimate connection between an acceleration and a force. Indeed, this law defines **force** as a phenomenon that can produce a change in an object's state of motion. This definition provides a simple way of recognizing forces in nature. In fact, we'll use the first law of motion extensively in this book to tell us how to recognize when a force, particularly a new kind of force, is acting.

Pool balls are an example of how uniform motion can suddenly change because of a force.

Inertia

We sometimes give the tendency of an object to remain in uniform motion—to resist changes in its state of motion—the special name **inertia.** In fact, Newton's first law of motion is sometimes called the law of inertia. Inertia is why your body presses back into your seat when your car accelerates forward, and why your body pushes forward against your seat belt when you hit the brakes.

Inertia is another one of the many terms in physics that's also often used in everyday speech. You might speak of the inertia in a company or government organization when you want to get across the idea that it is resistant to change. The physics definition is that inertia resists changes in motion.

Develop Your Intuition: Inertia

Have you ever been in a restaurant and had difficulty getting the ketchup out of a glass bottle? How could inertia help you get it out?

If you're like most people, you held the bottle upside down and shook it vigorously. When the ketchup came out with a plop, you were seeing inertia in operation—a successful application of Newton's first law of motion. Here's how it works.

When you held the bottle upside down and started moving it down, you gave everything in your hand—the bottle and its contents—downward velocity imparted by the force of your hand. When you suddenly stopped the motion of your hand, you caused a change in the state of motion of the bottle. The ketchup, however, is not slowed down by your hand, but only by the friction between itself and the walls of the bottle. Consequently, it obeys Newton's first law of motion and keeps moving in a straight line. "Plop!"

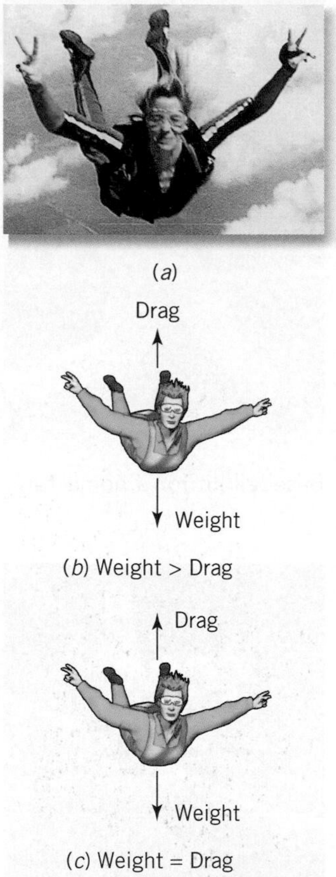

(a)

Drag

(b) Weight > Drag

Drag

Weight

(c) Weight = Drag

FIGURE 4-1. (a) A skydiver at terminal velocity falls with a constant speed. (b) Weight is greater than drag. Velocity is increasing, and there is acceleration downward. (c) Weight equals drag. Velocity is constant, and acceleration equals zero.

Balanced and Unbalanced Forces

One important aspect of the first law lies in the innocuous word "unbalanced." This word has to do with what happens when more than one force acts on an object. The unbalanced force on an object is often referred to as the *net force* acting on it. In the more complicated situations in which two or more forces act in different directions, the net force is sometimes called the *resultant force*.

To illustrate what we mean, imagine a skydiver jumping out of an airplane (see Figure 4-1a). We know that gravity acts on the skydiver, pulling her downward. But we also know that as she starts to fall she must push the air aside, and this push results in another force acting on her—a force engineers call *drag*, or *air resistance*. This force is directed upward (i.e., in the opposite direction to gravity) and it increases as she falls faster and faster. So as the skydiver falls, we can distinguish two distinct situations.

Unbalanced Force At the beginning of her fall, the skydiver is moving slowly, so the drag force is small (Figure 4-1b). Consequently, the downward force of gravity is much larger than the drag force. We define the *unbalanced force* on the skydiver as the difference between the downward and upward forces. At the start of the fall, then, this force is directed downward. This net force—the sum of all the forces acting on an object—is referred to as the *unbalanced force* in the first law of motion. According to this law, the skydiver should accelerate as she starts to fall—something we know to be true from everyday experience. She continues to accelerate as long as there is an unbalanced force acting.

Balanced Forces As the falling skydiver speeds up, the drag force grows. Eventually, the upward drag force equals the downward force of gravity (Figure 4-1*c*). At this point, the net force (gravity minus drag) is zero. There is no unbalanced force on the skydiver. According to the first law of motion, the balanced pair of forces means that the skydiver's acceleration goes to zero. Whatever velocity she had at the point where the net force went to zero is the velocity she has from that point on. This final speed is called the *terminal velocity*. For a skydiver falling toward the Earth the terminal velocity is typically about 100 miles per hour. Of course, once the parachute opens, the drag is greatly increased and the skydiver's velocity drops to just a few miles per hour.

Why doesn't a falling object stop moving when the net force goes to zero? According to the first law of motion, *any* change of motion requires the action of an unbalanced force. Stopping a falling object in midair requires that the object be accelerated (remember that a deceleration is just a negative acceleration). The object can't slow down unless some unbalanced force is acting on it, and since the net force on the object is zero there can be no change in its state of motion. Consequently, it just keeps moving along at the same speed. It keeps falling at the terminal velocity.

Another common example of opposing forces is connected to the phenomenon of friction. If you try to slide a heavy desk along a floor, you know that once you get the desk moving, you have to keep exerting a force to keep it moving. Stop pushing and the desk decelerates and stops. A desk moving across the floor at constant speed is not accelerating. (Remember that in order to accelerate, an object's velocity has to change.) This means that the net force acting on it must be zero. Since you are pushing and exerting a force (to the right, for example), there must be another force acting to the left. This is the force of friction, and it is generated whenever two surfaces slide past each other.

If you could see the contact between the bottom of the desk and the floor, you would see that both surfaces consist of microscopic peaks and valleys. When you move one surface across the other, the bonds that hold atoms and molecules together are constantly being broken and re-formed. You have to overcome these forces when you're pushing the desk.

Force of chair upward on you

Weight (Force of Earth downward on you)

FIGURE 4-2. Vector diagram of a person sitting in a chair and the forces acting on him.

Develop Your Intuition: Force Exerted by a Chair

Use Newton's first law of motion to describe what happens when you (1) sit in a chair that holds you up, and (2) sit in a chair that collapses under you.

In the first case, you don't accelerate; so, according to Newton's first law, the net force acting on you must be zero. Two forces act on you: the Earth exerts a downward force on you (your weight) and the chair exerts an upward force on you. Since the net force equals zero, these two forces must be equal in magnitude. The chair is strong enough to exert a force that exactly balances the force of gravity (Figure 4-2).

If, however, the chair is old and rickety, it might not be able to exert that much force. In this case, the net force is not zero and gravity will win. You will accelerate downward until you encounter an object (the floor, for example) capable of exerting an upward force equal to that exerted by gravity. Have you ever sat down in a chair that broke?

THE SECOND LAW

Newton's first law of motion tells you when a force is acting; the second law of motion tells you what the force does when it acts. The precise statement of the second law is:

> *When a net force acts on an object, the object accelerates in the same direction as that force. The acceleration is directly proportional to the net force and inversely proportional to the mass of the object.*

The **mass** of an object is simply the amount of matter contained in that object.

Newton's second law conforms to our everyday experience that it's easier to propel a bicycle than a car, easier to lift a child than an adult, and easier to deflect a ballerina than a defensive tackle. The second law of motion can be described as an equation.

1. In words:

> *The bigger the force, the greater the acceleration; the larger the mass, the smaller the acceleration.*

2. In an equation with words:

Force = Mass (in kilograms) × Acceleration (in meters per second^2)

3. In an equation with symbols:

$$F = m \times a$$

This equation, well known to generations of physics students, tells us that if we know the forces acting on a system of known mass, we can predict its future motion. Newton's second law supports our intuition that an object's acceleration is a balance between two factors. On the one hand, a net force causes an acceleration; therefore, the greater the net force, the greater the acceleration. The harder

A small force can cause a change in motion of a large object.

you throw a ball, the farther it goes; the harder you pedal a bike, the faster it goes. According to the second law of motion, the acceleration of an object is directly proportional to the net force applied.

On the other hand, the greater the object's mass—the more stuff you have to accelerate—the less effect a given force is going to have. A given force accelerates a golf ball more than a bowling ball, for example, and it takes twice as much force to start two chairs moving along the floor as it does to start one moving. In this way, Newton's second law of motion defines the balance between force and mass in producing an acceleration. Recasting the equation, we can see that acceleration is inversely proportional to mass:

$$\text{Acceleration} = \frac{\text{Force}}{\text{Mass}}$$

The Unit of Force

In the English system, force is measured in the familiar unit of the pound. In SI, the unit of force is called the **newton** (abbreviated N). Think about the second law's definition of force as mass times acceleration. A *newton* is defined as the force that accelerates a 1-kilogram mass at the rate of 1 meter per second per second.

$$1 \text{ N} = 1 \text{ kg-m/s}^2$$

For comparison, a 1-pound force is equal to about 4.45 N. See Looking at Force on page 86 for other examples of forces.

LOOKING DEEPER

When Forces Don't Act in the Same Direction

When more than one force acts on an object, but the forces don't act in the same direction, the recipe for finding the net force and resultant motion is a little more complicated. Any force can be represented as a vector—that is, as an arrow in which the direction of the arrow represents the direction of the force and the length of the arrow represents the strength, or magnitude, of the force. In Chapter 2 we learned how to add vectors that do not act in the same direction. When two forces act together, the resultant force is just the sum of the two vectors placed head to tail. In Figure 4-3 we show an example of such a situation. The two forces, shown as vectors A and B, act on a single object. On the right, we show the triangle whose sides are the vectors A and B, placed head to tail. The resultant force is shown as vector C, the third side of the triangle. This force has a magnitude equal to the length of C and points in the direction of C, as shown.

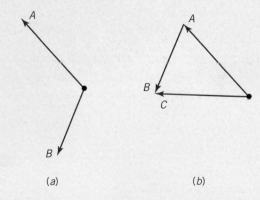

FIGURE 4-3. Vector addition of forces. (*a*) Forces *A* and *B* act on an object. (*b*) When vectors *A* and *B* are placed head to tail, the third side of the triangle is their sum, vector *C*.

If more than two forces are acting, you can find the resultant force for all of them by finding the resultant force for two, then finding the resultant force of that force with a third, then combining *that* resultant force with a fourth, and so on. Remember, no matter how many forces act, the object can accelerate in only one direction and the net effect must be the same as a single force.

Looking at Force

An apple weighs about 1 newton, or a quarter of a pound, roughly the same as a good-sized hamburger. That's a lot more than the force of attraction in the bonds of molecular DNA, about 10^{-7} N—although each molecule of DNA has about 100,000 such bonds. When it comes to large forces, consider the structures built to support roads, fly through the air, or control the flow of rivers. These are truly forces to reckon with.

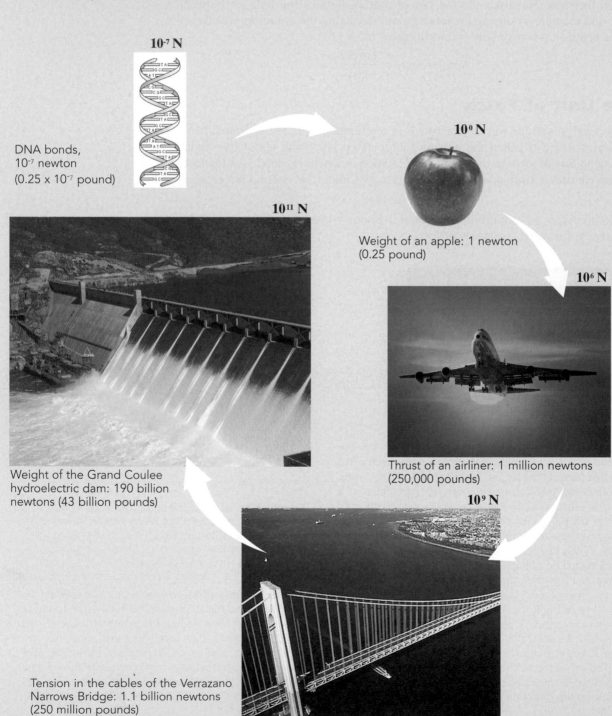

10^{-7} N

DNA bonds,
10^{-7} newton
(0.25×10^{-7} pound)

10^0 N

Weight of an apple: 1 newton
(0.25 pound)

10^{11} N

Weight of the Grand Coulee
hydroelectric dam: 190 billion
newtons (43 billion pounds)

10^6 N

Thrust of an airliner: 1 million newtons
(250,000 pounds)

10^9 N

Tension in the cables of the Verrazano
Narrows Bridge: 1.1 billion newtons
(250 million pounds)

THE THIRD LAW

Newton's third law of motion tells us that whenever a force is applied to an object, that object simultaneously exerts an equal and opposite force on whatever is applying the initial force.

For every action (force) there is an equal and opposite reaction (force).

For example, when you push on a wall it pushes back on you; you can feel the force on the palm of your hand. In fact, the force the wall exerts on you is equal in magnitude, but opposite in direction, to the force you exert on it. And the harder you push on the wall, the harder it pushes on you.

The third law of motion is perhaps the least intuitive of Newton's three laws. We tend to think of our world in terms of causes and effects, in which big or fast objects exert forces on smaller, slower ones. A pianist pushes down on the piano key to produce a note. A basketball player pushes down on the ball to bounce it. A meteorite streaks down from space and smashes into the surface of the Earth. What the third law tells us is that there is another way to look at each of these situations. The pianist does, indeed, exert a force on the key, but at the same time the key exerts an equal and opposite force on the pianist, bringing the motion of the finger to a stop (Figure 4-4). As the basketball player exerts a force on the ball, the ball exerts an equal and opposite force on his hand. (Figure 4-5). And even as the meteorite smashes into the solid Earth, the Earth exerts an equal and opposite force on the meteorite. The third law tells us that every time one object exerts a force on another object, the second object exerts an equal and opposite force on the first. Forces always come in pairs.

One important point about Newton's third law—a point that often causes a great deal of confusion—is that while it's true that forces always come in action–reaction pairs, *the two forces in this pair act on different objects.* The force exerted by the pianist's finger pushes on the key, but the force exerted by the key pushes on the pianist's finger. When you push on the wall, the force you exert acts on the wall. The force could, in principle, accelerate the wall (if you were pushing over a loose pile of bricks, for example). However, the force you exert on the wall cannot accelerate you because it does not act on you. The force exerted by the wall does act on you, and could cause you to accelerate. For example, if you bumped into the wall while running, you might bounce backward. Ouch!

FIGURE 4-4. A pianist presses down on the piano keys to make the notes sound; the keys press back up on the pianist's fingers.

Force of finger on piano key

Force of piano key on finger

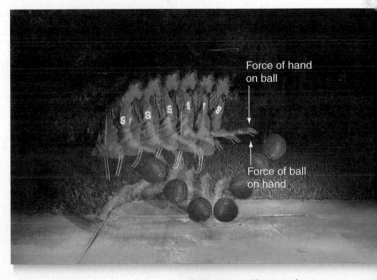

FIGURE 4-5. This strobe photo of a ball being dribbled shows that when you push on the ball, the ball exerts a force back on you.

Force of hand on ball

Force of ball on hand

Develop Your Intuition: A Horse and Cart

Here's a conundrum often used to bedevil beginning physics students: think about a horse and a cart. The horse exerts a forward force on the cart, but the cart exerts an equal and opposite backward force on the horse. How can the horse and cart possibly move forward?

The key to resolving this dilemma is to remember that Newton's second law of motion says that acceleration of an object results from the action of a net force *on that same object.* Take the horse, for example. It pulls on the cart, and therefore there is a backward force exerted on it by the cart. However, the horse also pushes backward against the Earth so there is a reaction force exerted on the horse by the Earth, and this force acts in the forward direction. If the horse pushes hard enough, the forward reaction force exerted on it by the Earth will be larger than the backward force of the cart, and the horse will move forward (Figure 4-6). The cart exerts a force on the horse and the horse exerts a force on the cart, but these two forces are not acting on the same object and so cannot be counted as part of the same net force causing an acceleration.

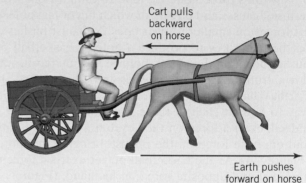

FIGURE 4-6. The horse pulls forward on the cart and the cart pulls backward on the horse. How can they move? Vector diagram of the horse shows only those forces acting on a single object.

Cart pulls backward on horse

Earth pushes forward on horse

If the surface is slippery (ice, for example), it's harder for the horse to push backward, so that the reaction force of the Earth is less, and the horse may not be able to move the cart. Have you ever tried to move a heavy object (a sled loaded with firewood, for example) on an icy surface and had your feet slip out from under you?

● NEWTON'S LAWS TAKEN TOGETHER

Isaac Newton's three laws of motion form a comprehensive description of all possible motions. Taken together, they give us a way of analyzing everything in the universe, from billiard balls rolling on a table to the track of a space probe. In many situations, all three laws have to be used together to analyze a particular situation.

This interplay of Newton's three laws of motion can be seen in a simple example. Imagine a child standing on roller skates holding a stack of baseballs. He throws the balls forward, one by one. Each time he throws a baseball, the first law tells us that he has to exert a force so that the ball accelerates. The third law then tells us that the baseball exerts an equal and opposite force on him. This force acting on the child, according to the second law, causes him to recoil backward.

While the example of the child and the baseballs may seem a bit contrived, it illustrates the exact principle by which rockets work. In a rocket engine, forces are exerted on hot gases, accelerating them out the tail end of the rocket. As with the child on skates, the first law tells us that the rocket exerts a force on the gas, the third law reveals that the gases exert an equal and opposite force on the rocket, and the second law quantifies the amount that this reaction force causes the rocket to accelerate. Every rocket, from Fourth-of-July fireworks to the space shuttle, works this way.

Develop Your Intuition: Rockets in Space

In the early days of rocketry, prestigious publications such as the *New York Times* argued that rockets could never work in space because there was no air for them to push against. If the exiting gases didn't push against any air, there would be no reaction push by the air against the rocket. How would you answer this argument using Newton's laws?

The fact is that a rocket works because of the reaction force exerted by exhaust gases against the rocket itself, not by the force the gases exert on the air (Figure 4-7). Thus, the rocket works whether there is air around it or not. The Apollo landings on and takeoffs from the Moon demonstrated convincingly that a rocket could indeed work in the absence of air.

FIGURE 4-7. The space shuttle *Discovery* rises from its launch pad at Cape Canaveral, Florida. As hot gases accelerate violently out the rocket's engine, the shuttle experiences an equal and opposite acceleration that lifts it into orbit.

THINKING MORE ABOUT

Newton's Laws of Motion: Isaac Newton Plays Volleyball

Newton's three laws of motion apply to every action of our lives. Think about a game of doubles volleyball at the beach. It's a beautiful sunny day with a light breeze off the ocean. The opposing team serves the ball low and hard, but your partner gives you a perfect pass. You set the ball high and outside, a foot off the net. She makes her approach, leaps high, and hits down the line for a side out.

During this one brief play, dozens of forces and motions took place. The ball changed direction and speed, players ran and jumped, sand

Newton's laws of motion apply everywhere, including the volleyball court.

grains shifted, and the wind blew while waves crashed in the background. Think about how Newton's laws of motion apply to these events.

The first law says that nothing happens without an unbalanced force. When a ball flies through the air or a person jumps high off the ground, a force must be involved. One everyday force is gravity, which causes the ball (and the players) to fall back to Earth. Contact forces, such as occur when the player's hand strikes the ball or when she leaps into the air by pushing against the sand, are also all around us. The origin of other forces, such as those that cause the wind to blow or waves

to crash onto the shore, are less obvious. But whatever the event, Newton's first law of motion tells us that a force must exist.

Newton's second law of motion allows us to calculate exactly how much force must be applied to achieve a given result (force equals mass times acceleration). A hard serve requires more force than a dink serve, just as a 36-inch vertical jump demands more force than a 12-inch jump. Volleyball players can't stop to perform complex mathematical calculations in the middle of their game. Nevertheless, our cumulative experience, living in a world of forces and motions, gives us a pretty good idea about how much force to use in a given situation. And the more you practice a sport, the better your intuition becomes about forces, motions, and Newton's second law.

Newton's subtle third law of motion comes into play throughout the volleyball game: all forces occur in pairs. As the player hits the ball, applying a force to propel it forward, the player's hand feels the force of the ball pushing back. (If you've ever received a painfully hard spike, you'll know about forces acting in pairs.) When you push down to jump high off the sand, the sand pushes with an equal and opposite force against you. As the wind blows against your hair, your hair simultaneously changes the path of the wind.

Some time when you're taking a walk, look at what goes on around you and think about how Newton's three laws of motion relate to what you see. What forces are involved when a car slows down, a leaf falls, or a squirrel runs up a tree? What is the unceasing interplay of forces and motions at work when you ride a bike, read a book, use a computer, or attend a concert? (For some other examples, look at Physics and Daily Life on p. 91. Can you identify the forces acting and the objects each force acts on? Can you identify the action–reaction pairs of Newton's third law?)

Summary

Isaac Newton combined the work of Kepler, Galileo, and others in his three **laws of motion.** The first law of motion states that nothing accelerates without an unbalanced **force** acting. Without an unbalanced force, objects remain at rest or in constant motion. **Inertia** is the tendency of an object to remain in uniform motion—to resist changes in its state of motion. The second law of motion quantifies net force (measured in **newtons**) in terms of the **mass** and acceleration of an object. The amount of acceleration is proportional to the force applied and inversely proportional to the mass. The third law of motion states that forces always act in equal and opposite pairs. Taken together, these three laws describe motions on Earth and in space.

Physics and Daily Life—Forces

Forces appear all around us all the time, although we may not be aware of them. Weight is a force and holding an object in the air also requires a force. And any time you change an object's motion—hit a volleyball, walk across the grass, raise your glass in a toast—you exert forces, while forces are exerted on you.

Force of ball on hand

Force of hand on ball

Weight

Force of ground on feet

Force of child on hands

Force of hands on child

Force of feet on ground

Force of ground on feet

Force of hand on cup upward

Force of cup on hand downward

Key Terms

force (measured in newtons) A phenomenon that can produce a change in an object's state of motion. (p. 81)

inertia The tendency of an object to remain in uniform motion—to resist changes in its state of motion. (p. 82)

mass The amount of matter contained in an object, independent of where that object is found. (p. 84)

newton (N) The SI unit of force that accelerates a 1-kilogram mass at the rate of 1 meter per second per second. (p. 85)

Newton's laws of motion Three laws that describe how any object in the universe behaves when acted on by any force; the *first law* states that a moving object will continue moving in a straight line at a constant speed, and a stationary object will remain at rest, unless acted on by an unbalanced force; the *second law* states that the acceleration produced on a body by a force is proportional to the magnitude of the force and inversely proportional to the mass of the object; the *third law* states that for every action (force) there is an equal and opposite reaction (force). (p. 80)

Key Equation

Force = Mass × Acceleration

Review

1. What is Isaac Newton's first law of motion? Give an everyday example of this law in action.

2. What is inertia? Give an everyday example of this phenomenon.

3. What is the force that causes a ball rolling on the floor to slow down and stop?

4. What is Isaac Newton's second law of motion? Give an everyday example of this law in action.

5. What is Isaac Newton's third law of motion? Give an everyday example of this law in action.

6. What is a force? What is the unit of force?

7. What is an unbalanced force? How does it relate to the net force?

8. How can you recognize the action of an unbalanced force?

9. According to Newton, what are the two kinds of motion in the universe?

10. How does Newton's third law apply to the blastoff of the space shuttle?

11. What is meant by an "equal and opposite force"?

12. Review how Newton's laws of motion come into play during a volleyball game.

13. Pick an everyday activity and discuss how Newton's three laws of motion come into play.

14. Think about your favorite sport. In what circumstances is being small an advantage? In what circumstances is being large an advantage? Why?

15. State Newton's second law of motion in your own words.

16. Why is mass significant in Newton's second law of motion?

17. How is inertia related to mass?

18. List some of the accomplishments of Sir Isaac Newton during the plague years of 1665 and 1666.

Questions

1. What pairs of forces act in the following situations?
    a. A pitcher throws a fast ball.
    b. A pencil rests on your desk.
    c. A car hits a tree.
    d. The wind pushes a sailboat.

2. A ball rolling along the floor doesn't keep moving in a straight line at constant speed. Why? What force is involved?

3. When you are moving up at constant speed in an elevator, there are two forces acting on you: the floor pushing up on you and gravity pulling down. Compare the strengths of these two forces. Explain your answer in terms of Newton's first law of motion.

4. You step into an elevator at the 10th floor and press the 2nd-floor button. The elevator accelerates downward briefly as it picks up speed. There are two forces acting on you: the floor pushing up on you and gravity pulling down. Compare these two forces while the elevator is accelerating downward. Explain your answer in terms of Newton's second law of motion.

5. A passenger on a Ferris wheel moves in a vertical circle at constant speed. Is she accelerating? Are the forces on her balanced? Explain.

6. A 28,000-pound jet airliner cruises at 500 miles per hour and an altitude of 35,000 feet. The forward thrust of the engines is 10,000 pounds. Assuming the plane maintains altitude and speed, what is the total air drag force pushing back on the plane? What is the total lift force pushing up on the plane? Explain.

7. A rocket blasting into space provides an example of Newton's third law of motion. Outline the forces acting on the rocket as well as on the propellants.

8. In order to slide a heavy desk across the floor at constant speed, you have to exert a horizontal force of 500 newtons. Compare the 500-newton horizontal pushing force to the frictional force between the desk and the ground. Is the frictional force greater than, less than, or equal to 500 newtons? Explain.

9. What does Newton's second law of motion tell us that the first law does not? Explain how Newton's first law of motion is actually a consequence of Newton's second law.

10. What force is involved when the following objects accelerate:
    a. A sailboat changes direction.
    b. A car turns a corner.
    c. A leaf falls from a tree.
    d. You turn a page of this book.

11. A race car drives along a straight track at constant speed. Are the forces acting on the driver balanced? Explain.

12. A heavily loaded freight train moves with constant velocity. Compare the net force on the first car to the net force on the last car. Explain.

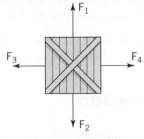

Questions 13, 14

13. In the figure, if the box is accelerating to the right, compare F_1 to F_2 and F_3 to F_4.

14. In the figure, if the box is accelerating upward and to the right, compare F_1 to F_2 and F_3 to F_4.

15. Five 10-N forces (represented by arrows in the figure) act on ropes connected to an iron ring in the directions shown.

Will the ring experience acceleration? If so, in what direction will it accelerate? Explain.

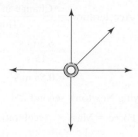

16. A picture hangs on the wall, supported by a wire connected to two corners of the picture frame, as shown. Draw two arrows that represent the two forces that the wires exert on the picture frame. Draw another arrow that represents the addition of those two forces.

17. Two 20-N forces and a 40-N force act on a hanging plant as shown. Will the plant experience acceleration? If so, in what direction will it accelerate? Explain.

20 N   20 N

40 N

18. When a rocket is launched, it carries a heavy load of fuel, which is burned during the ascent. How does the mass of the rocket change as it climbs up on its trajectory? Is the force needed to produce a given acceleration a few minutes after launch larger or smaller than the force needed to produce the same acceleration just off the launching pad?

19. In modern physics, we often talk about forces in terms of an exchange of particles between objects. To see how this might work, imagine the following situation. Two students are running side-by-side in a straight line to catch a train. One is carrying a heavy suitcase and halfway to the train he throws it over to his friend.

a. What forces are created as a result of the throwing of the suitcase?
b. What forces are created as a result of the friend catching the suitcase?
c. As a result of the change, are the students forced to deviate from straight line motion? Explain.

20. A bicycle rider accelerates from rest up to full speed on a flat, straight road. Compare the frictional force between the road and the tires pushing her forward to the air drag (and other frictional forces) pushing back: a. in the first few seconds of the ride, and b. after she has reached full speed.

21. When an object is moving in air, the air drag force is in the opposite direction to the velocity. A light foam ball is thrown up into the air. When is the net force on the ball the greatest: When it is moving up, when it is at the top of its trajectory, or when it is moving down? Explain.

22. A car is driving up a straight hill at a constant speed of 50 kilometers per hour. Is the net force on the car zero? A second car is driving over the crest of the hill at a constant speed of 50 miles per hour. Is the net force on the car zero?

23. When you kick a soccer ball, thus applying a force to the ball, what is the equal and opposite force? Which force is greater?

24. A female gymnast weighs 400 N. If she is hanging stationary from a high bar, what are the two forces acting on her? Compare the strength of those forces.

25. What is wrong with this statement: In a tug-of-war contest between two people, the person who pulls harder usually wins. (Assume the rope is light and it does not slip in either person's hands.)

Problem-Solving Examples

EXAMPLE 4-1

The Blue Whale

An adult blue whale, the largest animal on Earth, can have a mass as high as 150,000 kilograms (330,000 pounds). How much force must such a whale generate with its massive tail to achieve its swimming velocity of 30 kilometers per hour in 15 seconds?

REASONING: Newton's second law of motion says that force equals mass (in kilograms) times acceleration (in meters per second per second), so we need to know both the mass and the acceleration to calculate force. The mass is given as 1.5×10^5 kg. The acceleration is 30 kilometers per hour in 15 seconds, which must be converted to standard units of meters per second per second. To do this, first recognize that 30 kilometers equals 3×10^4 meters and 1 hour equals 3600 seconds.

$$30 \text{ kilometers/hour} = \frac{3 \times 10^4 \text{ meters}}{3600 \text{ seconds}}$$

$$= 8.3 \text{ m/s}$$

The change in velocity is thus 8.3 meters per second in 15 seconds.

$$\text{Acceleration} = \frac{\text{Change in velocity}}{\text{Time}}$$

$$= \frac{8.3 \text{ m/s}}{15 \text{ s}}$$

$$= 0.55 \text{ m/s}^2$$

Now, applying Newton's second law

$$\text{Force} = \text{Mass} \times \text{Acceleration}$$
$$= (1.5 \times 10^5 \text{ kg}) \times (0.55 \text{ m/s}^2)$$
$$= 82{,}500 \text{ newtons}$$

That's equivalent to almost 10 tons of force! ●

The Sum of Forces in Two Directions

EXAMPLE 4-2

A tree-removal team wants to direct the fall of a tree away from a house and power lines. To do this, they attach two ropes to the tree, as shown in Figure 4-8. The angle between the ropes is 90 degrees and the pull on each rope is 2000 N. What are the direction and magnitude of the net force acting on the tree?

REASONING: The net force is the vector sum of the two applied forces. We need to make a triangle to determine this resultant force. Two sides of the triangle are 2000 N long and the angle between them is 90 degrees. We then use the Pythagorean theorem to find the length of the hypotenuse of that triangle.

SOLUTION: The Pythagorean theorem tells us that if the lengths of two sides of a right triangle are a and b, then the length of the hypotenuse, c, is given by the equation

$$c^2 = a^2 + b^2$$

In this case, if we call the length of the resultant force F, we have:

$$F^2 = (2000)^2 + (2000)^2 = 8{,}000{,}000 \text{ N}^2$$

so that:

$$F = 2828 \text{ N}$$

Because the two sides of the triangle are of equal length, the resultant force is at an angle of 45 degrees from each of the ropes—halfway between them, in fact. ●

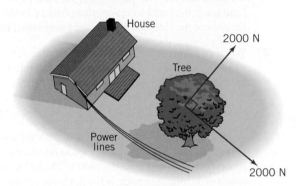

FIGURE 4-8. Pulling on ropes to deflect the fall of a tree. The net unbalanced force is the sum of the forces exerted on the ropes.

Problems

1. How much force must be applied during liftoff to accelerate a 20-kg satellite just enough to counter the Earth's gravitational acceleration of 9.8 m/s²?

2. Which of these objects has the greatest inertia: a mosquito, a VW bug, or an ocean liner? Justify your response by creating a mental experiment that uses Newton's second law, which can relate mass and inertia.

3. You are driving a car down a straight road at a constant 55 miles per hour.

   a. Are there any forces acting on the car? If so, list them.
   b. Is there a net force or unbalanced force acting on the car? Explain.

   c. If the car started to go around a long bend, still maintaining its constant speed of 55 miles per hour, is there a net force acting on the car? Explain.

4. John pushes a loaded wheelbarrow, which is initially at rest, with a constant horizontal force of 10 newtons. The mass of the wheelbarrow is 15 kg. Neglect friction forces.

   a. What is the constant acceleration of the wheelbarrow?
   b. If John pushes the wheelbarrow for 3 seconds, what distance does the wheelbarrow cover during this time?
   c. What is the speed of the wheelbarrow after 3 seconds?

5. Suzie (50 kg) is roller-blading down the sidewalk going 20 miles per hour. She notices a group of workers down the

walkway who have unexpectedly blocked her path, and she makes a quick stop in 0.5 seconds.

   a. What is Suzie's average acceleration in meters per second^2?

   b. What force in newtons was exerted to stop Suzie?

   c. Where did this force come from?

6. Margie (45 kg) and Bill (65 kg), both with brand new roller blades, are at rest facing each other in the parking lot. They push off each other and move in opposite directions, Margie moving at a constant speed of 14 ft/s. At what speed is Bill moving? (*Hint:* Recall from Newton's third law that Margie and Bill experience equal and opposite forces.)

7. Tracy (50 kg) and Tom (75 kg) are standing at rest in the center of the roller rink, facing each other, free to move. Tracy pushes off Tom with her hands and remains in contact with Tom's hands, applying a constant force for 0.75 seconds. Tracy moves 0.5 meters during this time. When she stops pushing off Tom, she moves at a constant speed.

   a. What is Tracy's constant acceleration during her time of contact with Tom?

   b. What is Tracy's final speed after this contact?

   c. What force was applied to Tracy during this time? What is its origin?

   d. What happened to Tom? If Tom moved, describe his motion, force, acceleration, and Tom's final velocity.

8. A fast-moving VW Beetle moving at 60 mph hit a mosquito hovering at rest above the road.

   a. Which bug (insect or VW) experienced the largest force?

   b. Which bug (insect or VW) experienced the greatest acceleration?

Investigations

1. Read a biography of Isaac Newton. Were all of his ideas about the physical world correct? What were his views on religion and mysticism?

2. Design an apparatus that could be used to determine the mass of an object in the weightlessness of space. (*Hint:* Apply Newton's second law of motion.)

3. Investigate how a sailboat can tack into the wind. What forces are involved? Is it possible for a sailboat to sail directly into the wind? Why?

4. Investigate the forces involved in launching the space shuttle into orbit. In the first seconds after liftoff, how much of the total force of the rocket engines is used to lift the fuel (as opposed to the shuttle itself)?

5. Investigate NASA's *Messenger* space probe, which will be launched in 2004 to orbit the planet Mercury. This complex mission will require more than 99 kilograms of fuel and other hardware for every pound of payload. What are the objectives of the mission? Why is so much mass of fuel required?

6. Propose a series of simple experiments that you might devise to test each of Newton's three laws in the laboratory. What measurements would you make? What would be the principal sources of error in these measurements?

 # WWW Resources

See the *Physics Matters* home page at **www.wiley.com/college/trefil** for valuable web links.

1. **www.physics.uoguelph.ca/tutorials/fbd/FBD.htm** The Free Body Diagram tutorial at the Department of Physics, University of Guelph.

2. **www-groups.dcs.st-and.ac.uk/~history/Mathematicians/Newton.html** A biographic site for Sir Isaac Newton.

3. **www-istp.gsfc.nasa.gov/stargaze/Snewton.htm** A NASA website summarizing Newton's Laws.

4. **www.physicsclassroom.com/Class/newtlaws/newtltoc.html** An animated Newton's Law's tutorial from physicsclassroom.com.

5

The Law of Universal Gravitation

KEY IDEA

Gravity is a universal attractive force that acts between any two masses.

PHYSICS AROUND US . . . In an English Garden

One of life's most familiar phenomena is gravity. If you drop a book, it falls. If you slip on the ice, you fall. Acorns fall from trees and rain falls from the sky. But have you ever looked up at the full moon on a clear night and wondered why it didn't fall? Almost 350 years ago the young Isaac Newton did just that.

According to Isaac Newton, it happened this way: he was walking in an apple orchard one fall day when he noticed an apple falling from a tree. From the first law of motion he knew that the change in the state of motion of the apple (from stationary to falling) had to have occurred because of the action of a force—a force known as gravity.

At the same time as he saw the apple fall, he noticed the Moon in the sky behind the tree. He knew that the Moon went around the Earth in a roughly circular orbit and if the laws of motion applied to the Moon, then the action of a force—some kind of force—was required to keep the Moon from flying off into space in a straight line. In a sense, Newton saw that the Moon is falling, just like the apple, but is also moving forward at the same time.

At this point, Newton asked a simple but very profound question: was it possible that the same force that caused the apple to fall also kept the Moon in its orbit? From this question came our modern understanding of the working of the solar system.

THE UNIVERSAL FORCE OF GRAVITY

Newton's three laws of motion don't say anything about the nature of the forces that act in the universe. Discovering the nature of those forces is a separate problem from understanding their effects. Much of the progress of science during and after Newton's time has been associated with the discovery and description of the forces that act in the world around us. Newton's second great contribution to science was to elucidate the nature of one of these forces.

Gravity is an attractive force that acts between any two objects in the universe. It is the most familiar (and insistent) force in our daily lives. It holds you down in your chair and keeps you from floating off into space. It guarantees that when we drop a ball or a book or a glass, they fall. The ancients knew the effects of what we call gravity, and Galileo and many of his contemporaries studied its quantitative properties. We discuss some of that work in connection with falling bodies in Chapter 3. It was Isaac Newton, however, who revealed the true universal nature of gravity.

Everyone knew that gravity pulled objects toward the Earth, but until Newton, most people assumed that gravity was local and operated only near the planet's surface. People believed that farther out, in the realm of the stars and planets, different rules applied to the turning of the celestial spheres. They would say that terrestrial gravity operated on the Earth and celestial gravity operated in the heavens, but that the two forces had little to do with each other. In their minds, there was no connection between a planet in orbit around the Sun and an apple falling toward the ground. Isaac Newton discovered that these two seemingly different kinds of gravity were, in fact, one and the same. In modern language, in a remarkable union of seemingly disparate elements, he unified earthly and heavenly gravity. This was another important step in simplifying our understanding of the universe; we can now explain a wide range of observations by applying one law instead of two.

Let's look at this problem in more detail (Figure 5-1). As we point out in Physics Around Us, the fact that the Moon moves more or less in a circle implies that some sort of force must be acting on it to keep it in orbit. The question that Newton asked had to do with the exact nature of that force. He knew that there was one force acting in the orchard—the force that caused the apple to accelerate as it fell to Earth. Newton's insight was that the force holding the Moon in its orbit could be the same as the force that made the apple fall—the familiar force of gravity. This force not only pulls the apple down, but it also extends out to the orbit of the Moon and keeps it from flying off in a straight line.

Eventually, Newton realized that the orbits of all the planets could be understood if gravity is not restricted to the surface of the Earth but is a force found throughout the universe. He formulated this insight (an insight that has been

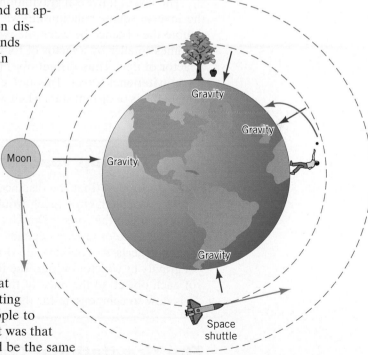

FIGURE 5-1. An apple falling, a ball being thrown, the space shuttle orbiting the Earth, and the orbiting Moon all display the influence of the force of gravity.

overwhelmingly confirmed by observations) in what is called **Newton's law of universal gravitation.**

1. In words:

Between any two objects in the universe there is an attractive force (gravity) that is proportional to the masses of the objects and inversely proportional to the square of the distance between them.

2. In an equation with words:

Force of gravity = Constant × $\dfrac{\text{First mass} \times \text{Second mass}}{\text{The square of the Distance between the masses}}$

3. In an equation with symbols:

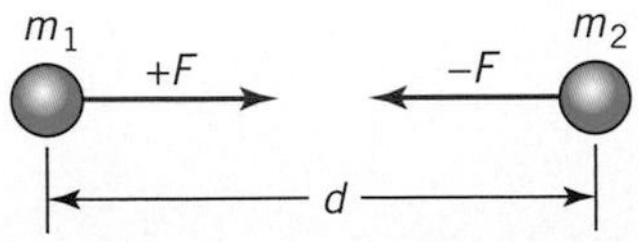

$$F = G \times \frac{m_1 \times m_2}{d^2}$$

where **G** is a number known as the **gravitational constant** (see next section).

m_1 $+F$ $-F$ m_2

d

FIGURE 5-2. Two masses separated by a distance. The force of gravity falls off rapidly as the objects get farther apart.

This law tells us that the more massive two objects are, the greater the gravitational force between them. If one of the masses doubles, then the force of gravity doubles. However, the greater the distance between them, the smaller the force is. The equation is illustrated in Figure 5-2.

The law of universal gravitation is the first example we have encountered of the inverse square relationship we discuss in Chapter 2. This means that if we double the distance between the two objects, the force of gravity between them becomes smaller by a factor of four. Triple that distance and the force drops by a factor of nine. Thus, distant objects have to be very massive to exert appreciable gravitational forces. This fact explains why we never consider the gravitational effects of distant stars when we talk about the orbits of planets in our solar system.

Develop Your Intuition: Moving the Sun

Suppose that the distance between the Earth and the Sun suddenly doubled. How much would the mass of the Sun have to increase in order to keep the force of gravity between the two the same?

The force of gravity depends inversely on the square of the distance between objects, so doubling the distance would cause a decrease in the force of gravity by a factor of four. The force of gravity depends directly on the mass of each object, so the mass of the Sun would have to be four times as big as it is now to compensate for the larger distance.

The Gravitational Constant

The equation for gravitational force incorporates the gravitational constant—the number, *G,* which is a constant of proportionality (see Chapter 2). In the equation for gravity, *G* expresses the exact numerical relation between the masses of two objects and the distance between them, on the one hand, and the gravitational force between them on the other. Henry Cavendish, a physicist at Oxford

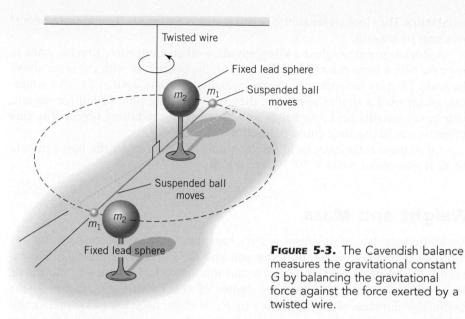

FIGURE 5-3. The Cavendish balance measures the gravitational constant G by balancing the gravitational force against the force exerted by a twisted wire.

University in England, first measured G in 1798 by using the apparatus shown in Figure 5-3. Cavendish suspended from a wire a dumbbell made of two small balls. Then he brought two large lead spheres near the balls. The resulting gravitational attraction between the small balls and the lead spheres turned the dumbbell until the twisted wire exerted a counterbalancing force strong enough to stop the rotation. By measuring the amount of twisting force (also known as a torque) on the wire, Cavendish measured the force on the dumbbells. This measured force, together with the known masses of the dumbbells, the masses of the heavy spheres, and knowledge of the distance between them, gave him the numerical value of everything in Newton's law of universal gravitation except G, which he then calculated using simple arithmetic. In metric units, the value of G is 6.674×10^{-11} m³/s²·kg. Most physicists think that this constant is universal, having the same value everywhere and at all times in our universe.

Ongoing Process of Science
The Value of *G*

It might surprise you to learn that the measurement of G is still of great interest to scientists around the world. It turns out that most fundamental physical constants, such as the mass of an electron or the electrical force between two charged particles, are known with great precision and accuracy, to many decimal places. But G is still only known, at best, to the fourth decimal place.

In November 1998 a group of 45 physicists met in London to celebrate the 200th anniversary of Cavendish's original experiment and to compare notes on several new determinations of the constant. All of these workers performed meticulous experiments: they enclosed their apparatus in a vacuum, they eliminated all stray magnetic and electrical fields, they checked and rechecked every aspect of their work.

Rival groups in France, Russia, and the United States used modern versions of the original Cavendish experiment, with weights suspended on a long wire or metal strips. They looked for subtle twisting effects and made their measurements over and over again.

A Zurich group weighed a kilogram mass on an exquisitely precise scale, in one case with a large mass below the scale, in the other case with the mass above the scale. The tiny difference in the two weights gave a measure of G. An American group tried a similar approach: they measured the time of fall for objects, with heavy weights held first below and then above the falling object. The tiny difference in falling time could be used to calculate G.

All of these techniques yield similar values; for the record, the best estimate for G is now about 6.674×10^{-11} m³/s²-kg. ●

Weight and Mass

Recall that the law of universal gravity says that there is a force between *any* two objects in the universe. Between you and the Earth, for example, there exists a force proportional to your mass and the (much larger) mass of the Earth. The distance between you and the center of the Earth is the radius of the Earth. This distance, which we denote by R_E, is about 6400 km (4000 miles); this is the number you would put in for the distance if you were calculating the force with which you are attracted to the Earth.

The gravitational attraction between you and the Earth would accelerate you downward if you weren't standing on the ground. As it is, the ground exerts an equal and opposite force to cancel gravity, a force you can feel in the soles of your feet.

An ordinary bathroom scale makes use of this interplay of forces. Inside the scale is a spring or some other mechanism that, when compressed, exerts an upward force. This upward force exerted by the scale keeps you from falling. The size of this counterbalancing force registers on a display, allowing you to measure your weight.

Weight, in fact, is just the force of gravity on an object. Your weight depends on where you are: on the surface of the Earth you have one weight, on the surface of the Moon another, and in the depths of interstellar space you would weigh next to nothing. Your weight is related to your mass, which is the amount of matter in your body (see Chapter 4). However, weight is different from mass. In interstellar space your weight would be zero, but your mass would not change.

Develop Your Intuition: Weight on a Mountaintop

Do you weigh the same at sea level as you do on a mountaintop?

The law of universal gravitation says that the farther apart two objects are, the smaller is the gravitational attraction between them. On the mountaintop, you are farther from the center of the Earth than you are at sea level, so you would actually weigh slightly less at a higher altitude. (In Problem 5 you get a chance to work out exactly what your weight reduction would be.)

The Acceleration due to Gravity: *g*

If we denote the mass and radius of the Earth as M_E and R_E, and your own mass as *m*, then the force that the Earth is exerting on you right now is given by the equation of gravity:

$$\text{Force} = G \times \frac{\text{First mass} \times \text{Second mass}}{\text{Distance}^2}$$

$$\text{Your weight} = G \times \frac{\text{Earth's mass} \times \text{Your mass}}{\text{Earth's radius}^2}$$

$$= G \times \frac{M_E \times m}{R_E^2}$$

or, rearranging the terms:

$$\text{Your weight} = m \times \left(\frac{G \times M_E}{R_E^2} \right)$$

Note that this equation is in the form: Force = Mass × [something in square brackets]. If we compare this to Newton's second law, Force = Mass × Acceleration, we see that what's in square brackets must be the acceleration you would feel if gravity were the only force acting on you. This missing number is identical to the quantity we call *g* in Chapter 3. In other words, *g* is the acceleration due to gravity at the Earth's surface:

$$\text{Force} = \text{Mass} \times g = \text{Weight}$$

and

$$g = \frac{G \times M_E}{R_E^2}$$

This result is extremely important. For Galileo, *g* was a number to be measured, but whose value he could not predict. For Newton, on the other hand, *g* was a number that could be calculated purely from the size and mass of the Earth.

One way of keeping weight and mass distinct in your mind is to remember that weight can change from one place to another; for instance, apples would weigh less on the Moon than on Earth, as shown in Figure 5-4. The mass of the

APPLES   APPLES

FIGURE 5-4. The weight of an object is different on Earth and on the Moon, but its mass is the same.

apples however, which measures the amount of material in the apples, is the same everywhere in the universe.

The Earth's Gravity

Given that the mass of the Earth is 6×10^{24} kilograms, what is the acceleration due to gravity at the Earth's surface?

REASONING AND SOLUTION: To answer this question, we just have to put the Earth's mass and radius into the expression for g given previously.

$$g = G \times \frac{M_E}{R_E^2}$$

$$g = 6.674 \times 10^{-11} \text{ m}^3/\text{s}^2\text{-kg} \times \frac{6 \times 10^{24} \text{ kg}}{(6.4 \times 10^6 \text{ m}^2)^2}$$

$$= 9.8 \text{ meters/second}^2$$

This number is the same constant that Galileo and others measured. Notice that because we now understand where g comes from, we can predict the appropriate value of gravitational acceleration not only for the Earth, but also for any object in the universe, provided we know its mass and radius. ●

Weighty Matters

A cantaloupe has a mass of 0.5 kilograms. What does it weigh?

REASONING AND SOLUTION: We are given a mass and want to find its weight. To answer this question, we have to calculate the force of gravity exerted on the cantaloupe at the Earth's surface. The relation between mass and weight is:

$$\text{Weight} = \text{Mass} \times g$$
$$= 0.5 \text{ kg} \times 9.8 \text{ m/s}^2$$
$$= 4.9 \text{ kg-m/s}^2 = 4.9 \text{ newtons}$$

This value is the weight of the cantaloupe. Note that the kilogram is not a unit of weight, despite its popular use. ●

GRAVITY AND ORBITS

Of all the motions that Newton and his contemporaries wanted to understand, none were more fascinating than the stately, sweeping orbits of moons and planets. Today's scientists have applied the same orbital equations derived in Newton's time to send humans to the Moon and robotic landrovers to the surface of the red planet Mars.

Circular Motion in Terms of Newton's Laws

Let's begin our analysis of orbits by looking quantitatively at the orbit of the Moon (or a planet) from the point of view of Newton's laws of motion. The fact that moons and planets do not move in straight lines tells us, from Newton's first law, that there must be a force acting to accelerate them; that is, to change the direction of their motion. In Chapter 3, we saw that the acceleration, a_c, required to keep an object such as the Moon moving around in a circular path of radius r with a constant velocity v is

$$a_c = \frac{v^2}{r}$$

This quantity is the centripetal acceleration. Newton's second law tells us that the force needed to produce this acceleration equals the product of the mass of the object times its acceleration. So the force needed to keep the object moving in a circle is

$$F_c = ma_c = \frac{mv^2}{r}$$

The robot rover *Sojourner* explored the surface of Mars for several days in 1997 before breaking down from the harsh conditions.

An object stays in a circular path only as long as this force acts. If the force disappears, Newton's first law of motion tells us that the object moves off in a straight line. Think of swinging a weight on a string in a circle around your head. The force that keeps the weight circling is the tension in the string; you can feel that force in your fingers. However, if you let go of the string, the weight doesn't keep circling around, but it flies off in a straight line in whatever direction it happened to be going at the moment of release.

Discus throwers use exactly this kind of action (Figure 5-5). First they spin around so that the discus, held at arm's length, is moving in a circular path (becase the thrower's hand is exerting a force, pulling the discus in). Then they release the discus so that it flies out over the field in a straight line.

The force in circular motion is directed toward the center of the circle, and hence is sometimes called the **centripetal force** (which means center-seeking force). The action of the centripetal force is illustrated for a weight on a string in Figure 5-6. The natural tendency of the weight is to move off in a straight line, as shown at point *A*. To keep it from flying off, you have to exert a force to pull it back into the circle. It then wants to fly off again, so that you have to pull it back again, and so on. The tug you feel in your hand as the weight goes around

FIGURE 5-5. A discus moves in a circular path as long as the athlete holds on to it, but it moves in a straight line once it is released.

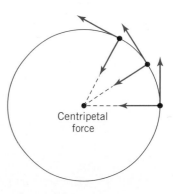

Centripetal force

FIGURE 5-6. Centripetal force acts to keep the weight moving in a circle at the end of the string.

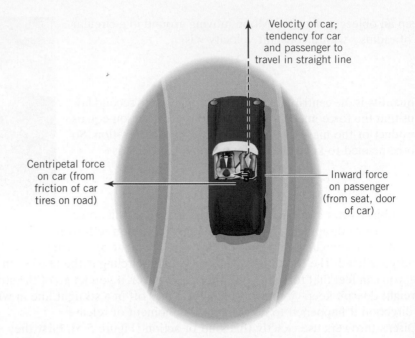

Velocity of car;
tendency for car
and passenger to
travel in straight line

Centripetal force
on car (from
friction of car
tires on road)

Inward force
on passenger
(from seat, door
of car)

FIGURE 5-7. When you're driving in a car moving around a curve, you feel the car seat and door of the car pushing you in toward the center of the circle.

is the reaction to your constantly pulling on the string, constantly tugging the weight back.

Another familiar example of centripetal force is a person sitting in a car as it goes around a curve (Figure 5-7). As the car starts into the turn, the passenger tends to move ahead in a straight line, following Newton's first law. The car seat or car door must exert a contact or frictional force to keep the passenger moving in a curve along with the car. This contact force is the centripetal force: the force that makes the passenger move in a circle. The passenger feels as if she is being thrown toward the outside of the car, but that is due to her inertia resisting the force pushing her toward the inside.

The Orbit Equation

Centripetal force is not a different kind of force; it simply describes any force that keeps an object moving in a circle. In the case of the circling weight, you can feel the force pulling on the string, so identifying centripetal force is simple. In the case of the Moon, however, the origin of the centripetal force isn't so obvious. Newton's insight was so important because he realized that gravity is the force that binds the solar system together by pulling the planets and moons back into their orbits. In fact, the basic equation that defines the orbit of a satellite—be it a planet, a moon, or the space shuttle—follows from this statement.

1. In words:

The centripetal force on a satellite in circular orbit is equal to the force of gravity exerted on that object.

2. In an equation with words:

$$\frac{\text{Satellite mass} \times \text{Satellite velocity squared}}{\text{Orbital distance}} = \frac{G \times \text{Satellite mass} \times \text{Central mass}}{\text{Distance squared}}$$

3. In an equation with symbols:

$$\frac{mv^2}{r} = \frac{GMm}{r^2}$$

where m is the mass of the satellite, M is the mass of the central body, r is the distance from the satellite to the center of the central body, and v is the speed of the satellite in its orbit.

This equation is simply an application of Newton's second law: force (the gravitational force, GMm/r^2) equals mass (m) times acceleration (v^2/r).

If we multiply both sides of this equation by r and divide both sides by m, we find that it reduces to

$$v^2 = \frac{GM}{r}$$

This orbit equation is a good one to examine because it tells us that for a given distance, r, between a satellite and its central body, there is one and only one speed, v, at which the satellite can move and remain in orbit (Figure 5-8). It also tells us that the speed at which a satellite has to move doesn't depend in any way on its mass. For example, a grapefruit orbiting the Earth at the same distance as the Moon would circle the Earth every 29 days, just as the Moon does. All modern satellites obey the orbit equation; for some examples of satellite applications, see Connection on page 106.

The force of gravity is not much different at the distance of a typical satellite orbit, such as the orbit of the space shuttle (6400 km + 200 km above Earth's surface), than it is where you are sitting ($r = 6400$ km). Why, then, does the shuttle stay in orbit? The reason is that enormous amounts of energy have been expended to get the shuttle moving very fast, so that its tendency to fly off in a straight line is just balanced by gravity. In that sense, the only reason you aren't in orbit at this moment is that you're not moving fast enough! How fast does the shuttle have to move? See Example 5-5 at the end of this chapter.

Apparent Weightlessness

We often see pictures of astronauts floating around in the space shuttle or the Space Station and we speak of them as being weightless. In fact, the force of the Earth's gravity at the orbit of the shuttle is pretty much the same as it is at

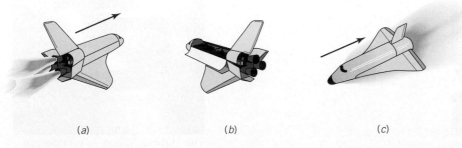

(a)   (b)   (c)

FIGURE 5-8. (a) The space shuttle reaches orbit by using its rocket boosters to achieve high acceleration. (b) Once in stable orbit, the space shuttle's speed is determined by the orbit equation; it does not use its engines at all. (c) To return to Earth, the shuttle fires its small thruster rockets to slow down, enabling Earth's gravity to pull it back down to the surface.

Physics and Modern Technology—Satellites

The physics of orbiting objects was first worked out by Kepler and Newton in the seventeenth century. Not until 1957 did advances in rocket technology and electronic instrumentation lead to a successful satellite launch. Today hundreds of satellites circle the Earth, from simple observation satellites with a camera to the International Space Station. However, the physics involved in all these orbits is the same.

With a navigation system in your car, you can locate your position on a road map of your neighborhood.

Long-distance cell phones receive messages relayed by a network of communications satellites.

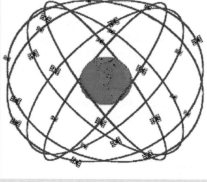

The Global Positioning System (GPS) uses 24 satellites to pinpoint the position of an object to within 15 meters or less, anywhere on the globe.

A weather satellite took this picture of a volcanic smoke plume rising from Mt. Etna in Sicily.

The Hubble Space Telescope can see astronomical objects from above Earth's atmosphere, greatly increasing its effectiveness.

the surface, so the astronauts aren't weightless. The force of gravity has not suddenly gotten smaller. Astronauts float in the shuttle because of the spaceship's acceleration as it moves around its orbit.

As seen from outside the spaceship, the only force acting is the Earth's gravity. Consequently, the spaceship is continuously pulled toward the Earth—otherwise it would escape into space. You can think of the spaceship (and its contents) as falling toward the Earth, even as it speeds along in its orbit. The point is that everything in the ship is falling at the same rate. So if the astronaut steps on a scale, that scale is falling at the same rate he is, and his weight registers as zero. We achieve apparent weightlessness in the presence of gravity.

 ## Develop Your Intuition: Weight in an Elevator

When you're in an elevator, you usually feel heavier when the elevator starts up and lighter when it starts down. Why?

These shifts in apparent weight, which would actually register as changes in your weight on a scale, come about even though the force of the Earth's gravity is essentially constant throughout the elevator trip (Figure 5-9). When you start upward, the elevator floor is accelerated into your feet, exerting a force that accelerates you upward. By Newton's third law of motion, your feet exert an equal and opposite force on the floor—or on a scale, if you're standing on one. This extra force causes the spring in the scale to compress and the reading to increase. The reverse process causes a feeling of partial weightlessness when the elevator starts down.

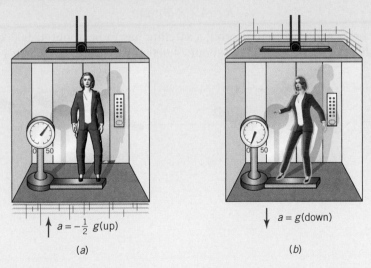

$a = -\frac{1}{2} g(\text{up})$

(a)

$a = g(\text{down})$

(b)

FIGURE 5-9. A person on a scale in an elevator feels (a) heavier while accelerating up, against gravity, and (b) lighter while accelerating down, with gravity.

Connection

Geosynchronous Orbits

Much of our modern communications system depends on relaying signals through satellites in orbit above the Earth. A particularly useful orbit is one in which the satellite moves just fast enough so that it appears to remain stationary above a point on the Earth.

Consider, for example, a satellite in orbit above the Earth's equator. If the satellite completes one revolution in 24 hours, that satellite appears to hover over the same spot on the Earth's surface, since that spot also completes a revolution in 24 hours. Such a satellite is said to be in *geosynchronous orbit*. What would the radius of the orbit have to be for the orbit to be geosynchronous?

If the satellite's orbit is located a distance R from the center of the Earth, then the satellite must travel a distance $2\pi R$ in 1 day. A day consists of 24 hours, each of which has 60 minutes of 60 seconds each. Consequently, the number of seconds in a day is

$$t = (24 \text{ hours/day}) \times (60 \text{ minutes/hour}) \times (60 \text{ seconds/minute}) = 86{,}400 \text{ s}$$

Now, the speed of the satellite must be

$$v = \frac{2\pi R}{t}$$

If we put this value of the velocity into the orbit equation, we find that

$$\frac{GM}{R} = \frac{4\pi^2 R^2}{t^2}$$

so that the radius of the orbit is

$$R^3 = \frac{GMt^2}{4\pi^2}$$

If we put in the mass of the Earth (6×10^{24} kg) and the value of t, we find that

$$R = 4.2 \times 10^7 \text{ m}$$

Thus, to be in geosynchronous orbit, a satellite has to be about 42,000 km from the Earth's center, or 36,000 km (about 24,000 miles) above the Earth's surface.

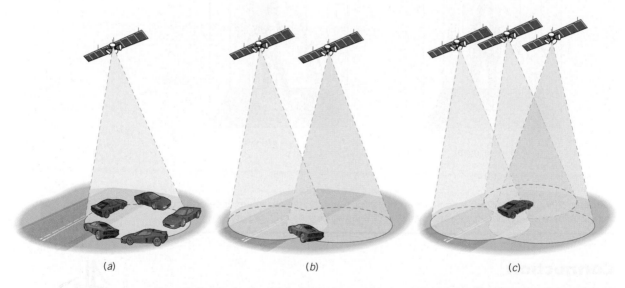

(a) (b) (c)

FIGURE 5-10. In the Global Positioning System (GPS), (*a*) a satellite identifies the receiver in a car as being somewhere on a circle. (*b*) When a second satellite identifies the same receiver as being on another circle, the car must be located at one of the two points where the circles intersect. (*c*) The third satellite determines exactly where the car is located.

This is a very high orbit. For reference, the space shuttle normally orbits a little over 100 miles above the surface.

The Global Positioning System (GPS), first developed by the United States Air Force, consists of 24 satellites placed in six orbits. These orbits are about 11,000 miles above the Earth's surface; they are not geosynchronous, so they move relative to the Earth's surface. Portable computers on your car, your plane, or even in your hand can pick up signals from three or more of these satellites and use them to determine your location on the Earth to within a few yards (Figure 5-10). ●

Physics in the Making
The Recovery of Halley's Comet

Of all celestial phenomena, none seemed more portentous and magical to the ancients than comets. These glorious lights in the sky, with their luminous sweeping tails, are not like the planets. They appear sporadically, and even in Newton's day appeared unpredictably. Yet even comets are subject to Newton's laws.

We associate the discovery of the orbital nature of comets with the British astronomer Edmond Halley (1656–1742). Halley led an adventurous (even swashbuckling) life before he settled down as Britain's Astronomer Royal. At various points in his life, he ran a diving company, captained a Royal Navy survey ship (facing down a mutiny in the process) and, if we are to believe the legend, used his growing reputation as an astronomer to travel around European capitals as a secret agent for his country. He was the first European astronomer to produce modern maps of the skies of the Southern Hemisphere, and his navigational maps of the Earth's magnetic field were used well into the nineteenth century.

(a)

(b)

Halley's comet: (a) on the Bayeux tapestry, showing the comet in 1066 A.D., and (b) in a telescope photo from 1986.

Halley was so wedded to the adventurous life that when he was proposed for a professorship at Oxford, one colleague wrote, "Mr. Halley expects [the professorship], who now talks, swears, and drinks brandy like a sea captain; so much that I fear that his ill behavior will deprive him of the advantage of this vacancy."

Nevertheless, he got the post at Oxford (the house he lived in is still there). About the time of Halley's appointment, astronomers were starting to think about explaining comets. Even though they knew about universal gravitation, they didn't have the mathematical tools to solve the problem of comets' elongated orbits. The reason for this difficulty is that, except for orbits that are nearly circular like those of the planets, the distance between a satellite and its central body varies considerably from one point to the next. Hence the gravitational force is not the same at each point around the orbit. The problem of deducing the shape of an orbit under these circumstances is a difficult one, but one that could be dealt with using the new mathematics of calculus, which was invented independently by Isaac Newton and the German mathematician Gottfried Leibniz.

In 1684, Halley visited Newton at Cambridge. Newton told him over dinner that according to his calculations, all bodies subject to a gravitational force would move in orbits shaped like ellipses. Bolstered by this information, Halley analyzed the historical records of some 24 comets. Knowing that the orbits had to be elliptical, he was able to use the observations to determine exactly the elliptical path along which each comet moved.

He found that three recorded comets—those that had appeared in 1531, 1607, and 1682—seemed to be following the same orbit. He realized that the sightings represented not three separate comets, but one comet that was appearing over and over again at intervals of about 75 or 76 years.

Predicting the next appearance of the comet wasn't as simple as you might think, because the gravitational effects of Jupiter and Saturn could change the period of the comet by several years. After some work, Halley predicted that the comet would reappear in 1758. On Christmas day 1758, an amateur astronomer in Germany sighted the comet coming back toward Earth. This so-called recovery of what is now known as Halley's comet marked a great triumph for the Newtonian picture of the world.

With characteristic aplomb, Halley (who had died in 1742) had had the last word on his prediction of the comet's return: "Wherefore if [the comet] should return again about the year 1758, candid posterity will not refuse to acknowledge that this was first discovered by an Englishman." ●

THINKING MORE ABOUT

The Clockwork Universe: Predictability

Newton bequeathed to posterity a picture of the universe that is beautiful and ordered. The planets orbit the Sun in stately paths, forever trying to move off in straight lines, forever prevented from doing so by the inward tug of gravity. The same laws that operate in the cosmos operate on Earth, and applying the scientific method led to the discovery of these laws. To an observer with Newton's perspective, the universe was like a clock, wound up and ticking along according to definite laws.

The Newtonian universe seemed regular and predictable. If you knew the present state of a system and the forces acting on it, the laws of

Whitewater is an example of a chaotic system, in which a small change in the initial position can produce a large difference in the outcome.

motion would allow you to predict its entire future. This notion was taken to the extreme by the French mathematician Pierre Simon Laplace (1749–1827), who proposed the notion of the *Divine Calculator.* His argument (in modern language) was this: if we knew the position and velocity of every atom in the universe and we had infinite computational power, then we could predict the position and velocity of every atom in the universe for all time. He made no distinction between an atom in a rock and an atom in your hand. According to the argument, everyone's movements are completely determined by the laws of physics to the end of time. You cannot choose your future. What is to be was determined from the very beginning.

While this argument raises many interesting questions, it has been rendered moot by two modern developments in science. One of these, the Heisenberg uncertainty principle (see Chapter 22), tells us that at the level of the atom it is impossible to know simultaneously and exactly both the position and velocity of any particle. (Heisenberg showed that any measurement of a particle's position alters its velocity, and vice versa.) Thus, you can never get all the information the *Divine Calculator* needs to begin working.

More recently, scientists working with computer models have discovered that there are many systems in nature that can be described in simple Newtonian terms but whose futures are extremely sensitive to initial conditions, making them, to all intents and purposes, unpredictable. These are called "chaotic systems," and the field of study devoted to them is called *chaos theory.*

Whitewater on a mountain stream provides a familiar example of a chaotic system. Imagine putting two chips of wood in water on the upstream side of the rapids. No matter how small you make the chips, or how close together they are at the beginning, those chips (and the water on which they ride) may be widely separated by the time they get to the end. If you knew the exact initial position of a chip and every detail of the waterway's shape and other characteristics with complete mathematical precision, you could, in principle, predict where the chip will come out downstream. But if there is the slightest uncertainty in your initial description, no matter how small, the actual position of the chip and your prediction will differ, often wildly. Every measurement in the real world has some uncertainty associated with it, so it is never possible to determine the exact position of the chip at the start of its trip. For all practical purposes, you cannot predict where it will come out even if you know all the forces acting on it.

The existence of chaos, then, tells us that there are some systems in nature in which the Newtonian vision of a completely predictable universe simply doesn't apply. Some important aspects of our lives, ranging from next week's weather to next year's health, are inherently unpredictable. Does science's inability to answer such important questions diminish its importance to society? In what ways does society prepare itself for the unpredictable aspects of the physical world?

Summary

Newton's law of universal gravitation describes **gravity,** the most prevalent force in our daily lives. At the Earth's surface, the gravitational force exerted on an object is called its **weight.** The same force that pulls a falling apple to Earth supplies the **centripetal force** that causes the Moon to curve around the Earth in its orbit. Indeed, the force of gravity (with the same **gravitational constant, G**) operates every-

where, with pairs of forces between every pair of masses in the universe. Newton's laws of motion, together with the law of universal gravitation, describe the orbits of planets, moons, comets, and satellites. They also allow us to derive Galileo's results and Kepler's laws of planetary motion (Chapter 3), thereby unifying the sciences of mechanics and astronomy.

Key Terms

centripetal force The force in circular motion, directed toward the center of the circle, that keeps an object following a curved or circular path. (p. 103)

gravitational constant (G) The exact numerical relation between the masses of two objects and the distance between them, on the one hand, and the gravitational force between them, on the other. (p. 98)

gravity The attractive force that acts between any two objects in the universe. (p. 97)

Newton's law of universal gravitation Newton's law that between any two objects in the universe there is an attractive force (gravity) that is proportional to the masses of the objects and inversely proportional to the square of the distance between them. (p. 98)

weight The force of gravity on an object. (p. 100)

Key Equations

$$\text{Force} = G \times \frac{\text{First mass} \times \text{Second mass}}{\text{Distance}^2}$$

$$\text{Force} = \text{Mass} \times g = \text{Weight}$$

$$(\text{Velocity of a satellite})^2 = G \times \frac{\text{Mass of central body}}{\text{Radius of orbit}}$$

Constants

$g = 9.8 \text{ m/s}^2$

$G = 6.674 \times 10^{-11} \text{ m}^3/\text{s}^2\text{-kg}$

Review

1. What similarity did Newton see between the Moon and an apple?

2. Why is gravity called a universal force?

3. State the law of universal gravitation.

4. Why is the gravitational constant, G, called a constant of proportionality? In Newton's equation for gravity, what is proportional to what?

5. How did Henry Cavendish determine the value of the gravitational constant, G?

6. What is the difference between weight and mass?

7. Does a bathroom scale measure weight or mass? Explain your answer.

8. What is centripetal force? Give an example of this force in action.

9. What supplies the centripetal force that keeps the planets in their orbits?

10. What is the relation between the velocity of a satellite, the radius of its orbit, and the mass of the central body?

11. According to the orbital equation, what factors determine the velocity of a satellite in orbit?

12. What is a geosynchronous orbit? Why are such orbits important to modern technology?

13. How did the work of Edmond Halley support Newton's theories?

14. What is a chaotic system?

15. What is the main idea of Newton's universal law of gravitation?

16. The gravitational constant is now known to 1 part in 10,000, yet physicists are still trying to measure this constant. Why?

17. Why did scholars of the sixteenth century distinguish between "terrestrial gravity" and "celestial gravity?"

18. What are the differences between the gravitational constant, *G,* and the acceleration due to gravity, *g*? Why is *g* not considered to be a universal constant?

19. Newton's equation for gravity incorporates an inverse square relationship between the force of gravity and the distance between two objects. What other familiar phenomena exhibit an inverse square relationship? (*Hint:* See Chapter 2.)

Questions

1. If this textbook is sitting on a table, the force of gravity is pulling it down. Why doesn't it fall?

2. Which of the following objects does not exert a gravitational force on you?

    a. this book   c. the nearest star
    b. the Sun   d. a distant galaxy

3. Two planets with the same diameter are close to each other, as shown in the figure. One planet has twice as much mass as the other planet. At which locations (A, B, C, or D) would both planets' gravitational force pull on you in the same direction? From among these four locations, where would you stand so that the force of gravity on you is a maximum i.e., at which point would you weigh the most?

4. Two iron spheres, of mass *m* and 2*m*, respectively, are shown in the figure. At which location (A, B, C, D, or E) would the net gravitational force on an object due to these two spheres be a minimum?

Questions 4, 5

5. Two iron spheres, of mass *m* and 2*m*, respectively, are shown in the figure. At which location (A, B, C, D, or E) would the net gravitational force on an object be at a maximum due to these two spheres?

6. If you moved to a planet that has the same mass as the Earth but twice the diameter, how would your weight be affected?

7. If you moved to a planet that has twice the mass of the Earth and also twice the diameter, how would your weight be affected?

8. The Earth exerts an 800-N gravitational force on a man. What gravitational force, if any, does the man exert on the Earth?

9. The environment in a satellite or space station orbiting the Earth is often referred to as a *weightless* environment. However, we have defined *weight* as the force of gravity on an object. Do you agree that objects on board orbiting satellites are *weightless?* Explain.

10. A bungee jumper feels weightless as she falls toward the Earth. Obviously the force of gravity has not disappeared simply because she has jumped off a high platform. What accounts for the weightless feeling people get when they fall freely?

11. The Earth's radius is about 3.7 times larger than the Moon's radius. If the Earth and the Moon had the same mass, which would have the greater acceleration due to gravity? Explain. Since we know that the Earth has a greater acceleration due to gravity, what does this tell you about the mass of the Earth compared to the mass of the Moon?

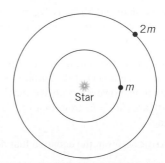

Questions 12, 13

12. Consider two planets of mass *m* and 2*m*, respectively, orbiting the same star in circular orbits. The more massive planet is twice as far from the star as the less massive planet. Which, if either, planet experiences a stronger gravitational attraction to the star? Explain.

13. Consider two planets of mass m and $2m$, respectively, orbiting the same star in circular orbits. The more massive planet is twice as far from the star as the less massive planet. Which planet is moving faster? Which planet has the shorter orbital period?

14. If our own Sun were twice as massive as it is, would the Earth have to move faster or slower in order to remain in the same orbit?

15. When Galileo first observed the four largest moons orbiting the planet Jupiter, he quickly determined the time it took for each moon to complete one orbit. Why won't this measurement allow us to determine the masses of the moons? Could such a measurement allow us to determine the mass of Jupiter?

16. How is weight related to mass?

17. How is Newton's law of gravitation related to Kepler's third law of planetary motion?

18. Does the Moon fall toward the Earth? Explain.

19. In *Star Trek* and other science fiction sagas, you often encounter a fictional device called a tractor beam, capable of pulling objects into the starship. Suppose that an object is falling under the influence of gravity and drag. In addition, imagine that a tractor beam on the ground is pulling the object down. If there is a limit to the force the tractor beam can exert, will the object still attain a terminal velocity?

20. If you fill a bucket partially with water and then swing it fast enough in a circle over your head, the water will stay in the bucket even when it is upside down. Since gravity is pulling the water down, why doesn't it spill out?

21. Why do you weigh less on the Moon than on Earth?

22. Would your mass change if you took a trip to the Space Station? Why or why not?

23. According to some nineteenth-century geological theories (now largely discredited), the Earth has been shrinking as it gradually cools. If so, would g have changed over geological time? Would G have changed over geological time?

Problem-Solving Examples

Football Player Mass

EXAMPLE 5-3

Suppose a football player has a mass of 120 kg. What is his weight in newtons? What is his weight in pounds?

REASONING AND SOLUTION: To find the weight in newtons, use the same equation as in Example 5-2.

$$\text{Weight} = \text{Mass} \times g$$
$$= 120 \text{ kg} \times 9.8 \text{ m/s}^2$$
$$= 1176 \text{ N}$$

From the *Table of Conversion Factors* in Appendix A we know that multiplying pounds by 4.45 will give newtons, so

$$\text{Weight (in pounds)} = \frac{\text{Weight (in newtons)}}{4.45}$$
$$= \frac{1176}{4.45} \text{ pounds}$$
$$= 264 \text{ pounds}$$

This weight shows the person to be large, but not unusually so for a football player. ●

Weight on the Moon

EXAMPLE 5-4

The mass of the Moon is $M_M = 7.18 \times 10^{22}$ kg and its radius R_M is 1738 km. If your mass is 60 kg, what would you weigh on the Moon?

REASONING: Once again we have to calculate the force exerted on an object at the surface of an astronomical body. This time, however, both the mass and the radius of the body are different from that of the Earth, while G is the same.

SOLUTION: From the equation that defines weight, we have

Weight = Force due to gravity

$$= \frac{G \times M_M \times 60 \text{ kg}}{R_M^2}$$

$$= \frac{(6.674 \times 10^{-11} \text{ m}^3/\text{s}^2\text{-kg}) \times (7.18 \times 10^{22} \text{ kg}) \times 60 \text{ kg}}{(1.74 \times 10^6 \text{ m})^2}$$

$$= 95 \text{ newtons}$$

This weight is about one-sixth of the weight the same object would have on the Earth, *even though its mass is the same in both places.* ●

The Space Shuttle

EXAMPLE 5-5

We can get an idea of how the orbit equation works by considering the space shuttle. A typical shuttle orbit is 200 kilometers (about 120 miles) above the Earth. How fast does a satellite at this distance have to move to stay in orbit?

SOLUTION: The equation for a satellite's speed, derived earlier in this chapter, is

$$v^2 = \frac{GM}{r}$$

As we saw earlier, the mass of the Earth is 6×10^{24} kg and its radius is 6.4×10^6 meters. The shuttle orbit, 200,000

($= 0.2 \times 10^6$) meters above the surface, therefore corresponds to a value of r of about 6.6×10^6 meters. If we put these values into the equation, we find that the square of the speed of the shuttle in its orbit is

$$v^2 = (6.674 \times 10^{-11} \text{ m}^3/\text{s}^2\text{-kg}) \times \frac{6 \times 10^{24} \text{ kg}}{6.6 \times 10^6 \text{ m}}$$

$$= 6 \times 10^7 \text{ m}^2/\text{s}^2$$

Taking the square root of this value, we see that the speed is

$$v = 7.8 \times 10^3 \text{ m/s}$$

This speed, equal to almost 8 km (5 miles) per second, is many times faster than commercial jet aircraft, which cruise at about 0.3 km (0.2 miles) per second.

One way of getting an idea of this speed is to ask how long it would take the shuttle to complete one orbit of the Earth. We can determine this time by rearranging the definition of speed into the form: time equals distance divided by speed. The length of the orbit is just the circumference of a circle of radius r, or $2\pi r$. Thus, the orbital period, or time for one revolution, is

$$\text{Orbital period} = \frac{2\pi r}{v}$$

$$= 2\pi \frac{6.6 \times 10^6 \text{ m}}{7.8 \times 10^3 \text{ m/s}}$$

$$= 5300 \text{ s}$$

$$= 89 \text{ minutes}$$

The space shuttle completes an orbit of the Earth every hour and a half. ●

Problems

1. What do you weigh in pounds? What do you weigh in newtons?

2. What is your mass in kilograms?

3. What would you weigh if the Earth were four times as massive as it is and its radius were twice its present value?

4. How long would our year be if our Sun were half its present mass and the Earth's orbit was in the same place that it is now?

5. How much would you weigh if you were standing on a mountain 200 km tall (*i.e.*, if you were standing still at about the altitude of a space shuttle orbit)? How much does this differ from your weight on the surface of the Earth? Would you be able to detect this weight difference on an ordinary bathroom scale?

6. **A.** Calculate the force of gravity on a 65-kg person in the following:

   a. at the surface of the Earth ($R = 6400$ km)
   b. at twice the Earth's radius
   c. at four times the Earth's radius

   B. Plot this gravitational force with distance. What pattern or relationship do you expect to obtain? Does your plot conform to your expectations?

7. Compare the gravitational force on a 1-kg mass at the surface of the Earth with that on the surface of the Moon ($M_M = 1/81.3$ mass of the Earth; $R = 0.27$ Earth radius).

8. How much less would you weigh on the top of Mount Everest than at sea level?

9. Calculate the weight in pounds and newtons of the following items:

   a. the Statue of Liberty (205 tons)
   b. a 40-ounce softball bat
   c. a solid rocket booster of the space shuttle (5.9×10^5 kg)

10. Calculate the weight in pounds and newtons of the three objects in Problem 9 if they were: a. on the surface of the Moon (see Problem 7); b. on the surface of Mars ($M = 0.11$ mass of the Earth; $R = 0.53$ Earth radius).

11. Calculate the speed and period of a ball tied to a string of length 0.3 meters making 2.5 revolutions every second.

12. Calculate the average speed of the Moon in kilometers per second around the Earth. The Moon has a period of revolution of 27.3 days and an average distance from the Earth of 3.84×10^8 meters.

13. Calculate the speed at the edge of a compact disc (radius = 6 cm) that rotates 3.5 revolutions per second.

14. Calculate the centripetal force exerted on the Earth by the Sun. Assume that the period of revolution for the Earth is 365.25 days and the average distance is 1.5×10^8 km.

15. The height of a mountain is limited by the ability of the atoms at the bottom to sustain the weight of the materials above them. Assuming that the tallest mountains on Earth are near this limit, how tall could a mountain be on the Moon? On Mars?

Investigations

1. In Chapter 1, we talk about astrology and whether a planet or star can influence our lives. Calculate the gravitational force on a newborn infant exerted by a star the size of the Sun 1 light year (9.5×10^{15} m) away. Compare it to the gravitational force exerted by a 100-kg physician 0.1 m away.

2. One objection that Copernicus's contemporaries raised to his theory was that if the Earth were really turning, we would all be thrown off the way that clay is thrown off a spinning potter's wheel. Use Newton's laws of motion and the law of universal gravitation to counter this argument.

3. In what sense is the Newtonian universe simpler than Ptolemy's? Suppose observations had shown that the two did equally well at explaining the data. Construct an argument you would make to say that Newton's universe should still be preferred.

4. If Kepler had been transported to another solar system, what would he have had to do in order to show that his laws applied there? What would Newton have had to do?

5. Some astronomers have proposed that Newton's law of gravitation may have to be modified over very large distances—that the gravitational "constant" varies over the immense scale of galaxies. What evidence do we have that gravitation is a universal force? How might you test this assumption? (*Hint:* Search the Internet for information on Modified Newtonian Dynamics or MOND.)

6. In what ways does gravity affect the form and function of living things? Relate this to both plants and animals.

7. Use the web to investigate some of the ongoing experiments to determine the value of G with greater precision and accuracy. Why is it so difficult to measure G to better than three decimal places, when most other physical constants are known to as many as 10 places?

8. Read a biography of Pierre Simon Laplace, who was one of history's most influential scientists. What were his major achievements? What major historical events occurred during his lifetime? How did his research influence his philosophical ideas?

 WWW Resources

See the *Physics Matters* home page at **www.wiley.com/college/trefil** for valuable web links.

1. **www.curtin.edu.au/curtin/dept/phys-sci/gravity/** An online gravity tutorial at the Department of Applied Physics, Curtin University of Technology.

2. **www.physics.purdue.edu/class/applets/NewtonsCannon/newtmtn.html** A humorous Java applet animating a woodcut from Newton's *Principia* that demonstrates satellite motion.

3. **www.physicsclassroom.com/Class/circles/circtoc.html** An animated tutorial from physicsclassroom.com discussing circular motion, planetary motion and universal gravitation.

4. **liftoff.msfc.nasa.gov/toc.asp?s=Satellites** Contains tutorials on types of satellites (including a section on geosynchronous satellites) and an extensive section on tracking current Earth-orbiting spacecraft live via the web.

6

Conservation of Linear Momentum

KEY IDEA

If no external forces act on a system, then the total momentum of the system is constant.

PHYSICS AROUND US . . . Dealing with Momentum

You get in your car, put your books on the seat next to you, and drive away. A squirrel runs across the road and you jam on your brakes to avoid hitting it. As you do so, the pile of books slides onto the floor of the car.

Later that day, in the cafeteria, you watch a fellow student hurrying across the floor with her tray. Stopping suddenly, she reaches out to grab her soft-drink cup to keep it from falling over and some of the drink sloshes over on her hand.

That night, you turn on your TV to watch a hockey game. The puck comes free and two players skate for

it at top speed. They collide just as they reach the puck and bounce off each other, taking themselves out of the action and leaving the puck for another player to pick up.

All of these occurrences (and countless more) illustrate a quantity called momentum, whose properties follow from Newton's laws of motion. And believe it or not, in addition to being involved in everyday events, momentum also governs such large-scale processes as the formation of planets and stars. As a result, momentum is one of the most important physical attributes of an object in motion.

LINEAR MOMENTUM

In Chapter 4 we introduced the property of inertia, which represents the tendency of a moving object to keep moving or a stationary object to remain stationary. Because of inertia, the only way to change an object's motion is to apply a force, as Newton's laws tell us. Everyday experiences provide all of us with an intuitive understanding of this concept. For example, we sense that a massive object such as a large train is very hard to stop, requiring a lot of force, even if it is moving slowly. You certainly wouldn't want to try and stop a train by standing in front of it!

At the same time, a small object moving very fast—a rifle bullet, for example—is also very hard to stop. Thus, our everyday experience seems to be telling us that the tendency of a moving object to remain in motion depends both on the mass of the object and on its speed. The greater the mass or the speed, the more difficult it is to stop the object or change its direction of motion.

Physicists encapsulate these notions in a quantity called **momentum** (plural, momenta). Momentum can be defined both in words and as an equation.

1. In words:

The momentum of an object is the product of that object's mass and velocity.

That means an increase in either mass or velocity increases momentum proportionally.

2. In an equation with words:

$$\text{Momentum} = \text{Mass} \times \text{Velocity}$$

3. In an equation with symbols:

$$p = m \times v$$

Here we use the letter p to denote momentum.

As we see in this chapter, momentum defined in this way is an extraordinarily useful concept in physics, and one that is deeply embedded in Newton's laws of motion. To begin, however, this definition leads to three important consequences.

1. Momentum is a vector quantity. Like velocity, momentum has both a magnitude *and* a direction. Thus, the rules for adding the total momenta of several objects are the standard rules for vector addition (see Chapter 2).

2. The definition of momentum matches our intuition about the tendency of objects to remain in motion. The equation tells us that the larger the mass or the greater the velocity, the greater the momentum.

3. The units of momentum are those of mass times velocity, or kg-m/s. There is no special name for this unit; it is simply written as a combination of the three basic units for mass, distance, and time.

Note that this product of mass times velocity is sometimes called **linear momentum** to distinguish it from *angular momentum,* which we'll discuss in the next chapter. In normal conversation, the term *momentum* by itself is understood to refer to linear momentum because we are referring to an object moving in a straight line. You can see other examples of momentum in Looking at Momentum.

Looking at Momentum

If you've ever swatted at a bee buzzing around your head, you know you can brush it away pretty easily. The momentum of a thrown baseball is about 1000 times greater than that, and you can certainly feel the impact when you catch it or hit it with a bat. A charging rhinoceros has about 1000 times more momentum than a baseball and could trample you flat. So just imagine the impact of a fully loaded oil supertanker, the biggest moving objects ever built, or the effect of a meteor crashing into the Earth.

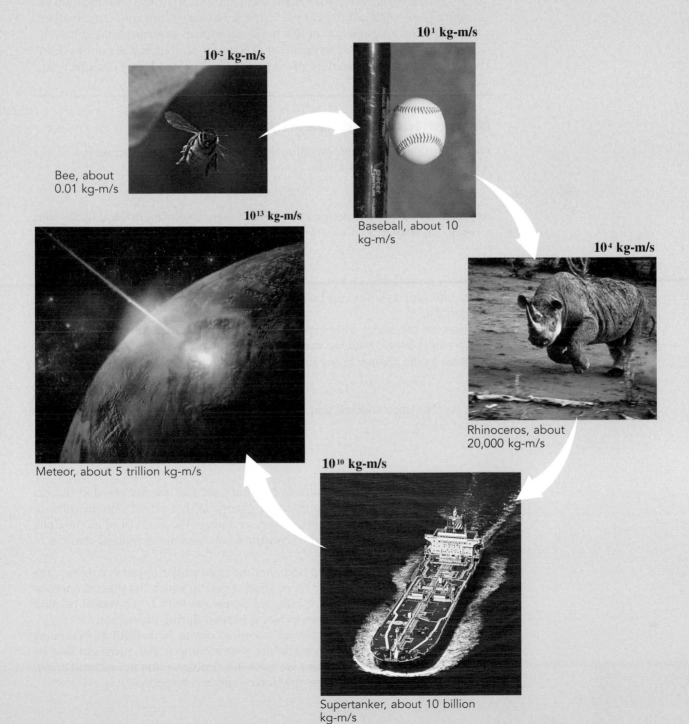

10^{-2} kg-m/s

Bee, about
0.01 kg-m/s

10^1 kg-m/s

Baseball, about 10
kg-m/s

10^{13} kg-m/s

Meteor, about 5 trillion kg-m/s

10^4 kg-m/s

Rhinoceros, about
20,000 kg-m/s

10^{10} kg-m/s

Supertanker, about 10 billion
kg-m/s

Develop Your Intuition: Billiards Momentum

Two billiard balls of equal mass m roll toward each other at equal speeds, v. What is the total momentum of the two balls?

Since the masses are equal and the speeds are equal, the magnitudes of the individual momenta of the two balls are equal. However, momentum is a vector quantity and has a direction. In this case, one ball has a momentum of magnitude mv directed toward the right and the other has a momentum of magnitude mv directed toward the left. The two momenta cancel each other out, so the total momentum of this particular system with two billiard balls is zero. In general, the total momentum of a group of objects does not have to be larger than the momentum of each individual object. It can even be zero, as in this case.

Momentum and Newton's Laws

We can learn more about momentum by looking at Newton's second law of motion. In equation form, it reads

$$F = m \times a$$

But since acceleration is defined to be the change in velocity divided by the time it takes that change to occur, this equation can be rewritten

$$F = m\frac{\Delta v}{\Delta t}$$

where the Greek letter capital delta, Δ, should be read "the change in" (see Chapter 2).

Now we can play a little mathematical trick. If the mass of the object in motion doesn't change, then the mass multiplied by the change in velocity must be the same as the change in the product of mass times velocity.

$$m\Delta v = \Delta(mv)$$

(Can you convince yourself that this is true?) In this case, we can write Newton's second law as

$$F = \frac{\Delta(mv)}{\Delta t}$$

In other words, this variation of Newton's second law tells us that the net force applied to an object is equal to the change in that object's momentum divided by the time it takes the momentum to change. The concept of momentum, then, is actually an integral part of Newton's laws and not a concept that has to be added to it.

Although we have assumed here that the mass of the object is constant, the equation is actually true for the more general case in which the mass changes as well. This situation may arise, for example, during the launch of a rocket because the mass of the rocket decreases as fuel is burned during the ascent.

This version of Newton's second law turns out to be helpful in examining more about momentum, as we see in the next section. It also turns out that by examining changes in momentum, we can solve problems that are difficult or impossible to solve by considering only forces and accelerations.

● IMPULSE

We can rearrange Newton's second law as written in the last section to get a better understanding of how forces act to change the momentum of a system.

1. In words:

> *The change in momentum is equal to the product of the net external force and the time during which it acts.*

2. In an equation with words:

> Change in momentum = Net force × Time interval

3. In an equation with symbols:

$$\Delta(mv) = F \times \Delta t$$

This form of Newton's second law tells us that a change in the momentum of a system is equal to the product of the net force that acts on it multiplied by the length of time that the force acts. In other words, a small force acting over a long time can produce the same change in momentum as a large force acting over a small time.

Physicists use the term **impulse** to refer to the product of the force multiplied by the time over which it acts. (This is another example of a common word in English given a very specific meaning in physics.) In terms of impulse, then, Newton's second law can be restated

> Impulse = Change in momentum

This version is often called the **impulse–momentum relationship.**

FIGURE 6-1. When a tennis ball is hit, the force of the impact often deforms the ball briefly before the ball breaks contact with the racquet.

Large Forces Acting for Short Times

Think of a tennis ball moving through the air while the player brings the racquet around to start his swing (Figure 6-1). The ball and the head of the racquet are in contact for only a fraction of a second, but during that short period the force is quite high. The total impulse, then, is large and the ball has a large momentum when it leaves the racquet. (See Example 6-2 in the Problem-Solving Examples, page 135.)

Small Forces Acting over Long Times

Think about a complementary case, in which a huge cruise ship enters a harbor and is nudged into place by a tugboat (Figure 6-2). The tugboat may push on the ship for several minutes, exerting a (relatively) small force over a long time. The direction of motion (and therefore the velocity) of the ship changes slightly as a result, but its mass is so large that even a tiny change in velocity produces a large change in momentum. To produce a large change in momentum, a large impulse is required, which is why the tugboat pushes for so long. (See Example 6-3 in the Problem-Solving Examples, page 135.)

FIGURE 6-2. A tugboat moves a much larger cruise ship by pushing against it steadily for a period of time, changing the ship's momentum bit by bit.

ADDING MOMENTA

Have you played a game of pool recently? If so, you remember hitting the white cue ball with the cue stick so that the cue ball collided with a colored ball, propelling the colored ball toward a pocket. Before the collision, the colored ball was stationary, while the cue ball was moving with a (more or less) constant velocity. After the collision, both balls were moving, although less swiftly than the cue ball had been moving before. You can see the same sort of behavior in many games; bowling, marbles, and croquet are examples. And, as we see in Chapter 9, such collisions are constantly taking place between the atoms that make up material objects. In all these cases, a transfer of momentum takes place between two colliding objects.

Total Momentum

Billiard balls and collections of atoms are examples of systems made up of many particles, each of which has a mass and a velocity. The **total momentum** P of such a system is defined as the sum of the momenta of all the objects in it. For the moment, let's talk only about objects that are all moving along the same line, either to the right or to the left.

CASE 1 ▪ Two balls moving in the same direction.
If the balls have masses m_1 and m_2 and velocities v_1 and v_2, respectively, and both are moving to the right, then the total momentum of the system is

$$P = m_1v_1 + m_2v_2$$

The total momentum of this system (Figure 6-3a) is, like the velocities, directed to the right.

CASE 2 ▪ Two balls moving in opposite directions.
If the balls have masses m_1 and m_2 and velocities v_1 and v_2, respectively, but the first is moving to the right and the second is moving to the left, then the total momentum of the system is

$$P = m_1v_1 - m_2v_2$$

In this case, the two momenta are in opposite directions (Figure 6-3b). In the special case in which the balls have equal momenta (for example, if they have the same mass and speed), then the total momentum of the system is zero.

CASE 3 ▪ More than two balls.
In this case (Figure 6-3c), the total momentum is the sum of all of the momenta of the balls. See that m_1 and m_2 are moving to the right, so their momenta are added, while m_3, m_4, and m_5 are moving to the left, so their momenta are subtracted.

$$P = m_1v_1 + m_2v_2 - m_3v_3 - m_4v_4 - m_5v_5$$

Internal Forces

Newton's laws tell us that when two billiard balls (or two atoms) collide, they exert forces on each other. If we call F_{12} the force that the first ball exerts on the

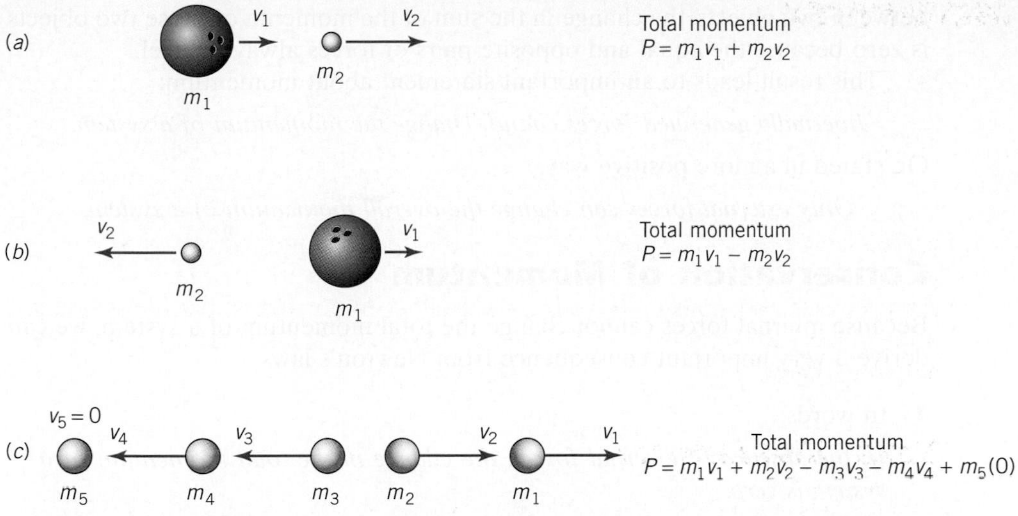

(a) v_1 v_2 m_2 m_1

Total momentum
$P = m_1v_1 + m_2v_2$

(b) v_2 m_2 v_1 m_1

Total momentum
$P = m_1v_1 - m_2v_2$

(c) $v_5 = 0$ v_4 v_3 v_2 v_1 m_5 m_4 m_3 m_2 m_1

Total momentum
$P = m_1v_1 + m_2v_2 - m_3v_3 - m_4v_4 + m_5(0)$

FIGURE 6-3. Balls of various masses move along the same line. When two balls are moving to the right (a), the total momentum is the sum of the individual momenta and is directed to the right. When all of the balls are not moving in the same direction (b, c), the total momentum is calculated by taking the difference between the momenta of the right-moving balls and of the left-moving balls.

second and F_{21} the force that the second ball exerts on the first, then Newton's third law tells us that

$$F_{12} = -F_{21}$$

We say that these kinds of forces are internally generated in the two-ball system. When we add up all the forces on the system, they cancel each other out. Thus, although each of the two balls feels an unbalanced force (and therefore accelerates), the entire system feels no net force.

Let's see how this works out in a simple example (Figure 6-4). Suppose we have two balls of mass m rolling toward each other with speeds v_1 and v_2. The impulse equation for the first ball says that

$$\Delta(mv_1) = F_{21}\Delta t$$

The equation for the second ball is

$$\Delta(mv_2) = F_{12}\Delta t$$

If we add these two equations together, we find that

$$\Delta(mv_1 + mv_2) = (F_{12} + F_{21})\,\Delta t = 0$$

The fact that the sum of the two forces is equal to zero follows from Newton's third law: these two forces must be equal and opposite.

A little thought should convince you that this result always holds no matter how many billiard balls or atoms are in a system. Whenever there is a collision

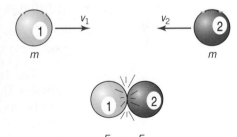

$$F_{12} = -F_{21}$$

FIGURE 6-4. Two balls of mass m roll toward each other with speeds v_1 and v_2. The internal forces of the collision sum to zero.

between two objects, the change in the sum of the momenta of those two objects is zero because the equal and opposite pairs of forces always cancel.

This result leads to an important statement about momentum:

Internally generated forces cannot change the momentum of a system.

Or, stated in a more positive way,

Only external forces can change the overall momentum of a system.

Conservation of Momentum

Because internal forces cannot change the total momentum of a system, we can derive a very important consequence from Newton's laws.

1. In words:

In the absence of external forces, the change in the total momentum of a system is zero.

2. In an equation with words:

The change in momentum of an isolated system equals zero.

3. In an equation with symbols:

$$\Delta P = 0$$

When physicists find a quantity that does not change during an interaction, they say that the quantity is "conserved." The conclusion we have just reached, therefore, is called the **law of conservation of momentum** and is stated as

If no external forces act on a system,
then the total momentum of that system remains the same.

In most practical situations, the law of conservation of momentum can also be written as

Initial momentum = Final momentum

or $$P_i = P_f$$

The law of conservation of momentum is of fundamental importance in physics. The outcome of almost every interaction between two or more particles or objects is determined in part by the conservation of momentum. As far as is known, whenever the conditions for momentum conservation have been satisfied, the law has never been violated.

THE NATURE OF CONSERVATION LAWS

Physicists have found several **conservation laws**—statements that a quantity is constant in nature—in addition to conservation of momentum. For example, we study conservation of angular momentum, energy, and electric charge in later chapters. Conservation laws are different in character from Newton's laws of motion and gravity, but they are just as fundamental, useful, and important. For example, the impulse–momentum form of Newton's second law says that if you apply a net force over an interval of time (an impulse), you cause a change in an object's momentum. Conservation of momentum says that if you don't apply an external force to a system, the total momentum of the system doesn't change—

momentum is conserved. That might sound like another way of saying the same thing, but there are important differences. In particular, the impulse/momentum form of Newton's second law says that some quantities change in an interaction; these changes are sometimes hard to measure. The principle of conservation of momentum identifies quantities that don't change in an interaction; these quantities are easy to measure.

To understand better what we can learn from a conservation law, imagine you are shooting a game of pool with a friend. You start out with 15 colored balls and a cue ball and you take turns hitting the cue ball into the other balls, trying to knock them into one of the six pockets. We impose a law of conservation of pool balls: no balls may be created or destroyed, but all 16 balls must stay on the table or in its pockets at all times.

After playing for 15 minutes, you count up the number of balls on the table and see that there are 5 balls plus the cue ball. You can't see any other balls, but you know there are 10 balls in the table pockets. Why? Because conservation of pool balls states 15 colored balls must be in the system at all times. You don't need to check each pocket on the table and count up the balls as long as you know the conservation law applies. The conservation law does not tell you the details of the game, such as the great shot you hit that sank 2 balls at once, but the law does keep track of the general state of the system.

Conservation laws can provide useful information about a system. Near the end of the game, your friend lines up a shot with only two colored balls left on the table (Figure 6-5a). You turn away for a minute and hear the loud smack of

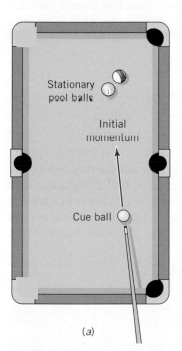 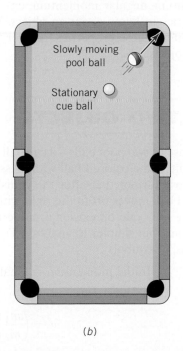 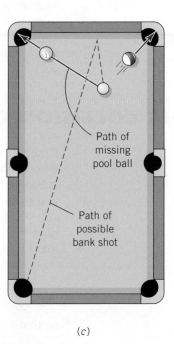

(a) (b) (c)

FIGURE 6-5. (a) The cue ball and two colored balls remain on a pool table. Your friend is about to shoot when you look away. (b) You hear the collision of the cue ball and look back to see the nearly stationary cue ball, and one colored ball moving slowly to the right. What happened to the other ball? (c) Conservation of colored balls tells you that the missing ball must be in one of the pockets, while conservation of momentum suggests that the ball was hit into the far left corner pocket, assuming that your friend didn't execute a difficult bank shot (dotted line) into the near left corner pocket!

the cue ball hitting the colored balls. When you turn back to the table, the cue ball is almost motionless on the table and one colored ball is rolling slowly toward the far right corner pocket (Figure 6-5*b*). What happened to the other ball? You can't see it, but you know from conservation of pool balls that since it's not on the table, it must be in one of the pockets. But which one? Here conservation of momentum applies. Once your friend hit the cue ball, no forces were exerted on the system of three balls. The cue ball started out fast, with a significant amount of momentum. The conservation law says that this momentum is still in the system: it can't be created or destroyed. So the missing ball must have gained momentum after being hit by the cue ball and must be in the far left corner pocket (Figure 6-5*c*). We examine the details of this transfer of momentum a little later in this chapter, but you don't have to calculate impulse or change of momentum to determine where the ball went. That's the beauty of conservation laws: you obtain information just from knowing that some quantity before an interaction is unchanged after the interaction.

Physicists often rely on conservation laws to determine the behavior of objects they can't observe directly. For example, they don't have to actually see atomic-scale particles to calculate forces and accelerations before and after a collision. Conservation laws apply to those particles, so we can still calculate information about them and about the interaction. It's not magic and it's not guesswork, it's simply applying known laws to a given situation.

Not all quantities in physics are conserved. There is no conservation of force or conservation of velocity, for instance. But those quantities that are conserved, such as momentum, angular momentum, energy, and electrical charge, are important properties to know about an object. Much current research in physics aims at determining these properties for everything from subatomic particles to astronomical objects.

THE COLLISION OF TWO OBJECTS

Consider the case of two pool balls—a cue ball with mass m_1 and velocity v_1 traveling to the right, and a colored ball with mass m_2 that is not moving before the collision. Suppose further that after the collision, the cue ball comes to a stop and the colored ball moves off to the right with velocity v_2 (Figure 6-6). (We are still considering only one-dimensional motion along a line here. Not a typical pool shot, perhaps, but simpler to analyze.) Then the law of conservation of momentum tells us, in symbols

$$\text{Initial momentum} = \text{Final momentum}$$
$$(m_1 \times v_1) + (m_2 \times 0) = (m_1 \times 0) + (m_2 \times v_2)$$

so that
$$v_2 = \left(\frac{m_1}{m_2}\right)v_1$$

This result matches our intuition, because if the masses of the two balls are equal, then the colored ball moves to the right with exactly the same velocity as the cue ball had before the collision.

We can gain more insight into the meaning of momentum by considering another example of a collision (Figure 6-7). Suppose that a speeding bullet (mass m_1, velocity v_1) and a powerful locomotive (mass m_2, velocity v_2 in the opposite

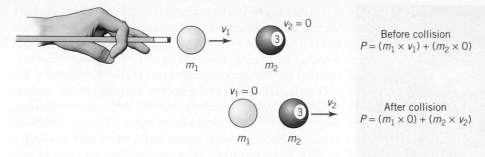

FIGURE 6-6. A cue ball with mass m_1 and velocity v_1 collides with a colored ball with mass m_2 that is not moving before the collision. Momentum is conserved.

direction) are approaching each other. The total momentum P of the system before impact is just

$$P = (m_1 \times v_1) - (m_2 \times v_2) = \text{Total momentum before collision}$$

The minus sign in this expression represents the fact that the bullet and the train are moving in opposite directions before the collision. Suppose, for the sake of argument, that after the collision the bullet recoils and is moving in the same direction as the train, but with velocity u_1, while the train keeps moving forward with velocity u_2 (where u_2 is not the same as u_1). The new total momentum is

$$P = (m_1 \times u_1) + (m_2 \times u_2) = \text{Total momentum after collision}$$

Conservation of momentum tells us that the total momenta before and after the collision have to be the same. Since the momentum of the bullet is reversed, the momentum of the train has to decrease and the train has to slow down (albeit by a very small amount).

You can, in fact, think of the collision as a transfer of momentum from the train to the bullet, with the train losing positive (say, rightward-directed) momentum and the bullet gaining this positive momentum as it reverses direction. In this case, the train with its larger mass has a momentum much larger than the bullet has, so it does not slow down very much. On the other hand, even a small amount of the train's momentum, when transferred to the bullet, produces a very large change in that object's momentum (in this case, it reverses the bullet's direction).

It's important to keep in mind that the law of conservation of momentum doesn't say that momentum can never change. The law just says that momentum won't change unless an outside force is applied. If a soccer ball is rolling across a field and a player kicks it, a force is applied to the ball as soon as the player's foot touches it. At that instant, the momentum of the ball changes, and that change is reflected in its change of direction and speed.

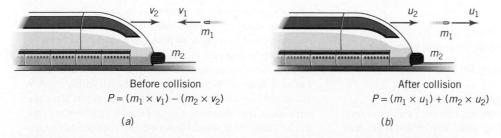

Before collision
$$P = (m_1 \times v_1) - (m_2 \times v_2)$$

(a)

After collision
$$P = (m_1 \times u_1) + (m_2 \times u_2)$$

(b)

FIGURE 6-7. (a) A bullet (mass m_1, velocity v_1) collides with a powerful locomotive (mass m_2, velocity v_2 in the opposite direction). (b) The train slows down a little bit, but the bullet completely reverses direction; however, total momentum is conserved.

The symmetrical patterns of exploding fireworks demonstrate conservation of momentum.

You may not have realized it at the time, but you were seeing the consequences of conservation of momentum the last time you watched a fireworks display. Think about a rocket arching up and, just to make things simple, imagine that the firework explodes just at the moment that the rocket is stationary at the top of its path, at the instant when its total momentum is zero. After the explosion, brightly colored burning bits of material fly out in all directions. Each of these pieces has a mass and a velocity, so each piece has some momentum. Conservation of momentum, however, tells us that when we add up all the momenta of the pieces right after the explosion, they should cancel each other out and give a total momentum of zero. For example, if there is a 1-gram piece moving to the right at 10 meters per second, there has to be the equivalent of a 1-gram piece moving to the left at the same velocity. Thus conservation of momentum gives fireworks their characteristic symmetrical starburst pattern.

LOOKING DEEPER

Collisions in Two Dimensions

Up to this point we have been talking about situations in which colliding objects move in one dimension only. However, conservation of momentum works in more complicated cases as well. For example, consider two billiard balls that make a glancing collision as shown (Figure 6-8). We can analyze this situation by recalling that momentum, like velocity, is actually a vector quantity.

When the two billiard balls approach each other at an oblique angle, their momenta can be represented by vectors, as shown in Figure 6-8a. As we discussed in Chapter 2, each of these vectors can be thought of as the sum of two components, one in the x-direction and one in the y-direction. These components are shown in the figure. When momenta are represented in this way, the conservation law can be stated

In the absence of external forces, momentum is conserved independently in the x- and y-directions.

We can examine this idea by looking at two situations.

CASE 1 ■ No external forces.
If the two billiard balls have velocities v_1 and v_2 before the collision and u_1 and u_2 after the collision (Figure 6-8b), then the statement tells us that

Momentum in x-direction before collision
    = Momentum in x-direction after collision

$$mv_{1x} - mv_{2x} = mu_{1x} - mu_{2x}$$

and

Momentum in y-direction before collision
    = Momentum in y-direction after collision

$$mv_{1y} - mv_{2y} = -mu_{1y} + mu_{2y}$$

CASE 2 ■ External force in one direction.
Another common situation occurs when an external force such as gravity acts in one direction, but no external force acts in the other. In this case, momentum is conserved in the direction in which no force acts, and Newton's second law describes the change in momentum in the other direction.

In the example of the exploding fireworks display, gravity is acting in the vertical y-direction and there is no external force in the horizontal x-direction. Consequently, momentum is conserved in the x-direction, so that any piece of material moving to the left has to be compensated by a piece of material moving to the right. In the up-and-down y-direction, however, gravity acts to pull the system back toward the ground, accelerating as it does so. What happens is shown in Figure 6-9. If you add up all the momenta at any moment after the explosion, they produce a vector that is exactly the same as the vector that would describe the rocket had the explosion never occurred.

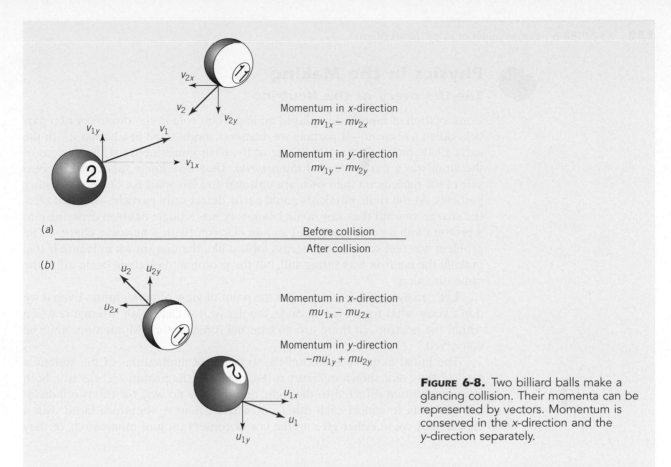

Momentum in x-direction
$mv_{1x} - mv_{2x}$

Momentum in y-direction
$mv_{1y} - mv_{2y}$

(a)

Before collision

After collision

(b)

Momentum in x-direction
$mu_{1x} - mu_{2x}$

Momentum in y-direction
$-mu_{1y} + mu_{2y}$

FIGURE 6-8. Two billiard balls make a glancing collision. Their momenta can be represented by vectors. Momentum is conserved in the x-direction and the y-direction separately.

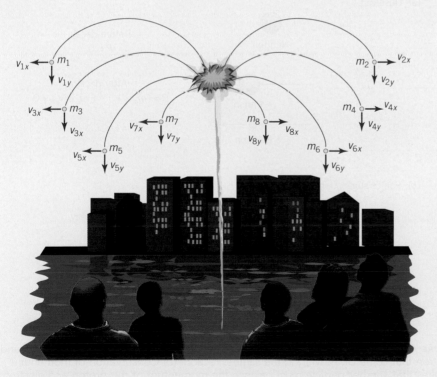

FIGURE 6-9. If you add up all the momenta of individual fragments at any moment after a fireworks explosion, they produce a vector that is exactly the same as the vector that would describe the momentum of the rocket had the explosion never occurred.

Physics in the Making
The Discovery of the Neutrino

Conservation of momentum played an important role in the discovery of a particle called a *neutrino*—a particle we discuss in more detail in Chapter 27. In the early 1930s, physicists knew that one of the basic constituents of the nucleus of the atom was a particle called the *neutron*. They also knew that neutrons outside of the nucleus, on their own, are unstable and fall apart (or decay) into other particles. At the time, physicists could easily detect only particles that had electric charge, so what they saw in the laboratory was a single neutron decaying into a proton (with a positive charge) and an electron (with a negative charge). The problem was that some of the decays looked like the one shown in Figure 6-10a: initially the neutron was sitting still, but the proton and electron came off in the same direction.

Let's analyze this situation from the point of view of momentum. Even if we don't know what forces act to cause the decay, it is clearly an internal reaction within the neutron, so there are no external forces acting. Momentum must be conserved.

The initial neutron is motionless, so the total momentum of the system is zero. In the event shown in Figure 6-10a, however, the proton and electron both have momentum directed to the right, and there is no way for the two individual momenta to cancel each other out. Physicists were therefore faced with a choice: they could either give up the law of conservation of momentum, or they

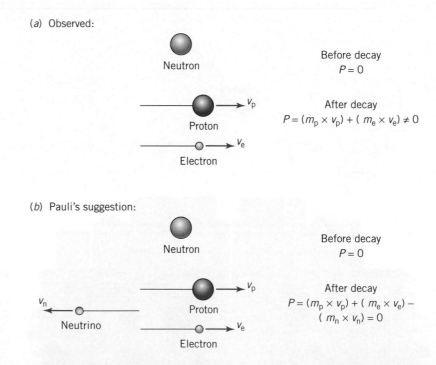

FIGURE 6-10. (a) The decay of a stationary neutron occasionally displays a proton and electron that come off moving in the same direction. (b) Conservation of momentum requires that another particle must balance the momentum of the proton and electron.

Physics and Daily Life–Momentum

Momentum depends on mass and velocity, so any time something moves, it has momentum. When you hit a volleyball, you give it momentum; when you walk across a lawn you have momentum—even moving your arm gives it momentum.

Momentum of struck ball

Momentum of rising body

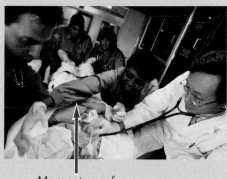

Momentum of moving arm

Momentum of walking child

could assume that another particle was emitted in the reaction, a particle they could not detect, which was carrying momentum directed to the left. With some reluctance, the German physicist Wolfgang Pauli took the latter route. He suggested that there was a particle (eventually called the neutrino, or "little neutral one") that, because it had no electrical charge, was invisible to experimenters, but that could balance the momentum of the electron and proton, as shown in Figure 6-10b. The hypothetical particle also solved other problems that had been encountered by experimenters.

For several decades, then, physicists accepted the existence of the undetectable neutrino because it allowed them to hold onto central laws of nature such as the conservation of momentum. When the neutrino was finally detected in 1956, this faith in the laws of nature was amply rewarded. ●

THINKING MORE ABOUT

Momentum: Why Isaac Newton Would Wear His Seat Belt

One of the authors (JT) knows a highway patrolman who makes a point about auto safety by saying, "I never pulled a dead man from behind a seat belt." Although some people do indeed die in car crashes while wearing seat belts, there is no question that seat belts greatly improve your chances of walking away from a crash. Given what you know about momentum and impulse, why should this be so?

Look at it this way: when you are sitting in a car, your momentum is your mass multiplied by the speed of the car. Call this quantity *P*. In a crash, the speed of the car—and of you—is rapidly reduced to zero. Therefore, the change in your momentum during the crash is *P*—your momentum before the crash minus your momentum after the crash. The impulse–momentum relationship tells us that a force must act to supply the impulse needed to bring about this change.

If you're not wearing a seat belt, you keep moving forward when the car stops, and your motion stops when your head hits the steering wheel or the windshield. Because these surfaces are hard, the time of the impact is short—a fraction of a second. Consequently, the force needed to stop you must be large. In addition, this force is applied to a small area of your skull, greatly increasing the pressure. Such a large focused force can be deadly.

If you're wearing a flexible seat belt, however, a smaller force is applied over a longer time interval to produce the same change in momentum. What's more, the broad, flat seat belt distributes that force over a much larger area of your body. The chances of injury are greatly reduced. This is one reason why seat belts save lives.

Given the physics of this situation, should states require that drivers and passengers wear seat belts? Is such legislation an intrusion on personal freedom? In what other ways do scientific principles influence our laws?

Other examples of momentum around us appear in Physics and Daily Life on page 131.

Summary

The **momentum** of an object is defined as the product of that object's mass and its velocity. Like velocity, momentum is a vector. In terms of momentum, Newton's second law says that the rate of change in the momentum of a system is equal to the net force.

Impulse is defined as the product of the net force and the time interval over which that force acts. Another way of stating Newton's second law, called the **impulse–momentum relationship,** is to say that the change in momentum of a system is equal to the impulse applied to it.

The **total momentum** of a system with several objects is defined as the sum of the momenta of all of those objects. Only external forces applied to a system can change its total momentum. Internal forces may change the momentum of one part of a system, but these changes always cancel out when they are added up over the entire system.

In the absence of a net external force, the total momentum of a system does not change. This statement, known as the **law of conservation of momentum,** is one example of a **conservation law.**

Key Terms

conservation law Statement that a quantity is constant in nature. (p. 124)

impulse The product of a force multiplied by the time over which it acts. (p. 121)

impulse–momentum relationship Restatement of Newton's second law, in which impulse equals the change in momentum. (p. 121)

law of conservation of momentum Statement that if no external forces act on a system, then the total momentum of that system remains the same. (p. 124)

linear momentum Another term used for *momentum* (the product of mass times velocity) when the object is understood to move in a straight line. (p. 118)

momentum The product of an object's mass and velocity. (p. 118)

total momentum The sum of the momenta of all the objects in a system. (p. 122)

Key Equations

Momentum = mass × velocity

Impulse = change in momentum
= net force × time (force is applied)

Review

1. What is momentum?

2. How is inertia related to linear momentum? Consider Newton's laws.

3. What does it mean to say that momentum is a vector quantity? Give an example of this.

4. What is an impulse? How is it derived from Newton's second law $F = ma$?

5. Which produces the greater impulse, a large force acting for a short time or a small force acting over a long time? Explain.

6. What is meant by the total momentum of a system? Give an example.

7. What effect do internally generated forces have on the total momentum of a system? Explain.

8. When adding up a system's total momentum, do we need to account for the direction of the individual particles that comprise the system? How so?

9. What do scientists mean when they say something is conserved?

10. What is the conservation of momentum?

11. Identify a physical quantity that is not conserved.

12. What happens when an external force is applied to a system? Is momentum conserved?

13. What is meant by a collision in two dimensions? Give an example.

14. In a two-dimensional collision, is momentum in the x-dimension conserved? How about the y-dimension?

Questions

1. Can a system of multiple objects moving in different directions have a total momentum of zero? How can this be?

2. Which has more momentum, an 18-wheel truck that is parked on a street or a mosquito buzzing around your ears?

3. A large truck that is moving to the right collides head-on with a stationary compact car. Which vehicle (if either) exerts the larger force on the other? Which force imparts a greater impulse? Assuming the collision forces are the only forces acting on the vehicles during the collision, which vehicle's momentum changes more? Compare the direction of the momentum change of the truck to the momentum change of the car.

4. Two balls are moving in opposite directions as shown in the figure. What is the direction of the total momentum of the system?

5. Two balls are moving in opposite directions as shown in the figure. What is the direction of the total momentum of the system?

6. What is the direction of the total momentum of the system of objects shown in the figure? Explain your answer.

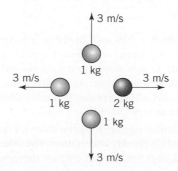

7. What is the direction of the total momentum of the system of identical tennis balls shown in the figure? Explain your answer.

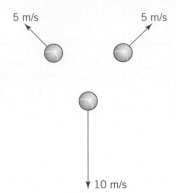

8. Two identical billiard balls, A and B, collide head-on. Before the collision, A is moving and B is stationary. After the collision, A is stationary and B is moving. Compare the speed of A before the collision to the speed of B after the collision. Explain how you reach your conclusion.

9. Which of the collisions in the figure are possible and which are impossible? The objects have identical masses. (*Hint:* Which collisions violate conservation of momentum?)

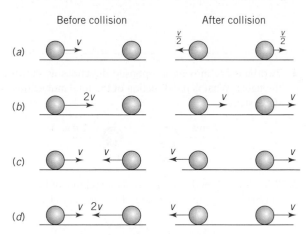

10. If you throw a ball high in the air, its momentum is apparently not conserved. First it moves up (upward momentum), then it stops (zero momentum), then it moves down (downward momentum). How come its momentum is not constant? Is there a system that includes a ball whose momentum is constant?

11. Two identical eggs are dropped from a height of 10 meters. One lands in the dirt and shatters. The other lands on a pillow and does not break. Compare the momentum change of each egg after each lands. In terms of collision time and forces, why does one egg break and the other does not?

12. Several years ago in New Orleans, a large ocean liner was unable to stop in time and crashed into a riverwalk dock, causing immense damage. Explain why a ship of such size would have such difficulty stopping, and describe the collision with the pier in terms of an impulse.

13. What are some reasons that you frequently see truck turnouts on mountain highways? (A truck turnout is a level or upward-sloping area alongside a steep downward highway where a truck can come to a stop and let its brakes cool.) Would truck brakes have more of a tendency to fail than passenger car brakes? Why? Consider momentum and Newton's laws.

14. Bungee jumpers use cords that are elastic. Explain in terms of impulse why a metal cord is *never* used in place of the elastic cord, even though it might be less likely to break. Explain this in terms of force, momentum, and impulse.

15. Why are people who have to jump from any appreciable height, such as parachuters or stunt people, taught to land with their knees bent and to roll on impact?

16. What is the purpose of a good 'follow-through' when swinging a golf club or a baseball bat?

17. Explain the theory behind air bags in cars. What are the advantages and disadvantages of their use? Why are airbags on occasion lethal to the occupant?

18. Explain to yourself in some detail just how conservation of momentum may be used to understand how a rocket moves. Can the motion of a rocket be completely explained by this conservation law? Why?

19. Conservation of momentum is really a consequence of Newton's second law. Explain the connection.

20. When new comets enter the solar system and move toward the Sun, they often display erratic motion, moving first one way and then another as large chunks of material evaporate in the heat. Use the concept of momentum to explain this behavior.

21. Modern fireworks displays include dramatic explosions in which flares perform spiral, twisting, or wiggling motions. How might these distinctive motions be produced?

22. While playing a game of pool, you line up a shot, strike the cue ball, and watch as it knocks the 9 ball into the corner pocket. How was momentum transferred in this situation? What happened to the momentum when the 9 ball entered the pocket? Was momentum conserved? If not, why not?

23. Car accidents are dangerous because the car may experience a very high acceleration (stopping fast in a head-on collision, for example). Consider a specific accident where a car slams into a wall and comes to a complete stop. Does the impulse imparted to a passenger depend on whether or not the passenger is wearing a seat belt? How come a seat belt makes the passenger safer?

Problem-Solving Examples

Large and Small Momenta

Compare the momentum of a baseball (mass 0.3 kg) with the momentum of a blue whale (mass 150,000 kg). Suppose the baseball is moving to the right at the speed of a good fastball (30 m/s). How fast would the whale have to be going to have the same momentum as the baseball?

SOLUTION: Momentum is defined as

$$p = m \times v$$

If the whale were swimming at the same speed as the baseball is thrown, the whale would have much more momentum. However, an object of large mass moving slowly can have the same momentum as an object of small mass moving quickly. How slowly would the whale have to move? Let's take a look at the numbers.

We can first find the momentum of the baseball by substituting the numbers for mass and velocity.

$$p = m \times v$$
$$= 0.3 \text{ kg} \times 30 \text{ m/s}$$
$$= 9 \text{ kg-m/s}$$

We are told that this is equal to the momentum of the whale. We also know the whale's mass and we want to find its velocity. We can rearrange the definition of momentum to solve for the speed v.

$$p = mv$$
$$v = \frac{P}{m}$$
$$= \frac{9 \text{ kg-m/s}}{1.5 \times 10^6 \text{ kg}} = 0.000006 \text{ m/s}$$

This is a *very* slow speed. It would take almost 50 hours for the whale to move 1 meter. ●

A Long Drive

What is the total impulse imparted to a 10-g golf ball that flies off the tee at 40 m/s (about 90 miles per hour)? Strobe photographs have indicated that in a golf drive, the club stays in contact with the ball for only about 1 millisecond (= 0.001 s). Assuming this time interval occurs for the given golf drive, how much force does the golfer exert?

REASONING: First we find the impulse by the impulse–momentum relationship. Then we use the results and the given time interval to calculate the force from the definition of impulse.

SOLUTION: By the impulse–momentum relationship, impulse equals the change in momentum, $\Delta(mv)$. In a golf drive the mass of the golf ball doesn't change, but its velocity increases, in this case from 0 to 40 m/s. The initial momentum of the golf ball is 0, because it is at rest on the tee. Therefore, the change in momentum is equal to the final momentum.

$$\text{Impulse} = \text{Change in momentum}$$
$$\Delta(mv) = m \times \Delta v$$
$$= 10 \text{ g} \times 40 \text{ m/s}$$
$$= 0.4 \text{ kg-m/s}$$

Once we know the impulse, we can use its definition to determine the net force applied to the ball for the given time interval.

$$\text{Impulse} = F \times \Delta t$$
$$F = \frac{\text{Impulse}}{\Delta t}$$
$$= \frac{0.4 \text{ kg-m/s}}{0.001 \text{ s}}$$
$$= 400 \text{ kg-m/s}^2 = 400 \text{ N} ●$$

Nudging a Cruise Ship

1. What is the impulse imparted to a cruise ship with mass 10^7 kg and velocity 5 m/s that is brought to rest by a tugboat?

   SOLUTION: Here the final momentum is 0, because the ship is brought to rest. The change in momentum thus equals the initial momentum.

   $$m \times v = 10^7 \text{ kg} \times 5 \text{ m/s}$$
   $$= 5 \times 10^7 \text{ kg-m/s}$$

2. If it takes 10 minutes to stop the ship, what is the average force exerted by the tugboat?

   SOLUTION: First we have to convert units from minutes to seconds.

   $$10 \text{ minutes} = 600 \text{ seconds}$$

From the impulse–momentum relationship, the change in momentum is equal to the impulse. This means that

$$\text{Impulse} = F \times \Delta t = \text{change in momentum}$$
$$F \times 600 \text{ s} = 5 \times 10^7 \text{ kg-m/s}$$

or

$$F = \frac{5 \times 10^7 \text{ kg-m/s}}{600 \text{ s}}$$
$$= 8.3 \times 10^5 \text{ kg-m/s}^2$$
$$= 8.3 \times 10^5 \text{ N}$$

This force is modest compared to the weight of the ship. It corresponds to the weight of an object with a mass of about 80,000 kg, which is less than 1% of the ship's mass. ●

Problems

1. Calculate the momenta of the following:
   a. A 200-gram rifle bullet traveling 300 m/s.
   b. A 1000-kg automobile traveling 0.1 m/s (a few miles per hour).
   c. A 70-kg person running10 m/s (a fast sprint).
   d. A 10,000-kg truck traveling 0.01 m/s (a slow roll).

2. Allison (30 kg) is coasting in her wagon (10 kg) at a constant velocity of 5 m/s. She passes her mother, who drops a bag of toys (5 kg) into the wagon.
   a. Do you expect the wagon to speed up or slow down? Why?
   b. What is the initial momentum of Allison and the wagon before her mother drops the toys in?
   c. What is the final momentum of Allison, the wagon, and the toys?
   d. What is the final speed of the wagon after Allison's mother drops the toys in?

3. What is the total momentum of each of the following systems?
   a. Two 1-kg balls move away from each other; one travels 5 m/s to the right, the other 5 m/s to the left.
   b. Two balls move away from each other, both traveling at 7 meters per second. One has a mass of 2 kg and the other has a mass of 3 kg.
   c. Two 1000-kg cars drive east; the first moves at 20 m/s, the second at 40 m/s.
   d. One of the 1000-kg cars moves west at 40 m/s, while the second moves east from the same starting point at a constant velocity of 30 m/s.

4. A 20-metric-ton train moves south at 50 m/s.
   a. At what speed must it travel to have twice its original momentum?
   b. At what speed must it travel to have a momentum of 500,000 kg-m/s?
   c. If there were a speed limit for this train as it traveled through a city, but not a weight limit, what mass in kilograms must be added to the train to slow it down to 20 m/s, while at the same time keeping the momentum the same as in part b?

5. **A.** Calculate the impulse imparted to the object in the following collisions.
   a. A 0.5-kg hockey puck moving at 35 m/s hits a straw bale, stopping in 1 second.
   b. A T-ball with a mass of 0.2 kg travels in the air at 15 m/s until it is stopped in the glove of a shortstop over a period of 0.1 seconds.
   c. A 12,000-kg tank moving at 4 m/s is brought to a halt in 2 seconds by a reinforced-steel tank barrier.

   B. What is the average net force exerted by these objects on the objects they collide with?

   C. Which is more important in determining the amount of damage an object sustains in a collision: the total momentum change (impulse) or the momentum change per unit time? Is the total area over which this force is applied important in determining how much damage is done?

6. A racing car with a mass of 1400 kg hits a slick spot and crashes head-on into a concrete wall at 90 km/hour, coming to a halt in 0.8 s. An ambulance weighing 3000 kg comes racing to the rescue, hits the same slick spot, and then collides with a padded part of the wall at 80 km/hr, coming to a halt in 2 seconds.
   a. What is the impulse exerted on each vehicle?
   b. What was the force exerted by each vehicle on the wall? What was the force exerted by the wall on each of the vehicles?
   c. What was the deceleration of each vehicle, from the time it contacted the wall to the time it completely stopped?

7. In Problem 6 in Chapter 4, you were asked to solve the following problem using the knowledge of Newton's laws that you had accumulated up to that point.
   Margie (45 kg) and Bill (65 kg), both with brand new roller blades, are at rest facing each other in the parking lot. They push off each other and move in opposite directions, Margie moving at a constant speed of 14 ft/s. At what speed is Bill moving?
   a. Use what you have now learned about momentum to answer this problem in a different way.
   b. Which method was easier for you to use to solve this problem, the Chapter 4 method or this one? How do the approaches compare? Are they really that different? Explain.

8. You are ice sailing in a boat on a very large, perfectly flat, piece of the Arctic. All of a sudden the wind dies and you cannot steer, but you have a boatload of frozen oranges that you have brought with you to eat. How can you try to stop yourself before your ice boat, heading for a deep crevice, goes over the edge? Use conservation of momentum to provide a solution.

9. Which object has a greater momentum, a 0.1-kg bullet traveling at 300 m/s or a 3000-kg truck moving at 0.01 m/s?

10. What is the total momentum of a two-particle system composed of a 1000-kg car moving east at 50 m/s and a second 1000-kg car moving west at 25 m/s?

11. Tony (60 kg) coasts on his bicycle (10 kg) at a constant speed of 5 m/s, carrying a 5-kg pack. Tony throws his pack forward, in the direction of his motion, at 5m/s relative to the speed of the bicycle just before the throw.

   a. What is the initial momentum of the system (Tony, the bicycle, and the pack)?

   b. What is the final momentum of the system immediately after the pack leaves Tony's hand?

   c. Is there a change in the speed of Tony's bicycle? If so, what is the new speed?

Investigations

1. The next time you are at an amusement park, go to the bumper cars. Observe what occurs in various collisions. What happens when a car occupied by a large heavy person hits a car with a very small person inside? What happens when cars with people of equal mass collide? Also, look at the effects of the angles of the collisions. Are your observations consistent with what you learned in this chapter?

2. Go to a pool hall or to a friend's house where there is a pool table. Set up different shots and observe whether momentum is conserved in these collisions. Ignore the various spins that may be applied to these balls as you try to further understand the nature of linear momentum and collisions. (The effect of spin is covered in the next chapter.)

3. Look up the masses and normal speeds for the balls used in several of your favorite sports. How much momentum do the balls typically have? What kind of impulses result when they are hit by or in turn collide with different solid objects such as bats, clubs, or human flesh?

4. Using the balls you studied in 3 above, design an experiment to measure the precise momentum of the different balls, along with the impulses generated in their normal use. Detail the methods that you could use to do this.

5. Research the history of weaponry. Examine the role momentum and impulse have played in the design and development of military machines and weaponry. Examine such objects as catapults, battering rams, cannonballs, tanks, and any others that you can think of. Has momentum been more important at different times? What types of trade-offs were involved in the design and manufacture of these machines?

6. To some physicists, particularly in the nineteenth century, conservation laws have represented more than just useful descriptions of nature. These scientists have seen a profound esthetic beauty and mathematical simplicity in such statements. James Prescott Joule, who helped to establish the law of conservation of energy (see Chapter 12), said, "Nothing is destroyed, nothing is ever lost, but the entire machinery, complicated as it is, works smoothly and harmoniously. . . . Everything may appear complicated in the apparent confusion and intricacy of an almost endless variety of causes, effects, conversions, and arrangements, yet is the most perfect regularity preserved—the whole being governed by the sovereign will of God." Imagine a universe in which momentum is not conserved and describe phenomena that might seem strange or different from our everyday experience. Is such a universe plausible? Is it less esthetic than our own?

7. Numerous serious scientific studies have been devoted to the threat posed to the Earth by collisions with asteroids. (It was just such a collision 65 million years ago that is believed to have caused the extinction of the dinosaurs.) Asteroids can be several kilometers wide traveling at speeds of hundreds of meters per second. Given what you know about momentum, what strategies do you think might be employed to save the Earth if a collision were found to be imminent?

 # WWW Resources

See the *Physics Matters* home page at **www.wiley.com/college/trefil** for valuable web links.

1. zebu.uoregon.edu/nsf/mo.html Animated Java simulation laboratories on the conservation of linear momentum from the Department of Physics, University of Oregon.

2. www.physicsclassroom.com/mmedia/momentum/cba.html A tutorial on impulse, momentum, and collisions from physicsclassroom.com including many animated problems and applets.

3. www.nhtsa.dot.gov/index.html Home of the US National Highway Traffic Safety Administration containing many resources regarding auto crash safety and testing.

7

Rotational Motion of an Object

KEY IDEA

In the absence of external torques, the angular momentum of any system is constant.

PHYSICS AROUND US . . . Spinning

An ice skater speeds up as she goes into a turn. Suddenly, she swings gracefully into a spin, both arms and one leg extended to the side. As she pulls her arm and leg in closer to her body, the rate of spin increases until, at the end, her features turn into a blur.

You are watching a news program at home when the anchor starts talking to a colleague half a world away. High above the Earth's atmosphere, a satellite relays the signals back and forth between the two reporters. Inside the satellite, meanwhile, a set of small spinning gyroscopes allow the onboard computers to keep track of which way the satellite's antennae are pointing so that the signals are aimed in the right direction.

It's winter. You wrap your coat tightly around you in the blustering wind. You know, however, that in six months you'll be walking around in shorts. You may even complain to a friend about how pleasant it would be if you could average the two seasons out, but you know it will never happen.

Amazingly, all three of these phenomena—skater, satellite, and seasons—are related to a quantity that physicists call *angular momentum*. All rotating objects have angular momentum, so in this chapter we also examine the physics of rotational motion.

● ROTATIONAL MOTION

So far we have discussed the motion of an object along a line, both straight lines and curved lines. We have not looked at any motion the object might have with respect to itself, such as spinning, as it moves along a line. However, real objects can spin around as they move, and they demonstrate different properties than objects that don't spin.

Many of the laws of mechanics for objects that move in a line have analogs for objects that spin. We see in this chapter that rotating objects have rotational speed analogous to linear speed and obey a rotational version of Newton's second law. We study rotational analogs for force and momentum as well. Much of the physics in this chapter will seem new and yet somewhat familiar.

All spinning objects display **rotational motion.** We can analyze this motion in terms of two properties: an axis of rotation and a rotational velocity.

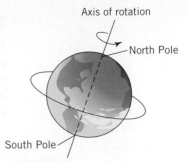

FIGURE 7-1. Each point on an imaginary line that passes through the center of the Earth, connecting the two poles, is stationary with respect to the Earth. Every other point in the planet turns around this axis of rotation.

Axis of Rotation

Imagine the Earth spinning in space. Everything on the surface is moving except for two points—the points we call the North and South Poles. These points remain stationary with respect to the object while everything else turns around them. You can imagine a line through the center of the Earth connecting the two poles (Figure 7-1). Each point on this line is also stationary with respect to the Earth, while every other point in the planet turns around it. This line, around which everything else rotates, is called the **axis of rotation** of the Earth. It is this axis on which an ordinary desktop globe spins.

Every rotating object has an axis of rotation. For example, the wheels of your car rotate around an axis at their center, along the car's axle. The Sombrero galaxy rotates around an axis perpendicular to its plane (Figure 7-2). Of course, your car's wheels turn many times each second, while the outer parts of a galaxy take hundreds of millions of years to make just one rotation. Nevertheless, the basic idea in each situation is the same. An overall rotation occurs around an axis, which itself remains stationary with respect to the object.

FIGURE 7-2. The Sombrero galaxy is a typical spiral galaxy, much like our own Milky Way, rotating about an axis through its center.

Develop Your Intuition: The Pizza Galaxy

A pizza maker rolls out the dough for the crust, then flips it spinning into the air to thin it out (and to show off) (Figure 7-3). What is the axis of rotation of the pizza dough while it's in the air?

Like the galaxy, the pizza dough spins (roughly) in a flat disk. At the center of the disk, an imaginary vertical line remains stationary with respect to the pizza dough. This line moves up and down with the dough, but no point on the line moves in a circle. This imaginary line through the disk of dough is the axis of rotation. Every rotating object has an axis of rotation, even if the axis is sometimes moving.

FIGURE 7-3. Pizza dough thins out into a circular disk as it spins, something like how a spiral galaxy forms.

Speed of Rotation

Period and Frequency The Earth spins about its axis of rotation about once every 24 hours. This daily cycle is one of the most familiar rotational motions for many of us. However, *any* rotational motion can be characterized by the time it takes for the object to make one complete rotation. This time is called the **period of rotation,** which is usually represented by *T*.

Another common way of talking about rotational motion is to specify the number of times an object completes a rotation in a given amount of time—a number called the **frequency of rotation,** which is usually represented by *f*. The frequency of Earth's rotation, for example, is 1 cycle per 24 hours or about 365 cycles per year.

Frequency can be defined in terms of any time period (a day, a year, etc.), but in many scientific applications frequency is defined as the number of rotations completed in 1 second. In the case of the Earth, only a small fraction of a rotation is completed in 1 second. In other cases (your car's wheels at high speeds, for example), many rotations may be completed in 1 second, so the frequency is some number larger than 1.

Rotational frequency is measured in a unit called the **hertz,** named after the German physicist Heinrich Rudolf Hertz (1857–1894), who discovered radio waves (see Chapter 19). A body that completes one rotation in 1 second has a frequency of 1 hertz (1 Hz).

Develop Your Intuition: How Often Does the Electric Current in Your Hair Dryer Change Direction?

Pick up any electric appliance in your house, such as a toaster or hair dryer, and find where its properties are specified. You will probably see a note that reads "60 Hz." What does this mean?

"Hz" is the abbreviation for hertz. This designation on the appliance means that it is intended for use with electric currents in which the direction of current changes 60 times each second (see Chapter 17), as in the United States. In Europe, the standard current is 50 Hz, which is one reason why it's not always possible to use American appliances in Europe.

Relation Between Period and Frequency Period and frequency are related to each other. An object that spins with a high frequency has a short period, while an object with a low frequency has a long period. An object that has a long rotational period (such as the Earth) completes only a small fraction of a revolution in a second, and hence has a low frequency. In fact, the relationship between the period and the frequency of a rotation is simple to state.

1. In words:

The longer the period, the lower the frequency, and vice versa.

2. In equations with words:

The period is equal to one divided by the frequency.

$$\text{Period} = \frac{1}{\text{Frequency}}$$

and

The frequency is equal to one divided by the period.

$$\text{Frequency} = \frac{1}{\text{Period}}$$

3. In equations with symbols:

$$T = \frac{1}{f} \quad \text{and} \quad f = \frac{1}{T}$$

Our Spinning Planet

What is the frequency of the Earth's rotation in hertz?

EXAMPLE
7-1

SOLUTION: We know that the period of the Earth's rotation is 1 day or 24 hours. We want frequency recorded in units of hertz, or cycles per second, so we first must calculate the number of seconds in 1 day. There are 60 seconds in a minute, 60 minutes in an hour, and 24 hours in a day. Consequently, the number of seconds in 1 day (the period of rotation) is

$$T = 60 \text{ s/min} \times 60 \text{ min/h} \times 24 \text{ h/day}$$
$$= 86{,}400 \text{ s/day}$$

Then, from the relationship between period and frequency,

$$\text{Frequency} = \frac{1}{\text{Period}}$$

$$f = \frac{1}{T} = \frac{1}{86{,}400} \text{ Hz}$$

$$= 1.16 \times 10^{-5} \text{ Hz} \ \bullet$$

Angular Speed Think about children on a moving carousel. The children on the inner horses, closer to the axis of rotation, travel a shorter distance during each rotation than children on the outer horses. Children on the inner horses must move at a slower speed than children on the outer horses. Yet we know that everyone on the carousel makes one circuit in the same amount of time. That shared aspect of rotational motion is called **angular speed.**

Imagine drawing a line that starts at the center of the carousel and passes through two children who are at different distances from the center before the rotation starts (Figure 7-4 on page 142). As the carousel turns, we can characterize the change in position of the children by the angle θ (the Greek letter theta), as shown. The point is that both children have moved through this same angle, even though they have traveled different distances along their respective paths.

1. In words:

Angular speed is the angle through which an object has moved about the axis of rotation divided by the time it takes it to go through that angle.

2. In an equation with words:

$$\text{Angular speed} = \frac{\text{Rotation angle}}{\text{Time}}$$

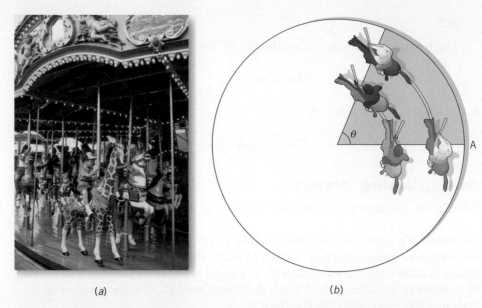

(a) (b)

FIGURE 7-4. (a) Children riding near the outside of a rotating carousel move faster than children near the center, but they all move with the same angular speed. (b) A line (A) from the center of a carousel passes through two children at different distances from the center. As the carousel turns, the change in position of the children is given by the angle θ, as shown. Both children have moved through this same angle, even though they have traveled different distances along their respective paths.

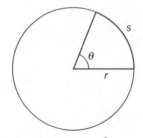

In radians, $\theta = \frac{s}{r}$

FIGURE 7-5. Angles involved in the measurement of angular speed are often given in terms of radians. If an object travels through an arc of length s (in meters) along a circle of radius r (also in meters), then the angle θ (measured in radians) is defined as $\theta = \frac{s}{r}$.

3. In an equation with symbols:

$$\omega = \frac{\theta}{t}$$

This definition of angular speed, ω (the Greek letter omega), is similar to the definition of linear speed as the distance traveled divided by the time it takes to travel that distance. Note, however, that while the two children on the carousel have the same *angular* speed, they have different *linear* speeds. In particular, the child farther from the center is moving faster.

It is customary to define the angle in the definition of angular speed in terms of a unit called the *radian*. If, as in Figure 7-5, an object travels through an arc of length s (in meters) along a circle of radius r (also in meters), then the angle θ (measured in radians) is defined as:

$$\theta = \frac{s}{r}$$

Coming Full Circle

EXAMPLE 7-2

If an object travels all the way around a circle, what is θ in degrees? In radians?

SOLUTION: A full circle is 360 degrees, so that is the value of θ in degrees. To get the value in radians, we note that if the object goes all the way around a circle, it will have traveled a distance equal to the circumference of the circle. Thus,

$$s = 2\pi r$$

From the definition of the radian, then, the angle through which the object travels is

$$\theta = \frac{s}{r}$$

$$= \frac{(2\pi r)}{r}$$

$$= 2\pi$$

Thus, 360 degrees is equal to 2π radians. ●

Angular Frequency   Each time a rotating object completes one revolution it traverses an angle of 2π radians. Sometimes it is useful to measure rotational motion in terms of the *angular frequency,* which is defined as the number of radians traversed in 1 second:

$$\text{Angular frequency} = 2\pi f$$

where f is the frequency of rotation in cycles per second, or hertz. Note that frequency f and angular frequency are measures of exactly the same physical phenomenon—namely, how rapidly an object rotates. The only difference is whether the units are cycles per second, or hertz (rotation frequency), or radians per second (angular frequency).

● TORQUE

Suppose you turned your bicycle upside down so that the wheels were in the air and you wanted to get the wheels spinning. What would you do? Most likely you would place your hand firmly on the wheel and move it downward—a process that would get the wheel turning. If you wanted the wheel to go faster, you would repeat the operation until you achieved the desired speed.

Let's look at this simple example from the point of view of the forces being applied (Figure 7-6). The axis of rotation is the axle that attaches the wheel to the bike. To get the wheel spinning, you have to apply a force that satisfies two criteria. First, the force must have a component that is tangent to the wheel; that is, in the plane of rotation and parallel to the edge of the wheel. Second, this force has to be applied some distance away from the axis. In this example the force is applied a perpendicular distance r away from the axis, where r is the wheel's radius. (This distance is sometimes called the "lever arm," a term that arises from the common use of torques in the operation of a simple lever.) A force applied in this way, satisfying these two criteria, is said to produce a **torque.**

If a tangential force F is applied a distance r from the axis of rotation of an object, then the torque τ (the Greek letter tau) is defined as

1. In words:

   The torque is the tangential force being applied times the perpendicular distance from the axis of rotation.

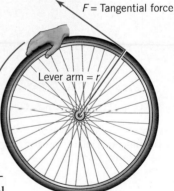

F = Tangential force

Lever arm = r

FIGURE 7-6. To get a wheel spinning about its axis, you have to apply a torque. The force that generates the torque must satisfy two criteria. First, the force must have a component that is tangent to the wheel; that is, in the plane of rotation and parallel to the edge of the wheel. Second, this force has to be applied some distance away from the axis. This distance is sometimes called the "lever arm," a term that arises from the common use of torques in the operation of a simple lever.

2. In an equation with words:

Torque equals tangential force times perpendicular distance.

3. In an equation with symbols:

$$\tau = F \times r$$

Torque plays a role in rotational motion analogous to the role of force in linear motion. If we want to change the motion of an object moving in a line, we have to apply a force. In the same way, if we want to change the way something is rotating, we have to apply a torque. In our example, we applied a torque to the bicycle wheel to start and then speed up its rotation.

It's important to remember that in order to produce a change in rotation, a force must have a component that is tangent (in a direction parallel to the edge of the wheel). If you grabbed the bicycle wheel and pushed inward toward the center or to the side (perpendicular to the plane of rotation), the wheel wouldn't start to spin. In these cases, you'd be applying a force without producing a torque.

Torque is needed to change the rotational motion of any object, not just a wheel. For example, when you open a door by pushing on the handle, you are exerting a force at a distance (the width of the door) from the axis of rotation (the hinges). The handle is placed at the edge of the door farthest from the hinges to increase the torque produced by a given force (your push). You probably know from experience that it's much harder to push a door open if you push on the side near the hinges.

The use of a torque underlies the operation of a simple lever. In this application, a beam is placed over a sharp edge, called the "fulcrum," with the length of the beam on one side of the fulcrum longer than the length on the other side (Figure 7-7). You apply a downward force to the long side of the beam in order to lift a load at the short side of the beam. Here the fulcrum serves as the axis of rotation for the beam. The load on the short side of the beam produces a torque in one direction that must be overcome by the torque in the opposite direction, produced by your applied force on the long side of the beam (your lever arm). Since your lever arm is longer, you can lift the load by exerting a smaller force than the weight of the load, but you must move the arm through a greater distance.

It is easier to open a door by pushing or pulling on the edge farthest from the hinges.

FIGURE 7-7. The use of a torque underlies the operation of a simple lever. A beam is placed over the fulcrum, with the length of the beam on one side of the fulcrum longer than the length on the other side. Applying a downward force to the long side of the beam allows you to lift a heavy object on the short side of the beam. The fulcrum serves as the axis of rotation for the beam.

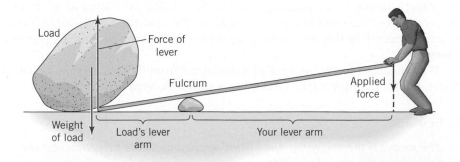

Braking

From Newton's laws of motion, it follows that any change in rotational motion, whether speeding up or slowing down, requires a force—a torque. When you are riding your bike and want to stop, you squeeze a lever attached to the handlebars. As you do this, two small plastic or rubber brake pads clamp down on the metal frame of the tire and the bike slows down (Figure 7-8). Where are the torques in this process?

The brake pads squeezing down on the metal frame generate a frictional force. This force acts in a direction tangent to the wheel and in a direction opposite to the direction of the wheel's motion. Because the frictional force is applied away from the axis of rotation, it produces a torque, in this case a torque that slows down the rotation rather than one that speeds it up.

The brakes in your car work the same way, although you can't see them in operation. When you put your foot on the brake pedal, tough ceramic pads clamp down on spinning metal parts of the wheel, producing a frictional torque that slows down the wheel.

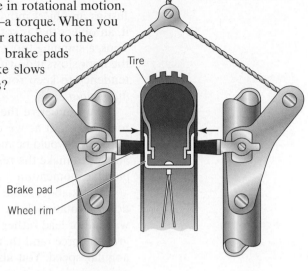

FIGURE 7-8. Brake pads exert a frictional force against the rim of a bicycle wheel to slow down the wheel's rotation and bring the bike to a stop.

Develop Your Intuition: Torque to the Rescue

When you change a tire on your car, it may happen that the nut holding the tire to the axle is rusted and difficult to turn. In this case, experienced mechanics sometimes put a length of pipe around the handle of the wrench, in such a way that the pipe extends farther out than the original handle of the wrench (Figure 7-9). Use the concept of torque to explain why this might be a useful thing to do.

The problem is that you can exert only so much force on the wrench and therefore produce only so much torque to turn the recalcitrant nut. However, torque depends on two factors: the amount of applied tangential force and the distance of that force from the axis of rotation. The pipe, in effect, makes the handle of the wrench longer—it increases r in the torque equation—and therefore allows you to exert a greater torque while applying the same force.

A word of caution: Sometimes this technique can generate more torque than you want. The authors have seen tire changers use such a long pipe that the steel nut was completely sheared off, requiring a trip to a mechanic to set things right.

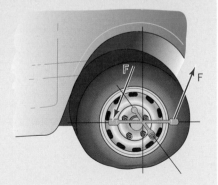

FIGURE 7-9. Tire wrench in use; an extender would increase the length of the lever arm so you could exert a larger torque.

● ANGULAR MOMENTUM

Just as an object moving in a straight line keeps moving unless a force acts on it, an object that is rotating keeps rotating unless a torque acts to make it stop. Thus, a spinning top will spin until the friction between its point of contact and the floor slows it down. A wheel will turn until friction in its bearing stops it. This tendency to keep rotating is often stated in terms of a quantity called *angular momentum*.

We can derive the formula for angular momentum step by step from Newton's laws, just as we derived the formula for linear momentum. However, the derivation would be more complicated, so we'll just point out some everyday examples to make the result seem reasonable. Afterward, we'll state a definition of angular momentum.

Think about some common experiences in which a torque speeds up or slows down a rotation. First, in the example of the bicycle wheel, imagine that the tire was full of lead rather than air. Common sense tells you that it would require a greater force (and therefore a greater torque) to spin the wheel up to the same angular speed. You also know from experience that if the radius of the wheel were twice as big, it would require a greater force (and therefore a greater torque) to get it rotating. The relation between torque and the change of rotation, then, must involve both the mass of the rotating object and the distance between the axis of rotation and the location of that mass.

Moment of Inertia

The quantity that describes the distribution of mass around an axis of rotation is called the **moment of inertia** (I). In general, the farther away the mass is from the axis, the greater the moment of inertia (Figure 7-10). For example, if you pick up a dumbbell with two weights on it, as shown, you have to apply a certain amount of torque to get it to rotate. If you move the weights out so that they are twice their original distance from the axis of rotation, it takes a greater torque to get the same amount of rotational speed—four times as much torque in this

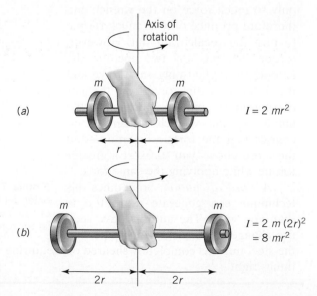

FIGURE 7-10. Moment of inertia (I) describes the distribution of mass around an axis of rotation. The farther away the mass is from the axis, the greater the moment of inertia.

(a) $I = 2\ mr^2$

(b) $I = 2\ m\ (2r)^2$
      $= 8\ mr^2$

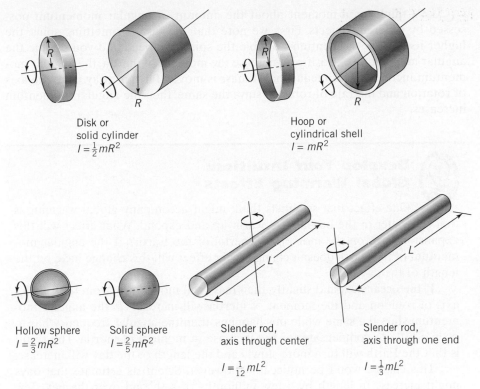

Disk or
solid cylinder
$I = \frac{1}{2}mR^2$

Hoop or
cylindrical shell
$I = mR^2$

Hollow sphere
$I = \frac{2}{3}mR^2$

Solid sphere
$I = \frac{2}{5}mR^2$

Slender rod,
axis through center
$I = \frac{1}{12}mL^2$

Slender rod,
axis through one end
$I = \frac{1}{3}mL^2$

FIGURE 7-11. Moments of inertia for several common objects rotating around various axes. All of these moments of inertia increase with the mass (m) of the object (double the mass and you double the moment of inertia), while the moment of inertia increases as the square of the dimension of the object (double the radius of a sphere, for example, and you increase its moment of inertia by a factor of four).

example. The torque must increase because moving the weights farther out increases the moment of inertia of the dumbbell.

In Figure 7-11, we show several common objects rotating around various axes, along with their moments of inertia. Notice that all the moments of inertia increase with the mass of the object—double the mass and you double the moment of inertia. Note also that the moment of inertia increases as the square of the dimension of the object—double the radius of a sphere, for example, and you increase the moment of inertia by a factor of four.

Definition of Angular Momentum

We are now ready to give a technical definition of **angular momentum.**

1. In words:

Angular momentum of an object depends on how the mass of the object is distributed and on its rate of rotation.

2. In an equation with words:

Angular momentum = Moment of inertia times angular speed.

3. In an equation with symbols:

$$L = I \times \omega$$

where L is the angular momentum and I is the moment of inertia.

Let's think for a moment about the amount of angular momentum possessed by different objects. First, we note that the faster something spins, the higher its angular momentum—double the spinning speed and you double the angular momentum. Also, if you increase the moment of inertia, the angular momentum increases. This means that if mass is moved farther away from the axis of rotation and the rate of rotation stays the same, then the angular momentum increases.

Develop Your Intuition: Global Warming Effects

One effect that scientists think might accompany global warming is that the water in the oceans will warm up and expand. What effect will this expansion have on the moment of inertia of the Earth? If the angular momentum of the Earth doesn't change, what effect will this change have on the length of the day?

If the oceans expand slightly, then mass will move away from the Earth's axis of rotation and the moment of inertia will increase. If the angular momentum stays the same while this happens, then the angular frequency has to drop a little to compensate for the increase in moment of inertia. The result is that the Earth will turn more slowly and the length of the day will increase.

This change won't be noticeable, however. Scientists estimate that days might increase in length by a few millionths of a second over the next few centuries.

Angular Momentum and Torque

Newton's second law describes the effect of a force on the rate of change of linear momentum (see Chapter 6). An analogous law for rotating bodies defines the effect of torque (a tangential force) on the rate of rotation (a change in angular momentum).

1. In words:

The rate of change in angular momentum of an object equals the net external torque on that object.

2. In an equation with words:

Net external torque = Change in angular momentum divided by
the change in time

3. In an equation with symbols:

$$\tau = \frac{\Delta L}{\Delta t}$$

The net external torque on an object is defined as the sum of the torques generated by all external forces acting on that object. Can you see the similarity between this equation and the equation in Chapter 6 for linear motion? In both cases, the force (or net external torque) on an object causes a change in momentum (or angular momentum) over a time interval, Δt.

CONSERVATION OF ANGULAR MOMENTUM

If a net external torque on a system is zero, the preceding equation tells us that the change in angular momentum must be zero. In other words, without the action of a force acting at a distance from the axis of rotation, the total angular momentum of a system cannot change. Like its counterpart, linear momentum, angular momentum is conserved in the absence of outside influences. This principle is known as the **conservation of angular momentum.**

1. In words:

If the net external torque is zero, the angular momentum of any system must stay constant over time.

2. In an equation with words:

The change in angular momentum of an isolated system equals zero.

3. In an equation with symbols:

$$\Delta L = 0$$

The consequences of the conservation of angular momentum that you're most likely to experience have to do with situations in which something happens to change an object's moment of inertia. In this case, the angular velocity must change as well so that the product of I and ω will stay the same. In most practical situations of this kind, you may find it helpful to express conservation of angular momentum in the form:

Initial angular momentum = Final angular momentum

or
$$I_i \times \omega_i = I_f \times \omega_f$$

(a)

A striking illustration of this point can be seen in figure skating competitions, as described in the Physics Around Us section. As the skater goes into the spin with her arms spread wide, her moment of inertia is high because an appreciable amount of mass (her arms) is located far from the axis of rotation (Figure 7-12a). As she pulls her arms in over her head, her moment of inertia drops (Figure 7-12b). Since no outside force is acting to affect the spin, her angular momentum must remain the same. The only way for this to happen is for her angular velocity (that is, her rate of spin) to increase.

We can make this same point by looking at the equation for angular momentum:

$$L = I \times \omega$$

If the skater pulls in her arms, her moment of inertia decreases. The only way the angular momentum L can stay the same is for her angular frequency ω to increase in a compensating way. So, for example, if I decreases by $\frac{1}{2}$, ω has to increase by a factor of 2 to keep L the same. Hence, her rotational speed doubles.

Can you can use this same reasoning to explain why the spin slows down when she puts her arms back out?

(b)

FIGURE 7-12. (a) Figure skater in a spin. (b) She can increase her rotation rate by pulling her arms and legs closer to her body, decreasing her moment of inertia.

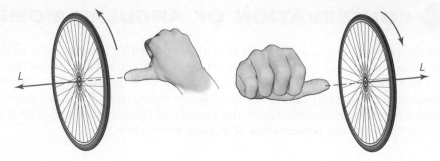

FIGURE 7-13. The right-hand rule for finding the direction of angular momentum for a spinning object. Curl the fingers of your right hand in the direction of the rotation; your right thumb will point in the direction of the angular momentum.

Footballs are thrown with a spiral motion to minimize wobble.

The Direction of Rotation

Like linear momentum, angular momentum has a direction. The so-called "right-hand rule" for finding this direction for a spinning object is simple: if you curl the fingers of your right hand in the direction of the rotation, your right thumb will be pointing in the direction of the angular momentum (Figure 7-13). For example, the angular momentum of the Earth points upward along the axis of rotation to the North Pole.

The conservation of angular momentum implies that *both* the size and direction of the angular momentum of an object remain fixed in the absence of torques. This fact explains why the orientation of an isolated spinning object, such as a football thrown in a spiral, is generally more stable than that of a nonspinning object, such as a football bouncing on the ground.

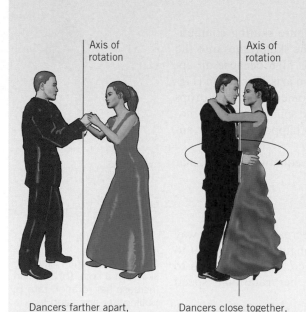

Dancers farther apart, larger moment of inertia, slower angular velocity

Dancers close together, smaller moment of inertia, faster angular velocity

Develop Your Intuition: Waltzers

Dancers doing a waltz normally hold each other at arm's length to avoid stepping on each other's feet. If they have to make a fast turn, however, you will often see them move together, with each partner putting his or her feet between or even behind those of the other. Why do they risk calamity this way?

The answer has to do with the conservation of angular momentum. Under normal circumstances, the couple's mass is located far from the axis of rotation (which is located between the dancers). By moving closer together, they reduce their moment of inertia and, like the ice skater, increase their angular velocity (Figure 7-14). This change in position helps them get through the turn, after which they can separate and slow down the rotation.

FIGURE 7-14. By moving closer together, dancers reduce their moment of inertia and increase their angular velocity. This change in position helps them get through fast turns, after which they can separate and slow down the rotation.

The Collapse of Stars

Stars often end their life cycle by undergoing a rapid collapse. The Sun, for example, which now has a radius of 7×10^8 meters, will collapse to a type of star called a "white dwarf," approximately the same size as the Earth. (The radius of the Earth is about 6.4×10^6 meters—less than a hundredth that of the Sun.) The Sun currently rotates about its axis once every 26 days. How will its rotation change when it becomes a white dwarf?

No outside forces will act on the Sun during its collapse, so its angular momentum must be conserved. As the Sun contracts to the size of the Earth, its moment of inertia will decrease dramatically because, as with a skater pulling in her arms, the mass will be located much closer to the axis of rotation. Consequently, the shrunken Sun's rate of rotation will have to increase dramatically to compensate. In other words, after the collapse, the Sun will rotate much faster than it does now.

We work out the numbers for this in Example 7-3 at the end of the chapter. The result turns out to be 3.1 minutes. In fact, some stars collapse to even smaller objects than white dwarves and they rotate hundreds of times *each second*.

Connection

Inertial Guidance Systems

The conservation of angular momentum plays an important role in the inertial guidance systems for navigation in airplanes and satellites. The idea behind such systems is very simple. A massive object such as a sphere or a flat circular disk is set into rotation inside a device in which very little resistance (that is, almost no torque) is exerted by the bearings. Once such an object, called a "gyroscope," is set into rotation, its angular momentum continues to point in the same direction, regardless of how the aircraft or rocket moves around it. By sensing the constant rotation and seeing how it is related to the orientation of the satellite, engineers can tell which way the satellite is pointed.

Toy gyroscopes and tops work in the same way (Figure 7-15). In the case of a spinning top, its weight starts to topple it over as it spins slightly away from perfectly vertical. However, when the weight is acting at a slight distance from the vertical axis, it produces a torque. This torque changes the angular momentum of the top; but what changes is the direction of the angular momentum, not its size. The angular momentum changes direction in such a way as to bring the top back to spinning vertically. However, the changing direction of angular momentum continues to cause a torque even as the top tries to get back to vertical. The result is that the top wobbles as it spins. The axis of rotation traces out the sides of an imaginary cone that gets larger as the top slows down from friction at the contact point.

Other spinning objects show this wobbling motion, which goes by the technical name "precession." You can certainly see the wobble in a thrown football if it is not thrown with a tight spiral. The Earth, too, wobbles in its motion around the Sun, as do all the other planets. ●

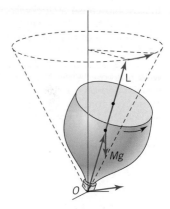

FIGURE 7-15. A toy top or gyroscope maintains its orientation as it spins. Inertial guidance systems rely on rapidly spinning gyroscopes that maintain orientation very accurately.

MOTION IN A PLANE WITH ROTATION

In Chapter 6 we learned how to analyze situations involving the momentum of an object moving in a straight line or in a plane under the influence of a force such as gravity. In this chapter, we have learned how to deal with a body that is rotating around an axis. The most general kind of motion involves a combination of these two types of motion. Think about the movements of a boomerang in flight or a Frisbee sailing through the air. In these cases, the flying objects are moving under the influence of gravity, but they are also rotating around an axis. We now have all the background we need to talk about these sorts of complicated motions.

Center of Mass

Imagine taking a ruler and balancing it on one finger. You know that the ruler will balance if you support it right in the middle. As shown in Figure 7-16, we can understand this situation in terms of the torques exerted on the ruler by gravity. The downward pull of gravity on the right-hand side produces a torque that, if it were the only force acting, would produce a clockwise rotation of the ruler. Similarly, the force of gravity acting on the left-hand side creates a torque that would produce a counterclockwise rotation. When the ruler is balanced on your finger, these two torques are equal and cancel each other out. The ruler sits stationary, not rotating at all. The point of support on which an object can be balanced like this is called its **center of mass** (or, sometimes, its **center of gravity**).

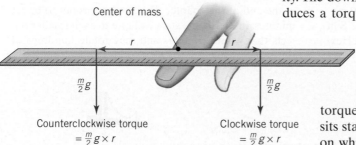

Center of mass

Counterclockwise torque
$= \frac{m}{2} g \times r$

Clockwise torque
$= \frac{m}{2} g \times r$

FIGURE 7-16. The downward pull of gravity on the right-hand side of a balancing ruler produces a torque that would cause a clockwise rotation of the ruler except for the force of gravity acting on the left-hand side that would cause a counterclockwise rotation. When the ruler is balanced on your finger, these two torques are equal and cancel each other out.

You can think of the center of mass of an object as being the average of the positions of the object's mass. For a regular geometrical object such as a cube or a sphere with a uniform mass distribution, the center of mass is at the geometrical center of the object. The center of mass of a smooth disk, for example, is at the exact center of the disk.

The center of mass of a system doesn't have to be at its geometrical center, however. Every object, no matter how complicated its shape or structure, has a single point where all the torques due to gravity cancel, a single point where it could be supported without rotation. For example, if one side of the ruler weighs more than the other side, the center of mass would be located more toward the heavy side.

Complex Motion

In Chapter 3 we saw that the discussion of motion in two dimensions could be greatly simplified by the law of compound motion, which allows us to break the problem up into a connected pair of simpler one-dimensional problems. In the same way, motions that include rotations can be broken up into a series of simpler problems. The analog of the law of compound motion for rotating objects is:

> *Motion that involves rotation can be thought of*
> *as the motion of the center of mass (treated as if the object*
> *were a single particle) plus rotation about the center of mass.*

Consider, as an example, a springboard diver performing a somersault in the air. Using the rule we have just stated, we could talk about her motion in two stages: (1) the motion of her center of mass and (2) the rotation around her center of mass. The motion of the diver's center of mass is simply that of a point particle under the influence of the force of gravity. As we saw in Chapter 3, the

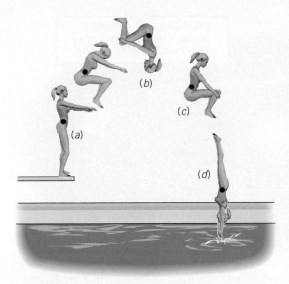

FIGURE 7-17. When the diver leaves the diving board (a) her body is vertical, but at the top of her arc (b), her hips are now horizontal. Thus the board applied a torque to her body when she jumped off, giving her angular momentum and causing her to rotate around her center of mass as the dive progresses. Once the diver is launched, there are no further torques acting on her. When she grabs her knees (c), which lowers her moment of inertia, she starts rotating faster to keep the angular momentum the same. When she straightens out again (d), the moment of inertia increases and the angular velocity drops.

path followed by such a particle is a parabola that rises to a peak and then falls, as shown in Figure 7-17.

To understand the motion around the center of mass, imagine traveling along the parabola with the diver. When she leaves the diving board, her body is vertical, as shown in Figure 7-17. At the top of her parabolic arc, her hips are now horizontal. When she reaches the water, she is again vertical but with her head down. In other words, the board applies a torque to her body when she jumps off, causing her to rotate around her center of mass as the dive progresses. At the beginning of the dive, she is rotating and therefore has a certain amount of angular momentum.

Once the diver is launched, however, there are no further torques acting on her, a fact that means that whatever angular momentum she had leaving the board remains constant. Thus, when she grabs her knees, an act that lowers her moment of inertia, she starts rotating faster to keep the angular momentum the same. This tucking action is what produces the spectacular rotation during the high part of the dive. When she straightens out again, the moment of inertia increases and the angular velocity drops. She enters the water cleanly.

A springboard diver always follows a parabolic path through the air.

Develop Your Intuition: Not a Swan Dive

How can a springboard diver enter the water so that there is no rotation about his center of mass? What would such a dive look like?

No matter what he does, the diver's center of mass will still move under the force of gravity through the same parabola. If he leaves the springboard without torque, however, his body must remain in a heads-up vertical position throughout the dive. There will be no rotation around his center of mass, and he will enter the water feet first. While such a dive is not likely to win an Olympic medal, he can make quite a splash by pulling his knees up with his arms to perform a "cannonball."

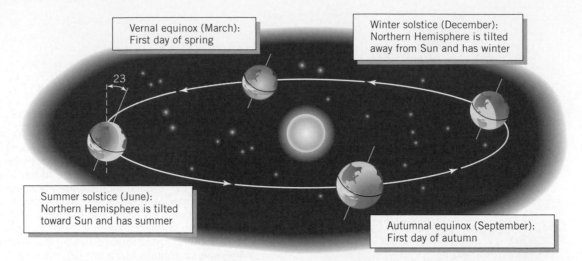

FIGURE 7-18. In the months of July and August, the Earth is tilted so that the Northern Hemisphere leans toward the Sun. During January and February, the Northern Hemisphere is tilted away from the Sun.

Connection

The Seasons

The continual progression of seasons on our planet is related to the conservation of angular momentum. As shown in Figure 7-18, the Earth's axis of rotation is tilted by an angle of 23 degrees to the plane of the Earth's orbit around the Sun. Over time scales of a few years, we can treat the Earth as if no significant torques act on it, so Earth's angular momentum must be constant. Both the size and direction of the Earth's angular momentum remain fixed, so that the direction of the axis remains the same over the course of a year. (The axis does change direction slightly due to precession, as we mentioned in the Connection section on inertial guidance systems, page 151, but this effect is small enough that we can neglect it in the present discussion.)

In the months of July and August, the Earth is tilted so that the Northern Hemisphere leans toward the Sun, as shown in Figure 7-18. (The tilt is oriented most directly toward the Sun at the summer solstice, around June 21 of each year.) This orientation means that more sunlight falls on each square foot of the Earth's Northern Hemisphere surface during this period than during January and February, when the Northern Hemisphere is tilted away from the Sun. The effect is similar to shining a flashlight on the floor from directly overhead or from an angle; the light is brighter when it covers the smaller area of surface from directly overhead (Figure 7-19). Earth's tilt is why the Northern Hemisphere experiences summer from June to August and why it experiences winter from December to February—*even though the Earth is closest to the Sun during those winter months.* We have seasons because the conservation of angular momentum ensures that the Earth's axis of rotation tilts in the same direction over the course of the year. ●

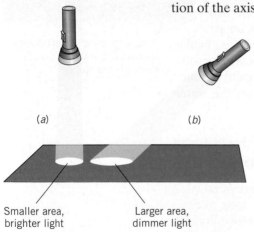

(a)   (b)

Smaller area, brighter light   Larger area, dimmer light

FIGURE 7-19. When a flashlight shines on the floor from directly overhead (a), the light is brighter and it covers a smaller surface area than when the light shines at an angle (b).

THINKING MORE ABOUT

Angular Momentum: The Spinning Solar System

The solar system, which includes the Earth, the Moon, the other planets, and lots of smaller objects that are gravitationally bound to the Sun, such as asteroids and comets, has a lot of angular momentum. The planets swing around the Sun in their stately orbits, while the Sun spins on its axis once every 26 days. But we know that, in the absence of an external force, angular momentum is conserved—it doesn't spontaneously increase or decrease. So where did all the solar system's angular momentum come from originally?

According to the most widely accepted theory, called the "nebular hypothesis," the Sun and planets formed almost 5 billion years ago from an immense, swirling, irregular cloud of dust and gas, called a "nebula," that extended across trillions of kilometers of space. Like any swirling cloud, the nebula had its own angular momentum—it was slowly spinning about its axis of rotation. Eventually, ever so slowly, the cloud began to spiral inward from the force of gravity due to the concentration of mass at its center (Figure 7-20). As the nebula became more dense and compact, it also began to spin faster. Ultimately, gravity pulled most of the dust and gas into the great central mass that we now call the Sun. Smaller clumps of matter called "planetesimals," which were orbiting too fast to fall into the Sun, formed the planets—Earth, Mars, Jupiter and so on.

The French mathematician and physicist Pierre Simon Laplace first proposed the nebular hypothesis in 1796. His work incorporated the earlier results of the German philosopher Immanuel Kant, who showed in 1755 that a contracting cloud of gas would form a disk in a plane perpendicular to the axis of rotation. Laplace went through calculations similar to those in the Looking Deeper box on the collapse of stars (see page 151), showing how the rotational rate of the Sun would speed up over time. However, he came up with a major problem. The calculations predicted that the Sun's period of rotation should be only a

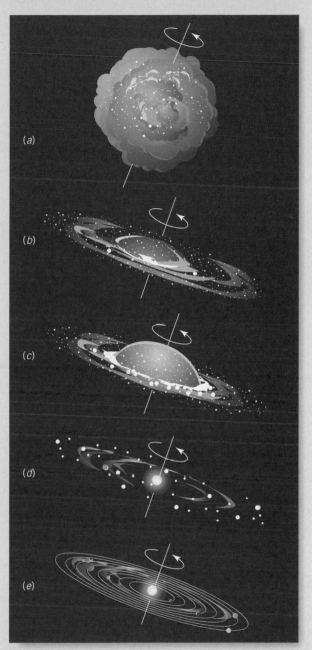

FIGURE 7-20. Astronomers today believe the solar system formed from a large cloud of gas and dust (*a*). As the cloud contracted under the influence of gravity, it started to spin faster and formed a flattened disk (*b*). Eventually, the Sun formed from the central part of the disk (*c*), while the outer parts formed into small rocky planetesimals (*d*) that over the course of time collided and grew into today's planets (*e*).

few hours, not a month. It seemed that the nebular hypothesis worked a little too well.

For a long time, the nebular hypothesis was discredited because of this result. Astronomers turned to other theories to explain the formation of the solar system, most of them based on various kinds of catastrophes. One idea was that a massive comet came so close to the Sun that it pulled out a long stream of material from the Sun, which eventually produced the planets. This theory was thrown out when astronomers learned that comets have nowhere near enough mass to cause such a disruption.

Another idea popular at one time was that the Sun was once part of a two-star or even three-star system. According to this idea, the system was unstable and eventually one of the stars collided with the Sun, causing the system to scatter apart and producing a stream of gas that became the planets. Beginning in the 1930s, physicists began to find problems with these catastrophe theories. For instance, calculations showed that a hot stream of matter from the Sun would dissipate, rather than condense to form planets. Other observations showed that the chemical makeup of the planets was not consistent with material pulled from the outer surface of the Sun, but must have formed under cooler conditions. This finding led astronomers to reconsider the nebular hypothesis, wondering what had gone wrong with Laplace's calculations.

Eventually, scientists realized a solution to the problem: the Sun has a strong magnetic field. As the early planets orbited the Sun, the Sun exerted magnetic forces on them, as well as gravitational forces. This magnetic force would have acted to sweep the planets along in faster orbits. However, by Newton's third law, the planets would have exerted a force back on the Sun, slowing its rotation. When physicists ran the calculations taking this into account, they found pretty close agreement with observations.

This history of a scientific theory is not unusual; it often happens that theories are abandoned because of a seemingly insurmountable problem, only to be resuscitated when a way to solve the problem is found. The important point to realize is that the success or failure of a scientific theory depends on how well it matches and explains testable observations.

Today, astronomers accept the nebular hypothesis as the most likely scenario for the origin of the solar system. The total angular momentum of the spinning Sun and orbiting planets, as well as the orientation of the axis of rotation of the entire solar system, is conserved from that swirling nebular cloud almost 5 billion years ago.

Given this acceptance, do you think that the original rejection of the nebular hypothesis was a failure of the scientific method? Do such changes of opinion suggest that scientists are fickle in the theories that they support? The physicist Richard Feynman once said "We are trying to prove ourselves wrong as quickly as possible, because only in that way can we find progress." What do you think he meant by that? How does that relate to the theory of solar system formation?

Summary

Rotational motion is exhibited whenever an object spins about an **axis of rotation,** which is the line about which the object turns. The **period of rotation** is the time it takes for a body to complete one entire cycle, and the **frequency of rotation** is the number of completed rotations per unit time. Frequency is customarily measured in **hertz,** which is defined such that 1 Hz corresponds to one complete rotation each second.

The **angular speed** of a rotating object is the angle through which the object rotates divided by the time it takes to go through that angle. The angular frequency measures the number of times a rotation goes through one radian each second.

When a tangential force is applied away from the axis of rotation, that force produces a **torque.** The magnitude of the torque is given by $\tau = rF$, where F is the tangential force and r is the distance from the axis to the point of application of the force.

The **moment of inertia** of an object measures the distribution of its mass. The more mass there is and the farther away it lies from the axis of rotation, the greater the moment of inertia. The **angular momentum** of a rotating body is defined as the product of its angular speed and its moment of inertia. The rate of change of angular momentum in any system is equal to the net external torque, a result that follows from the rotational form of Newton's second law. In the absence of a net external torque, the angular momentum of a rotating body cannot change—a result called the law of **conservation of angular momentum.**

The **center of mass** or **center of gravity** of an object is the average position of its mass and can be thought of as the point at which the object can be balanced without rotation. Motion that involves both rotation and ordinary movement through space can be broken down into two simple processes: motion of the center of mass dictated by Newton's laws of motion and rotation around the center of mass.

Key Terms

angular momentum The moment of inertia of a body, times its angular velocity. (p. 147)

angular speed The angle through which an object has moved about the axis of rotation, divided by the time it takes it to go through that angle. (p. 141)

axis of rotation The line through the center of an object, around which everything else rotates. (p. 139)

center of mass (center of gravity) The point of support on which an object can be balanced. (p. 152)

conservation of angular momentum If the net external torque is zero, the angular momentum of any system must stay constant over time. (p. 149)

frequency of rotation The number of times an object completes a rotation in a given amount of time. (p. 140)

hertz The unit of measure of frequency, corresponding to one complete rotation every second. (p. 140)

moment of inertia The quantity that describes the distribution of mass around an axis of rotation. (p. 146)

period of rotation The time it takes for an object to make one complete rotation. (p. 140)

rotational motion The spinning motion that occurs when an object rotates about an axis located within it, usually an axis through its center of mass. (p. 139)

torque The force applied perpendicular to a line from the axis of rotation, multiplied by the distance from the axis of rotation. (p. 143)

Key Equations

$$\text{Frequency of rotation} = \frac{1}{\text{Period of rotation}}$$

$$\text{Angular speed} = \frac{\text{Angle traversed}}{\text{Time it takes to traverse the angle}}$$

$$\text{Torque} = \text{Force applied} \times \text{Distance from the axis of rotation}$$

$$\text{Net external torque} = \frac{\text{Change in angular momentum}}{\text{Change in time}}$$

$$\text{Angular momentum} = \text{Moment of inertia} \times \text{Angular velocity}$$

Review

1. What is rotational motion? Give an example.
2. What is an axis of rotation? Give an example.
3. What is the period of a rotating body? In what units is the period described?
4. Define the frequency of rotation.
5. What is the period of the Earth's rotation? Its frequency?
6. What is the frequency in hertz of a disk that makes one complete revolution every second? Every 2 seconds?
7. Describe the relationship between frequency and period. What is the mathematical equation?
8. What is the angular speed of a rotating body?
9. How does angular speed differ from linear speed for a point on a rotating body? (Think in terms of actual displacement in meters vs. degrees.)
10. What is a radian?
11. For one complete rotation of a rotating body, what is the angular displacement in degrees? In radians?
12. Define the angular frequency of an object. What are its units?
13. What is a torque?
14. How can you increase the torque on an object without increasing the force applied?

15. Compare the role of torque in rotational motion to the role of force in linear motion. When you apply a torque to a rotating object, how does this affect the rate of rotation?

16. What is angular momentum?

17. How does the mass of an object affect its angular momentum?

18. What is a moment of inertia? How does the distribution of mass affect this?

19. What is the conservation of angular momentum? Give an example.

20. Compare and contrast angular momentum to linear momentum.

21. If angular momentum is conserved, does the moment of inertia have to stay the same? How about the angular speed? The product of these two quantities?

22. What is meant by the direction of angular momentum? Is the direction of angular momentum the same as the direction of an object's rotation about an axis? Compare this to the direction of linear momentum.

23. What is the center of mass of an object?

24. Is the center of mass always located at the geometric center of an object? Explain.

25. What is the law of compound motion for rotating objects?

26. How is the conservation of angular momentum responsible for the difference between summer and winter?

Questions

1. What is the frequency of the minute hand of a clock in hertz?

2. When you push on an object such as a wrench, a steel pry bar, or even the outer edge of a door, you are producing a torque equal to the force applied times the lever arm. At what angle to the lever arm should a force be applied to produce maximum torque and why?

3. A children's seesaw is essentially a plank balanced on a fulcrum. Explain its operation in terms of torque and angular momentum. What happens when one person is much heavier? Does it matter where on the seesaw each person sits? Explain.

4. From what you learned in this chapter, why was the invention of rifling in a long gun or cannon barrel so important? (*Rifling* is a series of screw-like grooves etched into the interior of a rifle barrel that imparts a spin to the bullet.)

5. Why does a helicopter have a tail rotor? Some of the largest helicopters have two rotors on top; do these two rotors spin in the same direction?

6. How does conservation of angular momentum affect the stability of a bike?

7. What are some of the reasons that people initially have a difficult time staying upright when they are learning to ride a bike? How does turning the bicycle wheel act to stabilize a cyclist when a bike is stationary?

8. How might a pole-vaulter pass over a 14-foot bar if she were only able to get her center of mass to reach 13.5 feet?

9. How would you describe the path of a flock of birds or a school of fish using the center of mass of the combined masses of the individuals in these populations? Why might you do this?

10. The Earth, the Sun, and most other objects in space are not uniform spheres. Instead, they tend to be much denser toward the center (the core) than on the outside (the crust).

For a body of a given mass and size, which will have the greater moment of inertia—a uniform sphere or a nonuniform sphere with a dense core? Why? (*Hint:* Think about the distribution of mass relative to its distance from the axis of rotation.)

11. If you were in a spaceship with an inertial guidance system (see page 151) and if an outside observer saw that the spaceship was rotating clockwise, what motion would you see in the gyroscope inside the ship?

12. Which planet has a longer period of rotation around the Sun, Mercury or the Earth? (*Hint:* Mercury is the closest planet to the Sun.) Which planet has a higher frequency?

13. While you're riding a bicycle (see figure), which has a higher rotation frequency, the front sprocket or the rear sprocket?

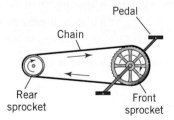

14. Five forces act on the outside of a wheel, as shown in the figure. Which of the five forces exert a torque about the center of the wheel?

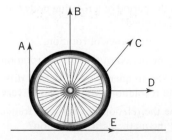

15. Consider two forces acting on the front tire of a bicycle wheel as the rider is braking: the road pushing up on the tire, and friction pushing back on the tire. Which of these forces exerts a torque about the center of the wheel? Explain.

16. Consider the simple dumbbell shown in the figure. It consists of two 10-kilogram spheres separated by a 2-meter light rod. Consider rotating it about two axes: one axis through the middle of the rod and one axis passing through one of the spheres. In which case will the dumbbell have a higher moment of inertia?

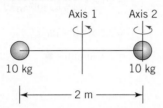

17. Consider the asymmetrical dumbbell shown in the figure. It consists of two spheres separated by a 2-meter light rod. Rotation about which axis (A, B, or C) involves the lowest moment of inertia?

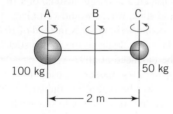

18. Consider the irregularly shaped object shown in the figure. Through which axis (A, B, or C) will the moment of inertia be greatest? Through which axis will it be least? Explain your reasoning.

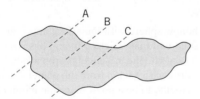

19. If everyone in the world moved to the equator, what would happen to the moment of inertia of the Earth? What would happen to the angular momentum of the Earth? What would happen to the angular speed of the Earth?

20. You have a ball tied to a string, and you spin it around in a horizontal circle. Your arm, sticking straight up in the air, defines the axis of rotation. Consider the force that the string exerts on the ball. Does that force exert a torque about the axis of rotation? Why or why not? Suppose you let some string slip through your fingers, allowing the ball to spin at a greater distance from the axis. What happens to the angular speed of the rotation? Explain.

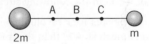

21. In the following figure, which location is most likely to be the center of mass of the dumbbell?

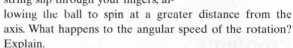

22. A 10-kilogram metal sphere is welded to a 10-kilogram metal disk as shown in the figure. Which location is most likely to be the center of mass of the object?

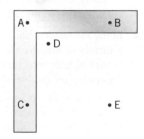

23. Which location is most likely to be the center of mass of the L-shaped object shown in the figure?

Problem-Solving Examples

EXAMPLE 7-3

The Sun's Collapse

In the Looking Deeper section, we discuss how stars spin faster as they collapse inward with time. We gave the starting and ending diameters of the Sun as 7×10^8 meters collapsing down to 6.4×10^6 meters. How fast will it be rotating when it becomes a white dwarf?

REASONING AND SOLUTION: The steps to follow for solving this problem are described in the Looking Deeper

section. From Figure 7-11, the moment of inertia of a uniform sphere is $(\frac{2}{5} M \times R^2)$. The Sun's angular frequency is $2\pi f$ or $2\pi/T$ where T is the period of rotation. Therefore, the angular momentum of the Sun today is:

$$\text{Angular momentum} = \frac{\text{Moment of inertia} \times 2\pi}{\text{Period}}$$

$$\text{Initial angular momentum} = \frac{(\frac{2}{5} M \times R_i^2)2\pi}{26 \text{ days}}$$

where R_i is the initial radius of the Sun before the collapse. The angular momentum of the Sun after the collapse will be:

$$\text{Final angular momentum} = \frac{\left(\frac{2}{5} M \times R_f^2\right)2\pi}{T}$$

where T is the (unknown) time it takes to do one rotation after collapse and R_f is the radius after collapse. However, because the angular momentum of the Sun is constant, by the law of conservation of angular momentum, we can set these two expressions equal:

Initial angular momentum = Final angular momentum

$$\frac{\left(\frac{2}{5} M \times R_i^2\right)2\pi}{26 \text{ days}} = \frac{\left(\frac{2}{5} M \times R_f^2\right)2\pi}{T}$$

Canceling the $\frac{2}{5}$, the mass of the sun, and 2π, we find

$$\frac{R_i^2}{26 \text{ days}} = \frac{R_f^2}{T}$$

so

$$T = 26 \text{ days} \times \left(\frac{R_f^2}{R_i^2}\right)$$

$$= 26 \text{ days} \times \frac{(6.4 \times 10^6 \text{ m})^2}{(7 \times 10^8 \text{ m})^2}$$

$$= 26 \times \left(\frac{4.1 \times 10^{13} \text{ m}^2}{4.9 \times 10^{17} \text{ m}^2}\right) \text{ days}$$

$$= 2.2 \times 10^{-3} \text{ days}$$

$$= 3.1 \text{ minutes} \bullet$$

Problems

1. **A.** What is the rotational speed in revolutions per second (hertz) of a CD in the following situations?

   a. The disc makes four revolutions in 48 seconds (this speed is much slower than that of a normal CD).
   b. The disc rotates six times in 240 seconds.
   c. The disc makes 1000 revolutions in 2 minutes.

   B. What are the velocities in radians per second for parts a, b, and c?

   C. What are the velocities in degrees per second for parts a, b, and c?

   D. How large a displacement in degrees occurs in each of parts a, b, and c if these discs spin for 30 seconds at the same speed? In radians?

2. In the mechanical and plumbing trades, many tools come in a variety of lengths. One of the reasons they are available in varying lengths is that different torques can be generated depending on the lengths of these tools.

   a. If you hold a 0.2-m wrench at its end and exert a force of 30 N, how much torque will you generate?
   b. If you use a 0.5-m wrench and exert a force of 45 N, how much torque can you generate?
   c. If a plumber needs to generate a torque of 160 N-m to unscrew a rusted pipe and can only generate a maximum force that day of 20 N because she has been out too late the night before, what length wrench is the smallest that she can use?
   d. What effect does the length of the handle of a wrench have on the torque that can be generated by it?
   e. To achieve the maximum torque from a given applied force to a lever arm such as a wrench, at what angle should the force be applied to the lever or wrench?

3. A meter stick is balanced perfectly on a fulcrum at the 0.5-m mark.

   a. If a 5-N weight is added to the very tip of the zero side of the meter stick, where do you need to place a 10-N weight to rebalance the stick?
   b. If you use an 8-N weight instead of a 10-N weight to balance the stick, where would you place it?
   c. If the second weight is 3 N instead of 10 N, where should you place it to balance the meter stick? Can the meter stick be balanced at all?
   d. If 1-kg and 2-kg masses are used in part a, where should these masses be placed for the stick to balance?
   e. What are the torques applied by each of the weights in parts a, b, and c?
   f. Repeat parts a–d with the fulcrum placed at 0.7 m.
   g. If the torques do not balance, what happens? Is angular momentum conserved?

4. What is the moment of inertia of each of the following objects?

   a. A hollow sphere with mass 5 kg and radius 0.5 m
   b. A solid ball that weighs 3 lb and has a radius of 1 foot
   c. A 200-kg satellite in a circular orbit around a small planet at a distance 5000 km from the planet's center (Consider the satellite to be a point mass; the moment of inertia of a single particle is MR^2.)
   d. A large truck tire of 0.75-m radius and mass 20 kg (assume all the mass is concentrated on the outer edge)

5. What is the angular momentum of the rotating objects in Problem 4 under the following circumstances?

   a. When the spheres in parts a and b rotate at 2 revolutions per second?
   b. When the spheres in parts a and b rotate at 1 radian per second?
   c. When the satellite in part c makes 1 revolution every 90 minutes?
   d. When the tire in part d spins at a rate of 1 revolution per second?

6. In what direction does the angular momentum vector point for the following situations (remember the right-hand rule)?

   a. A Ferris wheel spinning clockwise as you look at it

   b. A CD that spins counterclockwise as you look at it

   c. A bicycle wheel as the bike moves straight in a forward direction

   d. The left rear tire of a car moving straight backward in reverse

   e. The right rear tire of a car moving straight backward in reverse

7. If the direction of the angular momentum vector is pointed straight at you, in what direction does an object rotate?

8. Several children are playing on a merry-go-round in a park. Initially four of them, each weighing 20 kg, sit on the edge, 3 m from the center.

   a. If you neglect the weight of the merry-go-round, what is the initial angular momentum if it spins at a rate of 6 revolutions per minute?

   b. Not comfortable sitting on the edge of a spinning disk, the four children decide to walk to the center and sit halfway between the center and the edge, at 1.5 m. Will the angular velocity of the merry-go-round change? If so, what is the new angular velocity?

   c. Was angular momentum conserved when the children moved?

9. Astronomers know of collapsed stars called "pulsars" that rotate hundreds of times per second. The Sun (radius 7×10^8 m) now rotates once every 26 days. What would its radius have to be for it to rotate 100 times each second? Compare that radius to the size of the town you live in.

Investigations

1. Make a list of objects in your everyday life that rotate as part of their function. Think about the distribution of mass in these items and make notes on this. Based on your estimate of their mass distribution, rank the objects by how much torque is needed to cause them to experience an angular acceleration. Consider the distance the force is applied from the axis of rotation.

2. Research the development of gyroscopes. What are the principles behind them? What uses do they have?

3. Go to a pool hall or to a friend's house who has a pool table. Try putting different spins on the balls. Observe what happens to a spinning ball as it collides with another ball. Analyze this in terms of angular momentum. Is it conserved? Why or why not?

4. If you have the opportunity to work with a mechanic, plumber, or carpenter, ask to be shown how to use different hand tools and experiment with the effect of using tools of varying lengths. For example, try a long versus a short wrench, screwdriver, or hammer. Does your experience con-

cerning how much force is needed to do your chosen task comform with what you have learned in this chapter?

5. Using the library, the Internet, or both, research how the helicopter was developed. What were some of the problems that had to be solved to ensure stable flight, and how were they solved?

6. Investigate how a pitcher can throw a curve ball. How does the spin of the ball differ for the different kinds of curve balls?

7. Use the web to investigate the discovery of dark matter by Vera Rubin of the Carnegie Institution of Washington. How did her observations of the rotation of galaxies reveal that much of the mass of the universe is invisible?

8. Occasionally two galaxies collide so that hundreds of billions of stars coalesce into one giant galaxy. Use the web to investigate this process. What happens to the angular momentum of the new combined galaxy compared to the original two galaxies?

WWW Resources

See the *Physics Matters* home page at **www.wiley.com/college/trefil** for valuable web links.

1. **www.physics.brocku.ca/faculty/sternin/120/applets/CircularMotion/** An applet showing position, velocity and acceleration for uniform circular motion from the Department of Physics at Brock University.

2. **www.physics.uoguelph.ca/tutorials/torque/Q.torque.html** The Rotational Motion Tutorial at the Department of Physics, University of Guelph.

3. **web.hep.uiuc.edu/home/g-gollin/dance/dance_physics.html** Dedicated to the physics of dance for aficionados of both.

4. **www.windows.ucar.edu/tour/link=/cool_stuff/tour_evolution_ss_1.html** Discusses solar system evolution, including current theories of system formation.

8

Kinetic and Potential Energy

KEY IDEA

Kinetic and gravitational potential energy are interchangeable.
The total amount of energy in an isolated system is conserved.

PHYSICS AROUND US . . . The Roller Coaster

While visiting your local amusement park, you and your friends decide to take a ride on the roller coaster. You climb into a car, strap yourself in, and wait while an electric motor lifts the car slowly, overcoming the force of gravity that tries to pull you back down. Then you experience that breathless moment at the top, when you are poised to plunge downward but haven't quite started the scary descent.

With a whoosh of air (and perhaps a few screams from some riders), you plummet back toward Earth, careening faster and faster as you descend. At the last moment, the track changes direction and you start up again, repeating the sequence until the ride ends.

You may think of the ride as just good fun, but you have actually transformed yourself into a living demonstration of one of the most important features of nature—the transformation of energy from one form to another. This phenomenon underlies interactions throughout all of science, from the tiniest particles to the largest galactic clusters, from biology to chemistry to geology, from the sunlight that wakes you up in the morning to the electric lamp you turn off before going to sleep at night. The interchangeability of energy from one form to another is a powerful idea whose many implications we explore in the next several chapters.

WORK, ENERGY, AND POWER: WORDS WITH PRECISE MEANINGS

Whenever you ride in a car, climb a flight of stairs, or just take a breath, you use energy. At this moment, trillions of cells in your body are hard at work turning yesterday's food into the chemical energy that keeps you alive today. Energy in the atmosphere is generating sweeping winds and powerful storms, while the ocean's energy drives mighty currents and incessant tides. Meanwhile, deep within the Earth, energy in the form of heat is moving the very continent on which you are standing.

A turbulent sea is one of the greatest sources of energy on Earth.

Energy is all around you—in the ever-shifting atmosphere and restless seas, in simple bacteria and mighty redwood trees, in brilliant sunlight and shimmering moolight. Energy affects everything in the physical world, and the laws that govern its behavior are among the most important and overarching concepts in science.

In every situation where energy is expended you will find one thing in common. If you look at the event closely enough you will find that, in accord with Newton's laws of motion (Chapter 4), a force is exerted on an object to make it move. When your car burns gasoline, the fuel's energy ultimately turns the wheels of your car, which then exert a force on the road; the road exerts an equal and opposite force on the car, pushing it forward. When you climb the stairs, your muscles exert a force that pushes down on the stairs, enabling you to move upward against gravity. Even in your body's cells, forces are exerted on molecules in chemical reactions. Energy, then, is intimately connected with the application of a force.

In everyday conversation, we may speak of a small child's seemingly inexhaustible energy, of a song that sounds energetic, or of an athlete being energized. In physics, the term *energy* has a precise definition that is somewhat different from the ordinary meaning. However, to see what physicists mean when they talk about energy, we must first introduce the concept of *work*.

Work

In the lexicon of physics, we say that **work** is done whenever a force is exerted over a distance. When you picked up this book, for example, your muscles applied a force equal to the weight of the book over a distance of a foot or so. You did work (Figure 8-1).

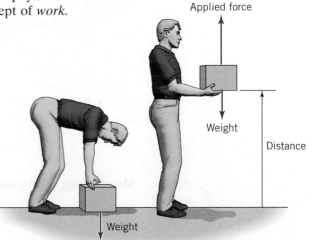

FIGURE 8-1. Work is done whenever a force is exerted over a distance.

This definition of *work* differs considerably from everyday use. From a physicist's point of view, if you accidentally drive into a stone wall and smash your fender, the wall does work because a force deformed the car's metal a measurable distance. On the other hand, a physicist would say that you haven't done any work if you spent an hour in a futile effort to move a large boulder, no matter how tired you got. Even though you have exerted a considerable force, the distance over which you exerted it is zero.

There's another way in which a force can do no work. Imagine carrying a briefcase as you walk along a flat street at a steady pace. In order to hold the briefcase, your hand exerts an upward force on it. But the briefcase moves in a direction perpendicular to that force, so that force does no work! You should keep in mind that when we talk about a force doing work, we really mean the part of the force that points in the direction of the motion. If there is no motion, or if the force is perpendicular to the motion, the work is zero.

It's also possible for work to be negative. For example, the gravitational force exerted on you as you climb a flight of stairs does negative work because you're moving up and the force is pulling down. If the force and the motion are in opposite directions, then the work done by that force is negative.

Physicists provide an exact mathematical definition to their notion of work.

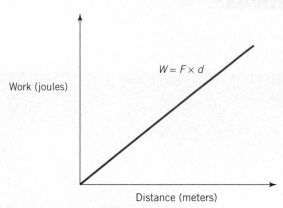

Work (joules)

Distance (meters)

FIGURE 8-2. The direct relationship between work and distance illustrated in graphical form.

1. In words:

Work is done whenever a force is exerted over a distance. The amount of work done is proportional to both the force and the distance.

2. In an equation with words:

Work (in joules) = Force (in newtons) $\times$ Distance (in meters)

where a joule is the unit of work, as defined next.

3. In an equation with symbols:

$$W = F \times d$$

The *direct relationship* between work and distance is illustrated in Figure 8-2. In practical terms, even a small force can do a lot of work if it is exerted over a long distance.

Using this equation, we can see that the units of work are equal to a force times a distance. In the metric system of units, in which force is measured in newtons and distance in meters, work is measured in newton-meters. This unit is given the special name **joule,** after the English scientist James Prescott Joule (1818–1889), one of the first people to understand the properties of energy. One *joule* (abbreviated 1 J) is defined as the amount of work done when you exert a force of 1 newton through a distance of 1 meter:

1 joule of Work = 1 newton of Force $\times$ 1 meter of Distance

Working Against Gravity

EXAMPLE
8-1

How much work do you do when you lift a 12-kilogram suitcase 0.75 meters off the ground (Figure 8-3)?

REASONING AND SOLUTION: We must first calculate the force needed to lift a 12-kilogram mass before we can determine the work done. This force is equal to the weight of the suitcase. From Chapter 5, we know that to lift a 10-kilogram mass against the acceleration of gravity (9.8 meters per second per second) requires a force given by:

$$\text{Force} = \text{Mass} \times g$$
$$= 12 \text{ kg} \times 9.8 \text{ m/s}^2$$
$$= 117.6 \text{ newtons}$$

Then, from the equation for work,

$$\text{Work} = \text{Force} \times \text{Distance}$$
$$= 117.6 \text{ newtons} \times 0.75 \text{ meter}$$
$$= 88.2 \text{ joules}$$

In North America, work is often measured in the English system of units (see Appendix A), where force is recorded in pounds and distance in feet. Work is thus measured in a unit called the *foot-pound* (usually abbreviated ft-lb), which corresponds to the work done in lifting a weight of 1 pound 1 foot upward against the force of gravity. ●

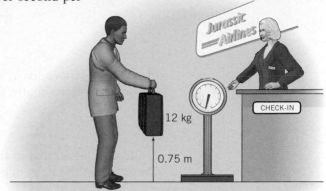

FIGURE 8-3. Lifting a suitcase off the ground requires you to do work against gravity.

Lifting Weights

It's not unusual for a professional athlete to be able to lift as much as 400 pounds (1780 newtons). Suppose that a football linebacker, lying on his back, pushes a 400-pound barbell from his chest to a position in which his arms are fully extended upward (Figure 8-4). (This action is called a "bench press" and gives rise to the sporting slang expression that an athlete can "press" a certain weight.) Estimate the amount of work he does in both SI and the English system.

EXAMPLE 8-2

REASONING: To calculate the work done, we need to multiply force times distance. The force in this problem is 400 pounds, but the distance isn't given. That means we have to estimate it. The athlete starts with the weight on his chest and keeps pushing until his arms are extended. This means that the distance over which the force is applied must be about the length of his arms. The length of a large man's arms is about 3 feet (in fact, the yard in the English system of units was originally defined as the distance from a man's nose to the end of his hand). So let's take the distance in the problem as 3 feet.

400 pounds

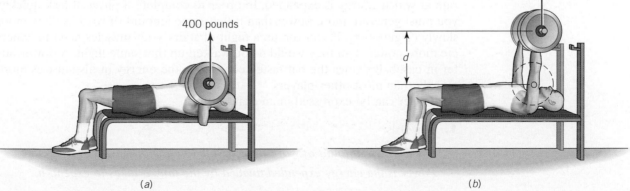

400 pounds

(a)

400 pounds

d

(b)

FIGURE 8-4. Athletes strengthen their muscles by working against gravity, for example by lifting weights. No pain, no gain!

SOLUTION: Work is force times distance, so

$$W = F \text{ (in pounds)} \times d \text{ (in feet)}$$
$$= 400 \text{ pounds} \times 3 \text{ feet}$$
$$= 1200 \text{ ft-lb}$$

Looking in Appendix A, we see that there are 1.356 joules for each foot-pound of energy. Consequently, this amount of work in SI is

$$W = 1200 \text{ ft-lb} \times 1.356 \text{ joules/ft-lb}$$
$$= 1627 \text{ joules}$$

Note once again that in this equation we can cancel units (in this case ft-lb) to be sure that the units of our answer are correct. ●

Compare the answers for Examples 8-1 and 8-2. Lifting a 12-kg suitcase 1.5 meter requires about 176 J of work; lifting a 400-lb barbell 3 feet (which is almost 1 meter) requires over 1600 J of work. Does this seem reasonable to you? (Note that 1 kilogram has a weight of about 2.2 pounds when $g = 9.8$ m/s^2.)

Energy

Energy is defined as the ability to do work. If a system is capable of exerting a force over a distance, then it possesses energy. The amount of a system's energy, which can be recorded in joules or foot-pounds (the same units used for work), is a measure of how much work the system might do. When you run out of energy, you simply can't do any work.

Energy is one of the most useful concepts in all of science, from both theoretical and practical viewpoints. Determining the energy of a system is often the key step in analyzing its future behavior, and we're all familiar with the importance of energy in our modern industrial society. We examine details of energy in the rest of this chapter and again in later chapters. Examples of energy in common situations appear in Physics and Daily Life on p. 168.

Power

Physicists define **power** as the rate at which work is done, or, equivalently, the rate at which energy is expended. In order to complete a physical task quickly, you must generate more power than if that same amount of work is done more slowly (Figure 8-5). If you run up a flight of stairs, your muscles need to generate more power than they would if you walked up that same flight. A power hitter in baseball swings the bat faster, converting the energy in his muscles more quickly than most other players.

Power can be expressed in an exact form.

1. In words:

Power is the amount of work done divided by the time it takes to do it.

Power is the energy expended divided by the time it takes to expend it.

2. In equations with words:

$$\text{Power (in watts)} = \frac{\text{Work (in joules)}}{\text{Time (in seconds)}}$$

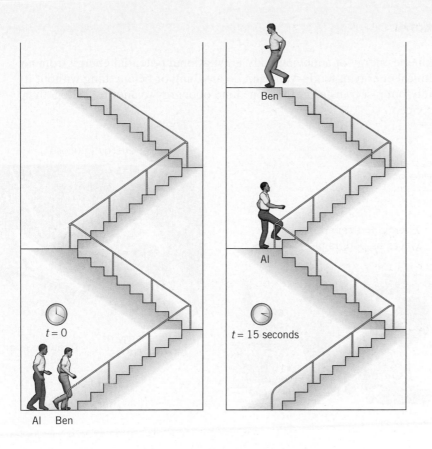

FIGURE 8-5. Power is the rate at which work is done, or, equivalently, the rate at which energy is expended. If Ben runs up two flights of stairs in the same time it takes Al to walk up one flight, then Ben expends twice as much power as Al.

or
$$\text{Power (in watts)} = \frac{\text{Energy (in joules)}}{\text{Time (in seconds)}}$$

where the *watt* is the unit of power, as defined next.

3. In equations with symbols:

$$P = \frac{W}{t} \quad \text{or} \quad P = \frac{E}{t}$$

In the metric system power is measured in **watts,** a unit named after James Watt (1736–1819), a Scottish inventor who helped to develop the modern steam engine that powered the Industrial Revolution. The watt, a unit of measurement that you probably encounter every day, is defined as the expenditure of 1 joule of energy in 1 second:

$$1 \text{ watt of Power} = \frac{1 \text{ joule of Energy}}{1 \text{ second of Time}}$$

When you change a lightbulb, for example, you look at the rating of the new bulb to see whether it's 60, 75, or 100 watts. This number provides a measure of the rate of energy that the lightbulb consumes when it is operating. Almost any electric hand tool or appliance in your home is also labeled with a power rating in watts. The unit of 1000 watts (corresponding to an expenditure of 1000 joules per second) is called a **kilowatt,** a commonly used measurement of electrical power. The English system, on the other hand, uses *horsepower,* which is defined as 550 foot-pounds per second (see Physics in the Making, page 169).

Physics and Daily Life—Energy

Energy is everywhere, whether kinetic energy of a moving body, gravitational potential energy from an object raised off the ground, chemical energy in food—you can't do anything or be anything without it. And it never disappears completely, but just transforms from one kind of energy to another. Altogether, a really useful and remarkable concept.

Kinetic energy of moving ball

Potential energy of body above ground

Energy and sound waves as people talk

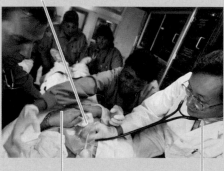

Chemical energy (glucose)

Kinetic energy of moving arm

Acoustic energy (sound)

The equation defining power as energy divided by time may be rewritten as:

$$\text{Energy (in joules)} = \text{Power (in watts)} \times \text{Time (in seconds)}$$

This equation says that if you know how much power you are using and how long you are using it, you can calculate the total amount of energy expended. The electric company calculates your electric bill in this way. The equation tells us that if you use 100 watts of power for 1 hour (by having a lightbulb turned on, for example), you have expended 100 watt-hours of energy, or one-tenth of a kilowatt-hour. This measurement of energy used appears on your electric bill.

The terms and units related to work, energy, and power are summarized in Table 8-1.

TABLE 8-1	Important Terms Used in Energy	
Quantity	**Definition**	**Units**
Force	Mass × Acceleration	Newtons
Work	Force × Distance	Joules
Energy	Ability to do work	Joules
Energy	Power × Time	Joules
Power	$\dfrac{\text{Work}}{\text{Time}} = \dfrac{\text{Energy}}{\text{Time}}$	Watts

Physics in the Making

James Watt and the Horsepower

James Watt devised the horsepower, a unit of power with a colorful history, so that he could sell his steam engines. Watt knew that the main use of his engines was in mines, where owners traditionally used horses to drive the pumps that removed water. The easiest way to promote his new engines was to tell the mining engineers how many horses each engine would replace. Consequently, he did a series of experiments to determine how much energy a horse could generate over a given amount of time. He found that, on average, a healthy horse can do 550 ft-lb of work every second over an average working day. Watt called this unit *horsepower* and rated his engines accordingly. We still use this unit (the engines of most cars and trucks are rated in horsepower), although these days we seldom build engines to replace horses. ●

TYPES OF ENERGY

Energy, the ability to do work, appears in many different kinds of physical systems, which give rise to many different kinds of energy. In this chapter we talk about only two of these categories—**kinetic energy,** which is energy associated with moving objects, and **gravitational potential energy,** which is a form of energy waiting to be released. We wait until Chapter 12, after we have discussed phenomena associated with heat, to talk about the many other categories of energy.

Kinetic Energy

Think about a cannonball flying through the air. When the iron ball hits a wooden target, the ball exerts a large force on the fibers in the wood, splintering them and pushing them apart, thereby creating a hole. The cannonball in flight clearly has the *ability* to do work because it is in motion. (If it were not in motion, it would remain on the ground and do no work at all.) This energy of motion is called *kinetic energy*. You can find countless examples of kinetic energy in the world around you. A swimming fish, a flying bird, a speeding car, a soaring Frisbee, a falling leaf, and a running child all have kinetic energy.

Our intuition tells us that two factors govern the amount of an object's kinetic energy. First, heavier objects have more energy than lighter ones: a bowling ball traveling 10 meters per second (a very fast sprint) carries a lot more kinetic energy than a ping-pong ball traveling at the same speed. In fact, kinetic energy is proportional to mass: double the mass and you double the kinetic energy (Figure 8-6a on page 170).

Second, the faster something is moving, the greater the force it is capable of exerting. A high-speed collision on the highway causes much more damage than a fender-bender in a parking lot. It turns out that an object's kinetic energy increases as the square of its velocity. A car moving 40 kilometers per hour has four times as much kinetic energy as one moving 20 km/h, while at 60 km/h a car carries nine times as much kinetic energy as at 20 km/h (Figure 8-6b on page 170). Thus, a modest increase in speed can cause a large increase in kinetic energy.

When a baseball breaks a window it demonstrates the existence of kinetic energy.

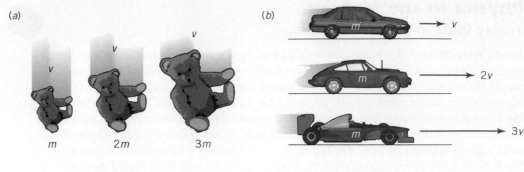

(a) Double the mass,
  double the kinetic energy

Triple the mass,
  triple the kinetic energy

(b) Double the speed,
  increase kinetic energy 4 times

Triple the speed,
  increase kinetic energy 9 times

FIGURE 8-6. (a) Double the mass and you double the kinetic energy. (b) Double the velocity and you increase kinetic energy by a factor of four.

These ideas, presented as equations, lead to the following definition.

1. In words:

Kinetic energy equals the mass of the moving object times the square of that object's velocity, multiplied by the constant $\frac{1}{2}$.

2. In an equation with words:

Kinetic energy (in joules) = $\frac{1}{2}$ × Mass (in kg) × [Velocity (in m/s)]^2

3. In an equation with symbols:

$$KE = \frac{1}{2}\,mv^2$$

We won't discuss where the constant $\frac{1}{2}$ comes from, but it must be included for the formula to be correct. Some examples of kinetic energy are shown in Looking at Energy, on p. 171.

Kinetic Energy Versus Momentum Although kinetic energy and momentum are both properties of moving objects and although both depend on the mass and the velocity of that object, they are quite different quantities. To make this point clear, here are two important differences between them:

1. Momentum is a vector, while kinetic energy is a positive scalar. In a system with two moving objects—two moving billiard balls, for example—the momenta of the two objects can cancel each other either partially or completely if they travel in opposite directions. The energies of the two objects, however, always add to each other to give the total energy of the system.

2. Momentum grows linearly with velocity, but kinetic energy grows as the square of velocity. Double the velocity of one of those billiard balls and the momentum doubles, while the kinetic energy increases by a factor of four. This difference is underscored by Example 8-5 at the end of the chapter in which the bowling ball has a greater momentum, while the baseball carries more kinetic energy.

Looking at Energy

Raindrops are wet, but they don't hurt you when they fall on your head. That's because one raindrop has very little energy; it takes a lot of raindrops to hurt someone. Large hurricanes have such energy that they can bring down buildings and wash people away; they are among the most energetic natural phenomena on Earth. Most of our experience falls between these extremes, but physicists study interactions with far less and far more energy than is familiar to most people. For instance, the impact of a large meteor (10-km in diameter) colliding with Earth, as probably happened 65 million years ago, would have released enough energy to destroy most life on the planet at that time.

10^{-4} J

Raindrop, 0.0001 joule

10^{-2} J

Pressing a computer key, 0.01 joule

10^{27} J

Impact of 10-km meteor striking Earth, 10^{27} joules

10^{6} J

Riding in a car at 55 mph, 600,000 joules

10^{19} J

Typical hurricane, 10^{19} joules

A balancing rock doesn't do any work as long as it stays in place. If it starts to fall over, however, you don't want to get in its way.

Potential Energy

Almost every mountain range in the country has a balancing rock, a boulder precariously perched on top of a hill so that it looks as if a little push would send it tumbling down the slope. If the balancing rock were to fall, it would acquire kinetic energy, and it would do work on anything it smashed into. This means that even though the balancing rock does no work while it is motionless, it still has the potential to do work. The boulder possesses energy just by virtue of having the potential of falling.

Energy that could result in the exertion of a force over a distance but is not doing so now is called "potential energy." In the case of the balancing rock, it is called *gravitational potential energy,* because it is the force of gravity that would cause the rock to move and exert its own force (its weight) on impact.

An object that has been lifted above the surface of the Earth possesses an amount of gravitational potential energy exactly equal to the total amount of work you would have to do to lift it from the ground to its present position.

1. In words:

The gravitational potential energy of any object equals its weight (the force of gravity exerted on the object) times its height above the ground.

2. In an equation with words:

$$\text{Gravitational potential energy (in joules)} = \text{Mass (in kg)} \times g \text{ (in m/s}^2) \times \text{Height (in m)}$$

where g is the acceleration due to gravity at the Earth's surface (see Chapter 3).

3. In an equation with symbols:

$$PE = mgh$$

In Example 8-1 we saw that if you lift a 12-kilogram suitcase 0.75 meters into the air, you do 88.2 joules of work and the suitcase acquires 88.2 joules of potential energy. This is the amount of work that would be done if the suitcase were allowed to fall, and it is the amount of gravitational potential energy stored in the elevated suitcase.

● THE CONSERVATION OF ENERGY

Interchangeable Forms of Energy

One of the most useful attributes of energy is that all of its many forms are interchangeable. Let's examine this property for the two forms of energy we have encountered so far—kinetic and gravitational potential energies.

A good way to understand this idea is to think about the roller coaster ride we discussed in the Physics Around Us section at the start of this chapter. While the car is being lifted to the top of the track for the start of the run, a force equal to the acceleration due to gravity (g) times the mass (m) of the car (plus the

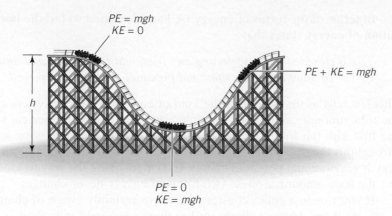

$PE = mgh$
$KE = 0$

$PE + KE = mgh$

$PE = 0$
$KE = mgh$

FIGURE 8-7. A roller coaster car has gravitational potential energy at the top of the hill and kinetic energy at the bottom of the hill. It has a mix of both kinds of energy at points in between.

passengers) is being applied over a distance h, where h is the height of the track. Thus, the work W done on the car is

$$W = mgh$$

This value is also the potential energy that the car has when it is resting at the top, ready to start down.

As the descent starts, two things happen: (1) the car starts to move and therefore acquires kinetic energy and (2) the car starts to lose height, so its potential energy begins to drop. By the time the car has reached the bottom of the first hill, it is at ground level, so its potential energy has been reduced to zero. At the same time, it is moving as fast as it is going to move, so its kinetic energy is at its maximum. On the way down, then, the potential energy has changed into kinetic energy—in other words, one form of energy has turned into another (Figure 8-7).

As the car starts up the second hill, this process goes on in reverse. As the car climbs, its potential energy increases while its kinetic energy drops. At the top of the second hill, when the car is momentarily stationary before its second swoop, all the kinetic energy it had at the bottom of the first hill has been converted back into potential energy. If we ignore losses due to friction (a subject we take up in Chapter 12), this back and forth transfer of energy between these two forms will go on forever.

We can summarize this result by saying that

> *Kinetic and gravitational potential energy are interchangeable.*

This statement is actually a special case of a more general rule, a rule that states that *all* forms of energy are interchangeable.

The Principle of Energy Conservation

Physicists always look for constants in their efforts to describe a changing universe. Is the total number of atoms or electrons in the universe constant? Is the total amount of electric charge fixed? We have seen in Chapter 6 that any statement that says that a quantity in nature does not change—that it is conserved—is called a *conservation law*. We have already seen how the conservation laws relating to linear and angular momentum help us understand the behavior of physical systems. If anything, the conservation law for energy is even more important.

In terms of the forms of energy we have discussed so far, the **law of conservation of energy** states that

In a closed system, neglecting any frictional forces, the total amount of kinetic and gravitational potential energy is conserved.

This law tells us that although the kind of energy in a given system can change, the total amount cannot. For example, when that roller coaster car starts down the first hill, the total amount of energy it has at the beginning is still there throughout the descent and the climbing of the next hill. Initially, all of this energy is gravitational potential energy. It changes along the way to kinetic energy, but the total amount of these two forms of energy never changes.

If you ride in a roller coaster car, you're certainly aware of changing speed as the car drops down or climbs up, but there's no direct indication of how much potential energy or kinetic energy you have along the way. There is no joule meter you can use to measure the potential energy of a balancing rock or the kinetic energy of a falling one. The fact that you can't see energy directly doesn't change the fact that it exists and can affect you.

The concept of energy allows us to look at the world so that we can analyze it. In fact, it is probably the single most successful tool ever devised for explaining how nature works and predicting its future behavior. Physicists often try to look at situations in several ways, each way giving a different understanding of what is actually happening. You can look at a roller coaster car in terms of force and acceleration, in terms of impulse and changes in momentum, or in terms of potential and kinetic energy. However, if you want to calculate the car's speed at the bottom of the hill, conservation of energy is by far the most useful way to analyze the problem. This situation often proves to be the case, which is why we spend several chapters exploring how we can use the idea of energy to understand various natural processes. The most important thing about conservation of energy is that it has been used by physicists for over 200 years and has never failed to work. Clearly, then, conservation of energy embodies a fundamental truth about the physical world, as valid as any concept known to humanity.

Energy of a Falling Body

The fact that the sum of kinetic and potential energies must be conserved gives us an easy way to analyze the fall of an object from a height. At the beginning of the fall, just before the object is released, its energy is all potential. If it is at a height h above the ground, then we saw earlier that its energy is *mgh* joules. As the object falls, this potential energy is gradually converted to kinetic energy, until, just before it hits the ground, the conversion is complete. The energy is now all kinetic and is given by the expression $\frac{1}{2}mv^2$. From this fact, we can deduce some important facts about falling bodies.

1. In words:

The kinetic energy at the end of a fall is equal to the potential energy at the beginning.

2. In an equation with words:

Initial potential energy = Final kinetic energy

3. In an equation with symbols:

$$mgh = \frac{1}{2} mv^2$$

If we cancel the mass from both sides of the equation, we find

$$gh = \frac{1}{2} v^2 \qquad \text{or} \qquad v = \sqrt{2gh}$$

In other words, the speed of the object at the end of the fall is independent of the object's mass.

Develop Your Intuition: How To View an Object Falling

Many thinkers, including brilliant scientists such as Aristotle, thought that heavier objects should fall faster than light ones and should therefore have a higher velocity at the end of the fall. How would you explain our preceding result to someone who argues that the heavier object has more oomph at the end, and therefore has to be moving faster?

It is true that the heavier object has more kinetic energy when it reaches the end of the fall and therefore makes a bigger impact. It is also true, however, that the heavier object had more potential energy at the beginning of the fall because a greater force had to be applied to lift it a distance *h*. The lighter object had less potential energy at the start and has less kinetic energy at the end. The effects of mass simply cancel out, and both fall at the same speed.

THE WORK-ENERGY THEOREM

Let's think again about the example of the roller coaster in the Physics Around Us section. We have answered a lot of questions about how it works but one question remains: how did it acquire that potential energy in the first place? We know that when the roller coaster car was lifted up, a force was exerted over a distance to overcome the downward pull of gravity. In the language of physics, work was done. In Chapter 12 we see that work is actually the result of the expenditure of different sorts of energy. In the case of the roller coaster it was electrical energy driving a motor, while in the case of a barbell being lifted it was energy in the muscles of the weight lifter. In this context, the lifting of the car or a weight is just one more example of energy being changed from one form to another. If we confine our attention to kinetic and gravitational potential energy, however, as we have done so far, then we can make one more statement about work and energy:

> *The total potential and kinetic energy of an object in a given state*
> *is equal to the work that was done to bring the object to that state.*

This statement is known as the **work-energy theorem.**

In practical situations, it often turns out that what you need to look at are the changes in kinetic and potential energy. The work-energy theorem applied to changes in energy gives this statement:

The work done on an object is equal to the sum of the changes in kinetic and potential energy.

LOOKING DEEPER

Collisions

Collisions between objects are an extremely important and constantly recurring theme in physics. Much of what we know about the nucleus of the atom (Chapter 26) and about elementary particles (Chapter 27) comes from studies of particle collisions at the subatomic level. The concepts of conservation of momentum and conservation of energy provide a useful way of analyzing any collision process.

Consider two objects of mass m_1 and m_2, moving initially with velocities v_1 and v_2. Suppose they collide and we want to know what happens. To do this, we call the final velocities u_1 and u_2; our task is to determine these velocities (Figure 8-8a). Conservation of momentum tells us that:

Initial momentum = Final momentum

$$m_1 v_1 - m_2 v_2 = -m_1 u_1 + m_2 u_2$$

while conservation of energy tells us that if there are no gains or losses of energy in the collision:

Initial energy = Final energy

$$\frac{1}{2} m_1 v_1^2 + \frac{1}{2} m_2 v_2^2 = \frac{1}{2} m_1 u_1^2 + \frac{1}{2} m_2 u_2^2$$

Here we have two equations and two unknown quantities to find: u_1 and u_2. The rules of algebra tell us that we can always find solutions to this kind of problem. If we know the initial velocities of the two objects, be they billiard balls or subatomic particles, we can always find the final state of the system. This is a good example of the way in which the Newtonian clockwork universe operates (see Chapter 5).

Let's take a simple case, in which the objects are equal in mass (call it m) and are approaching each other with equal (but oppositely directed) velocities of magnitude v. In this case, conservation of momentum tells us that

$$m(v - v) = 0 = m (u_1 + u_2)$$

so that

$$u_1 = u_2$$

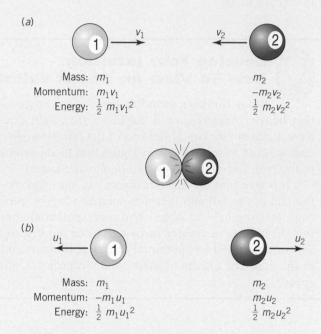

Mass: m_1 m_2
Momentum: $m_1 v_1$ $-m_2 v_2$
Energy: $\frac{1}{2} m_1 v_1^2$ $\frac{1}{2} m_2 v_2^2$

Mass: m_1 m_2
Momentum: $-m_1 u_1$ $m_2 u_2$
Energy: $\frac{1}{2} m_1 u_1^2$ $\frac{1}{2} m_2 u_2^2$

FIGURE 8-8. (a) A collision between two billiard balls is an example in which both momentum and energy are conserved. (b) If the two balls approach each other with the same speed before the collision, they rebound from each other with the same speed after the collision.

In other words, however fast the objects are moving after the collision, their velocities must be equal and opposite. Call the magnitude of the final velocity u. Then the energy equation tells us that if no energy is gained or lost in the collision,

$$\frac{1}{2} mv^2 + \frac{1}{2} mv^2 = \frac{1}{2} mu^2 + \frac{1}{2} mu^2$$

from which it follows that

$$u = v$$

In this collision, then, the two objects approach each other with the same speed, collide, and then move away from each other with the same speed that they had on approach (Figure 8-8b). If you think of two billiard balls colliding, you can see that this result is reasonable.

Develop Your Intuition: Adding Energy to the System

Consider a collision between billiard balls of equal mass that are traveling toward each other at equal speed v. At the moment of impact, an explosive cap on one ball explodes at the point of impact. How would the two equations in our analysis of collisions have to change for this situation?

All the forces associated with the cap are internal to the system, so they cannot affect the momentum of the system, which for our example remains zero. However, the cap does add energy to the system so that the energy equation now reads

$$\text{Initial energy} = \text{Final energy}$$

$$\tfrac{1}{2}mv^2 + \tfrac{1}{2}mv^2 + C = \tfrac{1}{2}mu^2 + \tfrac{1}{2}mu^2$$

where C is the amount of energy added to the system by the explosive cap. In this case, the final velocity, u, is higher than it would be without the cap, but the two balls still move away from each other at the same speed, back to back, so the final momentum is still zero.

THINKING MORE ABOUT

Simple Machines

The conversion back and forth from potential energy to kinetic energy is one of the most critical tasks of a technological society. To accomplish such energy conversions, humans have invented an extraordinary variety of *machines*, which are devices that change the direction or magnitude (or both) of an applied force. (This definition actually includes virtually all of the common tools, as well as more elaborate mechanical devices.) In the process, machines help us to apply a force over a distance—that is, to do work. In other words, machines help us convert between potential energy and kinetic energy in a wide variety of clever and useful ways. Three simple devices—the lever, the inclined plane, and the wheel and axle—lie at the heart of many familiar machines.

The lever Next time you watch a baseball game, notice how a power hitter can blast a baseball with one swing of the bat. The hitter achieves towering home runs by using a baseball bat—a beautiful example of the lever. A lever is simply a bar with a

support (called the "fulcrum"), as we described in Chapter 7. Used properly, this simple device can increase an applied force. In a baseball swing, the batter's pivoting body serves as the fulcrum, while the extended arms act as the lever arm (Figure 8-9). Everyday examples of levers include a

FIGURE 8-9. A view of a baseball hitter shows the lever effect of the pivoting body and extended arms.

claw hammer or crowbar, which can pry out a firmly lodged nail. The same principle comes into play when you use a tennis racquet or a sledge-hammer, which can increase the applied force on a struck object. Staplers, nutcrackers, fishing poles, and bottle openers are just a few other examples of levers in our daily lives.

The inclined plane Have you ever driven over a high mountain pass? You probably noticed how the road winds back and forth in sharp switch-backs. A switchback road is a simple machine called an inclined plane, which exchanges an increased travel distance for less effort (less power). Variations on the inclined plane include ramps and screws. The exact same principle also occurs in the wedge, which is used to cut and split

A mountain road with switchbacks is an example of an inclined plane. The car travels a longer distance but needs less power to climb the hill.

objects. Everyday examples of wedges include knives, scissors, axes, and your front teeth.

The wheel and axle The wheel and axle was one of the transforming technological inventions of human history. So common are wheels in our lives, that it's hard to imagine a society without this simple machine. The most obvious wheels in our lives are associated with vehicles—cars, bicycles, trains, and roller blades—in which wheels greatly reduce the friction of one object moving against another. But wheels appear in thousands of other devices, including clocks, fans, computer discs, ball bearings, gear trains, conveyor belts, and much more. In many of its everyday uses, including the steering wheel of your car, the capstan of a ship, pencil sharpeners, and valves, the wheel and axle may be thought of as a modification of a lever, in which the central axle acts like a fulcrum. To convince yourself of this idea, imagine common variants of water faucets, which can vary from simple levers to T-shaped handles or wheel-like valves.

Remarkably, many of the complex mechanical devices in our daily lives—cars, elevators, vending machines, and much more—are merely clever combinations of these three simple machines. Can you identify some of these simple machines in the devices you see around you? Some authorities classify the pulley, the wedge, and the screw as separate simple machines; try to see if you can recognize these elements as well. What do you think these simple machines all have in common?

Summary

Energy, which is measured in **joules,** is the ability to do **work**—the ability to exert a force over a distance. **Power,** which is measured in **watts** or **kilowatts,** is the rate at which energy is expended, that is, energy output per unit time.

Energy comes in several varieties. **Kinetic energy** is the energy associated with moving objects such as cars or cannonballs. **Gravitational potential energy,** on the other hand, is stored energy, ready for use; for example, the gravitational energy of dammed-up water. Energy can shift from one form to another, so that kinetic energy and gravitational potential energy are interchangeable. However, according to the **law of conservation of energy,** in a closed system, neglecting any frictional forces, the total amount of kinetic and gravitational potential energy is conserved.

The work needed to bring a system to a given state is equal to the sum of the kinetic and potential energies of the system. This statement, known as the **work-energy theorem,** relates the work done on a system to the total energy of that system.

Key Terms

energy The ability to do work. (p. 166)

gravitational potential energy The energy a body has by virtue of its position in a gravitational field. (p. 169)

joule The SI unit of work, corresponding to a force of 1 newton acting through 1 meter. (p. 167)

kilowatt The unit of 1000 watts (corresponding to an expenditure of 1000 joules per second). (p. 167)

kinetic energy The energy a body has by virtue of its motion. (p. 169)

law of conservation of energy The law that states that in a closed system the total amount of all forms of energy remains the same. (p. 174)

power The amount of work done divided by the time it takes to do it, or the energy expended divided by the time it takes to expend it. (p. 166)

watt The SI unit of power, defined as the expenditure of 1 joule of energy in 1 second. (p. 167)

work The product of the force exerted on an object times the distance over which it is exerted. (p. 163)

work-energy theorem The statement that the total potential and kinetic energy of an object in a given state is equal to the work that was done to bring the object to that state. (p. 175)

Review

1. What is the scientific definition of *work*? How does it differ from ordinary English use?

2. What is the definition of the *joule*? Why did scientists introduce this unit?

3. What is the difference between energy and power?

4. Is the kilowatt-hour a unit of energy or of power? How about kilowatt?

5. What is the difference between the watt and the horsepower?

6. What is the difference between the joule and the kilowatt-hour? Who uses which unit?

7. What is the definition of the *watt*? What is the relationship between the watt and the joule?

8. List some different kinds of energy. Explain how they differ from each other.

9. What factors determine the kinetic energy of a moving object?

10. Find something in your classroom or dorm room that possesses gravitational potential energy.

11. Look around your home and school. What objects in your everyday experience have the greatest potential energy?

12. What does it mean to say that different forms of energy are interchangeable?

13. What does it mean to say that energy is conserved?

14. What is the work-energy theorem? How does this theorem relate to the example of the roller coaster?

15. Give an example of the work-energy theorem in your home. Give another example at school.

16. What are the three simple machines? Give examples of each.

17. Identify some parts of an automobile and describe how they are used as simple machines.

Questions

1. A 50-pound crate is pushed across the floor by a 20-pound horizontal force. Aside from the pushing force and gravity, there is also a 50-pound force exerted upward on the crate and a 10-pound frictional force, as shown in the figure. Which of these forces does no work? Which does positive work? Which does negative work?

2. Two construction cranes are each able to lift a maximum load of 20,000 N to a height of 100 meters. However, one crane can lift that load in $\frac{1}{3}$ the time it takes the other. How much more power does the faster crane have?

3. As a freely falling object picks up downward speed, what happens to the power supplied by the gravitational force? Does it increase, decrease, or stay the same?

4. A pendulum swings left to right in the figure. At what locations in the pendulum's swing is the gravitational force

doing positive work? Negative work? No work? What is happening to the speed of the pendulum in each case?

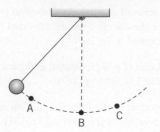

5. Where in the roller coaster ride shown in the figure is the gravitational force doing positive work? Negative work? No work? What is happening to the speed of the car in each case?

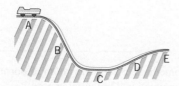

6. What kinds of energy are present in the following systems?

   a. Water behind a dam
   b. A swinging pendulum
   c. An apple on an apple tree
   d. The space shuttle in orbit

7. How do gravitational potential energy and kinetic energy shift in the following events?

   a. A baseball player hits a pop fly.
   b. A bungee jumper leaps off a high bridge.
   c. A meteor streaks down from space.
   d. An apple falls from a tree.

8. Identify which of the simple machines (the lever, the inclined plane, and the wheel and axle) are present in the following devices. (Note: Some devices incorporate more than one.)

   a. A toothbrush
   b. A fork
   c. A pizza cutter with a circular disk
   d. A saw
   e. A chisel
   f. A pencil sharpener

9. Which (if either) of the two objects shown has the greatest kinetic energy? Does it matter in which direction the objects are moving?

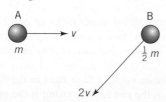

10. Where in the figure for Problem 4 does the pendulum have the greatest gravitational potential energy? Where does it have the greatest kinetic energy?

11. Where in the figure for Problem 5 does the roller coaster car have the greatest gravitational potential energy? Where does it have the greatest kinetic energy?

12. In the absence of air resistance, a falling rock gains kinetic energy and loses potential energy, with the total energy of the rock remaining constant. In the presence of air resistance, however, the rock eventually reaches terminal velocity. Now the kinetic energy is constant, but the potential energy continues to decrease as the rock falls toward the ground. What has happened to this missing energy?

13. According to the work-energy theorem, if work is done on an object, its potential and/or kinetic energy changes. Consider a car that accelerates from rest on a flat road. What force did the work that increased the car's kinetic energy?

14. Consider the block and tackle arrangement used to lift a 100-pound engine. What simple machine(s) is used in this arrangement? What force is necessary to hold up the engine? (*Hint:* How many ropes are actually supporting the weight?)

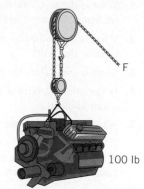

15. A 500-N crate needs to be lifted 1 meter vertically in order to get it into the back of a pickup truck. One option is to lift it directly up into the truck. Another option is to slide it up a frictionless inclined plane. Which method (if either) gives the crate more gravitational potential energy? What is the advantage of using the inclined plane?

16. List the several conservation laws that we have described thus far. In what ways are these laws similar?

17. Does the International Space Station have gravitational potential energy? Explain.

18. Is the total amount of gravitational and kinetic energy conserved in an open system? Why?

19. Many everyday devices incorporate more than one of the three simple machines—the lever, the inclined plane, and the wheel and axle. Identify the simple machine components in:

   a. A crowbar
   b. A pair of scissors
   c. A stapler
   d. A corkscrew

Problem-Solving Examples

Paying the Piper

EXAMPLE 8-3

A typical CD system uses 250 watts of electric power. If you play your system for three hours in an evening, how much energy do you use? If energy costs 8 cents a kilowatt-hour, how much do you owe the electric company?

REASONING AND SOLUTION: The given information consists of power and time, and you are asked to find an amount of energy. You can solve this problem by remembering the equation that relates energy used to amount of power:

$$\begin{aligned} \text{Energy} &= \text{Power} \times \text{Time} \\ &= 250 \text{ watts} \times 3 \text{ hours} \\ &= 750 \text{ watt-hours} \end{aligned}$$

Since 750 watts equals 0.75 kilowatt,

$$\text{Energy} = 0.75 \text{ kilowatt-hour}$$

The cost is

8 cents per kilowatt-hour × 0.75 kilowatt-hour = 6 cents. ●

Power Lifting

EXAMPLE 8-4

In Example 8-2 we talked about an athlete lifting a 400-pound weight. Suppose that, grunting and groaning, he lifts the barbell in 3 seconds. How much power is he expending in both English and SI units?

REASONING AND SOLUTION: Power is the work done divided by the time it takes to do it. Therefore, in the English system of units:

$$\begin{aligned} P \text{ (in ft-lb/s)} &= \frac{W \text{ (in foot-pounds)}}{t \text{ (in seconds)}} \\ &= \frac{1200 \text{ ft-lb}}{3 \text{ s}} \\ &= 400 \text{ ft-lb/s} \end{aligned}$$

But 1 horsepower is 550 ft-lb/s, so:

$$\begin{aligned} P \text{ (in hp)} &= \frac{400 \text{ ft-lb/s}}{550 \text{ ft-lb/s/hp}} \\ &= \frac{400}{550} \text{ hp} \\ &= 0.727 \text{ hp} \end{aligned}$$

In other words, for this very short period, the trained athlete is developing as much power as a small hand drill. In general, human beings can produce about $\frac{1}{4}$ hp over extended periods of time—a good deal less than a horse.

There are two ways to calculate the answer in SI units. One is to go to the table in Appendix A and find that one horsepower is equal to 745.7 watts. In this case

$$\begin{aligned} \text{Power (in watts)} &= 0.727 \text{ hp} \times 745.7 \text{ watts/hp} \\ &= 542.1 \text{ watts} \end{aligned}$$

The other way is to note from Example 8-2 that the athlete does 1627 joules of work in 3 seconds, so:

$$\begin{aligned} \text{Power (in watts)} &= \frac{1627 \text{ joules}}{3 \text{ seconds}} \\ &= 542.3 \text{ watts} \end{aligned}$$

The answers differ slightly due to rounding, but basically, they agree. ●

Bowling Ball Versus Baseball

EXAMPLE 8-5

What is the kinetic energy of a 4-kilogram (about 9-pound) bowling ball traveling down a bowling lane at 10 meters per second (22 miles per hour)? Compare this energy to that of a 250-gram (half-pound) baseball traveling 50 meters per second (110 miles per hour). Which object would hurt more if it hit you (that is, which object has the greater kinetic energy)?

REASONING AND SOLUTION: In this situation, the bowling ball has more mass but the baseball has more speed. The only way to really compare their energies is to substitute numbers into the equation for kinetic energy. For the 4-kg bowling ball traveling at 10 m/s,

$$\begin{aligned} \genfrac{}{}{0pt}{}{\text{Energy}}{\text{(in joules)}} &= \tfrac{1}{2} \times \text{Mass (in kg)} \times [\text{Velocity (in m/s)}]^2 \\ &= \tfrac{1}{2} \times 4 \text{ kg} \times (10 \text{ m/s})^2 \\ &= \tfrac{1}{2} \times 4 \text{ kg} \times 100 \text{ m}^2/\text{s}^2 \\ &= 200 \text{ kg-m}^2/\text{s}^2 \\ &= 200 \text{ joules} \end{aligned}$$

For the 250-gram baseball traveling at 50 meters per second,

$$\begin{aligned} \genfrac{}{}{0pt}{}{\text{Energy}}{\text{(in joules)}} &= \tfrac{1}{2} \times \text{Mass (in kg)} \times [\text{Velocity (in m/s)}]^2 \\ &= \tfrac{1}{2} \times 250 \text{ g} \times (50 \text{ m/s})^2 \end{aligned}$$

A gram is one-thousandth of a kilogram, so 250 g = 0.25 kg:

$$\text{Energy} = \frac{1}{2} \times 0.25 \text{ kg} \times 2500 \text{ m}^2/\text{s}^2$$
$$= 312.5 \text{ kg-m}^2/\text{s}^2$$
$$= 312.5 \text{ joules}$$

These results are summarized in Table 8-2. Even though the bowling ball is much more massive and has more momentum (mass × velocity) than a baseball, a hard-hit baseball carries more kinetic energy than a typical bowling ball because of its high velocity. ●

TABLE 8-2	Comparison of Kinetic Energy and Momentum			
	Mass (kg)	**Velocity (m/s)**	**Energy (J)**	**Momentum (kg-m/s)**
Bowling ball	4	10	200	40
Baseball	0.25	50	312.5	12.5

Gravitational Potential Energy

EXAMPLE 8-6

1. What is the gravitational potential energy of a 4-kilogram bowling ball 1 meter above the ground?

2. How high would a 250-gram baseball have to be held above the ground to have the same potential energy?

REASONING AND SOLUTION:

1. Apply the equation for potential energy to the 4-kg bowling ball 1 meter above the ground:

$$PE = mgh$$
$$= 4 \text{ kg} \times 9.8 \text{ m/s}^2 \times 1 \text{ m}$$
$$= 39.2 \text{ kg-m}^2/\text{s}^2$$
$$= 39.2 \text{ joules}$$

2. The second question asks for the height of a baseball in the case that:

$$(mgh)_{\text{bowling ball}} = (mgh)_{\text{baseball}}$$

Canceling g on both sides and inserting the known height of the bowling ball and the known masses of the two balls gives:

$$4 \text{ kg} \times 1 \text{ m} = 0.25 \text{ kg} \times h_{\text{baseball}}$$

Therefore,

$$h_{\text{baseball}} = \frac{4 \text{ kg} \times 1 \text{ m}}{0.25 \text{ kg}}$$
$$= 16 \text{ m}$$

The baseball would have to be held 16 meters (more than 50 feet) above the ground to hold the same amount of gravitational potential energy as the bowling ball. ●

Problems

1. How much work against gravity do you do when you climb a flight of stairs 3 meters high? Compare this work to the energy consumed by a 60-watt lightbulb in an hour. How many flights of stairs would you have to climb to equal the work of the lightbulb?

2. Would you rather be hit by a 1-kilogram mass traveling 10 meters per second, or a 2-kilogram mass traveling 5 meters per second?

3. Compared to a car moving at 10 miles per hour, how much kinetic energy does that same car have when it moves at 20 miles per hour? At 30 miles per hour? At 60 miles per hour? What do these numbers suggest to you about the difficulty of stopping a car as its speed increases?

4. A small air compressor operates on a 1.5-horsepower electric motor for 8 hours a day. How much energy is consumed by the motor daily? If electricity costs 10 cents a kilowatt-hour, how much does it cost to run the compressor each day? (*Note:* 1 horsepower equals about 750 watts.)

5. The joule and the kilowatt-hour are both units of energy. How many joules are equal to 1 kilowatt-hour?

6. Zak, helping his mother rearrange the furniture in their living room, moves a 50-kg sofa 6 meters with a constant force of 20 newtons. Neglecting friction,

   a. What is the work done by Zak on the sofa?
   b. What is the average acceleration of the sofa?

7. Georgie was pulling her brother (20 kg) in a 10-kg sled with a constant force of 25 newtons for one block (100 meters).

   a. What is the work done by Georgie?
   b. How long would a 100-watt lightbulb have to glow to produce the same amount of energy expended by Georgie?

8. A woman weight lifter can lift a 150-lb weight from the floor to a stand 3.5 feet off the ground. What is the total work done by the woman in ft-lb and joules?

9. The stair stepper is a novel exercise machine that attempts to reproduce the work done against gravity by walking up stairs. With each step, Brad (60 kg) simulates stepping up a distance of 0.2 meters with this machine. If Brad exercises for 15 minutes a day with a stair stepper at a frequency of 60 steps per minute, what is the total work done by Brad each day?

10. Calculate the amount of energy produced in joules by a 100-watt lightbulb lit for 2.5 hours.

11. Normally the rate at which you expend energy during a brisk walk is 3.5 calories per minute. (A calorie is the common unit of food energy, equal to 0.239 joules.) How long (in minutes) do you have to walk in order to produce the same amount of energy as in a candy bar (approximately 280 calories)?

12. How long (in minutes) do you have to walk to produce the same amount of energy as a 100-watt lightbulb that is lit for 1 hour? Refer to Problem 11.

13. You throw a softball (250 g) straight up into the air. It reaches a maximum altitude of 15 meters and then returns to you. (Assume the ball departed from and returned to ground level.)

    a. What is the gravitational potential energy (in joules) of the softball at its highest position?
    b. What is the kinetic energy of the softball as soon as it leaves your hand? (Assume that there are no energy losses by the softball while it is in the air.)
    c. What is the kinetic energy of the softball when it returns to your hand?
    d. From the kinetic energy, calculate the velocity of the ball.

14. Sleeping normally consumes 1.3 calories of energy per minute for a typical 150-lb person. How many calories are expended during a good night's sleep of 8 hours?

15. You leave your 75-watt portable color TV on for 6 hours during the day and evening, and you do not pay attention to the cost of this electricity. If the dorm (or your parents) charged you for your electricity use and the cost was $0.10 per kW-hr, what would be your monthly (30-day) bill?

16. While skiing in Jackson, Wyoming, your friend Ben (65 kg) started his descent down the bunny run, 25 meters above the bottom of the run. If he started at rest and converted all of his gravitational potential energy into kinetic energy,

    a. What is Ben's kinetic energy at the bottom of the bunny run?
    b. What is his final velocity?
    c. Is this speed reasonable?

17. Lora (50 kg) is an expert skier. If she starts at 3 m/s at the top of the lynx run, which is 85 meters above the bottom, what is her final speed if she converts all her gravitational potential energy into kinetic energy? What is her final kinetic energy at the bottom of the ski run?

18. The Moon has a mass of 7.4×10^{22} kg and completes an orbit of radius 3.8×10^8 m about every 28 days. The Earth has a mass of 6×10^{24} kg and completes an orbit of radius 1.5×10^{11} m every year.

    a. What is the speed of the Moon in its orbit? The speed of the Earth?
    b. What is the kinetic energy of the Moon in orbit? The kinetic energy of the Earth?

19. The current theory of the structure of the Earth, called plate tectonics, tells us that the continents are in constant motion. Right now, for example, the North American continent is moving at the rate of about 2 cm/year. Assume that the continent can be represented by a slab of rock 5000 km on a side and 30 km deep and that the rock has an average mass of 2800 kg/m³.

    a. What is the mass of the continent?
    b. What is the kinetic energy of the continent?
    c. Compare this to the kinetic energy of a jogger of mass 70 kg running at a speed of 5 m/s.

Investigations

1. Look at your most recent electric bill and find the cost of 1 kilowatt-hour in your area.

    a. Look at the back of your CD player or another appliance and find the power rating in watts. How much does it cost for you to operate the device for 1 hour?

    b. If you leave a 100-watt lightbulb on all the time, how much will you pay in a year of electric bills?
    c. If you had to pay $10.00 for a high-efficiency bulb that provided the same light as the 100-watt bulb with only 10 watts of power, how much would you save per year if you used the bulb for 4 hours each day?

 # WWW Resources

See the *Physics Matters* home page at **www.wiley.com/college/trefil** for valuable web links.

1. **www.vast.org/vip/book/HOME.HTM** The physics of roller coasters online.

2. **www.physicsclassroom.com/Class/energy/energtoc.html** Two useful animated lessons on work, energy, and power from physicsclassroom.com (includes discussions of physics of skiing and roller coasters).

3. **www.nu.ac.za/physics/1M2002/Energy%20work%20and%20power.htm** A site from the University of Natal discussing energy content of food.

4. **www.bodybuilding.com/fun/becker2.htm** Presents the physics underlying weight lifting, including gravity, friction, mechanical advantage, work, and power.

9

Atomic Structure and Phases of Matter

KEY IDEA

All matter consists of atoms, the building blocks of our world.

PHYSICS AROUND US . . . *Solid to Liquid to Gas*

You are walking along the street on a hot summer day, sipping an iced drink. Jostled by a passerby, you inadvertently spill some of the drink and an ice cube falls onto the sidewalk. In a few moments, the hard piece of ice begins to melt and is surrounded by a small puddle. Eventually, the ice cube disappears entirely, becoming a flat pool of water. One-half hour later, the water itself is gone, evaporated into the air.

A steelworker manipulates the controls of his console, tipping a huge bucket of red-hot molten steel into a mold. As the steel cools and solidifies, it is rolled into a sheet of tough, solid metal—perhaps a sheet that will be used to form the body of your next car.

You place a pot of water on the stove to boil water for coffee. A phone call distracts you and you forget all about it. Disaster is averted only when your roommate warns you that the water has all boiled off and the pot is about to burn.

All around us matter changes its physical form, from solids to liquids to gases and back again. The easiest way to understand these remarkable transformations is to realize that matter is made of invisibly small bits of material called *atoms*.

ENERGY, ATOMS, AND THERMODYNAMICS

In Chapter 8 we discuss the importance of the concept of energy in all areas of science, as well as in our daily lives. We take for granted the availability of energy in today's world. We just flick a switch or press a button and lights appear, computers start up, and we hear our favorite songs or see our favorite movies. Finding and securing energy resources has been one of the driving forces of global politics and economics for the past century or more and is still a vital concern for all nations. The by-products of energy consumption and their effects on the world environment have become a major concern as well.

It wasn't always like this. The ideas that energy could be measured and that changes in energy followed certain laws developed gradually during the late eighteenth and nineteenth centuries. The study of energy is known as **thermodynamics** and various physicists contributed to the laws of thermodynamics by the time the twentieth century dawned. However, there was no overall understanding as to why these laws worked. For instance, according to the first and second laws of thermodynamics, you can never build a perpetual motion machine, which would produce more energy than you put into it to make it run. Why not? Many physicists of the time would probably have answered, "Because the law says so, that's why not."

The great change in understanding the laws of energy came about from another fundamental idea in physics: that all matter consists of tiny particles called **atoms.** This idea has actually been around since the time of the ancient Greeks, over 2000 years ago. However, only in relatively recent times has it become clear that the behavior of matter at the atomic level actually determines what we observe and measure all around us. And one of the basic factors that most affects the behavior of atoms is energy.

The laws of thermodynamics deal with energy in the forms of heat, thermal energy, and work. However, before we discuss these ideas, in this chapter we introduce the atomic structure of matter and show how the energy of atoms affects whether a substance is solid, liquid, or gas. We show in later chapters that the atomic theory explains other properties of matter, from thermal expansion to strength of materials. With that discussion as background, we can explore what happens to a system of objects when the energy of the system changes.

ATOMS: THE BUILDING BLOCKS OF MATTER

In the first eight chapters we have examined everyday physical behavior of objects, including forces, which can cause objects to change their motion, and energy, which is necessary to exert a force over a distance. But what of physical objects themselves? What is the nature of matter, and why do objects display such an astonishing range of properties? To answer these questions we need to look at matter in much finer detail.

Imagine that you took a page from this book and cut it in half, then cut the half in half, cut half of that in half, and so on. Only two outcomes to this process are possible. If paper is smooth and continuous, there would be no end to this process, no smallest piece of paper that couldn't be cut further. On the other hand, you might come to a piece that could be divided no further, a smallest piece of matter. In which world do we live, a world where matter is continuous or a world in which there is a smallest piece? Finding the answer to this question has occupied many great minds over the course of more than 2500 years.

The Greek Atom

About 530 B.C. a group of Greek philosophers, the most famous of whom was a man named Democritus, gave this question some serious thought. Democritus argued (purely on philosophical grounds) that if you took the world's sharpest knife and started slicing chunks of matter, you would eventually come to a smallest piece—a piece that could not be divided further (Figure 9-1). He called this smallest piece the *atom*, which translates roughly as "that which cannot be cut." He argued that all material was formed from these atoms and that the atoms are eternal and unchanging, but that the relationships among atoms are constantly shifting. This argument provided a kind of intellectual bridge between two schools of thought among Greek philosophers, one of which argued that the fundamental reality of the world had to be eternal and unchanging, and the other that change was omnipresent in the universe.

It is important to realize that the Greek atomic theory is a part of philosophy and not part of science. For example, there is no place in it for observations and experiments. The only part of this historical episode that survives in modern science is the word *atom* itself, although even here, as we shall see in Chapter 21, it doesn't really describe the modern atom.

Elements

The beginning of modern atomic theory is generally attributed to an English meteorologist, John Dalton (1766–1844). In 1808, Dalton published a book titled *New System of Chemical Philosophy,* in which he argued that the new knowledge being gained by chemists about materials provided evidence, in and of itself, that matter was composed of atoms. Chemists knew that most materials can be broken down into simpler chemicals. If you burn wood, for example, you get carbon

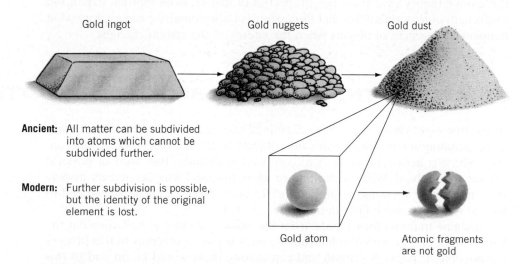

Gold ingot   Gold nuggets   Gold dust

Ancient: All matter can be subdivided into atoms which cannot be subdivided further.

Modern: Further subdivision is possible, but the identity of the original element is lost.

Gold atom   Atomic fragments are not gold

FIGURE 9-1. Repeatedly dividing a bar of gold, just like cutting paper repeatedly, produces smaller and smaller groups of atoms, until you come to a single gold atom. That single atom cannot be divided further by chemical means.

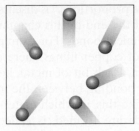
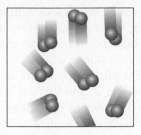
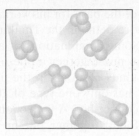

(*a*) Atoms   (*b*) Diatomic (two-atom)   (*c*) Triatomic (three-atom)
                              molecules   molecules

FIGURE 9-2. Atoms may be envisioned as solid balls that stick together to form molecules. Atomic models are often drawn with spheres, although we know that atoms are not solid objects. With this model, we can think of atoms as single spheres (*a*) and molecules as double spheres (*b*), triple spheres (*c*), or more.

dioxide, water, and all sorts of materials in the ash. If you use an electric current to break down water, you get the two gases hydrogen and oxygen. Dalton and his contemporaries also recognized that a few materials, called **elements,** could not be broken down into other substances. For example, you can heat wood to get charcoal (essentially pure carbon), but try as you might, you can't break down the carbon any further. Thus, chemists recognized that carbon is an element.

The hypothesis that we now call "atomism" was very simple. Dalton suggested that each chemical element corresponded to a species of indivisible objects called **atoms.** He borrowed this name from the Greeks, but other aspects of Dalton's theory were quite new. Two or more atoms stuck together form a **molecule**—the same term applies to any cluster of atoms that can be isolated, whether it contains two atoms or 1000. Molecules make up many of the different kinds of material we see around us. For example, a water molecule comprises one oxygen atom and two hydrogen atoms (thus, the familiar formula H_2O). In Dalton's view, atoms were truly indivisible—he thought of them as little bowling balls (Figure 9-2). In Dalton's world, then, indivisible atoms provide the fundamental building blocks of all matter.

ATOMS IN COMBINATION: COMPOUNDS AND MIXTURES

John Dalton saw chemical reactions, such as wood burning or a piece of metal dissolving in acid, as processes that involved the putting together and taking apart of various combinations of atoms. In fact, if you think of Dalton's atom as something like a bowling ball with Velcro patches scattered around its surface, you wouldn't be far off. Just as two bowling balls can stick together if their Velcro patches align, so too can two atoms come together to create a molecule. And just as you can separate the bowling balls if you apply enough force, so too can a molecule be broken down into its individual atoms. Today, as we see in Chapter 23, we refer to the matched "Velcro patches" that tie atoms together as *chemical bonds.* Materials such as water, consisting of molecules in which atoms are held together by chemical bonds, are called **compounds.**

Unlike the Greek philosophers, Dalton based his theory of the atom on a large amount of data from chemical experiments. One important law that was

Sand is a mixture composed of different kinds of light and dark grains. The white line gives the scale of 1 mm.

part of the basis for his work, called the "law of definite proportions," is discussed in more detail in the Thinking More About section at the end of this chapter.

Some combinations of atoms do not involve chemical bonds. For example, if you mix together a pile of iron filings with a pile of copper filings, there are two kinds of atoms in the final pile—copper and iron. Each bit of metal, however, has only one type of atom in it, and there is no bonding between the iron and copper bits. A collection of atoms or molecules like this, in which materials are found together but in which no chemical bonds are formed, is called a **mixture.** Sand on the beach is a common mixture, as are a cup of coffee with cream and sugar and even the air you breathe.

Physics in the Making
The Father of Modern Chemistry

One of the basic ideas on which Dalton's atomic theory rested was the principle of conservation of mass. This principle states that in a chemical reaction, there is no overall change in amount of mass—reactants may change their form and produce other substances, but mass is not destroyed. This is not an obvious concept; for example, if you burn a log in a campfire, producing smoke and ash, it certainly looks like mass has been lost. But if you trap all the smoke and weigh it along with the ash, you will find that the weight is the same as the initial log. The person who first proved this result, with experiments on combustion among other things, was Antoine Lavoisier (1743–1794).

Lavoisier was the first scientist to insist on the importance of quantitative measurements in chemical research. His textbook, *Traité Élémentaire de Chimie,* is the basis for modern chemical terminology and has been compared in its importance to chemistry with the influence of Newton's *Principia* in physics. Lavoisier disproved the accepted idea of combustion prevalent in the eighteenth century (the "phlogiston theory") and Dalton's work would not have been possible without Lavoisier's contributions.

Lavoisier was a member of the French nobility and the chief tax collector for an area of Paris. This made him a target of the French Revolution, and he was guillotined near the end of the Reign of Terror. ●

● ATOMIC MASS

In Dalton's theory, all atoms of the same species are identical, but atoms of different species have different masses. Measuring these masses was accomplished in the nineteenth century using a chemical principle first enunciated in 1811 by the Italian scientist Amedeo Avogadro (1776–1856). **Avogadro's principle** states that equal volumes of any gas at the same temperature and pressure must contain the same number of molecules.

To see how this fact was used to determine the relative masses of atoms, consider a simple electrolysis experiment in which an electric current is used to break up water molecules (Figure 9-3). In such an experiment, the volume of hydrogen gas produced (due to the number of hydrogen atoms) is twice the volume of oxygen (due to the number of oxygen atoms). This ratio of hydrogen to oxygen is expressed by the familiar chemical formula H_2O. Thus, half the weight of the hydrogen gas released from water, compared to the weight of the oxygen gas

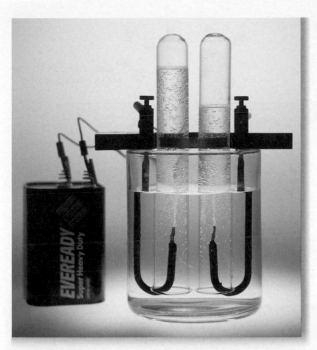

FIGURE 9-3. An electrolysis experiment on water shows that water separates into oxygen gas and hydrogen gas. The volume of the hydrogen is twice that of the oxygen, but the weight of the oxygen is eight times that of the hydrogen.

released, should yield the relative weights (and thus the relative masses) of the hydrogen and oxygen atoms.

In fact, when this experiment is done, the weight of the oxygen produced is about eight times the weight of the total amount of hydrogen, or 16 times the weight of half the hydrogen. If we define the mass of the hydrogen atom to be 1 atomic mass unit, then this result tells us that the oxygen atom has a mass 16 times that of hydrogen, or 16 atomic mass units.

Using similar techniques, we can determine the other atomic masses relative to hydrogen. Thus, the atomic mass of carbon is about 12, iron is 56, silver is 108, and gold is 197. When we speak of the mass (or weight) of an atom, we are referring to masses measured in this way, relative to hydrogen with atomic mass 1. (For technical reasons, today the atomic mass unit is defined in terms of carbon, not hydrogen, but the mass of carbon is still 12 units and the mass of a hydrogen atom is still 1 unit.)

PHASES OF MATTER

Once we understand that all materials are made from atoms, we can explain the kinds of transformations discussed in the Physics Around Us section at the beginning of this chapter. A given material may be made from certain groups of atoms or molecules, but how those atoms or molecules are arranged can lead the material to have very different properties.

For example, water always consists of molecules containing two hydrogen atoms and one oxygen atom. Water can appear in its familiar liquid form, but it can also be solid ice or it can be gaseous steam. Whatever its form, however, it is still made up of the same molecules. What changes is the effectiveness of the forces that hold those molecules together, due to the different energies of the molecules in water, ice, or steam.

FIGURE 9-4. Water occurs commonly in three different states—solid, liquid, and gas. These three states of matter differ in the organization of their molecules.

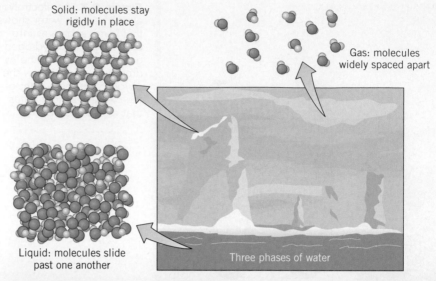

Solid: molecules stay rigidly in place

Gas: molecules widely spaced apart

Liquid: molecules slide past one another

Three phases of water

The three forms that water can take are examples of **phases of matter.** The three most common phases of matter are solid, liquid, and gas (Figure 9-4).

The Solid Phase

A **solid** is defined as a material that has a definite shape and volume and that is sufficiently rigid to counteract a force imposed on it. For example, if you squeeze this book, it doesn't collapse but exerts a force to counter your squeezing. This counter-force is what you feel in your fingers.

In Figure 9-5 we show one way that atoms can be combined to form a solid. Atoms in many solids are arranged in a regular, repeating array and they are held together by interatomic forces called *chemical bonds.* We discuss the exact nature of these forces in Chapter 23, but for the moment the easiest way to think about chemical bonds is to imagine that the atoms are held together by stiff springs, as shown in Figure 9-5*c*.

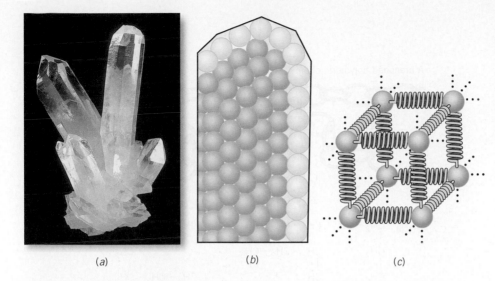

(a)   (b)   (c)

FIGURE 9-5. (a) A solid such as this quartz crystal has sharp, definite edges, reflecting the internal arrangement of its molecules. (b) A three-dimensional drawing of the crystal's regular arrangement of molecules. (c) Atoms are held together by bonds that behave like stiff springs.

A regular, repeating structure such as this is called a *crystal*. At the atomic level, solids from diamonds to table salt are arranged in this orderly way. You can think of a crystal as being formed from countless trillions of tiny boxes, each no more than a billionth of an inch on an edge, each in contact with several other identical boxes and containing exactly the same pattern of atoms (Figure 9-6). The key point is that atoms in a solid are locked into place and do not easily move around from one location to another. If you push on a crystal from the outside, the "springs" that hold the atoms together are compressed, and, by Newton's third law, exert a force that pushes back. This counter-force is how solids retain their shape when acted on by an outside force.

It is important to realize that crystals are only one way that atoms can arrange themselves into solids. In contrast to crystals, *glasses* include solids with predictable local atomic environments for most atoms, but no long-range order to the atomic structure (Figure 9-7). For example, in most common window and bottle glass, silicon and oxygen atoms form a strong three-dimensional framework with each silicon atom surrounded by four oxygen atoms. If you were placed on any atom in a glass, chances are you could predict the neighboring atoms. Nevertheless, glasses have no regularly stacked boxes of structure. Travel more than two or three atoms distant from any starting point and there is no way that you could predict whether you'd find a silicon atom or an oxygen atom.

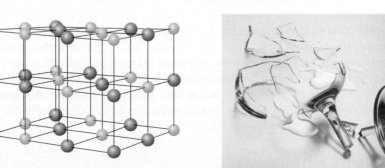

FIGURE 9-6. A crystal structure can be thought of as stacks of tiny boxes, each with the same pattern of atoms, arranged in a regular array.

FIGURE 9-7. Glass is not arranged in a regular pattern of atoms; it does not form geometrical crystals.

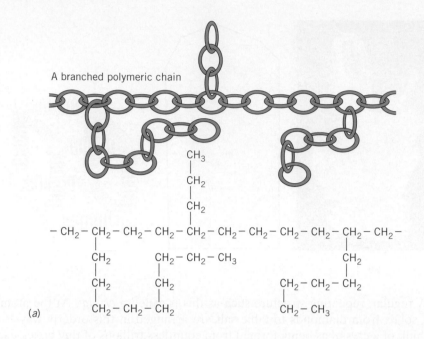

A branched polymeric chain

$$CH_3$$
$$|$$
$$CH_2$$
$$|$$
$$CH_2$$
$$|$$
$$-CH_2-CH_2-CH_2-CH_2-CH_2-CH_2-CH_2-CH_2-CH_2-CH_2-CH_2-$$
$$| \qquad\qquad\qquad\qquad | \qquad\qquad\qquad\qquad\qquad\qquad |$$
$$CH_2 \qquad\qquad CH_2-CH_2-CH_3 \qquad\qquad CH_2$$
$$| \qquad\qquad\qquad\qquad | \qquad\qquad\qquad\qquad\qquad\qquad |$$
$$CH_2 \qquad\qquad\qquad CH_2 \qquad\qquad\qquad CH_2-CH_2-CH_2$$
$$| \qquad\qquad\qquad\qquad | \qquad\qquad\qquad\qquad\qquad\qquad |$$
$$CH_2-CH_2-CH_2 \qquad\qquad\qquad\qquad\qquad\qquad CH_2-CH_3$$

(a)

An unbranched polymeric chain

$$CH_2-CH_2-CH_2-CH_2-CH_2-CH_2-CH_2-CH_2-CH_2-CH_2-CH_2-CH_2$$

(b)

FIGURE 9-8. (a) Branching and (b) unbranching polymers form from small molecules, just as long chains can form individual links.

Another kind of noncrystalline solid is a polymer. *Polymers* are extremely long and large molecules that are formed from numerous repeating smaller molecules like links of a chain. The atomic structure of these materials is often linear, with predictable repeating sequences of atoms along the polymer chain as well as branching side groups of atoms (Figure 9-8). Common polymers include numerous biological materials, such as animal hair, plant cellulose, cotton, spider webs, and clotted blood. Even DNA and RNA are polymers.

Polymers include all *plastics,* which are synthetic materials formed primarily from petroleum. A ubiquitous form of solid in our modern world, plastic features long chainlike molecules that are interlocked together like tangled strands of spaghetti in a bowl. Although almost unknown one-half century ago, plastics have become our most versatile commercial materials, providing an extraordinary range of uses: thin flexible sheets for lightweight packaging, dense castings for durable machine parts, strong fibers for clothing and rope, colorful moldings for toys, and many others. Plastics serve as paints, inks, glues, sealants, foam products, and insulation. New tough, resilient plastics have revolutionized many sports with products such as high-quality bowling and golf balls and durable helmets for cycling, football, and ice hockey, not to mention a host of completely new products from Frisbees to roller blades.

The Liquid Phase

A **liquid** is a material that maintains a constant volume, but assumes the shape of its container. Other than water, mercury, and a few biological fluids, few liquids occur naturally on Earth (at ordinary temperatures and pressures). Water, by far the most abundant liquid on the Earth's surface, is a dynamic component of geological change. In addition, water-based solutions are essential to all known forms of life.

In contrast to solids, atoms or molecules in a liquid do not stay in one place, but are free to slide over one another. You can picture the atoms or molecules in a liquid as being something like a container full of dry sand grains. The sand fills whatever volume it is poured into because individual sand grains can roll over one another. If the sand is poured into another container, the pile adjusts to conform to the new shape. By the same token, the relatively weak forces that hold the molecules together in a liquid suffice to keep the total volume of the liquid constant, but do not lock individual atoms in place. These forces are the same as those in a solid, but the molecules in a liquid have more energy and move around faster than molecules in a solid. As a result, the intermolecular forces are less effective at holding the molecules in place.

Connection
Surface Tension

Every molecule in a liquid has a weak attraction to every other molecule next to it. However, molecules at the surface of a liquid do not have any other molecules of the liquid above them, only below them. As a result, surface molecules experience a net force pulling them inward toward the rest of the liquid. This force is known as surface tension.

Surface tension in water is not a very large force, but it's enough to support the weight of light insects (e.g., water striders) or even a paper clip. Soap bubbles are round because the surface tension of the soapy water tries to minimize the area of the bubble around the volume of air inside; it turns out that a sphere has less surface area for a given volume than any other shape. In your lungs, the many tiny sacs (alveoli) where oxygen is absorbed by the bloodstream and carbon dioxide is released are also spherical. If these sacs had the same surface tension as water does, air pressure would not be enough to expand them. However, the alveoli are coated with a mucous lining that has a lower surface tension, enabling them to expand and fill up with air.

Surface tension is also the property of liquids that enables them to expand up into a thin tube, or capillary; the phenomenon is called "capillarity." You can see this behavior whenever you see water absorbed by the tiny spaces between fibers of a paper towel, but it is much more widespread than that. Capillarity has a role in the flow of blood through the smallest veins and arteries of the body (the capillaries) and is also involved in the transport of water and sap from the roots of trees to their upper branches. Can you see why surface tension is a topic of great importance in biology? ●

The Gaseous Phase

A **gas** is a material that retains neither its shape nor its volume, but expands to fill any empty container in which it is placed. The atoms or molecules in a gas are relatively far apart from one another and interact primarily by collisions. Thus, if gas is introduced into a container, the atoms rush outward until they encounter a wall, bounce off it, and return to collide with other atoms or molecules.

When a molecule of gas bounces off the walls of a container, Newton's first law of motion tells us that the wall must have exerted a force on the molecule. After

(a) Hot air balloon being filled with gas. (b) The balloon rises because its average density is less than that of air.

(a)

(b)

FIGURE 9-9. Gas molecules move constantly inside a tire. When they bounce off the inner walls of the tire, they cause an internal pressure.

all, that law tells us that, if an object with mass (such as a molecule) accelerates (by changing its direction of motion), then a force must have been applied to it. Newton's third law tells us that the molecule must also exert an equal and opposite force on the container. It is these collisions of countless molecules with the side of an automobile's tire, for example, that we perceive as air pressure (Figure 9-9).

The most common gas in our experience is the Earth's atmosphere, which is a mixture of oxygen and nitrogen, along with minor amounts of a few other gases (Figure 9-10). Even though you can't see the air around you, you can feel the effects of the gas molecules when the wind blows and exerts a force on you.

The term *fluid* is often used to refer collectively to both liquids and gases—materials that can change their shapes because individual atoms and molecules are free to change their positions.

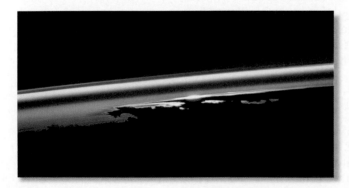

Ionosphere
Stratosphere (20 mi)
Troposphere (11 mi)
Earth's surface

FIGURE 9-10. Earth's atmosphere is a thin shell surrounding the solid and liquid parts of the planet. The air we breathe is mostly in the troposphere; clouds generally mark the boundary between the troposphere and the higher stratosphere.

Connection

The Fourth Phase of Matter

At extreme temperatures such as those of the Sun, high-energy collisions between atoms may begin to break the atoms down into even smaller particles called "electrons" and "nuclei" (see Chapter 26). Extreme heat can strip off electrons, creating a *plasma,* in which positive nuclei move about in a sea of electrons. Such a collection of electrically charged atoms is something like a gas, but displays unusual properties not seen in other states of matter. For example, plasmas are efficient conductors of electricity. They are too hot and reactive to be confined in any normal container, but can be confined in a strong magnetic field, or "magnetic bottle."

A plasma is the least familiar phase of matter to us, yet more than 99.9% of all the visible mass in the universe exists in this form. Not only are most stars composed of a dense hydrogen- and helium-rich plasma mixture, but several planets, including the Earth, have regions of thin plasma in their outer atmospheres. ●

CHANGES OF PHASE

In all of the discussion so far of phases of matter, we have omitted one very important fact: atoms and molecules are constantly in motion, even in solids and liquids. As we see in Chapter 11, the hotter a material is, the faster those atoms or molecules move.

Take a crystalline solid as an example of this point. Atoms are locked into place by the "springs" we call chemical bonds, but each atom wiggles around its equilibrium position in the solid. On the atomic scale, think of the solid as being in a constant state of microscopic jiggling and shaking, like a bowl of Jell-O during an earthquake. As the solid heats up, this atomic jiggling gets more and more violent until, when it gets hot enough, the atoms simply tear loose from their moorings and start to move around over each other. At this point, the material changes from a solid to a liquid (Figure 9-11). We say that the material has melted.

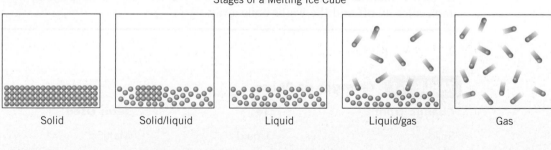

FIGURE 9-11. At the molecular scale, molecules in a solid gain energy as the temperature rises, eventually breaking free of their rigid locations and moving around in a liquid (melting). If the temperature continues to rise, the molecules of liquid gain enough energy to leave the substance altogether, becoming a gas (boiling).

Carbon dioxide in the solid phase (called "dry ice") does not melt into a liquid but goes directly to the gas phase, a process known as sublimation.

This is precisely what happens to the water molecules in an ice cube when it is left out at room temperature and changes into water.

This transformation from solid to liquid is an example of what scientists call a **change of phase:** a process by which a material, without changing its constituent atoms or molecules, changes their arrangement. A variety of familiar changes of phase are listed in Table 9-1.

Develop Your Intuition: Boiling Water Molecules

What does a boiling pot of water look like from an atomic point of view?

The water molecules move around, as already described, and are held in the liquid by relatively weak forces. As the motion of the molecules becomes more and more violent, and as the molecules start moving faster and faster, eventually the fastest molecules acquire enough kinetic energy to tear loose from the surface of the liquid. They fly off into the air as steam, leaving the liquid behind. As boiling continues, molecules from the middle of the pot of water have enough energy to escape the liquid; this is when bubbles form throughout the liquid and the entire pot is in turmoil. Eventually, if the source of heat is maintained (by leaving the pot on the stove, for example), all the molecules fly off and the water boils away.

TABLE 9-1 Changes of Phase

Change from	To	Term Used
Solid	Liquid	Melting
Liquid	Solid	Freezing
Liquid	Gas	Boiling or evaporation
Gas	Liquid	Condensation
Solid	Gas	Sublimation
Gas	Solid	Deposition

Atoms: Are They Real?

For much of the nineteenth and early twentieth centuries, a mild philosophical debate took place among scientists over the question of whether atoms are real or whether they are merely a useful mathematical construct. With advances in technology and in theoretical understanding over the past 3 centuries, increasingly convincing evidence has mounted for the reality of atoms. Here are some examples of this growing body of evidence.

1. **The behavior of gases** The Swiss physicist Daniel Bernoulli (1700–1782) realized that if atoms are real, they must have mass and velocity, and thus kinetic energy. He successfully applied Newton's second law of motion (force equals mass times acceleration) to atoms to explain the behavior of gases under pressure. Doubling the number of gas particles, or halving the volume, doubles the rate of collisions between the gas particles and the confining walls of its container. This increase in turn doubles the pressure, which equals the force per unit area. Increasing the temperature increases the average velocity of the gas particles and also results in an increase of pressure. Thus the idea of atoms of a gas is consistent with the observed behavior of gases.

2. **Chemical combinations** John Dalton advanced the atomic theory based in part on the law of definite proportions—an empirical law that states that for any given compound, elements combine in a specific ratio of weights. For example, water is always 8 parts oxygen to 1 part hydrogen by weight and carbon dioxide is always 12 parts carbon to 32 parts oxygen by weight. Furthermore, when two elements combine in more than one way, the ratios of weights for the two compounds is a small whole number. Thus, 12 pounds of carbon can combine with either 32 pounds of oxygen or 64 pounds of oxygen. These sorts of regularities are easy to understand in terms of the atomic theory—for example, there are always two hydrogen atoms for every one oxygen atom in water, and oxygen has a mass 16 times that of hydrogen.

3. **Radioactivity** The discovery in 1896 of radioactivity, by which individual atoms emit radiation (see Chapter 26), provided a compelling piece of evidence for the atomic theory. Certain materials called "phosphors" emit a brief flash of light when hit by this radiation. In 1903, upon seeing the irregular twinkling caused by this effect, even the most vocal skeptics of the atomic theory had to reconsider.

4. **Brownian motion** Brownian motion is an erratic, jiggling motion observed in tiny dust particles or pollen grains suspended in water. In 1905, Albert Einstein (1879–1955) demonstrated mathematically that such motions must result from forces—the forces due to random collisions of atoms. Einstein realized that any small object suspended in liquid would be bombarded constantly by moving atoms. At any given moment, purely by chance, more atoms hit one side of the particle than the other. The object is pushed toward the side with fewer collisions. A moment later, however, more atoms strike another surface, and the object changes direction. Over time, Einstein argued, these atomic collisions produce precisely the sort of erratic motion that you can see through a microscope.

Einstein used the mathematics of statistics to make several testable predictions about how fast and how far the suspended grains would move, based on the hypothesis that the motion was due to collisions with real atoms. French physicist Jean Baptiste Perrin (1870–1942) published careful measurements of Brownian motion in 1909—results that agreed with Einstein's calculations and thus convinced many scientists of the reality of atoms.

Note that, in spite of the variety of evidence for atoms, to this point all this evidence was indirect. Matter was observed to behave *as if* it were made of atoms, but atoms, themselves, had not been directly observed.

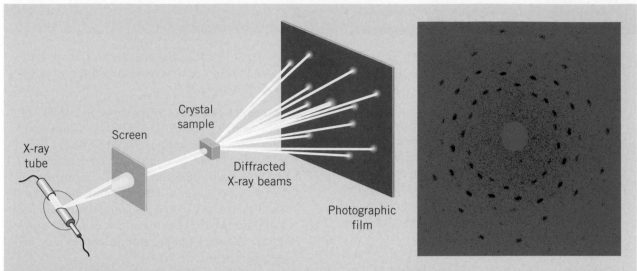

FIGURE 9-12. The regular atomic structure of a crystal can be revealed by the pattern of X rays that diffract off the crystal.

5. **X-ray crystallography** X-ray crystallography, developed in 1912, convinced any remaining skeptics by demonstrating the sizes (about 10^{-10} meter) and regular arrangements of atoms in crystals. X rays (see Chapter 19) are deflected by crystals in symmetric patterns that reveal the underlying regular atomic structure (Figure 9-12). X rays can't bounce off hypothetical ideas, so these images were further proof that atoms are real physical objects.

6. **Atomic-scale microscopy** In 1980, the first photograph of an individual atom was taken at the University of Heidelberg in Germany. This image was produced by an instrument called a "scanning tunneling microscope," which detects tiny flows of electrons in a microscopic needle placed next to a solid surface. Now, observational studies of individual atoms are undertaken around the world (Figure 9-13).

   When in the chain of historical events would you have been willing to believe that atoms are real?

When Dalton explained the existence of elements? When Einstein explained Brownian motion? When you were shown a picture such as the one in Figure 9-13? Never? What does it take to make something "real"? And, finally, does it make a difference to science whether or not atoms are real?

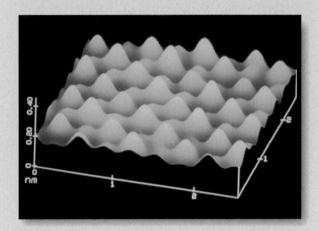

FIGURE 9-13. Atomic microscope image of individual atoms.

Summary

All matter is formed from tiny objects called **atoms.** The concept of the atom arose in Greek philosophy, but modern atomic theory, based on the observed chemical behavior of materials, began with John Dalton in the early nineteenth century. Chemical **elements** are materials that cannot be broken down into simpler substances by chemical means; each element corresponds to a different type of atom. Atoms combine to form **molecules,** which make up chemical **compounds,** held together by chemical bonds. Atoms or molecules that are found together but that are not bound together chemically are called **mixtures.** The relative masses of atoms can be determined through the use of **Avogadro's principle,** which says that equal volumes of gas at the same temperature and pressure contain the same number of atoms or molecules.

Atoms and molecules can exist in many **phases of matter.** The most common phases are **solid, liquid,** and **gas.** A solid maintains its shape and resists outside forces. Atoms in a solid are locked into place by interatomic (or intermolecular) forces. Liquids maintain their volume but not their shape, and their atoms are free to move around. Gases retain neither shape nor volume, and atoms or molecules in gases interact primarily through collisions.

Changes of phase, including melting, boiling, and freezing, result from changes in the intensity of atomic vibrations that occur when heat is added to or removed from a material.

Key Terms

atom The tiniest particle of matter that retains the chemical properties of an element. (p. 186)

Avogadro's principle The statement that equal volumes of any gas at the same temperature and pressure contain the same number of gas molecules. (p. 188)

change of phase A process by which a material, without changing its constituent atoms or molecules, changes their arrangement. (p. 195)

compound A material that is made up of two or more elements. (p. 187)

element A substance that cannot be broken down into other substances by chemical means. (p. 187)

gas A material that retains neither its shape nor its volume, but expands to fill any container in which it is placed. (p. 193)

liquid A material that maintains a constant volume, but assumes the shape of its container. (p. 192)

mixture A combination of two or more substances in which each substance retains its own chemical identity. (p. 188)

molecule Two or more atoms bound together by electric forces (chemical bonds). (p. 187)

phases of matter The different forms that matter can take; solid, liquid, and gas are the most common. (p. 190)

solid A rigid material that has a definite shape and volume. (p. 190)

thermodynamics The study of heat and energy. (p. 185)

Review

1. Explain Dalton's concept of an atom. How did it differ from the view of the Greeks?

2. Why is the atomic theory of Democritus considered to be philosophy and not science?

3. What is the difference between an atom and a molecule? Give an example of each.

4. How is a bowling ball with Velcro patches on it a good analogy to an atom? If two of these balls are stuck together, what does each component (ball, patches, and groups of such balls stuck together) represent? In what ways is this a bad analogy?

5. What is a compound? How does this differ, if at all, from a molecule?

6. From the perspective of a scientist, what is the critical difference between the Greeks' atomic theory and Dalton's theory?

7. What is a mixture? What role do chemical bonds have in a mixture? In a compound?

8. How do the masses of atoms of the same element compare to one another? Are the masses of atoms of different elements the same or different from one another?

9. What is Avogadro's principle? How was this principle used to determine the relative masses of atoms?

10. What is an atomic mass unit? How is it related to the mass of a hydrogen atom?

11. What are the three phases of matter discussed in this chapter? Do the molecules of a substance change as matter changes to another phase? Do the forces that bind molecules together change? Explain.

12. Define the solid phase of matter. Give an example.

13. What are the similarities and differences between ordinary glass and the structures we know as crystals? Explain.

14. How does heating a solid eventually lead to a liquid phase?

15. What is a polymer? A plastic? How are the molecules arranged in these materials?

16. Classify a crystal, a glass, and a plastic according to the range and direction of their repeating atomic structures.

17. What is a liquid? How are the molecules arranged in this phase of matter?

18. What is a gas? How are molecules and atoms arranged in a gas?

19. What is air (gas) pressure? What does pressure have to do with collisions and Newton's laws? Explain.

20. What are the two most common gases in our atmosphere?

21. What is meant by a change of phase? How does energy affect the change of phase of a substance?

22. What phase changes take place in the following processes: melting, freezing, boiling, condensation, sublimation?

23. Identify three examples of changes of state (other than in water) that occur in your everyday experience.

24. How did Bernoulli's work support the hypothesis that atoms were indeed real and not imaginary constructs?

25. What is the law of definite proportions? Why is it that when two elements combine in more than one way, the ratios of weights for the two compounds are usually small whole numbers?

26. How did the discovery of radioactivity support the reality of the atom? What is Brownian motion and how does it support atomic theory?

27. X-ray crystallography provided some of the first direct evidence of the existence of atoms. How did it provide this evidence?

28. When was the first photograph taken of an atom, and how was this done?

Questions

1. How many atoms are there in a water molecule? How many elements are there in a water molecule?

2. A carbon atom and an iron atom are moving at the same speed. Which atom has more kinetic energy?

3. Fullerenes are large molecules of carbon containing at least 60 carbon atoms. Discovered in 1985, fullerenes take a roughly spherical shape. The carbon atoms are arranged in a way that makes them look like soccer balls. Compare the mass of a C_{60} molecule to the mass of a single gold atom.

4. Based on what you have learned so far, should the techniques used to separate the individual components in a mixture be very different from the methods used to isolate the components that make up compounds? How so?

5. Describe the changes of state for a simple water molecule that goes from a solid to a liquid to a gas from the perspective of the forces that it experiences from its neighboring molecules.

6. Describe the changes of state for a simple water molecule that goes from a solid to liquid to a gas from the perspective of its average kinetic energy.

7. Do you think that there are factors other than temperature that might influence whether a phase change takes place? Specifically, how might the pressure that a material is subjected to influence whether it goes from a liquid to a gas or from a solid to a liquid?

8. What causes the Brownian motion of dust particles? Why aren't larger objects such as baseballs affected by this phenomenon?

9. Two gas-filled tanks have the same volume, temperature, and pressure. They are identical in every way except that one is filled with oxygen (O_2) gas and the other is filled with nitrogen (N_2) gas. Compare the number of gas molecules in each container. Which container weighs more?

10. A 1-liter tank contains 1,000,000,000 oxygen (O_2) molecules and 1,000,000,000 helium (He) atoms. Another tank contains 1,000,000,000 helium (He) atoms. The gases in the tanks have the same pressure and temperature. What is the volume of the tank that contains only helium?

11. Ammonia is a liquid that consists of molecules of (NH_3) (one nitrogen atom with three hydrogen atoms attached). Suppose ammonia is separated into nitrogen (N_2) gas and hydrogen (H_2) gas. If 1 liter of nitrogen is produced, what volume of hydrogen is produced?

12. Diamond and graphite are both solids composed of only carbon atoms. Since all carbon atoms are chemically identical, what do you think accounts for the vastly different properties of graphite and diamond?

13. Which have more average kinetic energy: the molecules in 10 grams of ice or the molecules in 10 grams of steam?

14. If the number of gas atoms in a container is doubled, the pressure of that gas doubles (provided the temperature and the volume of the container remain the same). Explain why the pressure increases in terms of the molecular motion of the gas.

15. Atmospheric pressure is approximately 15 pounds of force per square inch. How much force does the air exert on this side of this page of the book (approximately 100 square inches)? Why isn't it extremely difficult to turn the page?

16. How come crushed ice melts so much faster than an equal mass of ice cubes? (*Hint:* Think about making crushed ice by breaking ice cubes into small pieces. In order to melt ice, heat has to enter it. Where does heat enter the ice?)

Problems

1. The mass of a hydrogen atom is 1.67×10^{-27} kg.

   a. Calculate the weight of a hydrogen atom near the Earth's surface. (Recall from Chapter 5 that $W = m \times g$, where W is the weight, m is the mass, and g is the acceleration due to gravity, 9.8 m/s^2.)

   b. How many hydrogen atoms are there in 1 pound of hydrogen gas?

   c. Suppose that every person in the world (about 6 billion people in all) were employed as an atom counter. Each person would work a 40-hour week and be able to count one atom per second. How long would it take for 6 billion people to count the hydrogen atoms in 1 pound of hydrogen?

2. In gaseous form, oxygen consists of O_2 molecules and hydrogen consists of H_2 molecules. Suppose that instead of H_2 molecules, gaseous hydrogen consisted of H atoms. If this were the case, how many liters of hydrogen gas would be produced for each liter of oxygen gas when water, H_2O, was separated by an electric current?

3. Aluminum electroplating is a process by which aluminum is coated onto a metal object. The object is submerged in a liquid solution containing Al_2O_3 molecules. An electric current breaks up these molecules into oxygen gas and aluminum atoms. The aluminum is attracted to the object to be coated and forms a thin aluminum film on its surface. If a car bumper needs to be plated with 300 grams of aluminum using this electroplating process, what mass of oxygen gas is produced? [Assume masses of 27 atomic mass units (amu) for each aluminum atom and 16 amu for each oxygen atom.]

4. You blow up an ordinary party balloon with air until it has a diameter of 6 inches. Your friend blows up another balloon with helium gas until it has a diameter of 12 inches. Air consists mostly of O_2 and N_2 molecules, while helium gas consists of He atoms. Assume the pressure in each balloon is the same.

   a. Compare the number of helium atoms to the total number of O_2 and N_2 molecules.

   b. Air is about 80% nitrogen and 20% oxygen. Which balloon weighs more and by how much? (*Hint:* Imagine that there are 80 helium atoms in the helium balloon. Calculate the mass, in atomic mass units, of this amount of helium, and then compare to the mass of the corresponding number of oxygen and nitrogen molecules.)

Investigations

1. Which chemical elements do you encounter in more or less pure form in your daily experience? What are the properties of these elements and how are they used?

2. Chromatography is one of the most powerful tools a chemist has to separate out different components within mixtures. It can be used to separate everything from different colored pigments from plants to the various proteins in a cell. Explore the history of chromatography. What are the various types of chromatography and what principles are used to isolate individual components in often complex mixtures.

3. If you had never been told about atoms in science class, what observations might convince you of their existence? What might provide alternative hypotheses to the atomic theory of matter?

4. Investigate the history of extracting metal from metal ores. What techniques are involved in separating out and isolating the metal from compounds, such as iron oxide or bauxite (aluminum oxide)? How do these methods differ from methods that separate out mixtures?

5. In the Thinking More About section on page 197, we present six different significant pieces of evidence that support the once uncertain idea that atoms are indeed real. Pick one of these six pieces of supporting evidence and research the evidence in more detail by taking a closer look at one of the original experiments.

6. PVC, polyvinyl chloride, is a material that is essentially inert in its solid form and ubiquitous in daily life. Investigate the controversy surrounding its production and use by noting the arguments for it and against it, paying attention not only to the solid PVC plastic, but to the vinyl chloride gas used to make it as well. Does the evidence seem to indicate this is a dangerous product to you, or does it seem relatively safe?

7. Amedeo Avogadro, who proposed Avogadro's principle of gas volumes, also estimated the number of atoms that make up a given mass of material. Investigate the history of Avogadro's number and some of the dozen or more ways that scientists have attempted to measure this number.

8. Solids, liquids, and gases are the three common phases of matter in our daily lives, but another phase of matter—plasma—is far more abundant in the universe. (See the Connection on page 195.) Investigate plasmas to discover their unique properties, their distribution in the universe, and their possible technological applications.

WWW Resources

See the *Physics Matters* home page at **www.wiley.com/college/trefil** for valuable web links.

1. **http://cst-www.nrl.navy.mil/lattice/index.html** Information about and representations of 242 crystal lattice structures.

2. **http://chemed.chem.purdue.edu/genchem/topicreview/bp/ch2/mixframe.html** A review of elements, compounds and mixtures at Purdue University.

10

Properties of Matter

KEY IDEA

Properties of a material depend on the way in which its atoms are put together.

PHYSICS AROUND US . . . Giant Animals

Sometimes it's fun to watch old science fiction movies. You know the kind we mean, the ones in which giant grasshoppers attack Chicago, or Tokyo is menaced by a new kind of monster. In these old films, before computer technology made possible the construction of truly frightening monsters, the director's choices were either to make models of dinosaurs and giant gorillas or to scale up other sorts of animals—ants, for example, or spiders.

But could an ant with six skinny legs really become 20 feet tall? Could its outer skeleton support all that extra weight? Ever since Galileo's work in the seventeenth century on the properties of materials, scientists have known that the answer is no.

Today, this bit of scientific lore is well established among movie makers, and the old-fashioned, impossible monsters are seldom seen. That shouldn't stop you from enjoying the old movies, however.

OUR MATERIAL WORLD

Humans value different materials because of their diverse useful properties. Strong superglue, shatterproof glass, colorful dyes, resilient plastics, long-lasting paints, lightweight alloys, soft fabrics, flexible wire, efficient insulation, safe preservatives, effective drugs—the list goes on and on. Why do different materials show such an amazing range of properties?

Many of the everyday properties of materials can be understood if we think about the behavior of the atoms from which they are built. Conversely, once we understand how different kinds of atoms behave under various conditions, we can design materials that have the properties we're looking for. The list of materials in the previous paragraph contains only synthetic products; the ability to manufacture these substances springs from an understanding of how atoms interact. (See Physics and Modern Technology on page 205.) In many respects, understanding the interplay of atomic behavior and the resulting properties of matter has completely changed the modern world. Many of the devices common today but unheard of 50 years ago (lasers, computers, spacecraft, cell phones, credit cards) would not have been possible without advances in semiconductors, plastics, and metal alloys—the materials of the new age.

In this chapter we examine three familiar material properties: the density of materials, their response to pressure, and their elasticity. We encounter these properties every day in countless ways. We shall see that behaviors as diverse as floating in water, keeping aircraft in the air, and constructing buildings that don't fall down are all related to atoms and the way they interact with one another.

DENSITY

Pick up a brick and a piece of wood of the same size and you'll immediately notice that the brick is heavier. There are two reasons for this difference. First, the atoms in the brick include species such as potassium and silicon, which are more massive than the carbon and oxygen atoms that make up most of the wood. Second, the arrangement of atoms is such that the atoms in the brick are, on average, packed more closely together than the atoms in the wood. As a consequence, the brick weighs more than the wood, even though the two objects are the same size.

The property of matter that indicates the amount of material packed into a given volume is called **density.**

1. In words:

Density depends both on the kind of atoms from which a material is made and on how closely they are packed together.

2. In an equation with words:

Density equals mass divided by volume.

3. In an equation with symbols:

$$\rho = \frac{m}{V}$$

where the Greek letter ρ (rho) is customarily used to denote density.

Physics and Modern Technology—Materials Science

The products and devices of today could not have been imagined as recently as 50 years ago. In large part that's due to advances in understanding how atomic structure affects the properties of a material. This knowledge has led to the development of new materials for visual display, sports, medicine, housewares, and everything in between.

Light-emitting diodes (LEDs) emit light when an electric current runs through them. They provide visual displays for hand-held telephones.

Dacron was originally developed as a synthetic clothing material but is also used today for synthetic blood vessels.

Titanium alloys provide lightweight metal parts of outstanding strength and durability for everything from airplanes to bicycles to golf clubs.

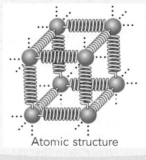

Atomic structure

Plastics made from petroleum by-products are used for everything from CD cases to styrofoam cups, keys to water bottles.

Liquid crystal displays (LCDs) allow light to pass through them when an electric current makes them align properly. You see them in calculators, watches, clock radios, and computer screens.

Teflon was discovered by accident but has become synonymous with nonstick cooking pans and utensils.

TABLE 10-1 Densities of Common Materials

Material	Density (grams/cubic centimeter)	Density (kilograms/cubic meter)
Solids		
Gold	19.3	19,300
Lead	11.3	11,300
Iron	7.9	7900
Diamond	3.5	3500
Aluminum	2.7	2700
Quartz sand	2.65	2650
Concrete	2.3	2300
Bone	1.8	1800
Ice	0.92	920
Wood (pine)	0.5	500
Liquids		
Mercury	13.6	13,600
Blood	1.05	1050
Water at 4°C	1.0	1000
Ethyl alcohol	0.79	790

Density is usually measured in units of grams per cubic centimeter or, in SI, in units of kilograms per cubic meter. (Note that a numerical value in kg/m^3 is exactly 1000 times larger than the value in g/cm^3.) A cubic centimeter of water at 4°C has a mass of 1 gram, so its density equals 1 g/cm^3. In this system of units, then, the density of any material tells you how much more, or less, massive a given volume of that material is compared to the same volume of water. The densities of some common substances are given in Table 10-1.

A Dense Metal

EXAMPLE 10-1

A cube of the platinum-like metal osmium measures 10 cm on a side and is found to have a mass of 22.6 kg. What is the density of osmium?

REASONING AND SOLUTION: Density is mass (in grams) divided by volume (in cubic centimeters). The mass is 22.6 kg, or 22,600 g, while the volume of the cube is given by

$$V = \text{Length} \times \text{Width} \times \text{Height}$$
$$= 10 \text{ cm} \times 10 \text{ cm} \times 10 \text{ cm} = 1000 \text{ cm}^3$$

Substituting these numbers:

$$\text{Density} = \frac{\text{Mass}}{\text{Volume}}$$

$$= \frac{22,600 \text{ g}}{1000 \text{ cm}^3}$$

$$= 22.6 \text{ g/cm}^3$$

In point of fact, osmium is the densest naturally occurring material at the Earth's surface. This metal is dense because it has massive atoms packed closely together. It is used in making hard, long-lasting metal alloys, including the metal used for some types of pen tips. ●

Develop Your Intuition: Density of Iron

What is the density of a 500-kilogram block of iron compared to an iron filing?

Density depends on a material's atomic structure, but does not depend on how much material there is. A 500-kilogram block of iron, in other words, has exactly the same density as an iron filing. The block's greater mass is accompanied by a greater volume, so when we divide mass by volume, we always get the same density.

 PRESSURE

The properties of liquids and gases are different from those of solids in some important respects. In this section we examine some concepts associated with liquids and gases, which together are often referred to as *fluids*.

When you go to a gas station, you may use a gauge to check the air pressure in your tires, and you may use a pump to inflate them. In atomic terms, what are you doing when you carry out these everyday operations?

Air is a gas and, as we showed in Chapter 9, molecules in the gas constantly strike the walls of the tire. The tire exerts a force on the molecules that changes their direction, and hence, from Newton's third law of motion, the molecules exert a force on the tire. This force causes pressure, which is defined as follows.

1. In words:

   Pressure *is a force divided by the area over which the force is exerted.*

2. In an equation with words:

   Pressure equals force divided by area.

3. In symbols:

$$P = \frac{F}{A}$$

Actually, in the case of a car tire, two pressures are involved—an outward pressure exerted by the air in the tire and an inward pressure exerted by the

When you measure the air pressure in a tire, you are actually measuring the difference between the pressure inside the tire and atmospheric pressure outside the tire.

atmosphere on the outside of the tire. By adding air to the tire, you increase the rate of collisions that contribute to the pressure inside the tire, so that the outward pressure exceeds the inward pressure and the tire expands. When you measure the pressure with a tire gauge, what you measure is the difference between these two pressures, a quantity called "gauge pressure." In a properly filled tire, the outward pressure from the air inside the tire is greater than the inward pressure from the air outside the tire; the material of the tire stretches to maintain that pressure difference.

The units of pressure in the English system are pounds per square inch (psi). This is the unit on an ordinary tire gauge. In SI, the unit of pressure is the pascal (Pa), which is defined to be a force of 1 newton exerted over an area of 1 square meter. The pascal is named in honor of Blaise Pascal (1623–1662), a French scientist and mathematician, who helped establish the study of statistics and did pioneering work in fluid mechanics.

Develop Your Intuition: Units of Pressure

Is a pressure of 1 Pa (a newton per square meter) bigger or smaller than a pressure of 1 psi (a pound per square inch)?

One meter is about 1 yard long and 1 yard is 36 inches. One square meter, therefore, contains approximately $(36)^2$, or over 1000, square inches. One newton, on the other hand, is the weight of approximately 0.1 kg, which is less than 1 pound. One pascal, then, represents a force smaller than 1 pound exerted over an area one thousand times larger than 1 square inch. Therefore, 1 pascal is much smaller than 1 psi. (In point of fact, everyday pressures are usually measured in a unit known as the megapascal, MPa, which is 1 million pascals.)

The Pressure–Depth Relationship

Water behind a dam, air in a balloon, mercury in a barometer, and lava in a volcano—in these and many other natural situations fluids exert pressure on their surroundings. Physicists have put considerable effort into understanding the relationship between a fluid's depth and the resulting pressure.

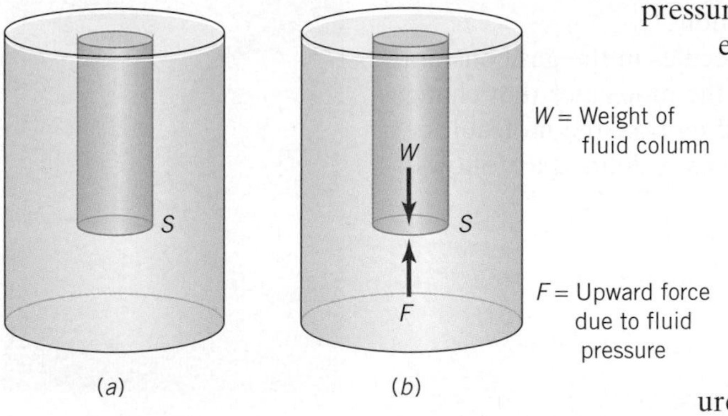

Imagine the situation shown in Figure 10-1, in which a column of fluid is isolated from its surroundings by a flexible membrane (think of the membrane as an oddly shaped balloon). The fluid might be water in a lake or ocean, the air outside your window, or oil in a hydraulic lift used by your auto mechanic to lift your car. Consider the bottom surface of the membrane, labeled S in the figure. This surface doesn't move, so by Newton's first law of motion, the forces on it have to balance. These forces are (1) the downward weight of the column of fluid above S and (2) the upward force due to the pressure of the fluid outside the membrane. (The fluid inside the membrane exerts an equal and opposite pressure, but that pressure acts on a different object, namely the fluid on the outside of the membrane.) In order for the surface to remain stationary, these forces must balance. Let's summarize this relationship.

W = Weight of fluid column

F = Upward force due to fluid pressure

(a)   (b)

FIGURE 10-1. Pressure–depth relationship. (a) A column of the fluid above an area S. (b) The weight of the fluid column must be balanced by the upward force associated with the pressure on S.

1. In words:

The weight of the fluid column must be balanced by the force associated with the upward pressure on S.

2. In an equation form:

The weight of the fluid column equals the fluid pressure times the area of S.

3. In an equation with symbols:

$$W = P \times A$$

where A is the area of the surface, W is the weight of the column, and P is the fluid pressure at the bottom of the column.

If we concentrate for the moment on liquids such as water and oil, we can simplify this equation. Many liquids are approximately *incompressible*—that is, their density does not change much with pressure. In this case, the density of liquid in the column in Figure 10-1 is the same throughout the column, so that the weight of liquid in the column is

$$W = \text{Density} \times \text{Volume} \times g$$
$$= \rho \times (A \times d) \times g$$

where d is the height of the column or, equivalently, the depth below the fluid surface of the bottom membrane S, and g is the acceleration due to gravity. (Remember that weight is equal to mass times g.) If we insert this expression for weight into the equation $W = P \times A$, the quantities A cancel on both sides of the equation, and we find

$$P \times A = \rho \times (A \times d) \times g$$
$$P = \rho \times d \times g$$

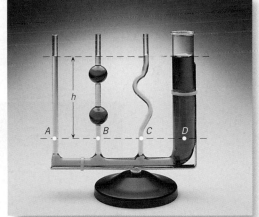

In other words, in a liquid, the pressure increases with depth—double the depth and you double the pressure.

Two points should be made about the pressure exerted by a fluid. First, although we have concentrated on the upward force due to pressure, in fact the pressure at any point in a fluid is the same in all directions because pressure is related to collisions of atoms or molecules, which also move in all directions.

The second point is that it makes no difference how much fluid is involved in our considerations. The water pressure 3 feet beneath the surface of a backyard swimming pool is exactly the same as the pressure 3 feet beneath the surface of the Atlantic Ocean (aside from a small difference due to the different densities of these two bodies of water.) For a given liquid, pressure depends on depth and nothing else (Figure 10-2). (This statement is true for a static fluid; that is, one with no motion. We look at moving fluids later in this chapter.)

FIGURE 10-2. The levels of water in a series of connected glass tubes are the same, independent of the shapes of the tubes.

Connection

The Design of Dams

A dam is a structure designed to hold back water flowing in a river. Typically, the water behind a dam forms a lake, and the sideways pressure exerted on the dam by the water piled up behind it increases with depth. The deeper the lake, the greater the pressure at the bottom of the dam, regardless of how much water is in the lake. This is why dams are usually much thicker at the bottom than at the top.

The earliest dams, built centuries ago, were made of packed earth and were intended to collect river water into a lake so the water could be used to irrigate fields. Over the years, the design and construction of dams evolved as engineers applied new ideas and new technology to these massive projects. Dams built across canyons were designed in a circular arch, curving away from the water; this design channeled the force of the water along the face of the dam to its ends, so the rocky sides of the canyon could help support the dam (Figure 10-3). The largest dams are built of concrete, a material that supports large pressures without cracking, The Hoover Dam at the Nevada–Arizona border is built in an arch and contains about 6.6 million tons of concrete; it was designed as a hydroelectric dam and supplies electricity to Las Vegas, among other areas.

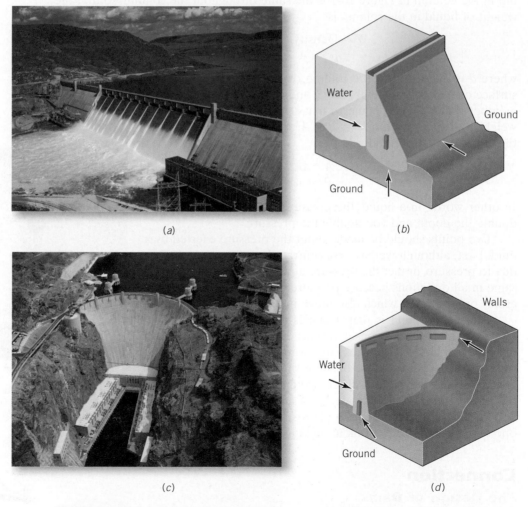

(a)

(b)

(c)

(d)

FIGURE 10-3. A big dam must withstand a great deal of pressure from the water against it. Gravity dams (a, b) and arch dams (c, d) are designed to transfer some of the forces due to that pressure to the ground and canyon walls where the dam is built.

The same principles apply to the design of large aquariums, whose windows need to be transparent while still holding back huge amounts of water. For example, the Monterey Bay Aquarium contains one of the largest windows in the world—54 feet long and 15 feet high—to offer visitors a view of the ocean.

Large tanks are often used to hold liquids in storage; these tanks must be strong enough to withstand the forces that develop from the pressure of the liquid. A disastrous example was the 1919 rupture of a metal tank used to store molasses in Boston. The tank was 90 feet high and 90 feet across, which meant it had to withstand forces at the bottom of over 2.5 million pounds. The flood of molasses released was initially 30 feet high, trapping and killing pedestrians and horses in its sticky ooze. The cleanup lasted for several weeks. ●

Connection
The Hydraulic Lift and Pascal's Principle

Our understanding of the physical behavior of fluids under pressure has led to numerous everyday practical applications. We now know that pressure is a property of atomic or molecular collisions. Therefore, if you increase the pressure of a static fluid in one place (by pushing on a plunger, for example), the collisions transmit that increase immediately to every part of the fluid. This immediate transmission of pressure to all parts of a fluid, known as **Pascal's principle,** explains the working of the hydraulic lift (Figure 10-4) that allows a mechanic to lift your car by exerting a small force.

The basic idea is that pushing down on the plunger shown on the left of Figure 10-4, increases the pressure throughout the fluid. If the piston on the right has a larger area than the plunger on the left, then the total upward force on that piston, which is equal to the pressure times the area of the plunger, is greater than the downward force exerted on the plunger on the left. This arrangement doesn't violate the law of conservation of energy because you have to move the plunger on the left through a greater distance. The work done—the product of force times distance—is the same for both plunger and piston.

Pascal's principle is the basis for all hydraulic systems and is involved wherever it is useful to produce a large force by exerting a small one. For example, the control

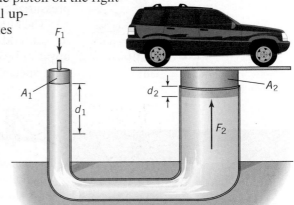

$$\text{Pressure}_1 = \text{Pressure}_2$$

$$\frac{F_1}{A_1} = \frac{F_2}{A_2}$$

$$F_2 = F_1 \times \frac{A_2}{A_1}$$

$$\text{Work}_1 = \text{Work}_2$$

$$F_1 d_1 = F_2 d_2$$

$$d_2 = d_1 \times \frac{F_1}{F_2}$$

Hydraulic mechanisms provide strong forces for controlling wing flaps on an airplane or plows on a bulldozer.

FIGURE 10-4. Mechanical lift in an auto repair shop. A small force moving a large distance is transformed into a larger force moving a smaller distance.

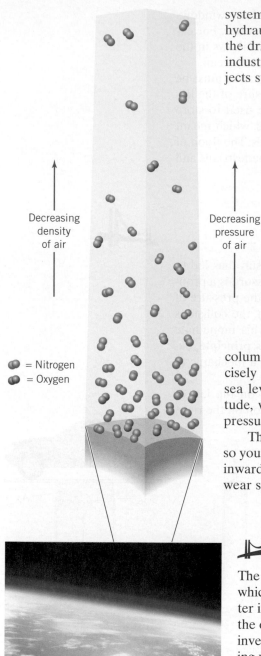

Decreasing
density
of air

Decreasing
pressure
of air

= Nitrogen
= Oxygen

FIGURE 10-5. Gas density in Earth's atmosphere. The pressure of the atmosphere at sea level is about 101,000 pascals (14.7 psi) and fades away to practically nothing at an altitude of 50 kilometers.

systems that operate the wing flaps and rudders on airplanes are usually hydraulic. Contemporary automobiles often use hydraulic systems to aid the driver in controlling both the brakes and the steering wheel. And in industry, large hydraulic presses are often used to shape metals into objects such as coins, office furniture, and machine parts. ●

Atmospheric Pressure

Like the liquid shown in Figure 10-1, gas in a large body such as Earth's atmosphere exerts a pressure that depends on the depth of the gas. Unlike liquid water or oil, however, the density of gas in a column above a point on the Earth's surface is not constant. Gas density is greater at the bottom of a column than it is higher up (Figure 10-5). As anyone who has hiked in the mountains can attest, the density of air at an elevation of 10,000 feet is considerably less than at sea level. By the time you get to about 100,000 feet, you are above 99 percent of the atmosphere.

Nevertheless, the air in the column above the Earth's surface weighs something, and that weight must be balanced by a pressure, just as in a liquid. If you hold out your hand at sea level and imagine drawing a square 1 inch on a side, then the column of air that extends from your hand into space will weigh precisely 14.7 pounds (or about 65 N). Thus, the atmospheric pressure at sea level is 14.7 psi. Atmospheric pressure drops off rapidly with altitude, which explains why airline passenger compartments have to be pressurized.

The human race evolved at the bottom of Earth's atmospheric ocean, so your body automatically exerts an outward pressure to counteract the inward pressure exerted by the atmosphere. This is why astronauts must wear special suits when they work in the vacuum of space.

Connection
The Barometer

The barometer is an instrument for measuring atmospheric pressure, which changes due to several different factors. A very simple barometer is shown in Figure 10-6. A long tube, open at one end and closed at the other, is filled with a dense liquid, such as mercury. The tube is then inverted into a bowl of the same dense liquid, with the closed end sticking up. The column of liquid in the tube starts to fall, but is opposed by the pressure of the atmosphere, pushing down on the surface of the liquid in the bowl. The level of liquid in the column can be calibrated to indicate normal atmospheric pressure. When atmospheric pressure drops, the weight of liquid that can be supported in the column drops, and the height of the liquid column falls. The reverse happens when atmospheric pressure rises.

Readings of atmospheric pressure—called "barometric pressure"—are one of the basic kinds of data for meteorology, the science of weather. In general, low barometric readings correspond to bad weather—clouds, rain, or snow—and high readings correspond to blue skies and fair weather. You've probably seen weather forecasts on television showing the progress of low-pressure systems and high-pressure systems as they follow the wind patterns of the atmosphere. The differences between high- and low-pressure systems are not large on a barometer, amounting to a few inches of height in the barometer's mercury column. (Normal barometric pressure is considered to be 30.00 inches of mercury. The lowest reading ever recorded in the northeastern United States, during the great 1938 hurricane, was 27.94 inches.) But these differences reflect changes in vast volumes of air in constant motion around Earth. Understanding these pressure readings and fitting them into the picture of temperature readings, measurements of wind speed and direction, and relative humidity form the basis for weather forecasting. ●

Buoyancy and Archimedes' Principle

Another intriguing property of fluids occurs because solids displace fluids. For example, when you get into a bathtub, the level of water in the tub rises. Every time you take a bath, you demonstrate what is now known as *Archimedes' principle,* named after the Greek scientist and mathematician Archimedes of Syracuse (287–212 B.C.). His principle deals with the force exerted by a fluid on an immersed object.

Think about a volume of material under the surface of a fluid, as shown in Figure 10-7. A volume of the fluid itself will not move but will be in equilibrium. Fluid pressure will exert force all around, greater at the bottom than at the top to support the weight of this volume of fluid. Now, if the molecules of a material, such as a rock or a piece of wood, are different from those of the fluid, the pressure around the object will still be the same, because the molecules of the fluid haven't changed. They will still exert the same forces on the object, including a net upward force. What has changed is the weight of the object. If the object's weight is greater than the upward force exerted by the fluid, the object sinks. If the object's weight is less than the upward force of the fluid, the object floats. This is the molecular basis of Archimedes' principle, which states that

> *The upward force exerted on an object immersed in a fluid*
> *is equal to the weight of the fluid that the object displaces.*

The force exerted by the fluid on the object is called the **buoyant force** and the effect is known as buoyancy. Buoyancy explains why it is easier to lift something—a child, for example—under water than in the air. The force needed to lift the child is her weight minus the upward buoyant force—the weight of the

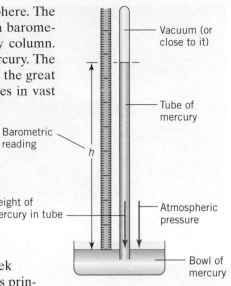

FIGURE 10-6. A barometer measures atmospheric pressure at a given time and place.

Fluid presses against arbitrary volume of fluid; force is greater at bottom of volume due to greater pressure

(a)

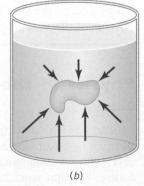

Fluid presses against object of same shape; pressure is still greater at bottom, providing net upward force

(b)

FIGURE 10-7. (a) A fluid exerts pressure on all sides of any arbitrary volume of the fluid; the pressure is greater at a greater depth. (b) For an object of the same shape immersed in the fluid, a net upward force (buoyant force) acts on the object.

A blimp exhibits buoyancy in the air.

An iceberg floats with 10% of its mass above the surface of the water because ice is only 90% as dense as water.

displaced water. For this reason, people undergoing physical therapy after a serious injury often spend time exercising their muscles while in water, where the force needed to support the body is less.

Archimedes' principle also explains why some objects float in water while others sink. If the weight of the object is greater than that of the displaced water, the net force on the object is downward and it sinks. On the other hand, if the weight of the object is less than the weight of the fluid it displaces, the upward buoyant force pushes the object to the surface. In fact, the buoyant force keeps pushing upward until the amount of material below the water line displaces just enough water to equal the object's weight. This fact means that objects with a density greater than that of water sink, whereas objects with a lower density than water float.

Archimedes' principle applies to all fluids, not just water. The buoyant force explains why blimps and hot-air balloons can travel through the air. Blimps are filled with lighter-than-air gases such as helium so that their average density (including the steel gondola containing passengers, crew, and engines) is less than that of the air. In a hot-air balloon, air is heated so it will expand and have a lower density than the surrounding air outside the balloon.

Archimedes' principle explains what people mean when they say, "That's only the tip of the iceberg." Ice has a density about 90% that of water. Thus, 90% of an iceberg's volume has to be hidden under the water's surface to produce a buoyant force equal to the iceberg's weight. What you see above water (the tip) is only 10% of what's actually there.

Develop Your Intuition: Can Steel Float?

We know that a piece of steel will sink if it's placed in water, so how can ships made of steel stay afloat?

The total amount of water displaced by a boat is the volume of the steel hull (or that part of the hull under water) *plus* the volume of air enclosed in that hull. The weight of the steel plus air is less than the weight of the water displaced, so the ship floats. However, if the enclosed air is replaced by water (think of the *Titanic*), the ship sinks.

Physics in the Making
Eureka

Legend has it that King Hieron II of Syracuse asked Archimedes to find out if a crown he had received was really pure gold. Archimedes had to do this, however, without damaging the crown in any way. He could weigh the crown, of course, but that wouldn't tell him if it were pure gold or just gold plate over a metal such as lead. Then, while getting into his bath one day, he saw the water in the tub rise and received an inspiration. He realized that if he put the crown under water, he could find its volume by seeing how much water it displaced. In modern language, he realized that he could measure both the mass and the volume of the crown and thus determine if its density matched that of gold.

Archimedes was so excited that he jumped out of his bath and raced through the streets to the palace, shouting, "Eureka!" ("I have found it!"). Different versions of the legend disagree as to whether he stopped to put his clothes on first. ●

Fluids in Motion: The Bernoulli Effect

One of the most remarkable and useful aspects of fluid properties is that fluids in motion exert pressures differently than fluids at rest. Daniel Bernoulli (1700–1782), a member of a family of contentious Swiss philosophers, mathematicians, and scientists, investigated the pressure exerted by fluids in motion. He found that he could summarize his results in a simple form:

> *The pressure exerted by a fluid on a surface decreases*
> *as the velocity of the fluid across the surface increases.*

We won't derive the equation that describes this **Bernoulli effect,** but it is simply an expression of conservation of energy for a moving fluid.

Bernoulli's principle explains in part how an airplane can stay aloft, even though it is heavier than the air it displaces. Next time you are at an airport, look at the wings of the planes you see there. You will notice that they are designed like the one shown in Figure 10-8, so that the upper part of the wing is more curved than the lower part. This shape is known as an "airfoil." When the plane starts to move down the runway, the air through which it moves has to travel faster over the top surface of the wing to catch up with the air moving under the bottom surface. By Bernoulli's principle, this means that the pressure on the top of the wing is less than it is on the bottom, so there is a net upward force, called **lift.** When the plane is moving fast enough, the lift force exceeds the weight of the plane and the plane lifts off the ground. Next time you fly, watch the wing during takeoff. On some aircraft, the wing is flexible enough so that you can actually see the tip moving upward just before the plane leaves the ground. (The

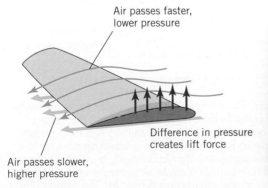

Air passes faster,
lower pressure

Difference in pressure
creates lift force

Air passes slower,
higher pressure

FIGURE 10-8. Bernoulli's principle. The pressure exerted by a fluid on a surface decreases as the velocity of the fluid across the surface increases.

Bernoulli effect is not the only principle involved in producing lift on an airplane. At least as important is Newton's third law. The shape of the wing also directs air downward as it passes the wing, giving the air a downward change in momentum. The reaction to this momentum change is an upward force acting on the plane.)

The airfoil shape appears in many different guises (Figure 10-9). For example, ski jumpers stretch far out over their skis to give their bodies an airfoil shape, helping produce a lift that takes them farther down the slope. The filled sails on a sailboat have the shape of a vertical airfoil, producing a horizontal force instead of a lifting force. Even many species of fish have evolved into airfoil-like shapes to produce more of a lifting effect as they swim through the water.

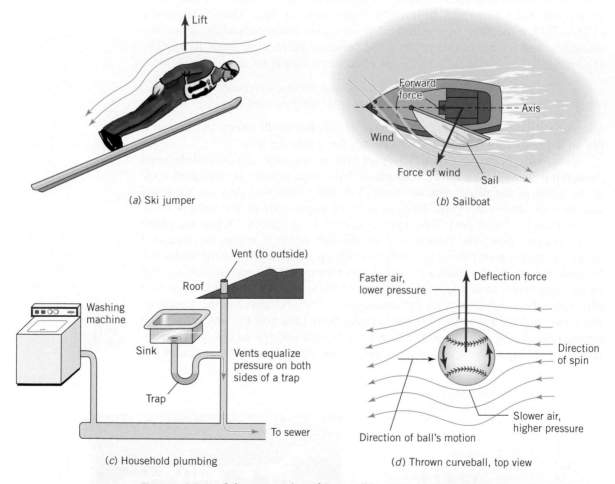

(a) Ski jumper

(b) Sailboat

(c) Household plumbing

(d) Thrown curveball, top view

FIGURE 10-9. Other examples of Bernoulli's principle. (a) A ski jumper turns his body into an airfoil to obtain vertical lift and jump farther. (b) A sailboat turns the force of wind into a pressure difference acting perpendicularly to the sail. (c) A trap in household plumbing maintains water in a pipe to prevent odors from traveling through the system. Flushing water could lower pressure by the Bernoulli effect, drawing water out of the trap, so vents must be provided to maintain outside air pressure on both sides of the trap, even with fast-running water in the connecting pipes. (d) The seams of a baseball move a layer of air around the ball as it spins, but the leading edge of the ball moves through the air faster, causing a sideways deflection.

THE IDEAL GAS LAW

The topics we have discussed in the last few sections—the pressure-depth relationship, Pascal's principle, Archimedes' principle, and the Bernoulli effect—apply to all fluids, whether liquid or gas. However, gases exhibit some unique behaviors that do not occur in liquids or solids.

During the eighteenth century, scientists made many experimental studies of the properties of gases. Two important results are:

1. If the pressure on a gas is held constant, the volume increases proportionally to the temperature (Charles's law).

2. If the temperature of a gas is held constant, the volume decreases proportionally with increases in pressure (Boyle's law).

Both of these laws (as well as others like them) conform with our everyday experience. If you squeeze a gas, its volume decreases (Boyle's law); if you heat a gas, it expands (Charles's law). Today, we summarize these kinds of results in a single equation, called the **ideal gas law.** As the name implies, this law does not apply exactly to any real gas, but it's a very good approximation for describing the behavior of all gases.

1. In words:

For a fixed amount of gas, the pressure, volume, and temperature of any gas are related to one another in a way consistent with Charles's and Boyle's laws.

2. In an equation with words:

For a fixed amount of an ideal gas, the product of pressure and volume divided by the temperature is a constant.

3. In an equation with symbols:

$$\frac{PV}{T} = \text{Constant} \qquad \text{or} \qquad \frac{P_1 V_1}{T_1} = \frac{P_2 V_2}{T_2}$$

where P, V, and T are the pressure, volume, and temperature of the gas; the subscripts refer to these properties for two different conditions of the gas.

Notice the following features of the ideal gas equation:

1. If the temperature is held constant, then the product of pressure and volume does not change. Consequently, if the pressure increases, then the volume must decrease, and vice versa. This is Boyle's law.

2. If the pressure is held constant, then the equation says that the volume is proportional to the temperature. This is Charles's law.

3. When we're using the ideal gas law, temperature must be measured on the Kelvin scale. We describe the Kelvin scale in more detail in Chapter 11; for now, you just need to know that temperatures on this scale are measured in kelvins. Kelvins are equal in *size* to degrees on the common Celsius scale, but the Kelvin scale begins at absolute zero (the lowest temperature possible), which is $-273°$ on the Celsius scale. The freezing point of water is $0°C$ or 273 kelvins, while the boiling point of water is $100°C$ or 373 kelvins.

> **Develop Your Intuition:**
> **Checking Tire Pressure**
>
> It is important to keep the tires on your car properly inflated. Why shouldn't you measure that inflation immediately after a long drive?
>
> The reason is that friction with the road heats the tire. Since V is approximately constant, the increase in T creates a corresponding increase in P. This effect means that the pressure you read from a hot tire will be higher than it would be if the tire were cold.

⬤ ELASTICITY

All around us, every moment of our lives, materials respond to stress. The floor of your room bends slightly as it supports your weight. Plastic grocery bags stretch as you cram them full of food. The arteries in your body swell from your blood pressure. These responses and many more are examples of *elasticity,* which is a term used to describe the way that materials change shape under the influence of external forces. Physicists have studied this fundamental material property for more than 3 centuries.

Hooke's Law

A thin wire provides a simple, one-dimensional example of elasticity. If an external force is applied to a wire—for example, by hanging a weight from it—the configuration of the wire's atoms changes in response. The "springs" between the atoms stretch and the wire gets longer. This stretching of the material is described by *Hooke's law,* named after the English scientist Robert Hooke (1635–1702).

1. In words:

   The harder you pull on something, the more it stretches.

2. In an equation with words:

   The change in length of a material is proportional to the applied force.

3. In an equation with symbols:

$$F = k\Delta L$$

   where ΔL is the change in length and k is a constant that varies from one material to the next. Once the force is removed, the wire returns to its original length.

A rubber band provides a familiar example of Hooke's law. Pull on a rubber band and it stretches, pull harder and it stretches more, as the law says—but only to a point. If you stretch a rubber band (or anything else) too far, it loses its elasticity and won't stretch any farther. In fact, it may even break. Once this *elastic limit* is reached, Hooke's law no longer applies. The internal arrangements of atoms and chemical bonds have been permanently changed. Bonds have been broken and the material is not the same as it was when the stretching began.

Tension and Compression

Another situation in which internal bonds between atoms become important occurs when materials are required to carry loads.
For example, the floor on which you are sitting right now could well
be held up by a series of wooden or steel beams or perhaps by a
concrete slab. Every time you drive over a bridge, your weight
(and the weight of your car) is supported by beams in
the bridge. From the tallest skyscraper to the hum-
blest cottage, ceilings and floors carry weight and
support loads, and it is the interatomic "springs"
that actually make these systems work.

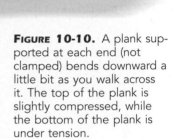

Take, for example, a plank supported at both
ends, as shown in Figure 10-10. You might come
across a plank like this being used as an impromptu
bridge on a wilderness trail or a construction site. When you stand on the plank, it
bows down in the middle, as shown. In this situation, the atoms along the top of
the plank are pushed together, compressing the interatomic bonds along the top
of the beam. We say that the top of the beam is under **compression.** At the same
time, the atoms along the bottom of the beam are pulled apart, causing the interatomic bonds there to stretch. The bottom of the plank is said to be under **tension.**
It is the combination of these forces—the springs along the top pushing and the
springs along the bottom pulling—that produces the force that supports your weight.

Over time, engineers have found that some materials, such as steel, support
tension better than other materials; this is why steel is used for suspension
cables in bridges and for other situations where tension is the main concern. Materials such as concrete support compression better than tension and are used
for the foundations of tall buildings, for dams, and for other compressive functions. Ultimately, the differences in behavior of these materials can be traced
back to the arrangements of their atoms and molecules.

FIGURE 10-10. A plank supported at each end (not
clamped) bends downward a
little bit as you walk across
it. The top of the plank is
slightly compressed, while
the bottom of the plank is
under tension.

Develop Your Intuition: Prestressed Beams

Concrete beams for highway bridges are often built so that they bow
upward in the middle before they are installed. This is called "pre-
stressed" concrete. Why is the beam made this way?

In this case, when the weight of a load is applied to the beam and it deforms downward, it will actually be straight. This minimizes the stress on the
beams when they carry the load.

Scaling

In Chapter 3 we talk about Galileo Galilei's contributions to experimental
physics. Galileo was an important figure in Italy, where he served as a consultant
to many important government offices. In particular, he was a consultant at the
Arsenal of Venice, the greatest naval shipyard of its time. People there were trying to build bigger ships, and they encountered a vexing problem: if they scaled
up the design of a successful smaller ship, there were all sorts of problems with
the result. Why wouldn't simple scaling work?

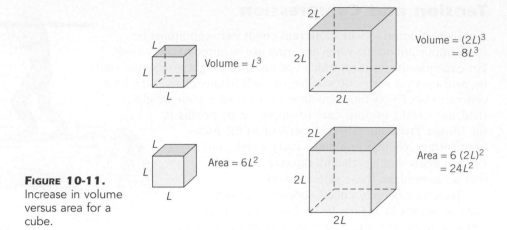

FIGURE 10-11.
Increase in volume versus area for a cube.

This question took Galileo into a study of the properties of materials and gave him an important insight into the relation between large and small structures. You can get a sense of his work by thinking about a simple problem. Imagine that you have a cube of material as shown in Figure 10-11 and you want to double its size. If the original cube is of length L on each side, then the doubled cube is of length $2L$.

This doubled cube is not as strong (for its size) as the original. Why not? Because the volume (and therefore the weight) of the doubled cube is eight times the weight of the original, but the surface area that must support that weight is only four times as big as the original. In other words, when we scale things up, the weight to be supported grows faster than the area available to support it.

Many consequences follow from this simple fact. One, relevant to the Physics Around Us section at the beginning of this chapter, is that an elephant is not only bigger than an ant, but it also looks different. The elephant's legs are much thicker in proportion to its body, for example. This difference arises because as the weight of an animal doubles, so must the cross-sectional area that carries the weight (in this case, the elephant's legs). The exact same principle applies to the size of columns that support tall buildings and the size of tires that support giant trucks.

THINKING MORE ABOUT

Modern Materials

Almost every material on Earth is easily classified as a solid, a liquid, or a gas, based on the arrangement and bonding of its atoms or molecules. But in the late nineteenth century, scientists synthesized an odd intermediate state of matter, called "liquid crystals." These materials are now used in many kinds of electronic devices, including the digital display of your pocket calculator.

The distinction between a liquid and a crystal is one of atomic-scale order: positions of atoms are disordered in a liquid, but ordered in a crystal. But think about what happens in the case of a liquid formed from very long and slender molecules. Like a box of uncooked spaghetti, in which the individual pieces are mobile but well oriented, these mol-

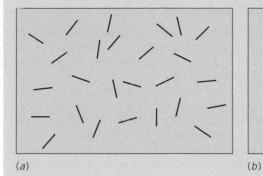

(a)   (b)

FIGURE 10-12. Schematic diagram of a liquid crystal. Under normal circumstances the elongated molecules are randomly oriented (a), but under special circumstances the molecules align in an orderly pattern (b).

ecules may adopt a very ordered arrangement even in the liquid form.

Under special circumstances (for example, when electricity passes near the liquid) all the molecules may align (Figure 10-12). This change in the arrangement of molecules causes a corresponding change in the physical properties of the liquid—a change that is used in a liquid crystal display.

Liquid crystals are also found in nature. Every cell membrane is composed of a double layer of elongated molecules called lipids (Figure 10-13). They separate the material inside the cell from its

surroundings just as effectively as if they were solid. Many scientists now think that these "lipid bilayers" originated in the primitive ocean as molecules similar to today's liquid crystals.

As with many other novel materials, liquid crystals were an unexpected laboratory discovery that eventually found important practical applications. Given such chance findings, what should be the relative importance of funding basic versus applied materials research? Should the federal government fund materials research or should corporate laboratories take the lead in these efforts?

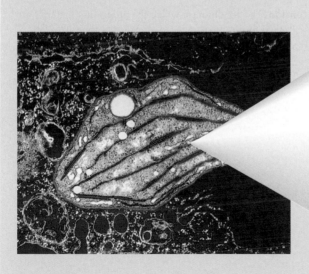

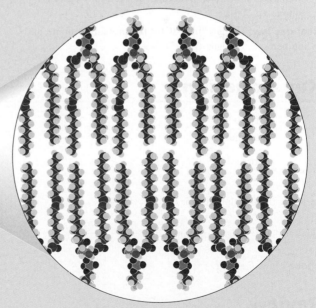

FIGURE 10-13. Cell membranes are composed of countless elongated lipid molecules that align into a bilayer, effectively separating the inside of a cell from the outside.

Summary

The **density** of a material, which relates to both the packing and the mass of a material's atoms, is defined as the mass of a sample of the material divided by its volume. **Pressure** is defined as a force divided by the area over which that force is exerted. Pressure in fluids is exerted by collisions between atoms or molecules and the container in which the fluid is held. The pressure in a fluid at any depth is determined by the weight of the fluid column above that depth—in a liquid, the pressure increases proportionately with depth. Changes in pressure are transmitted to all parts of a fluid, a rule known as **Pascal's principle.** Atmospheric pressure at sea level is approximately 14.7 psi or 0.1 MPa.

The upward force exerted on a body immersed in a fluid is equal to the weight of the fluid displaced (Archimedes' principle). This upward force is called the **buoyant force.**

The pressure on a surface decreases as the velocity of a fluid moving across that surface increases. This is called the Bernouilli effect and helps explain the phenomenon of **lift,** which enables airplanes to fly.

The behavior of the pressure, volume, and temperature of a gas is governed by the **ideal gas law,** which says that for a fixed amount of an ideal gas, the product of the pressure and volume is proportional to the temperature.

The change of length in a solid subjected to an outside force is proportional to the force applied (Hooke's law). Horizontal beams that support a load experience **compression** along their upper surfaces and **tension** along their lower surfaces. If an object is scaled up in size, the weight to be supported grows as the cube of the linear dimension, while the load-bearing surface grows only as the square.

Key Terms

Bernoulli effect The effect by which the pressure exerted by a fluid decreases as the fluid velocity increases. (p. 215)

buoyant force The upward force on an object due to the pressure of a fluid. (p. 213)

compression The condition in which the atoms of a material are squeezed closer together, due to an external force. (p. 219)

density The mass per unit volume of a substance; it is a measure of how much material is packed into a given volume. (p. 204)

ideal gas law The law that relates the pressure, volume, and temperature of a gas. (p. 217)

lift The net upward force on a wing due to the pressure difference between the top and the bottom of the wing. (p. 215)

Pascal's principle The statement that an increase of pressure of a static fluid in one place is transmitted immediately to every part of the fluid. (p. 211)

pressure A force divided by the area over which the force acts. (p. 207)

tension The condition in which the atoms of a material are pulled farther apart, due to an external force. (p. 219)

Key Equations

$$\text{Density} = \frac{\text{Mass}}{\text{Volume}}$$

$$\text{Pressure} = \frac{\text{Force}}{\text{Area}}$$

Pressure–depth relationship: $P = \rho \times d \times g$

Ideal gas law: $\dfrac{PV}{T} = \text{Constant}$ or $\dfrac{P_1 V_1}{T_1} = \dfrac{P_2 V_2}{T_2}$

Hooke's law: $F = k\Delta L$

Review

1. What is density? What are its units?

2. What two basic reasons account for the given density of a material?

3. Which is denser, a 2-foot square cube of balsa wood, or a 1-inch cube of the same wood? Explain.

4. What is pressure? What are its units?

5. How is the pressure at the bottom of a column of water related to its depth?

6. Why are dams much thicker at the bottom than at the top?

7. What is Pascal's principle? How is an increase in pressure in one part of a system transmitted to another part of a system?

8. How does the atmospheric pressure at sea level differ from the pressure at the top of a high mountain?

9. What does a barometer measure? How does it work?

10. What is Archimedes' principle?

11. How does a hydraulic lift work? Explain in terms of Pascal's principle.

12. What are some practical devices that utilize Pascal's principle, other than the hydraulic lifts mentioned in the text?

13. What is buoyancy? Explain in detail how an object's density determines whether or not it will float.

14. Use your own words to explain just how Archimedes was able to determine whether the king's crown was really gold.

15. What is Bernoulli's principle? How does this account for the lift that keeps an airplane aloft?

16. What is Boyle's law? What is Charles's law?

17. What is the ideal gas law?

18. Describe Hooke's law. What is the equation for it and what implications does it have for an object's breaking point?

19. What is meant by the elastic limit of an object? What happens to a material when stretched beyond this limit?

20. What is compression at the atomic level? What is tension? What forces are involved in each case?

21. Why can't you make something larger and stronger simply by scaling it up in size using the same proportions? Could an ant be made the size of an elephant while keeping its same proportions?

Questions

1. Is knowing only the density of a material enough to identify uniquely a material of unknown origin? Why or why not?

2. In everyday use, the word "dense" is often used interchangeably with the word "hard." In physics, density and hardness have completely different meanings. Which is denser, lead or diamond? Which do you think is harder?

3. Consider a cube of soft, spongy material. Which would result in a material with greater density, cutting out a piece of the cube that has one-eighth the volume (see figure), or compressing the cube until it has one-eighth the volume? Explain.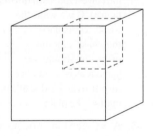

4. Consider two identical metal bottles that can be used to hold compressed gases. One is filled with air at atmospheric pressure, and the other is completely evacuated. Which bottle is heavier? Which bottle is denser? Explain.

5. Why are you weighed while submerged under water to determine your percentage of body fat?

6. Why do ice cubes float? Can you think of any possible ramifications for life in the oceans if this were not the case?

7. If you mixed oil and vinegar in one container, which would you expect to end up on top, and why?

8. Why does a plane extend flaps from its wings during take off and landing?

9. What happens when the angle of attack (the angle that a wing makes with the ground) increases as a plane climbs?

Is there an angle beyond which it becomes detrimental to the lift? Why might this happen?

10. Would the pressure at the bottom of a 3-foot holding tank be different if the tank held motor oil instead of water? Why or why not?

11. If you triple the depth of a column of fluid, what happens to the pressure at the bottom?

12. How does the pressure in a 3-foot-deep lake differ from the pressure in a 3-foot-deep hot tub 2 meters in diameter? Explain.

13. Why do some iron objects, such as iron ships, float when placed in water while other iron objects, such as nails, sink?

14. If the pressure on a gas in a flexible closed container is increased and the temperature remains constant, what happens to the volume of the gas? What has to be done to the gas while it is being compressed in order to maintain constant temperature?

15. Under constant pressure and with a constant amount of gas present, what happens to the volume of the gas if the temperature increases? Similarly, under the same conditions, what happens to the temperature if the volume of the gas is suddenly increased?

16. Which is likely to hurt more, having your bare foot stepped on by a 270-lb man wearing flat-soled loafers or having your foot stepped on by a 130-lb woman wearing high heels? Explain.

17. If you submerge a flexible air-filled balloon under water, what happens to the balloon's density? Why?

18. A flexible helium-filled party balloon is released in the atmosphere. As it gains altitude, what happens to the volume of the balloon? What about its density?

19. A helium-filled party balloon is released in the atmosphere. Imagine that the balloon is rigid, so that its volume cannot change. What happens to the buoyant force on the balloon as it gains altitude?

20. Suppose that a volleyball and a bowling ball are both completely submerged in water and have the same volume, as in the figure. (Of course, you would have to hold the volleyball beneath the water to keep it from popping up to the surface.) Which, if either, feels a greater buoyant force?

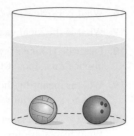

21. In the problem above, the volleyball is released and it floats up to the surface. The bowling ball, being denser than water, remains at the bottom (see figure). Which, if either, feels a greater buoyant force?

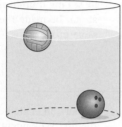

22. A 20-N rock hangs from a spring scale. The rock is lowered into a beaker of water that sits on another spring scale, but is not allowed to touch the bottom of the beaker (see figure). How do the readings on the two scales change?

23. Why do people who are rehabilitating from bone, muscle, and joint injuries often start their rehabilitation in pools? What role does buoyancy play?

24. How can huge cargo ships carry dense iron ore across the Great Lakes from the mines of Minnesota to various steel makers spread across the Midwest without the ships sinking from the weight of their loads? Why does such a ship ride higher in the water when empty?

25. Compare your ability to float in a very salty sea, such as the Great Salt Lake or the Dead Sea, to your ability to float in a fresh-water lake.

26. How does a baseball pitcher throw a curve ball? Explain how a ball curves in terms of the principles discussed in this chapter.

27. A fixed amount of helium gas is held inside a 1-liter container at a temperature of 25°C and atmospheric pressure. If the container expands to 2 liters without any change in temperature or amount of gas, what happens to the pressure and why?

28. Diamond is a hard transparent material made of only carbon atoms. Graphite is a black, soft material used to make pencil lead and is made of only carbon atoms. However, graphite and diamond have different densities. Explain how two materials made of identical atoms can have different densities.

29. In one scene in the movie *The Godfather II*, a solid gold phone is passed around a large table for everyone to see. Suppose the volume of gold in the phone was equal to the volume of a 10-centimeter cube of gold. Do you think such a phone could be casually passed around a table from hand to hand? Explain. (*Hint:* A 10-centimeter cube is 1 thousandth of a cubic meter. Look up the density of gold in Table 10-1.)

30. Helga says that although it's impossible to walk on a sea of water, it is possible to walk on a sea of mercury. She claims that if you step into a pool of mercury, you will only sink enough so that about half of your calf muscle is submerged. Ali disagrees with her statement. Who is right, Helga or Ali? Explain.

31. If you increase the temperature of a closed container of gas that has a fixed volume, the pressure inside will increase. The pressure on the walls of the container is due to the collisions of the gas molecules with the container wall. What must be happening to the gas molecules as the temperature is raised for them to exert a greater force on the walls?

32. A wedge-shaped piece of wood is floating in water with the widest part on the bottom and the narrowest part on top (see figure). If we want the wood to displace the least amount of water, should we leave it as is, should we turn it over, or doesn't it matter? Explain.

33. A 500-newton crate hangs from support beams as shown in the figure. State whether each beam is under compression or tension.

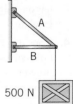

34. Suppose a small cube-shaped building is 10 meters on a side. What is the volume of this building? If it has a flat roof, what is the surface area of the five sides exposed to the outside? Answer the same questions for a building that is 5 meters on a side. Which building has a larger surface area to volume ratio? (The surface area to volume ratio is the surface area divided by the volume.)

35. Large things tend to have less surface area compared to their volume. Based on this fact, who is more likely to get cold in the winter, a fully grown man or a small child? Which is cheaper to heat in the winter, a single-family home or a living unit of the same size that is part of a large apartment building?

36. The average two-year-old boy is 36 inches (3 feet) tall and weighs 30 pounds. Suppose that when he is fully grown, he is 6 feet tall. If the rules of scaling apply, how much will he weigh when he is fully grown? Do you think the rules of scaling apply to growing humans?

Problem-Solving Examples

High-Pressure Footwear

A woman with a mass of 50 kg wears shoes with high heels. The square heels measure 0.5 cm on a side. At one point during her stride, all of her weight is on one heel. What is the pressure on the floor at that point?

SOLUTION: Pressure (in pascals, or newtons per square meter) equals force divided by area. The area of the heel is

$$A = 0.5 \text{ cm} \times 0.5 \text{ cm}$$
$$= 0.25 \text{ cm}^2$$
$$= 2.5 \times 10^{-5} \text{ m}^2$$

The force exerted is her weight, which is

$$F = m \times g$$
$$= 50 \text{ kg} \times 9.8 \text{ m/s}^2$$
$$= 490 \text{ N}$$

This means that the pressure is

$$P = \frac{F}{A}$$
$$= \frac{490 \text{ N}}{2.5 \times 10^{-5} \text{ m}^2}$$
$$= 2.0 \times 10^7 \text{ Pa}$$

This is a very high pressure, equal to about 3000 psi of floor. Even though the woman's weight is modest, the fact that it is applied over a small area means that the pressure is large. Engineers who design flooring materials know that they have to take effects like this into account so that high-heeled shoes don't damage the floor. ●

A Moving Air Parcel

The term "parcel" is used by meteorologists to refer to a volume of air with the same pressure and temperature throughout. Suppose a 1000-liter parcel of air at 1 atmosphere pressure and a temperature of 30°C rises up along the side of a mountain range. At the top, the pressure is now 0.75 atmosphere and the temperature is −10°C (cold enough for snow). What is the new volume of the parcel?

REASONING AND SOLUTION: Both the pressure and temperature have changed, so we need to use the ideal gas law for a fixed amount of gas. Remember that we have to use temperature in kelvins, so we first convert the given temperatures. The initial temperature is

$$T \text{ (in K)} = T \text{ (in °C)} + 273 = 30 + 273 = 303 \text{ K}$$

The final temperature is

$$T \text{ (in K)} = T \text{ (in °C)} + 273 = -10 + 273 = 263 \text{ K}$$

We solve the ideal gas law equation for the final volume, V_2, and then substitute the given data:

$$\frac{P_1 V_1}{T_1} = \frac{P_2 V_2}{T_2}$$

$$V_2 = V_1 \times \frac{P_1}{P_2} \times \frac{T_2}{T_1}$$

$$= 1000 \text{ L} \times \frac{1 \text{ atm}}{0.75 \text{ atm}} \times \frac{263 \text{ K}}{303 \text{ K}}$$

$$= 1157 \text{ L}$$

The decrease in pressure more than offset the decrease in temperature, and the air has expanded. ●

Problems

1. What is the weight of a column of water 5 feet high with a radius of 1 meter?

2. A perfectly spherical piece of metal is found at the bottom of a wishing well. If the mass of the object is 0.45 kg and the radius is 0.12 m, what is its density? If this object is pure gold, what is its mass? How much does it weigh?

3. Martin finds a piece of metal in a scrap yard and weighs it. Its mass is found to be 4740 kg and its volume is determined by immersion in water to be 0.6 cubic meters. What is the likely identity of this metal? What would the volume be of the scrap metal be if it had the same weight and were made of lead?

4. A water holding tank measures 100 m long, 45 m wide, and 10 m deep. Traces of mercury have been found in the tank, with a concentration of 60 mg/L. What is the total mass of mercury in the tank?

5. What is the mass of water required to fill a circular hot tub 3 meters in diameter and 1.5 meters deep? Do you think this will require special reinforcement of the floors if placed on the second floor?

6. A medic applies a force of 85 newtons to a 0.03-square-meter area that is bleeding. What is the pressure in pascals that she applies? What is the pressure in pounds per square inch? Is this a lot or just a little bit of pressure?

7. How much pressure in pascals is applied to the ground by a 104-kg man who is standing on square stilts that measure 0.05-m on each edge? How much pressure is this in pounds per square inch?

8. A column of water has a diameter of 2 meters and a depth of 10 meters.

    a. How much pressure in pascals is there at the bottom of the column?

    b. What is the weight of this column of water?

    c. What is the pressure if the column has a radius of 6 m? If it has a depth of 15 m?

    d. In what direction does this calculated pressure exist?

9. A dense plastic toy of mass 3 kg is floating just beneath the surface of a pond. What is the buoyant force on it?

10. A 2 × 4 (a piece of wood about 2 inches thick and 4 inches wide) from a nearby construction site floats near the shore of a lake. If it is floating in very calm water with half of its volume submerged and the other half of its volume just above the surface, what is the density of this 2 × 4? (Density of water = 1g/cm^3)

Investigations

1. Research the development of the modern submarine. How does a submarine control its buoyancy and hence its depth?

2. Often, local gyms and health clinics offer a body fat analysis in which they weigh you in water to get an estimate of your percentage of body fat. If you can, have this done to see what your percentage of body fat is. How does this method work? What is the procedure used and what principles explored in this chapter are fundamental to this technique?

3. Explore the problem of lift in airplanes. How did early airplane designers develop a wing that provided sufficient lift? What were some of the problems encountered and how were they solved?

4. Visit a construction site if you are able to and investigate what different types of materials are used and why they are used. What is built from concrete, what is built from steel and other metals and alloys, and what is built from wood and synthetic materials? How do the properties of these materials studied in this chapter affect their use? What are the trade-offs involved in terms of strength, cost, weight, and so forth, for the various materials you can see?

5. The *Alvin* was one of the first successful deep-sea diving submarines, exploring some of the deepest trenches in the Pacific. Investigate what were some of the obstacles in designing a vessel to withstand such depths. What materials were used to create a hull that was strong enough to withstand the pressure of the deepest trenches without being so heavy as to sink?

6. Muscles are remarkable for their elasticity. Use the Internet to investigate the biological materials that make up muscle fibers. What triggers muscle contractions? What physical changes do muscle materials undergo during this process?

7. Occasionally, large dams collapse, sometimes with catastrophic loss of life and property. Investigate the history of one such accident. What caused the dam to fail? What maintenance is required to keep the largest dams safe?

8. Investigate the design of a hydraulic lift near where you live or work. (You might find one as part of an elevator, at a service station, or on a piece of construction equipment). What fluid is used? How is the pressure generated? How much weight can it lift?

 # WWW Resources

See the *Physics Matters* home page at **www.wiley.com/college/trefil** for valuable web links.

1. **http://www.uncwil.edu/nurc/aquarius/lessons/buoyancy.htm** Simple lessons and activities regarding buoyancy and pressure, part of the physics of diving..

2. **http://www.walter-fendt.de/ph14e/buoyforce.htm** A Java applet for conducting a simple experiment for calculating the buoyant force applied to an object immersed in a liquid.

3. **http://www.walter-fendt.de/ph14e/hydrostpr.htm** A Java applet for conducting a simple experiment measuring hydrostatic pressure in liquids.

4. **http://www.allstar.fiu.edu/aero/Experiment1.htm** An experiment demonstrating the Bernoulli's effect.

5. **http://webphysics.ph.msstate.edu/javamirror/ntnujava/idealGas/idealGas.html** A Java applet simulating the relation between density, pressure, and partial velocity for an ideal gas.

6. **http://www.history.rochester.edu/steam/hero/** The "Pneumatics" of Hero of Alexandria, an ancient scientific treatise translated from the ancient Greek.

7. **http://zebu.uoregon.edu/nsf/piston.html** A simulation of the relationship between an ideal gas and pressure.

11

Heat and Temperature

KEY IDEA

Heat is a form of energy that can be transferred from one object to another because of a difference in temperature.

PHYSICS AROUND US . . . A Hot Summer's Day

The midday sun blazes as you walk across the blistered black asphalt toward your car. You wipe sweat from your brow, grab your keys, open the door, and feel a blast of hot air. As you slide inside, the seat sears your skin. The metal buckle of the seat belt burns. The suffocating heat assaults you.

But relief is at hand. The air conditioner kicks in, and within seconds refreshing waves of cool air wash over you. It feels so good!

Every day of our lives, in ways both big and small, heat flows around us. Heat flows from hotter objects such as door handles and seats to our cooler skin. Also, heat flows from our hot skin to the cooler air pouring out of the air conditioner. Little wonder, then, that scientists and engineers are fascinated by the movement of heat.

● TEMPERATURE

In Chapter 9 we explored different phases of matter—solids, liquids, and gases—as well as changes from one phase to another. In order to understand the causes of these dramatic transformations, we first have to examine the intertwined phenomena of heat and temperature. The differences between these two concepts were not clearly recognized until the early nineteenth century. However, understanding these differences led to important progress in our understanding of energy and its practical application in heat engines and machines.

Work, energy, inertia, momentum, force—over and over again in physics we encounter everyday words that have an exact scientific meaning. In this chapter we explore two more common terms, *temperature* and *heat,* that are often used interchangeably in day-to-day conversation, but which have very different scientific definitions. As you study this chapter, be sure to pay careful attention to the important distinctions between temperature and heat.

Temperature Defined

Temperature is one of the most familiar physical variables in our lives. You probably see or hear temperature measurements a dozen times a day. Weather reports document the current temperature and predict high and low temperatures for the coming week. You take your own body's temperature when you're feeling sick. And you set the temperature on your thermostat, your oven, and your refrigerator. As we'll see, temperature can be defined in different ways. But at the simplest level, **temperature** is a quantity that indicates how hot or cold an object is relative to a standard value. And, as we'll soon see, at a deeper level temperature is related to the speed at which the atoms and molecules in a substance are moving.

Temperature scales provide a convenient way to compare the temperatures of two objects. Several temperature scales have been devised over the centuries. All of these numerical scales are somewhat arbitrary, but every scale requires two easily reproduced temperatures for calibrating the measuring instrument (the thermometer). Two especially convenient standards are the freezing and boiling points of pure water. These reference temperatures are used today in the Fahrenheit scale (32°F and 212°F for freezing and boiling, respectively; see Figure 11-1) and the Celsius scale (0°C and 100°C for freezing and boiling, respectively). The Kelvin temperature scale, most commonly used in scientific research, also uses 100-degree increments between the temperatures of freezing and boiling water, but it defines 0 kelvin as **absolute zero,** which is the coldest possible temperature. Note that temperatures in the Kelvin scale are reported as kelvins (K), not as degrees Kelvin, for historical reasons we won't go into here.

Absolute zero is the temperature at which it is impossible to extract any energy at all from the vibrations of atoms or molecules. The temperature of absolute zero is approximately −273°C or −460°F. It turns out, therefore, that freezing and boiling of water occur at about 273 and 373 kelvins, respectively (Figure 11-1). We will come back to the idea of absolute zero in a little while; for now, we note that it is impossible to reach a temperature of absolute zero. Physicists have come very close, to

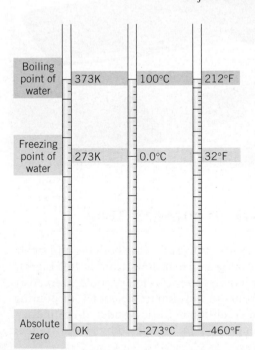

Boiling point of water — 373K — 100°C — 212°F

Freezing point of water — 273K — 0.0°C — 32°F

Absolute zero — 0K — −273°C — −460°F

FIGURE 11-1. The Fahrenheit, Celsius, and Kelvin temperature scales compared.

within one billionth of a degree. However, although an object can be at zero degrees or at a negative temperature on the Fahrenheit or Celsius scales, an object cannot achieve zero or a negative temperature on the Kelvin scale.

It's important to realize that while all of the conventional temperature scales involve the freezing and boiling points of water in some way, there is no reason that other fixed points could not be used. It's all just a matter of convenience. For example, until well into the twentieth century, brewers and distillers in parts of Europe used a temperature scale based on the freezing and boiling points of alcohol!

It's often necessary to convert from one temperature scale to another. American travelers, for example, often have to convert from degrees Celsius (used in most of the rest of the world) to degrees Fahrenheit. This conversion requires the following formula:

$$T \text{ (in °F)} = [1.8 \times T \text{ (in °C)}] + 32$$

The 1.8 in this formula reflects the fact that the Celsius degree is larger than the Fahrenheit degree by a factor of $\frac{9}{5}$ ($= 1.8$). There are 100 Celsius degrees between the freezing and boiling points of water, but 180 Fahrenheit degrees. Therefore the Fahrenheit degree is smaller. The 32 in the conversion formula reflects the fact that water freezes at 32°F but 0°C. To convert the opposite way, from Fahrenheit to Celsius, the formula is

$$T \text{ (in °C)} = \frac{T \text{ (in °F)} - 32}{1.8}$$

Scientists often convert between the Celsius and Kelvin scales, a simple process that just involves adding or subtracting 273:

$$T \text{ (in °C)} = T \text{ (in K)} - 273$$
$$T \text{ (in K)} = T \text{ (in °C)} + 273$$

Remember that all of these temperature scales measure the exact same phenomenon. They just use a different number scale to do so, much like measuring a distance in meters versus inches. Looking at Temperature on page 230 illustrates the range of temperatures physicists often deal with.

What To Wear?

You awake in Paris to a radio announcer forecasting a high temperature of 32°C. Should you wear gloves and an overcoat?

EXAMPLE
11-1

SOLUTION: Apply the equation to convert temperature from Celsius to Fahrenheit:

$$T \text{ (in °F)} = [1.8 \times T \text{ (in °C)}] + 32$$
$$= (1.8 \times 32) + 32$$
$$= 89.6°F$$

Looks like you won't need your overcoat today! ●

Thermometers

A **thermometer** is a device used to measure temperature. Most thermometers display temperature either digitally or on a numbered scale. Thermometers work by incorporating a material whose properties change with temperature. For example, many materials expand when heated and contract when cooled, and this

Looking at Temperature

Strange things happen to materials at very low and very high temperatures. We can't reach absolute zero, but we can come awfully close, to a billionth of a degree. Matter can exist in odd forms at these conditions, where atoms barely move at all. At 4 kelvins, helium becomes a liquid with unusual properties, such as spontaneously flowing up the walls of its container. At very high temperatures, solid rock melts and turns to lava; air conducts electricity, as in a lightning bolt; and hydrogen atoms fuse together, producing the energy in the Sun. Strange things indeed!

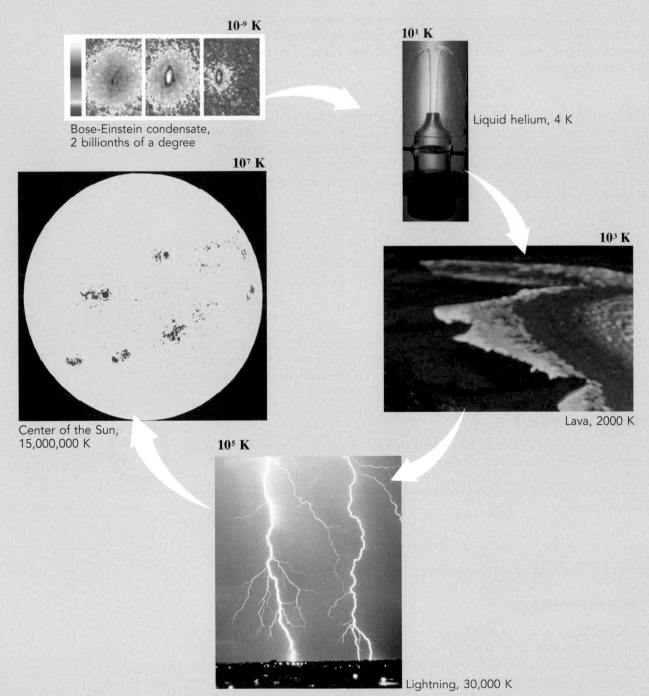

10^{-9} K

Bose-Einstein condensate,
2 billionths of a degree

10^1 K

Liquid helium, 4 K

10^7 K

Center of the Sun,
15,000,000 K

10^3 K

Lava, 2000 K

10^5 K

Lightning, 30,000 K

can be used to gauge temperature. In the old-style mercury thermometer, a bead of mercury expands with increasing temperature into a thin glass column; you read the height of the mercury against a scale marked in degrees. Many other (much safer) thermometers rely on changes in the electric properties of a temperature sensor, called a *thermocouple*.

One of the most visually intriguing types of thermometers, invented by Galileo Galilei in the early 1600s, uses changes in liquid density as a function of temperature. The Galilean thermometer consists of a large sealed flask with a liquid that changes density as it is heated. Suspended in this liquid are dozens of small numbered weights, each of a slightly different density. At low temperature, most of the weights rise to the top of the flask. As the temperature increases, the denser weights sink one by one to the bottom. The temperature is read simply as the lowest number on the weights that remain floating at the top of the thermometer.

The Atomic Basis of Temperature

We can use a thermometer to measure temperature, but what exactly is it that we are measuring? Temperature is actually a measure of the magnitude of the average speed of atoms and molecules, including their vibrations—that is, the kinetic energy of the particles that make matter. At lower temperatures a material's molecules have less kinetic energy than at higher temperature (Figure 11-2). On the other hand, if two materials are at the same temperature, then their atoms and molecules have the same kinetic energy. The important point here is that:

The higher an object's temperature, the faster its atoms or molecules move.

From this fact, it's clear that temperature is not a measure of the *total* kinetic energy of atoms and molecules in a substance. After all, if you double the amount of material, you double the total kinetic energy, even though the temperature remains the same.

This distinction between temperature and total molecular kinetic energy reveals an important point about using a thermometer. We use a thermometer to measure the temperature of a substance, such as a pot of water, the atmosphere,

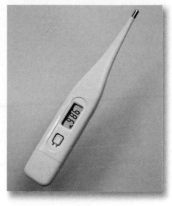

(a)

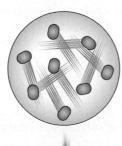

(b)

(a) A Galilean thermometer depends on the changes in density of a liquid with changes in temperature. (b) A modern digital fever thermometer depends on the changes in electrical properties of a crystal with changes in temperature.

Lower temperature, slower molecular speeds

Higher temperature, faster molecular speeds

FIGURE 11-2. Atomic view of temperature (cold and hot). Temperature measures the average kinetic energy of the molecules that make up an object.

or your mouth. It works like this: when you put a thermometer in your mouth, high-energy molecules in your body collide with lower-energy molecules in the thermometer. In the process, the molecules in the thermometer pick up energy and start to move faster. This process goes on until the molecules in the thermometer have the same average kinetic energy as those in your mouth, at which point no further change in temperature occurs. In other words, the thermometer is actually measuring its own temperature! The thermometer works perfectly well if the surroundings have much more mass (and thus more molecular kinetic energy) than the thermometer itself, because the thermometer can drain away some kinetic energy from that object without changing the object's temperature significantly. However, a thermometer can't be used to measure the temperature of an object such as a single drop of liquid, which is much less massive than the thermometer itself. In that case, the thermometer changes the temperature of the drop it's supposed to be measuring.

In the nineteenth century, scientists argued that there was a lowest possible temperature, called *absolute zero,* which would be reached when atoms came to a standstill. An atom, after all, cannot have a speed less than zero. With the advent of quantum mechanics (see Chapter 22), we recognized that atoms can never actually stop moving but that they can move into a state from which it is impossible to extract more energy. This is the modern definition of absolute zero, which is denoted as 0 K. Note that zero degrees in the Fahrenheit and Celsius scales does not correspond to any such milestone at the atomic level.

Thermal Expansion

Some important evidence for the link between temperature and atomic structure comes from the phenomenon called "thermal expansion." For example, think about the structure of a crystalline solid, as we described in Chapter 9. In such a solid, the atoms are separated from one another by springlike chemical bonds that vibrate about their equilibrium positions. As the temperature of the solid is raised, these vibrations become more energetic so that departures from the equilibrium positions become more pronounced. These motions, in turn, mean that each atom requires more room for its motion. As a result, the size of the crystal increases to accommodate this requirement. A similar argument could be given to show that most solids and liquids expand when heated.

This tendency for a heated object to expand is usually expressed in terms of a **coefficient of linear expansion,** which relates the length L of a material at one temperature to its length L' at another temperature.

1. In words:

The length of a solid object changes (usually expands) as temperature is raised.

2. In an equation with words:

The change in length is proportional to the change in temperature.

3. In an equation with symbols:

$$L' = L (1 + \alpha \Delta T)$$

where ΔT is the temperature change and α (the Greek letter alpha) is the coefficient of thermal expansion, a number that is different for different materials (Figure 11-3). This equation can also be written

$$\frac{\Delta L}{L} = \alpha \, \Delta T$$

where $\Delta L = L' - L$ is the change in length.

An important exception to this rule is water, which actually shrinks slightly when its temperature is raised from 0°C to 4°C, but expands for temperatures higher than 4°C. This fact is very important in nature because the shrinkage means that cold water is slightly more dense than warm water. As a result, when ponds of water begin to freeze over in the winter, the cold water at the surface sinks before turning to ice, allowing slightly warmer water to take its place. After all the water in the pond has reached 4°C, the water at the surface turns to ice first, since it floats on the denser 4°C water below it. The ice acts as further insulation for the water left in the pond. This makes it harder for ponds and lakes to freeze solid, which would be disastrous for the survival of fish and other aquatic organisms. Since biologists conjecture that life on Earth began in small pools and shallow lakes on the young planet, the thermal properties of water may have been an important factor in the early evolution of life.

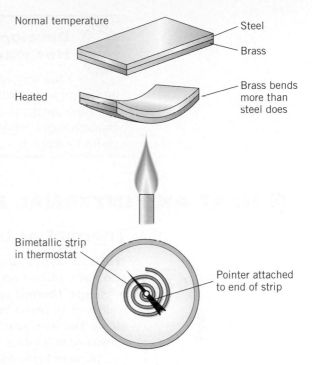

FIGURE 11-3. The thermostat in your home relies on the fact that different materials undergo different thermal expansions. At the heart of the thermostat is a bimetallic strip made of brass and steel. When heated, the brass expands more than the steel, causing the strip to bend. That bending, in turn, opens or closes an electrical switch that turns your furnace on or off.

A Bridge in the Summer

The coefficient of expansion for steel is 0.000056 per degree C. A particular bridge is 100 meters long when the temperature is 0°C. How long will that bridge be in the summer, when the temperature climbs to 40°C (over 100°F)?

EXAMPLE
11-2

SOLUTION: Putting the numbers into the thermal expansion equation, we find that the length will be

$$L' = L \,(1 + \alpha \, \Delta T)$$
$$= 100 \text{ m} \times [1 + (0.000056°\text{C}^{-1} \times 40°\text{C})]$$
$$= 100.22 \text{ m}$$

The length of a steel bridge, in other words, can change quite a bit from one season to the next—in this case by 22 cm. (almost 9 inches). Engineers take this fact into account by incorporating expansion joints into the bridge structure, like the one shown in the photo. As temperatures rise or fall these joints close together or open up. If the joints weren't there, the bridge might buckle on the first hot days of summer. ●

Expansion joints on a bridge allow the steel supports to expand in hot weather without buckling.

**Develop Your Intuition:
Hot Water and Bottle Caps**

One way of getting a hard-to-remove cap off a bottle is to hold it under hot water. Why does this work?

Both the metal cap and the glass bottle expand when heated, but the metal expands more. Under the hot water, the cap expands, becomes looser, and is easier to remove.

HEAT AND INTERNAL ENERGY

Thermal or Internal Energy

"Heat" is one of those words like "energy" that has an everyday meaning, but also has a precise scientific definition. Physicists distinguish between two kinds of energy. **Thermal energy,** sometimes called **internal energy,** is the energy that resides in an object because of the motion and interactions of its molecules and atoms. The term **heat** is reserved for energy that is transferred from one body or substance to another due to a temperature difference.

We now know that atoms and molecules, the minute particles that make up all matter (see Chapter 9), constantly move and vibrate and, therefore, possess kinetic energy. If molecules in a material move more rapidly, they have more kinetic energy and they are capable of exerting greater forces on each other in collisions.

If you touch an object whose molecules are moving with greater kinetic energy than the molecules in your hand, the collisions that occur between molecules will transfer some of that energy to the molecules in your hand. As a result, you perceive the object to be hot. By the same token, if an object feels cold, then the molecules in your hand have greater average kinetic energy than the molecules in that object. What we call "heat" in everyday speech, therefore, such as the heat of boiling water, is actually a transfer of thermal (internal) energy, which is the kinetic energy of wiggling atoms and molecules. Since heat is energy in transit from a substance at a higher temperature to another substance at a lower temperature, matter itself does not contain heat.

By contrast, all objects, whether scalding hot or frozen solid, have internal energy—the total energy of all the atoms and molecules. That internal energy represents the sum of all the molecular motions (vibrations, rotations, and other movements), as well as energies associated with the interatomic forces that hold groups of atoms and molecules together. Any time a substance absorbs or emits heat, the internal energy of that substance changes. Usually, a change in internal energy causes a corresponding change in the material's temperature. Under special circumstances, however, as when ice melts at 0°C or water boils at 100°C, the absorbed heat helps to break interatomic bonds while the temperature remains constant—a phenomenon called a *change of phase* (see Chapter 9).

Heat

The most familiar everyday attribute of temperature is that it tends to even out. A hot cup of coffee gradually cools, while a glass of cold water slowly warms up to room temperature. In each case, and in hundreds of other occurrences in your everyday life, some of a material's molecular kinetic energy is transferred from one substance to another as temperature evens out, a process that involves what we have defined as heat.

The transfer of energy by atomic collisions goes on all the time because atoms never sit still. If you've ever tried to warm a house during a cold winter day you've experienced this process. The thermal energy in the house gradually moves (as heat) to the cooler outside surroundings. Then if you turn off the furnace, the house begins to get cold. The only way to stay warm is to keep the furnace on.

Like a furnace, our bodies constantly convert chemical energy to thermal energy to maintain our core body temperature close to 98.6°F (37°C). Both your furnace and your body produce thermal energy on the inside, and that energy inevitably flows to the cooler outside as heat.

Heat is often measured in a unit called the *calorie*, which is defined as the amount of heat required to raise the temperature of 1 gram of water by 1 degree Celsius. The more familiar "Calorie" (with a capital *C*) is the unit used to measure stored energy in food and is equal to 1000 calories (with a lowercase *c*), or 1 kilocalorie. Since heat is a form of energy, it can also be measured using the SI unit joule or kilojoule. In the English system, heat energy is measured using the British thermal unit (BTU), which is defined as the amount of energy needed to raise the temperature of 1 pound of water by 1 degree Fahrenheit (Figure 11-4). Be sure to use the proper conversion factors when dealing with all these different common units for heat.

FIGURE 11-4. Comparison of heat units. One BTU is about the energy of a kitchen match.

Heat Capacity and Specific Heat

Heat capacity is a measure of how much heat an object can absorb. An object with a large mass has a higher heat capacity than an object of smaller mass. If we want to look at how much heat a particular kind of material can absorb, we need its specific heat. **Specific heat** is a measure of the ability of a material to absorb heat. It is defined as the quantity of heat required to raise the temperature of 1 gram of that material by 1°C. Water displays the largest specific heat of any familiar substance; by definition, 1 calorie is required to raise the temperature of 1 gram of water by 1°C. By contrast, you know that metals heat up quickly, so a small amount of heat can cause a significant increase in the metal's temperature.

Think about the last time you boiled water in a copper-bottom pot. It doesn't take long to raise the temperature of an empty copper pot to above the boiling point of water because copper, like most other metals, can't absorb much heat without having a large rise in temperature. In fact, 1 calorie of heat raises the temperature of a gram of copper by about 10°C. But water is a different matter; it must absorb 10 times more heat per gram than copper to raise its temperature by the same amount (Figure 11-5). Thus, even at the highest

FIGURE 11-5. Heat capacities of water and copper. The pot gets hot quickly, so most of the heat goes to boil the water.

stove setting, heating a pot of water to boiling can take several minutes. This ability of water to absorb or release large amounts of heat plays a critical role in the Earth's climate, which is moderated by the relatively steady temperature of the oceans.

Connection

The Heat Capacities of the Land and Oceans

Earth's climate is affected by many factors, but one important ingredient is the thermal energy stored in the oceans. During warm weather, sunlight falling on the ocean raises the water's temperature. Since water has a high specific heat and the ocean has a huge amount of mass, it takes a long time for the ocean to warm up and a correspondingly long time for it to cool off. Materials on land, however, have a lower specific heat and only a relatively thin layer of land surface is heated by the Sun. Thus, the land warms up and cools off much more quickly than the ocean.

Several regular features of Earth's weather are a direct result of the relatively high heat capacity of Earth's oceans. For example, the lands surrounding the Indian Ocean experience a seasonal wind pattern known as the *monsoon*. During the summer, the land on the Indian subcontinent warms up quickly while the ocean warms more slowly. The warm air over the land rises, drawing cooler air in from the ocean (Figure 11-6a). These winds carry a great deal of moisture after traveling over the ocean. When they rise over the heated land, they produce clouds and regular rainfall (the "rainy season").

On the other hand, in winter the ocean cools more slowly than the land. Thus, in winter the air over the ocean is warmer than the air over the land, and the wind pattern reverses. Warm air rises over the ocean, drawing in cooler, drier air from over the land (Figure 11-6b). The rains stop and a 6-month dry season begins. In this way, the weather experienced over a large part of our planet can be traced directly to the high heat capacity of water. ●

Weather patterns along an ocean coast are affected by the different heat capacities of land and water.

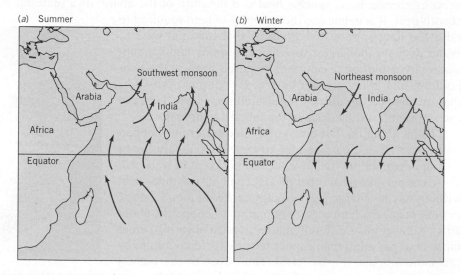

FIGURE 11-6. The Indian monsoon has a wet season and a dry season. (a) In summer, wet winds blow onshore toward the warmer land. (b) In winter, dry winds blow offshore toward the warmer ocean.

Develop Your Intuition: At the Beach

If you spend time along the beach, you have probably noticed that there is usually a breeze blowing out to sea in the early morning, but that the breeze calms down during the day and then blows in toward the land by late afternoon. Why does this happen?

At night, the land cools off. In the morning, then, the air over the ocean is warmer and it rises, drawing the cooler air from the land to replace it (Figure 11-7a). As the day passes, the land warms up, eventually becoming warmer than the sea. Then warm air rises over the land, and the (now relatively cooler) air from the ocean moves in to replace it (Figure 11-7b). The midday calm occurs when the two temperatures are about the same. Note that the same pattern of onshore and offshore breezes occurs on all coasts, but may happen earlier or later in the day, depending on the season and the location on Earth.

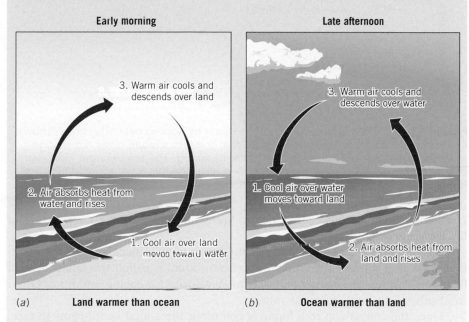

FIGURE 11-7. At the beach. (a) In the morning, the cool winds blow offshore; (b) by the afternoon and evening, the breeze has shifted to onshore.

Change of Phase

Specific heat gives us a good indication of how well any substance—gas, liquid, or solid—absorbs heat. However, if a change from one phase to another is involved, the situation is more complicated.

Suppose we take an ice cube from the freezer and transfer a small amount of heat energy to it each second. For a while, the added heat causes the temperature of the ice to rise, as shown in Figure 11-8. The molecules in the ice move faster and faster, but remain near their equilibrium positions. When the temperature reaches 0°C (32°F), however, another process comes into play. As heat energy flows into the ice, the bonds that hold the solid ice together begin to break.

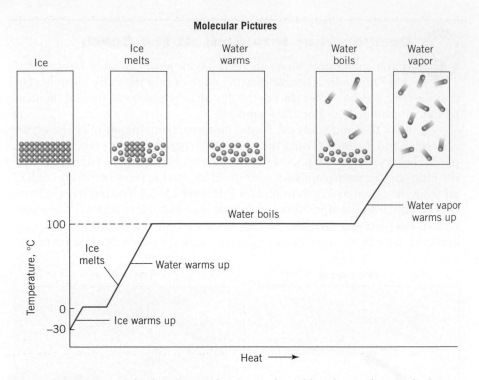

FIGURE 11-8. Ice is melted and turned to steam by adding heat. The graph shows the temperature versus heat added. Note that the temperature does not change while ice is melting, even though heat is being added. The temperature increases only after the ice has completely melted. A similar phenomenon occurs during the boiling of water.

Until enough energy has been added to break all the bonds, the temperature of the ice remains at 0°C.

The amount of heat needed to rearrange 1 gram of molecules from their configuration in a solid to the configuration in a liquid is called the **latent heat of fusion.** For water, this energy amounts to 80 calories for each gram of ice. Other solid substances have different values.

Once the conversion to a liquid is complete, the added heat begins to raise the temperature of the liquid water, as shown in the figure. (The fact that the slopes of the lines in Figure 11-8 are different for ice and liquid water is a reflection of the fact that solid ice and liquid water have different specific heats.) The molecules jitter faster and faster as the temperature rises. At 100°C (212°F), another phase transition takes place as the liquid water changes to water vapor. Once again, the liquid temperature remains constant while enough energy is added to the system to effect the change of phase. The amount of heat required to convert 1 gram of a liquid to a gas is called the **latent heat of vaporization** and is 540 calories per gram for water. Once all the water is in vapor form, there are no more changes of phase, and the normal increase of temperature with added heat goes on.

Note that the reverse processes—removing heat from steam to make liquid water and removing heat from liquid water to make ice—proceed in exactly the reverse order. The temperature remains fixed while enough energy is extracted from the system to allow the molecules to enter their new phase. Only when the change of phase has been completed does the temperature start to fall again.

Develop Your Intuition: Evaporation

People sweat, and the more active they are, the more they sweat. We've all seen pictures of athletes during a time-out with perspiration pouring off their bodies, soaking their uniforms. Why do people sweat? What purpose does it serve?

You can think about evaporation as a change of phase from liquid to gas. When water from your body's sweat glands comes through the pores of your skin to the surface, the water evaporates; that is, it turns from liquid water to water vapor. As it does so, the water absorbs heat from your body in an amount equal to the latent heat of vaporization. In other words, the evaporation of sweat from your skin acts to cool your body and keep your overall body temperature within normal bounds.

Many animals do not have sweat glands and cannot perspire from the skin. Instead, they must find other ways to cool their body temperature. For example, dogs pant heavily, allowing evaporation through the mouth and lungs.

In hospitals, patients running a high fever are sometimes given a rubdown with rubbing alcohol. Water has a higher heat of vaporization than alcohol, so it absorbs more heat per gram than alcohol does. However, alcohol evaporates very rapidly and so lowers the body temperature more quickly.

Physics in the Making

The Nature of Heat

What is heat? How would you apply the scientific method to determine its characteristics? That was the problem facing scientists 200 years ago.

They found that, in many respects, heat behaves like a fluid. It flows from place to place and seems to spread out evenly, like water that has been spilled on the floor. Some objects soak up heat faster than others, and many materials seem to swell up when heated, just like wood swells when it absorbs water. Thus, in the eighteenth century, after years of observations and experiments, many physicists mistakenly accepted the theory that heat is an invisible fluid, which they called "caloric." Supporters of the caloric theory of heat claimed that the best fuels, such as coal, are saturated with caloric, and they thought ice is virtually devoid of the substance.

Eventually, the caloric theory of heat was discarded as new observations failed to bear out the theory's predictions. In particular, the practical experience of machinists just did not support the idea of heat as a fluid. For instance, if heat is a fluid, then each object must contain a fixed quantity of that substance. However, Benjamin Thompson (later Count Rumford, 1753–1814), an eighteenth-century American who spent some time as a cannon maker, discovered that the amount of heat generated during cannon boring had nothing at all to do with the quantity of brass being drilled. Sharp tools, he found, cut brass quickly with minimum generation of heat, while dull tools made slow progress and produced prodigious amounts of heat. The amount of heat that could be generated seemed to be boundless. As long as the cannon borer was turned, it produced heat.

Thompson proposed an alternative hypothesis. He suggested that heat is nothing more than a consequence of the mechanical energy of friction, instead of an invisible fluid. He proved his point by immersing an entire cannon-boring

Benjamin Thompson (Count Rumford) helped disprove the caloric theory of heat by noting the transformation of mechanical energy into heat.

machine in water, turning it on, and watching the heat that was generated turn the water to steam. British chemist and popular science lecturer Sir Humphry Davy (1778–1829) further dramatized Thompson's point when he generated heat by rubbing two pieces of ice together on a cold London day. These demonstrations could not be explained by the caloric theory and marked the beginning of the end for that theory about the nature of heat. ●

The Mechanical Equivalent of Heat

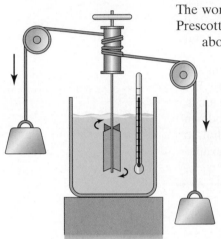

The work of Thompson, Davy, and others inspired the English researcher James Prescott Joule to devise an experiment to test the predictions of rival theories about the nature of heat. As shown in Figure 11-9, Joule's apparatus contained weights that were attached to ropes and lifted up. When the weight descended, the rope turned a paddle wheel that was immersed in a beaker of water. The weight had gravitational potential energy and, as it fell, that energy was converted into kinetic energy of the rotating paddle. The paddle wheel's kinetic energy, in turn, was transferred to internal energy of the agitated water molecules. As Joule suspected, the activity caused the water to heat up. Heat, he declared, is just another form of energy. In 1843, Joule even worked out how much heat was the equivalent of other kinds of energy. In modern units, Joule found that

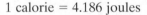

$$1 \text{ calorie} = 4.186 \text{ joules}$$

FIGURE 11-9. In his experiments, Joule demonstrated conclusively that heat is a form of energy. The kinetic energy of the turning paddle wheel heats the agitated water by a definite, measurable amount.

Joule's recognition of the equivalence of heat and mechanical energy was one of the great conceptual breakthroughs of physics. By disposing of the caloric theory once and for all, Joule showed that energy is the underlying concept governing all motions and interactions of objects. The study of energy—that is, *thermodynamics*—became one of the great fields of research for the rest of the nineteenth century. The practical benefits of this work included the improvement of the steam engine, the internal combustion engine, and electric generators—the sources of energy for modern industry and society. It is little wonder that the basic unit of energy was named for Joule.

● HEAT TRANSFER

Whenever two objects are at different temperatures, heat moves from the hotter to the cooler object. This movement of energy in the form of heat occurs all around us, all the time. You can't prevent heat from moving; you can only slow down its movement. In fact, scientists and engineers have spent many decades attempting to improve thermal insulation by studying the phenomenon known as **heat transfer**—the processes by which heat moves from one place to another. Heat transfer occurs by three mechanisms: conduction, convection, and radiation.

Conduction

Have you ever reached for a pan on a hot stove, only to burn your fingers when you grasped the metal handle? If so, you have experienced **conduction,** which is the movement of heat by atomic collisions.

Conduction occurs because of the motion of individual atoms or molecules. If a piece of metal such as a pot is heated at one end, the atoms at that end begin to move faster. When they vibrate and collide with atoms farther away from the heat source, they are likely to transfer kinetic energy to those atoms, so that those molecules begin moving faster as well. A chain of collisions occurs, with atoms progressively farther and farther away from the hot end moving faster (Figure 11-10).

To an outside observer, it appears that heat somehow flows like a liquid from the hottest part of the metal pot into the handle. There's nothing particularly mysterious about this process. Heat conduction is a result of collisions between vibrating atoms or molecules. When a fast-moving object collides with a slow-moving one, the fast object usually slows down and the slow object usually speeds up.

Steelworkers test a molten sample from the furnace. Because heat transfers from the hot end of the rod to the cool end by the process of conduction, the worker at the left must wear thick insulated gloves to hold the cool end of the rod.

When we pay our home heating bills, we are in large measure paying for the conduction of heat. The process works like this: In the winter the air inside the house is kept warmer than the air outside, so the molecules of air inside are moving faster than those outside. When these molecules collide with materials in the wall (a windowpane, for example), they impart some of their energy to the molecules in the wall. At that point, conduction takes place in the wall itself and the heat is transferred to the outside of the wall. There, the heat energy is transferred outdoors mostly by convection and radiation, which are processes that we will describe in a moment. In essence, your house becomes a kind of conduit: heat flows from the interior to the outside.

One way of slowing down the flow of heat out of a house is to add insulation to the walls or use thermal glass for the windows. Both of these materials are effective because of their low **thermal conductivity,** which is their ability to transfer heat energy from one molecule to the next by conduction. Have you ever noticed that a piece of wood at room temperature feels "normal," while a piece of metal at the same temperature feels cold to the touch? The wood and metal are at exactly the same temperature, but the metal feels cold because it is a good heat conductor; it moves heat rapidly away from your skin, which is generally warmer than air temperature. The wood, on the other hand, is a good *heat insulator;* it impedes the flow of heat and so it feels normal. You wouldn't want to live in a house made entirely of metal—we usually use relatively good insulators such as wood or masonry as primary building materials. The insulation in your home, especially in newer homes, is designed to have especially low thermal conductivity, so that heat transfer is slowed down (but never completely stopped). Thus, when you use special insulated windowpanes or put certain kinds of insulation in your walls, you make it more difficult for heat to flow outside, and thereby you use heat more efficiently. The result is that you need less energy to heat your house.

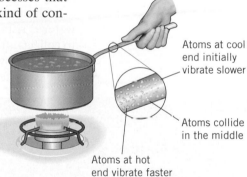

Atoms at cool end initially vibrate slower

Atoms collide in the middle

Atoms at hot end vibrate faster

FIGURE 11-10. Conduction transfers heat by the collision of atoms in a heated object.

Convection

Let's look carefully again at a pot of boiling water on the stove, as in Chapter 9. On the surface of the water you see a rolling, churning motion as the water moves and mixes. If you put your hand above the water, you feel heat. Heat has moved

FIGURE 11-11. Convection in a pot of water transfers heat by the bulk motion of the water.

from the water at the bottom of the pot to the top by **convection,** which is the transfer of heat by the bulk motion of a liquid or gas, as shown in Figure 11-11. Convection may be "forced"—for example, when a fan circulates cool air—or it may be driven by gravity, as in our pot of boiling water.

Water near the bottom of the pot expands as the flames heat it. Therefore, it weighs less per unit volume than the colder water immediately above it. A situation such as this, with colder, denser water above and warmer, less dense water below, is unstable. Under the force of gravity, the denser fluid tends to descend and displace the less dense fluid, which in turn begins to rise. Consequently, the warm water from the bottom rises to the top as shown, while the cool water from the top sinks to the bottom.

In convection, masses of fluid move in bulk and carry the fast-moving molecules with them. Heat moves by the actual physical motion of these masses of fluid.

Convection is a continuous, cyclic process as long as heat flows into the fluid. As cool water from the top of the pot arrives at the bottom, it is heated by the burner. In the same way, when the hot water gets to the top, its heat flows off into the air. The water on the top cools and contracts, while the water on the bottom gets hotter and expands. The original situation repeats continuously, with the less dense fluid on the bottom always rising and the more dense fluid on the top always sinking. This transfer of fluids results in a kind of rolling motion, which you see when you look at the surface of boiling water. Each of these regions of rising and sinking water is called a **convection cell.**

The areas of clear water that seem to be bubbling are the places where warm water is rising. The places where old bubbles (and scum, if the pot is dirty) tend to collect—the places that look rather stagnant—are where the cool water is sinking. Heat flows from the burner through the convection of the water and is eventually transferred to the atmosphere.

A heat island or city creates weather patterns inland. Warm air rises up and cooler air moves in to take its place.

Convection is a very efficient way of transferring heat. If you carefully study water in a pot on the stove you will notice that for a while the temperature at the surface of the water doesn't change appreciably. During this period, the heat transfers to the surface by the rather slow process of conduction through the water molecules. Eventually, when the temperature difference between the top and bottom becomes large enough, the water starts to move. At this point convection takes over and transfers the heat through the water.

Convection is a very common process in nature. From the small-scale circulation of cold water in a glass of ice tea to air rising above a radiator or toaster to large-scale motions of Earth's atmosphere and ocean currents, convection is at work. You may even have seen convection cells in operation in large urban areas. When you're in the parking lot of a large shopping mall on a hot summer day, you can probably see the air shimmer. What you are seeing is air, heated by the hot asphalt, rising upward. Some place farther away, perhaps out in the countryside, cooler air is falling. The shopping center with all its concrete is called a "heat island" and is the source of upward-rising air. It is the hot part of a convection cell, while the rest of the convection cell is the downward-flowing air elsewhere.

You may have noticed that the temperature in big cities is usually a few degrees warmer than in the outlying suburbs. Cities influence their own weather through the creation of convection cells. Rainfall is typically higher in cities than in the surrounding areas precisely because the warmer cities result in convection cells that draw in cool moist air from the surrounding areas.

Connection

Home Insulation

Today's homebuilders take heat convection and conduction very seriously. An energy-efficient dwelling has to stay warm in the winter and remain cool in the summer. A variety of high-tech materials provide effective solutions to this insulation problem.

Fiberglass, the most widely used insulation, is made of loosely intertwined strands of glass (Figure 11-12). It works by taking advantage of the fact that motionless air is an outstanding insulator. By trapping air into pockets, fiberglass minimizes the opportunities for conduction and convection of heat out of your home. Solid glass, by itself, is a rather poor heat insulator, but it takes a long time for heat to move along a thin, twisted glass fiber and even longer for heat to transfer across the occasional contact points between pairs of crossed fibers. Furthermore, a clothlike mat of fiberglass disrupts airflow and prevents heat transfer by convection. A thick continuous layer of fiberglass in your walls and ceiling thus acts as an ideal barrier to the flow of heat.

FIGURE 11-12. The interweaving of fibers in an ordinary piece of fiberglass, magnified in this photomicrograph, reduces heat transfer by convection and conduction.

If our houses were constructed with solid walls, then fiberglass would serve all our needs. However, windows pose a special problem. Have you ever sat near an old window on a cold winter day? Old-style single-pane windows conduct heat rapidly, as you can tell by putting your hand on such a window in winter. But how do we let light in without letting heat out? One solution is dual-pane windows with sealed, airtight spaces between the panes that greatly restrict heat conduction. In addition, builders employ a variety of caulking and foam insulation to seal any possible leaks around windows and doors, reducing drafts and resultant heat loss by convection. As a result, modern homes can be almost completely airtight (although some air has to be let in and out to prevent the house from becoming too stuffy). ●

Connection

Animal Insulation: Fur and Feathers

Houses aren't the only places where heat insulation is important. Birds and mammals maintain constant body temperatures despite the temperature of their surroundings, and both have evolved methods to control the flow of heat into and out of their bodies. Part of these strategies involve their natural insulating materials, including fur, feathers, and fat. Since most of the time an animal's body is warmer than the environment, insulation usually works to keep heat in.

Whales, walruses, and seals are examples of animals that have thick layers of fat to insulate them from the cold arctic waters in which they swim. Fat is a

Animals use feathers and fur for insulation.

poor conductor of heat, and it plays much the same role in their bodies as the fiberglass insulation in your attic.

Feathers also provide insulation; in fact, most biologists believe that feathers evolved first to help birds maintain their body temperature and were only later adapted for flight. Feathers are made of light, hollow tubes connected to each other by an array of small interlocking spikes. They have some insulating properties themselves, but their main effect comes from the fact that they trap air next to the body, and stationary air is an excellent heat insulator.

Birds often react to extreme cold by contracting muscles in their skin so that the feathers fluff out. This action has the effect of increasing the thickness (and hence the insulating ability) of the layer of trapped air. (Incidentally, birds need insulation even more than we do because their normal body temperature is 41°C or 106°F.)

Hair (or fur) is actually made up of dead cells similar to those in the outer layer of the skin. Like feathers, hair serves as an insulator and traps a layer of air near the body. In some animals (such as polar bears) the insulating power of the hair is increased by the fact that each hair contains tiny bubbles of trapped air. The reflection of light from these bubbles makes polar bear fur appear white—the strands of hair are actually translucent.

Hair grows from follicles in the skin, and small muscles allow animals to make their hair stand up to increase its insulating power. Over time, human beings have lost much of their body hair, as well as the ability to make most of it stand up. However, our mammalian nature is revealed in the phenomenon of goose bumps, in which muscles in the skin attempt to make the nonexistent hair stand up.

The main purpose of human clothing is to trap air near the body for insulation, just as animals and birds use fur or feathers. (This function of clothing was first recognized by Benjamin Thompson, who helped determine the nature of heat; see Physics in the Making on page 239.) Indeed, the earliest forms of clothing consisted of furs taken from animals hunted for that purpose. ●

Radiation

Everyone has experienced coming in on a cold day and finding a fire in the fireplace or an electric heater glowing red-hot. The normal reaction is to walk up to the source of heat, hold out your hands, and feel the warmth on your skin. But how does the heat move from the fire to your hands? It can't do so by conduction—it's too hard to transfer heat through the air that way. It can't be convection either, because you don't feel a hot breeze. The air in the room is almost stationary.

What you experience is the third kind of heat transfer—**radiation,** or the transfer of heat by *electromagnetic radiation,* which is a kind of wave that we discuss in Chapter 19. A fire, an electric heater, and the Sun all transfer heat in this form. This radiation travels like light from a hot source to your hand, where it is absorbed and converted into internal energy. You feel the warmth because of the energy that the radiation carries to your hand (see Figure 11-13).

Every object in the universe radiates energy. Under normal circumstances, as an object gives off radiant energy to its surroundings, it also receives radiant energy from the surroundings. Thus, a kind of equilibrium is set up and there is no net loss or gain of energy because the object is at the same temperature as its surroundings. It receives the same amount of energy as it radiates. However, if the object is at a higher temperature than its surroundings, it radiates more energy than it receives. Your body, for example, constantly radiates heat into its cooler surroundings. This energy can be detected easily at night with infrared goggles. You continue to radiate this energy as long as your body processes the food that keeps you alive.

Radiation is the only kind of energy that can travel through the emptiness of space. Conduction requires atoms or molecules that can vibrate and collide with each other. Convection requires atoms or molecules of bulk fluid, so that they can move. But radiation doesn't require anything in the environment to transport it; radiation can even travel through a vacuum. The energy that falls on Earth in the form of sunlight—almost all the energy that sustains life on Earth—travels through 93 million miles of intervening empty space in the form of radiation.

In the real world, all three types of heat transfer—conduction, convection, and radiation—occur constantly. Think about the heat generated by your body: heat conducts through your bones and teeth, heat convects as your blood circulates (an example of forced convection), and heat radiates from your skin into the cooler surroundings, where it is eventually absorbed. As your heat flows outward, it produces slightly higher temperatures in the surrounding air. The same three phenomena operate in a pot of boiling water. Heat moves through the metal bottom and sides of the pot by conduction, it moves through the boiling water by convection, and it moves from the sides and surface of the water by radiation. In fact, everywhere in the natural world, heat is constantly being transferred by these three mechanisms.

FIGURE 11-13. A fire transfers much of its energy by electromagnetic radiation.

Heat: Global Warming and the Greenhouse Effect

Heat constantly flows to Earth from the Sun, and it constantly flows outward from Earth into the cold blackness of space. Changes in the average temperature of Earth's surface arise from a complex combination of heat transfers that are not yet fully understood. But one possible source of change, called the greenhouse effect, is receiving a lot of attention.

Have you ever visited a greenhouse on a sunny winter day? It may be freezing cold outside, yet the temperature inside is much warmer. This contrast occurs because the Sun transfers heat by the radiation of visible light, which passes through the glass and warms up the interior. When the interior radiates away its own energy, it must do so by sending out infrared radiation. However, while the glass in a greenhouse is transparent to visible light, it is opaque to infrared radiation. The Sun's energy is thus trapped, warming the greenhouse until its temperature rises to the point where as much energy leaks out through the glass as comes in from the Sun.

In exactly the same way, Earth's atmosphere transmits the Sun's incoming radiation but some molecules in the air (particularly carbon dioxide) trap much of the Earth's heat as it flows away from the surface (Figure 11-14). These molecules play the same role for Earth that the glass does for the greenhouse, so the phenomenon has come to be known as the "greenhouse effect." Its effect is to warm the planet, just as the glass warms the greenhouse.

The greenhouse effect is vital to life on Earth. Without a layer of greenhouse gases, our planet would be a lifeless frozen ball. Modern concerns arise from the relatively rapid increase in atmospheric carbon dioxide during the past century—a consequence of burning fossil fuels such as coal, petroleum, and natural gas. Will this change in atmospheric composition cause a rise in global temperatures? Has such a change already begun?

The greenhouse effect is an area of continuing research in many fields of science. Physicists have examined the mechanisms of heat transfer in the atmosphere; chemists have studied which gases persist in the atmosphere and how they absorb heat; geologists have looked at rock and fossil formations to determine the history of Earth's temperature changes; and astronomers have simulated atmospheric models on other planets, such as Venus, which may have once had an accelerated greenhouse effect. Most experts now conclude that global warming is well under way and that the big question is not whether but how much global temperatures will rise over the next century. The best estimates are that the average global temperature will rise by a few degrees Celsius over that period.

If greenhouse warming is indeed happening, what sorts of things do you want to know about its possible consequences? How much in the way of consequences are you willing to accept in order to keep the convenience of using fossil fuels?

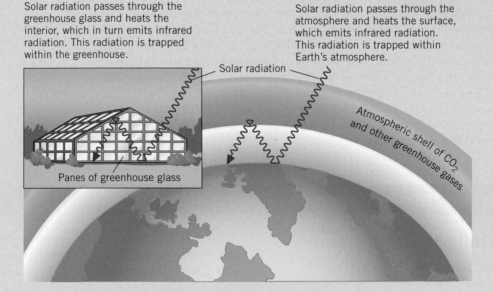

Solar radiation passes through the greenhouse glass and heats the interior, which in turn emits infrared radiation. This radiation is trapped within the greenhouse.

Solar radiation

Solar radiation passes through the atmosphere and heats the surface, which emits infrared radiation. This radiation is trapped within Earth's atmosphere.

Panes of greenhouse glass

Atmospheric shell of CO_2 and other greenhouse gases

FIGURE 11-14. The greenhouse effect. Just as the Sun's energy passes through the glass of a greenhouse and becomes trapped inside as heat, so too does the atmosphere act as a greenhouse to warm up Earth.

Summary

In everyday life, **temperature** is a quantity that indicates how hot or cold an object is relative to a standard value. All objects in the universe are at a temperature above **absolute zero;** thus, they hold some internal energy—the kinetic energy of the moving atoms. A **thermometer** is a device that measures temperature, according to a **temperature scale.**

Heat is a transfer of energy that moves from an object at a higher temperature to an object at a lower temperature.

Objects themselves contain **thermal energy** (also called **internal energy**), which is the kinetic energy of wiggling atoms and molecules. Most solids and liquids expand when heated; the amount of expansion is proportional to the **coefficient of linear expansion.**

   Specific heat is a measure of the ability of a material to absorb heat and is defined as the quantity of heat required to raise the temperature of 1 gram of that material by 1°C. An object's overall ability to absorb heat is measured by its **heat capacity.** When a change of phase is involved, the temperature of an object remains fixed while the bonds between its atoms and molecules are rearranged. The **latent heat of fusion** is the heat required to change 1 gram of a material

from a solid to a liquid, and the **latent heat of vaporization** is the heat required to change 1 gram of a substance from a liquid to a gas.

   There are three modes of **heat transfer** between two objects at different temperatures. **Conduction** involves the transfer of heat energy through the collision of individual atoms and molecules. **Thermal conductivity** measures the ability of substances to transfer heat. **Convection** involves the motion of a mass of fluid in a **convection cell,** in which atoms are physically transported from one place to another. Heat can also be transferred by **radiation,** which is electromagnetic energy that can travel across a room or across the vastness of space until it is absorbed.

Key Terms

absolute zero The lowest possible temperature, at which no energy can be extracted from atoms. (p. 228)

coefficient of linear expansion A quantity that relates the temperature change with the corresponding length change of a material. (p. 232)

conduction The transfer of heat due to atomic or molecular collisions. (p. 240)

convection Heat transfer due to the motion of a liquid or a gas. (p. 242)

convection cell A region of a fluid that is either rising or sinking due to the heat convection process. (p. 242)

heat The energy transferred from one body to another due to a difference in temperature between the two bodies. (p. 234)

heat capacity A measure of the change in temperature of an object on adding or removing heat; the amount of heat required to raise the temperature of the object by 1°C. (p. 235)

heat transfer The process by which thermal energy moves from one place to another. (p. 234)

latent heat of fusion The amount of heat required to change 1 gram of a solid material to a liquid when the solid is at its melting temperature; equivalently, the amount of heat that must be removed from 1 gram of liquid material to turn it

into a solid when the liquid is at its freezing temperature. (p. 238)

latent heat of vaporization The amount of heat required to change 1 gram of a liquid material to a gas when the liquid is at its boiling temperature; equivalently, the amount of heat that must be removed from 1 gram of a gaseous material to turn it into a liquid when the gas is at its condensation temperature. (p. 238)

radiation Heat transfer due to the emission and absorption of electromagnetic waves between two bodies at different temperatures. (p. 244)

specific heat The quantity of heat required to raise the temperature of 1 gram of a material by 1°C. (p. 235)

temperature A quantity that reflects how vigorously atoms or molecules are moving and colliding in a material. (p. 228)

temperature scale A standard of measurement for estimating temperature; familiar examples are the Fahrenheit and the Celsius scales. (p. 228)

thermal conductivity The ability of a material to transfer heat. (p. 241)

thermal energy or **internal energy** The energy of an object that results from the vibrations of individual atoms and molecules. (p. 234)

thermometer A device used to measure temperature. (p. 229)

Key Equations

Temperature conversions:

$T \text{ (in °F)} = [1.8 \times T \text{ (in °C)}] + 32$

$T \text{ (in °C)} = \dfrac{T \text{ (in °F)} - 32}{1.8}$

$T \text{ (in °C)} = T \text{ (in K)} - 273$

$T \text{ (in K)} = T \text{ (in °C)} + 273$

Review

1. What is temperature? Is it a relative or absolute quantity? Explain.

2. Describe three common temperature scales. What fixed points are used to calibrate them?

3. How do you convert degrees Fahrenheit to degrees Celsius? To kelvins?

4. What is meant by absolute zero? What happens to the movement of molecules at this temperature?

5. What is the numerical value of absolute zero in degrees Celsius? In degrees Fahrenheit?

6. In your own words, define temperature. How is absolute zero related to your definition of temperature?

7. What is the general principle behind the working of a thermometer? Give examples of different types of thermometers.

8. How did Galileo's thermometer work? How does the density of the liquid used affect the location of different weights suspended inside the liquid?

9. How does the kinetic energy of the atoms of a substance affect its temperature? If two objects have the same temperature, what can you say about the average kinetic energies of their molecules?

10. Can a thermometer be used to measure the temperature of something much smaller than itself? Why or why not?

11. What is the difference between temperature and heat?

12. What is a calorie? How many calories of heat are required to raise the temperature of 5 grams of water by 1°C?

13. How many calories make up 1 Calorie (capital C), the unit by which we measure the energy in our diets?

14. What is a BTU? How do you convert BTUs to calories?

15. What is the relation between the perception that something is hot and the motion of the molecules in it?

16. What is the difference, if any, between heat and internal energy?

17. How does the discovery of heat as a form of energy illustrate the scientific method?

18. What is specific heat? Is it the same for all materials?

19. What is the specific heat of water?

20. How does the relatively high specific heat of water affect Earth's climate? Specifically, what role does it play in the creation of monsoons?

21. How does the specific heat of water affect the winds blowing at different times of the day near large bodies of water?

22. Describe the historical process by which the notion of heat as a form of energy was developed.

23. What is a joule? What does it measure? How many calories make up 1 joule?

24. Identify the three ways that heat can be transferred and give examples of each.

25. What kind of heat transfer depends only on the collisions between individual atoms and molecules?

26. What kind of heat transfer depends on the bulk motion of large numbers of molecules?

27. By what process can heat be transferred across the vacuum of empty space?

28. Describe the kinds of heat transfer that occur when you cook a meal. Where does the heat energy come from? Where does it end up?

29. What is a fluid? Are all fluids liquid?

Questions

1. A glass of water sits on a table. The temperature of the water is the same as that of the glass. What can you say about the average kinetic energy of the water (H_2O) molecules compared to the silicon dioxide (SiO_2) molecules that make up the glass? Which is moving faster, the silicon dioxide molecules or the water molecules? (*Hint:* Look at the periodic table to determine the mass of a SiO_2 molecule versus a water molecule.)

2. A golf ball is dropped onto hard ground and, after a few bounces, comes to rest. Use the atomic basis of temperature to explain why the golf ball's temperature is slightly higher after it comes to rest.

3. A square hole is cut out of a piece of sheet metal, as shown in the figure. When the temperature of the metal is raised,

the metal expands. What happens to the size of the square hole? (*Hint:* Break up the piece of metal into eight smaller square pieces of sheet metal, then raise the temperature, then put them back together.)

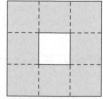

4. Suppose your gold wedding ring became stuck on your finger. Some home remedies suggest soaking your finger in ice water and then trying to remove the ring. If this remedy relies on thermal expansion effects, what does this tell you about the relative expansion coefficients of gold and your finger?

5. Two pieces of copper pipe are stuck together, as shown. One way to separate them is to run water at different tem-

peratures inside the inner pipe and over the outer pipe. Should the water running over the outer pipe be hotter or colder than the water running through the inner pipe? Explain.

6. Two thin strips of metal (A and B) are glued together at 0°C as shown in the figure. At 20°C they bend upward because the metals expand differently. Which metal, A or B, has a higher thermal expansion coefficient?

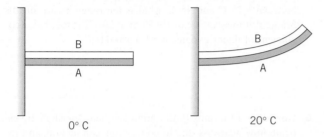

0° C   20° C

7. A mercury thermometer consists of a mercury-filled glass bulb that is connected to a narrow glass tube. Mercury thermometers are based on the thermal expansion of mercury: As the mercury expands, it rises up the tube. What can you say about the thermal expansion coefficient of mercury relative to the thermal expansion coefficient of glass? What would happen if their thermal expansion coefficients were the same?

8. A certain amount of heat is added to some water, and its temperature rises. The same amount of heat is added to a piece of aluminum with the same mass as the water. Compare the temperature change of aluminum to the temperature change of water.

9. If water had a lower specific heat, would your chances of enjoying a long, hot bath be greater or less? Explain?

10. Suppose a new liquid were discovered that is identical to water in every way except that it has a lower latent heat of vaporization. If you had to cook your pasta using either ordinary water or this new liquid, which would you choose and why?

11. Suppose a new liquid were discovered that is identical to water in every way except that it has a lower latent heat of fusion. Would it take a longer or shorter time to make ice out of this liquid in your freezer? Would this necessarily be a more desirable situation?

12. Suppose a new liquid were discovered that is identical to water in every way except that it has a lower specific heat. Consider taking a hot shower with this liquid. Would insulating the pipes from the hot water heater to the shower head be more or less important with this new liquid?

13. Why do some animals roll up into a ball when they are cold?

14. Why are feather beds warm, and why is goose down considered the best filling for a parka?

15. Imagine lying on a hot beach on a sunny summer day. In what different ways is heat transferred to your body? In each case, what was the original source of the heat energy?

16. Why do human beings wear clothes? Compare our behavior in this regard to other warm-blooded animals.

17. What is the difference between heat capacity and heat transfer?

18. Outline the three major modes of heat transfer. For each case, state if a medium is necessary, compare the motion of molecules in this medium, and relate the heat transfer to the presence or absence of a temperature difference in the medium.

19. The average specific heat of the human body is 83% of the specific heat of water. This value is higher than for most other solids, liquids, or gases. Why do you think the specific heat of the human body is closest to water?

20. Human beings must lose heat so their internal temperatures do not increase substantially above 37°C. The main mechanism for losing heat is sweating. Explain why this is an efficient mechanism to lose energy from the body.

21. Absolute zero, 0 K, is the lowest possible temperature. Temperatures below absolute zero do not exist. In terms of the molecular motion of a substance (a gas, for example), explain why there is a lowest temperature (absolute zero) but not a highest temperature.

22. Temperatures below absolute zero do not exist. However, suppose you are given an object that is right at absolute zero. Would it be possible to use this object to cool another object down to absolute zero?

23. Three identical potatoes are taken out of a hot oven to cool. The first is placed on the countertop. The second is wrapped in aluminum foil and placed inside a jar, and then the air is removed from the jar. The third potato is wrapped in aluminum foil and then placed on the countertop alongside the first. Can you place the potatoes in order of which will cool fastest?

24. One hundred grams of liquid A is at a temperature of 100°C. One hundred grams of liquid B is at a temperature of 0°C. When the two liquids are mixed, the final temperature is 50°C. What can you say about the specific heats of the two liquids? Explain your reasoning.

25. Two hundred grams of liquid A is at a temperature of 100°C. One hundred grams of liquid B is at a temperature of 0°C. When the two liquids are mixed, the final temperature is 50°C. Which material has a higher specific heat? Explain your reasoning.

26. On a very cold day you find that your key does not fit into your car door lock. Assuming this has happened because of thermal expansion effects, what can you say about the thermal expansion coefficient of your key relative to the thermal expansion coefficient of the lock?

Problems

1. Convert the following Fahrenheit temperatures to kelvins.

   a. 120°F   c. 11,500°F

   b. −40°F   d. −456°F

2. Convert the following Celsius temperatures to Fahrenheit.

   a. 300°C   c. 6,000°C

   b. −180°C   d. 40°C

3. Convert the following kelvin temperatures to Celsius.

   a. 80 K   c. 6000 K

   b. 300 K   d. 545 K

4. At what temperature is the Celsius and Fahrenheit value the same?

5. Convert 70°F to degrees Celsius and to kelvins.

6. The coefficient of linear expansion for a silver strip is $19 \times 10^{-6}/°C$. What is its length on a hot day when the temperature is 37°C if the strip is 0.20000 m long when it is −10°C?

7. If a 50-m steel footbridge experiences extreme temperatures between −15°C and 45°C, what is the range in size of this bridge if it measures exactly 50 m at 20°C? (Steel has a coefficient of linear expansion of 0.000011/°C)

Investigations

1. Research the daily high and low temperatures for the past week in a nearby big city and one of its surrounding smaller towns. On average, what is the difference in temperature? What causes this difference between the city and town?

2. Repeat Investigation 1, but compare the daily high and low temperatures for the past week in a coastal town with a nearby inland town.

3. Investigate the kind of insulation that is installed in your home. What could you do to improve your home's insulation?

4. Get an aluminum cup, a ceramic coffee mug, and a plastic drinking cup. Simply by feeling these objects, can you guess their relative thermal conductivities? What experiment could you perform to test your guess?

5. Visit a building supply store and look at the doors and windows they sell. What steps have manufacturers taken to reduce heat flow from homes?

6. How does a thermos bottle minimize heat transfer? Investigate how heat loss due to convection, conduction, and radiation is prevented by looking at the design of a thermos. Compare cheaper versions to the design of more expensive versions and explain the differences in terms of the concepts used in this chapter.

7. Boil a pan of water on your stove. Can you identify the convection cells in the pan? How hot does the water have to be before convection starts? (*Hint:* A small amount of food coloring can reveal the formation of convection cells.)

8. Investigate the history of temperature scales. Describe an obsolete temperature scale and its fixed points. Why was it abandoned? When was the Fahrenheit scale introduced, and what were its original fixed points?

 ## WWW Resources

See the *Physics Matters* home page at **www.wiley.com/college/trefil** for valuable web links.

1. **http://unidata.ucar.edu/staff/blynds/tmp.html** A tutorial on temperature from the National Center for Atmospheric Research.

2. **http://microgravity.grc.nasa.gov/combustion/index.htm** The NASA microgravity combustion science page describes the effects of microgravity on a candle flame, illustrating the effects of convection on combustion.

3. **http://www.psrc-online.org/classrooms/papers/coleman.html** A sample class on the physics of the greenhouse effect.

4. **http://jersey.uoregon.edu/vlab/Thermodynamics/** A simulation experiment on thermodynamic equilibrium of two ideal gases.

12

The First Law of Thermodynamics

KEY IDEA

The change in internal energy of a system is the difference between the heat added to the system and the work done by the system; thus mechanical energy is conserved.

PHYSICS AROUND US . . . The Great Circle of Energy

A few days ago you may have gone to your local gas station and filled up the tank of your car with gasoline. Have you ever wondered where that gasoline came from?

Hundreds of millions of years ago, a bit of energy was liberated from the incandescent core of the Sun. For thousands of years, that energy traveled outward to the Sun's surface. Then, in a mere 8 minutes, that energy made the trip through empty space to Earth in the form of sunlight, where it was absorbed by organisms known as algae floating on the warm ocean surface.

Through a process known as *photosynthesis,* the algae transformed the Sun's energy into the chemical energy of its complex molecules. Eventually the algae died and sank to the bottom of the ocean, where over millions of years those molecules were buried deeper and deeper. Under the influence of pressure and heat, the molecules were eventually transformed into energy-rich fossil fuel—petroleum.

Then, a short while ago, engineers pumped that petroleum with its stored chemical energy up out of the ground. At a refinery, the large molecules were broken down into gasoline and other compounds, and the gasoline was shipped to your town. The last time you drove you burned that gasoline, converting the stored energy into the engine's mechanical energy that moved your car.

What's remarkable about this story is that the mechanical energy and heat produced when your car burned that gasoline is exactly the same energy that was released by the Sun hundreds of millions of years ago. One of the great discoveries of physics is that energy can change forms many times over vast stretches of time, but the total amount of energy is constant.

251

THE MANY FORMS OF ENERGY

In Chapter 8 we introduced the notion of energy as the ability to do work or to exert a force over a distance. We discussed two kinds of energy—kinetic energy (or energy of motion) and gravitational potential energy (energy associated with being located at some height in a gravitational field). We saw that for these two types of mechanical energy (neglecting friction) two important principles hold. First, the energy of a system can change back and forth between these two forms, and, second, the total amount of kinetic and gravitational potential energy in an isolated system has to remain fixed.

In this chapter we examine many additional forms of energy other than mechanical energy. Despite this diversity in forms, however, the two general principles still hold: one form of energy can always be transformed into another, and the total amount always stays the same. As a result, we can extend conservation of energy to chemical systems and biological systems—to any system at all. In fact, in thermodynamics, we can define a *system* to be any collection of objects we're talking about, from engines to organisms, from molecules to galaxies. The principles of energy apply to all of them.

Heat as a Form of Energy

Energy, the ability to do work, appears in all natural systems and comes in many forms (Table 12-1). Two centuries ago scientists understood the behavior of kinetic and potential energy, but the nature of heat was far more elusive. We've seen that atoms and molecules—the tiny particles that make up all matter—move around and vibrate and, therefore, particles possess kinetic energy. Only other atoms and molecules experience the minuscule forces they exert, but that small scale doesn't make the force any less real. If molecules in a material move more rapidly, they have more kinetic energy and are capable of exerting greater forces on each other in collisions. If you touch an object whose molecules are moving fast, the collisions of those molecules with molecules in your hand exert greater force than if they were moving slowly, and you perceive the object to be hot. What we normally call *heat,* therefore, is simply a transfer of *thermal energy*—the kinetic energy of atoms and molecules. As we point out in Chapter 11, it's important to remember that the term *heat* should be used only to refer to thermal energy in the process of being transferred from one object to another due to a temperature difference.

TABLE 12-1	Kinds of Energy	
Potential Energy	**Kinetic Energy**	
Gravitational	Moving objects	
Chemical	Thermal	
Elastic	Sound and other mechanical waves	
Electromagnetic		
Mass		

Potential Energies

As we discussed in Chapter 8, potential energy is stored energy. In our daily lives we encounter many other kinds of potential energy in addition to the gravitational kind. Chemical potential energy is stored in the gasoline that moves your car, the batteries that power your radio, and the food you eat. All animals depend on the chemical potential energy of food, and all living things rely on molecules that store chemical energy for future use. In each of these situations, potential energy is stored in the chemical bonds between atoms (see Chapter 23).

Wall outlets in your home and at work provide a means to tap into electric potential energy, available to turn a fan or drive a vacuum cleaner. We see in Chapter 18 that electric potential energy depends on an electric field, just the way gravitational potential energy depends on a gravitational field.

In some situations, such as a tightly coiled spring, a flexed muscle, or a stretched rubber band, a different kind of energy is stored in the bonds between atoms. In these materials, work must be done to bring them to the energetic state we have described (for example, you have to exert a force over a distance to stretch a rubber band). This energy is stored in the chemical bonds and can be released when the situation is right (for example, when you let go of the rubber band). We say that these materials can contain elastic potential energy.

A refrigerator magnet holding a note in your kitchen carries magnetic potential energy. You have to exert a net force over a distance to remove the magnet from the refrigerator, and a net force is exerted over a distance on your hand when you put it back.

No matter which form the potential energy takes, however, in each case energy is stored, ready to do work.

(a)   (b)   (c)

Potential energy comes in many forms. (a) Explosives with a lighted fuse contain chemical potential energy. (b) A rock ready to fall possesses gravitational potential energy. (c) A drawn bow has elastic potential energy. All are examples of potential energy about to be converted into other forms.

Waves possess a form of kinetic energy.

Energy in Waves

Anyone who has watched surf battering a seashore knows that waves carry energy. In the case of water waves, large volumes of water are in motion and therefore possess kinetic energy. It is this energy that we see released when waves hit the shore. Other kinds of waves also possess energy (see Chapters 14 and 15). For example, when a sound wave is generated, molecules in the air are set in motion and the energy of the sound wave is associated with the kinetic energy of those molecules. When the energy of the wave reaches our ears, we hear sound. In Chapter 19 we will meet another important kind of wave, the kind associated with electromagnetic radiation, such as the radiant energy (light) that streams from the Sun. This sort of wave stores its energy in changing electric and magnetic fields.

Mass as Energy

The discovery that certain atoms, such as uranium, spontaneously release energy as they disintegrate—the phenomenon of radioactivity—led to the realization in the early twentieth century that mass is a form of energy. This principle is the focus of Chapter 28, but the main idea is summarized in Albert Einstein's most famous equation.

1. In words:

Every object at rest contains potential energy equivalent to the product of its mass times a constant, which is the speed of light squared.

2. In an equation with words:

$$\text{Energy (joules)} = \text{Mass (kg)} \times [\text{Speed of light (m/s)}]^2$$

3. In an equation with symbols:

$$E = mc^2$$

where c is the symbol for the speed of light, a constant equal to about 300,000,000 meters per second (3×10^8 m/s).

This equation, which has achieved the rank of a cultural icon, tells us that it is possible to change energy into mass and to change mass into energy. (Note: This equation does not mean the mass has to be traveling at the speed of light; the mass is assumed to be at rest.) Furthermore, because the speed of light is so great, the energy stored in even a tiny amount of mass is enormous.

Lots of Potential

EXAMPLE
12-1

According to Einstein's equation, how much potential energy is contained in the mass of a 1-gram grape sitting on your desk?

REASONING AND SOLUTION: Substitute the mass, 1 g, into Einstein's equation. Remember that a gram is 1 thousandth of a kilogram and the speed of light is a constant, 3×10^8 m/s.

$$
\begin{aligned}
\text{Energy (joules)} &= \text{Mass (kg)} \times [\text{Speed of light (m/s)}]^2 \\
&= 0.001 \text{ kg} \times (3 \times 10^8 \text{ m/s})^2 \\
&= 0.001 \times (9 \times 10^{16} \text{ kg-m}^2/\text{s}^2) \\
&= 9 \times 10^{13} \text{ joules}
\end{aligned}
$$

The energy contained in a mass of 1 gram is prodigious: almost 100 trillion joules, which is 25 million kilowatt-hours. The average American family uses about 1000 kilowatt-hours of electricity per month, so a single grape—if we had the means to convert its mass entirely to electric energy (which we don't)—could satisfy your home's energy needs for the next 2000 years!

In practical terms, Einstein's equation showed that mass could be used to generate electricity in nuclear power plants in which a few pounds of nuclear fuel is enough to power an entire city. ●

 THE INTERCHANGEABILITY OF ENERGY

You know from everyday experience that energy can transform from one kind to another. Plants absorb light streaming from the Sun and convert that radiant energy into the stored chemical energy of their cells. You eat plants and convert the chemical energy into the kinetic energy of your muscles—energy of motion that in turn can be converted into gravitational potential energy when you climb a flight of stairs, elastic potential energy when you jump in the air, or heat when you rub your hands together (Figure 12-1). The lesson from these examples is clear:

The many different forms of energy are interchangeable.

Bungee jumping provides a dramatic illustration of this rule (Figure 12-2, page 256). A bungee jumper climbs to a high bridge or platform, where an elastic cord is attached to an ankle. Then the jumper launches out into space and falls toward the ground until the cord stretches, slows the jumper down, and stops the fall.

From an energy point of view, a bungee jumper uses the chemical potential energy generated from food to walk up to the launching platform. The work that had been done against gravity to reach the launching platform provides the jumper with gravitational potential energy. During the long descent, the gravitational potential energy diminishes while the jumper's kinetic energy simultaneously increases. As the cord begins to stretch, the jumper slows down and kinetic energy is converted (gradually) to stored elastic potential energy in the cord. Eventually, the gravitational potential energy that the jumper had at the beginning is completely transferred to the stretched elastic cord. The cord then rebounds, converting some of the stored elastic energy back into kinetic energy and gravitational potential energy. All the time, some of the energy is also converted to heat, which increases the thermal energy of materials: thermal energy in the stressed cord, thermal energy due to friction on the jumper's ankles, and thermal energy in the air as it is pushed aside.

FIGURE 12-1. A pole-vaulter uses the energy of food to power his muscles for running. Kinetic energy of running is converted into elastic potential energy of the bent pole. That elastic potential energy converts to kinetic energy, giving the skilled vaulter sufficient gravitational potential energy to clear the bar.

FIGURE 12-2. Bungee jumping provides a dramatic example of energy changing from one form to another. Can you identify points at which the jumper has maximum kinetic energy? Maximum gravitational potential energy? Maximum elastic potential energy? Does the jumper ever have a combination of all three? What other kinds of energy are involved?

One of the most fundamental properties of the universe in which we live is that, theoretically, every form of energy on our list (Table 12-1) can be converted to every other form of energy.

Ongoing Process of Science
New Sources of Energy

The realization that energy can shift from one form to another drives many scientists to search for new energy-gathering technologies. For thousands of years humans have used sunlight to grow crops, wind to propel their ships, and waterfalls to turn their millstones. Energy from blowing wind, falling water, and brilliant sunlight is still all around us, just waiting to be used, if only we had the practical means to convert it cheaply and efficiently into electricity. Given society's incessant and growing demand for energy, research on new sources of energy continues to be a major scientific enterprise.

The National Renewable Energy Laboratory in Golden, Colorado, is one of the largest North American centers for this kind of research. This state-of-the-art facility, operated with about a quarter-billion tax dollars per year, is home to more than 1000 scientists, engineers, and support staff. Among the laboratory's most successful efforts is the development of novel photovoltaic materials, which are the increasingly familiar black wafers or foils that convert sunlight directly to electric energy in many objects, from roadside signs to pocket calculators.

A more exotic energy technology is being explored at the National Ignition Facility at California's Lawrence Livermore National Laboratory. There, a formidable array of 192 powerful lasers, each capable of blasting a hole through a cinder-block wall, is being constructed to focus on a BB-size pellet of hydrogen fuel. The resulting burst of energy will create, for a brief moment, a miniature sun that radiates intense heat and light as a portion of the hydrogen's mass is converted into energy (see Chapter 26). If all goes as planned, the lasers will fire in unison sometime early in the twenty-first century.

No one can predict how useful energy will be produced a century from now, but one thing is certain. Society will always need energy, and scientists will continue to ask how it can be harnessed. ●

ENERGY AND LIVING SYSTEMS: TROPHIC LEVELS

The concept of energy has been useful in areas of science other than physics. In chemistry, energy is one of the key criteria in determining whether a given reaction will take place. In biology, energy helps determine how an organism functions and how systems of organisms interact with the environment and with one another.

All of Earth's systems, both living and nonliving, transform the Sun's radiant energy into other forms. Just how much energy is available, and how is it used by living organisms? At the top of Earth's atmosphere, the Sun's incoming energy is about 1400 watts per square meter. To calculate the total solar energy we receive, we first need to calculate the cross-sectional area of the Earth in square meters (Figure 12-3a, page 258). The radius of the Earth is 6375 kilometers (6,375,000 meters), so the cross-sectional area is

$$\text{Area of a circle} = \text{Pi} \times (\text{radius})^2$$
$$= 3.14 \times (6{,}375{,}000 \text{ m})^2$$
$$= 1.28 \times 10^{14} \text{ m}^2$$

Thus the total power received at the top of Earth's atmosphere is

$$\text{Power} = \text{Solar energy per m}^2 \times \text{Earth's cross-sectional area}$$
$$= 1400 \text{ watts/m}^2 \times 1.28 \times 10^{14} \text{ m}^2$$
$$= 1.79 \times 10^{17} \text{ watts (or joules/second)}$$

Each second, the top of Earth's atmosphere receives 1.79×10^{17} joules of energy, but less than half that amount reaches the ground (Figure 12-3b). When solar radiation encounters the top of the atmosphere, about 25% of it is immediately reflected back into space. Another 25% is absorbed by gases in the atmosphere, and Earth's surface reflects an additional 5% back into space. These processes leave about 45% of the initial amount to be absorbed at the Earth's surface.

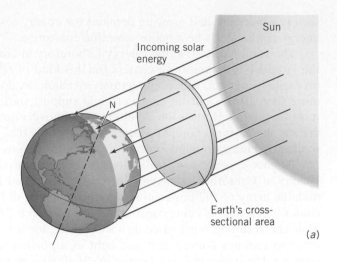

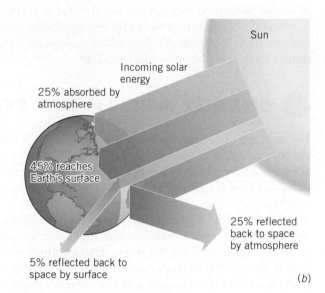

FIGURE 12-3. (a) The Sun illuminates Earth with a large amount of energy crossing through the Earth's cross-sectional area. (b) Much of this incoming energy does not reach Earth's surface.

All living systems take their energy from this 45%, but absorb only a small portion of this amount—only about 4% to run photosynthesis and supply the entire food chain. A much larger portion heats the ground and air or evaporates water from lakes, rivers, and oceans. This energy is all radiated back into space eventually.

To track the many changes of energy as it flows through the living systems of Earth, the concept of the food chain and its trophic levels is particularly useful. A **trophic level** consists of all organisms that get their energy from the same source (Figure 12-4). In this ranking scheme, all plants that produce energy from photosynthesis are in the first trophic level. These plants all absorb energy from sunlight and use it to drive chemical reactions that make plant tissues and other complex molecules. These plant tissues are subsequently used as energy sources by organisms in higher trophic levels.

The second trophic level includes all herbivores—animals that get their energy by eating plants of the first trophic level. Cows, rabbits, and many insects occupy this level. The third trophic level, as you might expect, consists of

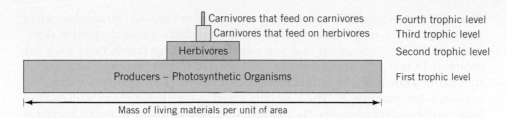

Carnivores that feed on carnivores — Fourth trophic level
Carnivores that feed on herbivores — Third trophic level
Herbivores — Second trophic level
Producers – Photosynthetic Organisms — First trophic level
Mass of living materials per unit of area

FIGURE 12-4. Living organisms are arranged in trophic levels according to how they obtain energy. The first trophic level consists of plants that produce energy from photosynthesis. In the higher trophic levels, animals get their energy by feeding on organisms from the next lowest level.

carnivores—animals that get their energy by eating organisms in the second trophic level. This third level includes such familiar animals as wolves, eagles, and lions, as well as insect-eating birds, blood-sucking ticks and mosquitoes, and many other organisms.

A few more groups of organisms fill out the scheme of trophic levels on the Earth. Carnivores that eat other carnivores, such as killer whales, occupy the fourth trophic level. Termites, vultures, and a host of bacteria and fungi get their energy from feeding on dead organisms and are generally placed in a trophic level separate from the four we've just described. (The usual convention is that this trophic level is not given a number because the dead organisms can come from any of the other trophic levels.)

Several species of animals and plants span the trophic levels. Humans, raccoons, and bears, for example, are omnivores that gain energy from plants and from organisms in other trophic levels, while the Venus flytrap supplements its diet with trapped insects.

Although you might expect it to be otherwise, the efficiency with which solar energy is used by Earth's organisms is very low, despite the struggle by all these organisms to use energy efficiently. For example, when sunlight falls on a cornfield in the middle of Iowa in August, arguably one of the best situations in the world for plant growth, only a small percentage of the solar energy striking the field is actually transformed as chemical energy in the plants. All the rest of the energy is reflected, heats up the soil, evaporates water, or performs some other function. It is a general rule that no plants anywhere transform as much as 10% of the solar energy available to them.

The same situation applies to trophic levels above the first. Typically, less than 10% of a plant's chemical potential energy winds up as tissue in an animal of the second trophic level that eats the plants. That is, less than about 1% (10% of 10%) of the original energy in sunlight is transformed into chemical energy of the second trophic level. Continuing with the same pattern, animals in the third trophic level also use less than 10% of the energy available from the second level.

You can do a rough verification of this statement in your supermarket. Whole grains (those that have not been processed heavily) typically cost about one-tenth as much as fresh meat. Examined from an energy point of view, this cost differential is not surprising. It takes 10 times as much energy to make a pound of beef as it does to make a pound of wheat or rice, and this fact is reflected in the price.

One of the most interesting examples of energy flow through trophic levels can be seen in the fossils of dinosaurs. In many museum exhibits, the most dramatic and memorable specimen is a giant carnivore—a *Tyrannosaurus* or *Allosaurus* with 6-inch dagger teeth and powerful claws. These impressive skeletons are illustrated so often that you might get the impression these finds are common. In fact, fossil carnivores are extremely rare and represent only a small fraction of known dinosaur specimens. Our knowledge of the fearsome *Tyrannosaurus,* for example, is based on only about a dozen skeletons, and most of those are quite fragmentary. By contrast, paleontologists have found hundreds of skeletons of plant-eating dinosaurs. This distribution is hardly chance. Carnivorous dinosaurs, like modern lions and tigers, were relatively scarce compared to their herbivorous victims. In fact, statistical studies of all dinosaur skeletons reveal a roughly 10:1 herbivore-to-carnivore ratio, a value approaching what we find today for the ratio of warm-blooded herbivores to warm-blooded carnivores, and much higher than the herbivore-to-carnivore ratio observed in modern cold-blooded reptiles. Many paleontologists cite this pattern as evidence that dinosaurs were warm-blooded.

THE FIRST LAW OF THERMODYNAMICS: HEAT, WORK, AND INTERNAL ENERGY

We have emphasized several times that energy is a fundamental concept of importance in all areas of science and technology. One reason why this is so is that conservation of energy applies in all these various forms and applications of energy. The general statement of energy conservation that applies to a system under consideration is called the *first law of thermodynamics.*

Before describing the first law, we must first think about the idea of a thermodynamic system. You can think of a **system** as an imaginary box into which you put some matter and some energy that you'd like to study. Scientists might want to study a system containing only a pan of water or perhaps a system consisting of a forest or even the entire planet Earth. Doctors examine your nervous system, astronomers explore the solar system, and biologists observe a variety of ecosystems. In each case, the investigation of nature is simplified by focusing on one small part of the universe.

If the system under study can exchange matter and energy with its surroundings—for example, a pan full of water that is heated on a stove and gradually evaporates—then it is an *open* system (Figure 12-5a). An open system is

FIGURE 12-5. (a) Boiling vegetables in a pot without a lid is an open system (heat and matter can flow in or out). (b) A closed Styrofoam cooler is an isolated system (no heat or matter can enter or leave the system).

(a)

(b)

like an open box where you can take things out and put things back in. Alternatively, if a system can exchange energy but not matter with its surroundings—for example, a tightly shut box made of heat-conducting material—the system is a *closed* system (Figure 12-5*b*). If the system can exchange neither matter nor energy with its surroundings—for example, a tightly shut box made of insulating material—the system is an *isolated* system. Earth and its primary source of energy, the Sun, together make a system that may be thought of for most purposes as isolated because there are no significant amounts of matter or energy being added from outside sources.

The first law of thermodynamics is a relationship among heat, work, and internal energy for a closed system. We've already defined "heat" as a transfer of energy into or out of a system and "internal energy" as the sum of the average kinetic energies of the particles of a system. For these purposes, work involves the motion of a system or part of a system in response to a net applied force. For example, a cylinder with a piston, as you can find in any internal combustion engine (in your car, say), does work when the gas in the cylinder expands, pushing the piston out. If you push the cylinder in, you're doing work on the system, instead of the system doing the work.

We're now ready to state the **first law of thermodynamics.**

1. In words:

In a closed system, the change in internal energy depends on the heat added to the system and the work done by the system.

2. In an equation with words:

Change in internal energy equals heat added minus work done (for a closed system).

3. In an equation with symbols:

$$\Delta U = Q - W$$

where U, Q, and W stand for internal energy, heat, and work, respectively.

As an example of the first law, consider the engine of your car. It has all kinds of gears and electrical connections; we won't worry about those. Instead, let's take as our system one cylinder with its piston. A mixture of air and gasoline enters the cylinder, is ignited by a spark, and expands, pushing the piston. We want to look at the closed system when the mixture is ignited and the gas is expanding—that is, when the heat is generated and the work is done. What the first law of thermodynamics says is that the work done by the piston is limited. The engine can do only as much work as there is energy available in the form of internal energy and added heat. Intuitively, this makes sense: you can't get energy out of nothing. To make the engine do work, you have to supply heat, in the form of burning fuel. Now, if you have an electric car, you can supply electric energy from a battery to make the engine do work. As far as the first law is concerned, that's fine, but the energy balance is still the same—you have to supply energy to get work done.

Energy is something like an economy with an absolutely fixed amount of money. You can earn it, store it in a bank or under your pillow, and spend it here and there when you want to. But the total amount of money doesn't change just because it passes through your hands. Likewise, in any physical situation, you can shuffle energy from one place to another. You can take it out of the account labeled "kinetic" and put it into the account labeled "potential;" you can spread it around into accounts labeled "chemical potential," "elastic potential," "heat," and so on. However, the first law of thermodynamics tells us that, in a closed system, you can never have more energy than you started with.

LOOKING DEEPER

Diet and Calories

The first law of thermodynamics has a great deal to say about the American obsession with weight and diet. Human beings take in energy with their food, energy we usually measure in Calories. (Note that the "Calorie" we talk about in foods is defined as the amount of energy needed to raise the temperature of 1 kilogram of water 1°C, a unit we call a "kilocalorie" in Chapter 11.) Let's consider our body as the system of interest. When a certain amount of energy is taken in, the first law says that only one of two things can happen to it: it can be used to do work and generate waste heat or it can be stored as internal energy. If we take in more energy than we expend, the excess is stored as chemical potential energy in the form of body fat. If, on the other hand, we take in less than we expend, energy must be removed from storage to meet the deficit, and the amount of body fat decreases.

Here are a couple of rough rules you can use to calculate how many Calories should be in your diet:

1. Under most circumstances, normal body maintenance uses up about 15 Calories per day for each pound of body weight.

FIGURE 12-6. Different kinds of physical activity use different amounts of energy.

2. You must consume about 3500 Calories to gain 1 pound of body fat.

Suppose you weigh 150 pounds. To keep your body weight constant, you have to take in

150 pounds × 15 Calories/pound
$$= 2250 \text{ Calories per day}$$

TABLE 12-2	Calories Burned in 10 Minutes of Exercise (150-lb person)
Activity	**Calories**
Volleyball	34
Walking (3 mph)	40
Bicycling (5.5 mph)	47
Calisthenics	49
Golf	54
Skating (moderate)	54
Walking (4 mph)	58
Tennis	68
Canoeing (4 mph)	70
Swimming (breaststroke)	72
Bicycling (10 mph)	81
Swimming (crawl)	87
Jogging (11 min-mile)	91
Handball/Racquetball	95
Skiing (downhill)	95
Mountain climbing	100
Skiing (cross-country)	108
Running (8 min-mile)	141

If you want to lose 1 pound (3500 Calories) in a week (7 days), you have to reduce your daily Calorie intake by

$$\frac{3500 \text{ Calories}}{7 \text{ days}} = 500 \text{ Calories per day}$$

Another way of saying this is that you have to reduce your Calorie intake to 1750 Calories—the equivalent of skipping dessert every day.

Alternatively, the first law says you can increase your energy use through exercise. Roughly speaking, to burn off 500 Calories you have to run 5 miles, bike 15 miles, or swim for 1 hour (see Figure 12-6 and Table 12-2). It's a whole lot easier to refrain from eating than to burn off the weight by exercise. In fact, most researchers now say that the main benefit of exercise in weight control has to do with its ability to help people control their appetites.

Develop Your Intuition: Working off Dessert

Your friend weighs 155 lb (mass of 70 kg) and doesn't want to gain more weight. However, he decides to celebrate an A on the physics test by having a hot fudge sundae with whipped cream (900 Calories) and then working off the calories by climbing up several flights of stairs. How high would he have to climb to burn off this dessert?

Fat in the body is a form of internal energy, so we want that to remain the same. From the first law, $\Delta U = 0 = Q - W$ or $Q = W$. That is, the work of climbing the stairs must equal the heat generated by eating the dessert, which we know is 900 Calories, or 900 kilocalories. We use the conversion factor that 1 kcal equals 4190 joules and get

$$Q = 900 \text{ kcal} \left(\frac{4190 \text{ J}}{1 \text{ kcal}}\right) = 3.77 \times 10^6 \text{ J}$$

This must equal the work done in climbing stairs, which equals mgh, where m is your friend's mass, g is the acceleration due to gravity, and h is the height we want to find (see Chapter 8). So we have

$$h = \frac{Q}{mg} = \frac{3.77 \times 10^6 \text{ J}}{(70 \text{ kg})(9.8 \text{ m/s}^2)}$$

$$= \frac{3.77 \times 10^6 \text{ J}}{686 \text{ N}} = 5500 \text{ meters (about 18,000 feet)}$$

He might be better off settling for a bowl of fruit after all.

Physics in the Making
Lord Kelvin and the Age of the Earth

The first law of thermodynamics provided physicists with a powerful tool for describing and analyzing the universe. Every isolated system, the law tells us, has a fixed amount of energy. Naturally, one of the first systems that scientists considered was the Earth and the Sun.

British physicist William Thomson (1824–1907), knighted in 1892 as Lord Kelvin, asked a simple question: How much energy could be stored in the Earth? Then, given the present rate at which energy radiates out into space, how old might the Earth be? Although simple, these questions had profound implications for philosophers and theologians who had their own ideas about Earth's relative

William Thomson, Lord Kelvin (1824–1907).

antiquity. Some Biblical scholars believed that the Earth could be no more than a few thousand years old. Most geologists, on the other hand, saw evidence in layered rocks to suggest the Earth was at least hundreds of millions of years old. Biologists also estimated that vast amounts of time were required to account for the gradual evolution of life on Earth. Who was correct?

Kelvin believed, as did most of his contemporaries, that the Earth had formed from a contracting cloud of interstellar dust. He thought that the Earth began as a hot body because impacts of large objects on it early in its history must have converted huge amounts of gravitational potential energy into heat. He used new developments in mathematics to calculate how long it had taken for the hot Earth to cool to its present temperature. He assumed that there were no sources of energy inside the Earth and found that the age of the Earth had to be much less than 100 million years. He soundly rejected the geologists' and biologists' claims of an older Earth because these claims seemed to violate the first law of thermodynamics.

Seldom have scientists come to such a bitter impasse. Two competing theories about the age of the Earth, each supported by seemingly sound observations, were at odds. The calculations of the physicist seemed unassailable, yet the observations of biologists and geologists in the field were equally meticulous. What could possibly resolve the dilemma? Had the scientific method failed?

The solution came from a totally unexpected source when scientists discovered in the 1890s that rocks hold a previously unknown source of energy, radioactivity, in which heat is generated by the conversion of mass. Lord Kelvin's rigorous age calculations were in error only because he and his contemporaries were unaware of this critical component of Earth's energy budget. The Earth, we now know, gains approximately half of its internal energy from the energy of radioactive decay. Revised calculations suggest an Earth several billions of years old, in conformity with geological and biological observations. ●

THINKING MORE ABOUT

Energy: Fossil Fuels

All life is rich in the element carbon, which plays a key role in virtually all the chemicals that make up our cells. Life uses the Sun's energy, directly through photosynthesis or indirectly through food, to form these carbon-based substances that store chemical potential energy. When living things die, they may collect in layers at the bottoms of ponds, lakes, or oceans. Over time, as the layers become buried, Earth's temperature and pressure may alter the chemicals of life into deposits of fossil fuels.

Fossil fuels, carbon-rich deposits of ancient life that can burn with a hot flame, have been the most important energy source during the past century and a half. Coal, oil (petroleum), and natural gas, the most common fossil fuels, are consumed in prodigious quantities around the world (Figure 12-7). They now account for approximately 90% of all energy consumed by industrial nations. In the United States alone, approximately 1 billion tons of coal and 2.5 billion barrels of petroleum are used every year.

Geologists estimate that it takes tens of millions of years of gradual burial under layers of sediments, combined with the transforming effects of temperature and pressure, to form a coal seam or petroleum deposit. Coal forms from layer upon layer of plants that thrived in vast ancient swamps, while petroleum represents primarily the organic matter once contained in algae, microscopic organisms that float near the ocean's surface. While these natural processes continue today, the rate of

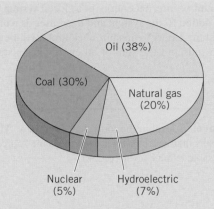

FIGURE 12-7. Sources of energy for industry. Note that most of our energy comes from fossil fuel.

coal and petroleum formation in Earth is only a small fraction of the fossil fuels being consumed. For this reason, fossil fuels are classified as nonrenewable resources.

One consequence of this situation is clear.

Humans cannot continue to rely on fossil fuels forever. Reserves of high-grade crude oil and the cleanest-burning varieties of coal may last less than 100 more years. Less efficient forms of fossil fuels, including lower grades of coal and oil shale (in which petroleum is dispersed through solid rock), could be depleted within a few centuries. All the energy now locked up in those valuable energy reserves will still exist, but in the form of unusable heat radiating far into space.

Given the irreversibility of burning up our fossil fuel reserves, what steps should we take to promote energy conservation? As a concerned citizen, would you vote to tax energy at a higher rate? Would you support government funding for research and development of new energy sources? Do you think alternative energy sources such as wind and solar power will be able to replace fossil fuels in 100 years? These questions all come back to the first law of thermodynamics: You can't get work without supplying energy.

About 90% of the energy in industrialized nations comes from fossil fuels such as petroleum.

Summary

Energy comes in several forms. Kinetic energy is the energy associated with moving objects such as cars or cannonballs. Potential energy, on the other hand, is stored, ready-to-use energy, such as the chemical energy of coal, the elastic energy of a coiled spring, the gravitational energy of dammed-up water, or the electric energy in your wall socket. Thermal energy or internal energy is a form of kinetic energy associated with vibrating atoms and molecules. Energy can also be transmitted by waves, such as sound waves or light waves. And early in the twentieth century it was discovered that mass is also a form of energy. All around us energy constantly shifts from one form to another, and all of these kinds of energy are interchangeable.

Photosynthetic plants in the first **trophic level** use energy from the Sun. These plants provide the energy for animals in higher trophic levels. Roughly speaking, only about 10% of the energy available at one trophic level finds its way to the next.

The most fundamental idea about energy, expressed in the **first law of thermodynamics,** is that it is conserved: the total amount of internal energy in a closed **system** increases as heat is added to the system and decreases as work is done by the system. Energy can shift back and forth between the different kinds, but the sum of all energy is constant.

Key Terms

first law of thermodynamics The law of physics that states the relationship between heat, work, and changes in the internal energy of a closed system. (p. 261)

system A collection of matter and energy that is controlled in such a way that its physical properties can be studied; systems can be open, closed, or isolated. (p. 260)

trophic level A level in the food chain hierarchy. All organisms that get their food from the same source belong to the same trophic level. (p. 258)

Key Equations

Energy associated with mass at rest (joules) = Mass (kg) $\times$ [Speed of light (m/s)]^2

First Law of Thermodynamics: $\Delta U = Q - W$

Review

1. Where did all the energy in the gasoline that powers your car come from? What was the role of photosynthesis in generating this energy? Where does the energy produced as waste heat in your car engine go to when your engine cools down? Is it lost completely?

2. What are some of the forms of energy discussed in this chapter? What are the two general principles that hold for all forms of energy?

3. What is heat? Describe the relationship between heat and kinetic energy.

4. What is thermal energy? What is the difference, if any, between thermal energy and heat?

5. What types of potential energy were mentioned in this chapter? What is the relationship between any potential energy and the ability to do work? (See Chapter 8 to review the scientific definition of *work.*)

6. How can mass be considered energy? Explain.

7. What famous equation describes the relationship between mass and energy?

8. What is the implication for the amount of energy stored in a small amount of mass, given the enormous value of the speed of light? How is it that just a few pounds of nuclear fuel can be enough fuel to provide the electric power of an entire city?

9. Are *all* forms of energy interchangeable? Explain.

10. How many energy conversions between different types of energies are there between the time a bungee jumper leaps off a tall bridge and the time he comes to a halt? What types of energy are involved? Is any energy actually lost?

11. Name some of the alternative energy sources being investigated by scientists that could perhaps meet future energy needs.

12. If only about 45% of the energy of the Sun reaches the surface of Earth, what happens to the other 55%?

13. What is the source of virtually all energy in the form of food and heat that we consume on this planet? Explain.

14. Think about your activities today. Pick one of them and identify the chain of energy that led to it. Where will the energy in that chain eventually wind up?

15. What are trophic levels? How much energy tends to be lost between levels and where does this energy go?

16. How efficient is a fourth-trophic-level predator in incorporating the energy of the Sun into its body tissue?

17. What does it mean to a scientist to say something is conserved? Give an example.

18. What is a system? How is this related to a conserved quantity such as energy?

19. What is the first law of thermodynamics?

20. Is energy ever completely lost and gone forever? Explain.

21. How old did Lord Kelvin estimate Earth to be? What was his reasoning behind the calculations? What was the flaw in his argument?

22. What is the ultimate origin of the energy in petroleum and coal? What are the implications for long-term energy use of continuing to rely on a high consumption level of fossil fuels?

Questions

1. Two glasses of water, A and B, have the same temperature but contain different volumes of water (see figure). In which glass are the water molecules moving faster? Which contains more thermal energy? Which requires more heat to increase its temperature by 1°C?

2. A wooden block is released from rest at the top of a frictionless inclined plane and slides down to the bottom, as shown in the figure. What conversions of energy are taking place as the block slides down the inclined plane? What would your answer be if there were friction between the block and the plane?

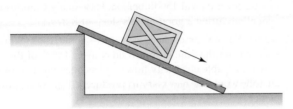

3. You use energy to heat your home. What ultimately happens to the energy that you pay for in your heating bill?

4. What happens on a hot summer day when the energy demand on your local power plant exceeds its energy output?

5. Describe how you might convert the elastic potential energy of a coiled spring into thermal energy. How might you convert thermal energy into gravitational potential energy?

6. Plants and animals are still dying and ending up at the ocean bottom today. Why then, do we not classify fossil fuels as renewable resources?

7. Some people say that you lose more Calories by eating celery than you gain. How could that be?

8. Many modern technologies are designed specifically to convert one kind of energy into another cheaply and efficiently. Photovoltaics, for example, convert light energy into electrical energy. Identify technologies that accomplish the following conversions:
   a. chemical potential energy into light
   b. electric potential energy into kinetic energy
   c. elastic potential energy into sound
   d. gravitational potential energy into kinetic energy
   e. kinetic energy into elastic potential energy
   f. mass into thermal energy
   g. thermal energy into chemical potential energy
   h. electric potential energy into chemical potential energy

9. Classify each of the following systems as either open or closed with respect to matter and with respect to energy. In each case do you have to qualify your answer in any way?
   a. your body
   b. a picnic lunch in a tightly shut Styrofoam cooler
   c. a picnic lunch spread out on a blanket
   d. your car
   e. Earth
   f. the Sun
   g. Earth and the Sun

10. A cylinder with a movable piston contains a gas, as shown in the figure. A weight is placed on top of the piston. When 100 joules of heat is added to the gas, the internal energy of the gas increases by 50 joules and the piston rises, doing 75 joules of work. Does this process violate the first law of thermodynamics? Explain.

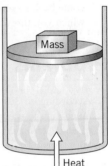

11. Suppose heat is added to a system, but the system does not expand. How much work is done by the system? The system could be anything, but for the sake of definiteness think of a gas in a rigid, closed container. Does the internal energy increase or decrease? Does the temperature increase?

12. A closed, rigid container contains an ice-water mixture at 0°C. Heat is added slowly and some, but not all, of the ice melts. How much work did the system do? Did the internal energy increase or decrease? Did the temperature increase?

13. Suppose you compress a gas, doing 100 joules of work on the gas. If 100 joules of heat is allowed to escape during the compression, what is the change in internal energy?

14. A metal spoon is dropped into a shallow pot of boiling water and its temperature increases to 100°C. The heat added to the spoon is almost exactly equal to the increase in the spoon's internal energy. How much work does the spoon do in this process? Explain by using the first law of thermodynamics.

15. Two pieces of metal are identical in every way, except that one has a much larger thermal expansion coefficient. If equal amounts of heat are added to both pieces of metal, which metal does more work on its surroundings? Which metal's temperature increases more? Explain.

16. Three identical blocks exchange heat in the following way (see figure): A transfers 500 joules of heat to B and B transfers 200 joules of heat to C. Suppose the blocks do not expand or contract, so no work is done. Which block's internal energy increased the most? Which block's internal energy decreased? Why is it impossible for block C to transfer heat to block A?

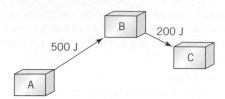

17. Suppose you squeeze an air-filled hollow rubber ball in your hand. Assuming no heat escapes, what happens to the internal energy of the air inside?

18. If you throw a Wham-O SuperBall against a wall, it will come back at you with approximately the same speed with which you threw it. If you throw it at a wall that's moving toward you, the ball will come back with a faster speed. If the wall is moving away from you, the ball will come back with a slower speed. Using the SuperBall example as an analogy, explain why a gas tends to heat up when it is compressed and why it tends to cool when it expands. What is the relationship between compression, expansion, and thermodynamic work?

19. Heat always flows from an object at higher temperature to an object at lower temperature. Is it also true that heat always flows from an object with higher thermal energy to one with lower thermal energy? Explain.

20. A microwave oven is capable of raising the temperature of foods to over 100°C, despite the fact that no part of the oven in contact with the food reaches anywhere close to 100°C. What can you conclude about the mechanism responsible for raising the temperature of things you heat in the microwave? (*Hint:* Look at the first law of thermodynamics. What are the two ways you can change the internal energy of a thermodynamic system?)

Problems

1. According to Einstein's famous equation, $E = mc^2$, how much energy is contained in a pound of feathers? A pound of lead? (*Hint:* You will first need to convert pounds into kilograms.)

2. Again using Einstein's equation, how much energy is contained in 0.001 g of matter? How about in 100 kg of matter?

3. Normally the energy expended during a brisk walk is 3.5 Calories per minute. How long (in minutes) do you have to walk in order to use up the Calories contained in a candy bar (approximately 280 Calories)? How long do you have to walk to produce the same amount of energy as a 100-watt light bulb that is lit for 1 hour? (*Note:* 1 Calorie = 1000 calories = 4186 joules.)

4. In order to lose 1 pound per week, you have to reduce your daily intake by 500 Calories per day. How long would you have to run in order to burn 500 Calories? (The Calories burned vary with the weight and intensity of the runner; assume you burn 7 Calories per minute.)

5. Sleeping normally consumes 1.3 Calories of energy per minute for a typical 150-lb person. How many Calories are expended during a good night's sleep of 8 h?

6. Above Earth's atmosphere we receive about 1400 W/m² of energy from the Sun. If you could convert 100% of this energy into usable electricity, how large a collecting area of solar cells would be necessary to produce a 1-gigawatt power plant?

7. In the Sun, 1 g of hydrogen consumed in nuclear fusion reactions produces 0.9929 g of helium; the other 0.0071 g of material is converted into other forms of energy.

   a. How much energy does this process produce in joules?

   b. How high could you raise the Mt. Palomar 5-m telescope (4.5×10^5 kg) with this energy? (*Recall:* Work = Force [Newtons] × Distance [meters]; also remember the definition of *force* from Newton's law in Chapter 4.)

   c. If you could convert 1 g of hydrogen into energy every second through nuclear fusion, the energy produced would be equivalent to how many 1-gigawatt power plants?

Investigations

1. What kind of fuel is used at your local power plant? What are the implications of the first law of thermodynamics regarding our use of fossil fuels? Our use of solar energy?

2. Investigate the history of the controversy between Lord Kelvin and his contemporaries regarding the age of the Earth. When did the debate begin? How long did it last? What kinds of evidence did biologists, geologists, and physicists use to support their differing calculations of Earth's age?

3. Further explore the structure of trophic levels by looking up the various efficiencies with which different types of organisms use the energy from the level below it. Pay attention to what types of organisms are more efficient than others. What is it that makes them more efficient? What effect does being warm-blooded versus cold-blooded have on the efficiency with which organisms use energy, if any? What trophic level(s) do humans usually occupy, and what is the efficiency of the conversion of solar energy into human tissue?

4. Investigate possible sources of energy as alternatives to petroleum and coal. Make a list of alternative sources that are either being explored by scientists or are currently available, and evaluate the pros and cons of using each. What is the economic cost of switching to them? Should economics play the dominant role in deciding whether to use them, or should other considerations take precedence?

5. Investigate technologies that are being developed to exploit wind power. How do wind turbines work? Where are the best locations in the United States for wind farms? Why might some environmentalists raise objections to the construction of these energy-producing facilities?

6. New technologies using high-frequency sound (ultrasound) can be used to promote chemical reactions (sonication) or to produce light (sonoluminescence). Investigate these energy-conversion phenomena and their possible uses in science and industry.

7. Several different dieting strategies have been advocated by experts in recent years. For example, some nutritionists recommend avoiding carbohydrates, whereas others restrict fats and sweets. Nevertheless, all diets rely to some extent on reducing the amount of chemical potential energy in the food you eat (Calories) that is converted to chemical potential energy in your body (fat). Investigate one of these diets from the standpoint of the first law of thermodynamics.

8. North America boasts vast deposits of oil shale, a rock that holds a significant amount of petroleum in its pores. The total petroleum reserves in oil shale are estimated to be more than 500 billion barrels—at least 20 times the known reserves in all the world's oil wells. Investigate the pros and cons of exploiting this potential fossil fuel resource.

9. Identify four sources of energy around us that are constantly being renewed. What sources of energy do we use that are not constantly renewed?

 # WWW Resources

See the *Physics Matters* home page at **www.wiley.com/college/trefil** for valuable web links.

1. **http://www.nu.ac.za/physics/1M2002/Energy%20work%20and%20power.htm** An illustration of the metabolic equivalence of work and energy content in food.

2. **http://www.npg.org/specialreports/bartlett_index.htm** A discussion of the long-term exhaustion of fossil fuel use for energy production.

3. **http://history.hyperjeff.net/statmech.html** An annotated timeline of thermodynamics and a related topic called "statistical mechanics."

13 | Entropy and the Second Law of Thermodynamics

KEY IDEA

Energy tends to transform spontaneously from more useful to less useful forms.

PHYSICS AROUND US . . . The Cafeteria

Next time you're in the cafeteria, as you go through the food line, think about how each course is prepared. Then try to imagine these processes in reverse. You can peel a piece of fruit, but you can't put it back together. It's easy to scramble eggs, but impossible to unscramble them. You can cook vegetables, but there's no way to uncook them. And once popcorn is popped, it can't be unpopped. Why is this so? Nothing in Newton's laws of motion or the law of gravity suggests that events work in only one way. Nothing that we have learned about energy, including the first law of thermodynamics, suggests that nature works in only one direction.

At the cafeteria you've probably noticed that foods and drinks that are very hot get cooler, while those that are very cold get warmer. A glass of ice water gradually gets warmer, while a plate of hot pasta gets cooler. Ice cream gradually melts, while hot fudge sauce hardens. These everyday events are so familiar that we don't give them a second thought, yet underlying the popping of popcorn and the cooling of a hot drink is one of nature's most subtle and fascinating laws—the second law of thermodynamics.

NATURE'S DIRECTION

The first law of thermodynamics states that the total amount of energy is constant, but it says nothing about the many ways that energy can shift from one form to another. In everyday life we experience severe limitations on energy transfers. Think again about the behavior of a hot bowl of soup. A hot bowl of soup becomes cooler, releasing heat to its surroundings, but it never spontaneously heats up further, absorbing additional heat from its surroundings. Either way, energy is conserved, yet one example is commonplace and the other seems to be impossible.

(a)

(b)

Another example: you know that gasoline burns in your automobile to produce heat and exhaust gases, which power the car. However, heat and exhaust gases never spontaneously combine to form gasoline. Again, the first law is obeyed either way, but nature seems to block some kinds of energy transfer.

Many familiar experiences can't be reversed. A pottery bowl falls to the ground and shatters into fragments. Your dorm room seems to get messy in the course of the week all by itself. And everyone and everything gets older—there is no turning back the clock (Figure 13-1). Evidently, there are restrictions on the flow of energy. Since the first law of thermodynamics does not explain these restrictions, we need a second law of thermodynamics.

STATEMENTS OF THE SECOND LAW

The influential British scientist and novelist C. P. Snow once called the second law of thermodynamics the scientific equivalent of the works of Shakespeare. That's a pretty strong statement. What exactly is this second law of thermodynamics, and why is it so important?

Throughout the universe, the behavior of energy is regular and predictable. According to the first law of thermodynamics, the total amount of energy is constant, although it may change from one form to another over and over again. Energy in the form of heat flows from one place to another by conduction, convection, and radiation. But the net energy flow goes in only one direction. Hot things tend to cool off; cold things tend to warm up; and an egg, once broken, can never be reassembled. These examples illustrate the concept of the second law of thermodynamics—one of the most fascinating and powerful ideas in science.

The **second law of thermodynamics** places restrictions on the ways heat and other forms of energy can be transferred and used to do work. We explore here three different statements of this law. The first and most intuitive of these statements is that heat does not flow spontaneously from a cold body to a hot body. The second statement follows from the first: you cannot construct an engine that does nothing but convert heat completely to useful work. The third, more subtle

(c)

FIGURE 13-1. Many events work in only one direction. (a) As a pool game begins, the cue ball contains all the kinetic energy, but this energy soon becomes distributed evenly through all the balls. (b) A building collapses, transforming the highly ordered structure into a disordered pile of rubble. (c) A flower, once cut, begins to wilt and decay.

statement of the second law is that every isolated system becomes more disordered with time. We consider these ideas in the following section of the chapter.

Although these three statements appear to be very different, they are actually equivalent. Given any one statement of the second law of thermodynamics, you can derive either of the other two as a consequence. For example, given the statement that heat flows only from hot to cold objects, a physicist can produce a set of mathematical steps that show that no engine can convert heat to work with 100% efficiency, which is another way of saying the second statement. In this sense, the three statements of the second law all say the same thing.

Heat Does Not Flow Spontaneously from a Cold Body to a Hot Body

The first statement of the second law of thermodynamics refers to the relative temperatures of two objects. If you take an ice cream cone outside on a hot summer afternoon, it melts. Heat flows from the warm atmosphere to the cold ice cream and causes its temperature to rise above the melting point. By the same token, if you take a cup of hot chocolate outside on a cold day, it cools as heat energy flows from the cup to the surroundings.

From the point of view of energy conservation, there is no reason why energy should necessarily work this way. Energy would be conserved if thermal energy stayed put or even if heat flowed from an ice cream cone to the warm atmosphere, making the ice cream colder than it was at the beginning. Our experience (and many an experimental confirmation) convinces us that heat flows spontaneously in only one direction—from hot to cold. It does not go the other way on its own. This everyday observation may seem trivial, but in this statement is hidden all the mystery of those changes that make the future different from the past: the directionality of time in the universe.

If you think about energy at the molecular level, the explanation of this version of the second law is easy to see. If two moving objects of the same mass collide and one of them is moving faster than the other, chances are that the slower object will speed up and the faster object will slow down. It's possible, but very unlikely, that events will go the other way. Thus, as we saw in the discussion of heat conduction, faster-moving molecules tend to transfer some of their energy to slower-moving ones. On the macroscopic scale, this process is seen as heat flowing from the warm regions to cold ones by conduction.

For the second law to be violated, the molecules in a substance would have to collide in such a way that slower-moving molecules slowed down even more, giving up their energy to faster molecules so they could go even faster. Our experience tells us that this situation doesn't happen. The second law of thermodynamics takes this experience and makes it into a general law of nature.

The second law does *not* state that it's impossible for heat to flow from a colder object to a hotter object. Indeed, you know that's exactly what happens in a refrigerator. When the refrigerator is operating, it removes heat energy from the colder inside to the warmer outside (Figure 13-2). You can verify this fact by putting your hand behind the refrigerator and feeling the warm air. The second law merely states that this action cannot take place *spontaneously,* of its own accord. If you wish to cool something in this way, you must supply energy. In fact, an alternative statement of the second law of thermodynamics could be, "a refrigerator won't work unless it's plugged in."

The second law of thermodynamics says that heat always flows from a warm body (the air) to a cold one (the ice cream).

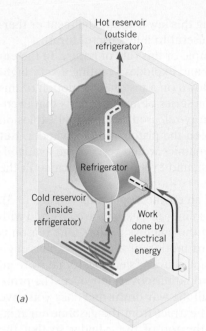

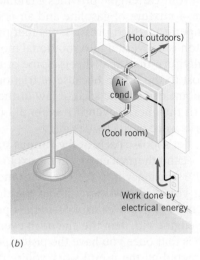

Hot reservoir
(outside
refrigerator)

Refrigerator

Cold reservoir
(inside
refrigerator)

Work
done by
electrical
energy

(a)

(Hot outdoors)

Air
cond.

(Cool room)

Work done by
electrical energy

(b)

FIGURE 13-2. (a) A refrigerator uses electric energy to remove heat from the inside cold reservoir and deposit it in the outside hot reservoir. (b) A window air conditioner works the same way to remove heat from a room and send it outside. In both cases the heat output is equal to the sum of the heat input plus the work done.

The second law doesn't tell you that you can't make ice cubes, only that you can't make ice cubes without expending energy. Paying the electric bill, of course, is another piece of our everyday experience.

You Cannot Construct an Engine That Does Nothing but Convert Heat Completely to Useful Work

The second statement of the second law of thermodynamics restricts the way we use energy. Recall that *energy* is defined as the ability to do work. This second statement of the second law tells us that whenever energy is transformed from heat to some other type of energy—from heat to electric energy, for example— some of that heat must be dumped into the environment and is unavailable to do work. The energy is neither lost nor destroyed, but it can't be used to make electricity to play your radio or fuel to drive your car.

Physicists and engineers use the term *efficiency* to quantify the percentage of useful energy an engine can deliver. **Efficiency** is the amount of work you get from an engine, divided by the amount of energy you put into it. In Chapter 12 we learned that different forms of energy are interchangeable and that the total amount of energy is conserved. According to the first law of thermodynamics, there is no reason why energy in the form of thermal energy can't be converted to electric energy with 100% efficiency. However, the second law of thermodynamics tells us that such a process isn't possible. The flow of energy in the form of heat has a direction. Another way of stating this law is to say that heat or thermal energy always transforms from a more useful to a less useful form.

Your car engine provides a familiar example of this rule of nature. In the engine, a mixture of gasoline and air is ignited and explodes, creating a very high-temperature, high-pressure gas that pushes down on a piston. The motion of this piston is converted into rotational motion of a series of machine parts that eventually turn the car's wheels. Some of the energy released in the initial explosion is lost to the piston because of friction between the piston and the cylinder. But in point of fact, the second law of thermodynamics would restrict the availability of the energy even if friction did not exist, and even if every machine in the world were perfectly designed.

Let's look closely at the various stages of an engine's operation (Figure 13-3). Why can't thermal energy in the exploding gasoline–air mixture be converted with 100% efficiency into the energy of motion of the engine's piston? One reason is that you can't just think about the downward motion of a piston—which engineers call the *power stroke*—in the operation of your car's engine. If that were all the engine did, then the engine in every car would turn over only once. The problem is that once you have the piston pushed all the way down, and once you have extracted all the useful work you're going to extract from the gasoline–air mixture, you still have to return the piston to the top of the cylinder so that the cycle can be repeated. (In actuality, the pistons in modern cars go down, up, and down again before they get back to the point where they can return energy

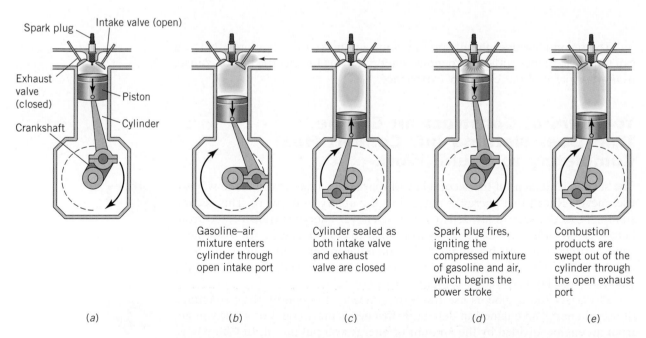

	Gasoline–air mixture enters cylinder through open intake port	Cylinder sealed as both intake valve and exhaust valve are closed	Spark plug fires, igniting the compressed mixture of gasoline and air, which begins the power stroke	Combustion products are swept out of the cylinder through the open exhaust port
(a)	(b)	(c)	(d)	(e)

FIGURE 13-3. The cycle of an automobile engine's piston. (a) The beginning of the intake stroke; (b) the middle of the intake stroke as a gasoline–air mixture enters the cylinder; (c) the beginning of the compression stroke; (d) the beginning of the power stroke when the spark plug fires, igniting the compressed mixture of gasoline and air; (e) the beginning of the exhaust stroke when combustion products are swept out. Note that each cycle involves two complete rotations of the crankshaft.

to the system.) In order to reset the engine to its original position so that more useful work can be done, some heat has to be dumped into the environment.

Ignore for a moment the fact that a real engine is more complicated than the one we're discussing. Suppose that all you had to do was to lift the piston up after the work had been done. The cylinder is full of air and, consequently, when you lift the piston up, the air is compressed and heated. In order to return the engine to the precise state it was in before the explosion, the heat from this compressed air has to be removed. In practice, it is expelled into the atmosphere as exhaust.

When the hot gas is absorbed into the atmosphere, the temperature of the atmosphere doesn't change significantly. After all, there's a lot more atmosphere than there is exhaust gas from a car. In such a situation, physicists call the atmosphere a "low-temperature reservoir," since it doesn't change its temperature. In a car engine, the hot gasoline–air mixture is constantly renewed at the same temperature; as a result, we can say it acts as a "high-temperature reservoir," losing some energy without lowering its temperature. The second law of thermodynamics says that any engine operating between two temperatures must dump some energy in the form of heat into the low-temperature reservoir. This heat is energy that flows through the engine, but that energy can't be used to do work. You can see how this works in the gasoline engine when the hot air produced in resetting the piston must be expelled as exhaust.

The consequence of this situation is that some of the energy stored in the gasoline can be used to run the car, but some must be dumped into the low-temperature reservoir of the atmosphere. Once that heat energy has gone into the atmosphere, it can no longer be used to run the engine. That energy simply dissipates and is no longer available. Thus, this version of the second law tells us that any real engine, or even an engine in which there is no friction, must waste some of the energy that goes into it. The engine works better if the difference between its hot input temperature and cold output temperature is as large as possible, but, no matter how large the difference, perfect efficiency is simply not possible.

This version of the second law of thermodynamics explains why petroleum reserves and coal deposits play such an important role in the world economy. They are high-grade and nonrenewable sources of energy that can be burned to produce very-high-temperature reservoirs. But no matter what we do, when these fossil fuels are burned to produce that high-temperature reservoir and generate electricity, a large portion of energy is then wasted.

Although the second law applies to engines that work in cycles, it does not apply to many other uses of energy. No engine is involved if you burn natural gas to heat your home or use solar energy to heat water, for example. In other words, burning fossil fuels or employing solar energy to heat your home directly can be considerably more efficient than using these fuels to generate electricity and then heating your home with electricity.

Develop Your Intuition: Materials for Jet Engines

Jet engines do not generally use simple steel or aluminum parts. Instead, they are made with special alloys and ceramic materials that can withstand very high temperatures, over 1000°C. Why is such high-temperature resistance necessary?

The exhaust gases of a jet engine are emitted to the atmosphere, so the output temperature cannot be adjusted. Therefore, for increased engine efficiency, the temperatures inside the engine should be as high as possible. Specially designed materials are needed to operate at these high temperatures without melting or becoming soft.

Efficiency

The second law of thermodynamics can be used to calculate the maximum possible efficiency of an engine. Let's say that the high-temperature reservoir is at a temperature T_{hot} and the low-temperature reservoir is at a temperature T_{cold} (where all temperatures are measured using the Kelvin scale). Then the maximum theoretical efficiency—the percentage of energy available to do useful work—of any engine in the real world can be calculated as follows:

1. In words:

Efficiency is obtained by comparing the temperature difference between the high-temperature and low-temperature reservoirs, with the temperature of the high-temperature reservoir.

2. In an equation with words:

Efficiency (in percent) =

$$\frac{\text{Temperature}_{hot} - \text{Temperature}_{cold}}{\text{Temperature}_{hot}} \times 100$$

3. In an equation with symbols:

$$e = \frac{T_{hot} - T_{cold}}{T_{hot}} \times 100\%$$

In a real machine, any loss of energy due to friction in pulleys, gears, or wheels will make the actual efficiency less than this theoretical maximum. This maximum possible efficiency is a very stringent constraint on real engines.

Consider the efficiency of a normal coal-fired generating plant. The temperature of the high-energy steam (the hot reservoir) is about 550 kelvins, while the temperature of the air into which waste heat must be dumped (the low-temperature reservoir) is around room temperature, or 300 kelvins. The maximum possible efficiency of such a plant is given by the second law as

$$\text{Efficiency (in percent)} = \frac{T_{hot} - T_{cold}}{T_{hot}} \times 100$$

$$= \frac{550 - 300}{550} \times 100$$

$$= 45.5\%$$

In other words, more than half of the energy produced in a typical coal-burning power plant must be dumped into the atmosphere as waste heat. This fundamental limit is independent of the engineers' ability to design the plant to operate efficiently. In fact, engineers have succeeded in making most generating plants operate within a few percent of the optimum efficiency allowed by the second law of thermodynamics. Figure 13-4 shows the maximum possible efficiency of an engine when the low temperature reservoir is at a temperature of 300 K.

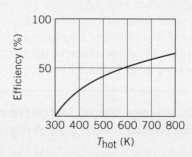

FIGURE 13-4. The maximum possible efficiency of an engine when the low temperature reservoir is the outside air at a temperature of 300 K.

Power Generation in Space

What improvement in efficiency might you obtain for a coal-burning power plant on the Moon, where the cold reservoir is 105 K?

SOLUTION AND REASONING: Apply the equation for efficiency.

$$\text{Efficiency (in percent)} = \frac{T_{hot} - T_{cold}}{T_{hot}} \times 100$$

$$= \frac{550 - 105}{550} \times 100$$

$$= 80.9\%$$

This is 35.4 percent higher than the maximum possible efficiency of a coal-burning power plant on Earth. On the Moon the efficiency of such a power plant would be much higher than on Earth because less energy is wasted as exhaust heat. ●

Steam Engines

What is the maximum efficiency of a steam engine that employs boiling water and dumps its waste heat into ice water?

SOLUTION AND REASONING: Again, apply the equation for efficiency, making sure to use temperature in the Kelvin scale (the temperatures of freezing and boiling water are 273 and 373 K, respectively).

$$\text{Efficiency (in percent)} = \frac{T_{hot} - T_{cold}}{T_{hot}} \times 100$$

$$= \frac{373 - 273}{373} \times 100$$

$$= 26.8\%$$

Today, even the best steam engines use less than one-third of the potential energy stored in coal or gas. But before these thermodynamic principles were understood, typical steam engine efficiencies were a meager 6%. Here's a clear case where the theoretical understanding of energy has led to dramatic benefits to society by showing that better efficiency was possible. ●

Every Isolated System Becomes More Disordered with Time

The third statement of the second law of thermodynamics, that every isolated system becomes more disordered with time, is in many ways the most profound. It tells us something about the order of the universe itself. Consequently, this idea of increasing disorder in the universe is perhaps the most familiar way of looking at the second law.

Order and Disorder To understand what the third statement of the second law means, you have to understand what a physicist means by the terms "order" and

Highly ordered
solid – low probability
(a)

This change
tends to be
spontaneous

Disordered liquid –
higher probability

An improbable way
for bricks to fall

A more probable way
for bricks to fall

(b)

FIGURE 13-5. Highly ordered, regular patterns of objects are less likely to occur than disordered, irregular patterns.

"disorder" (Figure 13-5). An ordered system is one in which some number of objects, be they atoms or automobiles, are positioned in a completely regular and predictable pattern. For example, atoms in a perfect crystal or automobiles in a perfect line are highly ordered systems. A disordered system, on the other hand, contains objects that are randomly situated, without any obvious pattern. Atoms in a gas or automobiles after a multicar pile-up on the freeway are good examples of more disordered systems.

We can devise countless simple experiments to illustrate the phenomenon of increasing disorder. For instance, take a deck of cards and put all the cards in order, by suit and by number. Then shuffle the deck of cards and see what happens. It becomes disordered—never the other way around. You can shuffle these cards a million million times, and chances are that they will not become reordered. In fact, even if you deal out only five cards, the chances of having a winning hand in poker are pretty slim (see Table 13-1).

Another simple experiment involves carefully layering two different colors of marbles or candy in a jar. Gently shake the jar and see what happens. The colors quickly become mixed up—never the other way around. You could shake the jar for a million years, and chances are the two colors would not separate into layers.

Highly ordered arrangements of cards are much less likely than disordered arrangements.

TABLE 13-1	Probabilities of Types of Poker Hands Dealt in the First Five Cards	
Type of Hand		**Probability**
No pair		1 in 2
One pair		1 in 2.5
Two pair		1 in 21
Three of a kind		1 in 47
Straight		1 in 255
Flush		1 in 509
Full house		1 in 694
Four of a kind		1 in 4165
Straight flush		1 in 72,193
Royal flush		1 in 649,740
Note: Number of possible poker hands = 2,598,960		

Yet another version of this simple experiment can be done with food coloring and water. Plop a drop of food color into a jar of water. Gradually, the molecules of the food coloring disperse through the water. The same thing is true of perfume or cologne: you put it on in the morning and by evening it's all gone, dispersed into the air. The molecules of perfume never spontaneously coalesce into a droplet.

This behavior of molecules, which spontaneously disperse, also applies to the distribution of velocities of particles in a gas. Imagine what happens when you mix two reservoirs of gas that are initially at different temperatures; for example, one gas at 0°C and a second gas at 100°C. The separation of gas particles into two distinct populations is a more ordered state, just like the separation of red and white marbles in a jar or the separation of pure water and a drop of food coloring. It's also an inherently unstable situation. As the gases mix, the temperature averages out—the system becomes more disordered.

Entropy Defined **Entropy** is a measure of the disorder in a physical system. In terms of entropy, the statement of the second law reads:

> *The entropy of an isolated system remains constant or increases.*

In other words, any system left to itself will change in the direction of the most disordered state. Without careful chemical controls, atoms and molecules tend to become more intermixed; without careful driving, automobile traffic also tends to become more disordered.

The example of food coloring and water reveals how such a process works. In the most likely situation, when molecules of water and food coloring are randomly mixed, the entropy is maximized. In the much less probable case of food coloring molecules coalescing in the water, the entropy is lower because the molecules are more ordered. Another way of saying this is that systems tend to avoid states of high improbability.

Probabilities in Nature The definition of entropy as a measure of disorder may seem a bit fuzzy, but the Austrian physicist Ludwig Boltzmann (1844–1906) placed it on a firm quantitative footing in the late nineteenth century. Boltzmann was born in Vienna and studied at the University of Vienna, where he spent most of his professional life as professor of theoretical physics. He was said to be an imposing man of physical strength combined with sensitivity and humor. But he also suffered from severe bouts of depression, and it was during one of those spells that he took his own life in 1906. Boltzmann used probability theory to demonstrate that, for any given configuration of atoms, the mathematical value of entropy is related to the number of possible ways you can achieve that configuration.

To get a feel for this idea, consider three orange balls numbered 1, 2, and 3, and three green balls numbered 4, 5, and 6. Ask yourself how many different ways there are to arrange these six balls in a row (Figure 13-6). There are six different possibilities for the location of the first ball, then five possibilities for the second, four for the third, and so on. If you multiply that out, you get

$$6 \times 5 \times 4 \times 3 \times 2 \times 1 = 720$$

It turns out that there are 720 ways to arrange these six numbered balls in a row—720 different possible configurations.

Food dye disperses in water; it never stays clumped together or goes back to being a single drop.

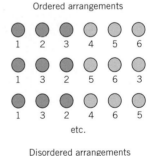

Ordered arrangements

Disordered arrangements

FIGURE 13-6. Ordered and disordered arrangements of six numbered balls (three orange and three green).

Now, how many of those arrangements have the ordered state with three orange balls followed by three green balls? There are exactly six different ways to arrange three orange balls ($3 \times 2 \times 1$), and then six different ways to arrange the three green balls. Altogether, that's 6×6, or 36, different configurations with three orange balls followed by three green balls out of 720 total configurations. Only 5% of all possible arrangements (36/720) are ordered in this way. All of the remaining 684 configurations are different. So, by a 19:1 margin, these other arrangements are much more probable, because there are many more ways to achieve a disordered state than an ordered one.

You can repeat this exercise for other numbers of balls (see Problems 5–7 at the end of the chapter). For a sequence of ten balls (five orange and five green), it turns out that there are more than 3.6 million different configurations, but only 120 of those sequences have five orange followed by five green balls. That's only 0.003% of all possible arrangements!

As the number of objects increases from 6 to 10 to trillions of trillions (as we find in even a few grams of atoms), the fraction of arrangements that is highly ordered becomes infinitesimally small. In other words, highly ordered configurations are improbable because almost every possible configuration is disordered.

The second law of thermodynamics and the behavior of entropy can thus be traced, ultimately, to the laws of probability. If you hold a glass of water in your hand, for example, its atoms are moving at more or less the same average velocity. It is extremely unlikely that the atoms will ever arrange themselves in such a way that one atom moves very fast in a collection of very slow ones. Over the course of time, any unlikely initial state like this evolves into a more probable state.

The concept of probability explains a number of paradoxes that puzzled scientists around the turn of the twentieth century. For example, it's possible that all the air molecules in the room in which you are sitting could suddenly rush over to one side of the room, leaving you in a perfect vacuum. You don't worry about this happening because you know that this event is highly unlikely. In fact, the probability is so low as to make it extremely unlikely you would see it happen even if you waited the entire lifetime of the universe.

Decreasing Entropy   While systems tend to become more disordered, the second law does not require *every* system to approach a state of lower order. Think about water, a substance of high disorder because water molecules are arranged at random. If you put water into a freezer, it becomes an ice cube, a much more ordered state in which water molecules have formed a regular crystalline structure. By placing water in the freezer you have caused a system to evolve to a state of higher order. How can this ordering be reconciled with the statement that isolated systems become more disordered?

The answer to this paradox is that this statement refers *only* to systems that are isolated. The refrigerator in which you make the ice cubes is not an isolated system because it has a power cord plugged into the wall socket that is ultimately connected to a generating plant. The isolated system in this case is the refrigerator, the generating plant, and their immediate environments. The second law of thermodynamics says that in this particular isolated system, the total entropy must increase. However, it does not say that the entropy has to increase in all the subparts of the system.

In this example, one part of the system (the ice cube) becomes more ordered, while another part of the system (burning fuel and the surrounding air at the generating plant) becomes more disordered. All that the second law requires is that the amount of disorder at the generating plant be greater than the amount of order at the ice cube. As long as this requirement is met, the second law is not violated. In fact, in this particular example the disorder at the generating plant greatly exceeds any possible order that could take place inside the refrigerator.

Physics in the Making
The Heat Death of the Universe

Some intellectuals viewed the nineteenth-century discovery of the second law of thermodynamics as a gloomy event. The prevailing philosophy of the time was that life, society, and the universe in general were on a never-ending upward spiral of progress. Darwin's 1859 publication of *On the Origin of Species,* which proposed that more complex forms of life could evolve from less complex forms, reinforced this particular notion of an ever-improving world. In this optimistic climate, the discovery that the energy in the universe was being steadily and irrevocably degraded was difficult for nineteenth-century scientists and philosophers to accept.

In fact, they felt that the second law inevitably meant that all the energy in the universe would eventually be degraded into waste heat and that everything in the universe would eventually be at the same temperature. They called this depressing end the "heat death" of the universe, and they saw it as the ultimate effect of the laws of thermodynamics. For example, the hero of the famous story *The Time Machine,* written by H. G. Wells in 1895, travels to a far distant future in which the Sun and the stars have all burned themselves out and the cold, dark Earth remains. Today, we realize that even if the universe does end in frozen desolation, such a demise will not occur for many hundreds of billions of years. ●

 ## CONSEQUENCES OF THE SECOND LAW

The Arrow of Time

We live in a world of four dimensions. Three of these dimensions define space and have no restrictions on the directions that you can travel. You can go east or west, north or south, and up or down in our universe. But the fourth dimension, time, behaves differently. Time has direction; we can never revisit the past.

Take one of your favorite home movies or just about any videotape and play it in reverse. Chances are that before too long you'll see something silly— something that couldn't possibly happen that will make you laugh. Springboard divers fly out of the water and land completely dry on the diving board. From a complete stop, golf balls roll along the ground and then fly off toward the tee. Ocean spray coalesces into smooth waves that recede from shore. Most physical laws, such as Newton's laws of motion or the first law of thermodynamics, say nothing about the direction of time. The motions predicted by Newton and the conservation of energy are independent of time—they work just as well if you play a video forward or backward.

The second law of thermodynamics is different: it takes into account a *sequence* of events. For example, heat flows from hot to cold, fuels burn to produce waste heat, the disorder of isolated systems never spontaneously decreases, and we all must get older. Time has a direction. We experience the passage of events as dictated by the second law. Scientists cannot answer the deeper philo-sophical question of why the arrow of time goes in only one direction, but through the second law they can describe how the effects of that directionality come about.

The tendency of all systems to change from an improbable to a more prob-able state accounts for this directionality of time that we see in the universe. From the point of view of the first law of thermodynamics, there is no reason that improbable situations can't occur. Fifteen slow-moving billiard balls have enough energy to produce one fast-moving ball. The fact that this situation doesn't occur in nature is an important clue as to how things work at the atomic level (see Figure 13-1).

Built-In Limitations of the Universe

The second law of thermodynamics has both practical and philosophical conse-quences. It poses severe limits on the ways in which humans can manipulate na-ture and on the way that nature itself operates. It tells us that some things cannot happen in our universe. In terms of energy, the first law says you can't get some-thing for nothing. The second law says you can't break even.

At the practical level, the second law tells us that if we continue to gener-ate electricity by burning fossil fuels or by nuclear fission, we are using up a good deal of the energy that is locked up in those concentrated nonrenewable re-sources. These limitations are not a question of sloppy engineering or poor design—they're simply built into the laws of nature. If you could design an en-gine or other device that extracted energy from coal and oil with higher effi-ciency than the second law allows, then you could also design a refrigerator that worked when it wasn't plugged in. Over the years, many inventors have tried to do just this, creating machines that supposedly could provide more work as out-put than they take in as heat or other energy. None of these so-called perpetual motion machines work this way and, according to the laws of thermodynamics, you simply cannot make such a machine. Some attempts have been quite ingen-ious, but there is always a flaw in the design that requires more energy than the machine can produce.

At the philosophical level, the second law tells us that nature has a built-in hierarchy of more useful and less useful forms of energy. The lowest or least use-ful state of energy is the reservoir into which all energy eventually gets dumped. Once the energy is in that reservoir, it can no longer be used to do work. For Earth, energy passes through the region that supports life, the *biosphere,* but is eventually lost as it is radiated into the black void of space.

Physics in the Making
America's First Theoretical Physicist

Most of the great names in American physics have been experimentalists. Start-ing with Benjamin Franklin (Chapter 16) and Benjamin Thompson (Chapter 11), American physicists were a practical lot, looking for ideas that would lead to spe-cific improvements in machines or would resolve particular problems. The ma-jor exception was one of the leading theoretical scientists of the nineteenth century, Josiah Willard Gibbs (1839–1903).

Gibbs came from an old New England family and lived quietly in New Haven, Connecticut, all his life. He started out as an engineer and was the first American ever to receive a Ph.D. in engineering. However, afterward he spent two years studying in Europe and became more interested in mathematical and theoretical physics. He was appointed Professor of Mathematical Physics at Yale in 1871, at the age of 32.

Gibbs studied how matter changes, in the most general ways. He developed the idea that is now called the "Gibbs free energy," one of the key concepts in determining whether a given chemical reaction will take place spontaneously. The Gibbs free energy combines two quantities: the first quantity is related to the chemical potential energy of the reactants and the second quantity is related to the entropy of the reactants. Gibbs showed that a reaction is determined by entropy considerations as much as by energy differences.

Despite the fundamental importance of his work, Gibbs was notoriously modest about himself. He published his work in the local *Transactions of the Connecticut Academy of Science,* which was seldom read by anyone outside the Yale community. However, Gibbs mailed copies of his articles to scientists around the world whose work might be affected by his own findings. Most other scientists could not follow his subtle and mathematically demanding work, but two of the great physicists of the time—James Clerk Maxwell and William Thomson, Lord Kelvin—recognized the brilliant work Gibbs was doing and gave lectures on his results to other European scientists.

Gibbs is now considered the founder of physical chemistry, establishing the physical principles that govern all chemical reactions. He also helped to establish the area of physics known as "statistical mechanics," which connects the laws of thermodynamics to the behavior of systems with large numbers of particles. It has been said that of all the great theories advanced in the nineteenth century, only the work of Gibbs continues to stand without serious changes after the new ideas of physics came along in the twentieth century. ●

THINKING MORE ABOUT

The Second Law: Evolution and Entropy

Entropy is such an important concept, so fundamental to the working of the universe, that it appears in many discussions that seem to have little to do with heat transfer. Economists, for example, use it to discuss disordered aspects of the economy, and communications engineers routinely use it to discuss the information (i.e., order) in radio and other transmissions. But perhaps the most interesting place where entropy crops up is in the intense debate about the origin of life on Earth.

Creationists, who believe that life appeared as the result of a single miraculous creation a few thousand years ago, point out that every life form is a highly ordered system—a system in which trillions of atoms and molecules must occur in exactly the right sequence. They argue that life could not possibly have arisen spontaneously without violating the second law. How, they ask, could a natural system go spontaneously from a disordered state of nonlife to the ordered state containing life?

This argument fails to take into account the fact that Earth is not itself an isolated system. The energy that drives living systems is sunlight, so that the isolated system that the second law speaks of comprises Earth *plus the Sun*. To make the evolution of life consistent with the second law, the increased order observed in living things must be offset by an even greater amount of disorder in the Sun. Once again, as with the earlier example of the ice cube, this requirement is easily met by the Sun and Earth taken together.

Science cannot yet describe in detail how life

arose, and some aspects of the process may never be known for sure. The scientific method, for example, is not an appropriate way to answer the question whether God was involved in the origin of life (see Chapter 1). However, science can help explain how the development of life, as a natural process, is consistent with the universal laws of thermodynamics.

Summary

The first law of thermodynamics promises that the total amount of energy never changes, no matter how you shift it from one form to another. The **second law of thermodynamics** places restrictions on how energy can be shifted. Three different but equivalent statements of the second law underscore these restrictions:

1. Heat does not flow spontaneously from a colder body to a hotter body.

2. It is impossible to construct a machine that does nothing but convert heat completely into useful work. That is, no engine can operate with 100% **efficiency.**

3. The **entropy** (measure of disorder) of an isolated system never decreases.

Key Terms

efficiency A measure of how much useful work you can get from an engine; it is equal to the work done by an engine divided by the heat input to the engine. (p. 273)

entropy A measure of the disorder in a system. (p. 278)

second law of thermodynamics The law of physics that places restrictions on the ways heat and other forms of energy can be transformed and used to do work. (p. 271)

Key Equation

$$\text{Efficiency (in percent)} = \frac{T_{hot} - T_{cold}}{T_{hot}} \times 100$$

Review

1. What is the first law of thermodynamics? What is meant by the directionality of energy flow? Does the first law of thermodynamics deal with the directionality of energy flow?

2. State the second law of thermodynamics in three different ways. Do they all say essentially the same thing?

3. Does the fact that heat does not flow spontaneously from a cold to a hot body violate the first law of thermodynamics in any way? Explain.

4. Give a molecular-level explanation of why heat does not flow spontaneously from a cold body to a hot body.

5. Why does a refrigerator that facilitates the flow of heat from a cool interior to a warm exterior not violate the second law? What must be supplied for this to happen?

6. If energy is defined as the ability to do work, can you construct an engine that does nothing but completely convert the energy in heat to useful work? Explain.

7. What is meant by the efficiency of an engine? Does the first law of thermodynamics preclude the construction of a 100% efficient engine? What about the second law?

8. Explain in terms of the motion of a piston why 100% efficiency in an engine is not possible.

9. What is the high-temperature reservoir in your car's engine? What is the low-temperature reservoir?

10. Why are petroleum reserves and coal deposits so important in producing a relatively efficient use of energy?

11. Does the second law apply to other uses of energy, such as heating your home with natural gas? Explain.

12. Identify three examples of the second law of thermodynamics in action that have occurred since you woke up this morning.

13. Why would a power plant in outer space be potentially very efficient?

14. What do physicists mean by the terms order and disorder?

15. Which type of system does the second law say the universe is going toward, a more ordered or a more disordered one?

16. What is entropy? What does the second law say about entropy?

17. What was Boltzmann's contribution toward the understanding of entropy?

18. Why are highly ordered configurations of objects improbable?

19. How do the laws of probability explain the behavior of entropy and the second law of thermodynamics? How likely is it that objects will be in an ordered state given the nearly infinite possible configurations available to systems with many objects in them? Explain.

20. Does the second law require *every* system to approach a state of lower order? Explain.

21. Give examples of ordered systems in nature. Give examples of disordered systems.

22. Do Newton's laws of motion have a direction in time? Why or why not?

23. Does the first law of thermodynamics have a direction in time? Why or why not?

24. What does the second law have to say about the directionality of time? How is probability involved?

25. Does the second law support a creationist view of the origin of the universe? Explain.

Questions

1. Why don't all the atoms in the room you're sitting in move to one side, leaving you in a vacuum?

2. Why are there big cooling stacks around nuclear reactors and coal-fired generating plants?

3. "Cogeneration" is a term used to describe systems in which waste heat from electric generating plants is used to heat nearby homes. Such systems achieve efficiencies much greater than 50%. Does cogeneration violate the second law? Why or why not?

4. Why is a perpetual motion machine impossible?

5. Seawater is full of moving molecules that possess kinetic energy. Could we extract this energy from seawater? Why or why not?

6. When ice freezes, water goes from a state of larger disorder to one with more order. Does this violate the second law of thermodynamics? Explain.

7. In the equation defining efficiency, why must you always use the Kelvin temperature scale? (*Hint:* Consider the effects of a temperature's sign and also of dividing by 0.)

8. Two systems contain vastly different amounts of internal energy. For the sake of definiteness, let's assume that system A contains 1,000,000 joules of internal energy and system B contains 100 joules of internal energy. Is it possible to say which direction heat will flow if these two systems are placed in thermal contact? Why or why not?

9. A cube of aluminum metal is placed in contact with a cube of copper metal. The average speed of the atoms in each metal is the same. Which way does heat flow? Explain. (*Hint:* Look at the periodic table to find the atomic masses of aluminum and copper. Which cube is at a higher temperature?)

10. During a complete cycle of an engine, the net internal energy change is 0. During that cycle, an amount of heat Q_{in} enters the engine, an amount Q_{out} leaves the engine, and

an amount of work W is done. The following table lists these quantities (in joules) for a variety of engines. Which of these engines violates the first law of thermodynamics (energy in equals energy out)? Which of these engines violates the second law of thermodynamics?

Engine	Q_{in}	Q_{out}	W
A	100	100	0
B	100	50	50
C	100	0	100
D	100	20	60
E	100	100	50

11. Imagine that it were possible to construct a reservoir at $-5\,K$ (below absolute zero). Suppose you ran an engine and used the $-5\,K$ reservoir as the cold reservoir. Why would such an engine violate the second law of thermodynamics?

12. An ice cube melts on the warm sidewalk on a hot summer day. What happens to the entropy of the ice cube? What happens to the entropy of the pavement? What happens to the entropy of the ice cube–sidewalk system?

13. You roll two six-sided dice. Why are you much more likely to roll a total of 7 than a total of 2?

14. A large parking lot contains 50 identical cars. Which is a higher entropy situation: when the cars are allowed to park anywhere, or when the cars are forced to park in between the lines in designated spaces? Explain in terms of the number of possible configurations of the system.

15. How many different ways are there to arrange five coins in a row if one is heads-up and the other four are tails-up (see figure)? What if two were heads-up and three were tails-up?

16. Two identical poker chips are placed on a board that is divided into a 2×3 checkerboard arrangement (see figure). If they are allowed to be on any two squares on the left half of the board, how many arrangements are there? What if they are allowed to be on any two squares on the board? What does this question have to do with entropy? (*Hint:* It may help to sketch all of the possible arrangements.)

17. Why do we say that the liquid state is more disordered than the solid state? Explain in terms of the number of possible arrangements of the atoms or molecules.

Problems

1. What is the theoretical efficiency of an engine that has a hot reservoir of 600 kelvins and a low-temperature reservoir of 300 kelvins?

2. If a steam engine has a high-temperature reservoir of 100°C and a low-temperature reservoir of 10°C, what is its maximum possible efficiency? What would be the efficiency if the low-temperature reservoir had a temperature of 278 K?

3. Calculate the maximum possible efficiency of a power plant that burns natural gas at a temperature of 600 kelvins, with low-temperature surroundings at 300 kelvins. How much more efficient would the plant be if it were built in the Arctic, where the low-temperature reservoir is at 250 kelvins? Why don't we build all power plants in the Arctic?

4. The Ocean Thermal Electric Conversion system (OTEC) is an example of a high-tech electric generator. It takes advantage of the fact that in the tropics, deep ocean water is at a temperature of 4°C, while the surface is at a temperature around 25°C. The idea is to find a material that boils between these temperatures. The material in the fluid form is brought up through a large pipe from the depths, and the expansion associated with its boiling is used to drive an elec-

trical turbine. The gas is then pumped back to the depths, where it condenses back into a liquid and the whole process repeats.

a. What is the maximum efficiency with which OTEC can produce electricity? (Hint: Remember to convert all temperatures to the Kelvin scale.)

b. Why do you suppose engineers are willing to pursue the scheme, given your answer in (a)?

c. What is the ultimate source of the energy generated by OTEC?

5. You have a collection of eight numbered balls; 1 through 4 are orange and 5 through 8 are green. How many different arrangements of these balls in a line are possible?

6. What percentage of those arrangements in Problem 5 have four orange balls followed by four green balls?

7. Repeat problems 5 and 6 for a collection of twelve numbered balls (six orange and six green).

8. In Problems 5–7, what happens to the probability of an ordered configuration as the total number of balls increases?

Investigations

1. Play a home movie or videotape backward. How many violations of the second law of thermodynamics can you spot?

2. Research the development of the steam engine. How efficient were the first steam engines and how was the efficiency eventually improved over time? Who were the people responsible for these improvements? Alternatively, do the same research for the automobile engine.

3. Spend a bit of time writing down all the ordered and disordered groups of objects or systems you see in a typical day. Write a very brief hypothesis for each group or system about the likelihood of the ordered and disordered items becom-

ing, respectively, disordered and ordered. Estimate the time within which this will occur. Check this diary at a later date to see if your hypotheses were correct. Are your expectations consistent with what you learned in this chapter?

4. Investigate the history of perpetual motion machines. What were some of the different types of machines imagined and built, and what were the problems associated with them?

5. Investigate and research the Carnot cycle, which is a more formal explanation of the version of the second law that says you cannot construct an engine that does nothing but convert heat completely to useful work. As part of your research, find and read a short biography of the French scientist Sadi Carnot, who was a key figure in the improvement of heat engines.

6. Investigate the design of waterwheels, which are machines that convert the gravitational potential energy of water into mechanical energy. What factors prevent such a wheel from operating with 100% efficiency?

 # WWW Resources

See the *Physics Matters* home page at **www.wiley.com/college/trefil** for valuable web links.

1. http://www.stirlingengine.com A discussion of the stirling engine.

2. http://electron4.phys.utk.edu/141/nov19/November%2019.html A discussion of entropy and ocean currents.

3. http://filebox.vt.edu/eng/mech/scott/index.html A collection of thermodynamic cycles, including steam engines and refrigerators.

4. http://www.taftan.com/thermodynamics/ Another collection of applied thermodynamics examples and descriptions, together with the laws of thermodynamics.

14

Vibrations and Waves

KEY IDEA

Waves carry energy from place to place without requiring matter to travel across the intervening distance.

PHYSICS AROUND US . . . A Day at the Beach

Few experiences are more relaxing than a day at the beach. The sight of waves washing ashore, the sound of good music, and the feel of the Sun's rays beating down help us forget about the pressure of exams and term papers. What might surprise you is that underlying all of these familiar experiences—ocean surf, sound, and the Sun's rays—is the phenomenon of waves.

Waves are all around us. Waves of water travel across the surface of the ocean and crash against the land. Waves of sound travel through the air when we listen to music. Some parts of the United States suffer from mighty seismic waves of rock and soil we know as "earthquakes." All of these experiences and many more are examples of waves in nature.

VIBRATIONS

The world around us is in constant motion. Ocean waves incessantly roll onto the shore, leaves flutter in the breeze, and pendulum clocks tick away the time. Even atoms and molecules wiggle constantly in their submicroscopic realm. All of these examples and countless more illustrate the phenomenon of **vibrations,** which are simply to-and-fro motions. You can watch the periodic vibrations of a child on a playground swing, feel the vibration of the ground as a train or heavy truck passes by, or hear the vibrations of a loud bell ringing the time. In all these instances, a back-and-forth motion occurs in time and space.

Vibration of a Pendulum

Have you ever watched the hypnotic swinging of a grandfather clock's pendulum? A pendulum, which is a weight suspended from a string or wire, swings back and forth with such regularity that for hundreds of years the most accurate clocks relied on their steady tick-tock motion.

The Italian physicist Galileo Galilei, who studied the behavior of falling objects (Chapter 3), also conducted pioneering experiments on pendulums. Galileo discovered that the time it takes for a pendulum to swing back and forth—its **period**—depends on its length: shorter pendulums swing faster (Figure 14-1a). Surprisingly, the pendulum's period does not depend on its mass or on the magnitude of the arc through which it swings (Figure 14-1b). You can try these experiments yourself by tying small weights to the end of a string.

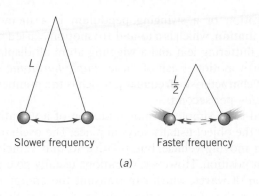

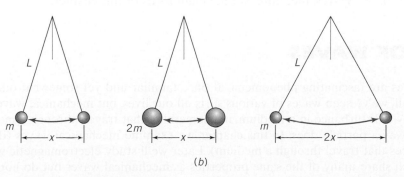

FIGURE 14-1. The periods of pendulums. (a) Shorter pendulums swing faster (higher frequency). (b) Pendulums of the same length swing at the same frequency, independent of their mass or amplitude.

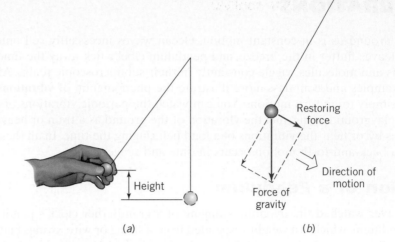

FIGURE 14-2. (*a*) Pulling a pendulum to the side increases the gravitational potential energy of the system. (*b*) Letting go of the pendulum initiates simple harmonic motion, with gravity providing the restoring force.

The steady motion of a pendulum is a consequence of gravity. You start the swinging by pushing the weight to the side, which causes it to gain gravitational potential energy (Figure 14-2*a*). When you let go, gravity acts as a *restoring force* that accelerates the weight back to swing through its central at-rest position (Figure 14-2*b*). In the absence of frictional forces, the pendulum would swing forever.

Simple Harmonic Motion

This steady vibration of a swinging pendulum is a memorable example of **simple harmonic motion,** which is a to-and-fro motion caused by a restoring force. A ringing bell, a fluttering leaf and a wiggling atom all display this kind of regular back-and-forth motion. Each of these *simple harmonic oscillators,* furthermore, has its own characteristic **frequency,** which is the number of back-and-forth vibrations, or *cycles,* per second.

The back-and-forth simple harmonic motion of a vibrating object, or oscillator, means that the object usually stays in place. The oscillator might be placed on a car or other moving vehicle, but, relative to that vehicle, it always returns to its equilibrium position. However, vibrations usually give rise to the remarkable phenomenon of waves, which can transmit the energy of the oscillator to other places. Waves, therefore, are the main focus of this chapter.

THE NATURE OF WAVES

Waves are fascinating phenomena, at once familiar and yet somewhat odd. After all, we've seen waves of various sorts all our lives, but mechanical waves are really a disturbance in a medium, not something that travels through a medium the way a particle does. In this chapter we examine mechanical waves (disturbances that travel through a medium). Later we'll study electromagnetic waves, which share many of the same properties as mechanical waves but do not need to travel in a medium. Scientists study waves by observing their distinctive behavior, particularly their unique ability to transfer energy without transferring mass. In this section we discuss some of the different properties of waves.

Energy Transfer by Waves

Energy can travel by two different carriers: the particle and the wave. For example, suppose you have a domino sitting on a table and you want to knock it over. This process requires transferring energy from you to the domino. One way to knock over the domino would be to take another domino (the "particle") and throw it. If you did this, then the muscles in your arm would impart kinetic energy to the thrown domino, which, in turn, would impart enough of that energy to the standing domino to knock it over (Figure 14-3a). In such a case, the energy is transferred through the motion of a solid piece of matter, as we have studied in Chapter 8.

Suppose, however, that you lined up a row of standing dominoes. If you knocked over the first domino, it would then knock over the second, which in turn would knock over the third, and so on (Figure 14-3b). Eventually, the falling wave of dominoes would hit the last one and you would have achieved the same result—energy has been transferred from you to the domino. In the case of the row of dominoes, however, no single object traveled from you to the most distant domino. You started a wave of falling dominoes and that wave knocked over the final one. **A wave,** then, carries energy from place to place without requiring matter to travel across the intervening distance. It's important to understand that the motion of the wave (from your hand to the final domino) is not the same as the motion of the individual dominoes in the row (which just fall over). As we shall see, these two kinds of motion—the motion of the wave and the motion of the medium—are general properties of all types of mechanical waves.

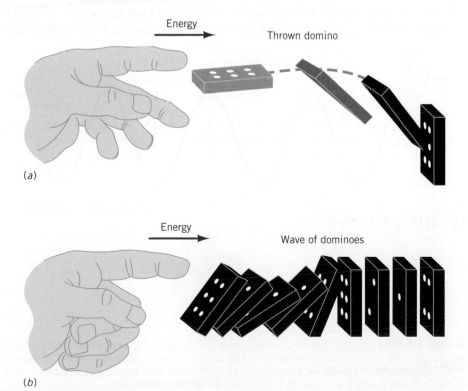

(a)

(b)

FIGURE 14-3. Two ways to knock over a domino. (a) You could throw another domino, thus imparting kinetic energy. (b) You could knock over a row of dominoes, causing a wavelike cascade.

Connection

Tsunamis

One of the most destructive waves known is the tsunami, which means "harbor wave" in Japanese. The term "tidal wave" is sometimes used, but it is a bit misleading because these waves have nothing to do with gravitationally generated tides. A tsunami is created when a geological event on the ocean floor—an undersea earthquake or a large landslide of rock off the flanks of a volcanic island—gives rise to a sudden change in the surface configuration of the sea. The effect is analogous to the wave pulse you can create by flipping one end of a rope while someone else holds the other end steady. Once the tsunami is generated, it can travel long distances—sometimes thousands of miles—and can cause great damage when it encounters land.

A word of warning: as a tsunami approaches a shore, the water is often pulled back toward the wave. As a result, someone standing on the shore suddenly sees the ocean recede, exposing the terrain of the bottom. If you are ever on shore and see this happen, get to high ground as fast as you can—the dangerous crest of the wave won't be far away! ●

On April 1, 1946, this tsunami struck Hilo, Hawaii. The man in the foreground was one of 159 people killed.

The Properties of Waves

Think about a familiar example of waves. You are standing on the banks of a quiet pond on a crisp autumn afternoon. There's no breeze and the pond in front of you is still and smooth. You pick up a rock and toss it out into the middle of the pond. As soon as the rock hits the water, a series of ripples move outward from the point of impact. In cross section, the ripples look like waves, as shown in Figure 14-4. You can use four measurements to characterize these ripples, each

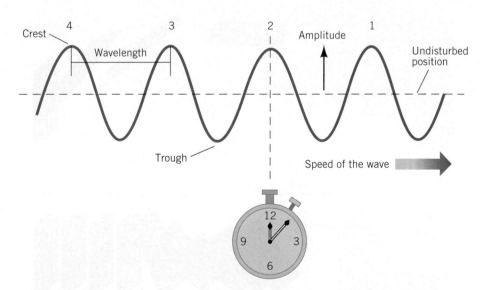

FIGURE 14-4. A cross section of a wave shows its characteristic properties: wavelength, amplitude, and speed. Successive wave crests are numbered 1, 2, 3, and 4. An observer at the position of the clock records the number of crests that pass by in a second. This is the frequency, which is measured in cycles per second, or hertz.

of which has an elevated *crest* (highest point of the wave) and a depressed *trough* (lowest point of the wave).

1. **Wavelength** is the distance between two adjacent crests—the highest points of neighboring waves. It's also the distance between two adjacent troughs or, in fact, between any two adjacent corresponding points of the wave. On a pond, the wavelength might be only a centimeter or two, while ocean waves may be tens or hundreds of meters between the crests. Wavelength is customarily represented by the Greek letter lambda (λ) and has units of distance (meters, feet, etc.).

2. The **frequency** of a wave (often represented by the letter f) is the number of wave crests that go by a given point every second. A wave that transmits one crest every second (completing one *cycle*) is said to have a frequency of 1 cycle per second or 1 **hertz** (abbreviated Hz); the units of hertz are 1/second. Small ripples on a pond might have a frequency of several hertz, while large ocean waves might arrive only once every few seconds. As with vibrations, the time for one cycle is the period of the wave, which is the reciprocal of the frequency.

3. *Velocity* is the speed and direction of the wave. Water waves typically travel a few meters per second, about the speed of walking or jogging, while sound waves travel about 340 meters (1100 feet) per second in air. (We shall see in Chapter 15 that the speed of sound can vary with temperature and altitude.)

4. **Amplitude** is the height of the wave crest above the undisturbed surface, such as the level of the calm pond before you tossed in the rock. In general terms, the amplitude is the maximum displacement of the wave from the undisturbed state of the medium, a definition that applies to all kinds of waves.

The Relationship Among Wavelength, Frequency, and Speed

A simple relationship exists among three fundamental wave measurements: wavelength, frequency, and speed. If we know any two of the three measurements, we can calculate the third using a simple equation.

1. In words:

   The speed of a wave is equal to the wavelength of the wave times the number of waves that pass by each second (frequency).

2. In an equation with words:

   Wave speed (in m/s) = Wavelength (in m) $\times$ Frequency (in Hz)

3. In an equation with symbols:

$$v = f \times \lambda$$

   where λ and f are the common symbols for wavelength and frequency.

This simple equation holds for all kinds of waves. In fact, it is not really a new formula but just the definition of *speed* (Speed = Distance divided by Time) applied to a wave.

FIGURE 14-5. Waves passing a sailboat reveal how wavelength, speed, and frequency are related. If you know the distance between wave crests (the wavelength) and the number of crests that pass each second (the frequency), then you can calculate the wave's speed.

To understand this relationship, think about waves on the water. Suppose you are sitting on a small sailboat, as shown in Figure 14-5, watching a series of wave crests passing by. You can count the number of wave crests going by every second (the frequency) and measure the distance between the crests (the wavelength). From these two numbers, you can calculate the speed of the wave. For example, if one wave crest arrives every 2 seconds (the period is 2, so the frequency is $\frac{1}{2}$) and the wave crests are 6 meters apart (the wavelength), then the waves must be traveling 6 meters every 2 seconds, which is a speed of $6 \times \frac{1}{2} = 3$ meters per second. You might look out across the water and see a particularly large wave crest that will arrive at the boat after four intervening smaller waves. You could then predict that the big wave is 30 meters away (five times the wavelength) and that it will arrive in 10 seconds (five times the period). That kind of information can be very helpful if you are plotting the best course for an America's Cup yacht race or estimating the path of potentially destructive ocean waves.

Seismic Waves

EXAMPLE
14-1

Earthquakes usually start from disturbances in Earth's crust, as huge landmasses bang up against one another. The kind of mechanical wave generated by an earthquake is called a *seismic wave*, and it plays a major role in how we study the structure of Earth's deep interior (see Thinking More About Waves at the end of this chapter). Suppose a seismic wave is detected passing through rock at a speed of 6 kilometers per second and a frequency of 2 crests per second. What is the wavelength of this wave?

REASONING AND SOLUTION: The relationship among wavelength, frequency, and speed applies to all waves. We know that

$$\text{Wave speed (in m/s)} = \text{Wavelength (in m)} \times \text{Frequency (in Hz)}$$

In this case, the speed is 6 kilometers per second, which equals 6000 meters per second, and the frequency is 2 cycles per second, which equals 2 hertz. We can rearrange the equation to solve for wavelength:

$$\text{Wavelength (in m)} = \frac{\text{Speed (in m/s)}}{\text{Frequency (in Hz)}}$$
$$= \frac{6000 \text{ m/s}}{2 \text{ Hz}}$$
$$= 3000 \text{ m}$$

These seismic waves, therefore, have a long wavelength—3 kilometers or about 1.8 miles. Nevertheless, they're much shorter than Earth's diameter and thus travel many times their wavelength before they die out. ●

KINDS OF WAVES

Waves can be classified according to the different motions of the medium through which the wave passes. The most common and important kinds of waves are transverse and longitudinal waves.

Transverse Waves

Imagine that a chip of bark or a piece of grass is lying on the surface of a pond when you throw in a rock. When the ripples go by, that floating object and the water around it move up and down, but they do not move to a different spot. At the same time, however, the wave crest moves in a direction parallel to the surface of the water. This means that the motion of the wave is different from the motion of the medium on which the wave moves—a point we made earlier about the wave in the row of dominoes. A wave like that on water, in which the motion of the medium (water in this case) is perpendicular to the direction of the wave, is called a *transverse wave*. Another example of a transverse wave is a wave transmitted along a stretched rope, string, or Slinky (Figure 14-6*a*).

You can observe (and participate in) this phenomenon if you ever go to a sporting event in a crowded stadium where fans "do the wave." Each individual simply stands up and sits down, but the visual effect is of a giant sweeping motion around the entire stadium. In this way, waves can move great distances, even though individual pieces of the transmitting medium hardly move at all.

Longitudinal Waves

Not all waves are transverse waves, like those on a rope. Sound is another form of wave, a wave that moves through the air and other media. When you talk, for example, your vocal cords move air molecules back and forth. The vibrations of these air molecules set the adjacent molecules in motion, which set the next group of molecules in motion, and so forth. A spherical sound wave moves out from your mouth, consisting of molecules of the air in motion, producing regions of alternately higher pressure (or compression) and lower pressure (or rarefaction). If you could see this wave, it would look very much like the ripples on a pond. The only difference is that in the air the wave crest that is moving out is not a raised portion of a water surface, but a denser region of air molecules. As a sound wave moves through the air, gas molecules vibrate forward and back in the same direction as (along) the wave. Sound, therefore, is a *longitudinal wave* (Figure 14-6*b*).

We explore more properties of sound waves in the next chapter. Suffice it to say that this longitudinal motion is very different from the transverse wave of a ripple in water, in which the water molecules move perpendicular to the direction of the waves. Note that in either longitudinal or transverse waves, the energy always moves in the direction of the wave, whether along a line like a rope, across a surface like a pond, or in all directions like a sound wave from your mouth. What is different is the direction of motion of the particles of the medium.

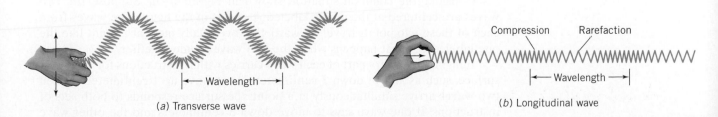

(a) Transverse wave

(b) Longitudinal wave

FIGURE 14-6. Two different kinds of waves. (*a*) in a transverse wave, the medium moves perpendicular to the direction of the wave. (*b*) In a longitudinal wave, the medium moves in the same direction as the wave.

> ### Develop Your Intuition: Doing the Wave
>
> Let's look again at the "wave" that fans in a large stadium sometimes do during ball games. Watching from a distance, you see a steady motion around and around the entire stadium. In this transverse wave, the individual people are the particles of the medium that move up and down, while the wave moves round and round. But think for a moment. Could fans in a stadium ever produce a longitudinal wave?
>
> In a longitudinal wave, the fans (the transmitting medium in this case) would have to move in the same direction as the wave. One way to accomplish this would be to have everyone lean quickly to the right as the wave crest arrives, then slowly straighten up again. It might not look as impressive as the usual "wave," but it would be longitudinal.

INTERACTIONS OF TWO OR MORE WAVES

Have you ever noticed how many different sources of sound you can hear at one time? For example, you can be driving in your car with the radio on and talking to your friends in the back seat at the same time, while still hearing the sounds of traffic outside. All of these different sound waves reach your ears at the same time, but you can recognize each one individually. How does that happen? To understand how our ears can perform this seemingly miraculous task, we need to consider how waves interact with one another.

Interference

One of the most striking features of waves occurs when waves from different sources overlap each other. This phenomenon, called **interference,** describes what happens when waves from two different sources come together at a single point. For instance, imagine that in our example of the pond, you throw two pebbles instead of one. In this case, two sets of ripples move out over the surface of the pond. One set is centered at the spot where the first pebble fell, while the other is centered where the second pebble fell. When the ripples overlap, we have two waves coming together from different sources. In this case each wave interferes with the other, and the height of the observed wave at every point is the sum of the heights (positive or negative) of the interfering waves.

Consider the common situation shown in Figure 14-7a. Suppose the two waves are centered at the points labeled A and B in the figure. The waves from each of these two points travel outward, and eventually meet at a point like the one labeled C. What happens when the two waves come together?

Imagine that every part of each wave carries with it instructions for the water surface, such as "move down 3 centimeters," or "move up 1 centimeter." When two waves arrive simultaneously at a point, the surface responds to both sets of instructions. If one wave says to move down 3 centimeters and the other wave says to move up 1 centimeter, the result is that the water surface moves down 2 centimeters.

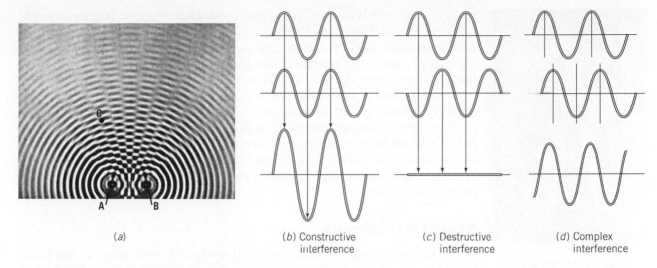

(a)   (b) Constructive
interference

(c) Destructive
interference

(d) Complex
interference

FIGURE 14-7. Examples of interference. (a) Two ripples start at points A and B; waves travel outward and eventually meet at a point like the one labeled C. The results can be (b) constructive interference, (c) destructive interference, or (d) more complex interference.

Each point on the surface of the water, then, moves a different distance up or down, depending on the instructions that are brought to it by the waves from point A and point B. One possible situation is shown in Figure 14-7b. Two waves, each carrying the command "go up 1 centimeter," arrive at a point together. The two waves act together to lift the water surface to the highest possible height it can have. This is a phenomenon called "constructive interference," or reinforcement. On the other hand, you could have the situation shown in Figure 14-7c, in which the two waves arrive at the point in such a way that one is giving the instruction to go up 1 centimeter and the other to go down 1 centimeter. In this case, the water surface does not move at all. This situation is called "destructive interference," or cancellation.

Intermediate cases, in which the two waves arrive with the crest of one displaced from the crest of the other, as in Figure 14-7d, result in a wave whose height at each point is somewhere between the sum of the crests and zero. Thus, there is a smooth transition between constructive and destructive interference.

Iridescence

A beautiful consequence of interference can be seen on paved roads on sunny afternoons. If you look carefully at the oil slicks on the pavement, you often see a phenomenon called "iridescence"—a rainbow of colors on the dark oil surface. In this case, two light waves are interfering with one another. One wave is the sunlight that bounces off the top of the oil film. The other wave is the sunlight that goes through the oil film and bounces off the bottom. These two waves may interfere constructively or destructively. As we see later, different wavelengths of light correspond to different colors. Light from different parts of the oil slick, then, produces different colors and creates the iridescent rainbow display.

When you clean the windows of your car on a warm sunny day, you often see a similar colorful phenomenon. As you wipe the squeegee along the glass, an iridescent sheen of colors appears. The colors seem to move around and then fade.

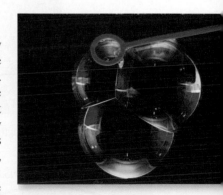

Example of iridescence in soap bubbles.

Example of iridescence in a peacock's tail feathers.

What's going on? What you are seeing is another effect of interference. The squeegee creates a thin film of water on the windshield. Two light waves—one reflecting from the top of the water and the other reflected from the glass—come together at your eye to produce colors. But as the Sun evaporates the water, the thickness of the film changes, and the wavelength corresponding to constructive interference changes as well. You see this as a change in color. When the film evaporates completely, the interference (and hence the display of colors) disappears.

Iridescence also explains many brilliant colors in nature. Peacock feathers reflect light off different layers of the feather structure; the resulting interference effects create the shimmering colors seen in the male bird. Hummingbirds, butterflies, and some kinds of fish all appear in bright, shimmering colors due to interference of light.

Standing Waves

A special kind of wave interaction, called a standing wave, plays an important role in many phenomena, from producing the musical note of a trumpet or organ to explaining the nature of the atom. The easiest way to understand the mechanics of a standing wave is to think about starting one on a rope attached to a tree. You hold the rope in your hand and shake it up and down. As a result, a wave pulse travels down the rope to where it is tied to the tree. At that point, the wave is reflected and moves back down the rope toward your hand. If, in the

FIGURE 14-8.
Examples of standing waves on a vibrating string of length L. (a) The longest wave, wavelength 2L; (b) the next longest wave, wavelength L; (c) the third longest wave, wavelength $\frac{2L}{3}$.

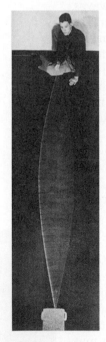

Wavelength = 2L

Wavelength = L

Wavelength = $\frac{2}{3}L$

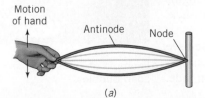

(a)

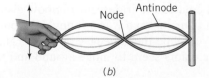

(b)

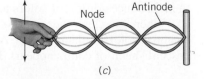

(c)

meantime, you keep shaking your hand and sending waves down the rope, those waves start to interfere with the waves coming back. If you time your hand movement just right, you can create a characteristic pattern of oscillation on the rope, as shown in Figure 14-8. This distinctive pattern is called a *standing wave* because the pattern does not move along the rope, but appears to simply stand in place.

Figure 14-8*a* shows the longest standing wavelength that can be sustained on the rope. If the length of the rope is L, the wavelength of this standing wave is $2L$. This is not, however, the only standing wave you can create. If you shake your hand faster (exactly twice as fast, in fact), you can get a pattern such as that shown in Figure 14-8*b*. If you shake the rope faster still, you can get the pattern shown in Figure 14-8*c*. In the last two patterns, there are places on the rope called *nodes* that do not move at all. The places where the rope moves the maximum distance are called *antinodes*. The wavelengths of the waves shown are L and $\frac{2}{3}L$, respectively. These types of standing waves play an important role in producing musical tones, as we see in Chapter 15.

 ## THE DOPPLER EFFECT

Once waves have been generated, their motion is independent of the source. It doesn't matter what kind of rock produces a wave in a pond; once produced, all such waves behave exactly the same way. This statement has an important consequence that was analyzed in 1842 by Austrian physicist Christian Johann Doppler (1803–1853). This consequence is called the Doppler effect in his honor. The **Doppler effect** describes the way the frequency of a wave appears to change if there is relative motion between the wave source and the observer.

Let's take sound as an example. Figure 14-9*a* shows the way a sound wave looks when the source is stationary relative to a listener—when you listen to your radio, for example. In this case everything sounds "normal."

FIGURE 14-9. The Doppler effect is illustrated by comparing (*a*) a stationary source of sound and (*b*) a moving source of sound.

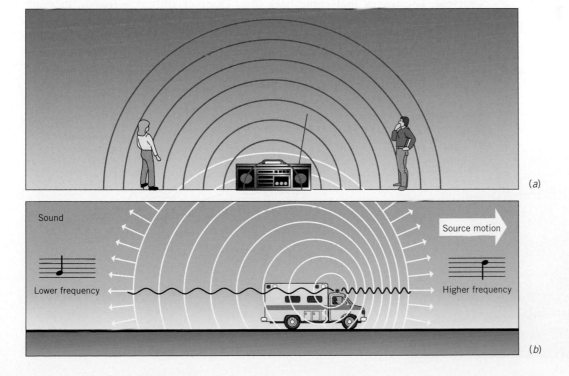

(*a*)

(*b*)

However, if the source of sound—a speeding ambulance, for example—is moving relative to the listener, a different situation occurs (Figure 14-9*b*). Periodically, a pulse of high pressure (the sound wave) moves away from the moving source of the sound wave and travels outward in a sphere, centered on the spot where the source was located when that particular pulse was emitted. By the time the source is ready to emit other pulses, it has moved. Thus the second sound-wave sphere emitted is centered at the new source location. As the source continues to move, it emits sound waves centered farther and farther from the original source—to the right in the figure—producing a characteristic pattern as shown.

To an observer standing in front of the source, the wave crests appear to be bunched up. To this observer, the frequency of the wave is higher than it would be from the same source if it were stationary—a classic example of the Doppler effect. In the case of a sound wave, this means that the sound is higher-pitched. On the other hand, if you are standing behind the moving source, the distance between crests is stretched out and it appears to you as if the wave has a lower frequency. If the wave is sound, then the pitch of the sound is lower.

You probably have heard the Doppler effect. Think of standing on a highway while cars go by at high speeds. Engine noise appears very high-pitched as a car approaches and drops in pitch as the car passes. This effect is particularly striking at automobile races, where cars are moving at very high speed. This sort of change of pitch was, in fact, the first example of the Doppler effect to be studied. Scientists hired a band of trumpeters to sit on an open railroad car and blast a single long, loud note as the train whizzed by at a carefully controlled speed. Musicians on the ground determined the pitches they heard as the train approached and as it receded, and they compared those pitches to the actual note the musicians were playing. The same sort of bunching up and stretching out of crests can happen for any wave, including light.

FIGURE 14-10. Doppler radar can measure the speed of water droplets in a tornado, moving toward the radar receiver on one side of the rotating funnel and away on the other side. In this way, the radar can determine wind speed.

The Doppler effect also has practical applications. For example, police radar units send out a pulse of electromagnetic waves that is reflected by the metal in your car. The waves that come back are "Doppler shifted"—that is, they are changed in frequency according to the Doppler effect. By comparing the frequencies of the wave that goes out and the wave that comes back, the speed of your car can be deduced. Meteorologists use a similar technique when they employ Doppler radar to measure wind speed and direction during the approach of potentially damaging storms (Figure 14-10).

(a)

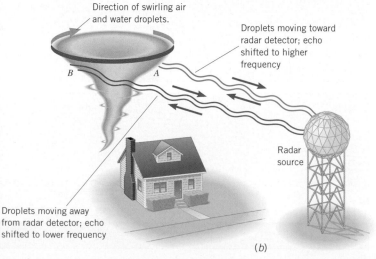

Direction of swirling air and water droplets.

Droplets moving toward radar detector; echo shifted to higher frequency

B   *A*

Radar source

Droplets moving away from radar detector; echo shifted to lower frequency

(b)

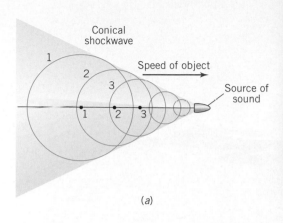

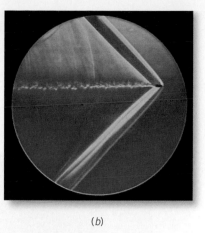

(a)

(b)

Figure 14-11. (a) When the speed of a sound source is greater than the speed of the sound waves, then the crests pile up on one another, forming a cone with the source at its apex. The edge of this cone is a region of high pressure (a shock wave) that is built up from the sound waves emitted by the speeding object. (b) A high-speed bullet that moves faster than the speed of sound creates a sonic boom.

Connection

Shock Waves and the Sound Barrier

According to the Doppler-effect relationship, the faster a source of sound is moving toward an observer, the higher the observed frequency of the source. But what happens when the source is moving so fast that its speed equals or even exceeds the speed of sound itself?

If you look back at Figure 14-9, you'll see that the crests of the emitted waves get closer together in the direction that the sound's source is moving. The faster the source moves, the closer together these crests become. When the speed of the source is the same as the speed of the sound waves, the crests pile up on one another, forming a cone with the source at its apex (Figure 14-11a). The edge of this cone is a region of high pressure, built up from the sound waves emitted by the speeding object. If the source is moving faster than the speed of sound, we call this edge a "shock wave." When a jet plane moves faster than the speed of sound, it creates a shock wave that moves along with the plane, extending down to the ground itself. We hear this high-pressure wave as it passes, causing the loud noise known as a "sonic boom." All objects that move faster than the speed of sound—supersonic aircraft, high-speed bullets, and even the tip of a bullwhip—create this boom (Figure 14-11b).

When experimental jet planes that could approach the speed of sound were first being built, they often shook from their own shock waves. Several attempts to go faster than sound were tragically unsuccessful because planes went out of control from this shaking. Pilots referred to a "sound barrier" that had to be broken for the plane to go faster than sound. There is indeed a force pushing against a plane as it approaches the speed of sound. As the plane pushes sound waves ahead of it, forming the shock wave, the waves push back against the plane by Newton's third law. The high-pressure shock wave causes water droplets in the air to condense suddenly, forming a fog behind the plane. Photos sometimes show planes emerging from this fog as if they were in fact breaking through some kind

FIGURE 14-12. When a plane exceeds the speed of sound, a high-pressure shock wave causes water droplets in the air to condense suddenly, forming a fog behind the plane. Photos sometimes show planes emerging from this fog as if they were breaking through some kind of physical barrier.

of physical barrier (Figure 14-12). But the sound barrier is simply the reaction force of the shock wave. ●

INTERACTIONS OF WAVES WITH MATTER: RESONANCE

Because waves carry energy, they are capable of affecting any matter with which they interact. We know that waves crashing into the shore eventually wear away the hardest rocks. Similar transfers of wave energy occur when light waves from lasers cut through solid metal or when a piercing sound shatters glass. The examples of water and light waves are fairly straightforward, involving little more than the wave having its effect through the brute-force delivery of energy. The example of a high-pitched sound breaking a wine glass, however, involves a more interesting process, one that physicists call **resonance.**

Imagine that you are pushing a child on a swing, and you are exhorted to "go higher." You know from experience that the way to increase the amplitude of the swing is to time your pushes so that you are pushing (thus delivering energy to the swing) just when the swing is at the highest point of its backward arc. By carefully timing your pushing, you can get the swing to go high enough to satisfy the most demanding child. In the language of physics, we say that you are timing your pushes to coincide with the *resonant frequency* of the swing.

Every physical system has one or more characteristic times or periods with which it interacts with outside influences. In the case of the swing, the characteristic time it takes for the swing to return to the high point of its arc is perhaps one swing every 3 or 4 seconds. In solid materials, the time it takes for atoms to vibrate back and forth in response to an outside force is typically less than a

The collapse of the Tacoma Narrows Bridge on November 7, 1940. The main span, 2800 feet long and 39 feet wide, began to oscillate on a day with strong, sustained winds.

millionth of a second. The *resonant frequency of the system* is defined to be 1 divided by the characteristic time during which the system responds.

The lesson of the swing, then, is that if energy is delivered to a system at the resonant frequency, the system is able to absorb large amounts of that energy. In the case of the swing, that means that the amplitude of the arc increases. In the case of a glass, if the crests of the sound wave arrive at just the right times, the glass experiences bigger and bigger oscillations until it shatters.

Perhaps the most famous example of the power of resonance took place in the state of Washington in 1940. The Tacoma Narrows Bridge had been completed and had been in operation for just four months when a strong wind came up that just happened to amplify the bridge's natural resonant frequency. The bridge had not been built with enough support to keep the roadway steady in such a wind. The bridge first began to sway and then to move in surges of ever increasing amplitude until it broke apart in spectacular fashion. The Tacoma Narrows Bridge has become a cautionary tale told to every beginning engineering student.

One of the authors (JT) remembers how those engineering majors would turn pale at football games at the University of Illinois. Students sat on an upper deck in the stadium, and, at crucial points in the game, they would start clapping and stamping their feet in unison. At these junctures, you could see the railing at the front of the upper deck begin to move rhythmically up and down. The engineering students were probably remembering those classroom films of the Tacoma Narrows Bridge collapse.

Resonance is not always destructive. For example, radio tuners are designed to have a variable resonant frequency. When you tune in a radio station, you're setting the tuner to allow one particular frequency of radio waves to be resonant. The radio electronics amplify that frequency so you can hear it while the electronics block out all other frequencies. When you turn to a different station's frequency, the new frequency becomes the resonant frequency. This is how the tuner selects just the frequency you want to hear from among all the radio waves broadcast in your listening area.

Develop Your Intuition: Avoiding Disaster

It is standard procedure in military units for soldiers to "break step" when they come to a bridge. Why is this practice followed?

Soldiers marching in step deliver precisely timed impacts to the ground. If those impacts happen to match a bridge's resonant frequency, it is possible that a repeat of the Tacoma Narrows Bridge disaster could occur.

Better safe than sorry!

THINKING MORE ABOUT

Waves: Seismic Waves and Earth's Interior

Scientists have tried for hundreds of years to learn the causes and behaviors of earthquakes, which have caused death and destruction throughout recorded human history. Today we have a pretty good understanding of why and how they occur. However, along the way, scientists found that you can't study earthquakes without first understanding waves, and the study of waves in the Earth led to our first glimpses of Earth's interior structure.

As we discussed in Example 14-1, earthquakes generate seismic waves, which are waves that travel through the Earth. Instruments to detect these waves and measure their properties are called seismographs. Using these instruments, geologists of the nineteenth century found that there are two different kinds of seismic waves (Figure 14-13). Those that reach the seismograph soonest after an earthquake are called primary waves, or P waves. P waves are longitudinal waves, like sound waves, and can travel through solid or liquid rock. Seismographs also detect other waves, called secondary waves or S waves. These events are transverse waves, like ripples on the surface

FIGURE 14-13. Two different kinds of seismic waves, longitudinal P waves and transverse S waves, travel through Earth. Geologists use a seismograph to detect these waves.

This damage was caused by a strong earthquake that struck Kobe, Japan, in January 1995.

of a pond. S waves can travel through solid rock, but not molten rock.

Scientists observed the pattern of seismic waves that were detected by seismographs around the world. Then they correlated their observations with the location of the initiating earthquake (called the earthquake's "epicenter"). After some years, they noted that S waves were never detected in a broad zone on the side of the Earth opposite from the location of the earthquake (Figure 14-14). In contrast, P waves were detected on the side of Earth opposite the earthquake, but not in either of two zones located at particular angles from the earthquake's location.

The Croatian Andrija Mohorovicic and the German Beno Gutenberg first worked out the explanation for these patterns of wave propagation through the Earth in the early twentieth century. Mohorovicic showed that all seismic waves are dispersed in a regular fashion when they encounter a boundary between Earth's outer crust and an inner layer of dense rock (the "mantle"), about 20 miles below Earth's surface—a boundary now known as the "Moho discontinuity." Guten-

berg created a series of mathematical models of the Earth to see which model could best explain the observed pattern of wave detection. It turned out that the observed pattern fit a model of the Earth with a dense solid inner core surrounded by a molten outer core, all centered within the mantle. The S waves cannot travel through this molten part of the Earth, so they never reach the far side; in effect, the outer core casts a shadow for the S waves. The P waves can travel through the liquid outer core, but bend when they enter the core and again when they leave it. These bends in their path keep them from penetrating to the zones shown in the figure.

Better instrumentation has allowed succeeding generations of geophysicists to refine this model of Earth's deep interior and to determine the depth of boundaries between inner core, outer core, and mantle. But Gutenberg's basic model, deduced by trial and error, is still correct. How does this trial and error approach fit into the scientific method described in Chapter 1? Is it essential for a scientist to have a fully formed hypothesis in order to make important discoveries?

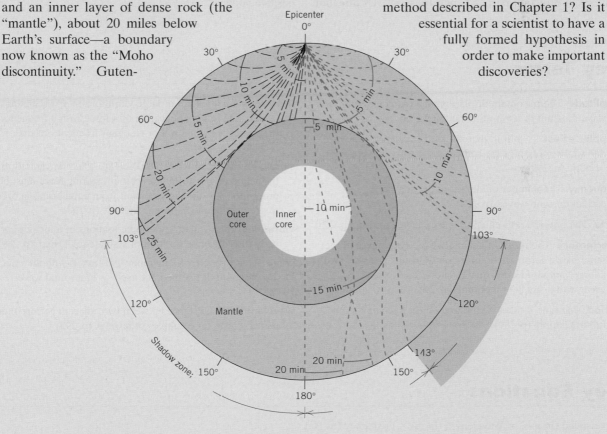

FIGURE 14-14. Seismic waves passing through Earth take a variety of paths. The speeds of P (longitudinal) and S (transverse) waves differ depending on the type of rock, its temperature, and the pressure. Also, S waves cannot travel through regions of molten rock.

Summary

Vibrations are back-and-forth motions of an object in time and space. The time required for a single oscillation is the **period.** A steadily swinging pendulum is an example of **simple harmonic motion,** which is a to-and-fro motion caused by a restoring force.

Waves, caused by vibrations, are a means of transferring energy across a distance. Mechanical waves move in a medium, but the movement of the wave is not the same as the movement of the medium. Waves are characterized by several quantities: **wavelength** (the distance between crests of the wave), **frequency** (the number of waves that go by a point each second), **amplitude** (the maximum height of the wave), and speed. Frequency is measured in a unit called the **hertz** (Hz); 1 Hz corresponds to one crest passing a point each second. These quantities are related by the equation:

$$\text{Speed} = \text{Frequency} \times \text{Wavelength}$$

Waves can be transverse (in which case the medium moves in a direction perpendicular to the wave's direction of motion) or longitudinal (in which case the medium moves in the same direction as the wave).

When two waves arrive simultaneously at a point, they interfere. The resulting disturbance is the sum of the disturbances from each wave by itself. **Interference** can be constructive (in which case the waves reinforce one another) or destructive (in which case they cancel one another out). Standing waves are disturbances that exist in a medium but do not travel. Standing waves can be thought of as the result of interference between outgoing and reflected waves.

Waves emitted by a moving source appear to have a higher frequency if the source is moving toward the observer and a lower frequency if it is moving away—a phenomenon known as the **Doppler effect.**

When waves strike an object, they can have a large effect if the wave crests arrive at intervals corresponding to the object's internal resonant frequency. This **resonance** effect can cause severe damage to structures but also has many positive applications.

Key Terms

amplitude The maximum displacement from equilibrium of a wave medium or a vibrating body. (p. 293)

Doppler effect A shift in the observed frequency of a wave due to the motion of the source of the wave, the observer, or both. (p. 299)

frequency The number of vibrations per second in oscillations and waves. (p. 290)

hertz The physical unit equal to one cycle per second. (p. 293)

interference The result of having two or more waves in the same place; constructive interference results in a wave with a larger amplitude; destructive interference results in a wave with a smaller amplitude. (p. 296)

period The time it takes a vibrating object to complete one full oscillation. (p. 289)

resonance The buildup of large vibrations in an oscillating system, due to the application of a force with a frequency that matches the natural or resonant frequency of the oscillator. (p. 302)

simple harmonic motion The oscillatory motion caused by the action of a restoring force; the restoring force must be proportional to the displacement from equilibrium in order to produce simple harmonic motion. (p. 290)

vibrations The oscillating, periodic motions of a medium or body forced from a position of stable equilibrium. (p. 289)

wave A disturbance of a medium that travels without a net displacement of the medium itself, as in a sound wave. (p. 291)

wavelength The distance between two adjacent corresponding points (crests, for example) of a wave. (p. 293)

Key Equations

Wave speed (in m/s) = Wavelength (in m) × Frequency (in Hz)

$$\text{Wavelength (in m)} = \frac{\text{Speed (in m/s)}}{\text{Frequency (in Hz)}}$$

1 hertz = 1 cycle/second

Review

1. What is a wave?

2. How can a wave carry energy? Give an example.

3. What is meant by simple harmonic motion? Explain how a weight on a spring or a pendulum represents this. How is this like a wave?

4. What is meant by the crest of a wave? What is meant by the trough?

5. What is the wavelength of a wave? What is its amplitude?

6. What is the frequency of a wave or harmonic oscillator? What is the period? How are these two quantities/concepts related to one another?

7. What is the commonly used unit of frequency? What are its units?

8. What is the velocity of a wave? How is it related to the frequency?

9. What is a transverse wave? Give an example.

10. What is a longitudinal wave? Give an example.

11. How would you measure the wavelength of a longitudinal wave on a stretched spring? In terms of the motion of a single coil of the spring, what are the period, frequency, and amplitude of the wave?

12. What does it mean to say that two waves interfere?

13. What is the difference between constructive and destructive interference?

14. What happens when two waves completely interfere with one another destructively? Constructively? Explain in terms of an ocean wave.

15. What is meant by a standing wave?

16. What is a node? What is an antinode?

17. How many nodes are there in a standing wave that is two wavelengths long? Three wavelengths? One wavelength?

18. What is the Doppler effect? Give an everyday example.

19. How does a police or fire siren sound as it approaches you? As it leaves you? What is happening to the wavelength and frequency of the sound waves as this occurs?

20. What is resonance? How could it contribute to the shattering of a glass or the destruction of a bridge?

21. What is the difference between resonance and interference?

22. If there were an earthquake off the coast of Alaska of a very large magnitude, how could that energy be transferred over very large distances? What type of waves in what different mediums might be involved?

Questions

1. Jim and Gina are swinging on adjacent, equal length swings at the school playground. Jim weighs about twice as much as Gina. Who, if either, will take less time to swing back and forth? What, if anything, will change if Jim swings while standing on the seat of his swing?

2. The magnitude of the arc through which a pendulum swings does not affect its period. If this were not true, the construction of a pendulum clock would be much more difficult. Why?

3. Two waves that travel through the same medium are sketched in the figure. Which one has the longer wavelength? Which one has the smaller amplitude? Which one has the higher frequency? Which one has the shorter period?

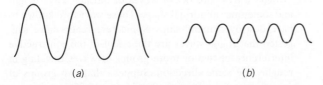

(a)   (b)

4. Two waves have the same speed. The first has twice the frequency of the second. Compare the wavelengths of the two waves.

5. A single wave pulse on a string is created by wiggling the end up and down once. The wave travels to the right (see figure). What kind of wave is this: transverse or longitudinal? What direction is the string moving at points A, B, and C in the figure? (*Hint:* Think about what the wave would look like a short time later.)

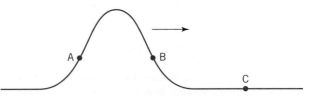

6. What happens to a piece of driftwood in a lake with waves? In what direction should it move as a result of the waves? Given this, how is energy transferred by a wave?

7. "Doing the wave" is a common activity in large football stadiums. Is the "wave" an example of a transverse or longitudinal wave? Explain.

8. **A.** In the elementary grades, some teachers illustrate a wave with the following activity. The students stand in a straight line side by side, shoulder to shoulder, with their hands at their sides. When the first student is

tapped on the shoulder by the teacher, that student then touches the next student at her side, who in turn then touches the next student down the line until the final student is touched.

a. This is an example of which type of wave? Explain.
b. If the pupils were holding hands, would the time for the wave to reach the last student be longer or shorter than in the first example? Why?
c. If the students were standing at arm's length, instead of side by side, would the time it takes for the wave to reach the last student be shorter or longer than in the first example? Why?
d. Using this information, can you relate the speed of a longitudinal wave to the density of the medium?

B. Instead of touching, suppose that the students jump up off the floor and that the next pupil cannot jump until the previous student is on the floor again. In this case, which type of wave do you have? Explain.

9. Why do waves break as they approach the shore?

10. What effect does the medium through which a wave moves have on the speed of transmission?

11. Two waves on a string travel toward each other as shown in the figure. They are 6 meters apart and traveling at 3 meters per second. Sketch the situation 1 second later. Sketch the situation 2 seconds later.

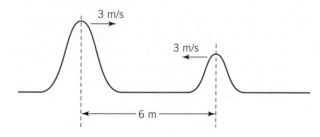

12. Two waves on a string travel toward each other as shown in the figure. They are 6 meters apart and traveling at 3 meters per second. Sketch the situation 1 second later. Sketch the situation 2 seconds later.

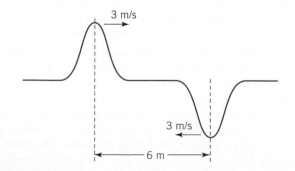

13. Two wave pulses travel toward each other as shown in the figure. (These pulse shapes are unrealistic for waves on a string but that does not matter for this problem.) They are moving at 1 meter per second. Sketch the situation 2 seconds later. Sketch the situation 3 seconds later. Sketch the situation 5 seconds later.

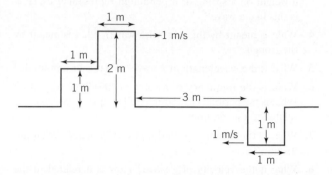

14. How many nodes and antinodes are there in the standing wave pattern shown?

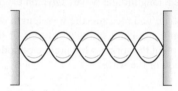

15. A police siren emits a 500-Hz tone. If the police car is chasing you, but catching up, how does that affect the sound you hear? What if you are pulling away?

16. What type of unique information might a Doppler radar give you that ordinary radar would not?

17. Suppose your car desperately needs shock absorbers, so that the coil or leaf springs are essentially supporting the weight of the car. You are driving down a long stretch of road that has equally spaced bumps in it. At one particular speed, but not others, your car bounces violently in reaction to driving over the bumps. What is happening?

18. If the speed of a wave doubles while the wavelength remains the same, what happens to the frequency?

19. "Rogue waves" are ocean waves of unusual wave height and sometimes abnormal shape. These waves are known to have destroyed many ships. Some researchers have suggested that rogue waves are the result of the constructive interference of two or more groups of waves traveling in roughly the same direction. Suppose that two groups of waves have wave heights of 15 feet and 25 feet, respectively, and are traveling in the same direction. A rogue wave near these groups of waves is observed to have a height of 50 feet. Does this support the theory that rogue waves are created by interference?

Problem-Solving Example

**EXAMPLE
14-2**

Standing Waves

Suppose the speed of a wave on a given piece of rope is 6 m/s and the rope is 3 m long. How often do you have to move your hand to produce each of the three standing waves shown in Figure 14-8?

REASONING AND SOLUTION: The frequency with which you shake your hand is the frequency of the wave you send out. To solve this problem we have to remember that for any wave the wavelength, frequency, and speed are related by

$$\text{Speed} = \text{Frequency} \times \text{Wavelength}$$

Since we know the speed of the wave (6 m/s) as well as the wavelengths of the three waves ($2L$, L, and $\frac{2}{3}L$, where

$L = 3$ m), it's simply a matter of using this equation. For the longest wave, the wavelength is 6 m (twice the length of the 3-meter rope), so that

$$6 \text{ m/s} = \text{Frequency} \times 6 \text{ m}$$

$$\text{Frequency} = \frac{6 \text{ m/s}}{6 \text{ m}}$$

$$= 1/\text{s or 1 Hz}$$

In other words, to obtain the longest possible standing wave, you have to shake your hand once a second. For the other two standing waves, the wavelengths are 3 and 2 m, respectively, so the frequencies are $\frac{6}{3} = 2$ Hz and $\frac{6}{2} = 3$ Hz. ●

Problems

1. You are pushing your little sister on a swing and in 1.5 minutes you make 45 pushes. What is the frequency (in hertz) of your swing pushing efforts?

2. Andrea was watching her brother in the ocean and noticed that the waves were coming in on the beach at a frequency of 0.33 Hz. How many waves hit the beach in 15 s?

3. Andrea asked her brother to take a 6-ft floating raft out of the water near the wave-swept shore. Using this raft as a measuring tool, she estimated that the wavelengths of these particular ocean waves were about 9 ft. How fast are these surface ocean waves if the frequency remains 0.33 Hz?

4. If an ocean wave passes a stationary point every 3 seconds and has a velocity of 12 m/s, what is the wavelength of the wave? Can you tell what the amplitude of the wave is from this information?

5. A standing wave on a rope 2-m long has two nodes.

   a. How many wavelengths does it have? If it has three nodes? If it has five nodes?

   b. Draw the standing waves in part (a) with two, three, and five nodes. Label the nodes and wavelengths.

   c. What wavelengths can create standing waves?

Investigations

1. Investigate how energy travels through Earth in an earthquake. What exactly is a seismic wave? Trace the development of seismography through history.

2. Write down a list of different phenomena in the physical world that can be represented by waves. Can anything with a simple, periodic, repeating characteristic be modeled with and described by waves? How about any object that moves in a circle at a constant speed through time? A pendulum? Something that just vibrates back and forth?

3. Next time you are near a lake or large body of water, stop and examine the waves. Can you estimate the amplitude and wavelength of the waves that day? Write down some notes on what might have caused these waves. Is there any debris floating in the waves? If so, how does it move? Investigate as many of the wave phenomena discussed in this chapter as you can. Organize your observations into a notebook, noting consistencies and inconsistencies with what you have learned in this chapter.

4. If you have access to a wave tank (usually a small tank or old aquarium) use it to further explore the nature of waves. Pay special attention to wave interference and the reflection of waves off the side of the tank.

5. Explore the history of tsunamis, or tidal waves. How often and where do they strike? What would be the effect of an impact of a large asteroid in the middle of the ocean in terms of these waves? Have popular film and the media overexaggerated this possible occurrence?

6. Investigate further just how engineers take resonance into account when designing structures. What specifically is done to avoid problems like the collapse of the Tacoma Narrows bridge?

7. Locate an old-fashioned Slinky, a coiled spring that has been used mostly as a child's toy. Use it to generate transverse and longitudinal waves and investigate as many of the wave phenomena discussed in this chapter as possible.

8. Investigate the design and use of atomic clocks, which rely on the precise and regular vibrations of individual atoms to measure time. What atoms are used in such a clock? How is the time unit of 1 second defined?

9. Investigate the technology of laser interferometry, which relies on waves of light energy that bounce off a surface.

How is this technique used to measure extremely short distances?

10. Modern seismometers can measure ground vibrations as small as a footstep. Investigate the design of such an instrument. How are tiny motions of the ground amplified?

 # WWW Resources

See the *Physics Matters* home page at **www.wiley.com/college/trefil** for valuable web links.

1. **http://www.falstad.com/ripple/** A Java applet simulation of a ripple tank, allowing all kinds of 2-D wave experiments and demonstrations with reflectors, lenses, etc.

2. **http://www.phy.ntnu.edu.tw/java/waveSuperposition/waveSuperposition.html** An animated Java demonstration of superposition of waves.

3. **http://users.erols.com/renau/harmonics.html** A very nice standing waves applet.

15 | Sound

KEY IDEA

Sound is a longitudinal wave that travels through a solid, liquid, or gaseous medium.

PHYSICS AROUND US . . . The Concert

Imagine yourself at a rock concert. As the house lights dim and the laser show starts, the first loud chords reach your ears. You notice several different instruments—guitars, drums, and synthesizers, not to mention the human voice—each contributing its distinctive part to the overall dynamic sound.

Or perhaps your tastes run more to symphonic music. A classical orchestra features an even greater variety of instruments, from the fat seven-foot-tall double basses to the slender piccolo, which measures less than 1 foot in length.

In diverse cultures across the globe, human inventiveness has led to an astonishingly rich array of string, wind, and percussion instruments, each with its own characteristic sound. What exactly is sound, and how can musical instruments produce such a range of sounds that are pleasing to our ears?

311

PROPERTIES OF SOUND

As we have seen in Chapter 14, sound is a longitudinal wave that travels through a medium such as air, water, or rock. When you talk, for example, your vocal cords vibrate in the surrounding air. This vibration produces a series of slightly higher and slightly lower air pressures, which move through the space surrounding you. Thus, sound is a longitudinal pressure wave. When a higher-pressure part of the wave strikes your friend's eardrum, the eardrum is pushed in. When a lower-pressure part of the wave strikes the eardrum, it moves back out. From this motion of the eardrum, your friend can detect the information transmitted by your sound wave.

In Figure 15-1a we show a snapshot of what that sound wave looks like after it has left your mouth. The dots represent molecules of air, and you can see that in some places these particles are packed together more tightly, while in others they are scattered more thinly. It is these successive regions of compression and rarefaction that travel through the air and make up the sound wave. If we take another snapshot a moment after this one (Figure 15-1b), we see that the whole pattern has moved outward, with the amount of movement determined by the speed of the wave.

You can make an analogy between the sound wave shown in Figure 15-1 and the ripples on water we analyzed in Chapter 14 by graphing the pressure versus distance along the wave's line of motion. In this case, we get a graph such as the one shown in Figure 15-2a. The high pressures correspond to the regions where air molecules are packed more tightly together, and low pressures correspond to regions where they are scattered more thinly. This graph looks just like a cross-sectional view of a ripple on the water. It even behaves like a ripple in that, if you

make a pressure-versus-distance graph for the second pattern (Figure 15-1b), it will look like the graph of Figure 15-2b. In other words, the pressure graph associated with the sound wave moves along just as if it were a wave on water.

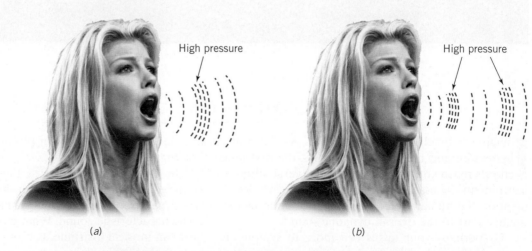

(a)   (b)

Figure 15-1. (a) Your voice creates a distinctive pattern of compressions and rarefactions in air. (b) As you speak, the sound waves move away from your mouth.

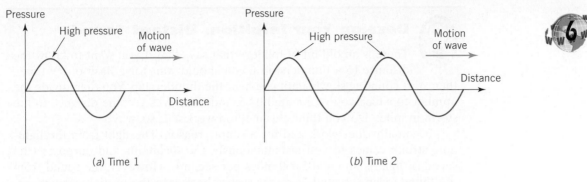

FIGURE 15-2. (*a*) The sound of your voice can be plotted on a graph as a distinctive pattern of air pressure. (*b*) As you speak, the pattern moves through the air.

The Speed of Sound

The speed of sound in air varies with the temperature of the air. At 0°C, the speed of sound is 331 m/s, while at 20°C (close to normal room temperature), the speed is a bit faster—about 344 m/s. This temperature effect on the speed of sound has an intriguing consequence. Have you ever noticed how much easier it is to hear distant voices and other sounds at night? During the daytime, upper levels of the air are generally cooler than those near the ground, so during the day sound waves tend to bend upward because the upper air levels slow them down (Figure 15-3*a*). At night, just the opposite effect occurs: sound waves tend to stay closer to the ground, and sound seems to travel farther (Figure 15-3*b*). This bending of sound waves is an example of a more general phenomenon known as *refraction*, which we will study in more detail in Chapter 20.

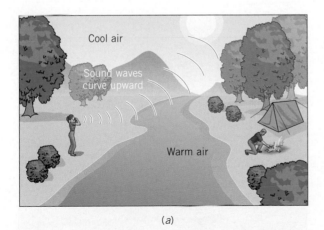

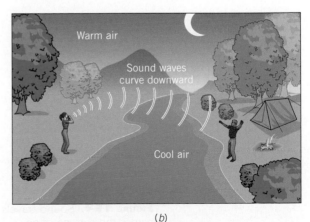

FIGURE 15-3. Sound waves bend when passing through layers of air at different temperatures. (*a*) During the daytime, upper layers of air are cooler, so sound waves tend to bend upward. (*b*) At night, upper layers of air tend to be warmer, so sound waves bend downward. That's why it's often easier to hear sounds from a distance at night.

Develop Your Intuition: Distant Sounds

There is an old bit of folklore that says that if you want to know the distance to a thunderstorm, you should wait for a flash of lightning, then start counting slowly until you hear the thunderclap. You then divide the number to which you've counted by 5, and that gives you the distance to the storm in miles. Do you think this method works? If so, why?

It actually does work and for a simple reason. The light from the lightning stroke comes to you instantaneously, for all intents and purposes (the speed of light is about 186,000 miles per second). However, the sound from the thunderclap (created when air rushes back into the partial vacuum created by the lightning) travels much more slowly and arrives much later. If you count slowly ("one one-thousand, two one-thousand, three one-thousand . . ."), you have estimated the number of seconds it took the sound from the storm to reach you. Sound travels at about 1000 feet per second, and there are about 5000 feet in a mile (5280, to be exact). Thus, dividing by 5 gives you an estimate of how many 5000-foot lengths (i.e., how many miles) the sound traveled before it reached you.

Note that to apply this method using the metric system, just divide the number of seconds by 3 to get the approximate distance in kilometers (1 km ~ 3250 feet).

Echoes

EXAMPLE
15-1

Have you ever visited an echo lake? You can find them all across North America, from Maine to California. Imagine standing on the shore, shouting "Hello!" and hearing the distinct echo from the far shore exactly 4 seconds later. What is the distance L across the lake?

REASONING AND SOLUTION: An echo is a sound wave that travels across a distance, bounces off some surface, and comes back to your ears. At an echo lake,

FIGURE 15-4. Sound travels across the surface of an echo lake, bounces off the far side, and comes back to its source. The distance across the lake can be measured by the time it takes for the echo to return.

Path of outgoing sound wave, distance L

Path of returning echo, distance L

the sound must travel across the lake and back, for a total distance of 2L (Figure 15-4). Recall the equation for distance in terms of speed and time:

$$\text{Distance (m)} = \text{Speed (m/s)} \times \text{Time (s)}$$

In this problem,

$$2L = 340 \text{ m/s} \times 4 \text{ s}$$
$$= 1360 \text{ m}$$

Therefore,

$$L = 1360/2 \text{ m} = 680 \text{ m}$$

In this way you can determine that the echo lake is about 2000 feet across just by standing on the shore and shouting. ●

The Nature of Sound Waves

Think about your many everyday experiences with sound. From your own observations, you can recognize some important attributes of sound:

1. **Sound travels in media other than air.** Have you ever been in a city and heard the sound of a subway deep beneath your feet? Or have you ever heard recordings of whale songs? These experiences should convince you that sound travels in media other than air—through water and solid materials, for example. How else could you hear your neighbor's stereo through the dorm walls? In general, air is a rather poor conductor of sound, and sound waves travel much faster and farther in water and in rock than they do in air.

2. **There are ways of generating sound waves other than using vocal cords.** Think about all the many ways that you can produce a sound. You can clap your hands, pluck a guitar string, set off a firecracker, or sand a piece of wood. Indeed, any event that causes a compression–rarefaction process in a medium can generate a sound wave. A hammer hitting a nail, for example, makes a sound because it compresses air around it (in this case, the hammer produces a sound pulse rather than a continuous wave). In a stereo system, a flexible diaphragm (often made of paper) in the speaker moves back and forth in response to electrical signals and thus causes increases and decreases in the air pressure in front of the speaker, similar to the sound wave caused by your vocal cords.

3. **Sound waves have all the properties of waves discussed in Chapter 14.** Unlike waves on the surface of water, sound waves are invisible. Nevertheless, they exhibit the full range of wavelike properties. As we have seen, sound waves have speed (about 340 m/s in air), and they have frequency and amplitude, as we describe in more detail later. Sound waves beautifully illustrate the Doppler effect, as you've experienced whenever a fast-moving vehicle whizzes by. (Indeed, as we mention in Chapter 14, the Doppler effect was first investigated using sound waves.) Finally, as we shall see, sound waves exhibit interference—a fact that is critical in the design of auditoriums.

Hearing and the Human Ear

The human ear, shown in Figure 15-5, is a complex organ that allows us to detect sound waves. To accomplish this feat, the ear senses the rapid changes in air pressure associated with sound and transmits these changes to the brain. First, as noted earlier, the membrane we call the eardrum vibrates in response to the pressure of the incoming sound wave. This vibration, in turn, is transmitted through a series of small bones to another membrane located on the cochlea, a coiled tube in the inner ear. The vibrations of this second membrane cause, in their turn, vibrations in the fluid inside the cochlea. The inside of the cochlea is formed in such a way that sounds of different frequencies have maximum effect at different places, stimulating different hair cells lining the walls of the cochlea. The movements of these hairs stimulate auditory nerves, which carry to the brain the signals you interpret as sound.

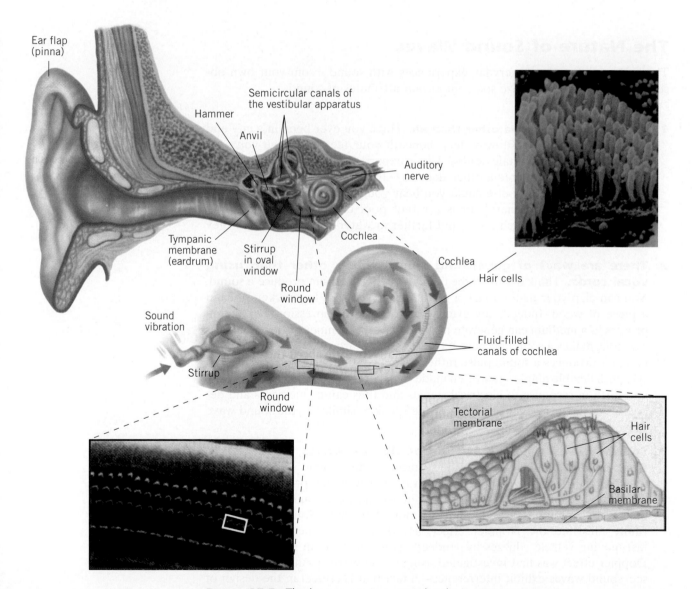

FIGURE 15-5. The human ear is a marvelously sensitive instrument.

Humans use sound to communicate, as do many other animals. But some animals have refined the use of sound to a much higher level. In 1793, the Italian physiologist Lazzaro Spallanzani did some experiments that established that bats use sound to locate their prey. First Spallanzani blinded some of the bats that lived in the cathedral tower in Pavia and then turned them loose. Weeks later, those bats had fresh insects in their stomachs, proving that they did not need sight to locate food. Similar experiments with bats that were deafened, however, showed that they could neither fly nor locate insects.

Today, we understand that bats navigate by emitting high-pitched sound waves—frequencies as high as 150,000 Hz, which is much higher than can be heard by humans. They then listen for the echo of those waves bouncing off other objects. By measuring the time it takes for a pulse of sound waves to go out, be reflected, and come back, the bat can determine the distance to surrounding objects, particularly the flying insects that constitute its diet (Figure 15-6). This process is called *echolocation*. Typically, a bat can detect the presence of an insect up to 10 meters away. In addition, the bat can use the Doppler effect to tell whether the target is moving, detecting the slight difference in frequency in an echo from an insect moving away from the bat (slightly lower frequency) or moving toward it (slightly higher frequency).

By contrast, some animals (elephants, for example) routinely use sound waves that are at the lowest range of human hearing, in the 20–40 Hz range, to communicate with one another over long distances. For example, humans

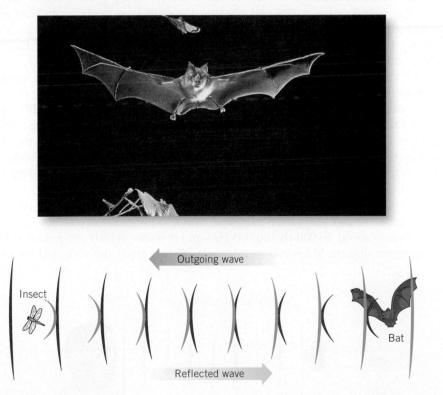

FIGURE 15-6. A bat emits a sound wave, which reflects off its target. By sensing the time it takes the sound to go out and back, the bat can tell how far away the target is.

experience the mating call of the female elephant more as a vibration than as sound, but the call attracts bull elephants from many miles away.

Whales, dolphins, and porpoises use echolocation as a navigation tool in the ocean, much as bats do in air. Sometimes, however, the sounds they emit are in the audible range for humans. Perhaps the most famous sophisticated uses of sound by animals are the songs of the humpback whales, which have been used in many commercial recordings. The function of these distinctively haunting songs remains a mystery. It appears, however, that all humpback whales in a wide area of ocean (the southern Atlantic, for example) sing similar songs, although some individual whales may leave out parts. Furthermore, the songs seem to change year by year, and whales in a given area are likely to change their songs together. We certainly have a lot to learn about how animals use sounds to communicate.

Connection
Applications of Ultrasound

We've mentioned that a bat emits sound waves with frequencies as high as 150,000 Hz. Sound waves with such high frequencies, above the 20,000-Hz limit of human hearing, are often called ultrasound. To see why such high-frequency sound waves can be very useful, think again about the bat searching for prey. Bats hunt insects, which are pretty small objects. To locate such a small object with a sound wave, you need a wave with a very small wavelength; a wave with a large wavelength would not reflect but would just keep going past the tiny insect. But sound waves with small wavelengths must have high frequencies since sound travels at a constant speed in air and speed equals wavelength times frequency. In other words, the advantage of ultrasonic frequencies is that they can reflect off very small objects.

Ultrasound is used in many different modern devices. For example, cameras with automatic range finders use beams of ultrasound to determine the distance to the object they are aimed at and set the camera's focus appropriately (Figure 15-7). Probably the most well-known application of ultrasound is in medical diagnosis. Pulses of sound at ultrasonic frequencies can travel through the body and partially reflect from every boundary between surfaces within the body. The reflections can be put together to form an image of organs within the body, such as the heart or a fetus, without any invasive surgery (Figure 15-8).

Ultrasonic frequencies are also used in sonar, which is a word derived from the phrase "sound navigation ranging." Sonar equipment sends out beams of ultrasound (about 20,000 to 100,000 Hz) under water, which are reflected from underwater objects. Ships can use the information from the reflected beams to

FIGURE 15-7. An autofocusing camera emits a pulse of ultrasound and measures the time for the echo to return.

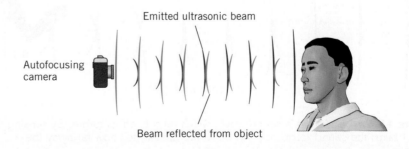

Emitted ultrasonic beam

Autofocusing camera

Beam reflected from object

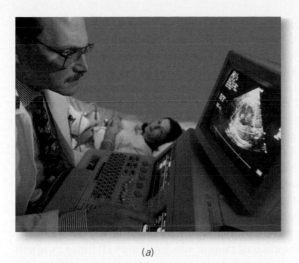
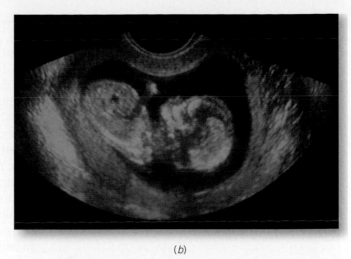

(a)   *(b)*

FIGURE 15-8. (*a*) The ultrasound emitter is placed on a pregnant woman's abdomen. (*b*) An image of the fetus is produced.

navigate around shallow reefs or detect sunken vessels, submarines, or even schools of fish. The physics involved is much the same as in echolocation: determining the location of objects by the use of sound instead of light. ●

INTENSITY AND FREQUENCY

The two attributes of sound that are most obvious to us in our everyday experience are its intensity (loudness) and its frequency (pitch). Both of these characteristics are related to physical properties of the sound wave.

Intensity and Loudness

Sound *intensity* is a measure of the energy of a sound wave and is a quantitative term that refers to a specific physical measurement: the energy carried by a wave through an area per unit of time. The intensity varies as the square of the wave's amplitude. Thus, the greater the difference in pressure between the regions of compression and rarefaction of the sound wave, the higher is its intensity. Intensity is closely related to the more subjective word **loudness,** which refers to how the sound is perceived by us. For example, a sound that seems loud and out of place in a quiet room may be almost unnoticeable when it's heard on a busy street corner, even though it has the same intensity.

The intensity of a sound is usually expressed in a unit called the **decibel (dB),** which is actually $\frac{1}{10}$ of a unit called the "bel," a unit named after the American scientist Alexander Graham Bell, who invented the telephone. (Remember from Chapter 2 that the prefix deci- means $\frac{1}{10}$.) For reasons of convenience, and because of long historical precedent, scientists and engineers use the decibel instead of the bel.

The decibel unit describes the relative intensity delivered by the sound wave. An incoming sound intensity of 10^{-12} watts per square meter—a sound that is

TABLE 15-1 Decibel Ratings	
Total deafness may occur	>160 dB
Jet plane taking off nearby	150 dB
Jackhammer	130 dB
Threshold of pain	120 dB
Rock concert; siren	110 dB
Harmful ranges for humans	>90 dB
Food blender; screaming baby	90 dB
Busy street traffic	70 dB
Normal conversation	60 dB
Buzzing mosquito	40 dB
Average whisper	30 dB
Pin dropped on a hard surface	20 dB
Rustling leaves	10 dB
Softest sound a human can hear	0 dB

barely audible to the human ear—is called 0 on this scale. A sound with 10 times that intensity is 10 dB, a sound with 100 (10^2) times that energy is 20 dB, a sound with 1000 (10^3) times that intensity is 30 dB, and so on. In Table 15-1 we give some familiar sounds and their decibel ratings.

This measure of sound intensity is an example of what physicists call a logarithmic scale. Each increase by a factor of 10 in the intensity associated with the sound wave results in an increase of 1 in the bel scale (and thus an increase of 10 in the decibel scale). While this convention might seem strange at first, there are many other examples of logarithmic scales in science. Earthquake energies, for example, are measured in a scale in which an earthquake of magnitude 5 releases 10 times as much energy as an earthquake of magnitude 4 and 100 times the energy of an earthquake of magnitude 3. The brightness of stars is also recorded as a logarithmic magnitude. And on a more fanciful level, the "warp drive" on ships in the *Star Trek* series also measures the speed of the ship on a logarithmic scale. In all these cases, a small range of numbers expresses a much larger range of magnitudes, which is the main advantage of a logarithmic scale.

Frequency and Pitch

Pitch, in contrast to loudness, is related to the frequency of a sound wave. The higher the frequency, the higher the pitch of the sound seems to be. Human hearing is typically sensitive to sounds in the frequency range from about 20 to 20,000 Hz, while some animals can sense sounds as low as 15 Hz (dogs) or as high as 150,000 Hz (bats).

It is useful to note that the speed of sound in air is essentially the same for all wavelengths. We can relate the frequency and wavelength of sound by the familiar equation

$$\text{Speed} = \text{Frequency} \times \text{Wavelength}$$

This equation tells us that the higher the frequency of a wave is, the shorter its wavelength, while the lower the frequency of a wave is, the longer the wavelength. Because of this relationship, we can say that

High-frequency sounds correspond to short wavelengths, and low-frequency sounds correspond to long wavelengths.

You can observe the consequences of this relationship between pitch and wavelength in the higher-pitched sounds emitted by smaller musical instruments or animals. For example, think about the shrill yapping of a Chihuahua versus the throaty growl of a Great Dane.

INTERFERENCE OF SOUND WAVES

Like all other waves, sound waves from two different sources interfere with one another when they come together. One interesting example of this phenomenon involves interference of sound waves in an auditorium. Occasionally, an auditorium is built in such a way that almost no sound can be heard in certain seats (so-called dead spots), while other seats receive unusually intense sound. This unfortunate situation results when two waves—for example, one directly from the stage and one bouncing off the ceiling or wall—arrive at those seats in just such a way as to cause significant destructive or constructive interference. One of the main goals of acoustical design of auditoriums, a field that relies on complex computer modeling of sound interference patterns, is to avoid such problems.

The next time you're in a modern auditorium, take a look around. You will probably see large blocks of brightly colored fabric that look like wall and ceiling decorations. These structures, called "sound baffles," are placed carefully by acoustic engineers to absorb sound at crucial spots in the room, thereby preventing a sound wave that would normally be reflected from traveling farther and causing the kind of destructive interference just described. When a new auditorium is built, there is often a period of "tuning" while devices like this are moved around and adjusted for optimal performance.

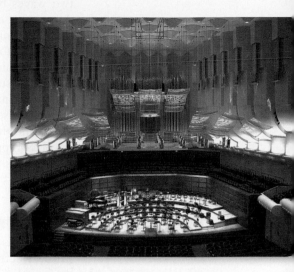

Davies Symphony Hall in San Francisco. This concert hall underwent a major renovation to improve its acoustics. Note the many sound baffles on the ceiling, hanging from the ceiling, and on the side walls.

Two concert halls that are famous for their excellent acoustical properties are the symphony halls in Boston, Massachusetts, and Vienna, Austria. Both of these halls were built in the nineteenth century, before architects were familiar with sound baffles and other devices for adjusting interference of sound waves. However, the decorative tastes of the time led the builders of these halls to include many statuettes and ornate wall decorations around the outer walls of the hall. These surfaces break up the sound waves just as effectively as modern techniques do, helping to produce a clear rich sound without dead spots.

Beats

Another interesting phenomenon occurs when two sound waves of almost the same frequency interfere with one another. As shown in Figure 15-9, if the two waves start out interfering destructively with one another, then with each cycle

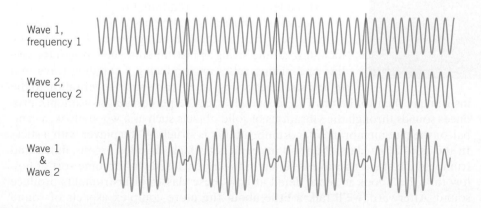

FIGURE 15-9. The origin of beat patterns. When two pitches of nearly the same frequency interfere, you hear a sequence of beats.

the crest of one arrives a little bit later than the crest of the other. Eventually the two waves begin to interfere constructively. But this constructive interference doesn't last either, because the falling behind of the crest continues until we recover the initial destructive interference. In this cycle, from one destructive interference to the next, one more crest of the higher-frequency wave has arrived at the listening point than the number of crests from the lower-frequency wave.

The result of this situation is that you hear a slow, pulsing sound when the two waves interfere. These pulses, called *beats,* result from the successive increases and decreases of intensity in the combined wave. When two frequencies combine to produce beats, the beat frequency equals the higher frequency minus the lower. Symphony musicians learn to avoid beats by adjusting the pitch of their instruments to play in tune.

THE SOUND OF MUSIC

Music, whether pop, jazz, or classical, consists of a pleasing succession of pitches. Any pitch (i.e., frequency) is possible, but musical pitches are usually selected from a specific sequence called a *scale.* In Western music, for example, the "well-tempered" scale consists of a sequence of pitches, each of which is the twelfth root of 2, or about 1.06, times the frequency of the next lower note. This relationship is derived from a 12-note scale; the thirteenth note has twice the frequency of the first note and thus sounds an *octave* higher. Look carefully at a piano keyboard and you'll see a 12-note pattern of white and black keys repeated over several octaves. Other cultures, such as Japan and India, use different musical scales, which can sound quite unusual to Western listeners.

Different-sized instruments in a New Orleans jazz ensemble play in different ranges. The large tuba on the left plays low notes, while the smaller trumpet at the center plays in a higher range.

Musical Instruments

The production of sound involves setting up a wave in the air. A wave pulse can be created by a single event such as clapping your hands or snapping your fingers. However, to set up a continuous sound, such as that produced by a musical instrument or a human voice, it is necessary to set up a standing wave that produces many pulses of the sound wave.

Three large classes of traditional (as opposed to modern electronic) musical instruments differ from one another in how they produce standing waves. The first class, including all the stringed instruments, such as guitars, violins, and pianos, relies on producing a standing wave in a tightly stretched string, which transmits the vibration to the instrument, then to the air. The second class, called percussion instruments, produces sounds through the vibrations of solid objects such as a wood block, a cymbal, or the taut membrane of a drumhead that is struck by the player with a stick. In the third large class, the wind instruments, which includes organs, flutes, and trumpets, the standing wave is set up in the air enclosed within some sort of hollow tube. Let's look in more detail at how these classes of instruments produce sound. Afterward, we'll talk a little about the more complex aspects of sound that make one instrument sound different from another.

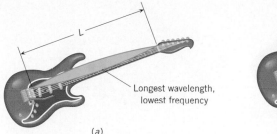
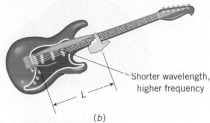

(a) Longest wavelength, lowest frequency

(b) Shorter wavelength, higher frequency

L

L

FIGURE 15-10. (a) An open guitar string vibrates at its lowest possible pitch. (b) A guitar player varies the pitch of her instrument by shortening the string with her fingers.

Stringed Instruments Think about what happens when you pluck a guitar string. Your finger pulls the stretched string to the side and lets it go. The string is then free to vibrate from side to side at its natural vibration frequencies. The lowest-pitched sound (corresponding to the longest wavelength and lowest vibration frequency) that can exist on the guitar string is shown in Figure 15-10a. This is the sort of standing wave pattern we discussed in Chapter 14. This vibration frequency produces the musical pitch of the note that we hear. The plucked string produces a sound by vibrating the instrument which in turn creates pressure waves in the air—traveling sound waves that match the frequency of the standing wave. These sound waves, in turn, travel at the speed of sound in air and come to your ear, allowing you to hear the plucked string.

Musicians who play stringed instruments can change the pitch of a stretched string in several ways. The frequency of the longest wave on a stretched string depends on three factors: the mass of the string, how tightly the string is stretched, and the length of the string. You can see all three of these factors come into play on a guitar or violin. First, the lower-pitched strings are always thicker and thus more massive compared to the higher-pitched strings. Then, to tune each string, musicians turn pegs on their instruments to tighten or loosen the strings. Finally, the musician's fingers press the strings down against the fingerboard to change the effective length of each string, producing the desired musical patterns of notes.

Pianos and harps employ a different strategy to change notes. Each of dozens of strings is pretuned to a specific pitch (88 different pitches in the modern piano, for example). Each pitch is sounded by striking a different piano key or by plucking a different string on the harp. The musician cannot change the effective length of a piano or harp string; the pattern of notes is created by the sequence of different strings that are struck.

The strings of a harp are fixed in length; the performer produces different notes by plucking different strings.

Percussion Instruments Perhaps the earliest forms of music were made by rhythmically striking objects that produce a pleasing sound. Percussion instruments, including all manner of drums, gongs, bells, and rattles, rely on the natural vibration frequencies of solid objects. Once struck, these objects vibrate at a characteristic set of frequencies, producing their characteristic sounds. Many instruments in the percussionist's vast arsenal of instruments, including ratchets, chimes, cymbals, and wood blocks, produce a single distinctive sound. In addition, a variety of mallet instruments, including xylophones, marimbas, and

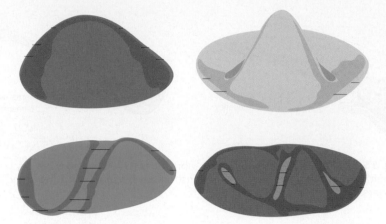

FIGURE 15-11. Drumheads vibrate with several different two-dimensional harmonics, which are analogous to those of a vibrating string. These four computer images show such vibrations with vertical exaggeration.

vibraphones, have a keyboard-like arrangement of dozens of individual wood or metal blocks, each sounding a specific pitch when struck.

Other percussion instruments can be tuned—for example, by changing the tautness of a drumhead. Next time you go to a symphony concert, watch as the timpanist carefully tunes the large copper drums. A drum (as well as a speaker in a stereo system) works in essentially the same way as a guitar string, except that in these cases a vibration is set up in a stretched membrane—a two-dimensional object, in contrast to a string, which has only one dimension (Figure 15-11).

Wind Instruments The third important class of traditional musical instruments, the wind instruments, rely on producing standing waves of a fixed pitch in an enclosed column of air. Wind instruments have existed since prehistoric times and now include a bewildering variety of forms, including all sorts of instruments that you blow into (saxophones, trumpets, flutes, oboes), as well as such mechanical instruments as pipe organs and even automobile horns. In order to understand how wind instruments work, it's important to distinguish the very different behavior of tubes with ends that are open versus those that are closed.

1. **At a closed end** Sound is a longitudinal wave, so the air molecules move in the same direction as the wave. Since the molecules can't move into the solid wall that closes the tube, the wave amplitude must be zero at the wall. In the language of Chapter 14, there must be a node in the standing wave at a closed end.

2. **At an open end** The sound wave must be at maximum amplitude at the open end of the tube—an antinode.

With this background, we can identify two different configurations for wind instruments. First, there are instruments (such as the clarinet or oboe) that are open at one end (the bell) and closed at the other (the mouthpiece), with a sound wave as shown in Figure 15-12a. In this case, there is a node at one end and an antinode at the other end, so the longest possible standing wave is of wavelength $4L$ (since only $\frac{1}{4}$ of a wavelength fits in the tube). Second, there are instruments (such as the flute and some organ pipes) in which the tube is open at both ends, as shown in Figure 15-12b. Here there are antinodes at both ends and the longest

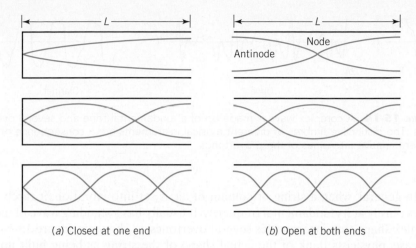

(*a*) Closed at one end   (*b*) Open at both ends

FIGURE 15-12. Wind instruments produce sound when a column of air vibrates in a tube. The two varieties of wind instruments rely on (*a*) a tube that is closed at one end or (*b*) a tube that is open at both ends.

standing wavelength is $2L$ (since only $\frac{1}{2}$ of a wavelength fits in the tube). Note that you could have a tube that is closed at both ends, but then the sound would have no way to leave the tube and could not be heard—rather an unfortunate situation for a musical instrument.

All modern wind instruments have the ability to change pitches by changing the length of the air column, and most of them employ one of three common mechanisms to do so. The trombone employs a long slide that changes the entire length of the instrument. Other brass instruments, including trumpets and horns, have valves that add or subtract fixed lengths of tubing. And all woodwind instruments, including flutes, saxophones, clarinets, and oboes, feature tubes lined with numerous keys and holes; opening and closing these holes also alters the effective length of the tube.

Fundamentals, Overtones, and Harmonics

A wave of a single frequency would be a rather unusual and haunting sound. Most of the sounds we hear are much more complex than that. To understand how complex sounds are generated, let's go back to the example of the vibrating guitar string. Think for a moment of all of the standing waves that could fit on the string, some of which are shown in Figure 15-13. In addition to the wave of wavelength $2L$ (a wave that is called the *fundamental,* or first **harmonic**), waves of wavelength $L, \frac{2}{3}L, \frac{1}{2}L,$ and so on can also fit on the string. Each of these higher harmonics or **overtones,** were it acting alone, would produce a specific frequency, corresponding to a specific pitch, with shorter and shorter wavelengths corresponding to higher and higher pitches.

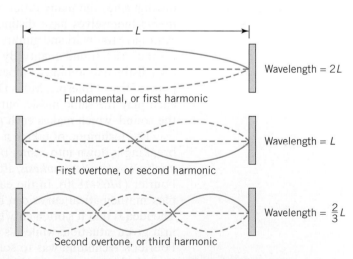

FIGURE 15-13. The first three harmonics of a vibrating string.

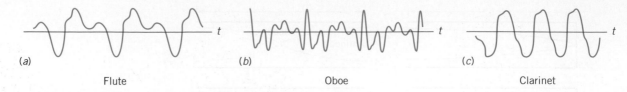

(a)	(b)	(c)
Flute	Oboe	Clarinet

FIGURE 15-14. A complex wave is made up of a fundamental tone and several overtones. The distinctive timbres of different musical instruments are a consequence of different relative intensities of these overtones.

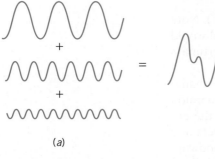

(a)

FIGURE 15-15. Adding waves. (a) The first three harmonics. (b) The combination of these waves into a complex sound wave.

In general, when a string or column of air is set into vibration, it rarely vibrates purely at its fundamental frequency. It usually has a standing wave of more complex shape that incorporates several overtones, as shown in Figure 15-14. In this case, physicists think of the actual shape of the string as being built up by combining standing waves of the various harmonics, using the standard techniques of wave interference we discuss in Chapter 14. For example, in Figure 15-15a we show three waves—the first three harmonics. In this example, the amplitudes of the second and third harmonics are $\frac{1}{2}$ and $\frac{1}{3}$ of the fundamental, respectively, but in principle they could be as large (or as small) as we liked. If we add these three waves together, we get the complex wave shown in Figure 15-15b. Thus, by adding together only three of the possible waves that can fit on the string, we can create a complex wave that looks like none of the three. If we keep adding overtones, we can build up any shape on the string that we want. This process, in turn, means that when we pluck the string, the resulting sound wave is a complex mix of several harmonics.

Have you ever wondered why a guitar sounds so different from a flute or piano, even when they play the same melody? The complex mix of fundamental and overtones, which musicians call the quality or *timbre* of the sound, is what distinguishes the many different musical instruments. In some cases, the instruments themselves have distinctive timbres. Violins by the great Italian master Antonio Stradivari and guitars by the American firm of Gibson are highly prized for their rich sound quality. By the same token, different musicians produce their own distinctive sounds. Compare the very different timbres of the great jazz trumpeters Dizzy Gillespie, Miles Davis, and Wynton Marsalis, for example. They can each play the same music, but each has a slightly different mix of overtones in the sound, which makes each player sound slightly different.

The technique of taking a complex shape, such as the wave on a string, and breaking it down into a sum of simple, single-frequency waves, is called *Fourier analysis* or *Fourier synthesis,* after the French mathematician Jean Baptiste Joseph Fourier (1768–1830). In the early nineteenth century, he showed that any complex mathematical curve can be broken down into a sum of simple harmonics. The idea is that it's often easy to analyze the behavior of a single-frequency wave in a given situation, while it's quite difficult to deal with more complex waves. Fourier's technique was to solve the problem for a simple wave and then add the solutions together to the get the solution for the more complex shapes. Scientists quickly adopted his technique to analyze all sorts of situations, from heat flow in engines to the way that a skyscraper sways in the wind. This is also the

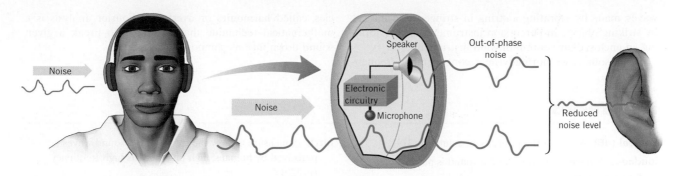

FIGURE 15-16. Noise-canceling headphones break down an incoming sound pattern into simple wave components and then generate new wave components that are out of phase with the incoming ones. The resulting sound is greatly reduced.

basis for noise-reducing headphones (Figure 15-16), which break down incoming noise into simpler wave components. They then generate new waves that are exactly opposite to the original sound pattern, causing destructive interference and effective cancellation of the noise.

Sound: Acoustic Versus Electronic Instruments

Electronic musical instruments, such as the familiar synthesizer, rely on wave generators to build up musical sounds by adding together many waves of different frequencies—a process that is closely related to Fourier analysis. Almost any sound, from a bass drum to a guitar to the human voice, can be duplicated by this technology. Synthesized music is remarkable for its range of effects, its versatility, and its amplified power.

Nevertheless, many musicians find greater satisfaction in traditional acoustic instruments, where the player has a more direct role in producing the good vibrations. After all, a violinist sweeps her bow across the strings to set the strings into vibration, a trumpet player takes a deep breath and buzzes his lips to produce a clear high note, and a percussionist strikes the cymbals and drums to make their dramatic sounds. Musicians can feel as well as hear the vibrations of these instruments; they can control the musical sound directly by the actions of their bodies. Many musicians crave this intimate physical feedback.

Should the difference between electronic and acoustic music matter? If a synthesizer can duplicate exactly the radiant tone of a Stradivarius violin or the rhythmic intensity of a rocker's set of drums, should the musician care about how the sound is produced? Should the audience?

Summary

Sound waves are longitudinal waves that travel through media such as air and water. The wave consists of alternate regions of high and low pressure in the medium. Sound is used by many animals for communication, navigation, and locating prey.

The intensity or **loudness** of a sound, measured in units of watts per square meter or **decibels** respectively, depends on the square of the amplitude of the wave. The **pitch** depends on the frequency, with higher frequencies corresponding to higher pitches. Sound waves exhibit interference, a fact that is important in the science of acoustics.

Musical sounds are normally produced by standing

waves, made by vibrating a string in stringed instruments, by striking objects in percussion instruments, or by vibrating an enclosed air column in wind instruments.

The sounds we normally hear contain many frequencies, called **harmonics** or **overtones.** Fourier analysis is a mathematical technique that allows us to break a given sound down into its component harmonics.

Key Terms

decibel (dB) A unit of the intensity of sound. (p. 319)

loudness A measure of how loud a sound is perceived by humans. (p. 319)

overtones (harmonics) A series of oscillations in which the frequency of each oscillation is a integral multiple of the fundamental frequency. (p. 325)

pitch A measure of how the frequency of sound waves is perceived by humans; high pitch means high frequency. (p. 320)

Review

1. Why is sound considered a longitudinal wave?

2. How does the compression and rarefaction of molecules result in sound?

3. What is the speed of a sound wave? How and why does this vary with temperature?

4. How can you tell the distance you are from a thunderstorm simply from seeing the flash of lightning and hearing the sound of the thunder? Explain.

5. What is an echo? How can it be used to tell the distance across an echo lake?

6. Can sound travel in media other than air? If so, what other media and how does this occur?

7. What are some of the other ways sound can be generated other than through the use of your vocal cords?

8. What wave properties does sound have? Explain.

9. How does the human ear sense the rapid changes in air pressure associated with sound?

10. How do bats locate their prey? What methods have been used to demonstrate this?

11. Give examples of how animals other than bats use sound.

12. What is meant by the intensity of a sound? How does it differ from loudness? What wave property is responsible for this?

13. What is the unit of relative loudness and who was it named after? What is the intensity in watts per square meter of a decibel value of zero?

14. What does it mean to say that the decibel system is a logarithmic system? What are some other examples of logarithmic scales used in science?

15. Do short sound wavelengths correspond to high or low frequencies? Long wavelengths?

16. At the extremes of human hearing, what strange qualities does sound perception take on?

17. How can destructive wave interference affect sound in a poorly designed auditorium? Explain exactly how this can occur.

18. How are good modern auditoriums designed so as to avoid problems with interference? Give an example.

19. What are beats and how do musicians avoid them?

20. What is a musical scale? An octave?

21. To get continuous sound out of a voice or instrument, why do you need to set up a standing wave?

22. What are the three main classes of traditional musical instruments, and how does each produce sound?

23. What three factors does the frequency of a stretched string depend on?

24. How do percussion instruments make use of the natural vibration frequencies of solid objects to produce their sounds? How do percussionists tune their instruments?

25. How do wind instruments work? Why must there be a wave node at the end of a closed-tube instrument? Why must there be an antinode at the end of an open-tube instrument?

26. How do musicians physically change the length of the air columns in their instruments and thus change the standing wave generated? How do brass instruments and woodwinds each do this?

27. What is a harmonic, or overtone? How is it produced and why is it desirable?

28. What is the timbre of a sound?

29. What is Fourier analysis and how is it used to analyze sound?

30. How does an electronic musical instrument differ from a traditional acoustic instrument?

Questions

1. In what ways are sound waves similar to water waves? How are they different?

2. If a tree falls in a forest, what kinds of waves are created? Where did the energy that produced those waves come from?

3. If an atomic bomb were detonated on the surface of the Moon, how long would it be until we heard the explosion here on Earth?

4. Runners line up side by side at the starting line of a road race. About how long would the starting line have to be for there to be a one-second delay between the sound of the starting gun reaching the closest runner and the farthest runner from the gun? Are roads usually this wide?

5. A longitudinal wave in a Slinky is a nice representation of a sound wave. Suppose a Slinky is cut into pieces and those pieces are connected by rigid rods, as shown in the figure. If the mass of the rods were the same as the mass of the missing links, how would the speed of the waves in this new Slinky compare to those in the original one? Explain your answer.

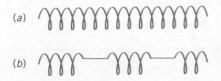

6. Why do you think that sound moves faster in water or rock than in air? What might be happening at a molecular level that accounts for this?

7. Do you think that sound travels slower or faster at very high altitudes than at sea level? Or does elevation make absolutely no difference in the speed of a sound wave? How would sound travel in outer space?

8. Why do you think that sound travels faster in warmer temperatures? Why does the elasticity or "springiness" of air increase with higher temperature?

9. What physical features make an echo lake produce echoes? What other types of spaces produce echoes?

10. Two stereo speakers are plugged into the same jack and emit pure 500-Hz tones. The speakers are separated by a distance of a few meters (see figure). As you walk along line A, describe what you hear. As you walk along line B, you hear alternating loud and soft sounds. Explain why this happens. What would change if you doubled the frequency and then walked along line B again?

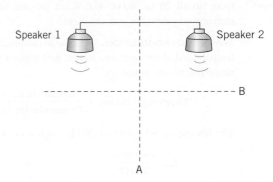

11. Two speakers are plugged into the same jack and emit a pure tone of 1000 Hz. You walk along a line connecting them (see figure). When you are between them, the loudness of the sound fluctuates up and down. When you are not between the speakers and are walking away from them, the loudness does not fluctuate but gradually decreases. Explain the difference.

12. A guitar and a flute are in tune with each other. Explain how a change in temperature could affect this happy situation.

13. A pure tone with frequency 500 Hz is played through two stereo speakers plugged into the same jack. As you walk around the room, you notice that the loudness of the sound alternates from loud to soft repeatedly. What is happening? Would anything be different if a 1000-Hz sound wave were used instead?

14. For most vibrating systems, the amplitude of vibrations does not affect the frequency as long as the amplitude is not too high. Explain why this fact makes the construction and use of musical instruments possible.

15. When a tuning fork of frequency 256 Hz vibrates alongside a piano string, beats are heard. The string is tightened slightly and the beats go away. Was the original frequency of the string greater or less than 256 Hz? Explain.

16. Lower-pitched strings on guitars and pianos often have copper wire wound around them. This wire does not make the string stronger or change the tension of the string. What purpose does this extra wire serve?

Problem-Solving Examples

EXAMPLE 15-2

The Limits of Human Hearing

The normal human ear can hear sounds at frequencies from about 20 to 20,000 Hz. What are the longest and shortest wavelengths you can hear?

REASONING AND SOLUTION: We have to calculate the wavelength needed for both the lowest and highest frequency, according to the equation

$$\text{Wavelength (in m)} = \frac{\text{Speed (in m/s)}}{\text{Frequency (in Hz)}}$$

The lowest audible note, at 20 Hz, requires a wavelength

$$L = \frac{340 \text{ m/s}}{20 \text{ Hz}}$$

$$= 17 \text{ meters (about 50 feet long)}$$

Similarly, the highest audible note, at 20,000 Hz, is produced by a wavelength

$$L = \frac{340 \text{ m/s}}{20,000 \text{ Hz}}$$

$$= 0.017 \text{ meters (about two-thirds of an inch)}$$

At these extremes of human hearing, sounds take on a strange quality. The highest-frequency notes have a shrill, piercing quality, while at the lowest frequencies we don't so much hear the notes as feel them as a kind of low rumble. Listen carefully to your surroundings to see if you can identify everyday sources of these highest and lowest pitches. ●

EXAMPLE 15-3

The Limits of Human Hearing Revisited

Of all the traditional musical instruments, a large pipe organ has the widest range of pitches. Given that the human ear can hear sounds at frequencies from about 20 to 20,000 Hz, what are the longest or shortest open-ended organ pipes you are likely to see? (Treat organ pipes as columns of air open at both ends.)

REASONING AND SOLUTION: In Example 15-2, we have seen that these two frequencies correspond to wavelengths of 17 m (about 50 feet) and 0.017 m (about $\frac{2}{3}$ inch), respectively. Open organ pipes producing these notes need to be about half these wavelengths—approximately 8.5 and 0.009 meters, respectively. Most large pipe organs have pipes ranging from about 8 meters to less than 0.05 meter in length. Next time you have the chance, visit a church or auditorium with a large pipe organ and look at the vari-

ety of pipes. Not only are there many different lengths, but there are also many distinctive shapes, each sounding like a different instrument. ●

A mighty pipe organ.

Problems

1. Rank the following media from slowest to fastest in terms of the speed with which a sound wave travels through them: air at 40°C, steel, air at 0°C, water.

2. Rosa and Jon were asked by their physical science teacher to determine the speed of sound. While walking to their dormitories after class, Jon clapped his hands, which Rosa and Jon heard a moment later as an echo. The echo bounced off a building that was 300 ft away. They knew that they could not measure the brief time for a single clap to return, so they had a brilliant idea. Jon clapped and then started to clap as soon as he heard the echo, and he then continued this synchronized clapping so that Rosa could measure the frequency. Rosa counted 56 of Jon's claps in one-half minute.

   a. What is the frequency of Jon's clapping?
   b. What is the speed of sound as determined by Rosa and Jon?
   c. How does this speed compare with the speed of sound at room temperature (about 340 m/s)?

3. Rosa and Jon decided to test their results, so they walked an additional 100 ft away from the wall. Using their calculated sound speed in Problem 2, answer the following.

   a. Predict their new clapping frequency. Will it be greater than, equal to, or less than the frequency determined in Problem 2?
   b. What is the new time between successive claps?
   c. How many claps will Jon have to make in one-half minute?

4. Upon seeing a flash of lightning, Giselle counts 7 seconds until she hears the rumble of thunder. Assuming that the flash travels to her instantaneously, use the estimation method given on page 314 to determine how far away the lightning struck. If the temperature was 20°C, what was the distance using the known speed of sound at that temperature? How does this distance compare to the estimated values?

5. Anna was on vacation and came across an echo lake. Wanting to know how far she had to swim to get across the lake to the other side, she yelled across "Hello!" Assume that the temperature is 20°C and the speed of sound is 344 m/s.

   a. If 5 seconds later she heard her own echo, estimate the distance in feet, miles, and meters across the lake.
   b. If she had heard the echo in 3 seconds, what would the distance be?

6. A. Assume that the speed of a sound wave produced by an elephant at 20°C is 344 m/s and the frequency is around 25 Hz.

   a. What is the wavelength of this wave?
   b. Can you tell the amplitude or the intensity/loudness of the sound from this information? Why?
   c. Do you expect the wavelength of sound to be shorter or longer for higher-frequency sounds? Why?
   d. Consider the compression and rarefaction pattern of a sound wave and identify each of the following terms in the context of the elephant's sound wave: wavelength, frequency, and amplitude.

6. B. Repeat the calculations in part a for a high-frequency sound of 150,000 Hz that might be produced by a bat in order to confirm your answer in part c. Would a human hear this sound?

7. Use the data in Table 15-1 for the following problem. The decibel system is a logarithmic scale that measures the intensity of a sound relative to a 0-decibel level that corresponds to a sound wave with intensity 10^{-12} watts/m².

   a. How many times more intense is a food blender than a normal conversation? Than a pin dropped on a soft surface?
   b. Compare the intensity of sound of a jet at takeoff to that of an average loud rock concert.
   c. How much more intense is the sound of a jet compared to a normal conversation? Compared to a sound with a decibel value of zero?

8. A dog can hear sounds in the range from 15 to 50,000 Hz. What are the wavelengths that correspond to these sounds at 20°C?

9. Draw two sound waves together on the same graph. First draw the cases of destructive and constructive interference when the waves have the same frequency. Also, draw a case in which the waves have different frequencies. Is the interference constructive, destructive, or a mix of the two in this case?

10. If the length of a flute open at both ends is 0.4 m, what is the longest possible wavelength that could be produced by this instrument?

11. Calculate the fundamental frequency of a 4-meter organ pipe that is open at both ends. Calculate the same frequency for a pipe that is open at one end and closed at one end. Assume that the temperature is 20°C and the speed of sound is 344 m/s.

12. A stringed instrument has a maximum string length of 0.3 m and is tuned so that a wave travels along the string at 120 m/s.

   a. What is the string's fundamental frequency at this length and wave speed?
   b. What are some of the overtones or harmonics associated with this fundamental frequency?
   c. Is each of the fundamental waves and its overtones a standing wave?
   d. Do overtones enrich a sound or take away from it? What role does wave interference play in producing the final sound you hear? Explain your answer.

Investigations

1. Whales communicate over distances of thousands of miles. Investigate how they do this. What kind of wave is used? What is its speed? Its frequency? Its wavelength? Listen to a recording of humpback whale sounds.

2. Go to several local and campus concert halls and rate the quality of the acoustics for each one you are able to visit. What characteristics do the halls with good acoustics share? What are some of the obvious design additions that improve the sound? Do some initial research into the design of these structures before going so you know what to pay attention to in terms of design.

3. How do submarines use sound and sonar to navigate and to detect other ships? Research the history and development of these sound detection and sound suppression technologies. Evaluate the progress made so far, and find out if scientists see any further improvements on the horizon.

4. Go to a church with a pipe organ and listen to a concert, and if possible take a closer look at the pipes. How many are there and what shapes and sizes do they come in? Are they open- or close-ended? Interpret the sounds in terms of what you learned in this chapter. Which pipes produce high-pitched sounds and which ones low-pitched sounds? How is the volume or intensity controlled?

5. A symphony concert is a good place to hear a variety of sounds. Watch how the musicians tune their instruments before they play. What does each player do? Why is it effective?

6. Investigate the wave generators used in the production of electronic music. How do they do this, and what is the actual physical generator of sound?

7. Go to your local stereo store and compare stereo systems. What are the differences in sound from different systems? Do systems in the same price range always sound the same? If not, how are they different and why is this the case? Similarly, what is different in terms of sound generation in some of the cheaper versus the high-end systems?

8. How does a hearing aid work? What site(s) in the ear does it target?

9. How does a car muffler work? Find out just how it suppresses engine sound. What decibel levels are heard with and without it?

10. Pick up a guitar sometime and, as you play around with it, note how the thickness of the string affects the sound as you play or pluck away. Also note how the length of the string affects the sound as you hit different frets by pressing down the string at different places. Try adjusting the tuning by changing the tightness of the strings and note how a tighter versus a looser string changes the sound. Would changing the actual composition of the strings make a difference?

11. What types of animals use sound to communicate? What are some organisms that do not use sound to communicate? What might be some of the evolutionary advantages and disadvantages of using sound?

12. What are the types of instruments that were available to "primitive" or premodern cultures? How do they compare to modern instruments?

 WWW Resources

See the *Physics Matters* home page at **www.wiley.com/college/trefil** for valuable web links.

1. **http://asa.aip.org/sound.html** Interesting sounds from the Acoustical Society of America.

2. **http://www.silcom.com/~aludwig/musicand.htm** An equation-free description of the physics of sound, with animations.

3. **http://www.phy.ntnu.edu.tw/java/sound/sound.html** A visible/audible Java applet that demonstrates Fourier synthesis.

4. **http://hyperphysics.phy-astr.gsu.edu/hbase/sound/soucon.html#c1** The physics of sound concept maps.

5. **http://www.earaces.com/anatomy.htm#Cross%20Section%20Of%20Ear** The anatomy of the human ear.

6. **http://www.phys.unsw.edu.au/PHYSICS_!/SPEECH_HELIUM/speech.html** The physics of human speech.

7. **http://www.geocities.com/Vienna/3941/index.html** Essays on physics of brass wind instruments, concentrating on the trumpet.

8. **http://www.kettering.edu/%7Edrussell/Demos.html** An extensive collection of acoustics and vibrations animations.

9. **http://www.harmony-central.com/Guitar/harmonics.html** The physics of guitar strings for musicians.

16 | Electric and Magnetic Forces

KEY IDEA

Electricity and magnetism are fundamental forces of nature.

PHYSICS AROUND US . . .
Late for Work at the Copy Center

Imagine waking up on a cold, dry winter day and discovering you're late for work. You rush to throw on your clothes and comb your hair with rapid, vigorous strokes. Looking in the mirror, you realize that the folds of your shirt are sticking together and your hair is standing on end. When you finally get to work, you try to photocopy an important report, only to find the copies sticking to each other, slowing your efforts. Static electricity has struck.

Imagine walking home from work on a hot, humid summer day, listening to a radio station on your headphones. The sky is dark and you hear the distant rumble of thunder. The radio has a lot of static. Suddenly, a jagged lightning bolt slices the horizon. You

decide to take the subway and hurry to get home before the thunderstorm. You swipe your train pass through the turnstile, but it takes a couple of tries before it registers. Static electricity has, quite literally, struck again.

Believe it or not, the force that causes your clothes to stick together and your hair to stand on end in the cold, dry winter is exactly the same force that causes lightning in the hot, humid summer. It is the force of static electricity. The force that enables your radio to play and the card reader to operate is a different force, but it is affected by electricity. This is the force of magnetism. We examine both of these forces in this chapter.

333

NATURE'S OTHER FORCES

According to Newton's laws of motion, nothing accelerates without a force. However, the law of gravitational force that Newton described cannot explain many everyday events. How does a refrigerator magnet cling to metal, defying gravity? What makes a compass needle swing around to the north? How could gravity make static cling wrinkle your shirt or lightning shatter an old tree? These phenomena point to the existence of underlying forces that are different from gravity.

The forces we are talking about have been known (and even used) by people for a long time. They go by the names "electricity" and "magnetism." In this chapter, we explore the properties of these forces. In the next chapter, we look at one of the most amazing facts about them—the fact that they are connected to one another, despite their apparent differences.

Newton may not have thought much about electricity and magnetism, but he did give us a method for studying them: First, observe natural phenomena and learn how they behave. Then, organize those observations into a series of natural laws. Finally, use those laws to predict future behavior of the physical world. This is the process we have called the *scientific method*.

In particular, we find Newton's first law of motion (see Chapter 4) to be very useful in our investigation of nature's other forces. According to this law, whenever we see a change in the motion of any material object, we know that a net force has acted to produce that change. Thus, whenever we see such a change and can rule out the action of known forces such as gravity, we can conclude that the change must have been caused by a hitherto unknown force. We use this line of reasoning to show that electric and magnetic forces exist in the natural world.

STATIC ELECTRICITY

The modern understanding of static electricity began in the eighteenth century with a group of scientists in Europe and North America who called themselves "electricians." These researchers were fascinated by the many curious phenomena associated with nature's unseen forces. Their thoughts were not focused on practical applications, nor could they have imagined how their work would transform the world.

Various phenomena related to electricity have been known since ancient times. The Greeks knew that if you rub a piece of amber with cat's fur and then touch other objects with the amber, those other objects repel one another. The same thing happens, they found, if you rub a piece of glass with silk: objects touched with the glass repel one another. On the other hand, if you bring objects that have been touched with the amber near objects touched with the glass, they attract one another. Objects that behave in this way are said to possess **electric charge,** or to be "charged."

The force that moves objects toward and away from one another in these simple demonstrations was named **electricity** (from "electro," the Greek word for amber). In these simple experiments, the electric charge doesn't move once it has been placed on an object, so the force is also called **static electricity.**

The electric force is clearly different from gravity. Unlike the electric force, gravity is never repulsive: when a gravitational force acts between two objects,

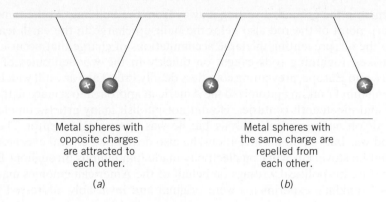

FIGURE 16-1. There are two kinds of electric charges, designated positive and negative. (a) Opposite charges attract, while (b) like charges repel.

it always pulls them together. The electric force, on the other hand, can attract some objects toward one another and push other objects apart. The electric force, furthermore, is vastly more powerful than gravity. A pocket comb charged with static electricity easily lifts a piece of paper against the gravitational pull of the entire Earth.

Today, we understand that the properties of the electric force arise from the existence of two kinds of electric charge (Figure 16-1). We say that objects touched by the same source, be it amber or glass, have the same electric charge and are repelled from one another. On the other hand, an object touched with amber has a different electric charge than a second object touched with glass. This difference is reflected in their behavior—they attract one another. We can summarize this behavior by saying

Like charges repel each other, unlike charges attract each other.

When the girl touches the electrically charged sphere, her hair becomes electrically charged as well. Individual hairs repel one another and thus stand on end.

Physics in the Making
Benjamin Franklin and Electric Charge

The most famous North American "electrician" was Benjamin Franklin (1706–1790), one of the pioneers of electrical science as well as a central figure in the founding of the United States of America. Franklin began his electric experiments in 1746 with a study of electricity generated by friction. Most scientists of the time thought that electric effects resulted from the interaction of two different electric fluids. Franklin, however, became convinced that the transfer of a single electric fluid from one object to another could explain all electric phenomena. He realized that objects could have an excess or a deficiency of this fluid, and he applied the names negative and positive to these two situations.

Following this work, Franklin demonstrated the electric nature of lightning in June 1752 with his famous (and extremely dangerous) kite experiment. A mild lightning stroke hit his kite and passed along the wet string to produce sparks and an electric shock. Not content with acquiring theoretical knowledge, Franklin followed his discovery of the electric nature of lightning with the invention of the lightning rod, a metal rod with one end in the ground and the other end sticking up above the roof of a building. The rod carries the electric charge of

Lightning rods allow charge accumulated in the ground during thunderstorms to leak harmlessly into the air. If the charge becomes great enough, lightning can be attracted to the metal rod and the electricity run harmlessly into the ground. Benjamin Franklin first introduced lightning rods such as this.

lightning into the ground, diverting it away from the building. At the same time, the sharp point of the rod also helps the built-up charge in the earth leak gently into the air, preventing the large accumulations of charge that occur in lightning strikes. Lightning rods caught on quickly in the wooden cities of North America and Europe, preventing countless deadly fires. They are still widely used.

Benjamin Franklin epitomized the American approach to science in the eighteenth and nineteenth centuries. He did not publish many articles on electrical theory or on mathematical analysis, but he was a practical inventor. The lightning rod was but one of his inventions; he also developed bifocal eyeglasses and the Franklin stove. His work on electricity made him famous throughout Europe years before his political writings on behalf of the American colonies made him famous. Franklin's experiments were original and invariably addressed key issues in the science he was investigating. His practical approach made him the pioneer of American engineers and inventors, whose successors included Joseph Henry (the man behind the electrical transformer), Alexander Graham Bell (the telephone), Thomas Edison (the lightbulb, the phonograph, moving pictures) and the Wright Brothers (the airplane). ●

The Movement of Electrons

We now understand that there are two kinds of electric charge, called *positive* and *negative* after Franklin's terminology. This fact enables us to explain observations of both the attractive and repulsive behavior of charged objects. As we have seen in Chapter 9, all objects are made up of minute building blocks called *atoms*. As we'll see in Chapter 21, atoms are made up of still smaller particles, many of which have an electric charge. In the modern view, negatively charged electrons move around a heavy, positively charged nucleus located at the center of every atom. Electrons and the nucleus have opposite electric charges, so an attractive force exists between them. This force in atoms plays a role similar to that played by gravity in keeping the solar system together. Most atoms are electrically neutral because the positive charge of the nucleus cancels the negative charge of the electrons. (Again, we stress that, in speaking of electric charge in terms of electrons, we are applying modern ideas. Benjamin Franklin didn't even know about atoms, much less electrons and nuclei.)

Electrons, particularly those in the outer orbits, tend to be rather loosely bound to their parent nucleus. These electrons can be removed from the atom and, once removed, can move freely (until attracted to some other positive charge). When electrons are pulled out of a material, they no longer cancel the positive charges in the nuclei. The result is a net excess of positive charge in the object, and we say that the object as a whole has acquired a positive electric charge. Similarly, an object acquires a negative electric charge when extra electrons are pushed onto it. This happens when you run a comb through your hair on a dry day: electrons are knocked onto the comb from your hair, so the comb acquires a negative charge. Simultaneously, your hair loses electrons, so individual strands become positively charged.

During a thunderstorm, the same transfer of charge occurs on a much larger scale as wind and rain disrupt the normal distribution of electrons in clouds. When a negatively charged cloud passes over a tall tree or tower, the violent electric discharge called lightning may result from the attraction of positive charges on the ground and negative charges in the cloud. (Note that in the case of lightning, both the positive and negative charges move.)

Although historical investigations of electric charge tended to concentrate on somewhat artificial experiments, we have come to know that electrically charged particles play important roles in many natural systems. For example, virtually all the atoms in the Sun have lost electrons and thus are positively charged. Atoms that have lost or gained electrons from their normal neutral state are called **ions.** Ions may be positive (the atom has lost electrons) or negative (the atom has gained electrons). In all advanced life forms, including human beings, ions routinely move into and out of cells to maintain the processes of life. As you read these words, for example, positively charged potassium and sodium ions are moving across the membranes of cells in your optic nerve to carry signals to your brain.

Some Facts About Electric Charge

If we think of the electric charge on an object as being the sum of the electric charges on particles in that object's atoms, then several interesting points follow.

1. **Objects can acquire an electric charge through friction,** which allows electrons to move from one body to another. In the examples already discussed, electrons move either to or from an object being rubbed, depending on the details of the materials involved. If electrons move out, the object becomes positively charged; if electrons move in, it becomes negatively charged.

2. **Objects can acquire a charge through the process of induction.** Suppose you have a metal ball, as shown in Figure 16-2, and you bring a negatively charged object near it. The negatively charged electrons are repelled and move to the far side of the ball. If you give them a way of moving away from the metal ball (for example, by connecting a wire between the ball and the ground, as shown), the electrons leave the ball. If you now disconnect the wire and then take the original negatively charged object away, you will find that the ball has a positive charge (because it has lost some of its electrons) even though it was never touched by any charged object. This process is known as charging by induction.

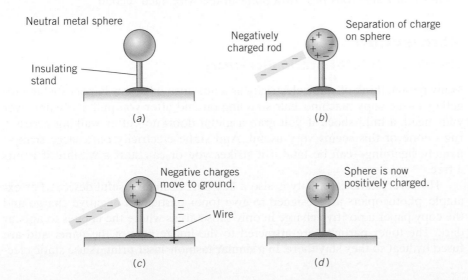

FIGURE 16-2. A hollow metal sphere illustrates the process of charging by induction. (a) The sphere is initially neutral. (b) If a negatively charged rod is brought close to the sphere, charges redistribute on the sphere's surface. (c) Negative charge can be drained off the far side of the sphere. (d) The result is a positively charged sphere, even though the rod did not touch it.

Develop Your Intuition: Charging by Induction

How can you use induction to produce a negatively charged metal ball? In this case, you connect the wire to the metal ball and bring up a positively charged object. Electrons in the metal ball move toward the positively charged object, leaving the far side of the ball with a positive charge. Electrons from the ground are then attracted to that positive charge and flow into the ball. Remove the wire and the ball will have acquired a negative charge because of those excess electrons.

3. **The total amount of electric charge in a closed system remains the same.** If electric charge is really carried as discrete particles inside the atom and if in the normal course of affairs these particles are not created or destroyed, then an important consequence follows. If you think of these particles as being like two colors of marbles in a jar, then there are many things you can do—you can move one or both colors around, put them in other jars, put some back in the original jar, and so on. However, no matter what you do, the total number of marbles of each color remains the same. In just the same way, the total number of positive and negative charges in a closed system stays the same, which means that the net charge on that system does not change. In the language we introduced in Chapter 6, the electric charge of a system is conserved, a result that is known as the law of **conservation of electric charge.**

4. **The basic unit of electric charge is the coulomb.** Like other physical quantities, the amount of electric charge on an object can be measured and assigned a unit. In the case of electric charge, that unit is called the *coulomb* (C) after a French scientist whose work we will describe in a moment. The easiest way to think of the coulomb is to say that if you have a pile of 6.25×10^{18} electrons, the magnitude of their total charge is 1 coulomb. This may seem like a very large number of electrons, but remember that electrons are very small. For reference, if you turn on an ordinary reading lamp, about this many electrons pass by a point in the wire each second.

Connection

The Good Side of Static Electricity

Many people think of static electricity as a nuisance or worse. Papers sticking together in the copy machine, hair standing on end after you pull a sweater over your head, a mild shock if you grab a metal doorknob after walking across a rug—none of this seems very useful. And static electricity on a large scale—namely, lightning—can be fatal if it strikes you or can start a wildfire if it hits a tree.

However, static electricity is also a key part of many useful devices. For example, photocopiers are designed to give toner particles a negative charge and the copy paper a positive charge in only those areas where the copy is to appear dark. The toner particles are attracted to the image part of the paper and are fused by heat so they stay there. In a similar fashion, laser printers use static electricity to attract toner particles to those places where the laser has hit on the paper being printed (Figure 16-3).

Large electrostatic precipitators are used in industry to remove particles of soot and dust from waste gases released to the environment. The tiny particles are attracted to a charged grid as the exhaust fumes pass through the smokestack; it is then a simple matter to turn off the grid and remove the accumulated soot without having it become air pollution.

On a more basic level, atoms in molecules are held together in chemical bonds by electrostatic forces. In fact, atoms consist of positive protons and negative electrons held together by their electric force of attraction. It's hard to think of a more fundamental force than that! ●

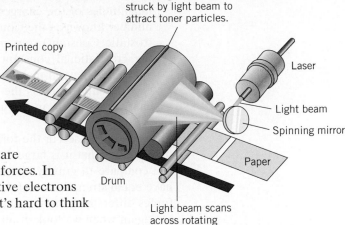

FIGURE 16-3. In a laser printer, light from a laser beam scans across a rotating drum, affecting the ability of the drum to hold a charge. A rod then applies a charge of static electricity to only those areas of the drum struck by the light. These areas attract toner particles, which are then pressed against the paper to be printed.

 ## COULOMB'S LAW

Electricity remained something of a mild curiosity until the mid-eighteenth century, when scientists began applying the scientific method to investigate it. One of the first tasks was to develop a precise statement about the nature of the electric force. The French scientist Charles Augustin de Coulomb (1736–1806) was most responsible for this work. During the 1780s, at the same time the United States Constitution was being written by Benjamin Franklin and others, Coulomb devised a series of experiments in which he passed different amounts of electric charge onto objects and then measured the force between them.

Coulomb observed that if two electrically charged objects are moved farther away from one another, the force between them gets smaller, just as with gravity. In fact, if the distance between two objects is doubled, the force decreases by a factor of four—the familiar 1/distance^2, or inverse-square relationship, that we have seen in the law of universal gravitation in Chapter 5. Coulomb also discovered that the size of the force depends on the product of the charges of the two objects—double the charge on one object and the force doubles, double the charge on both objects and the force increases by a factor of four, and so on.

After repeated measurements, Coulomb summarized his discoveries in a simple relationship known as **Coulomb's law:**

1. In words:

The force between any two electrically charged objects is proportional to the product of their charges divided by the square of the distance between them.

2. In an equation with words:

$$\text{Force (in newtons)} = k \times \frac{\text{First charge} \times \text{Second charge}}{(\text{Distance between them})^2}$$

3. In an equation with symbols:

$$F = k \times \frac{q_1 \times q_2}{d^2}$$

where the distance d between the two charges is measured in meters; the magnitudes of the charges q_1 and q_2 are measured in coulombs; and k is a number known as the Coulomb constant (also known as the Universal Electrostatic Constant) a number that plays the same role in electricity that the gravitational constant G plays in gravity. In SI units, k has the value 9.00×10^9 newton-meter^2/coulomb^2. This number, like G, can be determined experimentally and is the same for all charges and all separations of those charges anywhere in the universe.

As you can see in the following example, the most striking thing about the number k is that it is large. In fact, now that we have a force in nature that we can compare to gravity, the first thing we notice is that the two forces do not have equal strengths: one (electricity) is much stronger than the other. Nature's forces differ from one another—this is a feature of the world we live in, as we see again when we look at all the fundamental forces in Chapter 27.

Electric and Gravitational Forces Compared

EXAMPLE 16-1

The simplest atom is hydrogen, in which a single electron circles a single positively charged particle known as a *proton* (Figure 16-4). The masses of the electron and proton are 9×10^{-31} kg and 1.7×10^{-27} kg, respectively. The charge on the proton is 1.6×10^{-19} C, and the charge on the electron has the same magnitude but is negative. A typical separation of these two particles in an atom is 10^{-10} m. Given these numbers, what are the values of the electric versus the gravitational forces of attraction between these two particles?

REASONING AND SOLUTION: We need to apply Newton's equation for gravitational force (which requires two masses and a distance) and Coulomb's equation for electric force (which requires two charges and a distance).

For gravity, the force between an electron and a proton is

$$F \text{ (in newtons)} = \frac{G \times \text{Mass}_1 \text{ (kg)} \times \text{Mass}_2 \text{ (kg)}}{\text{Distance}^2 \text{ (m)}}$$

$$= \frac{(6.7 \times 10^{-11} \text{ m}^3/\text{kg-s}^2) \times (1.7 \times 10^{-27} \text{ kg}) \times (9 \times 10^{-31} \text{ kg})}{(10^{-10} \text{ m})^2}$$

$$= 1.0 \times 10^{-47} \text{ N}$$

FIGURE 16-4. The gravitational attraction between an electron and a proton in a hydrogen atom (*a*) is many orders of magnitude smaller than the electrical attraction between these two charged particles (*b*).

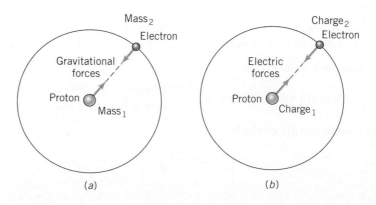

(a)   (b)

For electricity, the force between these two charged particles is

$$F \text{ (in newtons)} = \frac{k \times \text{Charge}_1 \text{ (C)} \times \text{Charge}_2 \text{ (C)}}{\text{Distance}^2 \text{ (m)}}$$

$$= \frac{(9 \times 10^9 \text{ N-m}^2/\text{C}^2) \times (1.6 \times 10^{-19} \text{ C}) \times (1.6 \times 10^{-19} \text{ C})}{(10^{-10} \text{ m})^2}$$

$$= 2.3 \times 10^{-8} \text{ N}$$

From this calculation we can see that, in the atom, the electric force (2.3×10^{-8} N) is many orders of magnitude (or factors of 10) larger than the gravitational force (1.0×10^{-47} N). This is why, in our discussion of the atom in subsequent chapters, we ignore the effects of gravity completely. ●

Polarization

Because of the inverse square nature of Coulomb's law, it is possible for an object to exert an electric force even though it has no net electric charge. For example, in some molecules the internal structure is such that the positive charges are located on one side of the molecule and the negative charges on the other, as shown in Figure 16-5. In this case, the total charge on the molecule is 0. However, if we put an electric charge at a point near the molecule, such as the point labeled A in Figure 16-5b, then an electric force appears on the molecule. This net electric force occurs because the force exerted by the charge at A on the negative charges on the molecule is greater than the force exerted on the positive charges, simply because the negative charges are closer to point A. This effect, known as *polarization,* is common in nature. In fact, it occurs whenever the charge within a molecule is not distributed symmetrically.

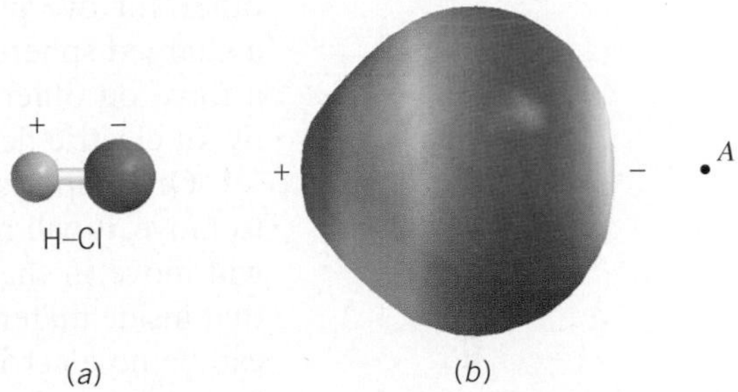

You see the results of polarization every day in the behavior of ordinary water. Water molecules, which are composed of two hydrogen atoms linked to one oxygen atom, have a negatively charged side and a positively charged side. When you sprinkle the compound sodium chloride (ordinary salt crystals) into water, the positively charged sodium atoms and negatively charged chlorine atoms interact with the positive and negative sides of water molecules. These forces enable the salt to dissolve in the water.

FIGURE 16-5. (a) A polar molecule such as HCl (hydrogen chloride, which dissolves in water to form hydrochloric acid) is electrically neutral, but it has regions that are more positive and regions that are more negative. Such a molecule will align itself with an electric field. (b) These electrically charged regions can be represented by colors—red for negative and blue for positive.

The Electric Field

Imagine that you have an electric charge sitting at a point. The charged object could be a piece of lint, an electron, or one of your hairs. If you brought a second charged object to a spot near the first, the second object would feel a force. If you then moved the second object to another spot, it would still feel a force, but the force would, in general, have a different magnitude and point in a different direction than at the first spot. In fact, the second charged object would feel a force at every point in space around the first object.

You can make a picture that represents this fact, as in Figure 16-6. The arrow at each point around a positively charged object represents the force that would be felt by a second, tiny, positively charged object if that second object

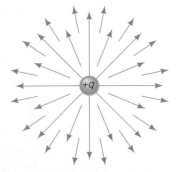

FIGURE 16-6. An electric field surrounding a positive charge, +q, may be represented by arrows radiating outward. Any charged object that approaches +q experiences a greater electric force the closer it gets. Positively charged objects are repelled, while negatively charged objects are attracted.

were brought to the point in space where the arrow originates. The collection of all the arrows that represent these forces is called the **electric field** of the original charged object. We can think of every charged object as being surrounded by such a field, as shown in the figure. Notice that the electric field is defined as the force that would be felt by a positive charge if that charge were located at a particular point, so that the field is present even if no other charge is actually there.

Technically, the electric field at a point is defined as the force that would be felt by a +1-coulomb charge if it were brought to that point. The field is usually drawn so that the direction of the arrow corresponds to the direction of that force and the length of the arrow corresponds to its magnitude. Figure 16-7 shows the shape of the electric field for two equal and opposite charges near one another, for two plates with equal and opposite electric charges on them, and for a charged sphere. In each case, the fact that the charge on the object can exert a force on other charges means that we can envision the object as surrounded by an electric field that permeates space.

One important point about electric fields is that if a positive charge is free to move, it will move in the direction indicated by the field. (A negative charge will move in the direction opposite to that indicated by the field.) This means that inside materials such as metals, in which there are many free electrons, there can be no electric field at all. If there were, the electrons would move until they had changed the charge configuration in such a way as to cancel the field. This phenomenon, known as *shielding,* illustrates an important difference between the electric and gravitational fields (see Chapter 5). Nothing can shield the gravitational force—Earth pulls on you whether you're inside your car or not. The metal casing of the car, however, can shield you from electric fields generated on the outside. This is why the safest place to be in a lightning storm is inside a metal container such as a car. The charges in the lightning bolt may move around the car's surface, but they (and their attendant fields) can't establish a stable field

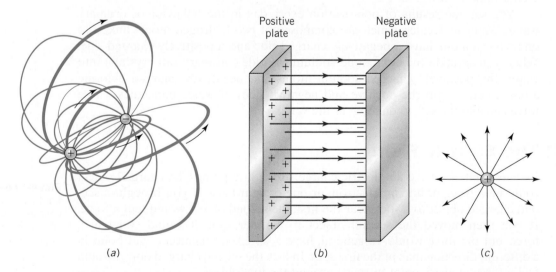

FIGURE 16-7. Electric fields may be represented as lines connecting positively and negatively charged objects. A positively charged object will move in the direction of the field lines. (*a*) Two equal and opposite charges near one another. (*b*) Two flat plates with equal and opposite electric charges. (*c*) A charged sphere.

within the shielding provided by the metal surfaces. For the same reason, airplanes can fly in bad weather without putting the passengers at risk of lightning strikes.

 MAGNETISM

Just as electricity was known to the ancient philosophers, so too was the phenomenon we call *magnetism*. The first known **magnets** were naturally occurring iron minerals. If you bring one of these minerals (a common one is called "magnetite," or "lodestone") near a piece of iron, the iron is attracted to it. You have undoubtedly seen experiments in which magnets were placed near nails, which jumped up and hung from the magnets.

Nails are electrically neutral, so electrical attraction doesn't make the nails move. Similarly, gravity cannot cause the nails to jump up. The fact that the nails behave in this way tells you that there must be yet another force in nature, a force different from both electricity and gravity. The simple experiment of picking up a nail with a magnet illustrates beyond a shadow of a doubt that there is a **magnetic force** in the universe—a force that can be identified and described by the same methods we used to investigate gravity and electricity.

Whereas electricity remained a curiosity until well into modern times, magnetism was put to practical use very early. The compass, invented in China and used by Europeans to navigate the oceans during the age of exploration, is the first magnetic device on record. A sliver of lodestone, left free to rotate in a horizontal plane, will align itself in a north-south direction. We use compasses so often these days that it's easy to forget how important it was for early travelers to know directions, particularly travelers who ventured in sailing ships out of sight of land.

In the late sixteenth century, the English scientist William Gilbert (1544–1603) conducted the first serious study of magnets. Although revered in his day as a doctor (he was physician to both Queen Elizabeth I and King James I), his most lasting fame came from his discovery that every magnet can be characterized by what he called *poles.* If you take a piece of naturally occurring magnet and let it rotate, one end of the magnet points north and the other end points south. These two ends of the magnet are called **poles** and are given the labels **north** and **south.**

In the course of his research, Gilbert discovered many important properties of magnets. He learned to magnetize iron and steel rods by stroking them with a lodestone. He discovered that hammered iron becomes magnetic and that the iron's magnetism can be destroyed by heating. He realized that Earth itself is a giant magnet, a fact that, as we shall see, explains the operation of the compass.

Gilbert also documented many of the most basic aspects of the magnetic force. He found that if two magnets are brought near one another so that the north poles are close together, a repulsive force develops between the magnets and they are forced apart. The same thing happens if two south poles are brought together. However, if the north pole of one magnet is brought near the south pole of another magnet, the resulting force is attractive. In this respect, studies of magnetism seem to mimic the eighteenth-century studies of static electricity. William Gilbert's results can be summarized in a simple statement:

> *Every magnet has two poles; like magnetic poles repel each other, while unlike poles attract each other.*

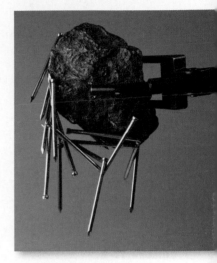

The common mineral magnetite is a natural magnet.

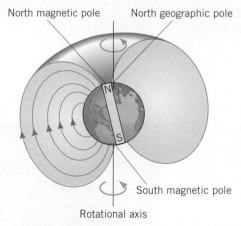

FIGURE 16-8. A compass needle and Earth. Any magnet will tend to twist because of the forces between its poles and those of Earth. Note that Earth's north and south magnetic poles don't quite line up with Earth's axis of rotation.

Once you know that a magnet has two poles, you can understand how a compass behaves. Earth itself is a giant magnet, with one pole in northern Canada and the other pole in Antarctica. If a piece of magnetized iron (for example, a compass needle) is allowed to rotate freely, one of its poles will be attracted to and twist around toward Canada in the north, and the other end will point to Antarctica in the south (see Figure 16-8). There is a confusing matter of nomenclature we need to address here. The first people who used magnetic compasses understandably painted an "N" on the pole that pointed north. Given that we now know that opposite magnetic poles attract each other, we have two options: (1) We can call the pole labeled "N" the north pole of the compass, in which case it is the south magnetic pole of the Earth that is located in Canada. Alternatively, we can say that the north magnetic pole of the Earth is located in Canada and that the "N" on the magnetic compass stands for the "north-seeking pole," with the understanding that this is actually the magnet's south pole. Since many people (the authors included) find it more reasonable to keep north poles, either magnetic or geographic, in the Northern Hemisphere, the second option is the one most used. Either procedure, however, is correct and consistent with what we know about magnets.

The Magnetic Field

Just as the electric force can be represented in terms of an electric field, so too can the magnetic force be represented in terms of a magnetic field. If a small compass needle is brought near a magnet, as shown in Figure 16-9, the forces exerted by the magnet twist the needle around. (In the language of Chapter 7, we say that the forces exert a torque on the compass needle.) In general, the needle's direction is different at different locations around the magnet.

We can imagine mapping out a magnetic field by using the procedure shown in Figure 16-9. We lay down a large magnet and then bring a compass needle, as shown, to its north pole. The compass aligns itself, pointing toward the magnet's north pole. After the needle has settled down, we bring up another compass and align it nose to tail with the first, then bring up a third, a fourth, and so on until we have a line of compasses stretching from the north to the south pole of the original magnet, as shown. We draw a line through these compasses and then repeat the operation for other initial orientations of the first compass. This results in a series of lines around the magnet, as shown in Figure 16-10*a*.

Just as we can imagine any collection of electric charges as being surrounded by an electric field—imaginary field lines—we can imagine every magnet as being surrounded by an imaginary set of lines such as those in the figure. These lines are drawn so that if a compass were brought to a point in space, the needle would turn and point along the line passing through that point. The number of lines in a given area is a measure of the strength of the forces exerted on the compass. Collections of lines that map out the directions in which compass needles would point are called **magnetic field lines.** They help us visualize the shape of the **magnetic field** around a magnetized object.

The magnetic field shown in Figure 16-10 is a particularly important one because it is the field associated with a bar magnet with a north and a south pole, called a *dipole field,* and we encounter this sort of field often in the natural world.

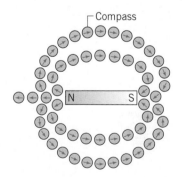

FIGURE 16-9. Magnetic field lines curve from the north pole to the south pole of a magnetic dipole. Small compass needles placed in this field will align with these field lines.

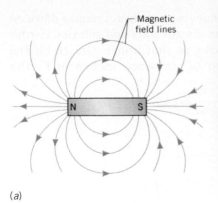

Magnetic field lines

(a)

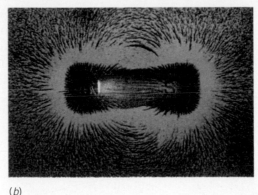

(b)

FIGURE 16-10. (*a*) A bar magnet and its magnetic dipole field. (*b*) Iron filings placed near a bar magnet align themselves along the field.

Isolated Magnetic Poles

All magnets found in nature have both north and south poles—you never find one without the other. Even if you take an ordinary bar magnet and cut it in two, you don't get a north and a south pole in isolation. Instead, you get two small magnets, each with a north and a south pole (Figure 16-11). If you take each of those halves and cut it in half, you will continue to get smaller and smaller dipole magnets. In fact, it seems to be a general rule of nature that

There are no isolated magnetic poles in nature.

In the language of physicists, a single isolated north or south magnetic pole would be called a "magnetic monopole." Although physicists have conducted extensive searches for monopoles, no experiment has yet found unequivocal evidence for their existence. In the next chapter we see why.

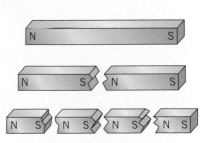

FIGURE 16-11. Cut magnets. If you break a dipole magnet in two, you get two smaller dipole magnets, not an isolated north or south pole.

 MAGNETIC FORCES ON CHARGED PARTICLES

When a particle that carries an electric charge moves through a region containing a magnetic field, a force is exerted on the particle. The magnitude of the force depends on the strength of the magnetic field, as well as on the particle's velocity and charge. The force is always exerted in a direction perpendicular to the direction in which the particle is moving, and is greatest when the particle moves perpendicular to the magnetic field. The force becomes zero when the particle moves in the same direction as the field.

The direction of the force is given by a *right-hand rule*. (*A word of caution:* There are several right-hand rules in physics—this is just one of them.) It works like this: Point the index finger of your right hand in the direction of the particle's motion and the middle finger of your right hand in the direction of the magnetic field. Then the force is in the direction of your extended thumb for a positively charged particle and in the opposite direction for a negatively charged one.

For example, in Figure 16-12 we show a positively charged particle entering a region where the magnetic field is pointing into the plane of the paper. Applying the right-hand rule tells us that the force on the particle is directed upward, toward the top of the page, so that the particle follows the curved path shown.

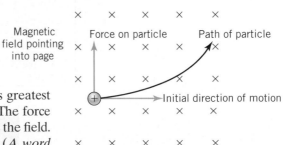

Magnetic field pointing into page

Force on particle

Path of particle

Initial direction of motion

FIGURE 16-12. A positively charged particle enters a region where the magnetic field is pointing downward into the paper. The right-hand rule tells us that the force on the particle is directed toward the top of the page, so that the particle follows the curved path shown.

Fast-moving charged particles called *cosmic rays* are always raining down on Earth. Some come from the Sun, others from distant stars and galaxies. Earth's magnetic field deflects some of these particles, as shown in Figure 16-13. The curved paths of some rays describe a kind of corkscrew motion in Earth's

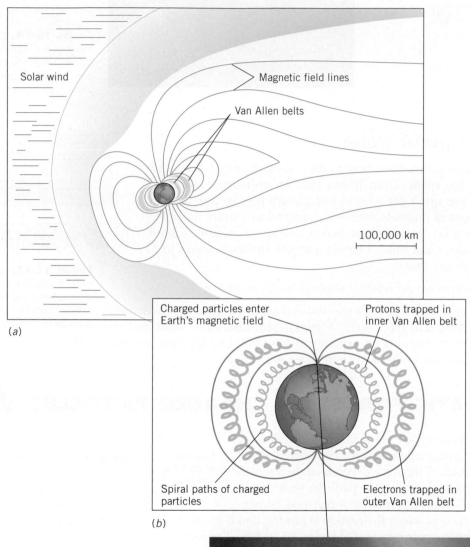

Solar wind

Magnetic field lines

Van Allen belts

100,000 km

(a)

Charged particles enter Earth's magnetic field

Protons trapped in inner Van Allen belt

Spiral paths of charged particles

Electrons trapped in outer Van Allen belt

(b)

FIGURE 16-13. Earth's magnetic field deflects the constant rain of fast-moving charged particles called cosmic rays. Many rays adopt a corkscrew motion in Earth's magnetic field, producing the van Allen belts. Because the magnetic field lines converge near the poles, the particles are reflected back toward the equator and are then reflected back again at the other pole. The glow caused by the collisions of these particles is called the aurora borealis.

(c)   Aurora caused by charged particles entering atmosphere

magnetic field, producing what are called the *Van Allen belts* (named after the American physicist James Van Allen, who discovered them in 1958). These two rings consist of particles that travel corkscrew paths up and down between Earth's poles. Because the magnetic field lines begin to converge near the poles, the particles are reflected back toward the equator and are then reflected back again at the other pole.

When Earth's magnetic field is distorted (as it often is when massive numbers of particles are thrown outward from the Sun), charged particles in the van Allen belts actually enter the atmosphere, producing the beautiful displays known as the "northern lights," or *aurora borealis*, and "southern lights," or *aurora australis*.

Connection
Television and Computer Screens

You don't need to live in the far north or south of the world and view the auroras in order to see the effects of magnetic forces on charged particles. All you have to do is turn on your television or computer.

Picture tubes work by shooting beams of electrons onto a coated screen (Figure 16-14*a*). The electrons are emitted by an electron gun that is basically a wire heated to high temperatures. Electrically charged plates accelerate the electron beams to high speeds directed toward the screen, which is coated with chemical phosphors that glow when the electrons strike them. The beams are guided across the screen from side to side and from top to bottom (Figure 16-14*b*) while the internal electronics adjust the beam intensities to match the incoming signal, producing dots of the right strength and color to create the desired picture.

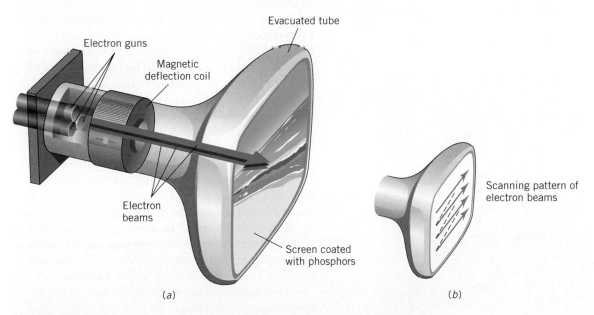

(*a*)   (*b*)

FIGURE 16-14. (*a*) A picture tube works by beaming electrons from a heated wire onto a coated screen that glows when the electrons strike it. The beam passes through the center of several wire coils that produce a changing magnetic field while internal electronics adjust the beam intensities to match the incoming signal and produce the desired picture. (*b*) This magnetic field deflects the beams back and forth from top to bottom across the entire screen.

In order to produce a smoothly moving image on the screen, the electron beams sweep across the entire screen 30 times per second. How can the beams be guided so accurately and so rapidly at the same time? The answer is magnetic fields. The beams pass through the center of several coils placed in front of the electron guns. The coils produce a changing magnetic field that deflects the beams as needed, back and forth from top to bottom.

In recent years, newer picture screens use liquid crystals instead of electron guns. These screens can be flat for portable computers or wall-mounted televisions. However, electron-beam tubes are still used in oscilloscopes for medical and electronic diagnosis. ●

THINKING MORE ABOUT

Magnetic Navigation

Many living things in addition to humans use Earth's magnetic field for navigation. Scientists at the Massachusetts Institute of Technology demonstrated this ability in 1975 when they were studying a single-celled bacterium that lived in the ooze at the bottom of nearby swamps. They found that the bacteria incorporated about 20 little chunks of the mineral magnetite into their bodies. These chunks were strung out in a line, in effect forming a microscopic compass needle.

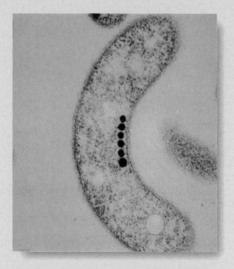

Grains of iron minerals in this bacterium allow it to tell up from down.

Because Earth's magnetic field dips into Earth's surface in the Northern Hemisphere and rises up out of the surface in the Southern Hemisphere, the Massachusetts bacteria have a built-in up and down indicator. This internal magnet allows the bacteria to navigate down into the nutrient-rich ooze at the bottom of the pond. Interestingly, related bacteria in the Southern Hemisphere follow the field lines in the opposite direction to get to the bottom of their ponds.

Since 1975, similar internal magnets have been discovered in many animals. Some migratory birds, for example, use internal magnets as one of several cues to guide them on flights thousands of miles in length. In one case—the Australian silvereye—there is evidence that the bird can sense the magnetic field of Earth through a process involving the modification of molecules normally involved in color vision.

Over the past few decades, people have built large communications and military detection systems that produce very strong electric and magnetic fields. Think, for example, of military radar systems or of systems designed to allow communications with submarines. When these systems are transmitting, migrating birds sometimes become disoriented by these fields and lose their ability to navigate.

How much attention should be paid to the effects of electric and magnetic fields on wildlife that we create? If human beings need to have a particular technical capability, how much risk to wildlife is acceptable?

Summary

The forces of **electricity** and **magnetism** are quite different from the universal gravitational force that Newton described in the seventeenth century. Nevertheless, Newton's laws of motion provided scientists with a way to describe and quantify a range of intriguing electromagnetic behavior.

Static electricity is the interaction of **electric charges,** such as electrons and protons. An excess of electrons imparts a **negative charge,** while a deficiency causes an object to have a **positive charge.** An **ion** is an atom that has either too many or too few electrons. The transfer of electrons between objects causes phenomena such as lightning, static cling, and the small sparks produced when you walk across a wool rug on a cold winter day. The total electric charge of a closed system stays the same. This is the law of **conservation of electrical charge.**

Objects with like charge experience a repulsive force, while oppositely charged objects attract one another. These observations are quantified in **Coulomb's law,** which states that the magnitude of the electrostatic force between any two objects is proportional to the product of the charges of the two objects and inversely proportional to the square of the distance between them.

Other scientists, investigating the very different phenomenon of magnetism, observed that every **magnet** has a **north pole** and a **south pole** and that magnets exert forces on one another. No matter how many times a magnet is divided, each of its pieces has two poles—there are no isolated magnetic poles. Like magnetic poles repel one another, while opposite poles attract one another. A compass is a needle-shaped magnet that points to the poles of Earth's magnetic field.

Electric and **magnetic forces** can be described in terms of **electric** and **magnetic fields**—imaginary lines that indicate the directions of forces that would be experienced in the vicinity of electrically charged or magnetic objects.

Charged particles traveling through a magnetic field experience a force that causes them to be deflected. The direction of deflection is different for positive or negative charges.

Key Terms

conservation of electric charge The physical law that states that the total amount of charge in the universe does not change. (p. 338)

Coulomb's law The law of physics that defines the force between electrically charged objects. (p. 339)

electric charge The property of an object or particle that quantifies its response to electric and magnetic phenomena; charges are either **positive** or **negative,** so named because their effects tend to cancel one another out. (p. 334)

electric field A measure of how much electric force an electric charge would experience at a particular location in space; it is defined as the force per unit charge. (p. 342)

electric force The force that one electric charge exerts on another. (p. 339)

electricity A general term used to define the presence and or motion of electric charges. (p. 334)

ion An atom that has lost or gained electrons, thus acquiring an electric charge. (p. 337)

magnet Material or object that possesses a magnetic property such that it is attracted to or repelled from another magnet. (p. 343)

magnetic field A measure of the effect that a magnet has on the space surrounding it, or the effect that would be felt by a magnet if it were at a particular location in space. (p. 344)

magnetic field lines Imaginary lines in space that give the direction of a compass needle. (p. 344)

magnetic force The force that a magnet exerts on other magnets or on moving electric charges. (p. 343)

poles (north and **south)** The two ends of a magnet; one is called the *north pole* and one is called the *south pole*. (p. 343)

static electricity Electric charges that are at rest; also used to describe the force due to electric charges that are at rest. (p. 334)

Key Equations

1 coulomb = Magnitude of charge on 6.25×10^{18} electrons

$$\text{Electrostatic force} = k \times \frac{\text{First charge} \times \text{Second charge}}{(\text{Distance between charges})^2}$$

Universal Electrostatic Constant (Coulomb constant): $k = 9.00 \times 10^9 \ \frac{\text{N-m}^2}{\text{C}^2}$

Review

1. How do we know that there is such a thing as an electric force?

2. How do two electrically charged objects behave when brought near one another?

3. How can the movement of negative charges, such as electrons, produce a material that has a positive charge?

4. What is static electricity? What are some examples of it in everyday life?

5. When you walk across a rug to open a door, you can get a shock when you reach out toward the doorknob even if you don't actually touch it. Why is this so?

6. Compare the electric force to the force of gravity. Which is stronger? Do both depend on mass?

7. Explain how eighteenth-century scientists thought of electricity in terms of fluids. Does this make sense?

8. What were Benjamin Franklin's contributions to the understanding of electricity?

9. How does a lightning rod work?

10. What do we mean, at the atomic level, when we say that something is electrically charged?

11. How can objects acquire an electric charge through friction?

12. What type of charge results when electrons are removed from a particular material? When they are added to a material?

13. How do objects acquire charge through induction? Do the objects need to contact one another physically for this to happen? Explain.

14. What is meant by the conservation of electric charge?

15. What is the basic unit of electric charge? How many electrons are equivalent to 1 unit of charge?

16. What is Coulomb's law? How does the strength of attraction or repulsion of objects vary with the distance between them?

17. What is k, the Coulomb constant? How is this determined? Does it have the same value everywhere?

18. How can an object exert an electric force even if it has no net electric charge? What is meant by polarization?

19. What is an electric field? When does such a field exist?

20. What happens to free electrons in metals under the influence of an electric field?

21. What is shielding? What is the difference between an electric and a gravitational field with respect to shielding?

22. How do you know there is such a thing as a magnetic force?

23. What was the important result of William Gilbert's work on magnetism? How is this similar to the electric force?

24. What is a magnetic field?

25. What type of magnetic fields are found in nature?

26. What is a monopole? A dipole?

Questions

1. There is an old saying that lightning never strikes the same place twice. Given what you know about electric charge, is this statement likely to be true? Why?

2. If you double the distance between two charged objects, how does this affect the electric force between them? What if you triple the distance?

3. If you double the charge on one of two charged objects, how does the force between them change?

4. Three small spheres carry equal amounts of positive electric charge. They are equally spaced and lie along the same line, as shown. What is the direction of the net electric force on each charge due to the other two charges?

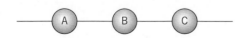

Questions 4, 5

5. Three small spheres carry electric charge. They are equally spaced and lie along the same line, as shown. They all have the same amount of charge, but sphere A and C are positive and sphere B is negative. What is the direction of the net electrical force on each sphere due to the other two spheres?

6. Four small charged spheres sit at the corners of a square, as shown in the figure. Sphere A is negatively charged and the other three have an equal amount of positive charge. Reproduce the figure and draw an arrow at sphere A that represents the net electric force on sphere A due to the other three charges. Repeat this for spheres B, C, and D. Which sphere has the greatest net force acting on it?

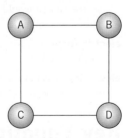

7. Two charged spheres sit near each other, as shown in the figure. They carry an equal amount of negative charge. What is the direction of the electric field created by these two charges at locations a, b, and c? (Point b is at the exact midpoint between spheres A and B.)

Questions 7, 8

8. Two charged spheres sit near each other, as shown in the figure. Sphere A is negative and sphere B carries an equal amount of positive charge. What is the direction of the electric field created by these two charges at locations a, b, and c?

9. Two spheres carry the same amount of positive charge. Reproduce the figure and draw an arrow that represents the direction and strength of the electric field at a, b, c, and d.

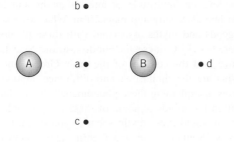

Questions 9, 10

10. Two spheres carry equal and opposite amounts of electric charge. Sphere A is positive and sphere B is negative. Reproduce the figure and draw an arrow that represents the direction and strength of the electric field at a, b, c, and d.

11. Static cling makes your clothes stick together. What is actually happening in your dryer to generate it? Does this make sense given what you have learned in this chapter?

12. The Greeks had a legend that there was an island in the Mediterranean Sea made entirely of lodestone. They used this story as an argument that ships should not be built with iron nails. How does this argument work? Are there other reasons for not building ships with iron nails?

13. Object A and object B are initially uncharged and are separated by a distance of 2 meters. Suppose 10,000 electrons are removed from object A and placed on object B, creating an electric force between A and B. Is this force attractive or repulsive? If an additional 10,000 electrons are removed from A and placed on B, how much does the electric force change?

14. Object A and object B are initially uncharged and are separated by a distance of 1 meter. Suppose 10,000 electrons are removed from object A and placed on object B, creating an attractive force between A and B. If an additional 10,000 electrons are removed from A and placed on B and the objects are moved so that the distance between them is increased to 2 meters, how much does the electric force between them change?

15. The magnetic field at the equator points north. If you throw a positively charged object (for example, a baseball with some electrons removed) to the east, what is the direction of the magnetic force on that object?

16. The magnetic field at the equator points north. If you throw a negatively charged object (for example, a baseball with some extra electrons) to the east, what is the direction of the magnetic force on that object?

17. A charged particle is located on the right side of a bar magnet and is moving to the right, as shown. If the particle is being deflected in such a way that its path is curving out of the page, is the particle negative or positive?

18. Two identical bar magnets are aligned as shown. What is the approximate direction of the magnetic field created by this arrangement at locations a, b, and c?

19. A small bar magnet pulls on a larger one with a force of 100 newtons. What is the magnitude of the force the larger one exerts on the smaller one?

20. A charge of +1 coulomb is placed at the 0-cm mark of a meter stick. A charge of −1 coulomb is placed at the 100-cm mark of the same meter stick. Is it possible to place a proton somewhere on the meter stick so that the net force on it due to the charges at the ends is 0? If so, where should it be placed? Explain.

21. A charge of +1 coulomb is placed at the 0-cm mark of a meter stick. A charge of +4 coulombs is placed at the 100-cm mark of the same meter stick. Is it possible to place a proton somewhere on the meter stick so that the net force on it due to the charges at the ends is 0? If so, where should it be placed? Explain.

22. Identify five objects in your room that would not be possible without discoveries in electromagnetism.

Problems

1. Based on electric charges and separations, which of the following atomic bonds is strongest? [*Hint:* You are interested only in the relative strengths, which depend only on the relative charges and distances.]

a. A +1 sodium atom separated by 2.0 distance units from a −1 chlorine atom in table salt

b. A +1 hydrogen atom separated by 1.0 distance unit from a −2 oxygen atom in water

c. A +4 silicon atom separated by 1.5 distance units from a −2 oxygen atom in glass

2. Assume that in interstellar space the distance between two electrons is about 0.1 cm.

a. Is the electric force between the two electrons repulsive or attractive?
b. Calculate the electric force between these two electrons.
c. Calculate the gravitational force between these two electrons. Is this an attractive or repulsive force?

d. Which force is greater and by how many orders of magnitude?

3. Repeat Problem 2 for two protons at the same distance from one another.

4. Assume that you have two objects, one with a mass of 10 kg and the other with a mass of 15 kg, each with a charge of $−3.0 \times 10^{-2}$ C and separated by a distance of 2 meters. What is the electric force that these objects exert on one another? What is the gravitational force between them?

Investigations

1. Read the novel *Frankenstein* (or see the classic 1931 movie with Boris Karloff and Colin Clive). Discuss the ideas about the nature of life that are implicit in the story. Does it represent a realistic picture of scientific research? Why?

2. How does a copying machine work? What role does electricity play in the transfer of an image from the pages of a book to the copied final product?

3. Explore how electric eels generate electric shocks.

4. How does an electroencephalograph (EEG) work? How does it differ from an electrocardiograph (EKG)?

5. Many kinds of living things, from bacteria to vertebrates, incorporate small magnetic particles. Investigate the ways in which living things specifically use magnetism.

6. Explore the multitude of ancient myths and legends regarding electricity and magnetism. What were some of the legends and myths associated with these phenomena that were held by Chinese, the Hindus, various Near Eastern cultures, and the cultures of the early Greeks and Romans? What are the similarities and differences among these cultures in explaining these phenomena? Try to place yourself within the minds of some of these early peoples. How rational would these myths seem to you if you lived in their times? Similarly, can you identify any beliefs held today about either electricity or magnetism that are not fully supported by mainstream scientists employing the scientific method? How should such questions be investigated?

WWW Resources

See the *Physics Matters* home page at **www.wiley.com/college/trefil** for valuable web links.

1. **http://www.gel.ulaval.ca/~mbusque/elec/main_e.html** Exploring electric fields. Field plots for user-defined arrangements of charge.

2. **http://www.fi.edu/franklin/rotten.html** The life, times, and experimental history of Benjamin Franklin.

3. **http://www.aip.org/history/electron/** A site celebrating the 100th anniversary of the discovery of the electron.

4. **http://www.ee.umd.edu/~taylor/frame1.htm** A gallery of famous scientists and personalities from the historical study of electricity and magnetism.

5. **http://thorin.adnc.com/~topquark/fun/JAVA/electmag/electmag.html** A history and function of the mass spectrometer.

6. **http://www.geo.mtu.edu/weather/aurora/** A site containing breathtaking photographs of and the physics underlying the aurora borealis or "northern lights."

7. **http://pwg.gsfc.nasa.gov/istp/earthmag/demagint.htm** A collection of websites discussing Earth's magnetic field, celebrating the 400th anniversary of the publication of *De Magnete* by William Gilbert.

17 | Electromagnetic Interactions

KEY IDEA

Electricity and magnetism are two aspects of the same force—the electromagnetic force.

PHYSICS AROUND US . . . The Electrical World

During the next hour or so, look around you and notice all the electric equipment that affects your life. Are you in a building? Electrons flowing through copper wires in the walls are what make the lights go on. They supply power to the outlets on the walls—the places where you plug in everything from stereos to vacuum cleaners. Are you listening to your Walkman or using a laptop computer? Then you are using the energy stored in batteries to create a flow of electrons to run your equipment. Drying your hair? The fan is driven by an electric motor. Starting your car? The battery turns the engine over when you use your ignition key.

It would be hard to imagine the modern world without electricity. Yet all of these electronic devices (and countless more) spring from a few discoveries made by basic researchers in the mid-nineteenth century. In this chapter, we examine these discoveries. Many of them involve the close relationship between electric and magnetic forces, which only becomes apparent when we look at electric charges in motion. We are no longer looking just at static electricity; now we are studying the complete range of electromagnetic phenomena.

FORCES ON MOVING CHARGES

In our everyday experience, static electricity and magnetism seem to be two unrelated phenomena. After all, what could be more different than static cling and refrigerator magnets? Yet scientists in the nineteenth century, probing deeper into the electric and magnetic forces, discovered remarkable connections between the two—a discovery that transformed every aspect of technology.

In Chapter 16 we examined the forces acting between two or more stationary charges. It turns out that moving electric charges experience very different forces in addition to the force of static electricity. These additional forces are magnetic. There's a beautiful symmetry at work here: moving electric charges give rise to magnetic fields, and changing magnetic fields give rise to electric fields. In this chapter we discuss these effects one at a time and then show how they can be brought together into one description of the **electromagnetic force.**

Despite the fact that people had known about both electricity and magnetism for millennia, their fundamental connection could not be discovered until scientists had at their disposal a mundane device that we use every day without thinking—the storage battery. The battery enabled scientists to have a reliable source of electricity for use in experiments. As our understanding of electricity and magnetism grew, the importance of batteries for supplying electric energy in applications became evident.

BATTERIES AND ELECTRIC CURRENT

Although we encounter static electricity in our everyday lives, most of our contact with electricity comes from moving charges. In your home, for example, negatively charged electrons move through wires to run all of your electric appliances. A flow of charged particles is called an **electric current.**

Until the work of the Italian scientist Alessandro Volta (1745–1827), scientists could not produce persistent electric currents in their laboratories and therefore knew little about them. As a result of his investigations into the work of Luigi Galvani (see Physics in the Making on page 355), Volta developed the first **battery,** a device that converts stored chemical energy in the battery materials into kinetic energy of electrons flowing through an outside wire.

The first batteries were crude affairs that featured alternating disks of two different metals, such as lead and silver, in salt water. We now use much more sophisticated batteries to start our cars and run all sorts of portable electric equipment, but the principle is the same. Your car battery, a reliable and beautifully engineered device that routinely performs for years before it needs replacing, is also made of alternating plates of two kinds of material (lead and lead oxide) immersed in a bath of dilute sulfuric acid. When the battery is being discharged, the lead plate interacts with the acid, producing lead sulfate (the white crud that collects around the posts of old batteries) and some free electrons. This process causes free electrons in an external wire to travel toward the other plates, where they interact with the lead oxide and sulfuric acid to form more lead sulfate. The electrons flowing through the outside wire are what enable you to start your car.

A woodcut engraving of Volta's early metal disk battery shows alternating disks of two metals connected to containers of salt water. These two containers served as the positive and negative terminals of this primitive battery.

When the battery is completely discharged, it consists of plates of lead sulfate immersed in water, a configuration from which no energy can be obtained. Running a current backward through the battery, however, runs all the chemical reactions in reverse and restores the original configuration. We say that the battery has been recharged. Once this is done, the whole cycle can proceed again. In your car, the generator constantly recharges the battery whenever the engine is running, so it is always ready to use.

We use many other kinds of batteries in our everyday life—small ones to power miniature electronic devices, larger ones to run our laptop computers. Some of these batteries, like the one in your car, can be recharged, while others we simply use, recycle, and replace. All of them, however, provide us with a source of electric current. It was this feature of Volta's invention that led to the discovery of the connection between electricity and magnetism.

Physics in the Making
Luigi Galvani and Life's Electric Force

Scientists of the eighteenth century discovered remarkable links between life and electricity. Of all the phenomena in nature, none fascinated these scientists more than the mysterious "life force" that allowed animals to move and grow. An old doctrine called vitalism held that this force is found only in living things and not found in the rest of nature. Luigi Galvani (1737–1798), an Italian physician and anatomist, added fuel to the debate about the nature of life with a series of classic experiments demonstrating the effects of electricity on living things.

Galvani's most famous investigations employed an electric spark to induce convulsive twitching in amputated frogs' legs—a phenomenon not unlike a person's involuntary reaction to a jolt of electricity. Later he was able to produce a similar effect simply by poking a frog's leg simultaneously with one fork of copper and one of iron. In modern language, we would say that the electric charge and the presence of the two metals in the salty fluid in the frog's leg led to a flow of electric charge in the frog's nerves, a process that caused contractions of the muscles.

Galvani, however, argued that his experiments showed that there was a vital force in living systems, something he called "animal electricity," that made them different from inanimate matter. This idea gained some acceptance among the scientific community, but provoked a long debate between Galvani and the Italian physicist Alessandro Volta. Volta argued that Galvani's effects were caused by chemical reactions between the metals and the salty fluids of the frogs' legs. In retrospect, both of these scientists had part of the truth. Muscle contractions are indeed initiated by electric signals, even if there is no such thing as animal electricity, and electric charges can be induced to flow by chemical reactions.

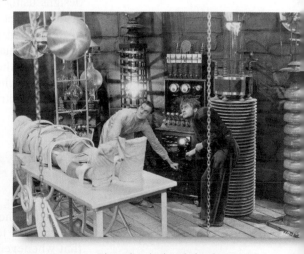

The idea behind the legend of Frankenstein may have been suggested by early experiments on animal electricity.

The controversy that surrounded Galvani's experiments had many surprising effects. On the practical side, as we discuss in this chapter, Volta's work on chemical reactions led to the development of the battery and, indirectly, to our modern understanding of electricity. However, the notion of animal electricity proved a great boon to medical quacks and con men, and for centuries various kinds of electric devices were palmed off on the public as cures for almost every known disease.

Finally, in a bizarre epilogue to Galvani's research, other researchers used batteries to study the effects of electric currents on human cadavers. In one famous public demonstration, a corpse was made to sit up and kick its legs by electric stimulation. Such unorthodox experiments helped inspire Mary Shelley's famous novel, *Frankenstein.* ●

Connection
Batteries and Electric Cars

Modern urban areas are troubled by pollutants introduced into the air by the burning of gasoline in cars and trucks. A possible solution to this problem is the introduction of battery-powered electric cars. In these cars, chemical energy stored in batteries, rather than in gasoline, provides the motive power.

Battery-powered electric cars store chemical energy in batteries rather than in gasoline.

Many electric vehicles, such as golf carts, are already in use, but much more powerful batteries are needed for cars traveling on modern highways. The most difficult technical problem standing in the way of developing practical electric cars is the fact that batteries actually store rather small amounts of energy for their weight. For example, for the kind of lead-acid battery that you use to start your car to supply as much energy as a gallon of gasoline, you would need a battery weighing 3500 newtons (800 pounds). Because the car has to move the batteries as well as its normal load, this heavy weight leads to problems.

Perhaps the most severe problem is the comparatively short distance an electric car can go before the batteries need to berecharged, a distance called the car's range. Many first-generation electric cars have effective ranges of only 65–80 kilometers (40–50 miles), although engineers are steadily improving the performance of these cars. Two types of advanced batteries—nickel-cadmium (the type you probably use in your laptop computer) and other designs using nickel and other metals—deliver about twice the power of lead-acid batteries. Before long, many of us may be driving electric cars with ranges of more than 100 miles.

The primary advantage of electric cars is that they don't emit exhaust in the area where they are driven. We should keep in mind, however, that the electricity used to charge the batteries was most likely generated by nuclear or coal-burning electrical plants somewhere outside the city. You can't get energy for free! ●

● MAGNETIC EFFECTS FROM ELECTRICITY

In the spring of 1820, a strange thing happened during a physics lecture in Denmark. The lecturer, Professor Hans Christian Oersted (1777–1851), was using a battery to demonstrate some properties of electricity. By chance he noticed that whenever he connected the battery to a circuit (so that an electric current began to flow through the wire), a nearby compass needle began to twitch and turn. When he disconnected the battery, the needle went back to pointing north. Connecting the battery in the opposite direction, and thus reversing the direction of current flow, caused the compass needle to swing in the opposite direction. This accidental discovery led the way to one of the most profound insights in the history of science. Oersted had discovered that electricity and magnetism—two forces that seem as different from one another as night and day—are in fact intimately related to one another. They are two sides of the same coin.

In subsequent studies, Oersted and other physicists established that whenever electric charge flows through a wire, a magnetic field appears around that wire. A compass brought near the wire will twist around until it points along the direction of the magnetic field. This observation leads to an important experimental finding in electricity and magnetism:

The motion of electric charges creates magnetic fields.

The magnetic field produced by a current in a wire, as shown in Figure 17-1, is in the form of concentric circles with the current at the center. Thus, if a compass in front of the wire points to the right, a compass at a corresponding point behind the wire will point to the left. The direction of the magnetic field, in fact, is given by a simple mnemonic known as a *right-hand rule*. For the wire shown in Figure 17-1, the rule works like this: You put the thumb of your right hand in the direction of the current. (For historical reasons, we assume the current consists of positive charges. If the current consists of electrons, you point your thumb in the opposite direction to the electron flow.) When you have done this, the fingers of your right hand are curling in the direction of the magnetic field (that is, your fingers point in the same direction that a compass needle would if it were in the same spot). Note that this right-hand rule, which gives the direction of a magnetic field, is different from the right-hand rule given in Chapter 16, which gives the direction of the force on a moving particle and the rule for angular momentum (Chapter 7).

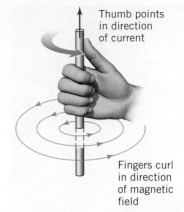

Thumb points in direction of current

Fingers curl in direction of magnetic field

FIGURE 17-1. The right-hand rule states that when the thumb of the right hand points in the direction of a positive electric current, the fingers curl in the direction of the resultant magnetic field.

> ## Develop Your Intuition: Magnetic Field of a Straight Current-Carrying Wire
>
> Suppose you align a wire in part of a circuit in a straight line so that electrons are moving from east to west. If you place a compass above the wire, in what direction will the needle point? What happens if you place the compass below the wire?
>
> The compass needle will point in the direction of the magnetic field of the wire, so we use the right-hand rule to determine the direction of that field. If electrons are flowing from east to west, then the direction of a *positive* current is from west to east, so point your thumb to the east. If you then look along the wire from east to west, your fingers are curling around the wire in a counterclockwise direction. If you place a compass above the wire, it will point south; if you place the compass below the wire, it will point north.

Like all fundamental discoveries, the discovery of this law of nature had important practical consequences. Perhaps most important, it led to the development of the **electromagnet,** a device composed of many coils of wire that produces a magnetic field whenever an electric current flows through the wire. Almost every electric appliance in modern technology uses this device.

The Electromagnet

Electromagnets work on a simple principle, as illustrated in Figure 17-2. If an electric current flows in a loop of wire, then a magnetic field is created around the wire, just as Oersted discovered in 1820. That magnetic field has the shape

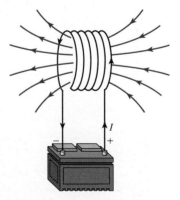

FIGURE 17-2. A schematic drawing of an electromagnet reveals the principal components—a loop of wire and a source of electric current. When current goes around the loop, a magnetic field forms through it.

Electromagnets, which can be turned on and off, are ideal for moving scrap iron at a junkyard.

sketched in the figure, a shape familiar to you as the dipole magnetic field shown in Figure 16-10. The direction of the field is given by the right-hand rule.

The important point about the magnetic field associated with a loop of wire is that it has exactly the same shape as the field of a permanent magnet. You could, in fact, imagine replacing the loop of wire by a bar magnet and this replacement would not affect the magnetic field in the region.

We can, in other words, create the equivalent of a magnetized piece of iron simply by running electric current around a loop of wire. The stronger the current (i.e., the more electric charge we push through the wire), the stronger is the magnetic field. However, unlike a bar magnet, an electromagnet can be turned on and off. To differentiate between these two sorts of magnets, we often refer to magnets made from materials such as iron as "permanent magnets."

Electromagnets are used in all sorts of practical ways, including buzzers, switches, and electric motors. In each of these devices a piece of iron is placed near the magnet. When a current flows in the loops of wire, the iron is pulled toward the magnet. In some cases, the electromagnet can be used to complete a second electric circuit by pulling an iron switch closed. As soon as the current is turned off in the electromagnet, a spring pushes the iron back and the current in the second circuit also shuts off.

Connection
The Electric Motor

Look around your room and try to count the number of electric motors that you use every day. They appear in fans, clocks, disk drives, VCRs, CD players, hair dryers, electric razors, and dozens of other familiar objects. Electromagnets are crucial components in every one of these electric motors.

The simplest **electric motors,** as shown in Figure 17-3, employ a pair of permanent magnets and a rotating loop of wire inside the poles of the magnets. Let's say the current in the rotating loop is directed so that when the loop is oriented as shown in Figure 17-3a the north pole of the electromagnet lies near the north pole of the permanent magnet and the south pole of the electromagnet lies near the south pole of the permanent magnet. The repulsive forces between like poles cause the wire loop to spin. As the loop gets to the position shown in Figure 17-3b, attractive forces between unlike poles make the electromagnet continue to spin in the same direction. When the electromagnet gets to the position shown in Figure 17-3c, the current in the coil reverses, changing the poles of the electromagnet. Now the south pole of the electromagnet lies just past the south pole of the permanent magnet and the north pole of the electromagnet lies just past the north pole of the permanent magnet. The repulsive forces between like magnetic poles act to continue the rotation (Figure 17-3d). The complex apparatus in Figure 17.3 shows how, in practice, an electric current can be moved from a wire to a rotating shaft.

This simple diagram contains all the essential features of an electric motor, but most electric motors are much more complex. Typically, they have three or more different electromagnets and at least three permanent magnets, and the alternation of the current direction is somewhat more complicated than we have indicated. By artfully juxtaposing electromagnets and permanent magnets, inventors have produced an astonishing variety of electric motors: fixed-speed for the second hand

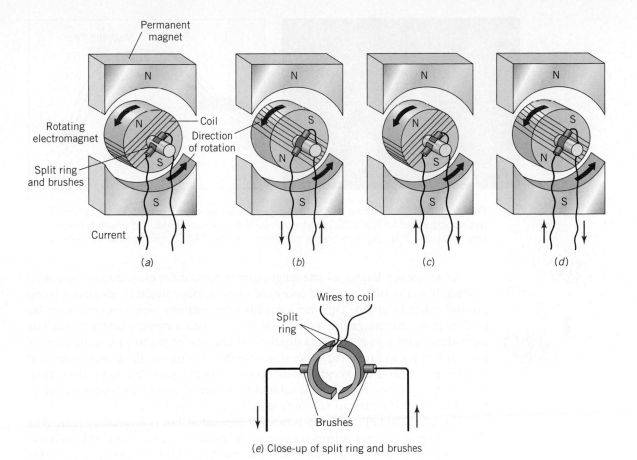

Permanent magnet

Rotating electromagnet

Coil

Direction of rotation

Split ring and brushes

Current

(a)   (b)   (c)   (d)

Wires to coil

Split ring

Brushes

(e) Close-up of split ring and brushes

FIGURE 17-3. An electric motor. The simplest motors work by placing an electromagnet that can rotate between two permanent magnets. (a) When the current is turned on, the north and south poles of the electromagnet are repelled by the north and south poles of the permanent magnet. (b) The north and south poles of the electromagnet are attracted to the south and north poles, respectively, of the permanent magnet. Rotation continues in the same direction. (c) As the electromagnet rotates halfway around, the current direction is switched, reversing the poles of the electromagnet. The poles are now back in the same orientation as in part a. (d) Rotation continues in the same direction, ready for another reversal of current direction. (e) Close-up of split ring and brushes.

of your clock, variable-speed for your food processor, reversible motors for power screwdrivers and drills, and specialized motors for many industrial uses. ●

Magnetic Fields in Nature

The universe holds countless magnetic fields, including the Sun's powerful field, Earth's field that is responsible for the working of a compass, and the magnets on your refrigerator. As we have seen, whether large or small, every magnetic field is created by the motion of electric charges. At the atomic scale, for example, an electron in motion around an atom constitutes a current, similar to the circle of wire we discussed above. The only difference is that the current in the atomic loop consists of a single electron going around and around a nucleus, while the current in the wire consists of many electrons moving around and around in a much larger loop.

 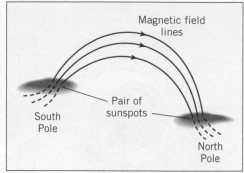

FIGURE 17-4. The surface of the Sun experiences strong magnetic fields. Loops of gas superheated to 2,000,000 kelvins reveal the magnetic field that arises between two sunspots. This negative image was taken by the *TRACE* solar probe in 1999.

As we see in Chapter 24, the magnetism in permanent magnets can be traced ultimately to the summation of countless current loops made by electrons going around orbits in atoms. This fact explains why ordinary magnets can never be broken down into magnetic monopoles. If you break a magnet down to one last individual atom, you still have a dipole field because of the atomic-scale current loop. If you try to break the atom down further, the dipole field disappears and there is no magnetism except that associated with the particles themselves. Thus magnetism in nature is ultimately related to the arrangement and motion of electric charges, rather than to anything intrinsic to matter itself.

The magnetic field of Earth is more complicated, but is ultimately related to the swirling of the liquid iron core near the planet's center. And, although the actual mechanism is slightly different, the magnetic field associated with a star such as the Sun can ultimately be attributed to the swirling of the plasma in the Sun's interior (Figure 17-4).

Ongoing Process of Science
Explaining Magnetic Field Reversals

If the magnetic fields of bodies such as Earth and the Sun were steady and unchanging, scientists would have little trouble explaining them. But these bodies display striking variations in their magnetic fields. The Sun's magnetic field reverses itself (that is, the north and south pole interchange) every 11 years. More amazing, the geological record shows that a similar thing happens on Earth. The magnetic field stays in the same direction for a long period of time, but the reversal itself happens quickly. Over a period of several thousand years, Earth's magnetic field diminishes in strength and vanishes, only to reappear again with the poles reversed. Geophysicists have documented dozens of such reversals over the last 100 million years. The last one occurred about 740,000 years ago and many scientists think we're due for another field reversal soon.

We now happen to live in a period when our planet's south magnetic pole is in Antarctica. But sometime in the future, probably within the next few hundred thousand years, the position of Earth's magnetic poles will flip. How can that be? To try to explain the complex behavior of flipping magnetic poles, scientists have suggested that convection cells in the liquid iron core of the planet occasionally disrupt the orderly rotation of the fluid. It is fair to say, however, that we do not yet have a detailed understanding of the behavior of the magnetic field of our own planet.

ELECTRIC EFFECTS FROM MAGNETISM

Once Oersted and others demonstrated that magnetic effects arise from electricity, it did not take long for scientists to realize that electric effects arise from magnetism. British physicist Michael Faraday (1791–1867) is the person who is usually associated with this discovery.

Faraday's crucial experiment took place on August 29, 1831, when he placed two coils of wire—in effect, two electromagnets—side by side in his laboratory. He used a battery to pass an electric current through one of the coils of wire, and he watched what happened to the other coil. Astonishingly, even though the second coil of wire was not connected to a battery, a strong electric current developed. We now know what happened in Faraday's experiment: as the current increased in the first loop of wire, it produced an increasing magnetic field in the neighborhood of the second loop. This changing magnetic field, in turn, produced a current in the second loop by means of a process called **electromagnetic induction** (see Figure 17-5). An identical effect was observed when Faraday

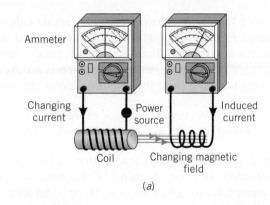

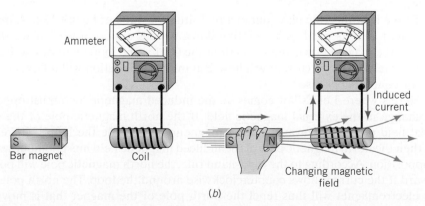

FIGURE 17-5. Electromagnetic induction. (a) When a changing current flows in the circuit on the left, a current is observed to flow in the circuit on the right, even though there is no battery or power source in that circuit. (b) Moving a magnet into the region of a coil of wire causes a current to flow in the circuit, even in the absence of a battery or other source of power.

waved a permanent magnet in the vicinity of his wire coil—he produced an electric current without a battery.

A simple law can summarize Michael Faraday's research:

Changing magnetic fields produce electric fields.

The "induced" electric field can be used to generate an electric current in a circuit.

When a magnetic field passes through a loop of wire, three actions can produce a current in that wire.

1. We can change the strength of the field (by moving the source of the field closer to or farther away from the loop, for example).
2. We can change the size of the loop.
3. We can change the orientation of the loop (by tilting it, for example).

Any one of these changes, or most combinations of them, can cause a current to flow in a circuit. Note, however, that it is the *change* that produces the current. Current does not spontaneously flow through a loop sitting in an unchanging magnetic field. Loosely speaking, if the amount of magnetic field penetrating the loop changes, then there will be an induced current in the loop.

The direction of the current flow is given by a rule called *Lenz's law,* after the Russian scientist Emil Lenz (1804–1865). To understand the rule, we have to go back to Oersted's discovery and remember that once a current starts flowing in a loop, it produces its own magnetic field, which is a different one from the magnetic field being imposed from the outside. Lenz's law states that

The direction of induced current is such
that it opposes the change that produces it.

Lenz's law is actually a statement of conservation of energy, because an induced current that reinforced the original changing magnetic field would lead to unlimited electric energy. As we know by now, there is no such thing as unlimited energy.

To see how this rule plays out in a real situation, look at Figure 17-6. A magnet is held with its north pole pointing down toward a loop of wire. If we start to move the magnet down, the strength of the field at the loop increases, so Faraday's law tells us that a current will flow. But in which direction will it flow, clockwise or counterclockwise?

Here's where Lenz's law comes in: the induced magnetic field must oppose the change in the external magnetic field. If the north magnetic pole of the external field points downward and the magnetic field inside the loop is increasing, then the north magnetic pole of the induced magnetic field must point upward in opposition. According to the right-hand rule, the north magnetic pole will point upward if the current flows counterclockwise around the loop. The north pole of this electromagnet will thus repel the north pole of the magnet that is moving downward—in effect, opposing the motion of that magnet.

By the same token, if the south magnetic pole moves downward toward the loop of wire, then Lenz's law tells us that the induced current will flow in a clockwise direction so that south magnetic poles will be opposed to each other. Thus,

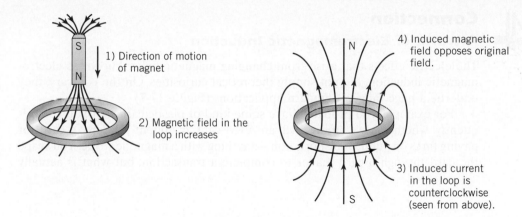

1) Direction of motion of magnet

2) Magnetic field in the loop increases

4) Induced magnetic field opposes original field.

3) Induced current in the loop is counterclockwise (seen from above).

FIGURE 17-6. Lenz's law. (1) A magnet held with its north pole pointing down moves toward a loop of wire. (2) As a result, the strength of the field at the loop increases. (3) Faraday's law indicates that a current will flow, while Lenz's law states that the direction of induced current is such that it opposes the change that produces it. According to the right-hand rule, a current flowing counterclockwise around the loop produces a magnetic field with a north pole that points upward at the center of the loop. (4) This induced magnetic field will repel the north pole of the magnet that is moving downward, in effect opposing the motion of that magnet.

once we set a train of events in motion by starting to move the external magnet, the current in the loop sets up its own magnetic field to oppose that motion and produce a new electric current.

Develop Your Intuition: Direction of Induced Current

Suppose you orient a bar magnet with the north pole pointing toward a loop of wire, just as in Figure 17-6, but now you move the magnet away from the loop instead of toward it. Is a current induced in the loop? If so, in what direction does it flow?

An electric field is induced by a changing magnetic field, whether increasing or decreasing, so a current *is* induced by moving the magnet away from the loop. To determine its direction, you can work through the following sequence of steps: The magnetic field through the loop is decreasing, so the direction of current should act to increase the field. This means establishing a south pole above the loop, trying to attract the north pole of the bar magnet and increase the magnetic field. To get a south pole above the loop, current must flow clockwise around the loop, as seen from above, the opposite direction from that shown in the figure.

Note that because the induced magnetic field of a coil opposes the magnetic field of the moving magnet, you have to exert a force to move the magnet near the coil. This force does work and this work is the source of the energy of the induced magnetic field. Once again, energy does not come from nothing; you have to supply energy to get energy.

Connection
A World of Electromagnetic Induction

The electric effects that arise from changing magnetic fields, known as electromagnetic induction, are not simply theoretical curiosities. On the contrary, they underlie a huge range of modern applications (Figure 17-7).

For example, you probably have some kind of plastic card that you use frequently, whether it's a credit card, an ATM card, a student ID card, a card for paying fares on public transportation—anything with a magnetic stripe. You swipe the card through a card reader to complete a transaction, but what is actually

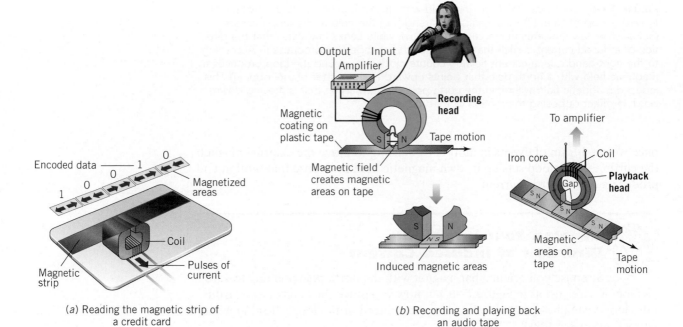

(a) Reading the magnetic strip of a credit card

(b) Recording and playing back an audio tape

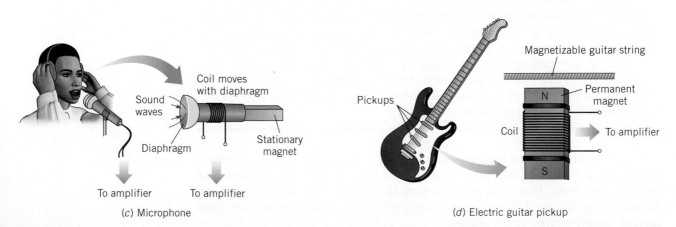

(c) Microphone

(d) Electric guitar pickup

FIGURE 17-7. Examples of electromagnetic induction.

happening when you do that? Information is encoded on the magnetic stripe, in a pattern of tiny north and south poles associated with small grains of magnetic material embedded in the card. The information is usually your name, an account number, and an expiration date. When you move the card through the reader, the magnetic areas induce currents in the reading head, transmitting the information for verification.

The same idea is involved in a cassette deck (Figure 17-7b). When music is recorded on magnetic tape, the recording head becomes an electromagnet and creates a series of tiny magnetic poles, or areas, on the tape. The pattern of north and south poles follows the pattern of alternating current in the electromagnet, which follows the pattern of sound waves in the original music. When you play back the tape, the magnetized areas on the tape induce currents as they pass through a magnetic field in the playback head, allowing the recorder to recreate the original pattern of sound. Thus, the recording is made by using the magnetic effects of electric currents, and the recording is played back by using the electric effects of magnetic fields.

Several types of microphones work with a moving coil that induces an electric current as it moves back and forth with the vibrations of the sound (Figure 17-7c). The pickups of an electric guitar work the same way, with current in a coil moving in time with the metal strings of the guitar (Figure 17-7d). The strings have to be metal so they can be magnetized by a magnet in the pickup, which can then detect the vibrations of the string as oscillations in a magnetic field.

Metal detectors at airport security checkpoints also depend on electromagnetic induction. As you walk through the magnetic field of the detector, nothing happens. But if you have a piece of metal on you, it interacts with the field and induces a current in the detector. ●

Electric Generator

Figure 17-8 illustrates the **electric generator,** or dynamo, a vital tool of modern technology that demonstrates electromagnetic induction. Place a loop of electric wire with no batteries or other power source between the north and south poles of a strong horseshoe magnet. As long as the loop of wire stands still, no current flows in the wire. However, as soon as we begin to rotate the loop, a current flows in the wire. This current flows in spite of the fact that there is no battery or other power source in the wire.

From the point of view of the electrons in the wire, any rotation changes the orientation of the magnetic field. The electrons sense a changing magnetic field and hence, by Faraday's findings, move and form a current in the loop. If we spin the loop continuously, a continuous current flows in it. In most electric generators the current flows in one direction for half of the rotation and then flows in the opposite direction for the other half of the rotation. This vitally important device, the electric generator, followed immediately from Faraday's discovery of electromagnetic induction.

In an electric generator, some source of energy—such as water passing over a dam, steam produced by a nuclear reactor or coal-burning furnace, or wind-driven propeller blades—turns a shaft attached to the coils of wire. In your car, the energy to turn the alternator coils in a magnetic field comes from the gasoline that is burned in the motor. In every generator, the rotating shaft links to coils of wire that spin in a magnetic field. Because of the rotation, electric

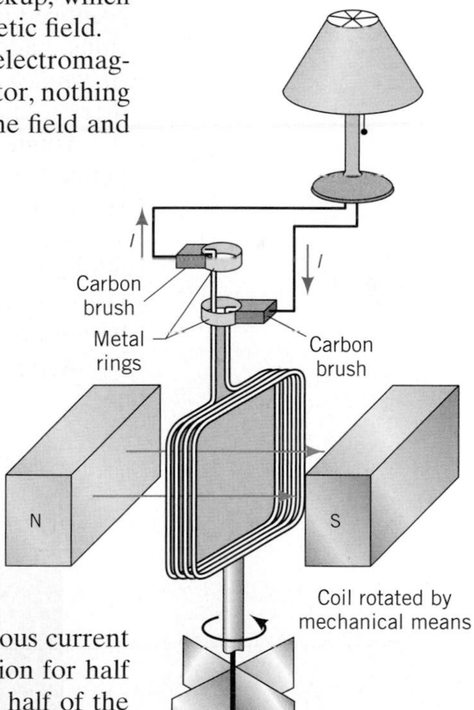

FIGURE 17-8. An electric generator. As long as the loop of wire rotates, the amount of magnetic field penetrating the loop changes and current flows in the wire.

current flows in the wire and can be tapped off onto external lines. Almost all the electricity used in the United States is generated in this way.

You may have noticed a curious fact about electric motors and generators. In an electric motor, electric energy is converted into the kinetic energy of a spinning shaft, while in a generator, the kinetic energy of a spinning shaft is converted into electric energy. Thus motors and generators are, in a sense, exact opposites in the world of electromagnetism.

Because the current in the generating coils flows first one way and then the other, it does the same thing in the wires in your home. This kind of current, the kind used in household appliances and cars, is called **alternating current (AC)** because the direction keeps alternating. In contrast, chemical reactions in a battery cause electrons to flow in one direction only and produce what is called **direct current (DC).**

Physics in the Making

Michael Faraday

Michael Faraday, one of the most honored scientists of the nineteenth century, did not come easily to his profession. The son of a blacksmith, he received only a rudimentary education as a member of a small Christian sect. Faraday was apprenticed at the age of 14 to a London book merchant, and he became a voracious reader as well as a skilled bookbinder. Chancing upon the *Encyclopaedia Britannica,* he was fascinated by scientific articles, and he determined then and there to make science his life.

Young Faraday pursued his scientific career in style. He attended a series of public lectures at the Royal Institution by London's most famous scientist, Sir Humphry Davy, a world leader in physical and chemical research. Then, in a bold and flamboyant move, Faraday transcribed his lecture notes into beautiful script, bound the manuscript in the finest tooled leather, and presented the volume to Davy as his calling card. Michael Faraday soon found himself working as Davy's laboratory assistant.

After a decade of work with Davy, Faraday had developed into a creative

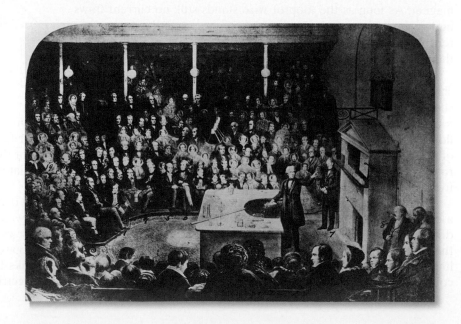

Michael Faraday (1791–1867) lecturing before an audience in London.

scientist in his own right. He discovered many new chemical compounds, including liquid benzene, and enjoyed great success with his lectures for the general London public at the Royal Institution. He is usually credited with being the first to develop the idea of electric field lines as a way of visualizing forces due to an isolated charge. However, his most lasting claim to fame was a series of classic experiments through which he discovered electromagnetic induction—a central idea that helped link electricity and magnetism. Two different units in electricity have been named after him, the faraday and the farad, as a tribute to his contributions to physics. No less an authority than Albert Einstein noted that, just as in mechanics Galileo performed the classic experiments that eventually led to Newton's laws of motion, so Faraday performed the experiments in electricity that eventually led to Maxwell's equations. ●

Maxwell's Equations

Electricity and magnetism are not distinct phenomena at all, but are simply different manifestations of one underlying fundamental entity—the **electromagnetic force.** In the 1860s, Scottish physicist James Clerk Maxwell (1831–1879) realized that the four very different statements about electricity and magnetism that we have discussed constitute a single coherent description of electricity and magnetism. These four mathematical statements have come to be known as Maxwell's equations because he was the first to realize their true import by manipulating the mathematics to make important predictions—predictions that we discuss in detail in Chapter 19. For reference, these four fundamental laws of electricity and magnetism—Maxwell's equations—are as follows:

James Clerk Maxwell (1831–1879).

1. Coulomb's law: like charges repel, unlike charges attract, with a force that has an inverse-square dependence on the distance between the charges.
2. There are no magnetic monopoles in nature.
3. Magnetic phenomena can be produced by electric effects.
4. Electric phenomena can be produced by magnetic effects.

These equations, like Newton's laws of motion and the laws of thermodynamics, summarize the behavior of an entire aspect of the universe. The three sets of laws taken together described everything in the universe known to scientists at the end of the nineteenth century. Together, they make up what is usually known as *classical physics*—that is, the physics that describes normal-sized objects moving at normal speeds. In the twentieth century, the theory of relativity (which describes objects moving near the speed of light) and quantum mechanics (which describes the world of the atom) were added to the physicist's repertoire. These two subjects (which we discuss later) are often referred to as *modern physics*.

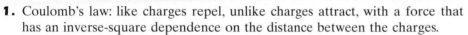

THINKING MORE ABOUT

Electromagnetism: Basic Research

It's hard to imagine modern American society without electricity. We use it for transportation, communication, heat, light, and many other necessities and amenities of life. Yet the men who gave us this marvelous gift were not primarily concerned with developing better lamps or modes of transportation. In terms of the categories we introduced in Chapter 1, they were doing basic research. Galvani and Volta, for example, were drawn to the study of electricity by their studies of frog muscles that contracted by jolts of electric charge. Volta's first battery was built to duplicate the organs found in electric fish. Scientific discoveries, even those that bring enormous practical benefit to humanity, can come from unexpected sources.

What does this tell you about the problem of allocating government research funding? Can you imagine trying to justify funding Galvani's experiments on frogs' legs to a government panel on the grounds that it might lead to something useful? Would a nineteenth-century government research grant designed to produce better lighting systems have produced the battery (and, eventu-ally, the electric light), or would it more likely have led to an improvement in the oil lamp? How much funding do you think should go to offbeat areas (on the chance that they may produce a large pay-off), as compared to projects that have a good chance of producing small but immediate improvements in the quality of life?

While you're thinking about these issues, you might want to keep in mind Michael Faraday's response when he was asked by a political leader what good his electric motor was. He is supposed to have answered, "What good is it? Why, Mr. Prime Minister, someday you will be able to tax it!"

Summary

Nineteenth-century scientists discovered that the seemingly unrelated phenomena of electricity and magnetism are actually two aspects of one **electromagnetic force.** Hans Oersted found that an **electric current** passing through a coil of wire produces a magnetic field. The **electromagnet** and **electric motor** were direct results of his work. Michael Faraday discovered the opposite effect of **electromagnetic induction** when he induced an electric current by placing a wire coil near a changing magnetic field. Faraday's work led to the first **electric generator,** which produces an **alternating current (AC). Batteries,** on the other hand, develop a **direct current (DC).**

James Clerk Maxwell realized that the many independent observations about electricity and magnetism constitute a complete description of electromagnetism.

Key Terms

alternating current (AC) Electric current that changes direction periodically in a circuit. (p. 366)

battery A device that uses stored chemical energy as an electric power source. (p. 354)

direct current (DC) Electric current that is always in the same direction in a circuit. (p. 366)

electric current A flow of charged particles. (p. 354)

electric generator A device that uses electromagnetic induction to produce electricity. (p. 365)

electric motor A device that does work by running electric current through coils of wire in the presence of permanent magnets. (p. 358)

electromagnet A device that creates a magnetic field by running electric current through coils of wire. (p. 357)

electromagnetic force The term that describes the combined effect of the electric and magnetic forces. (p. 354)

electromagnetic induction The effect in which a changing magnetic field causes electric current in a loop of wire. (p. 361)

Review

1. What is electric current?

2. What is a battery? How does a car battery work?

3. How can a battery be recharged?

4. Make a list of five everyday items that depend on battery power. In each case, what alternative sources of energy might you use?

5. What led to the discovery of the connection between electricity and magnetism?

6. What was vitalism?

7. Explain the difference between Galvani and Volta's view of the vital or electric force in living beings. Who was right?

8. How does an electric car work? What are some of the advantages and disadvantages of electric cars?

9. What happens when a current starts to flow through a wire near a compass? What does this tell us about the connection between electricity and magnetism?

10. How did Oersted show that that magnetic fields can indeed be created by the motion of electric charges?

11. What is meant by the phrase "direction of the magnetic field"?

12. What is the right-hand rule? How is this used to determine the direction of a magnetic field? Where else in this text have you seen a right-hand rule used?

13. What is an electromagnet? How is this similar to a bar magnet? How is it different?

14. How does a simple electric motor work? Give some examples of these motors in ordinary life.

15. What is alternating current, and of what importance is it in the operation of a simple electric motor?

16. How are magnetic fields created in nature? Give an example.

17. How is an electron circling an atomic nucleus similar to an electric current going through a coiled wire? Is each a true current? Explain.

18. What is the nature of a magnetic field at the atomic level? What role do electrons play in the creation of this field?

19. What is the present thinking as to what causes Earth's magnetic field? What are some of the problems with this theory? Explain.

20. How did Faraday show that electric effects arise from magnetism?

21. Describe three ways that current can be made to flow when a magnetic field passes through a loop of wire.

22. Which is more significant in Question 20, the magnetic field itself or the change in the magnetic field? Explain.

23. How do we know the direction of an induced current? Explain.

24. What is Lenz's law?

25. How does an electric generator work? How do magnets and coils of wire combine to produce the electricity generated?

26. What are the similarities and differences between a simple electric motor and an electric generator?

27. What is alternating current? Give an example.

28. What is direct current? Give an example.

29. What are the four Maxwell equations?

30. What do Maxwell's equations tell us about the unity of electricity and magnetism? Explain.

31. Was Faraday's classic experiment on electromagnetic induction an example of pure research or applied research?

Questions

1. The figure represents two long, straight, parallel wires extending in a direction perpendicular to the page. The current in the left wire runs into the page and the current in the right wire runs out of the page. What is the direction of the magnetic field created by these wires at locations a, b, and c? (b is at the exact midpoint between the wires.)

2. The figure represents two long, straight, parallel wires extending in a direction perpendicular to the page. The current in both wires flows out of the page. What is the direction of the magnetic field created by these wires at locations a, b, and c? (b is at the exact midpoint between the wires.)

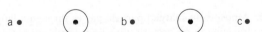

3. An electric current runs through a coil of wire as shown in the figure. A permanent magnet is located to the right of the coil. If the magnet is free to rotate, will it rotate clockwise or counterclockwise? Explain.

4. Suppose you have a Frisbee with a copper wire glued around its outer circumference. When you throw the Frisbee correctly, it maintains a constant orientation with the ground; if you throw it incorrectly, it will wobble. In which of these cases, if either, will a current be induced in the copper wire due to Earth's magnetic field?

5. Suppose you are in a location where the magnetic field of Earth points north and is horizontal to the ground. If a circular wire is rotated as shown in the figure (the axis of rotation is along the north-south direction), will there be an induced current in the wire? Explain. What if the axis of rotation were in the east-west direction?

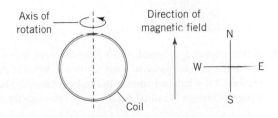

6. If you only had a magnet and a coil of wire, how might you generate a small amount of electricity in that coil?

7. If you took an electric motor and turned it by hand, what do you think would happen in the coils of wire?

8. What is the underlying basis of the magnetic field in a magnetized piece of iron?

9. Why can't you find a magnetic monopole by taking an atom apart?

10. What would be the effect of using direct current to power the simple electric motor described in this chapter?

11. Explain just how motors and generators can be considered opposites in the realm of electromagnetism. How are they similar and how do they differ?

12. If a current were flowing clockwise around a loop placed on your desk, which direction would the resulting magnetic field be inside the loop?

13. A rectangular piece of wire is moving to the right as shown. It passes through a region where there is a magnetic field pointing into the page, as shown (magnetic field indicated by the shaded region). When the loop of wire is in the position shown in the figure, is there an induced current in the loop? Explain.

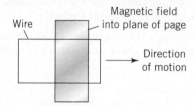

Wire

Magnetic field into plane of page

Direction of motion

14. A rectangular piece of wire is moving to the right as shown. It passes through a region where there is a magnetic field pointing into the page, as shown (magnetic field indicated by the shaded region). When the loop of wire is in the position shown in the figure, is there an induced current in the loop? If there is an induced current, does it run clockwise or counterclockwise? Explain.

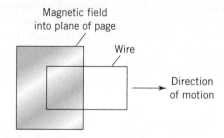

Magnetic field into plane of page

Wire

Direction of motion

15. A bar magnet is dropped, north pole down, so that it falls through a circular piece of wire, as shown. What is the direction of the induced current in the loop (A or B) before it passes through the wire? After it passes through the wire?

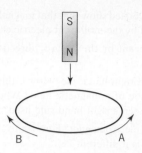

16. A square loop of copper wire is moving to the right in a uniform, downward-pointing magnetic field, as shown. What is the direction of the force on an electron moving to the right in this magnetic field? Use your answer to explain why there is no induced current in this square current loop.

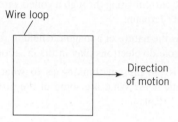

Wire loop

Direction of motion

17. If you could wrap a wire around Earth's equator at an altitude of 200 kilometers and run an electric current through it in the westerly direction, would this have the effect of reinforcing or canceling Earth's natural magnetic field at Earth's surface?

18. A long straight wire is aligned north-south and carries current in the northerly direction. What is the direction of the magnetic field created directly below the wire? What is the direction of the magnetic field created directly to the right of the wire?

19. A long straight wire is aligned north-south and carries current in the northerly direction. What is the direction of the magnetic field created directly above the wire? What is the direction of the magnetic field created directly to the left of the wire? If a proton is traveling north directly above the wire, what is the magnetic force on the proton due to the wire?

20. A positively charged particle passes through a laboratory traveling in an easterly direction. There are both electric and magnetic fields in the room and their effects on the charged particle cancel. If the electric field points upward, what must be the direction of the magnetic field?

Problems

1. The electric field at a point in space is defined as the force per unit charge at that point in space. That is, it is the force that would be exerted on a 1-coulomb charge if one were at that point in space. Therefore, we can write the electric field E of a charge q at a distance d from that charge, experienced by a charge $Q = 1$ coulomb, as

$$E = \frac{F}{Q} = k\frac{q}{d^2}$$

The electric field has a direction such that it points toward negative charges and points away from positive charges. Suppose you rub a balloon in your hair and it acquires a static charge of -3.0×10^{-9} coulombs.

   a. What are the units of electric field?
   b. What is the strength and direction of the electric field created by the balloon at a location 1 meter due north of the balloon?
   c. What is the strength and direction of the electric field created by the balloon at a location 2 meters above the balloon?
   d. Your hair acquired an equal amount of positive charge when you rubbed the balloon on your head. What is the strength and direction of the electric field created by your head, at the location of your feet, 1.5 meters down?

2. The strength of Earth's magnetic field at the equator is approximately equal to $B = 50 \times 10^{-6}$ T, where B is the customary symbol for magnetic field and T stands for tesla, the unit of magnetic field. The force on a charge q moving in a direction perpendicular to a magnetic field is given by $F = qvB$, where v is the speed of the particle. The direction of the force is given by the right-hand rule, as described in

Chapter 17. Suppose you rub a balloon in your hair and your head acquires a static charge of 3.0×10^{-9} coulombs.

   a. If you are at the equator and driving west at a speed of 30 meters per second, what is the strength and direction of the magnetic force on your head due to Earth's magnetic field?
   b. If you are at the equator and driving north at a speed of 30 meters per second, what is the strength and direction of the magnetic force on your head due to Earth's magnetic field?
   c. If you are driving east, how fast would you have to drive in order for the magnetic force on your head to equal 200 newtons (probably enough to knock you over)?

3. In the laboratory, you have arranged to have a magnetic field that is pointing north with a strength of $B = 0.5$ T an electric field that points downward with a strength of $E = 6.0 \times 10^6$ N/C. An electric charge with a magnitude $q = 10^{-9}$ C passes through the laboratory. The force on the charge due to the electric field is given by $F = qE$. The force on the charge due to the magnetic field is given by $F = qvB$, where v is the speed of the particle. The direction of the magnetic force is given by the right-hand rule, as described in Chapter 16.

   a. What direction would the charge have to travel in order that the charge passes through the room undeflected? Neglect the gravitational force. (*Hint:* What direction does the charge have to travel so that the direction of the electric force is opposite to the direction of the magnetic force?)
   b. What is the strength of the electric force?
   c. How fast would it have to travel so that it passes through the room undeflected?

Investigations

1. Read Jules Verne's *The Mysterious Island*. Do you think it is possible for castaways to build an electric generator as described in the book?

2. Take apart an old electric razor or other small motor-driven appliance and dissect the motor. How many permanent magnets are inside? How many separate coils of wire?

3. Calculate the cost per hour to run a large radio, or boom box, with batteries. Then calculate how expensive the same radio would be to run for 1 hour when plugged into a household electric outlet, using the electric rates in your area. Is there a large difference?

4. How long is the average commute between home and work for people in your area? Might electric cars be of use in the future?

5. Research the local power plants in your area. How are they the same? How do they differ? Examine the fuels used, the actual generators themselves, and the pollution produced, as well as the types of technology employed to reduce this pollution.

6. Further explore the history of the development of the electric car. Who have been the main developers of this technology, and how are obstacles to the efficiency of these cars being overcome? From your research, when do you think you will see such cars on the road in large numbers, if ever?

 **WWW Resources**

See the *Physics Matters* home page at **www.wiley.com/college/trefil** for valuable web links.

1. **http://ippex.pppl.gov/interactive/electricity/** A website of the Internet Plasma Physics Education Experience (IPPEX).

2. **http://physics.uwstout.edu/staff/scott/animate.html#faraday** Animations illustrating Faraday's law of electromagnetic induction.

3. **http://physicsed.buffalostate.edu/SeatExpts/EandM/motor/index.htm** A discussion of how to construct and analyze the "world's simplest electric motor."

4. **http://www4.ncsu.edu/~rwchabay/emimovies/** A series of movies illustrating electric and magnetic interactions, including the right-hand rule.

18 | Electric Circuits

KEY IDEA

An electric circuit is a closed path through which electrons move to do work.

PHYSICS AROUND US . . . How Electricity Does Work

You press a button on your food processor and the machine springs into action, slicing up ingredients for supper. You flip a wall switch and the room is flooded with light. You click a button on your radio or TV and sit back to enjoy the programming. You get in your car and turn the ignition switch, and the engine surges into life, ready to take you to your destination.

All these experiences and countless more involve the movement of electric charges around a closed path—what we call an *electric circuit*. Although you probably never think about them, circuits are very much a part of your life. In fact, electric power is accessible in a large part of the world today, so people can use electric circuits in the home or at work. Circuits are essential for being able to use electric energy, which is the major form of energy in common use. In this chapter we look at the basic elements of circuits that make them such an important part of contemporary society.

● BASIC ELEMENTS OF ELECTRIC CIRCUITS

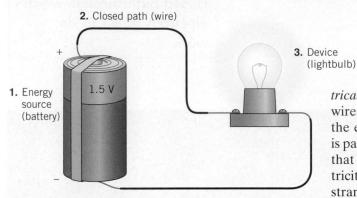

2. Closed path (wire)

3. Device (lightbulb)

1. Energy source (battery)

1.5 V

+

−

FIGURE 18-1. Every circuit consists of three parts: (1) a source of energy such as a battery; (2) a closed path, usually made of metal wire, through which the current can flow; and (3) a device such as a lightbulb that uses the electric energy.

Most people come into contact with electric phenomena through electric circuits in their homes and cars. In its most basic form, an **electric circuit** is an unbroken path of material that carries electricity; we call this material an *electrical conductor.* In most cases, the conductor is a metal wire or cable, typically made of copper. For example, the electric light that you are using to read this book is part of an electric circuit that begins at a power plant that generates electricity many miles away. That electricity continues through power lines (usually made of strands of aluminum) into your town and is distributed on overhead or underground wires until it finally arrives where you live. There, a circuit includes the light bulb and copper wires that run through the walls of your home.

Every circuit consists of three parts: a source of energy such as a battery; a closed path, usually made of metal wire, through which the current can flow; and a device such as a motor or a lightbulb that uses the electric energy (Figure 18-1). It is important to remember that a circuit must provide a path for current to go from the source to the operating device as well as a return path from the device back to the source. If any part of the circuit is incomplete (an *open circuit*), the device will not work.

Measuring Electric Currents

A good way to think about electric circuits is to draw an analogy between electrons flowing through a wire and water flowing through a pipe (Figure 18-2). In the case of water, we use two quantities to characterize the flow: the amount of

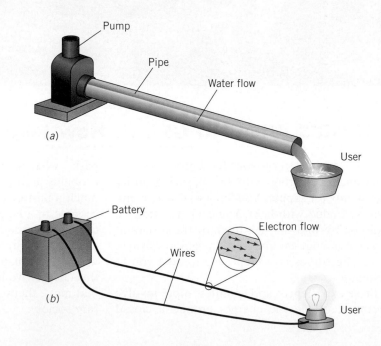

Pump

Pipe

Water flow

User

(a)

Battery

Electron flow

Wires

User

(b)

FIGURE 18-2. Electrons flowing through a wire are analogous to water flowing through a pipe. In the case of water (a), the flow is characterized by the amount of water that passes a point each second and the pressure difference that makes the water flow. In an electric circuit (b), the flow of electrons (current) is measured in amperes and the electric potential (voltage) is measured in volts.

water that passes a point each second and the pressure difference that makes the water move. For example, a city water system may send water through large pipes and push that water (i.e., generate pressure) with a large pump. In small towns, on the other hand, water is stored in a tall tower and the pull of gravity on the water is used to generate pressure—the taller the tower, the greater the pressure in the pipes.

The numbers we use to describe the flow of electrons in an electric circuit can be thought of in an analogous way. The quantity that is analogous to the volume of water flowing through the pipe is the number of electrons that flow through the wire, called the *electric current.* Current is measured in a unit called the **ampere** or **amp,** named after French physicist Andre-Marie Ampere (1775–1836). One amp corresponds to a flow of 1 coulomb (the unit of electric charge) per second past a point in the wire:

1 amp of current = 1 coulomb of charge per second

Alternatively, 1 amp of current corresponds to 6.25×10^{18} electrons moving past a point in the wire each second. Typical household appliances use anywhere from about 1 amp (a 100-watt bulb) to 40 amps (an electric range with all burners and the oven blazing away).

The quantity that is analogous to pressure is called the **electric potential.** (Note that the word "potential" does not denote an energy, as it does in Chapters 8 and 12.) You can think of the device that generates this potential—either a battery or an electric generator—as a kind of pump that supplies energy to the electrons that flow through the circuit. The electric potential of an energy source is measured in a unit called the **volt** (V), which was named after Alessandro Volta, the Italian scientist who invented the chemical battery. In ordinary use, we often refer to the electric potential as the **voltage** of the energy source in a circuit.

A water tower uses gravitational potential energy to generate pressure; the taller the tower, the greater the pressure in the pipes.

As we have said, you can think of voltage in circuits in much the same way you think of water pressure in your plumbing system. The power of the water pump or the height of the water tower produces water pressure. In an electric circuit, more volts mean more oomph to the current, just as more water pressure makes for a stronger flow of water. Typically, a new flashlight battery produces 1.5 volts, a fully charged car battery produces about 12 volts, and ordinary household circuits operate on either 120 or 240 volts. (See Looking at Voltage, page 376.)

Wires through which the electrons flow are analogous to pipes carrying water: the smaller the pipe, the harder it is to push water through it. Similarly, it is harder to push electrons through some wires than others. The quantity that measures how hard it is to push electrons through wires is called **electrical resistance,** and it is measured in a unit called the **ohm,** after the German physicist Georg Ohm (1789–1854).

To understand the physical basis for resistance, think about electrons moving through a copper wire. Every so often, one of the moving electrons collides with a copper atom. In that collision, the electron, on average, loses energy and the atom gains energy. As a result of these collisions, some of the electrons' energy is converted into heat. The wire gets warmer, and the electric energy converted into heat has to be replaced by the power source in order to keep the electrons moving along the wire.

In real materials, such as copper, collisions happen so often, and the electrons change direction so often, that the actual speed of an electron in the wire (a quantity known as the *drift velocity*) is actually very low—often a fraction of a centimeter per second. However, even though the electrons themselves move

Looking at Voltage

Your body runs on electricity; nerve impulses are basically electrical in nature. However, the voltages involved are rather small, less than one-tenth the voltage of a standard flashlight battery. Our modern society runs on electrical energy that exceeds the body's voltage levels by over 1000 times. Really large voltages are generated in power plants and carried by transmission lines; these voltages can be very dangerous.

10^{-1} V

Nerve impulse, 0.1 V

10^{0} V

Flashlight battery, 1.5 V

10^{6} V

Transmission lines, 500,000 V

10^{2} V

Wall outlet, 120 V

10^{4} V

Power plant, 25,000 V

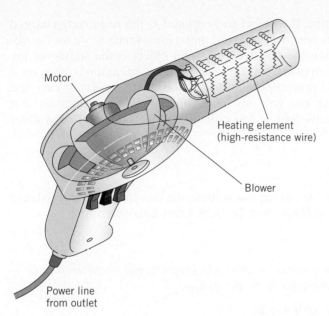

Motor

Heating element
(high-resistance wire)

Blower

Power line
from outlet

FIGURE 18-3. A hair dryer uses high-resistance wire to generate heat.

very slowly, electric signals travel through the wire at almost the speed of light. This situation is analogous to a wave on water, which can move quickly even though the water molecules do not.

Greater electrical resistance in wires and other conducting materials means that more of the electrons' energy is converted into heat. For example, ordinary copper wire has a low resistance, which explains why we use it to carry electricity around our homes. On the other hand, toasters, space heaters, and hair dryers contain high-resistance wires so that they produce large amounts of heat when current flows through them (Figure 18-3). In transmission lines, it is important that as much energy as possible gets from one end of the line to the other; thus electric power companies use very thick low-resistance (high-efficiency) wires.

Connection

Electrolysis and Electroplating

Not all electric circuits involve electrons flowing along wires. For example, transistors and integrated circuit chips, which are part of all modern electronic equipment, use both positive and negative charge carriers moving along paths etched into thin layers of crystalline silicon. As another example, consider the large electroplating industry.

Electrolysis employs a reaction opposite to what takes place in batteries. A battery uses a chemical reaction to produce an electric current; electrolysis uses electric current to drive a chemical reaction that normally wouldn't occur. Michael Faraday first worked out the principles of electrolysis. The idea is that two objects are placed in a solution of ions and connected to a strong current supply. Electrons are stripped from one object, leaving that object with a positive charge, and are supplied to the second object, giving it a negative charge. The ions in the solution are attracted toward the two objects, with negative ions moving to the positively charged object and positive ions moving toward the negatively charged object. The moving ions complete a circuit, allowing current to flow through the entire apparatus.

Chrome electroplating is a technology based on electrolysis.

In electroplating, the object to be plated is the negatively charged object and attracts positively charged metal ions, which settle on the object's surface. Often the positively charged object is the source of the ions; it slowly dissolves as the process continues. Chromium is often used in this way to add protection and decoration to steel parts of cars and motorcycles. Silver and gold are also common plating materials that give a rich decorative look at less cost than objects made of solid silver or gold. ●

Ohm's Law

The close relationship among the voltage, the current, and the resistance of a circuit is called **Ohm's law.** In 1826, Ohm discovered that

1. In words:

The current in a circuit is inversely proportional to the resistance and directly proportional to the voltage.

2. In an equation with words:

Current (in amps) = Voltage (in volts) divided by Resistance (in ohms)

or   Voltage = Current × Resistance

3. In an equation with symbols:

$$I = \frac{V}{R} \quad \text{or} \quad V = IR$$

where I is the standard symbol used to denote electric current.

Ohm's law tells us that, if we want to increase the amount of current flowing in a circuit, we can do one of two things: we can increase the voltage, or we can lower the resistance (Figure 18-4). If you think of our analogy with the flow of water in a pipe, this makes sense. You can increase the flow of water by increasing the pressure (for example, by using a more powerful pump) or by installing a

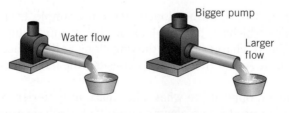

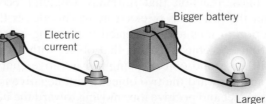

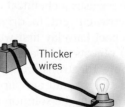

FIGURE 18-4. Ohm's law says that an increase in current can be accomplished by an increase in voltage or a decrease in resistance. This behavior is analogous to water flowing in a pipe; greater water flow results from either a bigger pump or a wider pipe.

larger-diameter pipe. In the same way, in an electric circuit you can increase current by increasing voltage (for example, by using a more powerful pump) or by lowering the resistance (for example, by using a larger-diameter copper wire).

Ohm's law does not apply to all materials, but it does apply to metals used in electrical wiring and to many other materials found in simple electric circuits. However, it does not work for the semi-conductor materials we will discuss in Chapter 25.

You can understand the behavior of lightning in terms of electric circuits (Figure 18-5). In a thunderstorm, collisions between particles in the clouds produce a buildup of negative charge at the bottom of the cloud, which attracts a corresponding buildup of positive charge in objects on the ground underneath the cloud. This buildup creates a voltage between cloud and ground. The lightning stroke is the electric current that runs between the two when the voltage is high enough. Lightning normally strikes tall objects such as buildings and trees because they're closer to the thunderstorm and the positive charge on the ground tends to accumulate in these tall objects. When the potential difference between clouds and ground becomes large enough, the air becomes ionized and electricity passes through it. As we have mentioned earlier, the lightning rod invented by Benjamin Franklin uses this principle by allowing the lightning to flow through a low-resistance bar of metal instead of through the building.

Buildup of negative charge at bottom of thundercloud

Positive charge accumulates in objects beneath the cloud

FIGURE 18-5. Lightning occurs when collisions between particles in a thundercloud produce a buildup of negative charge at the bottom of the cloud, which attracts a corresponding buildup of positive charge in objects on the ground underneath the cloud. The lightning stroke is the electric current that runs between the two when the voltage is high enough.

 ## ELECTRIC POWER

In Chapter 8 we saw that power, measured in watts, is the rate at which work is done or, equivalently, the rate at which energy is expended. This concept is important in the design and application of electric circuits.

The *load* in any electric circuit is the "business end"—the place where useful work gets done. The filament of a lightbulb, the heating element in your hair dryer, and coils of wire in an electric motor are typical loads in household circuits. The power used by the load depends both on how much current flows through it and on the voltage. The greater the current or voltage, the more power is used. A simple equation allows us to calculate the amount of electric power used.

1. In words:

The power consumed by an electric appliance is equal to the product of the current flowing through it and the voltage across it.

2. In an equation with words:

Power (in watts) = Current (in amps) times Voltage (in volts)

3. In an equation with symbols:

$$P = I \times V$$

TABLE 18-1	Terms Related to Electric Circuits		
Term	**Definition**	**Unit**	**Plumbing Analog**
Voltage	Electric pressure	volt	Water pressure
Resistance	Resistance to electron flow	ohm	Pipe diameter
Current	Flow rate of electrons	amp	Flow rate of water
Power	Current times voltage	watt	Rate of work done by moving water

This equation tells us that both the current and the voltage have to be large for a device to consume high levels of electric power. Table 18-1 summarizes some key terms about electric circuits.

Starting Your Car

EXAMPLE
18-1

When you turn on the ignition of your automobile, your 12-volt car battery must turn a 400-amp starter motor. How much power is required to start your car?

REASONING AND SOLUTION: In order to calculate electric power, we need to multiply current times voltage. We apply the equation for electric power:

$$\text{Power (watts)} = \text{Current (amps)} \times \text{Voltage (volts)}$$
$$= 400 \text{ amps} \times 12 \text{ volts}$$
$$= 4800 \text{ watts} = 4.8 \text{ kilowatts}$$

Most early automobiles were started by a hand crank, which might have required 100 watts of power, a reasonable amount for an adult to exert. Modern high-compression automobile engines require much more starting power than could be generated by one person. That's why your car has a powerful starting motor. ●

Connection

Capacitors

It often happens that you don't need much energy to keep an electric device running, but you do need an occasional jolt of energy for some particular function. For example, a light meter on a camera requires only a small amount of current from a battery, but if you want to use a flash attachment for taking pictures at night, you need a good-size burst of energy for the flash. The circuit device that enables you to store large amounts of electric energy and then use it when needed is called a *capacitor*.

In its simplest form, a capacitor consists of two parallel conducting plates kept a small distance apart. When placed in a circuit with a battery or other voltage source, one plate accumulates positive charge and the other plate accumulates negative charge. The charge stays on the plates; it can't cross the insulating gap between them. In this way you can build up a good deal of electric potential energy in the capacitor. When you want to tap this energy, you just close a switch on some other part of the circuit connected to the capacitor and the charge immediately leaves the plates as a current in this new part of the circuit.

Capacitors come in different sizes and can store different amounts of charge. The ratio of stored charge to voltage across the capacitor is called *capacitance* and is measured in units called "farads" (after Michael Faraday). Modern capacitors use a thin layer of insulating material, such as paper, wax, or plastic, between thin conducting sheets, such as metal foil. Then the whole device is wrapped up into a tight cylinder.

Capacitors are common circuit components for supplying energy in repeated bursts instead of a continuous stream. Think of windshield wipers set to an intermittent sweep. Capacitors are also used for equipment that needs more energy to start up than to continue operation, such as TV sets and automobiles. (The large capacitor in a car is often called a condenser because it "condenses" electric energy into a larger concentration than the battery can supply.) Capacitors are also used to shield delicate circuit components from large currents that might damage them; the large current runs up a charge in the capacitor instead. Surge protectors also use capacitors to protect electronic equipment from large bursts of current triggered by lightning or a power outage. ●

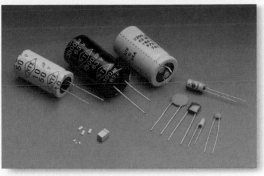

A variety of capacitors.

Resistive Loss

Ohm's law and the equation for electric power provide us with a very simple way of calculating the power dissipated in the resistance of an electric circuit. Consider, as in Figure 18-6, a resistance of R ohms (the bulb filament) in a circuit through which a current I is flowing. According to Ohm's law, the voltage drop across the resistor is

$$V = IR$$

But the power equation tells us that the power being dissipated in the resistor is

$$P = IV$$

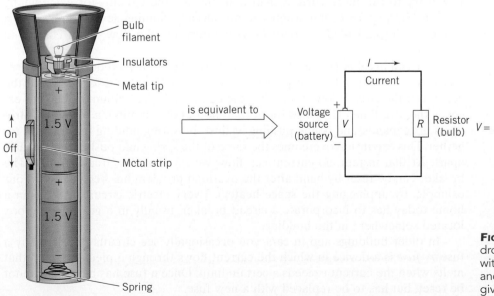

FIGURE 18-6. The voltage drop V (in volts) in a circuit with resistance of R (in ohms) and current I (in amps) is given by Ohm's law: $V = IR$.

This is the power that moving electrons transform into heat in the resistor. In this example, the resistor could be a useful appliance such as a toaster or a lightbulb, but it could also be another part of a circuit, such as a length of copper wire.

Combining these two equations, we find that the amount of energy dissipated as heat each second in a resistor in any electric circuit is

$$P = I^2R$$

This equation tells us that if we double the current through a wire or circuit element, we *quadruple* the amount of heat generated (or, equivalently, the amount of energy lost). This fact has important implications for the design of the nation's power grid, as we'll see in Example 4.

CIRCUIT SAFETY

Electric circuits are all around us. Because these circuits can release a great deal of electric energy, we have to be aware of the fact that they can be dangerous. Two different sorts of dangers are associated with electricity: the generation of fires, and electric shock. Each of these dangers is met in a different way in modern electric systems.

Overloading Circuits

Because current flowing through a wire generates heat, there is always the danger that a wire will overheat and ignite materials around it, causing a potentially lethal fire. Consequently, electric codes are written so that the size of the wire in every circuit in a building is set so that it can carry the maximum allowed current without overheating. This is why the wires that carry high currents (such as those that run to an electric stove) have to have a much larger diameter than the familiar cords that power ordinary electric lamps.

In general, most household circuits are designed to carry 15 amps of current. If you try to run more current than that through the circuit—for example, by plugging in a space heater when you are already running several lights and a stereo—a device called a *circuit breaker* interrupts the flow of all current in the circuit.

A **circuit breaker** (Figure 18-7) is a somewhat complicated device that works like this: A spring holds two pieces of metal, called contacts, together. All the current in the circuit flows through these contacts. If the amount of current exceeds the safe limits (15 amps for normal household circuits), a bimetallic strip heats up, expands, and bends, pushing against the spring holding the contacts together. This expansion overcomes the force of the spring and pushes the contacts apart. At that instant, no current can flow; we say that the circuit is *open*. The breaker can be reset by hand after the overload problem has been fixed (in our example, by unplugging the space heater). Every electric circuit installed in a home today has to incorporate a circuit breaker, usually in a gray electric box located somewhere in the building.

In older buildings and in cars, you occasionally see circuits protected by a fuse. A *fuse* is a device in which the current flows through a piece of metal that melts when the current exceeds a certain limit. Once a fuse has blown, it cannot be reset, but has to be replaced with a new fuse.

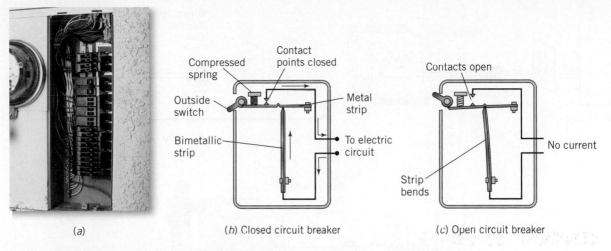

FIGURE 18-7. A circuit breaker opens when electric current exceeds a critical value. This safety feature prevents a circuit from overheating.

Develop Your Intuition: Tripping a Circuit Breaker

You have a 100-W lightbulb on in your room, along with your 360-W stereo receiver and a 1600-W electric heater, all plugged into the same 120-V outlet. The circuit is protected by a 20-amp circuit breaker. Suppose you decide to plug in your 650-W hair dryer to the same circuit; will you trip the circuit breaker when you turn on the hair dryer?

The question is, how much current does each appliance draw? The sum must be less than 20 amps for the circuit breaker to stay closed. The current for each appliance equals its power use divided by the voltage:

$$\text{Lightbulb} \quad \frac{100\ \text{W}}{120\ \text{V}} = 0.8\ \text{amp}$$

$$\text{Stereo} \quad \frac{360\ \text{W}}{120\ \text{V}} = 3.0\ \text{amp}$$

$$\text{Heater} \quad \frac{1600\ \text{W}}{120\ \text{V}} = 13.8\ \text{amp}$$

$$\text{Hair dryer} \quad \frac{650\ \text{W}}{120\ \text{V}} = 5.4\ \text{amp}$$

Without the hair dryer, the circuit is carrying a current of 17.6 amps, primarily because of the heater. Adding another heat-generating appliance such as a hair dryer will be too much; the total current of 23 amps will exceed the rating of the circuit breaker and it will open.

Electric Shock

Another danger involves situations in which human beings inadvertently make themselves part of an electric circuit. As outlined in the Connection: The Propagation of Nerve Signals on page 385, the human nervous system is electrical

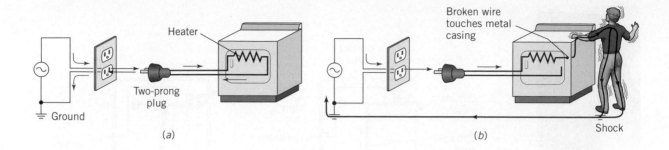

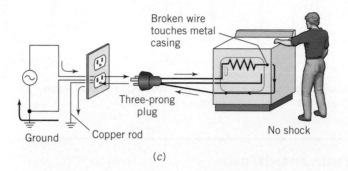

FIGURE 18-8. (*a*) Electric appliances that use a two-pronged plug may cause an electric shock. (*b*) If the insulation on wires inside a metal appliance rubs off and you touch the appliance casing, current will try to flow through your body into the ground. (*c*) A three-pronged plug alleviates this problem.

in nature and can be damaged and even destroyed by flows of current through the body.

A common situation is shown in Figure 18-8. The insulation on wires inside the metal casing of an appliance has, over time, been rubbed off, so that the metal frame is in electrical contact with the wire. This means that the entire metal casing is at 120 V. If you touch the casing (Figure 18-8*b*), current will try to flow through your body into the ground. As shown in the following examples, the amount of current (and the damage to you) depends on the resistance that your body offers to that current.

A Little Shock

If you are standing on a dry floor and your skin is dry when you touch the appliance frame that has damaged wires, the resistance of your body could be as high as 100,000 ohms. How much current can flow through your body?

REASONING AND SOLUTION: Ohm's law allows us to calculate the current when we know both the voltage (120 volts) and resistance (100,000 ohms). Solving for the current,

$$I = \frac{V}{R}$$

$$= \frac{120 \text{ volts}}{100,000 \text{ ohms}}$$

$$= 0.0012 \text{ amp}$$

This is a small current. You will probably feel a little tingle, but not much else. ●

A Big Shock

Now suppose you touch the appliance frame that has damaged wires when you are wet, perhaps when you are standing in a puddle of water or (much worse!) a bathtub. Water is a good conductor because of the ions it contains. In this case, your resistance could be quite low, perhaps only 1000 ohms. Now how much current flows?

SOLUTION:

$$I = \frac{V}{R}$$

$$= \frac{120 \text{ volts}}{1000 \text{ ohms}}$$

$$= 0.12 \text{ amp}$$

This large current is enough to kill you. In fact, muscle spasms may occur when about 0.015 amp flows through the body, and currents of 0.07 amp can stop the heart from beating if they last long enough.

The bottom line: ***Never*** touch electric devices when you are in contact with water! ●

Grounding

To counter the danger from electric shock around the home, modern electric circuits are grounded. Look at a wall receptacle and you'll note that there are usually three holes to accommodate an electric plug. The two flat prongs of a plug carry the normal current, but the third round prong carries an extra wire that connects the electric device to the ground (Figure 18-8c). If it happens that one of the "hot" wires touches the casing of the device, as in Examples 18-2 and 18-3 a very low-resistance path is opened up. Potentially dangerous current is carried from the high-voltage wires directly into the ground. When a low-resistance path such as this has opened up, we say that we have a **short circuit.**

As a result of the short circuit, large amounts of current begin to flow and the circuit breaker trips, preventing the current from flowing in the circuit until the problem is repaired. This process of **grounding** a circuit protects people by ensuring that the circuit is taken out of operation as soon as a potential for delivering electric shock develops (Figure 18-7c).

Connection

The Propagation of Nerve Signals

All of your body's movements, from the beating of your heart to the blinking of your eyes, are controlled by nerve impulses. Although nerve signals in the human body are electrical in nature, they bear little resemblance to the movement of electrons through a wire. Nerve cells of the type illustrated in Figure 18-9, called neurons, form the fundamental element of the nervous system. A neuron consists of a central body with a large number of radiating filaments at each end. These filaments connect one nerve cell to many others. The long filament that carries signals away from the central nerve body and delivers those signals to other cells is called the axon.

The membrane surrounding the axon is a complex structure, full of channels through which atoms and molecules can move. When the nerve cell is resting, positively charged molecules tend to remain outside the membrane, while

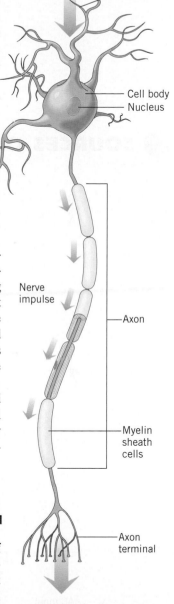

FIGURE 18-9. A nerve cell (neuron) consists of a central body and a number of filaments. The dendrites receive incoming signals and the axon conducts outgoing signals away from the cell body. The myelin sheath allows the nerve signal to move faster.

negatively charged molecules remain inside. However, when an electric signal triggers the axon, the membrane is distorted and, for a short time, positive charges (mainly sodium atoms) pour into the cell. When the inside becomes more positively charged the membrane changes again and positive charges (this time mainly potassium) move back outside to restore the original charge. This charge disturbance moves down the filament as a nerve signal. When the signal reaches the end of one of the filaments, it is transferred to the next cell by a group of molecules called *neurotransmitters* that are sprayed out from the end of the upstream cell, and received by special structures on the downstream cell. The reception of neurotransmitters initiates a complex and poorly understood process by which the nerve cell decides whether or not to send a signal down its axon to other cells. Thus, although the human nervous system is not an ordinary electric circuit, it does operate by electric signals. ●

● SOURCES OF ELECTRIC ENERGY

As we have seen in Chapter 17, there are two primary ways of converting other forms of energy into electric energy. The oldest technology, first developed by Alessandro Volta, is the battery, which uses chemical potential energy to move electrons through a wire. The battery produces current in which the electrons always flow in the same direction, which we call *direct current* (DC).

The other device, associated with Michael Faraday, is the electric generator. In a power plant using a generator, stored energy in some form (gravitational potential energy of falling water, for example, or stored chemical energy in coal) is used to turn a shaft between large magnets, producing electric current by electromagnetic induction. Current produced by a generator flows in one direction for half of a rotation of the shaft and in the other direction for the other half. This we call *alternating current* (AC).

We use both kinds of current in our daily lives. When you turn on a portable CD player, boot up your laptop computer, or start your car, you use direct current generated by batteries. When you plug your stereo into a wall outlet or turn on an electric stove, you use alternating current generated at a distant power station.

This interplay between direct and alternating current in our lives makes two kinds of devices, called *rectifiers* and *transformers,* very important.

Rectifiers

A **rectifier** is an electric device that converts AC into DC. Rectifiers are important because many types of electronic equipment, including cell phones, stereos, and laptop computers, are built so that their internal circuits run on direct

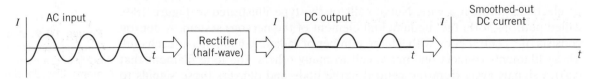

FIGURE 18-10. A simple rectifier acts as a one-way gate for electric current. Alternating current entering from the left is converted into a series of bumps, each of which flows in the same direction. Other electronic devices are used to smooth out the bumps and produce the steady DC current shown on the right.

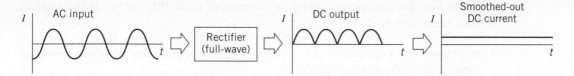

FIGURE 18-11. In a modern rectifier, special electronics flip the voltage on the current that is going the "wrong" way. The result is a system in which the incoming AC is converted into a series of contiguous pulses, all with current moving in the "right" direction, and then smoothed out.

current. However, when you plug them into the wall outlet or charge their batteries you bring in alternating current. The first thing you have to do to make the device operate is to convert AC to DC.

We discuss the details of how a common rectifier works at the atomic level in Chapter 25. For the moment, however, think of a rectifier as a one-way gate for electric current—a gate that opens when electrons move to the right, for example, but closes when they try to move back to the left. The effect of such a device is shown in Figure 18-10. The alternating current is shown by the graph on the left, in which the current is first negative, then positive, then negative, and so on. This corresponds to electrons in the circuit moving first one way, then the other way, then back to the first way, and so on. This current enters from the left (AC input) and is converted into a series of bumps (DC output). In each of these bumps the current is always flowing in the same direction. After this step, other electronic devices are used to smooth out the bumps and produce the steady DC current shown on the right.

Actually, in a real rectifier, we do not throw away half of the incoming electric energy, as in this simple example. Instead, by use of clever electronics, engineers flip the voltage on the current that is going the "wrong" way. The result is a system such as that shown in Figure 18-11, in which the incoming AC is converted into a series of contiguous pulses, all with current moving in the "right" direction, and then smoothed out.

Transformers

Engineers often need to change the voltage associated with AC current. For example, when electricity is generated at a power plant, it is usually sent through wires over long distances. As we see in Example 18-4, there is less loss due to resistance if a small current is sent at a high voltage than if a larger current is sent at a low voltage. For this reason, it is advantageous to increase or decrease the voltage of electric current flowing through a wire. The **transformer** is the device that carries out this task.

Transmitting Electric Power

Suppose you have a power source (such as a generator) that delivers 1 kilowatt of electric power. This power has to be sent over a long electric wire that has a resistance of 5 ohms. Calculate the energy dissipated as heat in the resistor if the power is sent in the form of

EXAMPLE
18-4

a. 10 amps at 100 volts
b. 0.1 amp at 10,000 volts

(Note that the total power, given by the current times the voltage, is the same in both these situations.)

REASONING AND SOLUTION: As previously shown, the power dissipated in a resistor is given by the expression

$$P = I^2R$$

All we have to do, then, is substitute the currents in the two cases:

a. Current = 10 amps, so

$$P = (10 \text{ amps})^2 \times 5 \text{ ohms}$$
$$= 500 \text{ watts}$$

b. Current = 0.1 amp, so

$$P = (0.1 \text{ amp})^2 \times 5 \text{ ohms}$$
$$= 0.05 \text{ watt}$$

In other words, in the first case almost half the power from the generator is lost in heating the wire, while in the second case almost none is lost. This result shows why electric power is always transmitted at very high voltages and low currents. ●

A simplified picture of a transformer is shown in Figure 18-12*a*. A loop of wire, called the "primary coil," is connected to an alternating-current power source. Situated on top of this wire loop is another loop, called the "secondary coil," which is connected to wires leading away from the device. When AC current flows through the primary coil, it creates a varying magnetic field. This oscillating field, in turn, causes a change in the magnetic field enclosed in the secondary coil. By electromagnetic induction, a voltage *V* is induced in the second coil. If the second coil is part of a circuit, the induced voltage causes current I_{out} to flow in the circuit. This result makes sense because, assuming no losses, the energy flowing into the transformer must be the same as the energy flowing out. Note that the secondary coil is not connected to a power source; it is linked to the primary coil only by the changing magnetic field. The only link between the two coils, in fact, is through the varying magnetic field.

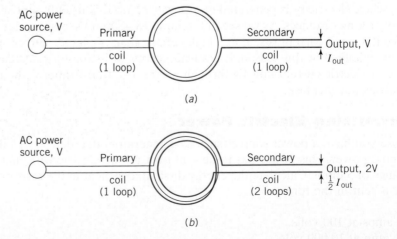

FIGURE 18-12. (*a*) A simplified picture of a transformer with one loop of wire in both the primary and secondary coils. (*b*) In a transformer in which the primary coil has one loop of wire and the secondary coil has two loops, half the current and twice the voltage flows in the secondary coil.

In this simple example, the current and voltage in the secondary coil are exactly the same as that in the primary coil (ignoring losses due to resistance in the wires). Suppose, however, that we have a situation like that shown in Figure 18-12b, with the secondary coil consisting of two loops of wire. In this case, voltage V is induced in each of the two loops in the secondary coil, producing a total voltage of $2V$. Here the energy flowing in (per unit of time) is $I_{in}V$, and this must be the same as the energy flowing out:

$$I_{in}V = I_{out}\,(2V)$$

$$I_{out} = \frac{I_{in}}{2}$$

Thus the output current is half of the original current.

Adding more loops to the secondary coil of a transformer further increases the secondary voltage and decreases the secondary current. This kind of transformer, in which the secondary voltage is greater than the primary, is called a *step-up transformer.* You can apply the reasoning we've just used to a transformer that has more coils in the primary than in the secondary. In this case, the voltage in the secondary coil is less than that in the primary coil and the current is correspondingly higher—a situation known as a *step-down transformer.* Both kinds of transformers are used routinely in the electric industry. Step-up transformers at power stations provide high voltages for transmission lines, while step-down transformers in your local neighborhood reduce the voltages to manageable levels for delivery to homes (Figure 18-13).

In general, if the primary coil of a transformer has N_1 loops and the secondary coil has N_2 loops, then the relations between the primary and secondary currents and voltages are

$$I_{secondary} = \left(\frac{N_1}{N_2}\right) I_{primary}$$

and

$$V_{secondary} = \left(\frac{N_2}{N_1}\right) V_{primary}$$

(a)

(b)

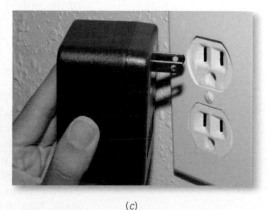
(c)

FIGURE 18-13. (a) A step-up transformer converts current from power generators to high voltage for long-distance transmission. (b) A step-down transformer lowers voltage from power lines for home use. (c) A step-down transformer lowers voltage from home power outlets for use in delicate electronic devices.

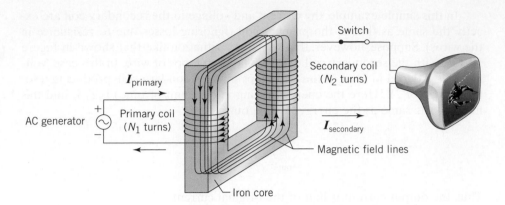

FIGURE 18-14. A typical transformer has both the primary and secondary coils wrapped around an iron core.

A typical transformer is shown in Figure 18-14. Both the primary and secondary coils are wrapped around an iron core, as shown. The core, in essence, traps the magnetic field produced by the primary and leads it around so that it all passes through the secondary coil. It is the presence of the iron that makes transformers heavy—a feature you may have noticed if you've ever picked one up.

The Power Grid

Transformers are routinely used to help bring electric power from distant generating plants to cities. First, in order to send the electricity over long distances with minimum loss, the output of a generator is stepped up to hundreds of thousands of volts. This high-voltage current provides the power that runs through the transmission lines that you see snaking across the open countryside near generating plants.

When this power comes into an urban area, the voltage is stepped down to 600 volts to be carried around the city on overhead lines. Finally, on a power pole near your home, another transformer steps the voltage down further, to the 240 volts that comes into your house. You can often see these final transformers on power poles in residential areas—they look like small garbage cans at the top of the pole.

Physics in the Making

The War of AC Versus DC

Nationwide systems of electric power supply cannot happen overnight. The planning and funding of such an enormous project takes years. The development of electric power as a replacement for natural gas and oil began in the 1880s, but it was not a smooth process.

The great leap forward in the clamor for available electric power came with Thomas Edison's improvement of the electric lightbulb, which he patented in 1880. Edison ran special trains to bring 3000 people to his laboratory in Menlo Park, New Jersey, to see his demonstration of hundreds of lamps in the streets and shops of the neighborhood, all run by a single central generator. News that Edison had solved the problem of electric lights caused stocks of gas companies to drop sharply worldwide.

Edison formed a company to build generators and offer electric power to businesses and neighborhoods in New York. He advocated direct current and claimed it was safe and easy to use. However, around 1890, competition arrived from the engineer and inventor George Westinghouse. Westinghouse had recognized the economic value of using transformers with alternating current to transmit power at low currents (high voltages) and had formed a partnership with the engineer Nikola Tesla to build practical power supplies of AC. Westinghouse Electric began offering power to businesses and factories at lower costs than Edison could match.

During most of the 1890s the war raged between AC and DC. Edison played up every fatal electric accident as due to high-voltage transmission. The state of New York introduced electrocution by alternating current as a means of capital punishment at that time, and Edison's friends tried to make AC a synonym for danger and death. In retaliation, Tesla invented an electric chair that worked on DC, a device still in use today.

Eventually, Westinghouse's financial arguments, as shown in Example 18-4, were too persuasive to ignore. Today he is considered the main figure in the worldwide use of AC for electric power. ●

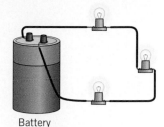

(a) Series circuit

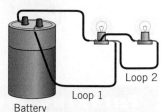

(b) Parallel circuit

FIGURE 18-15. (a) In a series circuit, two or more loads are linked along a single loop of wire. (b) In a parallel circuit, different loads are situated on different wire loops.

PARALLEL AND SERIES CIRCUITS

Common household circuits come in two different types, depending on the arrangement of wires and loads. In **series circuits** (Figure 18-15a), two or more loads (a series of lightbulbs, for example) are linked along a single loop of wire. In **parallel circuits** (Figure 18-15b), by contrast, different loads are situated on different wire loops. Both types of circuits are used in every household (Figure 18-16).

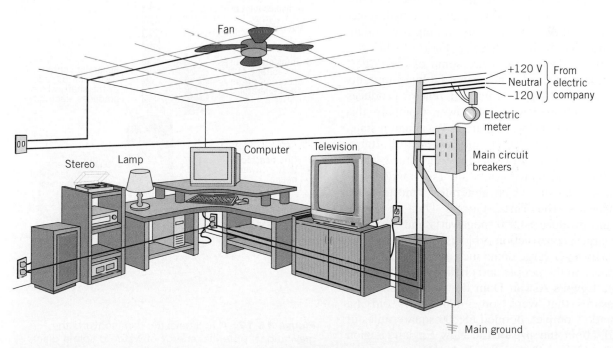

FIGURE 18-16. A typical household electric system has several parallel circuits, each with a number of lights and appliances or outlets in series. Each of these circuits has its own circuit breaker, which will trip if the circuit becomes overloaded.

The basic idea is that in a series circuit the same current flows through each circuit element. The voltage drop across each element, then, is given by Ohm's law and depends on the resistance of the element. In a parallel circuit, on the other hand, the voltage drop across each element is the same, and the amount of current flowing through each element can be different.

The differences between these two types of circuits can become obvious around Christmastime. Many older strands of Christmas lights are linked by a single series circuit. If any one light burns out, the entire strand goes dark because the electric circuit is broken. It can be a frustrating experience trying to find the one bad bulb. Most modern light strands, on the other hand, feature several parallel loops, each with just a few lights. So today, if one light burns out, only a few bulbs along the strand go dark.

THINKING MORE ABOUT

Hydroelectric Dams

Electric generating plants come in various types. Some run on fossil fuel, such as oil, coal, or natural gas. Others depend on nuclear reactors, geothermal vents, or the heat from solar energy. All of these power plants use heat to change water to steam, and the steam turns the turbines that rotate magnetic coils, inducing electric currents. However, the largest power plants are hydroelectric, using the energy of water falling over a dam to turn huge turbines directly (Figure 18-17).

Hydroelectric dams are some of the largest structures on Earth. They generate huge amounts of electricity. A typical nuclear reactor produces about 1200 megawatts of power. Compare this with the Grand Coulee Dam in Washington state, which generates 10,000 megawatts; the Itaipu Dam on the border of Paraguay and Brazil, which produces 12,600 megawatts, and La Grande Complex in Canada, which generates almost 16,000 megawatts. The Three Gorges Dam in China should produce 84,000 megawatts and is one of the largest construction projects ever undertaken.

However, large dams such as these have huge impacts on the people and environment of the region. Egypt's Aswan Dam flooded archeological treasures that were some 4000 years old. La Grande Complex flooded 68,000 square miles of land, about the size of the New England region of the United States. The Itaipu Dam cost over

$18 billion and was partly responsible for Brazil's economic problems in the 1990s. The Three Gorges Dam has flooded one of the greatest scenic attractions in all of China and has forced the relocation of over 1 million people. Major dams such as these turn rivers into lakes in landscapes that are not designed for stable lakes.

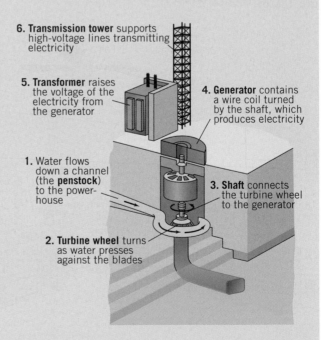

FIGURE 18-17. A hydroelectric dam converts the gravitational potential energy of water to spin a giant turbine that generates electric energy.

Is the availability of cheap electricity worth the cost of relocating people and wildlife? How can you determine the balance of benefits versus costs, both financial and otherwise? What would you think if plans were announced to build a hydroelectric plant near your home?

Summary

An **electric circuit** is an unbroken path of conducting material and contains a source, a closed conducting path, and a load. The quantity of charge flowing per second (the current) is measured in **amperes** or **amps,** the **electric potential** in **volts,** and the **electrical resistance** in **ohms.** When electrons flow through a wire, some of their energy is converted into heat. **Ohm's law** tells us that the **voltage** drop across any resistance is given by the product of the current times the resistance. The power dissipated in a circuit is given by the product of the voltage times the current.

Circuit breakers protect property by creating an open circuit when the current in a circuit exceeds a certain value. **Grounding** circuits protects people from electric shock. A short circuit is a low-resistance path from the positive to the negative side of a voltage source and may cause overheating of circuits.

A **rectifier** is a device that converts alternating current to direct current, and it is often used in modern electronic devices. A **transformer** is a device that changes the current and voltage of alternating current through the action of electromagnetic induction. Transformers play an important role in distributing electric power over long distances.

In a **parallel circuit,** all circuit branches have the same voltage. In a **series circuit,** the same current flows through all circuit elements.

Key Terms

ampere (amp) The physical unit used to define a quantity of electric current. (p. 375)

circuit breaker A device that prevents too much current from flowing through it by creating an open circuit. (p. 382)

electric circuit An unbroken path of material that carries electricity. (p. 374)

electric potential The quantity in electric circuits that is analogous to pressure in water flowing through pipes; it is the potential energy per unit charge across a region in a circuit. (p. 375)

electrical resistance (measured in **ohms**) The quantity that defines how hard it is to run electric current through an object. (p. 375)

grounding The process of connecting an element of a circuit to the ground to provide a safe path for the current in the case of an overload. (p. 385)

ohm The physical unit that defines the amount of electrical resistance. (p. 375)

Ohm's law The relationship among voltage, current, and resistance in a circuit. (p. 378)

parallel circuit A circuit, or part of a circuit, that consists of two or more loads linked together along different loops of wire. (p. 391)

rectifier An electric device that converts alternating current to direct current. (p. 386)

series circuit A circuit, or part of a circuit, that consists of two or more loads linked together along a single loop of wire. (p. 391)

short circuit A low-resistance path that causes potentially dangerous amounts of current in a circuit. (p. 385)

transformer A device that converts a high voltage to a low voltage (step-down transformer), or a low voltage to a high voltage (step-up transformer). (p. 387)

volt The physical unit used to quantify the amount of electric potential. (p. 375)

voltage (measured in **volts**) Synonymous with *electric potential.* (p. 375)

Key Equations

Power dissipated in a circuit element: $P = IV$

Voltage drop across a resistor (Ohm's law): $V = IR$

Power dissipated in a resistor: $P = I^2R$

Review

1. What is an electric circuit? What are its three key parts?

2. What are some of the similarities between electrons flowing through a wire and water flowing through a pipe? What are some of the differences?

3. What does an ampere measure? Where do you run into this term in your everyday experience?

4. What is electric potential?

5. What does the volt measure? Where do you run across this term in your everyday experience?

6. How is voltage like the pressure in a city water system? Explain.

7. What does the ohm measure? Where do you run across this term in your everyday experience?

8. What is the physical basis of resistance? Explain.

9. What is meant by the drift velocity of an electron? Is this a very fast velocity compared to the speed of the electric signal itself? Explain.

10. What is the relationship between the heat generated by electrons flowing in a wire and the resistance of the particular wire? How does this relationship affect the design of appliances such as toasters and space heaters?

11. What is the relationship between the voltage across a circuit, the current through it, and the resistance of the wire that composes the circuit?

12. What are two things that can be done to increase the flow of current in a circuit? How is this analogous to the water flow in a pipe?

13. How is lightning like an electric circuit? Explain.

14. What is the relationship between the power an appliance consumes, the voltage across it, and the current through it?

15. If we double the amount of wire through a wire or a circuit element, how much do we increase the heat generated? What equation demonstrates this?

16. What are two of the principal dangers of electric circuits?

17. Why do wires that carry large currents have much larger diameters than wires that carry power to ordinary electric lamps?

18. What is the purpose of a circuit breaker? How does it work?

19. What are the similarities and differences between a circuit breaker and a fuse?

20. Why should you *never* touch an electric device when you are in contact with water? Explain this in terms of Ohm's law.

21. What do you think are some of the more common causes of short circuits in household electric appliances? How might they be prevented?

22. What does grounding an electric circuit mean? How is this done?

23. What is a short circuit? How does grounding an electric device help decrease the danger of a short circuit?

24. What are some similarities and differences between the electric circuits in your home and the transmission of nerve impulses in your body?

25. What is direct current? Give an example of it.

26. What is alternating current? How does it differ from direct current?

27. What are the advantages and disadvantages of alternating current?

28. What is a rectifier? Name some ordinary devices that use a rectifier. Why is it needed and how does it work?

29. What is a transformer and why is it used? Give some examples of its use.

30. How does a transformer work? What is the purpose of the primary coil, and what is the purpose of the secondary coil?

31. Are the primary and secondary coils of a transformer directly linked? Explain.

32. Why is electric power transmitted at very high voltages?

33. What are the differences between a step-up and a step-down transformer? When is each used?

34. What is the relationship between the number of primary and secondary loops and the current going in and out of a transformer? What is the relationship between the voltages and the number of coils?

35. What do we mean by a series circuit? Give an example.

36. What is a parallel circuit? Give an example. In what ways does it differ from a series circuit?

37. What are some of the advantages of direct current? What are the disadvantages?

38. What are some common examples of series circuits? Parallel circuits? In what situations might one be preferable to the other?

Questions

1. What is the difference between electric current and electric charge?

2. Suppose 100,000 electrons per second pass a particular point in a wire. Is this a large or small current compared to one ampere? Explain.

3. Why should you never seek the shelter of a tree if you are caught outside in a thunderstorm?

4. Suppose a very large number of ants are confined to the inside of a long cardboard tube. Half of the ants are red and half are black. The red ants have acquired a positive charge by donating some of their electrons to the black ants. Thus, the black ants have acquired an equal and opposite amount of negative charge. Let the tube be aligned so that we can distinguish between the right and left ends of the tube. In which of the following situations is there an electric current? If there is an electric current, state its direction.

a. The black ants are moving to the right and the red ants are not moving.

b. The black and red ants are moving in the same direction at the same speed.

c. The black ants are moving to the left and the red ants are moving to the right.

d. None of the ants is moving.

e. The black ants are not moving, but half of the red ants are moving to the right and half are moving to the left.

5. Consider the system of three water pipes connected at a junction, as shown in the figure. You measure 5 gallons per minute flow in pipe A toward the junction and 10 gallons per minute flow in pipe B toward the junction. What is the water flow in pipe C?

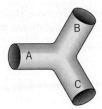

Questions 5, 6

6. Consider the system of three water pipes connected at a junction, as shown in the figure. You measure 5 gallons per minute flow in pipe A away from the junction and 10 gallons per minute flow in pipe B toward the junction. What is the water flow in pipe C?

7. Consider three wires connected at a junction, as shown in the figure. You measure 100 electrons per second flowing in wire A toward the junction and 200 electrons per second flowing in wire B toward the junction. What is the electron flow in wire C?

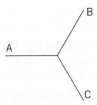

Questions 7, 8

8. Consider three wires connected at a junction, as shown in the figure. You measure 200 electrons per second flowing in wire A toward the junction and 300 electrons per second flowing in wire B away from the junction. What is the electron flow in wire C?

9. Why do you think a thicker wire has less resistance than a thinner wire made of exactly the same material?

10. A copper wire carries 1 amp of electric current. Do the electrons flowing in the wire give it a negative charge? Explain.

11. How do the length, diameter, and temperature of a copper wire affect its resistance?

12. A water tank empties itself by draining water out of the bottom through a pipe network. The figure shows two possible configurations of pipes leading out of the bottom. In each case, the pipe has the same diameter and there are two identical constrictions in the pipe. (A constriction is simply a location where the pipe diameter is smaller.) In which case will the water tank drain in less time? Explain.

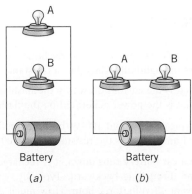

(a)   (b)

13. The figure represents two possible ways to connect two lightbulbs to a battery. All the bulbs are identical. In which case will the total current running through the battery be greater? Explain.

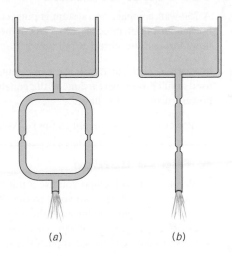

Questions 13, 14

14. The figure represents two possible ways to connect two lightbulbs to a battery. Bulbs A are identical and have less resistance than bulbs B, which are also identical. Which bulb has the most current running through it? Explain.

15. Does a transformer work with direct current? Why or why not?

16. How does a three-way lightbulb work?

17. Even though copper conducts electricity very well, it does have some resistance. How would the resistance of a 1-meter-long thick copper wire compare to the resistance of a 1-meter-long thin copper wire? Explain.

18. Why do you think strings of holiday or party lights are wired in parallel, not in series?

19. Which has more resistance: a 50-watt lightbulb or a 100-watt lightbulb? Explain.

20. Two lightbulbs are wired in series and connected to a 12-volt battery. What happens to the current through the battery if a third bulb is added in series? What happens to the power dissipated by the bulbs?

21. Two lightbulbs are wired in series and connected to a 12-volt battery. What happens to the current through the battery if a third bulb is wired in parallel with the other two bulbs? What happens to the power dissipated by the bulbs?

Problem-Solving Examples

The Power of Sound

EXAMPLE 18-5

A 250-watt compact disc system is plugged into a normal household outlet rated at 120 volts. How much current flows through the stereo at full power?

REASONING AND SOLUTION: We can calculate the current used at full power by rearranging the equation that relates power, current, and voltage:

$$\text{Power (watts)} = \text{Current (amps)} \times \text{Voltage (volts)}$$

We can manipulate this equation to find the current:

$$\begin{aligned}\text{Current} &= \frac{\text{Power}}{\text{Voltage}} \\ &= \frac{250 \text{ watts}}{120 \text{ volts}} \\ &= 2.08 \text{ amps}\end{aligned}$$

This current is slightly more than the current that flows through two 100-watt lightbulbs. ●

A Trip to Europe

EXAMPLE 18-6

While on a trip to Europe you use a transformer to convert European voltage, 240 volts, to the 120 volts typical of North American appliances. If the primary coil of your transformer has 40 loops of wire, how many loops must be in the secondary coil to accomplish this transformation?

REASONING AND SOLUTION: In this case we know the primary voltage (240 V), the secondary voltage (120 V), and N_1 (40). To determine N_2 we have to rearrange the transformer equation:

$$V_{\text{secondary}} = \left(\frac{N_1}{N_2}\right) V_{\text{primary}}$$

so

$$\begin{aligned}N_2 &= N_1 \left(\frac{V_{\text{primary}}}{V_{\text{secondary}}}\right) \\ &= 40 \left(\frac{240}{120}\right) \\ &= 80 \text{ loops}\end{aligned}$$ ●

Problems

1. In a simple DC circuit for a flashlight, a 1.5-V battery powers the lightbulb, which has a resistance of 2 ohms.

   a. What is the current drawn by the flashlight?
   b. What is the power generated by the flashlight?

2. In a typical household 100-W bulb, the current drawn is about 1 amp. What is the resistance of a 100-W lightbulb?

3. When a car battery runs down, it is recharged by running current through it backward. Typically, you might run 5 amps at 12 volts for 1 hour. How much energy does it take to recharge a battery?

4. A single electron has a charge of 1.60×10^{-19} C. How many electrons does it take to produce 1 ampere of current if they all pass a specific point in 1 second?

5. A typical 1.5-V alkaline D battery is rated at 3.5 amp-hours.

   a. What is the power that can be expended by the battery?
   b. What is the total energy stored by this battery?

6. A standard flashlight has two 1.5-V D batteries arranged so that their voltages add together, with a 2-W bulb. These batteries are rated at 3.5 amp-hours each.

   a. What is the total power that can be expended by the batteries?
   b. What is the total energy stored in the flashlight batteries?
   c. How long can the flashlight keep the bulb lit at its rated power of 2 W?

7. Giselle and Anna decided to impress their physical science professor with a simple experiment investigating the resis-

tance of a 100-W lightbulb. They used a VARIAC (which provides a variable voltage to a circuit) in a series circuit with one 100-W lightbulb. They measured the current and voltage, which are listed in the following table.

Voltage (volts)	Current (amps)
120	0.81
100	0.72
80	0.62
60	0.51
40	0.40
20	0.23

a. Calculate the resistance of the lightbulb for each voltage setting.

b. Does the resistance of the 100-W bulb increase or decrease with the voltage? With the current?

c. From your personal experience, can you predict whether the temperature of the filament of the lightbulb increases or decreases with the voltage?

d. Using this information, speculate on the reason(s) the resistance of the lightbulb changes.

8. Most household circuits have fuses or circuit breakers that open a switch when the current in the circuit exceeds 15 amps. Will the lights go off when you plug in an air conditioner (1 kilowatt), a TV (250 watts), and four 100-watt lightbulbs? Why?

9. Energy-efficient appliances are important in today's economy. Suppose that a lightbulb gives as much light as a 100-watt bulb, but consumes only 20 watts while costing $2.00 more. If electricity costs 8 cents per kilowatt-hour, how long will the bulb have to operate to make up the difference in price?

10. An energy-efficient air conditioner draws 7 amps in a standard 120-volt circuit. It costs $40 more than a standard air conditioner that draws 12 amps. If electricity costs 8 cents per kilowatt-hour, how long would you have to run the efficient air conditioner to recoup the difference in price?

Investigations

1. Explore the beginnings of the electric generation and transmission utility industry. How was the choice made to use alternating current over direct current to transmit power? Was there any disagreement over which to use and, if so, what were the arguments for and against each?

2. Make an inventory of all your electric appliances. How many watts does each use?

3. Most household circuits have fuses or circuit breakers that open a switch when the current in the circuit exceeds 15 amps. How many of the appliances that you listed in Investigation 2 could you run on the same circuit without overloading it?

4. Examine your most recent electric bill. How much power did you use? How much did it cost? Is there a discount for electricity used at off-peak hours? Examine your living place

for all the locations where you use electricity. Plan a strategy for reducing your electric bill by 10% next month. You can reduce consumption by turning off lights and appliances when not in use, installing lower-wattage bulbs, or using electricity during low-rate times.

5. How many kilowatts of electric power does a typical commercial power plant generate? How much electricity does the United States use each year? Is this amount going up or down?

6. Identify the major electric circuit components in your automobile. Which require the greatest power?

7. What materials other than copper might be used as common wire for electricity? What are the advantages and disadvantages of other such materials?

WWW Resources

See the *Physics Matters* home page at **www.wiley.com/college/trefil** for valuable web links.

1. **http://www.physics.uoguelph.ca/tutorials/ohm/index.html** DC circuits tutorials from the University of Guelph, Canada.

2. **http://www.ee.umd.edu/~taylor/frame1.htm** A gallery of electromagnetic personalities, including Ampere, Ohm, Franklin, and Volta.

3. **http://jersey.uoregon.edu/Voltage/** An animated circuit simulator teaching Ohm's law from the University of Oregon.

4. **http://micro.magnet.fsu.edu/electromag/java/transformer/index.html** A simple Java applet demonstrating a simple two stage transformer from Florida State University.

19 | The Electromagnetic Spectrum

PHYSICS AROUND US . . . Light All Around Us

You wake up when your clock radio goes off, flooding your bedroom with sound. Bright sunlight streams through the window as you jump out of bed, clean up, get dressed, and go down to make breakfast. In the kitchen, you pop a couple of slices of bread into the toaster, barely noticing how the heating coils glow a dull orange as they turn the bread brown. You also warm up a cup of coffee in your microwave oven. While you wait, you idly scan the notes held to your refrigerator by tiny magnets.

Did you realize that everything mentioned in this story, from the light of the Sun to the microwaves of your oven to the refrigerator magnets, are examples of the connection between electricity and magnetism? In particular, all these events involve electromagnetic waves.

ELECTROMAGNETIC WAVES

Physicists characterize waves by a mathematical expression called a "wave equation," which describes the movement of the medium for every wave, whether it's an ocean wave moving through water, a sound wave in air, or a seismic wave moving through Earth. Physicists have learned that whenever an equation of this (or some closely related) form appears, a corresponding wave is seen in nature.

Soon after Maxwell wrote the four equations that describe electricity and magnetism, he realized that some rather straightforward mathematical manipulation led to yet another equation, one that describes waves. The waves that Maxwell predicted are rather strange sorts of things, and we'll describe their anatomy in more detail later. However, the important point is that these are waves in which energy is transferred not through matter but through electric and magnetic fields. For example, it appears from the equations that whenever an electric charge is accelerated, one of these waves is emitted. Maxwell called them **electromagnetic waves** or **electromagnetic radiation.** An electromagnetic wave is a wave that incorporates electric and magnetic fields that fluctuate together; once it starts, the wave keeps itself going, even in a vacuum.

Maxwell's equations also predicted exactly how fast the waves would move—the wave velocity depends only on known constants such as the universal electrostatic (Coulomb) constant in Coulomb's law (see Chapter 16). These constants were known from experiments, and when Maxwell put the numbers into his expression for the velocity of his new waves, he found a very surprising answer. The predicted velocity of the wave turned out to be 300,000 kilometers per second (186,000 miles per second).

If you just had an "aha!" moment, you can imagine how Maxwell felt. The number that he calculated is the speed of light, which means that the waves described by his equations are actually the familiar (but mysterious) waves of **light.**

This result was astonishing. For centuries, scientists had puzzled over the origin and nature of light. Newton and others had discovered natural laws that describe the connections between forces and motion, as well as the behavior of matter and energy. But light remained an enigma. How did radiation from the Sun travel to Earth? What caused the light produced by a candle?

There is no obvious reason why static cling, refrigerator magnets, or the workings of an electric generator should be connected in any way to the behavior of visible light. Yet Maxwell discovered that light and other kinds of radiation are a type of wave that is generated whenever electric charges are accelerated.

Anatomy of an Electromagnetic Wave

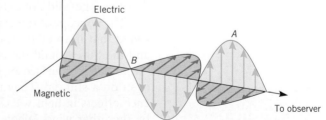

FIGURE 19-1. A diagram of an electromagnetic wave shows the interdependence of the changing electric field, the changing magnetic field, and the direction of the moving wave. *A* and *B* indicate points of maximum and minimum field strength.

A typical electromagnetic wave, shown in Figure 19-1, consists of electric and magnetic fields arranged at right angles to one another and perpendicular to the direction the wave is moving. To understand how the waves work, go back to Maxwell's equations that describe the behavior of electric and magnetic fields.

Recall that a changing electric field generates a magnetic field, while a changing magnetic field generates an electric field. So what happens when an electric charge such as an electron in a wire vibrates? The vibrating electric field produces a changing magnetic field, which in turn generates a changing electric field, back and forth. In other words, once one kind of field starts to change, it automatically changes the other kind of field, and that change, in turn, affects the original field.

Once you understand that the electromagnetic wave has this kind of ping-pong arrangement between electricity and magnetism, you can understand one of the most puzzling aspects of it—the fact that the wave can travel through a vacuum. Every other wave we have talked about is easy to visualize because the wave moves through a medium. We know that the motion of the wave is not the same as the motion of the medium, but in every other way the medium is there to give the wave tangible support. The electromagnetic wave is different. It is a wave that needs no medium whatsoever, but simply keeps itself going through its own internal mechanisms.

Electromagnetic waves, then, carry radiant energy, or radiation, created when electric charges accelerate. Once the waves start moving, however, they no longer depend on the source that emitted them.

Physics in the Making

The Ether

When Maxwell first proposed his idea of electromagnetic radiation, he was not prepared to deal with a wave that required no medium whatsoever. Previous scientists who had studied light, including such luminaries as Isaac Newton, assumed that light must travel through a hypothetical substance called the "ether" that permeates all space. The ether, they thought, served as the medium for light, and so Maxwell assumed that the ether provided the medium for his electromagnetic waves. In Maxwell's picture, the ether was a tenuous, transparent substance, perhaps like Jell-O, that filled all space. An accelerating charge shook the Jell-O at one point, causing electromagnetic waves to move outward at the speed of light.

The idea of an ether goes back to the ancient Greeks, and for most of recorded history scholars imagined that the vacuum of space was filled with this invisible substance. Not until 1887 did two U.S. scientists, Albert A. Michelson (1852–1931) and Edward W. Morley (1838–1923), working at what is now Case Western Reserve University in Cleveland, perform experiments that demonstrated that the ether could not be detected. This failure was interpreted to mean that the ether did not exist.

The concept of the experiments was very simple. Michelson and Morley reasoned that if an ether really existed, then the motion of Earth around the Sun and the Sun around the center of our galaxy would produce an apparent ether "wind" at the surface of Earth, much as someone riding in a car feels a wind even on a still day. They used very sensitive instruments to search for interference effects in light waves, effects that would result from the deflection of light by the ether wind. When their experiments turned up no such deflection, they concluded that the ether does not exist.

In 1907, Albert Michelson became the first U.S. scientist to win a Nobel Prize, an honor that recognized his pioneering experimental studies of light. ●

Light

Once Maxwell understood the connection between electromagnetism and light, his equations allowed him to draw several important conclusions. For one thing, because the speed of the electromagnetic waves depends entirely on the nature of the interactions between electric charges and magnetic fields, it cannot depend on the wavelength or frequency of the wave itself. Thus, all electromagnetic waves, regardless of their wavelength or frequency, have to move at exactly the same speed. This speed—the **speed of light**—turns out to be so important in science that we give it a special symbol, *c*. The speed of electromagnetic waves in a vacuum is one of the fundamental constants of nature.

For electromagnetic waves, the relation among speed, wavelength, and frequency takes on the familiar form

$$\text{Wavelength} \times \text{Frequency} = c$$
$$= 300{,}000 \text{ km/s} \ (= 186{,}000 \text{ miles/s})$$

In other words, if you know the wavelength of an electromagnetic wave, you can calculate its frequency, and vice versa.

Figuring Frequency

The wavelength of yellow light is about 580 nanometers, or 5.8×10^{-7} m. What is the frequency of a yellow light wave?

REASONING: We know that for all electromagnetic waves,

$$\text{Wavelength} \times \text{Frequency} = 300{,}000 \text{ km/s} = 3 \times 10^8 \text{ m/s}$$

We want to determine frequency, so rearrange this equation:

$$\text{Frequency} = \frac{3 \times 10^8 \text{ m/s}}{\text{Wavelength}}$$

SOLUTION: This equation reveals that for yellow light with a wavelength of 5.8×10^{-7} m,

$$\text{Frequency} = \frac{3 \times 10^8 \text{ m/s}}{5.8 \times 10^{-7} \text{ m}}$$
$$= 0.52 \times 10^{15} \text{ Hz}$$
$$= 5.2 \times 10^{14} \text{ Hz}$$

(Remember, 1 hertz equals 1 cycle per second.) In order to generate yellow light by vibrating a charged comb, you would have to wiggle it more than 5 hundred trillion (520,000,000,000,000) times per second. ●

The Energy of Electromagnetic Waves

Think about how you might produce an electromagnetic wave with a simple comb. Electromagnetic waves are generated any time a charged object is accelerated, so imagine combing your hair on a dry winter day when the comb picks up a static charge. Each time you move the comb back and forth, an electromagnetic wave traveling 300,000 kilometers per second is sent out from the comb.

If you wave the electrically charged comb up and down slowly, once every second, you create electromagnetic radiation, but you're not putting much energy into it. You produce a low-frequency, low-energy wave with a wavelength of about 300,000 kilometers. (Remember, each wave moves outward 300,000 kilometers in a second, which is the separation between wave crests.)

If, on the other hand, you could vibrate the comb vigorously—say at 300,000 times per second—you would produce a higher-energy, high-frequency wave with a 1-kilometer wavelength. By putting more energy into accelerating the electric charge, you wind up with more energy in the electromagnetic wave.

Visible light, the first example of an electromagnetic wave known to humans, bears out this kind of reasoning. A glowing ember has a dull red color, corresponding to relatively low energy. Hotter, more energetic fires show a progression of more energetic colors, from the yellow of a candle flame to the blue-white flame of a blowtorch. These colors are merely different frequencies, and therefore different energies, of light. High-frequency visible light corresponds to a blue color, whereas low-frequency visible light appears red.

Red light has a wavelength corresponding to the distance across about 7000 atoms or about 700 nanometers (a nanometer is 10^{-9} meter, about 40 billionths of an inch). Red light is the longest wavelength that the eye can see and is the least energetic of the visible electromagnetic waves. Violet light, on the other hand, has a shorter wavelength, corresponding to the distance across about 4000 atoms, or about 400 nanometers, and is the most energetic of the visible electromagnetic waves. All other colors have wavelengths and energies between those of red and violet. We explore this important relationship between frequency and energy in the next chapter.

THE PARTS OF THE ELECTROMAGNETIC SPECTRUM

A profound puzzle accompanied Maxwell's original discovery that light is an electromagnetic wave. Waves can be of almost any wavelength. Water waves on the ocean, for example, range from tiny ripples to globe-spanning tides. Yet visible light spans an extremely narrow range of wavelengths, only about 400 to 700 nanometers. According to the equations that Maxwell derived, electromagnetic waves could exist at any wavelength (and, consequently, any frequency) whatsoever. The only constraint is that the wavelength times the frequency must be equal to the speed of light. Yet when Maxwell looked into the universe, he saw visible light as the only example of electromagnetic waves. It was as if a splendid symphony was playing, ranging from the deep bass of the tuba to the sharp shrill of the piccolo, but you could hear only a couple of notes from a single violin.

In such a situation it would be natural to wonder what had happened to the rest of the waves. Physicists looked at Maxwell's equations, looked at nature, and realized that something was missing. The equations predicted that there ought to be more kinds of electromagnetic waves than visible light; waves that no one had seen up to that time; waves performing the waltz between electricity and magnetism, but with frequencies and wavelengths different from those of visible light. These as-yet-unseen waves would have exactly the same structure as the one shown for the electromagnetic wave in Figure 19-1, but they could have either longer or shorter wavelengths than visible light, depending on the acceleration of the electric charge that created them. These waves would move at the speed of light and would be exactly the same as visible light except for the differences in the wavelength and frequency.

Between 1885 and 1889, German physicist Heinrich Rudolf Hertz (1857–1894), after whom the unit of frequency is named, performed the first experiments that confirmed these predictions. He discovered the waves that we now know as radio waves. Since that time, all manner of electromagnetic waves have been discovered, from those with wavelengths longer than the radius of Earth to those with wavelengths shorter than the size of the nucleus of the atom. They

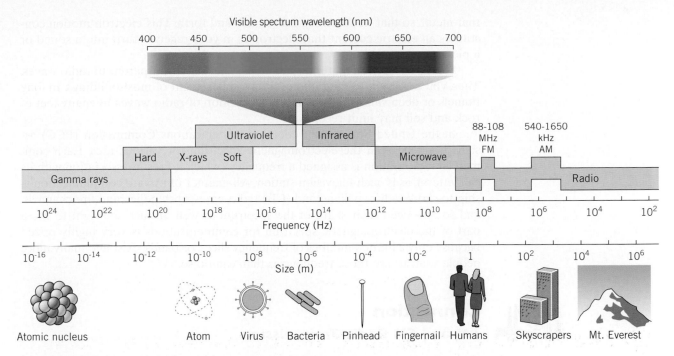

FIGURE 19-2. The electromagnetic spectrum includes all kinds of waves that travel at the speed of light, including radiowaves, microwaves, infrared radiation, visible light, ultraviolet radiation, X rays, and gamma rays. Note that sound waves, water waves, seismic waves, and other kinds of waves that require matter in order to move travel much slower than light speed.

include radio waves, microwaves, infrared radiation, visible light, ultraviolet radiation, X rays, and gamma rays. This entire collection of waves is called the **electromagnetic spectrum** (Figure 19-2). Remember that every one of these waves, no matter what its wavelength or frequency, is the result of an accelerating electric charge.

Radio Waves

The **radio wave** part of the electromagnetic spectrum ranges from the longest waves, those whose wavelength is longer than the size of Earth, to waves a few meters long. The corresponding frequencies, from roughly a kilohertz (1000 cycles per second, or kHz) to several hundred megahertz (1 million cycles per second, or MHz), correspond to the familiar numbers on your radio dial. There are various subdivisions of radio waves, but the most important fact about them is that, like light, they can penetrate long distances through the atmosphere. This characteristic makes radio waves very useful in communication systems.

Have you ever been driving at night and picked up a radio signal from a station 1000 miles away? If so, you have had firsthand experience of the ability of radio waves to travel long distances through the atmosphere. Pushing electrons back and forth rapidly in a tall metal antenna can produce a typical radio wave of the type used for communication. This acceleration of electrons produces outgoing radio waves, just as throwing a rock in a pond produces outgoing ripples. When these waves encounter another piece of metal (for example, the antenna in your radio or TV set), the electric fields in the waves accelerate electrons in that metal, so that its electrons move back and forth. This electron motion constitutes an electric current that electronics in your receiver turn into a sound or a picture.

Most construction materials are at least partially transparent to radio waves. Thus, you can listen to the radio even in the basement of most buildings. In long tunnels or deep valleys, however, the absorption of radio waves by many feet of rock and soil may limit reception.

In the United States, the Federal Communications Commission (FCC) assigns frequencies in the electromagnetic spectrum for various uses. Each commercial radio station is assigned a frequency (which it uses in association with its call letters), as is each television station. All manner of private communication—ship-to-shore radio, civilian band (CB) radio, emergency police and fire channels, and so on—need their share of the spectrum as well. In fact, the right to use a part of the electromagnetic spectrum for communications is very highly prized because only a limited number of frequency slices, or bands, exist and many more people want to use those frequencies than can do so.

Connection

AM and FM Radio Transmission

Radio waves carry signals in one of two ways: AM or FM. Broadcasters can send out their programs at only one narrow range of frequencies, a situation very different from music or speech, which uses a wide range of frequencies. Thus radio stations cannot simply transform a range of sound-wave frequencies into a similar range of radio-wave frequencies. Instead, the information to be transmitted must be compressed in some way on the narrow frequency range of the station's radio waves.

This problem is similar to one you might experience if you had to send a message across a lake with a flashlight at night. You could adopt either of two strategies. You could send a coded message by turning the flashlight on and off, thus varying the brightness (the amplitude) of the light. Alternatively, you could change the color (the frequency) of the light by passing blue or red filters in front of the beam.

Radio stations also adopt these two strategies (see Figure 19-3). All stations begin with a carrier wave of fixed frequency. This is the broadcast frequency of the station. AM radio stations typically broadcast at frequencies between about 530 and 1600 kHz, whereas the carrier frequencies of FM radio stations range from about 88 to 110 MHz.

The process called *amplitude modulation* (AM) depends on varying the strength (or amplitude) of the radio's carrier wave according to the sound signal to be transmitted (Figure 19-3a). Thus the shape of the sound wave is impressed on the radio's carrier wave signal. When this signal is received by your radio, its interior electronics recover the original sound signal and use it to run the speakers. This original sound signal is what you hear when you turn on your radio.

Alternatively, you can slightly vary the frequency of the radio's wave according to the signal you want to transmit, a process called *frequency modulation* (FM), as shown in Figure 19-3b. A radio that receives this particular signal can unscramble the changes in frequency and convert them into electric signals that run the speakers so that you can hear the original signal. TV broadcasts,

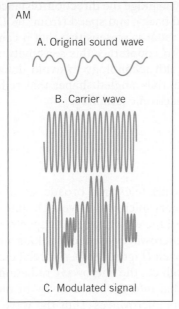
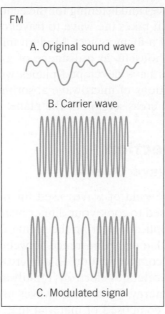

(*a*) Amplitude modulation

(*b*) Frequency modulation

FIGURE 19-3. AM (amplitude modulation) and FM (frequency modulation) transmission differ in the way that a sound wave (A) is superimposed on a carrier wave (B) of constant amplitude and frequency. The carrier wave can be varied, or modulated, to carry information (C) by altering its amplitude or its frequency.

which use carrier frequencies about a thousand times higher than FM radio, typically transmit the picture on an AM signal and the sound on an FM signal at a slightly different frequency.

Another important difference between AM and FM radio stations is that the lower-frequency waves of AM broadcasts may partially reflect off layers of the atmosphere, which allows them to be heard at great distances (especially at night). However, higher-frequency FM transmissions require line-of-sight reception, which restricts their range to about 50 miles. ●

Microwaves

Microwaves include electromagnetic waves whose wavelengths range from about 1 meter (a few feet) to 1 millimeter (1 thousandth of a meter, or about 0.04 inch). The longer wavelengths of microwaves travel easily through the atmosphere, like their cousins in the radio part of the spectrum, although rock and building materials absorb most microwaves. Therefore, microwaves are used extensively for line-of-sight communications. Most satellites broadcast signals to Earth in microwave channels, and these waves also commonly carry long-distance telephone calls and TV broadcasts. The satellite dish antennas that you see on private homes and businesses are designed primarily to receive microwave transmissions, as are the large cone-shaped receivers attached to the microwave relay towers found on many hills or tall buildings.

The distinctive transmission and absorption properties of microwaves make them ideal for use in aircraft radar. Solid objects, especially those made of metal, reflect most of the microwaves that hit them. By sending out timed pulses of

A stealth fighter relies on combinations of microwave-absorbing materials and angled shapes to reduce the apparent cross section of the plane.

microwaves and listening for the echo, you can judge the direction, distance (from the time it takes the wave to travel out and back), and speed (from the Doppler effect) of a flying object. Modern military radar is so sensitive that it can detect a single housefly at a distance of a mile. To counteract this sensitivity, aircraft designers have developed planes with stealth technology to avoid detection—combinations of microwave-absorbing materials, angled shapes that reduce the apparent cross section of the plane, and electronic jamming.

Connection

Microwave Ovens

The same kind of waves used for phone calls, television broadcasts, and radar can be used to cook your dinner in an ordinary microwave oven. Despite the different applications, from the point of view of the electromagnetic spectrum there is no fundamental difference between the microwaves used for cooking and those used for communication. In a microwave oven (Figure 19-4), a special electronic device accelerates electrons rapidly and produces the microwave radiation, which carries energy. These microwaves are guided into the main cavity of the oven, which is composed of material that scatters microwaves. Thus the wave energy remains inside the box until it is absorbed by something.

It turns out that microwaves are absorbed quickly by water molecules. This means that the energy used to create microwaves is carried by those waves to food inside the oven, where the energy is absorbed by water molecules inside the food and converted into heat. This absorption of microwave energy results

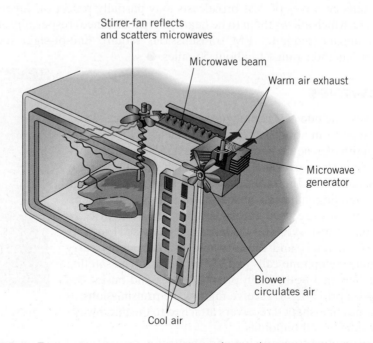

Stirrer-fan reflects
and scatters microwaves

Microwave beam

Warm air exhaust

Microwave
generator

Blower
circulates air

Cool air

FIGURE 19-4. Every microwave oven contains a device that generates microwaves by accelerating electrons. The walls scatter the microwaves until they are absorbed, usually by water molecules in the food.

in a very rapid rise in temperature and rapid cooking. In a microwave oven, then, food is heated from the inside, unlike an ordinary oven, where heat travels from the surface to the interior. Microwaves do not heat paper and glass, which don't contain many water molecules.

Metal is a good reflector of microwaves; a metal fan is often contained in a microwave oven to help scatter the beam around the oven. This explains why you don't want to wrap food in metal foil before putting it in the microwave oven; the foil would scatter the radiation and prevent it from reaching the food inside and would probably overload the oven's circuits in the process. ●

Infrared Radiation

Infrared radiation includes wavelengths of electromagnetic radiation that extend from 1 millimeter down to about 1 micron (10^{-6} meter, or less than 1 ten-thousandth of an inch). Our skin, which absorbs infrared radiation, provides a crude kind of detector. You feel infrared radiation when you put your hands out to a warm fire or over the cooking element of an electric stove. Infrared waves are what we feel as radiant heat.

Warm objects emit infrared radiation, and this fact has been used extensively in both civilian and military technology. Infrared detectors are used to guide air-to-air missiles to the exhaust of jet engines in enemy aircraft, and infrared detectors are often used to "see" human beings and warm engines at night. Similarly, many insects (such as mosquitoes and moths) and other nocturnal animals (including opossums and some snakes) have developed sensitivity to infrared radiation; thus they can see in the dark.

Infrared detection is also used to find heat leaks in homes and buildings (Figure 19-5). If you take a picture of a house at night using film that is sensitive to infrared radiation, places where heat is leaking out show up as bright spots on the film. This information can be used to correct the loss and thus conserve energy. In a similar way, Earth scientists often monitor volcanoes with infrared detectors. The appearance of a new hot spot may signal an impending eruption.

FIGURE 19-5. A photograph using infrared film reveals heat energy escaping from houses. This false-color image is coded so that white is hottest, followed by red, pink, blue, and black.

Develop Your Intuition:
Uses of Infrared Beams

Have you ever been to a public restroom in an airport or office building and found plumbing that works automatically? Water comes on in the sink as soon as you put your hands under the faucet and turns off as soon as you take your hands away. But there's no visible light beam anywhere, so how does the system know you're there?

What you're not seeing are beams of infrared radiation (Figure 19-6a). A fiber-optic cable inside the faucet emits a beam with a wavelength of 850 nanometers, which is not visible to the human eye. A second fiber-optic cable receives the beam reflected from the bottom of the sink. When you put your hand in the path of the beam, a sensor notes that a change in intensity of the reflected beam has occurred and it initiates electronic switches that turn on the water. When you take your hand away, the sensor returns to normal and the switches close.

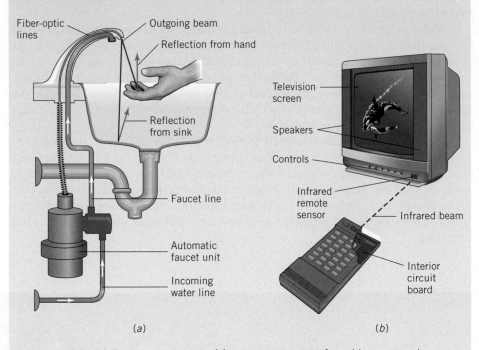

(a)   (b)

FIGURE 19-6. (a) Faucets in many public restrooms use infrared beams to detect a person's hands. Water comes on when you put your hands under the faucet and turns off when you take your hands away. (b) Your TV remote control sends a pattern of infrared signals to a sensor on the TV set.

Does this seem like a fairly exotic application of infrared beams? There's a far more common application closer to home: the TV remote control (Figure 19-6b). Every function on the remote is coded into a microchip. When you press a button, the chip activates a light-emitting diode that sends a pattern of infrared signals to a reader on the TV set. The pattern is usually repeated five times per second to make sure the message is received.

Visible Light

All of the colors of the rainbow are contained in **visible light,** whose wavelengths range from red light at about 700 nanometers down to violet light at about 400 nanometers (Figure 19-7). From the point of view of the larger universe, the visible world in which we live is a very small part of the total picture (see Figure 19-2).

Our eyes distinguish several different colors, but these portions of the electromagnetic spectrum have no special significance except in our perceptions. In fact, the distinct colors that we see—red, orange, yellow, green, blue, and violet—represent very different-size slices of the electromagnetic spectrum (Figure 19-8). The red and blue portions of the spectrum are rather broad, spanning more than 50 nanometers of wavelengths; we thus perceive many different wavelengths as red or blue. In contrast, the yellow part of the spectrum is quite narrow, encompassing wavelengths from only about 570 to 590 nanometers.

Why should our eyes be so sensitive to such a restricted range of the spectrum? The Sun's light is especially intense in this part of the spectrum, so some biologists suggest that our eyes evolved to be especially sensitive to these wavelengths in order to take maximum advantage of the Sun's light. Our eyes are ideally adapted for the light produced by our Sun during daylight hours. Animals that hunt at night, such as owls and cats, have eyes that are more sensitive to infrared wavelengths, radiation that makes warm living things stand out against the cooler background.

FIGURE 19-7. A glass prism separates white light into the visible spectrum.

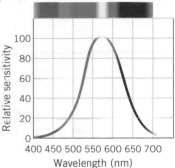

FIGURE 19-8. Humans perceive the visible light spectrum as a sequence of color bands. The relative sensitivity of the human eye differs for different wavelengths. Our perception peaks near wavelengths that we perceive as yellow, although the colors we see have no special physical significance.

(a)

(b)

(c)

A variety of chemical reactions produce light energy.

Develop Your Intuition: Stars of Different Colors

Not all stars in the sky have the same color. For example, the star Betelgeuse, located in the shoulder of the constellation Orion, is distinctly red, while the star Rigel, located in Orion's foot, emits blue light. Which of these stars is at a hotter temperature?

Blue light has a shorter wavelength (higher frequency) than red light, and higher frequencies mean higher energies. So Rigel emits light of higher energy than does Betelgeuse and is the hotter star. In fact, most young stars emit blue light, indicating their high temperatures. Near the end of their lives, their temperatures decrease and their light turns redder. Betelgeuse is classified as a red giant star, meaning it is nearing the last stages of its lifetime. Nevertheless, since stars last for billions of years, Betelgeuse still has a long time to go.

Connection
The Eye

The light detector with which we are most familiar is one we carry around with us all the time—the human eye. Eyes are marvelously complex organs, turning incoming electromagnetic radiation into images through the use of a combination of physical and chemical processes (Figure 19-9).

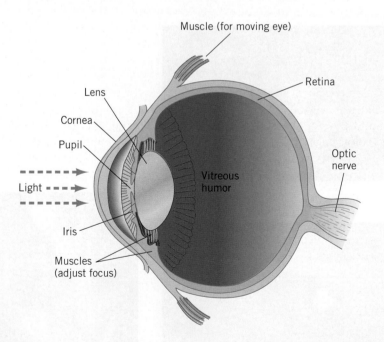

FIGURE 19-9. A cross section of the human eye reveals the path of light, which enters through the protective cornea and travels through the colored iris. The pupil is the aperture through which light passes and changes in size to control the amount of light entering the eye. Muscles move the eye and change the shape of the lens, which focuses light onto the retina, where the light's energy is converted into nerve impulses. These signals are carried to the brain along the optic nerve.

Light waves enter the eye through a clear lens whose thickness can be changed by a sheath of muscles around its edges. The direction of the waves is changed by refraction in the lens so that they are focused at receptor cells located in the retina at the back of the eye. There the light is absorbed by two different kinds of cells, called rods and cones (the names come from their shapes, not their functions). The rods are sensitive to light and dark, including low levels of light; they give us night vision. Three kinds of cones, sensitive to red, blue, or green light, allow us to see colors.

The energy of incoming light triggers complex changes in molecules in the rods and cones, initiating a series of reactions that eventually lead to a nerve signal that travels along the optic nerve to the brain. ●

Ultraviolet Radiation

At wavelengths shorter than visible light, we begin to find waves of high frequency and therefore high energy and potential danger. The wavelengths of **ultraviolet radiation** range from 400 nanometers down to about 100 nanometers (the size of 100 atoms placed end to end) in length. The energy contained in longer ultraviolet waves can cause a chemical change in skin pigments, a phenomenon known as tanning. This lower-energy portion of the ultraviolet is not particularly harmful by itself.

On the other hand, shorter-wavelength (higher-energy) ultraviolet radiation carries more energy—enough that this radiation can damage skin cells, causing sunburn and skin cancer in humans. The wave's energy is absorbed by cells and can cause extensive damage to DNA, the critical molecule that transfers biological information from one generation of cells to the next. The ability of ultraviolet radiation to kill living cells is used by hospitals to sterilize equipment and kill unwanted bacteria.

The Sun produces intense ultraviolet radiation in both longer and shorter wavelengths. Fortunately, our atmosphere (and particularly the ozone layer in the atmosphere) absorbs much of the harmful short wavelengths and thus shields living things. Nevertheless, if you spend much time outdoors under a bright Sun, you should protect exposed skin with a sunblocking chemical, which is transparent (colorless) to visible light but absorbs harmful ultraviolet rays.

The energy contained in both long and short ultraviolet wavelengths can be absorbed by atoms, which in special materials may subsequently emit a portion of that absorbed energy as visible light. (Remember, both visible light and ultraviolet light are forms of electromagnetic radiation, but visible light has longer wavelengths, and therefore less energy, than ultraviolet radiation.) This phenomenon, called fluorescence, provides the black-light effects so popular in stage shows and nightclubs. We examine the origins of fluorescence in more detail in Chapter 21.

X Rays

X rays are electromagnetic waves that range in wavelength from about 100 nanometers down to 0.1 nanometer, smaller than a single atom. These high-frequency (and thus high-energy) waves can penetrate several centimeters into most solid matter, but are absorbed to different degrees by all kinds of materials. This fact allows X rays to be used extensively in medicine to form visual images of bones and organs inside the body. Bones and teeth absorb X rays much

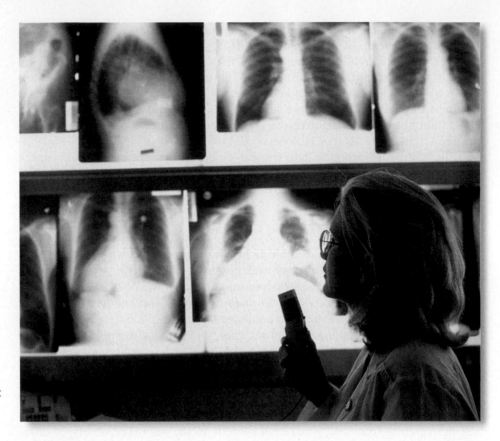

A physician examines medical X rays. Internal structures are revealed because bones and different tissues absorb X rays to different degrees.

more efficiently than skin or muscle, so a detailed picture of inner structures emerges. X rays are also used extensively in industry to inspect for defects in welds and manufactured parts.

The X-ray machine in your doctor's or dentist's office is something like a giant lightbulb with a glass vacuum tube. At one end of the tube is a tungsten filament that is heated to a very high temperature by an electric current, just like in an incandescent lightbulb. At the other end is a polished metal plate. X rays are produced by applying an extremely high voltage—negative on the filament and positive on the metal plate—so electrons stream off the filament and smash into the metal plate at high velocity. The sudden deceleration of the negatively charged electrons releases a flood of high-energy electromagnetic radiation—the X rays that travel from the machine to you at light speed.

Ongoing Process of Science
Intense X-Ray Sources

X rays have become supremely important in many facets of science and industry. X-ray crystallographers use beams of X rays to determine the spacing and positions of atoms in a crystal, physicians use X rays to reveal bone fractures and other internal injuries, and many industries use X rays to scan for defects in manufactured products. However, many potential applications, such as structural studies of very small crystals or scans of unusually large manufactured products, are unrealized because of the relatively low intensity of conventional X-ray sources.

A major effort is now under way to develop new, more powerful X-ray sources. One such facility, the Advanced Photon Source (APS) near Chicago, Illinois, generates intense X-ray beams 1 billion times stronger than conventional sources by accelerating electrons in a circular path. (Remember, electromagnetic radiation is emitted when charged particles are accelerated.) Scientists converge on the APS from around the world to study the properties of matter. Eventually, an X-ray laser (see Chapter 21) might produce even more powerful X-ray beams, although such technology is now only a dream. ●

Gamma Rays

The wavelengths with the highest energies in the electromagnetic spectrum are called **gamma rays.** Their wavelengths range from slightly less than the size of an atom (about 0.1 nanometer, or 10^{-10} meter) to less than 1 trillionth of a meter, or 10^{-12} meter. Gamma rays are normally emitted only in very high-energy nuclear and particle reactions (see Chapters 26 and 27), and they are not as common on Earth as the other kinds of electromagnetic radiation that we have talked about.

Gamma rays have many uses in medicine. Some types of medical diagnosis involve giving a patient a radioactive chemical that emits gamma rays. If that chemical concentrates at places where bone is actively healing, for example, then doctors can monitor the healing by locating the places where gamma rays are emitted. The gamma-ray detectors used in this specialized form of nuclear medicine are both large (to capture the energetic waves) and expensive. Doctors also use gamma rays for the treatment of cancer in humans. In these treatments, high-energy gamma rays are directed at tumors or malignancies that cannot be removed surgically. When the gamma-ray energy is absorbed in those tissues, the tissues die and the patient has a better chance to live.

Gamma rays are also studied in astronomy because many of the interesting processes going on in our universe involve bursts of very high energy and, hence, the emission of gamma rays. The Compton Gamma Ray Observatory, a satellite

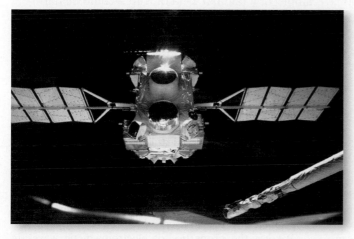

(a)

(b)

(a) NASA's Compton Gamma Ray Observatory, which operated from 1991 to 1999, detected exploding stars and active galaxies. (b) Some of the most energetic events ever observed in the universe, such as this gamma-ray-emitting star in a distant galaxy, were recorded by this satellite.

launched in 1991, documented many such energetic events, including exploding stars and active distant galaxies. In 1999, the Compton's guidance system failed, so in June 2000 it was allowed to fall into the atmosphere and burn up.

THINKING MORE ABOUT

Electromagnetic Radiation: Is ELF Radiation Dangerous?

Maxwell's equations tell us that *any* accelerated charge emits waves of electromagnetic radiation, not just radiation with frequencies of millions or billions of hertz. In particular, the electrons that move back and forth in wires to produce the alternating current in household wiring generate electromagnetic radiation. Every object in which electric power flows, from power lines to toasters to computers, is a source of this weak, extremely low-frequency (ELF) radiation.

For more than a century, human beings in industrialized countries have lived in a sea of weak ELF radiation, but until recently no questions were raised about whether that radiation might have an effect on human health. In the late 1980s, however, a series of books and magazine articles created a minor sensation by claiming that exposure to ELF radiation might cause some forms of cancer, most notably childhood leukemia.

Scientists tended to downplay these claims because the electric fields most residents experience due to power lines are a thousand times weaker than those due to natural causes (such as electrical activity in nerve and brain cells). They also pointed out that age-corrected cancer rates in the United States (with the exception of lung cancer, which is caused primarily by smoking) have remained constant or dropped over the last 50 years, even though exposure to ELF radiation has increased enormously. They also questioned the statistical validity of some studies: more detailed analyses of results did not demonstrate the connection between ELF radiation and disease. In 1995, the prestigious American Institute of Physics reviewed the scientific literature on this subject and concluded that there is no reliable evidence that ELF radiation causes any form of cancer, and most funding for research in this area was cut off.

This situation is typical of encounters at the border between science and public health. Preliminary data indicate a possible health risk but do not prove that the risk is real. Settling the issue by further study takes years, while researchers carefully collect data and weigh the evidence. In the meantime, people have to make decisions about what to do. In addition, as in the case of ELF radiation, the cost of removing the risk is often very high.

Suppose you are a scientist who has shaky evidence that some common food—bread, for example, or a familiar kind of fruit—could be harmful. What responsibility do you have to make your results known to the general public? If you stress the uncertainty of your results and no one listens, should you make sensational (perhaps unsupported) claims to get people's attention?

Summary

The motion of every wave can be described by a characteristic wave equation. James Clerk Maxwell recognized that simple manipulation of his equations pointed to the existence of **electromagnetic waves** or **electromagnetic radiation,** consisting of alternating electric and magnetic fields that can travel through a vacuum at the **speed of light.** This discovery solved one of the oldest mysteries of science, the nature of **light.** Although visible light was the only kind of electromagnetic radiation known to Maxwell, he predicted the existence of other kinds with longer and shorter wavelengths.

Soon thereafter a complete **electromagnetic spectrum** of waves was recognized, including **radio waves, microwaves, infrared radiation, visible light, ultraviolet radiation, X rays,** and **gamma rays.**

We use the properties of electromagnetic waves in countless ways every day—in radio and TV, heating and lighting, microwave ovens, tanning salons, medical X rays, and more. Much of science and technology during the past 100 years has been an effort to find new and better ways to produce, manipulate, and detect electromagnetic radiation.

Key Terms

electromagnetic spectrum The entire collection of electromagnetic waves, from the shortest wavelength (gamma rays) to the longest wavelength (radio waves) and everything in between. (p. 403)

electromagnetic wave or **electromagnetic radiation** A wave that incorporates electric and magnetic fields that oscillate together. (p. 399)

gamma rays Electromagnetic waves with the shortest wavelength; they are usually emitted in nuclear reactions. (p. 413)

infrared radiation Electromagnetic waves with wavelengths in the range of about 1 micron to 1 millimeter; they are what we feel as radiant heat. (p. 407)

light Electromagnetic radiation detectable by the human eye, with wavelengths from about 400 to 700 nanometers in length. (p. 399)

microwaves Electromagnetic waves with wavelengths between about 1 millimeter and 1 meter. (p. 405)

radio waves Electromagnetic waves with the longest wavelength, used to broadcast radio signals. (p. 403)

speed of light, c The speed at which all electromagnetic waves travel when in a vacuum, 3.00×10^8 m/s. (p. 401)

ultraviolet radiation Electromagnetic waves with wavelengths from about 100 nanometers to 400 nanometers; they are known to cause sunburn. (p. 411)

visible light The specific frequencies of electromagnetic waves that can be detected by the human eye; they include all colors of the rainbow. (p. 409)

X rays Electromagnetic waves with wavelengths from about 0.1 nanometer to 100 nanometers. (p. 411)

Key Equations

1 hertz = 1 cycle/second

For light: Wavelength (m) $\times$ Frequency (Hz) $= c$

Constant: Speed of Light: $c = 300,000$ km/s $= 3 \times 10^8$ m/s

Review

1. Through which medium does an electromagnetic wave move? Explain.

2. Can an electromagnetic wave move through a complete vacuum? How?

3. How fast do electromagnetic waves travel? What was the significance of the discovery that light also moved at the same speed?

4. If a source of electromagnetic waves stops emitting waves, will the waves that have already been emitted be affected in any way?

5. What was meant by the "ether"? What prompted the assumption that it existed?

6. What did Michelson and Morley's experiment show about the medium that was thought to carry electromagnetic radiation?

7. What value is c, the speed of light? [*Hint:* Don't forget the units.]

8. What is the relationship among the frequency, the wavelength, and the speed of light?

9. What is the relationship between the frequency of an electromagnetic wave and its energy?

10. Which is hotter, the blue flame of a furnace or the orange flame of a campfire? Explain.

11. What is the range of wavelength of visible light in nanometers? How long is a nanometer? Write this using a decimal point.

12. What is meant by the term "electromagnetic spectrum"?

13. What is the significance of accelerating charges to the electromagnetic spectrum? What role did Hertz's discovery of radio waves play in the understanding of this spectrum?

14. How is a radio wave produced? What role does an antenna play in this?

15. What is the difference between AM and FM radio? How is each of these types of waves produced?

16. Identify some of the differences and similarities between radio waves and microwaves. Which has a longer wavelength? Which is more energetic?

17. Identify three common uses of microwaves.

18. How does a microwave oven work? What role does water play in facilitating its function? Why does a glass or paper container that holds the food inside such an oven not get hot even while the food does?

19. How do we perceive infrared radiation? Which of our five senses best detects it?

20. What is the difference between red light and yellow light? What is it that determines a difference in the color of light?

21. What quality of light do the rods within your eye detect? Do the cones detect?

22. Why is short-wave ultraviolet light more damaging to your skin than long-wave ultraviolet light?

23. What is an X ray and how is it generated in a simple machine in your doctor's office?

24. Why are X rays used for medical diagnosis? What other wavelengths of electromagnetic radiation are used in medicine?

25. How much energy does a gamma ray have relative to other parts of the electromagnetic spectrum? What is its wavelength and frequency relative to these other parts of the spectrum?

26. What are some uses of gamma rays?

27. What kinds of electromagnetic radiation can you detect with your body?

28. What is ELF radiation?

29. Summarize the arguments for and against ELF radiation with regard to human health.

Questions

1. Compare an electromagnetic wave to a wave on a pond. What are the similarities and differences?

2. In what ways do sound waves differ from radio waves? In what ways are they similar?

3. Suppose a sound wave and a light wave have the same frequency. Which one has the longer wavelength?

4. White light is a combination of all frequencies of electromagnetic waves in the visible spectrum. In a vacuum, all frequencies of light travel at the same speed. Suppose for a moment that lower frequencies traveled slower than higher frequencies. Would a distant star look any different? Explain. If that star suddenly disappeared, what would be the color of the last light that you would see from the star?

5. Why would walking down a flight of stairs be very hazardous if our eyes detected only infrared light? (*Hint:* What does the amount of infrared light emitted by an object indicate?)

6. An object that looks white when exposed to sunlight reflects all colors of light. What does a white object look like when it is exposed to red light? What does a red object look like when it is exposed to blue light?

7. Compare the frequency, speed, and wavelength of microwaves versus visible light.

8. Compare the frequency, speed, and wavelength of radio waves versus ultraviolet light.

9. Which has more energy, visible light or ultraviolet light? What determines the energy of electromagnetic waves?

10. A person is just as likely to get sunburned on a cloudy day as on a sunny day. Does this evidence support the hypothesis that ultraviolet light, not visible light, causes sunburn? Explain.

11. What is the primary difference between a radio wave and a sound wave? What is the difference between a radio wave and a light wave?

12. If someone asked you to prove that electromagnetic waves can travel in a vacuum, what would you say?

13. What is the difference between a gamma ray and an infrared ray?

14. Your friend proposes an experiment to measure the speed of light using items she has collected in her garage. Why should you be suspicious that this experiment may not work?

Problems

1. Radio and TV transmissions are being emitted into space, so *Star Trek* episodes are streaming out into the universe. The nearest star is 9.5×10^{17} meters away. If civilized life exists on a planet near this star, how long will they have to wait for the next episode?

2. What is the frequency of the wave used by your favorite radio station? What is the wavelength of that station's radio waves? If the station is 50 km away, how long does it take for the radio waves to reach you from the station?

3. The FM radio band in most places goes from frequencies of about 88 to 108 MHz. How long are the wavelengths of the radiation at the extreme ends of this range?

4. The AM radio band in most places goes from frequencies of about 535 to 1610 KHz. How long are the wavelengths of the radiation at the extreme ends of this range?

5. What are the frequency and wavelength of a microwave from a typical microwave oven? Does this have any implications for how these ovens are constructed? Why or why not?

6. A. What is the range of wavelengths in nanometers that make up the following?

    a. red light   c. orange light

    b. green light   d. blue light

  B. What is the range of each of these in meters?

7. If an X ray has a wavelength of 5 nanometers, what is its frequency?

8. Which has greater energy, an X ray with a wavelength of 90 nm or one with a wavelength of 2 nm? Explain.

9. If the frequency of an electromagnetic wave is 10^6 Hz what is the wavelength in nanometers? In meters? What type of electromagnetic wave is this?

10. Repeat Problem 9 for an electromagnetic wave with a frequency of 10^{21} Hz.

Investigations

1. Visit a local hospital and see how many types of electromagnetic radiation are used on a regular basis. From radio waves to gamma waves, how are the distinctive characteristics of each portion of the electromagnetic spectrum used at the facility?

2. What frequencies of electromagnetic radiation, if any, do police, fire, and medivac personnel in your community use for emergency communications? What are the corresponding wavelengths of these signals? What organizations allocate and monitor these frequencies?

3. In large metropolitan areas, a license to broadcast electromagnetic waves at an AM frequency may change hands for millions of dollars.

  a. Why is electromagnetic "real estate" so valuable? Investigate how frequencies are divided up and who regulates the process. Should individuals or corporations be allowed to own portions of the spectrum or to buy and sell pieces of it?

  b. Currently the only portions of the electromagnetic spectrum that are regulated by national and international law are the longer wavelengths, including radio waves and microwaves. Why are the shorter wavelengths, including infrared radiation, visible light, ultraviolet radiation, and X rays, not similarly regulated?

4. Different colors represent different wavelengths of electromagnetic radiation. Investigate the process by which the human eye detects color, as well as the means by which the brain interprets color. Do all mammals see in color? How do we know?

5. Investigate which portions of the electromagnetic spectrum from sunlight reach the surface of Earth. What happens to the other wavelengths?

 ## WWW Resources

See the *Physics Matters* home page at **www.wiley.com/college/trefil** for valuable web links.

1. http://imagine.gsfc.nasa.gov/docs/science/know_l1/emspectrum.html A discussion and tutorial on the electromagnetic spectrum from NASA's education outreach resources.

2. http://lectureonline.cl.msu.edu/~mmp/applist/Spectrum/s.htm A Java applet illustrating the electromagnetic spectrum.

3. http://hamjudo.com/notes/cdrom.html Electromagnetic mayhem and experiments at home with your own microwave oven.

4. http://www.mcw.edu/gcrc/cop/static-fields-cancer-FAQ/toc.html Electromagnetic fields and human health issues from the Medical College of Wisconsin.

5. http://webphysics.ph.msstate.edu/javamirror/ntnujava/emWave/emWave.html An animation that describes the propagation of electromagnetic waves.

20 | Classical and Modern Optics

KEY IDEA

Mirrors, lenses, and other optical devices alter the paths of electromagnetic waves by scattering, transmission, and absorption.

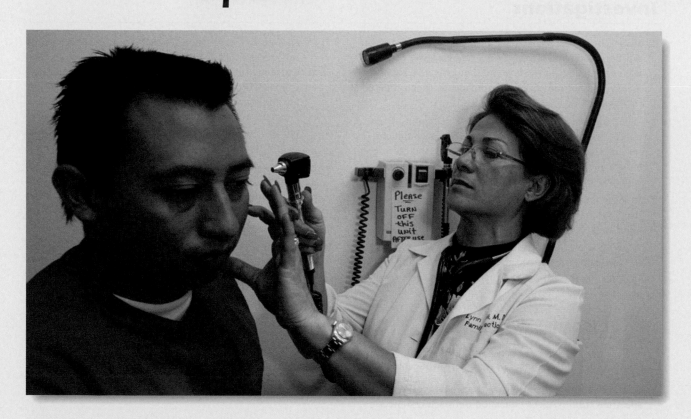

Physics Around Us . . . A Day in the Life

You pass your physical checkup easily this year. The X rays of the knee ligaments you tore last year show they have healed well. You won't need arthroscopic surgery after all. Examination of your blood under a microscope indicates you have a normal range of red and white cells. Your doctor puts on his reading glasses to sign your form, saying you're in excellent health. As you put on your clothes, you smile at yourself in the mirror; no health problems for you this year.

That night you celebrate the good news by heating up a pizza in the microwave and watching a movie on TV. The new satellite dish is working fine and reception is great. In the middle of the movie you get a call from a friend spending a semester abroad in Europe. Her voice is carried to you on cables laid across the floor of the Atlantic Ocean.

Many of the devices you've seen or used today—the X-ray machine, the arthroscopic surgery instrument, the microscope, the doctor's glasses, the mirror, the microwave, the satellite dish, the telephone cables, even your eyes themselves—involve applications of the principles of optics, both classical and modern.

ELECTROMAGNETIC WAVES AND MATTER

Interactions between electromagnetic waves and matter encompass some of the most important phenomena in physics and daily life. We rely on the entire electromagnetic spectrum, from radio waves to gamma rays, in countless ways—in cooking, communications, space exploration, medical diagnosis, surgery, and the full range of scientific and artistic pursuits (see Physics and Daily Life—Optics, page 420). Interactions between light and matter provide the foundation for what we know about the natural world. This chapter examines some of these interactions.

Recall some of the distinctive properties of electromagnetic waves. In a vacuum or in a uniform medium (and in the absence of a large mass), electromagnetic waves travel in straight lines. In a vacuum, all of these waves travel with the same speed, the *speed of light,* designated by *c.* As we have seen in Chapter 19, there is a reason for these sorts of similarities. All electromagnetic waves have basically the same structure (crossed electric and magnetic fields) and they differ from one another only in wavelength and frequency. This similarity means that although visible light is the most familiar electromagnetic wave, every device that we use to control the light we see (the lenses in eyeglasses, for example, or the mirror in your bathroom) has an analogous device to control other parts of the spectrum.

The challenge facing researchers is to devise means to produce, detect, and change the direction of electromagnetic waves in every part of the spectrum. Devices such as lenses and mirrors that are used to alter and control electromagnetic radiation are collectively called *optical devices* or **optics.** (We should note that the term *optics* is used to refer both to the field of physics devoted to the behavior of electromagnetic radiation and to the apparatus used to study and control that behavior.) Originally, the term *optics* was applied only to devices that work on visible light. That's hardly surprising since, for most of the history of science, light was the only electromagnetic wave known. Today, the term *optics* has been broadened to include the entire spectrum, so we can speak of "X-ray optics" or "microwave optics."

The interactions between electromagnetic waves and matter make it possible to build a wide variety of optical devices. Radiation interacts with matter in three principal ways:

1. The radiation can be scattered from the material's surface.

2. The radiation can be absorbed by the material.

3. The radiation can be transmitted through the material, often changing direction in the process.

All these processes are familiar from our everyday experience with light.

In describing the way that light and other kinds of electromagnetic radiation move through devices, it is often convenient to visualize the direction in which the wave is moving rather than the wave itself. For example, if, as shown in Figure 20-1, a series of crests and troughs are moving to the right, then we can represent the motion of the wave by following the path of a particular point on a wave crest, as shown. In this way, a single line replaces the entire crest and trough structure of the wave. The line that traces the motion of the wave is called a **ray.** In much of this chapter we are concerned with tracing rays of light and other kinds of electromagnetic radiation through various kinds of optical systems.

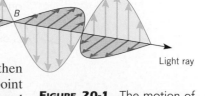

FIGURE 20-1. The motion of a light wave can be represented as a line that traces the motion of the wave, called a *ray.*

Our lives are surrounded by electromagnetic waves, from light to radio to infrared. The laws of **optics**—reflection, absorption, and refraction—are so common that we hardly notice them; they are part of the way things are. If you look around you, you can find them in operation everywhere.

Eyes absorb visible light

Food absorbs infrared radiation

Electric grills give off low-frequency radiation

Color images are created from pixels in computer screen

Telephone signals are transmitted by total internal reflection

The laser light is reflected, allowing the scanner to read the bar code

● SCATTERING

When waves encounter material bodies, they often scatter in many different directions. For example, when a water wave on a lake encounters a rock sticking out of the surface, the wave produces an outgoing circular wave centered on the rock. That circular wave, moving in a direction different from the incoming wave, is called the **scattered** wave.

In just the same way, when electromagnetic waves encounter obstacles, they scatter as well. Most familiar surfaces scatter light in many different directions—a process called *diffuse scattering* (Figure 20-2a). White surfaces are particularly efficient at scattering visible light diffusely. In the same way, the interior of your microwave oven is specially designed to scatter microwaves efficiently so that they reach all parts of the food being heated.

Waves scattered by a rock produce circular waves.

Perhaps the most important optical device that scatters light is the **mirror,** which relies on a scattering process known as **reflection.** When a beam of parallel light rays encounters a smooth mirrored surface, the scattered light rays are also parallel (Figure 20-2*b*). As shown in Figure 20-3, we define the *angle of incidence* as the angle that the direction of the incoming radiation makes with a line drawn perpendicular to the surface. Similarly, the *angle of reflection* is the angle that the direction of the reflected radiation makes with that same perpendicular. A simple rule relates these two angles:

The angle of incidence equals the angle of reflection.

For a beam of light encountering a smooth mirrored surface, this rule explains how it is that we see a reflected image when we look at the mirror. Light rays from neighboring points on an illuminated object (such as your face) travel in all directions. However, consider two rays from neighboring points on your face that take parallel paths to the mirror. If these parallel light rays strike the mirror at a particular angle, then they are reflected in parallel light rays at the same angle. Because the light rays travel in parallel bundles, when they arrive at your eye they allow you to see an undistorted image of the object being reflected. An ordinary flat mirror works this way.

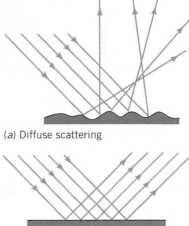

(*a*) Diffuse scattering

(*b*) Reflection

FIGURE 20-2. (*a*) Diffuse scattering of light rays off an irregular surface. (*b*) Reflection of light rays off a flat mirror surface.

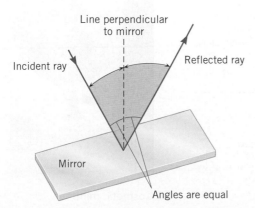

FIGURE 20-3. A light ray strikes a mirror at the angle of incidence, which is the angle that the direction of the incoming radiation makes with a line drawn perpendicular to the surface. Similarly, the angle of reflection is the angle that the direction of the reflected radiation makes with that same perpendicular. The angle of incidence equals the angle of reflection.

Develop Your Intuition: Mirror Image Reversal

Why is it that when you look at your face in the bathroom mirror, the image is reversed (that is, the right side of your face appears on the left side of the image, and vice versa)?

Because it is reflected, the light that comes to your eyes from the left side of your face appears to come from a point behind the right side of the mirror. The analogous situation occurs for light from the right side. Thus, the image appears reversed. Convince yourself of this by holding up your right hand in front of a mirror and seeing how its image appears.

We can learn several important lessons from the simple process of light's reflection from a mirror. When you look at a mirror image, there's nothing to tell your eye that it is not seeing light coming directly from the illuminated object, rather than being reflected from a mirror. Consequently, your brain interprets the incoming light signals as indicating the position of a real object. Thus, to you it looks as if the object is located behind the plane of the mirror. This tendency of

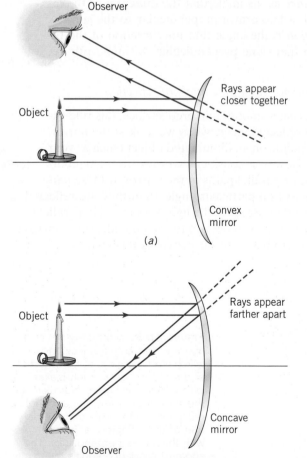

Store observation mirror

Cosmetic mirror

FIGURE 20-4. Curved mirror surfaces. (*a*) A *convex* mirror, which bows outward, makes an object appear smaller than it really is. This kind of mirror is often used in stores to survey large areas of the store from one place. (*b*) A *concave* mirror bows inward and makes an object appear larger than it really is. This kind of mirror is used for shaving mirrors and cosmetic mirrors.

the eye to see objects by tracing back along straight light rays plays an important role in understanding several optical illusions we discuss later in the chapter.

Some useful mirrors are not plane surfaces, but are curved in some way. For example, in Figure 20-4a we show a person in front of a *convex mirror*—one that bows outward. In this case, we can still consider rays from neighboring points on the object (the candle flame) that travel in parallel lines to the mirror, but they strike at points where the directions of the perpendiculars to the mirror surface are different, as shown. The angle of incidence, in other words, is different for the two rays and so are the angles of reflection. Once the rays have left the mirror, they no longer travel in parallel lines. When your eye receives these rays, it traces them back along the direction of travel and (mistakenly) assumes that the two neighboring points on the object are closer together than they really are. You see the object as being smaller than it really is. This kind of mirror is often used in stores to survey large areas of the store from one place.

A mirror curved the other way—a *concave mirror*—produces exactly the opposite effect. As shown in Figure 20-4b, someone standing close to the mirror sees adjacent points farther apart than they really are and hence sees an object as larger than it really is. This kind of mirror is used for shaving mirrors or cosmetic mirrors.

Some concave mirrors have another interesting property. If parallel rays of light fall on a concave mirror that is shaped like a parabola, all the reflected rays pass through a single point, as shown in Figure 20-5. The point at which all the rays are focused is called the *focal point* of the mirror, and the distance between the mirror and the focal point is called the *focal length*.

Parabolic mirrors are useful in many situations. It often happens that we have a weak signal—a faint light source, for example, or a weak microwave signal beaming down from a satellite. In this situation, we need a way to concentrate that signal and make it strong enough for our sensors to detect. If we let the signal fall on a parabolic mirror, all the radiation that falls on the mirror is brought to the focal point and the signal at that point is much stronger than the unreflected signal.

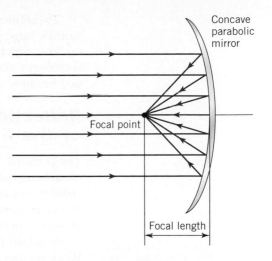

FIGURE 20-5. A concave parabolic mirror focuses incoming parallel rays to a focal point. The distance between the mirror and the focal point is the focal length.

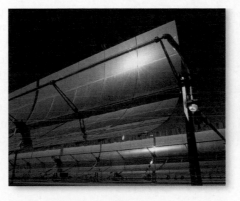

(a)   (b)   (c)

A variety of devices employ parabolic mirrors, including (a) radio telescopes, (b) solar energy collectors, and (c) satellite TV receivers.

Parabolic mirrors are the basis for many important devices. Most of the world's large astronomical telescopes, for example, are built around parabolic mirrors, which collect and focus the light of distant stars and galaxies. The astronomers' slang for these telescopes—"light buckets"—tells us that they are valued because they can collect large amounts of light and focus it.

Reflection in Other Parts of the Electromagnetic Spectrum

The principle of scattering for visible light can be applied to other parts of the electromagnetic spectrum as well. For example, radio telescopes (which monitor radio waves emitted by distant sources) also rely on parabolic dishes that reflect and concentrate radio waves at the focal point, where a receiver measures them. However, in this case the radio waves are reflected from a mirror made of metal.

Satellite dishes, which detect microwaves, operate on the same principle. Weak microwave signals sent from satellites are reflected by the parabolic surface of the dish and focused at one point. Next time you see one of these dishes, look for the receiver suspended above the dish, precisely at the focal point.

Infrared and ultraviolet radiation can also be reflected off mirrorlike surfaces. In fact, you may have noticed a curved metal reflector at the back of a space heater.

Connection
Active Optics

Large telescope mirrors capable of collecting large amounts of light pose special technical problems for engineers. For one thing, a mirror made from a single block of glass is very heavy. Not only does this weight make it difficult to move, but gravity causes the glass to sag, destroying the parabolic shape. Other effects, such as changing temperature and wind buffeting, also make the use of big mirrors difficult.

One way around these sorts of problems, exemplified by the 10-meter Keck telescopes in Hawaii, is called *active optics*. The mirrors of these telescopes are not made from solid blocks of glass, but are actually 36 interlocking hexagonal pieces, each part of a large parabola. The pieces are mounted on hydraulic supports that can change the position of the mirror. Twice each second, sensors around the segments report to a computer about the segment's position, and each segment is adjusted to compensate for distortions caused by wind, temperature, and gravity. In this way, the mirror retains its parabolic shape, even when the telescope is moving.

All of the largest modern land-based telescopes now use similar technology. ●

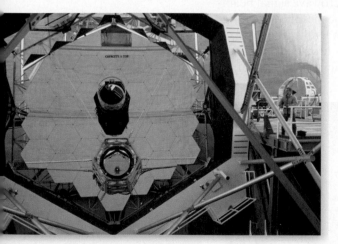

The Keck Telescopes in Hawaii employ multiple mirrors.

Fiber Optics

When light moves from a more dense to a less dense medium (from glass to air, for example), the ray bends away from a line perpendicular to the boundary surface. This is a process called *refraction*. If the light approaches the surface at a large enough angle, this bending can make light skim along a material's surface,

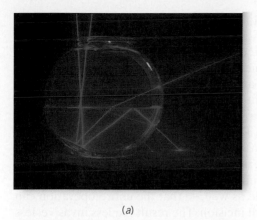

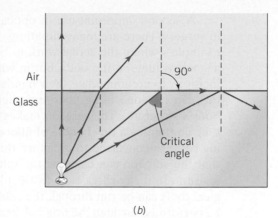

(a)   (b)

FIGURE 20-6. Total internal reflection occurs when the angle of incidence becomes larger than a critical value. In this case, the light is not able to leave the material but is reflected back inward.

as shown in Figure 20-6. If the angle of incidence becomes larger than this critical value, the light is not able to leave the material but is reflected back inward. This phenomenon, known as *total internal reflection,* is the basis for modern fiber-optic technology.

The basic working principle of **fiber optics,** shown in Figure 20-7, is that a beam of light enters a long glass fiber at one end, but traveling at such an angle that each time it reaches the glass–air surface, it is reflected back into the glass. Thus, whatever enters one end of the fiber comes out the other end.

Today, in the most efficient optical fibers, the density of the glass fiber is made to vary from the long central axis to the outside so that the light actually follows the kind of wavy path shown in Figure 20-7a. In this way optical fibers can transmit almost 100% of their light from one end to the other.

Fiber optics finds critical uses in many modern technologies. As we have seen in Chapter 19, the wavelength of visible light is quite short—only a few hundred billionths of a meter. This short wavelength makes it possible to pack a lot more information into a light signal of a given length than into much longer radio waves or waves traveling through copper wires. Consequently, if we send signals via light waves in optical fibers, we can pack a lot more information into each fiber and carry a much heavier load. A typical copper wire, for example, can transfer tens of thousands of phone conversations, while an optical fiber can carry upward of half a million. The first commercial optical-fiber phone line was installed in downtown Chicago in 1977; today the majority of long-distance phone calls are transmitted in this way.

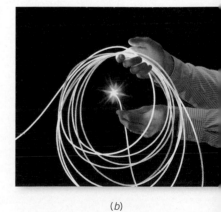

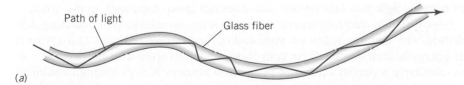

(a)

FIGURE 20-7. (a) In an optical fiber, a beam of light enters a long glass fiber at one end, traveling at such an angle that each time it reaches the glass–air surface, it is reflected back into the glass. In modern optical fibers, the material is designed in such a way the the index of refraction of the glass is different near the edge than it is at the center. The light is bent gently as it approaches the edge rather than being reflected. Optical fibers can reflect almost 100% of a light beam from one end to the other. (b) Optical fibers used in microsurgery.

A second important use of optical fibers has been in medicine, particularly in surgery. There are many situations—removing torn cartilage from a joint, for example—where the actual work to be done in the surgical procedure involves a small volume in the body, but in which surgeons used to have to make large incisions to accommodate their hands and scalpels. Today, microsurgery such as this can be done using fiber-optic light sources to illuminate and observe the body's structures. The surgeon makes a small incision in the body and inserts a tube containing a tiny bundle of fibers. The light traveling out through the twisting cable is converted to a picture that can be shown on a TV-like monitor, so that the surgeon can examine the interior of the body without making a large incision. In many cases, such as the joint surgery already mentioned, small surgical tools can be run through the cable as well, so that the entire operation can be performed through the original small incision. The result is a less invasive, less traumatic experience for the patient and usually leads to a much more rapid recovery than conventional surgery. So common has this sort of procedure become that it is often referred to by the slang term "Band-Aid surgery" because after it is over only an ordinary Band-Aid is needed to cover the wound.

ABSORPTION AND TRANSMISSION

When electromagnetic waves travel through matter, they do not behave in the same way as they do in a vacuum. The waves, after all, consist of electric and magnetic fields, and matter consists of atoms made from electrically charged particles such as electrons. These electrons can absorb energy from the wave—for example, the wave's electric fields can accelerate the electrons. Depending on how the atom is put together and the way in which it is bonded to other atoms in the material, several things can happen.

The energy that the electron absorbs from the wave may be converted to many different forms, since all forms of energy are interchangeable (see Chapter 12). For example, the energy from the wave might eventually find its way into kinetic energy of atoms in the material. In this case, the electric and magnetic energy in the wave is reduced and the thermal energy in the material increases. The intensity of the electromagnetic wave diminishes as it passes through the material and the temperature of the material increases.

In some materials, all the energy in the wave is converted into thermal energy or other forms of energy and the wave simply disappears. In this case, we call the process **absorption.** Materials that absorb or scatter electromagnetic radiation that falls on them are said to be **opaque**—that is, they are materials that do not allow the radiation to pass through them. A sheet of metal, for example, is opaque to light because light that falls on one side does not make it through to the other.

Absorption of electromagnetic radiation is an essential step in detecting that radiation. The metal antenna on your radio, for example, absorbs radio waves; that energy is then amplified by your radio and converted into the sounds you hear. Similarly, a dentist's piece of X-ray film absorbs X rays through chemical reactions that ultimately produce lighter and darker regions on the film. Those light and dark areas correspond to parts of your teeth that absorb greater and lesser numbers of X rays and thus reveal cavities. But you don't need to rely on modern technology to make this point. Go outside on a sunny day and you'll find that your skin is an excellent absorber of both infrared radiation (which you feel as heat) and ultraviolet radiation (which gives you a sunburn).

Electromagnetic waves can undergo a very different fate than absorption when encountering matter. It is possible that the wave, even though it loses energy, actually emerges from the other side of the material. In this case we call the process **transmission** and say that the material is **transparent.** A pane of ordinary window glass is an example of a material that is transparent to visible light because light from the outside easily passes through so that you can see what is on the other side.

Think about everyday phenomena that use electromagnetic radiation and you'll realize that many materials must be transparent to these waves. For example, your radio works inside your house or school; that means that walls, floors, carpeting, and windows must be transparent to radio waves. Earth's atmosphere is transparent to a wide range of electromagnetic radiation, including most wavelengths of radio waves, microwaves, and visible light. Fortunately for us, the atmosphere effectively absorbs most harmful ultraviolet radiation and X rays.

Having made the distinction between transparent and opaque materials, however, we have to make a couple of points:

1. It is possible for a material to be transparent for one wavelength of radiation, but opaque for another. Ordinary window glass, for example, readily transmits radiation at the wavelengths of visible light but is opaque to radiation in the infrared. This effect explains the operation of a greenhouse, by which the Sun's radiation warms the interior of a greenhouse, which then emits infrared radiation. Much of that heat energy is prevented from escaping quickly back into space because the glass is opaque to that wavelength. A similar phenomenon is the basis for the greenhouse effect (see Chapter 11).

2. Many intermediate situations are possible between complete transparency and complete opacity. A pane of clean, high-quality window glass is almost completely transparent, but when it is dirty more of the incoming light is absorbed. It's still more or less transparent, but not as transparent as it was. This fact, after all, is why we routinely clean the windshields of our cars. Thus, it's better to think of transparent and opaque as two extremes on a continuum, with most materials falling somewhere in between.

● REFRACTION

When light is transmitted through a material substance, its path and speed may change in significant ways. These changes in direction and speed are called **refraction,** and they form the basis of countless optical devices, including eyeglasses, cameras, microscopes, and binoculars.

Because of the fact that electromagnetic radiation interacts with atoms when it passes through different materials, the radiation moves through these materials more slowly than it does in a vacuum. An analogy (although not an exact one) may help you think about why an electromagnetic wave travels more slowly in matter than in a vacuum. Imagine that two travelers arrive at an airport on the East Coast of the United States, both bound for the same airport on the West Coast. One takes a direct cross-country flight; the other changes planes, first in Chicago and then in Denver, before coming to his destination. While both travelers move at exactly the same speed while they are in the air, the one on the direct flight clearly arrives at the destination more quickly. An outside observer would therefore say that the traveler on the direct flight was moving faster (had a higher velocity) than the one who changed planes.

In just the same way, the individual waves discussed here all move at the same speed as they would in a vacuum between atoms, but the net effect of the interference process is, on average, to slow the wave down.

Index of Refraction

If the speed of an electromagnetic wave in a particular material is v, then the **index of refraction,** n, of that material is given by the following expression:

1. In words:

> *The index of refraction of a material is the ratio of the speed of the wave in a vacuum divided by the speed of the wave in that material.*

2. In an equation with words:

$$\text{Index of refraction} = \frac{\text{Speed in vacuum}}{\text{Speed in material}}$$

3. In an equation with symbols:

$$n = \frac{c}{v}$$

Since light always travels more slowly in materials than in a vacuum, the index of refraction is always a number greater than 1. A few typical values are given in the following table.

Material	n
Air	1.0003
Water	1.33
Ethyl alcohol	1.36
Crown glass	1.52
Table salt	1.53

EXAMPLE
20-1

Light in the Water

How fast does light travel in water?

SOLUTION: In a vacuum, light travels at a speed of 3×10^8 m/s. In water, the index of refraction is 1.33, as listed in the table. Then the definition of the index of refraction tells us that

$$1.33 = \frac{3 \times 10^8 \text{ m/s}}{v}$$

where v is the speed of light in water. This means that

$$v = \frac{3 \times 10^8 \text{ m/s}}{1.33}$$

$$= 2.25 \times 10^8 \text{ m/s}$$

This value is three-fourths of light's speed in a vacuum. ●

The fact that radiation travels at different speeds in different materials means that the direction of a wave's motion changes when it passes through a boundary between one material and another. This effect is the basis for refraction.

Another analogy may help you understand refraction. Imagine horses racing across an open meadow, with each horse running at exactly the same speed, so that they stay neck and neck as they move. You can think of the line of horses as marking the crest of a wave.

Now suppose that, as in Figure 20-8, there is a marsh along one edge of the field—a marsh in which the horses get bogged down and move more slowly. As each horse enters the marsh, it slows down, while the horses in the open field keep running at their original pace. The result, as shown, is that the line wheels around as more and more horses enter the marsh. Eventually, the horses are neck and neck again, but the line is moving in a different direction.

In just the same way, an electromagnetic wave entering a medium with a high index of refraction wheels around and moves in a different direction. The easiest way to picture the behavior of such a wave is to imagine each bit of the wavefront emitting its own little wave, with these waves undergoing interference to reconstruct the wave front farther along. As shown in Figure 20-9, while the wave is in a medium with a low index of refraction (i.e., a medium in which the speed of light is high), the wavefront moves forward because of the interference of the wavelets. When the wave reaches the second material, however, the wavelets generated by that part of the wavefront travel more slowly. Just like the line of horses encountering the marsh, the wavefront wheels around and changes direction.

If we trace the ray corresponding to the wave encountering the boundary, we get a diagram such as that shown in Figure 20-9b. We can make two qualitative statements about the behavior of radiation at a boundary:

1. When a wave moves from a medium with a low index of refraction to one with a high index of refraction, its direction of motion moves closer to a line perpendicular to the surface.

2. When a wave moves from a medium with a high index of refraction to one with a low index of refraction, its direction of motion moves farther from a line perpendicular to the surface.

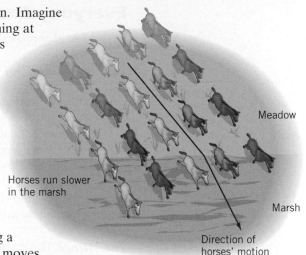

Horses run slower in the marsh

Meadow

Marsh

Direction of horses' motion

FIGURE 20-8. A line of horses traveling from a field to a marsh appears to change direction as the horses entering the marsh slow down. The same phenomenon is demonstrated by light rays traveling from a material of lower to higher index of refraction.

(a)

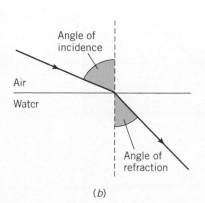

Angle of incidence

Air

Water

Angle of refraction

(b)

FIGURE 20-9. A light ray bends at the boundary of materials with higher and lower refractive indices.

Everyday Examples of Refraction

Let's look at three everyday examples of refraction.

The Swimming Pool You may have had the experience of standing next to a swimming pool and looking at a friend standing in the water. In this situation, you may have noticed that her legs appear to be shorter than they really are. This optical illusion is a consequence of refraction.

As shown in Figure 20-10, a light ray from the bottom of your friend's feet will be bent away from the perpendicular as it moves from water (high index of refraction) to air (low index of refraction). When this ray enters your eye, your brain assumes that the actual location of her foot can be obtained by tracing back along the line of the ray. Thus, to you, it appears that her feet are higher than they actually are.

The Mirage Have you ever had the experience of driving along a highway on a hot summer day and seeing a stretch of highway ahead of you that looks as if it were covered with water? If you have, you know that when you actually get to the "wet" spot, it turns out to be ordinary dry highway, while the "wet" spot has moved farther ahead. This phenomenon is an example of a *mirage*.

A mirage works like this: The air near the highway is heated and has a lower index of refraction than the cooler air higher up. Thus, light entering this air is bent away from the perpendicular, as shown in Figure 20-11. Because the air gets hotter the closer it is to the ground, this bending is a continuous process, since the light enters layers with successively lower indices of refraction.

Eventually, the direction of the incoming light is turned completely around,

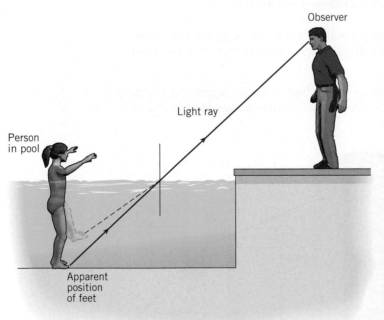

FIGURE 20-10. A light ray from the bottom of a swimming pool is bent away from the perpendicular as it moves from water (high index of refraction) to air (low index of refraction). To you it appears that the pool is shallower than it actually is.

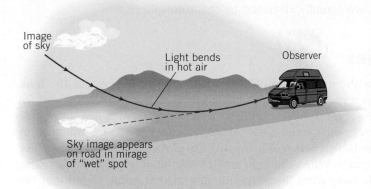

FIGURE 20-11. A mirage occurs when air near the ground is heated and has a lower index of refraction than the cooler air higher up. Thus, light entering this air is bent away from the perpendicular. What you see is light from the sky whose direction has been changed by refraction.

as shown. When the light rays enter your eye, you trace them backward and think that they are coming from a spot on the highway ahead of you. In actuality, you are seeing light from the sky—light whose direction has been changed by refraction.

Develop Your Intuition: A Green Mirage

If you are driving toward a forested mountain, the highway mirage may look green. Where do you suppose that light originated?

The green light originated on the mountainside and was refracted up to your eye. The green you see is actually the color of the leaves on the trees.

Twinkling Stars The fact that the index of refraction of air changes with temperature also explains another common experience, the twinkling of stars at night. Stars are very far away, so they can be thought of as point sources of light (because even though the stars are big, the distance to the stars is very much larger than the size of any star). As a ray of light makes its way through the atmosphere, it encounters currents of air of different temperatures and is refracted this way and that. Thus, the ray that finally comes to your eye in one instant has followed a circuitous course, as shown in Figure 20-12, and the star appears to be at the position labeled A. A ray coming through a split second later encounters a different pattern of air temperatures, however, so it follows a slightly different path and appears to be at the nearby position B. As a result of the movement of the atmosphere, then, the position of the star seems to shift around—a phenomenon that the eye interprets as twinkling.

Needless to say, twinkling has always been a problem for astronomers, and it explains why they put so many major observatories above the atmosphere, where refraction will not take place. The Hubble Space Telescope is an excellent example of this approach. Located over 100 miles above Earth's surface, the Hubble Space Telescope intercepts light rays before they have a chance to be refracted by the atmosphere. (Other satellite observatories, such as those for X rays

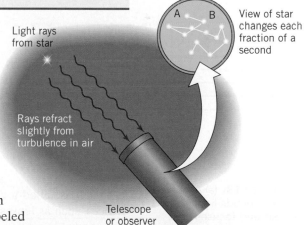

FIGURE 20-12. The twinkling of stars at night arises from the fact that the index of refraction of air changes with temperature. As a ray of light makes its way through the atmosphere, it encounters currents of air of different temperatures and is refracted this way and that. Your eye interprets this phenomenon as twinkling.

The Hubble Space Telescope.

and infrared radiation, are placed in orbit because those kinds of radiation are actually absorbed by the atmosphere.)

Lenses

Perhaps the most common and familiar optical device is the **lens,** which is a piece of transparent material designed to bend and focus light (Figure 20-13a). When a light ray from a distant source encounters the curved glass surface, it undergoes refraction and, as it enters the glass, it bends toward the perpendicular to the surface at that point. When the light ray leaves the glass on the other side, it bends away from the perpendicular at that point. (Note that because of the curved glass, the perpendiculars at the entry and exit points do not go in the same direction.)

An observer standing on the other side of the lens sees the light ray coming at an angle, as shown, and assumes that the light originated at a point above the top of the actual object. Thus, the observer sees the object as larger than it actually is. This basic principle is employed in numerous devices, including magnifying glasses, eyeglasses, and binoculars.

A lens like the one shown, which bends light rays toward its axis, is called a *converging lens.* If parallel rays of light from a distant source fall on the lens, then all the light is brought together at a single point, as shown in Figure 20-13a. The point at which all the light rays come together is called the *focal point* of the lens, and the distance between the focal point and the lens is called its *focal length.*

Another kind of device, shown in Figure 20-13b, is called a *diverging lens.* Rays of light entering this sort of lens are bent away from the central axis, so that someone looking through the lens sees what appears to be an object in back of the lens, as shown. An image such as this, where light rays do not actually originate on the object itself, is called a *virtual image.* A virtual image cannot be affected by other devices, such as lenses, farther along in the optical system.

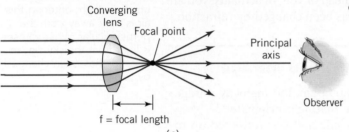

(a)

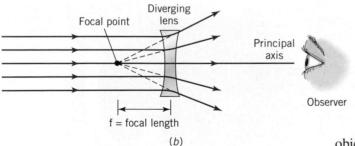

(b)

FIGURE 20-13. (a) A converging lens bends light toward its axis and focuses light to a focal point; the distance between the focal point and the lens is its focal length. (b) A diverging lens bends light away from the axis, so that someone looking through the lens sees what appears to be an object in back of the lens. Light rays moving along the axis in the center of the lens are not bent.

Connection
Telescopes and Microscopes

Lenses occur in many important scientific instruments. In Figure 20-14, we sketch two kinds of telescopes. In Figure 20-14a, we see the sort of parabolic collecting mirror we have already discussed. Telescopes like this, based on the principle of reflection, are called *reflecting telescopes.* Another design, shown in Figure 20-14b, focuses light through a lens—an arrangement known as a *refracting telescope.*

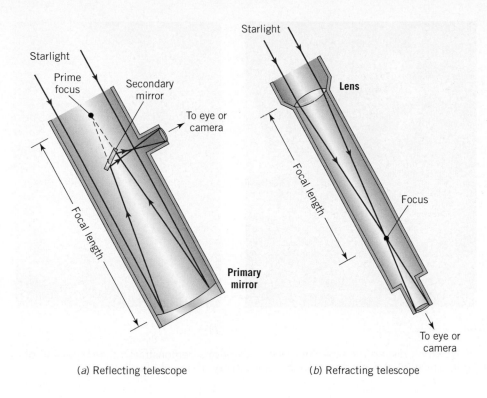

FIGURE 20-14. Two kinds of telescopes. (*a*) Telescopes based on the principle of reflection are reflecting telescopes. (*b*) Telescopes that focus light through lenses are refracting telescopes.

Because it is easier to build large mirrors than large lenses, the biggest astronomical telescopes—those intended to capture as much light as possible—are designed as reflectors. On the other hand, many small telescopes used by amateur astronomers and bird watchers are refractors.

The *microscope* is a device designed to magnify images. In a classical optical microscope such as the one shown in Figure 20-15, a strong light beam shines through a thin sample—a slide with biological tissue on it, for example, or a thin slice of rock—and is then focused through a series of converging lenses. The result is that the image is magnified, so that objects much too small to be seen with the naked eye become visible. ●

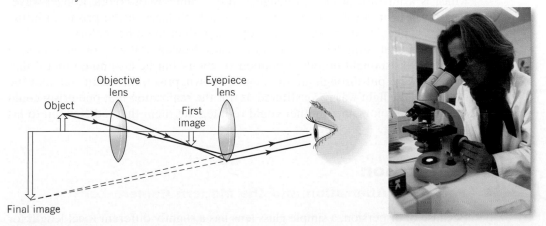

FIGURE 20-15. An optical microscope employs a sequence of lenses to magnify small objects.

(a)　　　　　　　　　　　　　　　(b)

(a) Newton's dispersion experiment with two prisms demonstrated the dispersion of light, which (b) also produces the brilliant display of a rainbow.

Dispersion

Thus far we have treated the index of refraction as a single number for a given material that is applicable to all wavelengths. If this were true, the world would be a far less colorful place than it is. Numerous optical phenomena, including majestic rainbows and the brilliant sparkle of diamonds, are the result of a process called *dispersion*.

In general, the index of refraction of materials does depend on the wavelength of the radiation. When visible light enters glass, for example, long wavelengths are bent less than short ones. This phenomenon, known as **dispersion,** explains why a glass prism or raindrops in the atmosphere can break up sunlight, which is a mixture of all wavelengths, into a rainbow of colors. Longer-wavelength red light, for example, is bent least at both faces of the prism, so in the end it has been bent through a different angle than the other colors.

It's an interesting historical fact that Isaac Newton not only observed that a prism breaks sunlight up into a rainbow of colors, but he also noted that if that rainbow were put through another, upside-down, prism, the result was that the original white light was reconstituted. It was the realization that one prism could separate the colors and another could re-integrate them that led Newton to his theory of color.

Connection
Chromatic Aberration and the Modern Camera Lens

Because of dispersion, a simple glass lens has a slightly different focal length for different colors. In practical terms, this difference means that if such a lens were put into a camera, you would see halos of different colors surrounding every object in a picture. This unwanted effect is known as *chromatic aberration.*

Multi-Layer Diffractive Optical Element (Conceptual Diagram)

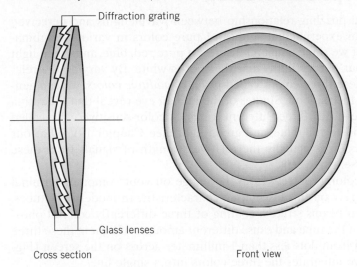

Cross section   Front view

Correction of Chromatic Aberrations by the Multi-Layer Diffractive Optical Element

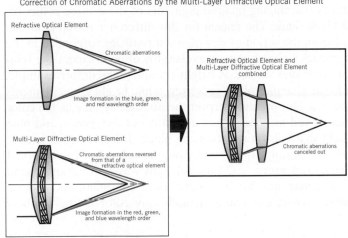

FIGURE 20-16. In modern cameras, chromatic aberration is corrected by placing a second lens made of a different glass behind the primary lens. A good camera lens today may have up to six nested lenses to correct for this and other kinds of aberrations.

In modern cameras, chromatic aberration is corrected by placing a second lens made of a different glass behind the primary lens, as shown in Figure 20-16. In fact, a good camera lens today may have up to six nested lenses to correct for this and other kinds of aberrations. ●

 # A WORLD OF COLORS

Color is one of the most familiar, yet most complex, optical phenomena. As we have seen in Chapter 19, the visible portion of the electromagnetic spectrum can be described simply in terms of the wavelength of the electromagnetic wave. The longest wavelengths correspond to red light, whereas the shortest wavelengths correspond to violet. All visible light consists of some combination of these wavelengths. However, it is important to realize that the phenomena our eyes and brains perceive as colors are related to wavelengths of light in a rather indirect way.

Human Perception of Color

The remarkable and puzzling relationship between wavelengths and perceived colors emerged from experimental studies of pure colors in various combinations. For example, if we shine equal intensities of pure red, blue, and green light on the same spot, our perception is that the spot is white. By varying the relative intensities of these three colors, called *primary additive colors,* we can generate every color of the rainbow. Studies of the human eye reveal that the retina in the back of the eye has three different kinds of color-sensitive cells, called "cones," that respond to red, blue, and green light (see Chapter 19). What our minds interpret as color is actually the relative strength of signals from these three kinds of light-sensitive cells.

Many familiar colored objects, from the image on your computer terminal to the comic pages, rely on this fascinating characteristic. In modern TV tubes, for example, electron beams strike a coating of three differently colored phosphors on the inside of the tube and cause different amounts of light in these three colors to be emitted from dots less than 1 millimeter across on the screen (Figure 20-17a). Your eye integrates the three colors into a single hue.

Although beams of red, green, and blue light can combine to produce white light, every painter knows that mixing opaque pigments of these three colors results in a dark, almost black shade. The reason for this difference is that paint acquires its brilliant color from the action of tiny pigment particles suspended in the liquid. These pigment particles absorb most wavelengths of light except the relatively narrow range of color that we see. A bright orange pigment, for example, typically absorbs light in the blue part of the spectrum, while bright green pigments absorb efficiently in the red. Our brains interpret the absence of blue as the color orange and the absence of red as green. Thus, a mixture of red, green, and blue paint absorbs all the light that falls on it, producing black, at least in principle.

It is possible to produce a range of colors by mixing pigments. The colors of paint that play the role of red, green, and blue for light waves are cyan (a kind of turquoise), yellow, and magenta. Mixing appropriate combinations of these three primary subtractive colors can make virtually any color of the rainbow (Figure 20-17b).

Connection

Four-Color Printing

A book such as this one, with many color pictures in it, is printed by a process called four-color printing. You might ask why we need four colors to make the pictures in a book or magazine when there are only three primary subtractive colors. The reason is a practical one. Although mixing the three colors together, as previously described, produces black in principle, in practice it is very difficult to get a pleasing shade of black in this way. Consequently, a fourth color, black, is added to the palette to produce color pictures that are pleasing to the eye. ●

The Physics of Color

All kinds of interactions between light and matter—scattering, transmission, and absorption—depend on the wavelength of the electromagnetic radiation. For example, a piece of green stained glass preferentially absorbs red wavelengths but is transparent to green wavelengths. Similarly, your favorite red sweater absorbs

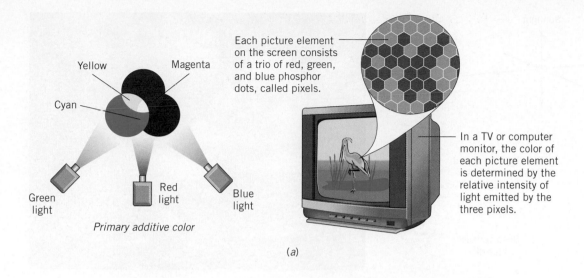

Each picture element on the screen consists of a trio of red, green, and blue phosphor dots, called pixels.

In a TV or computer monitor, the color of each picture element is determined by the relative intensity of light emitted by the three pixels.

Yellow Magenta

Cyan

Green light   Red light   Blue light

Primary additive color

(a)

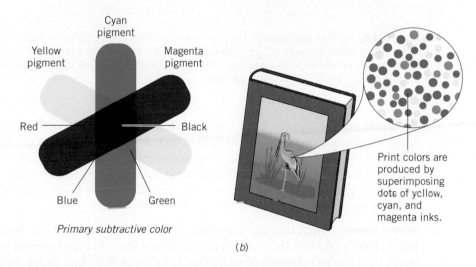

Cyan pigment

Yellow pigment   Magenta pigment

Red — — Black

Blue   Green

Primary subtractive color

Print colors are produced by superimposing dots of yellow, cyan, and magenta inks.

(b)

FIGURE 20-17. (a) The colors of many objects, including television screens, are produced by combining dots of the three primary additive colors—red, green, and blue. (b) Color printing combines dots of the three primary subtractive colors—cyan, yellow, and magenta.

green light but scatters red light. Every colored object absorbs, transmits, and scatters different wavelengths of light differently.

Let's consider three of the most familiar examples—blue sky, white clouds, and red sunsets—all of which are governed by the scattering of white light from the Sun. Two general rules govern the scattering of electromagnetic waves. First, if the object causing the scattering is much smaller than the wavelength of the radiation, then shorter wavelengths are scattered much more strongly than longer ones. Second, if the object causing the scattering is much larger than the wavelength of the radiation, then all wavelengths are scattered equally.

1. Blue sky Why does the daytime sky appear blue? As shown in Figure 20-18a, when you look at the sky in a direction away from the Sun, what you

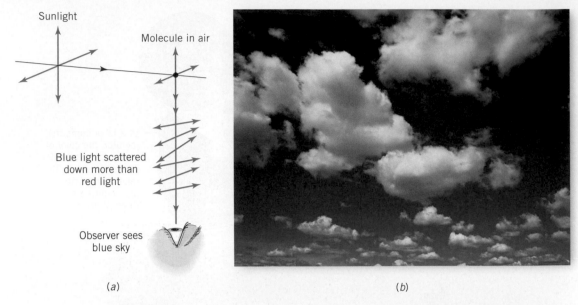

(a)

(b)

FIGURE 20-18. (a) The sky appears blue because short wavelengths (blue light) are scattered much more strongly off air molecules than are longer ones (red light). Thus, the light we see when we look away from the sun is blue. (b) Clouds appear white because they scatter all wavelengths of light.

are seeing is light that has been scattered from molecules in the atmosphere. Since the size of molecules (tenths of nanometers) is much less than the wavelength of visible light (hundreds of nanometers), we expect that the short wavelengths (blue light) are scattered much more strongly than longer ones (red light). Thus, the light we see when we look in this direction is blue, which explains the color of the sky.

2. **White clouds** When sunlight encounters clouds, it scatters from droplets of water (Figure 20-18b). These droplets have dimensions that are typically much larger than the wavelengths of visible light, although their size varies. The net result is that light of all colors is scattered equally from the cloud and the clouds appear white.

3. **Red sunsets** Light from the Sun is white—a collection of all visible wavelengths. As that light comes through the atmosphere, the blue light is scattered out of the beam (making the sky appear blue). What remains is light that is predominantly yellow, which explains the daytime appearance of the Sun. Toward sunset, however, the light from the setting Sun has to travel through much more of the atmosphere (Figure 20-19), so more scattering occurs. Once the blue has been removed from the beam, the yellow and green follow, leaving only the red light in the beam. This gradual filtering explains the appearance of the Sun at sunset.

You can see an interesting application of the rules for scattering by looking at the sky above a large city. Typically, the sky directly overhead looks blue, but near the horizon it is often a hazy white. The reason for this is that the sky above the city is normally full of tiny particles of dust and other material. Light

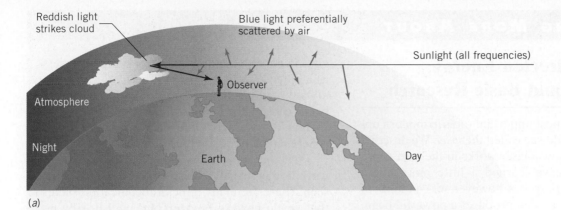

(a)

(b)

FIGURE 20-19. (a) Light from the Sun is white, but blue light is scattered out of the beam (making the sky appear blue). Toward sunset, however, the light from the Sun has to travel through much more of the atmosphere, so more scattering occurs. Once the blue has been removed from the beam, the yellow and green follow, leaving only the red light in the beam. (b) This gradual filtering explains the appearance of the Sun at sunset.

coming from the horizon, then, has travelled a long distance through these particles. It is a mixture of the normal blue light coming from atmospheric scattering and the many wavelengths resulting from scattering by the particles. As a result, the normal blue is washed out, and the sky appears a pale white.

THINKING MORE ABOUT

Optics: Directed Energy Weapons and Basic Research

One of the most important tools in modern optics is the device called the *laser.* We describe the details of how a laser works in the next chapter, after we have learned a little more about atoms. For our purposes, however, we can think of a laser as a device that produces a powerful beam of electromagnetic radiation, usually in the form of visible light.

You have undoubtedly seen lasers in use. They are often used as pointers in classroom lectures, for example—they produce a red spot on the screen—and they are used extensively to produce visual effects at rock concerts. They are also widely used as surgical tools in medicine, as we see in Chapter 21, and as cutting tools in industry, to name a few modern applications.

They also have a presence, although less benign, in science fiction. The phaser in the *Star Trek* series, for example, is a weapon modeled on the laser, as are countless clones in other movies and TV shows. The development of the laser as a weapon, however, is not just fiction. It has played a very real role in modern military research.

The basic idea behind what are called directed energy weapons is simple. An intense beam of electromagnetic radiation is directed toward a target, which absorbs the energy of the beam. If the beam is powerful enough, the target will be weakened—a metal target, for example, might actually melt. For this reason, some scientists and engineers have thought about using powerful lasers as a defense against missiles. The idea is to "illuminate" an incoming warhead with an intense beam, causing the metal casing to weaken enough so that the warhead will burn up in the atmosphere, much like a meteor.

Whether or not such a system can be made to work in a practical defense system remains to be seen. There are many difficult technical problems that would have to be overcome, and it is by no means clear that this would be the best system to develop. Nevertheless, the discussion about directed energy weapons illustrates an important point about basic research in physics and other sciences. When people at universities in the United States were developing the first lasers, they had no idea that these devices might someday improve surgical techniques in the nation's hospitals, much less be developed as weapons. In fact, one of the authors (JT) remembers hearing Arthur Schawlow, who shared the Nobel Prize for developing the laser, speculating that, possibly, it could be used to make a better device for erasing letters typed on paper (people still used typewriters rather than word processors in those days). He thought of his device primarily as a tool to do basic research in atomic physics, nothing more.

Every major discovery has applications far beyond what can be imagined at the beginning. Do you think scientists should consider what those uses might be before they begin research? Should there be government laws and regulations in this area? Why or why not?

Summary

Optics is the study of light and its interactions with matter. When electromagnetic radiation encounters matter, it can be **scattered, absorbed,** or **transmitted.** A **mirror** is an optical device that uses **reflection** to change the direction of light rays. Radio telescopes, radar receivers, satellite dishes, and X-ray telescopes are all examples of instruments that work by reflection.

A material that absorbs all the radiation of a given wavelength that falls on it is said to be **opaque,** while one that absorbs little of that radiation is said to be **transparent.**

The ratio between the speed of light in a vacuum and its speed in a material is called the **index of refraction** of that material. When radiation passes a boundary between different materials, its direction changes. The ray moves closer to the perpendicular to the boundary if the wave is moving from a medium of low to high index of refraction and away from it when moving from a medium of low to high index of refraction. This bending effect is called **refraction.**

The laws of refraction govern the operation of **lenses** and of optical instruments such as microscopes and some telescopes. This phenomenon is the basis of **fiber optics,** a technology that is revolutionizing both communications and medicine.

Different wavelengths of radiation often have slightly different indices of refraction in materials, a phenomenon known as **dispersion.** Dispersion causes a prism to split white light into its constituent colors.

The color of a single ray of light is determined by its wavelength. All colors can be made from combinations of red, green, and blue light, which are the three primary additive colors. All colors can also be made by mixing cyan, magenta, and yellow pigments, which are the three primary subtractive colors.

Key Terms

absorption The conversion of electromagnetic wave energy into thermal energy, resulting in a reduction (partial or complete) of the wave strength. (p. 426)

dispersion The phenomenon that different wavelengths of light refract different amounts when entering a medium. (p. 434)

fiber optics A technology that uses long and thin glass fibers to carry light great distances using the principle of total internal reflection. (p. 425)

index of refraction Defined for a specific material, the ratio of the speed of light in a vacuum to the speed of light in that material; it is a measure of how much light slows down and bends as it enters the material. (p. 428)

lens A piece of transparent material designed to bend and focus light. (p. 432)

mirror A device that scatters light by reflection. (p. 421)

opaque materials Materials that absorb or scatter electromagnetic radiation. (p. 426)

optics The branch of physics dealing with the manipulation and analysis of electromagnetic waves. (p. 419)

ray The path taken by a beam of electromagnetic radiation, a line drawn perpendicular to the wavefront. (p. 419)

reflection The return of light from a surface on which it falls. (p. 421)

refraction The change in direction and speed of a wave as it enters a different medium. (p. 427)

scattering The process of changing the direction (and sometimes the properties) of a wave as it encounters an obstacle. (p. 421)

transmission The process of a wave passing through a material (even though it may lose some of its energy). (p. 427)

transparent materials Materials that transmit electromagnetic waves. (p. 427)

Key Equation

$$\text{Index of refraction} = \frac{\text{Speed in vacuum}}{\text{Speed in material}}$$

Review

1. What is meant by the term "optics"?

2. Is optics strictly concerned with visible light? Explain.

3. What are three ways that electromagnetic radiation interacts with matter?

4. What is a ray? Why is this description of the movement of an electromagnetic wave so useful?

5. Is the concept of a ray used only in connection with visible light?

6. What does it mean to say that a wave scatters? What exactly is a scattered wave, and in what direction does it move?

7. What is the angle of incidence of an incoming ray of radiation? The angle of reflection?

8. How is the angle of incidence related to the angle of reflection?

9. Do objects look smaller or larger in a concave mirror? Why?

10. Similarly, do objects look smaller or larger in a convex mirror? Why?

11. What is a focal point? How must a mirror be shaped to produce this?

12. What is the focal length of a mirror?

13. Why are parabolic mirrors frequently used in astronomy? What is meant by the term "light bucket"?

14. Infrared cameras capture electromagnetic radiation in the infrared part of the spectrum. Are these cameras considered optical instruments? List at least five other optical devices that are not concerned with visible light.

15. What are some of the technical difficulties associated with the construction of large telescope mirrors? How does active optics help mitigate some of these difficulties?

16. How does an electromagnetic wave behave in a vacuum? In a material? Is the energy of the wave conserved in each case? If not, what happens to it?

17. Fill a drinking glass with water and look through the water. Describe the different ways that light is interacting with matter.

18. Identify a substance that
    a. absorbs radio waves
    b. scatters microwaves
    c. transmits visible light
    d. absorbs x-rays
    e. scatters infrared radiation

19. What is absorption? What occurs to make a material opaque to light or to other electromagnetic radiation?

20. What is transmission? Can a material be transparent for one wavelength of radiation and opaque for another? Explain.

21. Propose an experiment to test whether lead is transparent to radio waves.

22. Describe the phenomenon of refraction. How and why does this occur?

23. What is the index of refraction of a material? Explain this in terms of a ratio.

24. What happens to the direction of a ray when it passes from a material with a high index of refraction, through a boundary, to a material with a low index of refraction? Similarly, what occurs in a transition from a low to a high index of refraction?

25. How does refraction account for a mirage? Explain.

26. Why do stars twinkle? What is the reason so many telescopes are placed as high in the atmosphere as possible?

27. What is a converging lens? A diverging lens? How does each work?

28. What is the difference between a reflecting telescope and a refracting telescope? Where is each commonly used?

29. What type of lens is used in a microscope? How does this magnify the image of the specimen?

30. What is meant by the term "total internal reflection"? How does this principle explain the workings of fiber optics?

31. Give several examples of the use of fiber optics. What makes it such a valuable technology?

32. What is dispersion? How does this account for the colors generated by a prism?

33. Explain the phenomenon of chromatic aberration. How is this corrected in photographic lenses?

34. What happens when you shine equal intensities of red, blue, and green light on the same spot? What happens when you mix paints from these three colors together? Why are the outcomes different?

35. What is a primary additive color? A primary subtractive color?

36. Why is the sky blue?

37. Why are the clouds white?

38. Why are sunsets red?

39. How are most colors we see like many of the sounds we hear?

Questions

1. How does the smoothness of a mirror affect the clarity of the image you see? What is the difference between a set of parallel rays reflected off a very smooth mirror and the same rays reflected off a more bumpy mirror made of the exact same material?

2. Two large plane mirrors are aligned so that they are parallel and facing each other, as shown. A flashlight beam strikes the bottom mirror, as shown. Where will the flashlight beam hit the bottom mirror the second time?

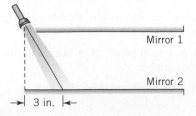

3. A baseball is sitting near a plane mirror, as shown. At which of the marked locations could an observer stand and see the image of the baseball?

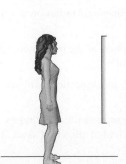

4. Sally stands in front of a plane mirror, as shown. Do you think she is able to see her feet in the mirror, or does she need a full-length mirror?

5. As you walk toward a full-length plane mirror, your image walks toward you. If your speed is 1 meter per second, what is the speed of your image?

6. A laser beam passes through the focal point of a concave parabolic mirror, labeled F in the figure. After the beam reflects off the mirror, which other point will the beam pass through?

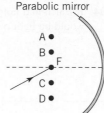

Parabolic mirror

7. In a nighttime infrared image of a heated house, the windows glow brightly. However, this book claims that glass is opaque to infrared radiation. Resolve this apparent dilemma. (*Hint:* Think about the difference between the heat conduction properties of the walls and the glass windows.)

8. The figure shows a beam of light striking a flat interface between air and glass. Which is the correct refracted beam, A, B, C, or D?

9. Diamonds have a very high index of refraction. How does this help to account for their sparkle? How does the cutting of diamonds into facets increase the sparkle you see?

10. Light passes through a triangular-shaped piece of glass as shown. Which is the correct emerging beam, A, B, or C? How would your answer change if the glass were submerged in a liquid that had the same index of refraction as the glass?

11. If the atmosphere did not scatter light, what would you see when you looked at the daytime sky? Explain.

12. If the atoms and molecules in the sky had about 10 times their present size, would you expect the daytime sky to be blue? If not, what color would you expect? Explain.

13. A man is standing by the lake looking at a piece of cork floating on the surface. At that moment, he notices a fish in the same line of sight as the cork. The figure shows the actual position of the fish. Where is the cork?

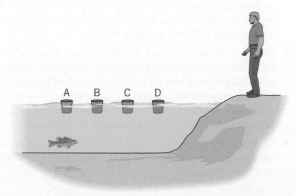

14. Why do people wear light-colored clothing in summer and dark-colored clothing in winter?

15. What is the frequency range of light seen when you look at a white-colored object? A black-colored one?

16. Why do stained-glass windows look gray from the inside of a church at night but bright from the outside?

17. For a house window to be as energy efficient as possible, which wavelengths of the electromagnetic spectrum should it be transparent to, and which wavelengths should it be more opaque to?

18. What kind of electromagnetic radiation can you detect with your body?

19. Why don't planets twinkle the way stars do? (*Hint:* How big do planets appear, as compared to stars?)

20. How does a concave mirror change the image seen as compared to a flat plane mirror? Is the image enlarged or reduced? Why does this image differ from a flat plane mirror?

21. What shape are the security mirrors often placed high in the corners of stores, and why are they shaped that way?

22. Some of the mirrors you might see in an amusement park make some parts of you seem large, while at the same time making other parts seem smaller. How is this accomplished?

23. Is a lens used to capture an image different from a lens used to project that image back onto a screen for viewing? Explain.

24. When you focus an optical instrument such as a camera or microscope, what are you actually doing to the lens and what happens to the focal point of the image?

25. Why is the image projected onto the back of the retina in your eye upside down? How can we see something as right side up if this is the case?

26. In movies that involve a character lost in the wilderness somewhere, you often see the hero vainly try to spear a fish in a river or tidal basin. Even if his aim is good, how should he aim to spear the fish? Due to refraction, is the fish actually located nearer or farther from where he sees it?

27. In many weapons systems, including aircraft, fiber optics is used to transmit information. Why might fiber optics be particularly useful in a military application such as a battlefield environment?

28. If you left a glass fiber-optic cable unshielded by any plastic covering, should the light still be able to travel through the cable? Explain.

29. A polar bear actually has translucent colorless fur and black skin. What are the benefits it derives from this and how is this similar to the workings of fiber optics?

30. Which parts of the electromagnetic spectrum, if any, scatter all wavelengths equally from atoms and molecules?

31. It sometimes happens that in cities located on a coast a hazy white layer can be seen over the city, but the layer is much less pronounced, or even absent, over the water. Explain why this should be so.

32. If you swim just below the surface of a pool and look straight up, you will see the sky. However, if you look at a glancing angle to the surface of the water, you will see a reflection of the bottom of the pool. What's going on?

33. In order to have a wide field of view, the passenger-side

mirror on an automobile or truck is curved outward (it is a convex mirror). Why is the driver-side mirror not usually convex?

34. If Earth had no atmosphere the days would be shorter (it would be light out for a shorter time each day). Why?

Problems

1. How fast does light travel through crown glass? Take the index of refraction of crown glass to be 1.52 and the speed of light to be 3×10^8 m/s.

2. If the speed of light through material Z is 2.5×10^8 m/s, what is this material's index of refraction?

3. Diamond has a high index of refraction at about 2.4, which helps account for its sparkle. How fast does light travel through a diamond? Using Problem 1, which material, diamond or crown glass, bends a light ray more as it passes from air into the respective material?

Investigations

1. Next time you see a rainbow after a sun shower, note the orientation of the rainbow in relation to the light from the Sun. Which direction is the Sun coming from in relation to the rainbow and in which direction does there seem to still be rain? Research the origin of rainbows. When do they occur, and what is the role of reflection and refraction in creating them?

2. Examine a microwave oven or, better yet, obtain an old broken oven that you can take apart. Locate the source of microwaves. Which materials in the oven transmit microwaves? Which ones scatter microwaves? Do you think any of the components absorb microwaves? Why?

3. The eye is a very sophisticated and remarkable instrument through which we interpret much of the physical world. How does it work? What is its structure and how and where does it focus the incoming light rays to form the images we see? In terms of the concepts discussed in this chapter, what happens when light comes into a normal healthy eye? How is the eye similar to a camera and how does it differ?

4. Continuing Investigation 3, find out what is actually happening when a person is nearsighted, farsighted, or has an astigmatism. Where is the focal point in each of these conditions in relation to where it ought to be if the eye were normal, and how is the lens of the eye misshapen so that this occurs? What types of lenses are used to correct these conditions, and how and where do these lenses refocus images onto the retina so that the focus is better? How does the surgical procedure "radial keratotomy" work to correct vision permanently?

5. Search the Web or the library for images created by the Hubble Space Telescope. Compare these images to images of the stars from the best terrestrial telescopes. How do they differ, if at all? Investigate the history of the Hubble Space

Telescope. What types of optical technologies are on board? What exactly plagued it in its early years? How was this initially corrected for? What have been some of its key discoveries, and how much of the universe has been seen with it?

6. Explore further the history of the development of fiber optics. What types of technological advances in terms of materials and optical instruments helped facilitate its widespread use? Can you find examples of its use not given in the chapter? How does fiber optics react to temperature variation and electrical interference, and how easy would it be to eavesdrop on such a communication system compared to other traditional communication mediums? How difficult is it to connect a fiber-optic cable to other cables and devices in telephone systems? Is this a problem? Finally, what is its future, as you see it?

7. Go into a classroom where a microscope is available, examine it, and use it. How many lenses are there and how do you focus them? What happens to the rays of light as they hit the object and then move through the microscope to your eye? What are the smallest images seen with a traditional optical microscope using traditional glass lenses? How does an electron microscope work? What are the smallest images seen to date and how are they seen?

8. Go to an amusement park nearby or a fair. Look for a Hall of Mirrors or some other place where they have a mirror exhibit. As you look into all the various mirrors displayed, notice the images reflected and try to explain to yourself what is happening to the light rays in each instance. How are the respective mirrors shaped to produce the effects seen?

9. Compare the workings of an infrared camera to an ordinary camera that captures visible light. What are the similarities and the differences?

 **WWW Resources**

See the *Physics Matters* home page at **www.wiley.com/college/trefil** for valuable web links.

1. **http://www.phy.ntnu.edu.tw/java/Rainbow/rainbow.html** A site that discusses and models the dispersion of light through raindrops, creating the curved spectrum known as a rainbow.

2. **http://www.phy.ntnu.edu.tw/java/Lens/lens_e.html** The classic thin lens and mirror simulator, this applet shows how lenses and mirrors make use of the laws of reflection and refraction to create various kinds of images.

3. **http://www.phy.ntnu.edu.tw/java/optics/prism_e.html** A demonstration of how refraction and reflection occur in a prism, including calculated intensities of reflected and refracted beams. The index of refraction is controllable.

4. **http://www.exploratorium.edu/light_walk/index.html** The famous walk through the physics of pinhole images from the San Francisco Exploratorium.

5. **http://www.phy.ntnu.edu.tw/java/image/rgbColor.html** An applet that allows the mixing of colored light, the mixing of pigments, and the use of filters to demonstrate color production and separation.

6. **http://school.discovery.com/lessonplans/interact/electromagneticspectrum.hml** A brief introduction to the electromagnetic spectrum from Discovery School.

21

Atomic Structure and Interactions

KEY IDEA

The arrangement of protons and electrons in atoms determines their chemical properties.

PHYSICS AROUND US . . . A Deep Breath

Take a deep breath. As you feel the air moving into your lungs, pause for a moment to think about what it is you are taking into your body. You probably know that some of the air you are taking in is made from atoms of a gas called oxygen. These atoms are taken into your bloodstream and carried to your cells, where they participate in chemical reactions that provide you with the energy you need to live. These reactions are similar to what happens when a piece of wood burns in a fire—atoms of oxygen combine with atoms of carbon to produce a molecule called carbon dioxide, which is in your breath when you exhale. This carbon dioxide, in turn, is taken in by plants, whose chemical processes produce the oxygen that you breath in. Atoms of oxygen, in other words, move around through the biosphere. They appear in combination with different sorts of atoms at different times, but the oxygen atoms themselves don't change.

Another part of the air you breathe in (most of it, actually) is made from a different kind of atom, called nitrogen. Unlike oxygen, nitrogen doesn't take part in combustion, either inside or outside the body. Isn't it amazing that these two atoms, mingled so closely together, can have such different properties?

The next time you're out in the country breathing in that fresh air, think about the fact that the unchanging atoms you are drawing in are part of an endless cycle of combination and recombination—a cycle that makes life on our planet possible.

TURNING INWARD

By the end of the nineteenth century, the industrialized world was feeling pretty confident. The shimmering prospect of endless progress permeated public thought—after all, this was the generation that had seen the telegraph remove barriers to communication, electricity light up their cities, and trolley cars move people around with unprecedented speed. There seemed to be no limit to what human society could accomplish.

The accomplishments of physics seemed to fit this pattern. Newton's laws of motion explained the motion of the solar system and other material objects. Maxwell's equations codified and linked the phenomena of electricity and magnetism (and, incidentally, led to the availability of commercial electric power). The laws of thermodynamics explained the behavior of heat and gave engineers a way of improving the steam engines that had powered the Industrial Revolution (although, to be honest, many improvements came from plain old-fashioned tinkering).

And yet . . . and yet . . . there were still mysteries. They seemed small things—footnotes in the triumphal march—but they had something in common. As the century turned, scientists began to see more new and puzzling things in the behavior of the smallest bits of matter. Eventually, they turned their attention inward, toward the atom and its constituents. In the end, these new phenomena led to an entirely new way of looking at nature.

A Bewildering Variety of Atoms

In Chapter 9 we discussed John Dalton's atomic theory and how it slowly gained acceptance during the nineteenth century. The idea that each chemical element corresponded to a different kind of atom and that atoms could combine to form molecules explained all the observations and rules that scientists had made. However, there were still doubters, scientists who recognized that the concept of atoms explained many observations, but were still not certain that they existed in reality.

One of the biggest problems about the atomic theory was that there was no obvious reason why different atoms should behave so differently from one another. For example, experiments showed that an atom of iron is about 56 times heavier than an atom of hydrogen, but nobody knew why. Suppose you assume that atoms of iron are simply bigger than atoms of hydrogen and that all the different elements are just slightly different in size (and mass). You could arrange the elements according to weight, from lightest to heaviest, so that each atom would be slightly heavier than the one next to it. An atom of nitrogen is only slightly heavier than an atom of carbon, and an atom of oxygen is only slightly heavier than an atom of nitrogen. But these three elements—carbon, nitrogen, and oxygen—are very different from one another. Carbon is a black solid, nitrogen is a clear gas that doesn't react strongly with most other elements, and oxygen is a clear gas that reacts vigorously with most other elements. Why should such a slight difference in size cause such large differences in behavior? Scientists concluded that there had to be more of an explanation than Dalton's simple atoms.

Discovering Chemical Elements

In the early 1800s, the list of known chemical elements was rapidly expanding, but contained only a few dozen entries. Today, the periodic table lists more than 110 elements, of which 92 appear in nature and the rest have been produced artificially. Most of the materials we encounter in everyday life are not elements but compounds of two or more elements bound together. Table salt, plastics, stainless steel, paint, window glass, and soap are all made from a combination of elements.

Nevertheless, we do have experience with a few chemical elements in our everyday lives:

- **Helium:** A light gas that has many uses in addition to filling party balloons and blimps. In liquid form, helium is used to maintain superconductors at low temperatures (see Chapter 24).
- **Carbon:** Pencil lead, charcoal, and diamonds are all examples of pure carbon. The differences among these materials have to do with the way the atoms of carbon are linked together, as we discuss in Chapter 24.
- **Gold:** A soft, yellow, dense, and highly valued metal. For thousands of years the element gold has been coveted as a symbol of wealth. Today it coats critical electrical contacts in spacecraft and other sophisticated electronics.

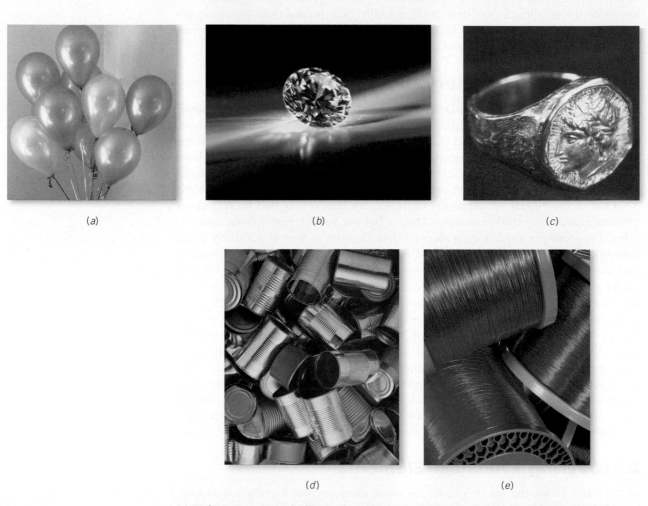

(a) (b) (c)

(d) (e)

(a) Helium in a party balloon; (b) carbon in diamonds; (c) gold in the tomb of King Tutankhamen; (d) aluminum in cans; and (e) copper in electrical wire.

- **Aluminum:** A lightweight metal used for many purposes, from overhead power lines to airplane parts and building construction. The dull white surface of the metal is actually a combination of aluminum and oxygen, but if you scratch the surface, the shiny material underneath is pure elemental aluminum.
- **Copper:** The reddish metal of pennies and pots. Copper wire provides a relatively inexpensive and efficient conductor of electricity.

Although we know of more than 90 different elements in nature, many natural systems are constructed from just a few. Six elements—oxygen, silicon, magnesium, iron, aluminum, and calcium—account for almost 99% of Earth's solid mass. Most of the atoms in your body are hydrogen, carbon, oxygen, or nitrogen, with smaller but important roles played by phosphorus and sulfur. And most stars are formed almost entirely from the lightest element, hydrogen. These differences in the behavior of the elements suggested that atoms might have a complex internal structure that made them different from one another.

THE INTERNAL STRUCTURE OF THE ATOM

Dalton's idea of the atom as a single indivisible entity was not destined to last. In 1897, English physicist Joseph John Thomson (1856–1940) unambiguously identified a particle called the **electron,** which has a negative electric charge and is much smaller and lighter than even the smallest atom known. Because there was no place from which a particle such as the electron could come, other than inside the atom, Thomson's discovery provided incontrovertible evidence for what some physicists had suspected for a long time: Atoms are not the fundamental building blocks of matter, but they are made up of particles that are smaller and more fundamental still. Table 21-1 summarizes some of the important terms related to atoms.

The Atomic Nucleus

The most important discovery about the structure of the atom was made by New Zealand–born physicist Ernest Rutherford (1871–1937) and his coworkers in Manchester, England, in 1911. The basic idea of the experiment is sketched in

TABLE 21-1	Important Terms Related to Atoms
Atom	The smallest particle that retains its chemical identity
Electron	A subatomic particle with negative charge and small mass
Nucleus	The small, massive central part of an atom
Proton	A positively charged nuclear particle
Neutron	An electrically neutral nuclear particle
Ion	An electrically charged atom
Element	A chemical substance made up of only one kind of atom
Molecule	Any collection of two or more atoms bound together

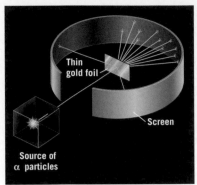

FIGURE 21-1. In Rutherford's experiment, a beam of alpha particles was scattered by atomic nuclei in a piece of gold foil. A lead shield protected researchers from the radiation.

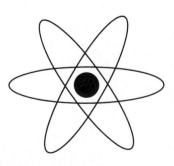

The Nuclear Regulatory Commission uses a highly stylized atomic model as its logo.

Figure 21-1. The experiment started with a piece of radioactive material—matter that sends out energetic particles (see Chapter 26). For our purposes, you can think of radioactive materials as sources of tiny subatomic bullets. The particular material that Rutherford used produced bullets that scientists had named *alpha particles,* which are positively charged particles thousands of times heavier than electrons. By arranging the apparatus as shown, Rutherford produced a stream of these subatomic bullets moving toward the right in the figure. In front of this stream, he placed a thin foil of gold.

The experiment was designed to measure something about the way atoms are put together. At the time, people believed Thomson's idea that the small, negatively charged electrons were scattered around the entire atom, more or less like raisins in a bun. Rutherford was trying to shoot "bullets" into the "bun" to see what happened. He expected the results to confirm Thomson's model of the atom.

Instead, what the experiment revealed was little short of astonishing. Almost all the subatomic bullets either passed right through the gold foil unaffected or were scattered through very small angles. This result is easy to interpret: it means that most of the heavy alpha particles passed through spaces in between the gold atoms and that the alpha particles that hit the gold atoms were only moderately deflected by the relatively low-density material in them. However, about one alpha particle in a thousand was scattered through a large angle; some even bounced straight back. What could possibly cause the relatively heavy alpha particles to rebound in this way? Rutherford said that it was like firing a heavy artillery shell at a piece of tissue paper and having it bounce back and hit you.

After almost two years of puzzling over these extraordinary results, Rutherford concluded that a large part of each atom's mass is located in a very small, compact object at the center, which he called the **nucleus.** About 999 times out of 1000 the alpha particles either missed the atom completely or went through the low-density material in the outer regions of the atom. About 1 time out of 1000, however, the alpha particle hit the nucleus and bounced backward through a large angle.

You can think of the Rutherford experiment in this way. If the atom were a large ball of mist or vapor with a diameter greater than a skyscraper and the nucleus were a bowling ball at the center of that sphere of mist, then most bullets shot at the atom would go right through. Only those that hit the bowling ball would ricochet back through large angles. In this analogy, of course, the bowling ball plays the part of the nucleus, while the mist is the domain of the electrons.

As a result of Rutherford's work, a new picture of the atom emerged, one that is very familiar to us. Rutherford described a small, dense, positively charged nucleus sitting at the atom's center, with lightweight, negatively charged electrons circling it, like planets orbiting the Sun. (The nucleus must have a positive charge because the positively charged alpha particles bounced back from the nucleus rather than becoming attracted to it.) Indeed, Rutherford's discovery has become an icon of the modern age, adorning everyday objects from postage stamps to bathroom cleaners. Thus, by trying to confirm one model of the atom, Rutherford wound up proposing an entirely different model. This interplay of model and experiment has been typical of modern physics. In the first half of the twentieth century, many ideas about atomic structure were proposed, tested, and further refined. We examine many of these ideas in the next few chapters.

Later on, physicists discovered that the nucleus itself is made up primarily of two different kinds of particles (see Chapter 26). One of these carries a positive charge and is called a *proton*. The other, whose existence was not confirmed until 1932, carries no electric charge and is called a *neutron*.

For each positively charged proton in the nucleus of the atom, there is normally one negatively charged electron associated with the atom. The electric charges of the electrons and the protons are of equal magnitude and so cancel out; thus atoms are normally electrically neutral. In some cases, atoms either lose or gain electrons. In this case, they acquire an electric charge and are called *ions*.

Why the Rutherford Atom Couldn't Work

The picture of the atom that Rutherford developed is intellectually appealing, particularly because it recalls to us the familiar orbits of planets in our solar system. However, we have already learned enough about the behavior of nature to know that the atom just described could not possibly exist in nature. Why do we say this?

We learned in Chapter 3 that an object traveling in a circular orbit is constantly being accelerated—it is not in uniform motion because it is continually changing direction. Furthermore, we learned in Chapter 19 that any accelerated electric charge must give off electromagnetic radiation, as described by Maxwell's equations. Thus, if an atom really fit the Rutherford model, the electrons moving in their atomic orbits would constantly give off energy in the form of electromagnetic radiation. This energy, according to the first law of thermodynamics, would have to come from somewhere (remember conservation of energy!), so as the electrons gave up their energy to electromagnetic radiation, they would gradually spiral in toward the nucleus. Eventually, the electrons would fall into the nucleus and the atom would cease to exist in the form we know.

In fact, if you put in the numbers, the life expectancy of the Rutherford atom turns out to be less than a second. Given the fact that many atoms have survived billions of years, since almost the beginning of the universe, this calculation poses a serious problem for the simple orbital model of the atom.

WHEN MATTER MEETS LIGHT

Almost from the start, the Rutherford model of the atom encountered difficulties. Some of the problems involved its violations of fundamental physical laws as we have described, whereas others were more mundane—the Rutherford model simply did not explain all the behavior of atoms that scientists knew about. Rutherford and his contemporaries knew that the planetary model of the atom was a step forward from Dalton's idea of featureless spheres, but they also knew that, in the usual way of scientific progress, more work was needed to refine the model further. The first decades of the twentieth century were a period of tremendous ferment in physics as people scrambled to find a new way of describing the nature of atoms.

Niels Bohr (1885–1962) with his five sons, including Aage Bohr. Both won Nobel prizes in physics.

The Bohr Atom

In 1913, Niels Bohr (1885–1962), a young Danish physicist working in England, produced the first model of the atom that avoided the kinds of objections encountered by Rutherford's model. The **Bohr atom** does not match well with our intuition about the way things ought to be in the real world, but it was the precursor of the modern view of the atom's internal structure.

The young Bohr was deeply immersed in studying how atoms interact with light and other forms of electromagnetic radiation. He knew that some new ideas were circulating in theoretical physics at the time—ideas that things in the world of the atom were different from the way they were understood in the familiar Newtonian world. In particular, he knew that physicists were suggesting that in the atomic world, energy comes in discrete bundles, called *quanta* (see Chapter 22). Bohr wondered what the consequences would be if the angular momentum of electrons circling the nucleus also came in discrete units. In this case, electrons circling the nucleus, unlike planets circling the Sun, could not maintain their positions at just any distance from the center. Bohr found that if their angular momentum could only have certain discrete values, then there were only certain positions—he called them "allowed energy levels" or "allowed orbits"—located at specified distances from the center of the atom in which an electron could exist for long periods of time without giving off radiation. (As we shall see in Chapter 22, in the modern view of the atom, the analogy between electrons and planets is no longer accepted as completely accurate, although it provides a rough approximation of reality.)

Bohr's picture of the atom (Figure 21-2) embraces the idea that the electron can exist at a specific distance r_1 from the nucleus, at a distance r_2, or at a distance r_3, and so on, each distance corresponding to a different electron energy level. As long as the electron remains at one of those distances, its energy is fixed. In the Bohr atomic model, the electron cannot ever, at any time, exist at any place between these allowed energy states.

One way to think about the Bohr atom is to imagine what happens when you climb a flight of steps. You can stand on the first step or you can stand on the second step. It's very hard, however, to imagine what it would be like to stand somewhere between two steps. In just the same way, an electron can be in the first allowed energy level or in the second one, but it can't be in between these allowed energy levels. In terms of energy, both the steps and the electrons in an atom may be represented by a simple pictorial description (Figure 21-3). Each time you change steps in your home, your gravitational potential energy changes. Similarly, each time an electron changes levels, its energy changes.

An electron in an atom can be in any one of a number of allowed energy levels, each corresponding to a different distance from the nucleus. You would have to exert a force over a distance to move an electron from one allowed

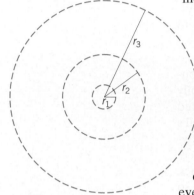

FIGURE 21-2. A schematic diagram of the Bohr atom showing the first three energy levels and respective distances (r_1, r_2, and r_3) from the nucleus.

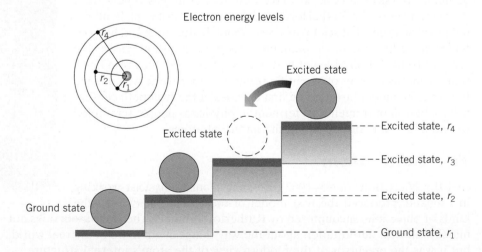

Electron energy levels

FIGURE 21-3. Stairs provide an analogy for the energy changes associated with electrons in the Bohr atom.

energy level to a higher level, just as your muscles have to exert a force to get you up a flight of stairs. Thus, the allowed energy levels of an atom occur as a series of steps, as shown in the figure. An electron in the lowest energy level is said to be in the *ground state,* while all energy levels above the ground state are called *excited states.*

Photons: Particles of Light

One major feature of the Bohr atom is that an electron in a higher energy level can spontaneously move down into an available lower energy level. This process is analogous to that by which a ball at the top of a flight of stairs can bounce down the stairs under the influence of gravity.

Assume that an electron is in an excited state, as shown in Figure 21-4. The electron can move to a lower energy state, but if it does, something must happen to the extra energy. Energy can't just disappear. This realization was Bohr's great insight. The energy that's left over when the electrically charged electron moves from a higher state to a lower state is emitted by the atom in the form of a single packet of electromagnetic radiation—a particle-like bundle of light called a **photon.** Every time an electron jumps from a higher to a lower energy level, a photon moves away at the speed of light. The energy of the photon, which is proportional to its frequency, is equal to the difference between the electron's initial and final energy levels.

The concept of a photon raises a perplexing question: Is light—Maxwell's electromagnetic radiation—a wave or a particle? We explore this puzzle at some length in Chapter 22, once we have learned more about the behavior of atoms.

The interaction of atoms and electromagnetic radiation provides the most compelling evidence for the Bohr atom. If electrons are in excited states and if they make transitions to lower states, then photons are emitted. If we look at a group of atoms in which these transitions are occurring, we see light or other electromagnetic radiation. Thus, when you look at the flame of a fire or the fluorescent light in the ceiling, you are actually seeing photons that have been emitted by electrons jumping between allowed states in that material's atoms.

Not only does the Bohr model give us a picture of how matter emits radiation, it also provides an explanation for how matter absorbs radiation. For example, start with an electron in a low-energy state, perhaps its ground state. If a photon arrives that has just the right amount of energy to raise the electron to a higher allowed energy level (the next step up), the photon can be absorbed and the electron is pushed up to an excited state (see Figure 21-4b). The absorption of light is a mirror image of light emission.

This picture of the interaction of matter and radiation is exceedingly simple, but two key ideas are embedded in it. For one thing, when an electron moves from one allowed state to another, it cannot ever, at any time, be at any place in between. This rule is built into the definition of an allowed energy level. This means that the electron must somehow disappear from its original location and reappear in its final location without ever having to traverse any of the positions in between. This process, called a **quantum leap** or **quantum jump,** cannot be visualized, but it is something that seems to be fundamental in nature—an

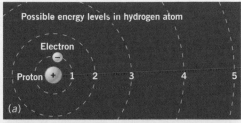

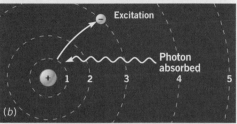

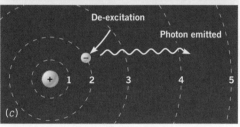

FIGURE 21-4. Electrons may jump between the energy levels shown in (a) and, in the process, (b) absorb or (c) emit energy in the form of a photon.

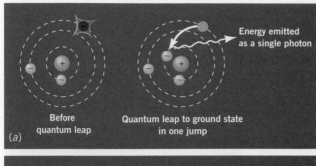

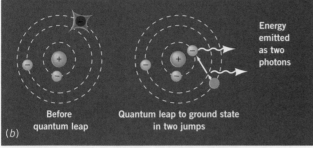

FIGURE 21-5. An electron can jump from a higher to a lower energy level in a single quantum leap (a) or by multiple quantum leaps (b).

example of the quantum weirdness of nature at the atomic scale that we discuss in Chapter 22.

The second key idea is that if an electron is in an excited state, it can, in principle, get back down to the ground state in one of several different ways. Look at Figure 21-5. An electron in the upper energy level can move to the ground state by making one large jump and emitting a single photon with a large amount of energy. Alternatively, it can move to the ground state by making two smaller jumps, as shown. Each of these smaller jumps emits a photon of somewhat less energy. The energies emitted in these smaller jumps are generally different from one another, but the sum of the two energies equals that of the single large jump. If we had a large collection of atoms of this kind, we would expect that some electrons would make the large leap while others would make the two smaller ones. Thus, when we look at a collection of these atoms, we would measure three different energies of photons.

This curious behavior of electron energy levels helps to explain the familiar phenomenon of fluorescence. Recall that the energy of electromagnetic radiation is related to its frequency. In fluorescence, the atom absorbs a higher-energy photon of ultraviolet radiation (which our eyes can't detect). The atom then emits two lower-energy photons, at least one of which is in the visible range. Consequently, shining ultraviolet black light on the fluorescent material, makes it glow with a bright, visible color.

(a) (b) (c)

The elements (a) sodium, (b) potassium, and (c) lithium impart distinctive colors to a flame. The color from each element corresponds to the frequency of the photons emitted by that element as its electrons change energy level from higher energy to lower.

A key point about the Bohr model is that energy is required to lift an electron from the ground state to any excited state. This energy has to come from somewhere. We have already mentioned one possibility: that the atom absorbs a photon of just the right frequency to raise the electron to a higher energy level. There are other possibilities, however. For example, if the material is heated, atoms will move faster, gain kinetic energy, and undergo more energetic collisions. In these collisions, an atom can absorb energy and then use that energy to move electrons to a higher state. This explains why materials often glow when they are heated—the glow occurs when electrons drop back down after being raised to a higher energy level.

Develop Your Intuition: Bright Lights! Big City!

All big cities have large downtown areas of restaurants, stores, and movie theaters—all places with big electric signs in a dozen different colors. Most large electric light displays are gas discharge tubes—commonly called neon signs—in which gas at a low pressure glows when an electric current passes through it. How do these electric signs appear in so many different colors?

The glow of a gas discharge tube is caused by changes in the electron energy states in the gas atoms. Those atoms absorb energy, their electrons move to higher energy levels, and then the electrons fall back to lower energy levels and release the energy as light. The most common gas in these tubes is neon, which produces photons of orange-red light. If you want a different color, you can use a different gas with different energy levels. Chemically inert noble gases are particularly popular in gas discharge tubes: argon glows bluish-green, krypton glows bluish-white, and xenon glows blue.

Bright neon lights decorate a city at night.

In fact, krypton emits such an intense light that it is commonly used for lighting airport runways. Other ways to produce bright colors with gas discharge tubes include mixing in other gases or coating the inside of the tube with fluorescent materials.

Gas discharge tubes are gradually being replaced with displays of light-emitting diodes (LEDs), which are solid crystals that also emit light when electrons fall back from excited energy levels. These remarkable materials are safer and they use less energy. However, the range of colors presently available is not nearly as wide as with gas discharge lights.

An Intuitive Leap

Bohr first proposed his model of the atom based on an intuition guided by experiments and ideas about the behavior of things in the subatomic world. In some ways, the Bohr model was completely unlike anything we experience in the macroscopic world; indeed, the model seemed to some a little bit crazy. It took two decades for scientists to develop a theory called *quantum mechanics* that showed why electrons can exist only within Bohr's allowed energy states and not in between them. We discuss this justification for the Bohr atom in Chapter 22, but remember that the justification occurred long after the initial hypothesis. Physicists accepted the Bohr model because it worked—it explained what they saw in nature and allowed them to make predictions about the behavior of real matter.

How could Bohr have come up with such a strange picture of the atom? He was, as we have said, guided by some of the early work that led to the theory of quantum mechanics (see Chapter 22). In the end, however, this explanation is unsatisfactory. Many people at the time studied interactions of atoms and light, but only Bohr was able to make the leap of intuition to his description of the atom. This insight, like Newton's realization that gravity might extend to the orbit of the Moon, remains one of the great intuitive achievements of the human mind.

SPECTROSCOPY

Whenever energy is added to a system with many atoms in it, electrons in some atoms jump to excited states. As time goes by, some of these electrons make quantum leaps down to the ground state, giving off photons as they do. If some of those photons are in the range of visible light, the source appears to glow.

You may not realize it, but you have looked at such collections of atoms all your life. Common mercury vapor streetlamps contain bulbs filled with mercury gas. When electric current passes through the gas, electrons move up to excited states. When they jump down, they emit photons that give the lamp a bluish-white color. Other types of streetlights, often used at freeway interchanges, use bulbs filled with sodium atoms. When sodium is excited, the most frequently emitted photons lie in the yellow range, so the lamps look yellow. Yet another place where you can see photons emitted directly by quantum leaps, as mentioned above, is in the vivid glowing colors of fluorescent objects, which are often used in black light displays at the theater.

Fluorescent minerals appear dull and ordinary under daylight, but glow with brilliant colors under ultraviolet light.

From these examples, we can draw two conclusions: (1) quantum leaps are very much in evidence in your everyday life, and (2) different atoms give off different characteristic photons. The second of these two facts is extremely important for physicists. If you think about the structure of an atom, the idea that different atoms emit and absorb different characteristic photons shouldn't be too surprising. Electron energy levels depend on the electrical attraction between the nucleus and the electrons, just as the orbits of the planets depend on the gravitational attraction between the planets and the Sun. Different nuclei have different numbers of protons, so electrons circling them are in different energy levels. In fact, the energies between allowed energy levels within atoms are different in each of the hundred or so different chemical elements. Because the energy and frequency of photons emitted by an atom depend on the differences in energy between these levels, each chemical element emits a distinct set of characteristic photons.

You can think of the collection of characteristic photons emitted by each chemical element as a kind of fingerprint—a set of wavelengths that is distinctive for that chemical element and none other. This feature opens up a very important application. The total collection of photons emitted by a given atom is called its **spectrum,** a characteristic pattern that can be used to identify chemical elements even when they are very difficult to identify by any other means. (Molecules also have individual spectra, but these are far more complicated than atomic spectra.)

In practice, the identification process works because light from the gaseous atoms is spread out after being passed through a prism (Figure 21-6). Each possible quantum jump corresponds to light at a specific wavelength, so each type

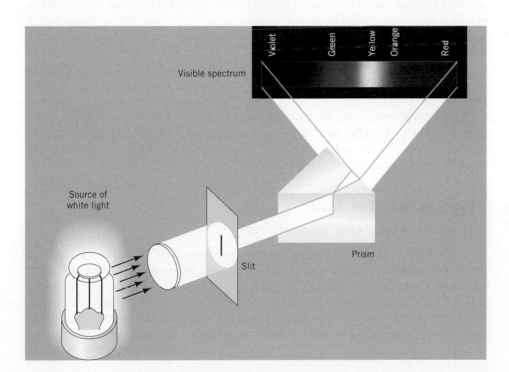

FIGURE 21-6. A glass prism spreads out the colors of the visible spectrum.

FIGURE 21-7. Line spectra, shown here for hydrogen, sodium, and neon, provide distinctive fingerprints for elements and compounds.

of atom produces a different set of lines, as shown in Figure 21-7. This spectrum is the atomic fingerprint.

The Bohr model suggests that if an atom gives off light of a specific wavelength and energy, then it also absorbs light at that wavelength. The emission and absorption processes, after all, involve quantum jumps between the same two energy levels, but in different directions. Thus, if white light shines through a material containing a particular kind of atom, then certain wavelengths of light are absorbed. When you observe that light on the other side of the material, then certain lines of color are missing. The dark areas corresponding to the absorbed wavelengths are called *absorption lines*. This set of lines is as much an atomic fingerprint as the set of colors that glowing atoms emit. And although the use of visible light is very common, these arguments hold for radiation in any part of the electromagnetic spectrum.

Spectroscopy has become a standard tool that is used in almost every branch of science. Astronomers use emission spectra to determine the chemical composition of distant stars, and they study absorption lines to determine the chemical composition of interstellar dust and the atmospheres of the outer planets. Spectroscopic analysis is also used in manufacturing to search for impurities on production lines and by police departments when conducting investigations to identify small traces of unknown materials.

Physics in the Making

The Bohr Atom and Spectroscopy

Bohr's model of the atom was pretty radical for its time. Max Planck had first proposed the idea of a quantum of energy in 1900, but even he wasn't sure if nature really worked this way. One of the few scientists to take the idea seriously was Albert Einstein, who proposed in 1905 that light itself consisted of quanta, which he called *photons*. But many physicists remained skeptical. Then along came Niels Bohr in 1913 saying that the very nature of atoms required quanta of energy and angular momentum. Why did physicists accept this strange idea in fewer than 10 years?

One of the great mysteries of the time, one of the mysteries alluded to at the beginning of this chapter, was the reason for spectroscopic patterns. Why should hydrogen, for example, emit light of just those wavelengths observed in its spectrum? Many scientists tried to find a formula for the pattern of wavelengths, without any luck. Finally, in 1885, a Swiss schoolteacher named Johann Balmer found a formula by trial and error that fit the wavelengths of the most prominent series of lines in hydrogen. Nobody knew why the formula worked, but it did.

Others got into the act. Hydrogen shows several series of lines in different parts of the electromagnetic spectrum, and each series is named after the person who discovered it: the Lyman series, the Paschen series, and so on. In 1890, the Swedish spectroscopist Johannes Rydberg found a formula that fit all these series. He had only to plug in simple integers such as 1, 2, and 3 to get Balmer's series, Lyman's series, or any other series. Rydberg's formula involved multiplying by a constant (now called "Rydberg's constant") with a value of 1.097×10^7 meter^{-1}. He didn't know why his constant had this value or why the formula worked, but it did.

When Bohr proposed his model of the atom, it explained how spectral lines are produced in the atom. He was able to derive a formula for the wavelengths emitted when electrons dropped from several excited levels in hydrogen to the ground state; the result matched Lyman's series. Bohr also derived a formula for the wavelengths emitted when electrons dropped from the excited levels in hydrogen to the first excited level; the results matched Balmer's series. All the series known for hydrogen turned out to fit particular transitions of electrons between allowed levels.

In all these formulas, Bohr had to multiply energies by a constant, but in his case it was a complicated factor involving the speed of light, the mass and charge of an electron, and other known constants. When Bohr plugged in these values and came up with a number for his constant, lo and behold, he got 1.097×10^7 meter^{-1}. Thus, Bohr showed how to calculate the Rydberg constant from more fundamental physical constants. This strong, direct confirmation of his theory went a long way toward winning acceptance of the Bohr model. (Bohr received the Nobel Prize for his work in 1922, one year after Einstein and four years after Planck.) ●

Connection
Spectra of Life's Chemical Reactions

In a classic set of experiments in the early 1940s, scientists used spectroscopy to work out in detail how large molecules called enzymes govern chemical reactions in living cells. In these experiments, a fluid containing the materials undergoing the chemical reactions was allowed to flow down a tube. As the fluid moved farther down the tube, the reaction progressed closer and closer to completion. By measuring spectra at different points along the tube, scientists were able to follow the changes in the behavior of electrons as the chemical reactions went along. In this way, part of the enormously complex problem of understanding the chemistry of life was unraveled.

More recently, scientists have begun to develop instruments that can use the principles of spectroscopy to identify pollutants emitted by automobile tailpipes as cars drive by. If they are successful, we will have a major new tool in our battle against air pollution and acid rain. ●

Physics in the Making
The Story of Helium

You have probably encountered helium, perhaps to inflate party balloons. Helium gas turns out to be a very interesting material, not only for its properties (it's less dense than air, so helium-filled balloons float up), but also because of the history of its discovery.

The word helium comes from "helios," the Greek word for Sun, because helium was first discovered by identifying a new set of spectral lines in light from the Sun, work done in 1868 by English scientist Joseph Norman Lockyer (1836–1920). Helium is very rare in Earth's atmosphere and before Lockyer's discovery scientists were not even aware of its existence. Following the discovery, there was a period of about 30 years when astronomers accepted the fact that the element helium existed in the Sun, but they were unable to find it on Earth.

The discovery of this hitherto unknown spectrum led to a very interesting problem. Could it be that there were chemical elements on the Sun that simply did not exist on our own planet? If so, it would call into question our ability to understand the rest of the universe, for the simple reason that if we don't know what an element is and can't isolate it in our laboratories, then we can never really be sure that we understand its properties. In fact, the existence of helium on Earth wasn't confirmed until 1895, when Sir William Ramsay identified its spectrum in a sample of radioactive material. ●

Connection
The Laser

The Bohr model provides an excellent way of understanding the workings of one of the most important devices in modern science and industry—the *laser*. The word "laser" is an acronym for *l*ight *a*mplification by *s*timulated *e*mission of *r*adiation. At the core of every laser is a collection of atoms—a crystal of ruby, perhaps, or a gas enclosed in a glass tube. The term "stimulated emission" refers to a process that goes on when light interacts with these atoms in a special way.

If an electron is in an excited state and one photon of just the right energy passes nearby, then the electron may be stimulated to make the jump to a lower energy state, thus releasing a second photon. By "just the right energy" for the first photon, we mean a photon whose energy corresponds to the energy gap between two electron energy levels in the atom.

What's so special about the photons emitted by the stimulated electron? Remember that light is a form of electromagnetic radiation that can be described as a wave. In a laser, the crests of all the emitted photon waves line up exactly with the crests of the first photon, and the signal is enhanced by constructive interference. In the language of physics, we say that the photons are "coherent." Thus, in stimulated emission you have one photon at the beginning of the process and two coherent photons at the end.

Now suppose that you have a collection of atoms where most of the electrons are in the excited state, as shown in Figure 21-8. If a single photon of the correct frequency enters this system from the left and moves to the right, it passes the first atom and stimulates the emission of a second photon. You then have two photons moving to the right. As these photons encounter other atoms, they, too, stimulate emission so that you have four photons. It's not hard to see that

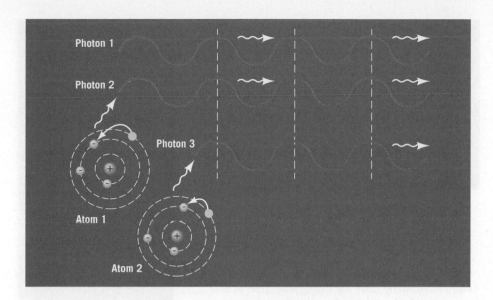

FIGURE 21-8. Lasers produce a beam of light when one photon stimulates the emission of other photons.

light amplification in a laser happens very quickly, cascading so that soon there is a flood of photons moving to the right through the collection of atoms. Energy added to the system from outside continuously returns atoms to their excited state—a process called "optical pumping"—so that more and more coherent photons can be produced.

In a laser (Figure 21-9), the collection of excited atoms is bounded on two sides by mirrors so that photons moving to the right hit the mirror, are reflected, and make another pass through the material, stimulating even more emissions of photons as they go. If a photon happens to be lined up exactly perpendicular to the mirrors at the end of the laser, it will continue bouncing back and forth. If its direction is off by even a small angle, however, it will eventually bounce out through the sides of the laser and be lost. Thus only those photons that are exactly aligned wind up bouncing back and forth between the mirrors, constantly amplifying the signal. Aligned photons traverse the laser millions of times, building up an enormous cascade of coherent photons in the system. Because only photons moving in exactly the right direction are amplified, the laser beam does not spread out very much, but stays tightly bunched. In the language of physics, we say that it is "collimated."

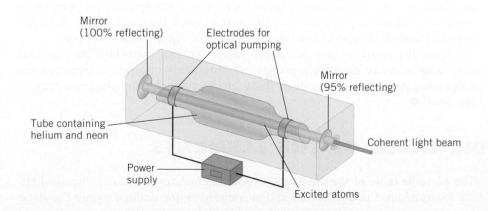

FIGURE 21-9. The action of a laser. Electrons in the laser's atoms are continuously "pumped" into an excited state by an outside energy source, and the beam of coherent photons is released when the electrons return to their ground state.

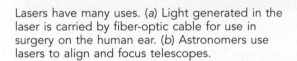

Lasers have many uses. (*a*) Light generated in the laser is carried by fiber-optic cable for use in surgery on the human ear. (*b*) Astronomers use lasers to align and focus telescopes.

(*b*)

The mirror from which the beam reflects is designed to be partially reflective—perhaps 95% of the photons that hit the mirror are reflected back into the laser. The remaining 5% of photons that leak out form the familiar laser beam, while the mirror at the other end reflects all the light that strikes it. Thus, the beam is made of intense, coherent light.

Laser beams have been applied in thousands of ways in science and industry since their development in the 1960s. Low-power lasers are ideal for optical scanners, such as the ones in supermarket checkout lines, and they make ideal light pointers for lectures and slide shows. The fact that the beam of light travels in a straight line makes the laser invaluable in surveying over long distances—for example, modern subway tunnels are routinely surveyed by using lasers to provide a straight line underground. Lasers are also used to detect the movement of seismic faults in order to predict earthquakes. In this case, a laser is directed across the fault, so that small motions of the ground are easily measured. Finely focused laser beams have revolutionized delicate procedures such as eye surgery. Much more powerful lasers that can transfer large amounts of energy are often used as cutting tools in factories, as well as implements for performing some kinds of surgery. The military has also adopted laser technology in targeting and range finders and in designs for futuristic energy-beam weapons.

From the point of view of science, lasers are also important because they enable us to make extremely precise measurements of atomic structures and properties. Almost every modern study of the atom depends in some way on the laser. ●

THE PERIODIC TABLE OF THE ELEMENTS

The **periodic table of the elements,** which systematizes all known chemical elements, provides a powerful conceptual framework for understanding the structure and interaction of atoms. Dmitri Mendeleev, the Russian scientist who developed the periodic table in the nineteenth century (see Chapter 1), assigned each element an integer called the *atomic number.* We now know that the atomic number corresponds to the number of protons in the atom, or, equivalently, if the atom is not charged, to the number of electrons surrounding the nucleus. If you arrange the elements as shown in Appendix C, with elements getting progressively heavier as you read from left to right and from top to bottom, as in a book, then elements in the same vertical column have very similar chemical properties.

Periodic Chemical Properties

The most striking characteristic of the periodic table is the similarity of the elements in any given column. For example, the far left-hand column of the table lists highly reactive elements called *alkali metals* (lithium, sodium, potassium, etc.). Each of these soft, silvery elements forms compounds (called salts) by combining in a one-to-one ratio with any of the elements in the seventh column (fluorine, chlorine, bromine, etc.). Water dissolves these compounds, which include sodium chloride, or table salt.

The elements in the second column (beryllium, magnesium, calcium, etc.) are metallic elements called the *alkaline earths,* and they too display similar chemical properties. For instance, these elements combine with oxygen in a one-to-one ratio to form colorless compounds with very high melting temperatures.

Elements in the far right-hand column (helium, neon, argon, etc.), by contrast, are all colorless, odorless gases that are almost impossible to coax into any kind of chemical reaction. These are called *noble gases* and they find applications when ordinary gases are too reactive. For example, helium is used to lift blimps because the only other lighter-than-air gas is the dangerous, explosive element hydrogen. Argon fills incandescent light bulbs because nitrogen or oxygen would react with the hot filament.

In the late nineteenth century, scientists knew that the periodic table worked—it organized the 63 elements known at that time and implied the existence of others—but they had no idea why it worked. Their faith in the periodic table was buttressed by the fact that, when Mendeleev first wrote it down, there were holes in the table—places where he predicted elements should go, but for which no element was known. The ensuing search for the missing kinds of atoms produced the elements we now call scandium (in 1876) and germanium (in 1886).

Why the Periodic Table Works: Electron Shells

With the advent of Bohr's atomic model and its modern descendants, we finally have some understanding of why the periodic table works. We now realize that the pattern of elements in the periodic table mirrors the spatial arrangement of electrons around the atom's nucleus—a concentric arrangement of electrons into *shells.*

The atom is largely empty space. When two atoms come near enough to one another to undergo a chemical reaction—such as a carbon atom and an oxygen atom in a burning piece of coal—electrons in the outermost shells meet one another first. These outermost electrons govern the chemical properties of materials, so we have to document the behavior of these electrons if we want to understand the periodic table.

In the process of studying atoms and their electrons, scientists discovered a curious fact. Electrons obey what is called the *Pauli exclusion principle,* which says that no two electrons in an atom can occupy the same state at the same time. (Note that the word "state" as used in this principle is not the same as an orbit or shell—there are normally many states in each shell.) One analogy is to compare electrons to cars in a parking lot. Each car takes up one space and once a space is filled, no other car can go there. Electrons behave just the same way. Once an electron fills a particular niche in the atom, no other electron can occupy the same niche. A parking lot can be full long before all the actual space in the lot is taken up with cars, because the driveways and spaces between cars must remain empty. So, too, a given electron shell can be filled with electrons long before all the available space is filled.

In fact, it turns out that there are only two spaces that an electron can fill in the innermost electron shell, which corresponds to the lowest Bohr energy level. One of these spaces corresponds to a situation in which the electron spins clockwise on its axis, the other to a situation in which it spins counterclockwise. When we start to catalog all possible chemical elements in the periodic table, we have element 1 (hydrogen) with a single electron in the innermost shell, and element 2 (helium) with two electrons in that same shell. After these two elements, if we want to add one more electron, it has to go into the second electron shell because the first electron shell is completely filled. (Note that this second shell is at a different energy than the first.) This situation explains why only hydrogen and helium appear in the first row in the periodic table.

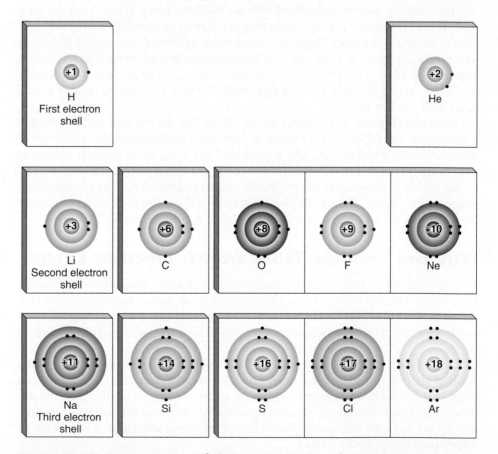

FIGURE 21-10. A representation of electrons in a number of common atoms.

Adding a third electron yields lithium, which is an atom with two electrons in the first shell and a single electron in the second electron shell. Lithium is the element just below hydrogen in the first column of the periodic table, because both hydrogen and lithium have a lone electron in their outermost shell.

The second electron shell has room for eight electrons, a fact reflected in the eight elements of the periodic table's second row, from lithium with three electrons to neon with ten. Neon appears directly under helium, and we expect these two elements to have similar chemical properties because both have a completely filled outer electron shell. In fact, both helium and neon are colorless, odorless, nonreactive gases.

Thus, a simple counting of the positions available to electrons in the first two electron shells explains why the first row in the periodic table has two elements in it and the second row has eight. By similar (but somewhat more complicated) arguments, you can show that the Pauli exclusion principle requires that the next row of the periodic table have 8 elements, the next 18, and so on (Figure 21-10). Thus, with an understanding of the shell-like structure of the atom's electrons, the mysterious regularity that Mendeleev found among the chemical elements becomes an example of nature's laws at work.

Develop Your Intuition: Predicting Chemical Formulas

The periodic table tells you how many electrons are in the outer shell of each element. We know that atoms are usually most stable (less likely to react chemically with other atoms) with a completely filled outer shell of electrons, which means eight electrons in most cases. Can you use that information to predict the formulas of simple compounds of the elements?

Elements at the left in the periodic table generally donate electrons to form a stable outer shell. Elements at the right in the periodic table generally accept electrons to complete a stable outer shell. Thus, an atom with one electron in its outer shell can combine with an atom with seven electrons in its outer shell, forming a compound with elements in a one-to-one ratio. Table salt, sodium chloride, has one atom of sodium for every atom of chlorine. Other alkali metals form similar compounds with halogens, such as potassium iodide or lithium fluoride. Atoms with two electrons in the outer shell can combine in one-to-one ratio with atoms containing six electrons in the outer shell. Examples are calcium oxide and magnesium sulfide.

Atoms can also form molecules containing three atoms, where two atoms containing one electron in the outer shell combine with an atom containing six electrons in the outer shell. This way you get potassium oxide, with two atoms of potassium for every atom of oxygen, and sodium sulfide, with two atoms of sodium for one of sulfur. You can also have molecules formed from one atom with two electrons in its outer shell and two atoms with seven electrons in the outer shell. Examples are calcium chloride, with two atoms of chlorine for every one of calcium, and magnesium fluoride, with two atoms of fluorine for every atom of magnesium.

See if you can figure out other combinations of atoms that might form molecules. How many atoms of sodium combine with one atom of nitrogen? How many atoms of aluminum combine with how many atoms of oxygen?

THINKING MORE ABOUT

Atoms: What Do Atoms Look Like?

Throughout this book you will find drawings of atoms. In this chapter we draw atoms as electrons in circular shells around a central nucleus. In Dalton's original work, atoms appear as little spheres in pictures of molecules. In other chapters in this book, atoms are portrayed as fuzzy clouds, waves, or even collections of dozens of smaller spherelike particles. So, what do atoms really look like?

Strictly speaking, we only see something when electromagnetic waves from the visible part of the spectrum enter our eyes. We are accustomed, however, to talking about other ways of "seeing." You cannot see X rays being absorbed by your teeth unless some intermediary system—a film or an electronic video monitor, for example—converts the X rays into a pattern that can be detected in the visible region of the electromagnetic spectrum. Similarly, astronomers often convert data from radio waves, infrared radiation, and other radiation into false-color images of distant objects. Scanning tunneling microscope pictures of atoms come from another such

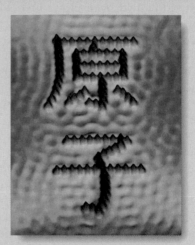

A scanning tunneling microscope image of individual iron atoms on copper forms the Japanese character for "atoms."

transformation—the amount of electric charge at a particular point on a material's surface is converted by electronics into peaks and valleys on a digital image.

So what does an atom look like? Is an X-ray picture of your teeth more real than the scanning tunneling microscope picture of the atom? Why or why not? Does the fact that we can't see atoms with our eyes mean that they don't exist?

Summary

All the solids, liquids, and gases around us are composed of about 100 different elements. Atoms, the building blocks of our chemical world, combine into groups of two or more; these groups are called molecules. Each atom contains a massive central **nucleus** made from positively charged protons and electrically neutral neutrons. Surrounding the nucleus are **electrons,** which are negatively charged particles that have only a small fraction of the mass of protons and neutrons.

Early models of the atom treated electrons like the planets orbiting around the Sun. Those models were flawed, however, because each electron, constantly accelerating, would have to emit electromagnetic radiation continuously. Niels Bohr proposed an alternative model in which electrons exist in various energy levels, much as you can stand on different levels of a flight of stairs.

Electrons in the **Bohr atom** can shift to a higher energy level by absorbing the energy of heat or light. Electrons can also drop into a lower energy level and in the process release heat or a **photon,** an individual electromagnetic wave. These changes in electron energy level are called **quantum leaps** or **quantum jumps. Spectroscopic** studies of the light emitted or absorbed by atoms—the atom's **spectrum**—reveal the nature of each atom's electron energy levels.

Each atom's electrons are arranged in concentric shells. When two atoms interact, electrons in the outermost shell come into contact. This shell-like electronic structure is reflected in the organization of the **periodic table of the elements,** which lists all the elements in rows corresponding to increasing numbers of electrons in each shell and in columns corresponding to elements with similar numbers of outer-shell electrons and thus similar chemical behavior.

Key Terms

Bohr atom A model of the atom proposed in 1913 by Danish physicist Niels Bohr; the Bohr atom revolutionized our understanding of physics. (p. 451)

electron A negatively charged fundamental particle; it is one of the primary building blocks of the atom. (p. 449)

nucleus A very small, dense, positively charged object at the center of every atom; nuclei are made up of protons and neutrons. (p. 450)

periodic table of the elements The systematic organization of the known chemical elements in terms of their electron configurations. (p. 462)

photon The quantum of electromagnetic radiation; a particlelike bundle of light. (p. 453)

quantum leap or **quantum jump** The process of changing location without having to traverse any of the positions in between; usually in reference to electrons changing energy levels in an atom. (p. 453)

spectroscopy The analysis of emission and absorption spectra of materials to determine their chemical properties. (p. 458)

spectrum The total collection of photons emitted by an atom; the distribution of the frequencies of photons emitted by a radiating system. (p. 457)

Review

1. What is an element? How many elements exist in nature?

2. What are most stars made of?

3. What three particles make up almost every atom? What are the major differences among these particles?

4. Review the basic components of Rutherford's experiment. What was it about the results of the experiment that led Rutherford to the conclusion that the atom has a nucleus?

5. In what ways is the structure of an atom like the solar system of planets orbiting around the Sun? In what ways is it different?

6. How does the nucleus differ from electrons?

7. What makes atoms neutral with respect to electric charge?

8. What is an ion?

9. Think of an analogy for the Rutherford experiment other than the cloud-of-mist-plus-bowling-ball we describe in the chapter. How does your analogy differ from that one? In what ways is it better or worse?

10. Why would an atom like Rutherford's model constantly give off energy, and how would this affect the electron orbit?

11. How does Bohr's model of the atom differ from that of Rutherford's? What is an allowed energy level?

12. What are quanta? How did this concept influence Bohr's thinking on atomic structure?

13. How is the Bohr atom similar to a set of steps?

14. What is the ground state of an electron? An excited state?

15. What is a photon?

16. What are you actually seeing when you look at a fire or the red-hot coil of an electric stove?

17. Describe the relationship between a photon and a quantum leap. What makes such a leap or jump occur?

18. Describe the alternative ways an electron in an excited state can return to its ground state. How much energy is emitted in each case as the electron returns to this ground state, and in what form is it emitted?

19. What is fluorescence?

20. Cite three examples of everyday objects with bright emission spectra.

21. Do different atoms give off different characteristic photons when excited? What is it about the structures of different elements that would lead to an element emitting photons of unique energies?

22. What is the spectrum of an element?

23. How is the emission spectrum of an element detected? What does each individual line stand for?

24. What is an absorption spectrum, and what is the specific technique used to capture it? What do the lines of this spectrum stand for?

25. Compare and contrast an emission spectrum with an absorption spectrum.

26. How might astronomers on Earth use spectroscopy to determine chemical elements that occur in stars?

27. How was helium discovered?

28. Describe the basic components of a laser. How does a laser work?

29. What is meant by optical pumping in a laser, and what specific role do mirrors play in helping to generate laser light? What is the 5% of reflected light that leaks out?

30. What is the atomic number of an element?

31. How was the periodic table of elements developed? How did the holes in the original table eventually help confirm its validity?

32. What does it mean to say the periodic table was useful because it worked? How does this relate to the scientific method? Describe an imaginary discovery that might have invalidated the periodic table.

33. In what ways are all the elements in a given column of the periodic table similar? In what ways are they different?

34. What is an electron shell? How many electrons are in the first shell of an atom? The second?

35. What is the Pauli exclusion principle?

36. How do electron shells help explain how the periodic table works?

37. What do atoms really look like?

Questions

1. The leaves of a tree are bright green. What do you think a leaf's absorption spectrum might look like?

2. Draw three different emission spectra that would appear red. (*Hint:* Think about different kinds of red objects, including a red laser, a red-hot coal, and a red sweater.)

3. Rutherford's experiment involved firing nucleus-sized bullets at atoms of gold. He found that one atom in 1000 bounced backward. Using the kind of simple pictures of the atom introduced in the chapter, speculate about how the experiment might have turned out if atoms were completely uniform in mass. What if electrons were more massive than the nucleus? (*Hint:* What happens when a bowling ball collides with a Ping-Pong ball?)

4. Advertisers often describe improvements in their products as a "quantum leap." Is this an appropriate use of the term? How big is a quantum leap?

5. Based on your knowledge of Newton's laws of motion, the laws of thermodynamics, and the nature of electromagnetic radiation, explain why the Rutherford model of the atom couldn't work.

6. When you shine invisible ultraviolet light (black light) on certain objects, they glow with brilliant colors. How can this behavior be explained in terms of the Bohr atom?

7. Why do different lasers have beams with different colors?

8. Space probes often carry compact spectrometers among their scientific hardware. What kind of spectroscopy might scientists use to determine the surface composition of the cold, outer planets that orbit the Sun? How might they use spectroscopy to determine the atmospheric composition of these planets?

9. Suppose a particular atom has only two allowable electron orbits. How many different wavelength photons (spectral lines) would result from all electron transitions in this atom?

In the following three questions the energy levels of the atom are represented by horizontal lines, where the vertical spacings between the lines are proportional to the energy differences between the levels.

10. Suppose an atom has three equally spaced energy levels, as shown. How many different photons would result from all possi-

ble electron jumps between levels? Which jump corresponds to the highest frequency photon?

11. Suppose an atom has three energy levels, as shown. How many different photons would result from all possible electron jumps between levels? Which jump corresponds to the longest wavelength photon?

12. An atom has four equally spaced energy levels, as shown. How many different photons would result from all possible electron jumps between levels? Which jump(s) corresponds to the longest wavelength photon? Which jump corresponds to the photon with the highest frequency?

13. Figure 21-4 shows five of the energy levels in the hydrogen atom. Consider three electron jumps: from level 5 to level 2, from level 4 to level 2, and from level 3 to level 2. These jumps give off red, blue-green, and violet photons. Which jumps correspond to which colors? Explain.

14. A beam of white light shines through a sample of cool hydrogen gas, as shown. Describe the light that comes out the other side. The hydrogen is then heated to a very high temperature; describe the light that comes out the other side.

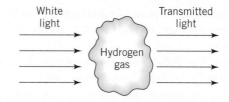

15. In his famous experiment, Rutherford fired alpha particles at a thin gold film. Most of the alpha particles went through the film and a very few bounced back. Suppose instead that about one-half the alpha particles bounced back and one-half went through. How would this have changed his conclusion about the structure of the atom?

16. In the process of fluorescence, an atom absorbs a photon of ultraviolet light and emits two or more photons of visible light. Is the reverse process possible? That is, it is possible for an atom to absorb a photon of visible light and emit photons of ultraviolet light?

17. A 100-watt lightbulb becomes warm and glows brightly enough to light a small room. On the other hand, a 100-watt laser can cut holes in steel and would not be effective at lighting a small room. What is it about the light coming from these two sources that accounts for these differences?

18. Silicon (Si) and nitrogen (N) are adjacent to carbon (C) on the periodic table. Si and C have many similar chemical properties but C and N do not. What accounts for this difference?

19. Explain why sodium chloride (NaCl) is such a stable compound.

20. Explain why two hydrogen (H) atoms combine with one oxygen (O) atom to form water (H_2O).

Problems

1. How many protons do the following elements have?
   a. Hydrogen (H)   d. Calcium (Ca)
   b. Carbon (C)   e. Iodine (I)
   c. Sulfur (S)

2. How many electrons does each of the elements in Problem 1 have when they are electrically neutral?

3. How many electrons do the following ions have?
   a. +1 of sodium (Na)   d. −1 of bromine (Br)
   b. −2 of oxygen (O)   e. +2 of calcium (Ca)
   c. +3 of iron (Fe)   f. +4 of lead (Pb)

4. How many protons do the ions in Problem 3 have? Does this differ from the number these elements would have if they were electrically neutral?

5. If a one-electron atom can occupy any of four different energy levels, how many lines might appear in that atom's spectrum?

6. If you were told that fluorine is an extremely reactive element (that is, it combines readily with other elements), what other elements could you guess were also extremely reactive? Why?

7. If you were told that argon (Ar) is an exceptionally unreactive element, what other elements could you guess were also extremely unreactive? Why?

Investigations

1. Investigate the history of the discovery of the chemical elements. What technological innovations led to the discovery of several new elements? Which was the most recent element to be discovered and how was it found?

2. Simple handheld spectrometers are available in many science labs. Use one to look at the spectra of different kinds of lightbulbs: an incandescent bulb, a fluorescent bulb, a halogen bulb, and any other kinds available to you. What differences do you observe in their spectra? Why?

3. Place pieces of transparent materials between a strong light source and the spectrometer mentioned in Investigation 2. Does the spectrum change? Why?

4. Why do colors look different when viewed indoors under fluorescent light than outdoors in sunlight? How might you devise an experiment to quantify these differences?

5. Investigate the variety of lasers that are currently available. What is the range of wavelengths available? How are different lasers used in medicine? In industry? In science?

WWW Resources

See the *Physics Matters* home page at **www.wiley.com/college/trefil** for valuable web links.

1. **http://www.chemistry.org/portal/a/c/s/1/acsdisplay.html?DOC=sitetools\periodic_table.html** A sophisticated periodic table of the elements site by the American Chemical Society. See especially the filling of electron shells.

2. **http://www.achilles.net/~jtalbot/history/** A large site devoted to the history of the laser, including rich links to laser physics applications, design, related phenomena, and newsworthy lasers.

3. **http://www.achilles.net/~jtalbot/data/elements/index.html** A site devoted to gas discharge spectra of the light elements.

4. **http://www.colorado.edu/physics/PhysicsInitiative/Physics2000/elements_as_atoms/index.html** An excellent site devoted to the quantum atom. This location within that site runs an animated tutorial with simulations that describes the physical properties of electrons within elements, followed by a tutorial on the periodic table.

5. **http://www.colorado.edu/physics/PhysicsInitiative/Physics2000/quantumzone/index.html** An excellent site devoted to the quantum atom. This location within that site runs an animated tutorial with simulations that starts with gas discharge spectra, leads through spectroscopy, the Bohr atom and electron energy levels.

6. **http://www.almaden.ibm.com/vis/stm/gallery.html** A scanning tunneling microscope image gallery from IBM. The art and beauty of atomic images.

7. **http://micro.magnet.fsu.edu/electromag/java/rutherford/index.html** A Java applet simulation of the Rutherford Experiment from Florida State University.

8. **http://www.aip.org/history/electron/** An online exhibit describing the discovery of the electron by J.J. Thompson from the American Institute of Physics.

22

Quantum
Mechanics

KEY IDEA

At the subatomic scale,
physical quantities are
quantized; any
measurement at that scale
significantly alters the
object being measured.

PHYSICS AROUND US . . . Quantum Mechanics in Your Life

Have you used a computer lately? How about a digital camera? A car radio? A calculator?
All these devices, and many more, operate according to the laws of a strange and wonderful area of physics known as *quantum mechanics,* which governs the behavior of individual atoms and subatomic particles. Take the digital camera as an example. At the heart of every digital camera is a plate of light-sensitive material called a "photoelectric device"—the same kind of material that converts the Sun's energy into electricity in a solar cell and measures brightness in a light meter. Light hitting this plate causes the release of individual electrons, which are detected by receivers in the camera. The pattern of electrons leaving the plate ultimately produces the picture you see. The interaction involved between light and electrons cannot be explained correctly by classical physics—you need quantum mechanics to understand it.

In the same way, the building block of every computer is a device known as a "transistor." As this device is turned on and off (a process we describe in more detail in Chapter 25), your computer takes in, processes, and outputs information in the form of individual electric charges. Again, the behavior of these charges cannot be described correctly by classical physics. The entire information revolution depends on the laws of quantum mechanics. In fact, all the everyday objects listed above, and many more, are practical consequences of the laws of the quantum world.

471

THE WORLD OF THE VERY SMALL

In Chapter 21 we saw that when an electron moves between energy levels and emits a photon, it is said to make a *quantum leap*. The term "quantum mechanics" refers to the theory that describes this process, as well as many other events at the scale of the atom. The word "quantum" comes from the Latin word for "bundle," and mechanics, as we have seen in Chapter 4, is the study of the motion of material objects. **Quantum mechanics,** then, is the branch of physics that is devoted to the study of the motion of objects that come in small bundles, or *quanta*. We have already seen that material inside the atom comes in little bundles—tiny bundles of matter we call *electrons* travel in orbits around another little bundle of matter we call the *nucleus*. In the language of physicists, the atom's matter is said to be *quantized*.

Electric charge is also quantized—every electron has a charge of exactly −1, while every proton has a +1 charge. We have seen that photons emitted by an atom can have only certain values of energy, so that energy levels in atoms, as well as the energy they emit, are quantized. In fact, inside the atom, in the world of the submicroscopically small, *everything* comes in quantized bundles.

Our everyday world isn't like this at all. Although we've been told since childhood that the objects around us are made up of atoms, for all intents and purposes we experience matter as if it were smooth, continuous, and infinitely divisible. For almost all everyday human activities, there's no advantage to knowing that the world is made of atoms. For example, when was the last time you had to think about matter in terms of atoms?

The quantum world is foreign to our senses. All the intuition that we've built up about the way the world operates—all the gut feelings we have about the universe—comes from our experiences with large-scale objects made up of apparently continuous material. If it should turn out (as it does) that the world of the quantum doesn't match our intuition, we shouldn't be surprised. We can't see or experience firsthand the world at the scale of the atom, so we have no particular reason based on everyday observations to believe that it should behave one way or the other.

This warning may not make you feel much better as you learn just how strange and different the quantum world really is, but it might help you come to intellectual grips with a most fascinating part of our physical universe.

Measurement and Observation in the Quantum World

Every measurement in the physical world incorporates three essential components:

1. A sample—a piece of matter to study
2. A source of energy—light or heat or kinetic energy that interacts with the sample
3. A detector to observe and measure that interaction

When you look at a piece of matter such as this book, you can see it because light bounces off the book and comes to your eye, which is a very sophisticated detector of electromagnetic radiation (see Chapter 19). When you examine a piece of fruit at the grocery store, you apply energy by squeezing it to detect if it feels ripe. You may listen to sound waves generated by a CD before you buy it—a process that involves having the CD interact with a laser beam.

Many professions employ sophisticated devices to make measurements. Air traffic controllers reflect microwaves (radar) off airplanes to determine their positions, oceanographers bounce sound waves (sonar) off deep-ocean sediments to map the sea floor, and dentists pass X rays through your teeth and gums to look for cavities. In our everyday world we assume that such interactions of matter and energy do not change the objects being measured in any appreciable way. Microwaves don't alter an airplane's flight path, nor do sound waves disturb the topography of the ocean's bottom. And although prolonged exposure to X rays can be harmful, the dentist's brief exploratory X-ray photograph has no immediate effect on the tooth. Our experience tells us that a measurement can usually be made on a macroscopic object—something large enough to be seen without a microscope—without altering that object because the energy of the probe is much less than the energy of the object.

A radar antenna sends out microwaves that interact with flying airplanes, are reflected, and are detected on their return. This allows air traffic controllers to keep track of where airplanes are in the sky.

The situation is rather different in the quantum world. If you want to "see" an electron, you have to bounce energy off it so that the information can be carried to your detectors. But nothing at your disposal can interact with the electron without simultaneously affecting it. You can bounce a photon off it, but in the process the electron's energy changes. You can bounce another particle off it, but the electron will recoil like a billiard ball. No matter what you try, the energy of the probe is too close to the energy of the electron being measured. The electron cannot fail to be altered by the interaction.

Many everyday analogies illustrate the process of measurement in the quantum world. For example, it's like trying to locate the position of a bowling ball by bouncing other bowling balls off it. The act of measurement in the quantum world poses a dilemma analogous to trying to discover if there is a car in a dark tunnel when the only means of finding out is to send another car into the tunnel and listen for a crash. With this technique you can certainly discover whether the first car is there. You can probably even find out where it is by measuring the time it takes the probe car to crash. What you *cannot* do, however, is assume that the first car is the same after the interaction as it was before. In the same way, nothing in the quantum world can be the same after the interaction associated with a measurement as it was before.

In principle, this argument applies to any interaction, whether it involves photons and electrons or photons and bowling balls. However, as we demonstrate in Example 22-1 later in this chapter, the effects of the interaction for large-scale objects are so tiny that they can be ignored, while in the case of interactions at the atomic level they cannot be ignored. This fundamental difference between the quantum and macroscopic worlds is what makes quantum mechanics quite different from the classical mechanics of Isaac Newton. Remember that every experiment, be it on planets or fruit or quantum objects, involves interactions of one sort or another. The consequences of small-scale interactions make the quantum world different, not the fact that a measurement is being made.

The Heisenberg Uncertainty Principle

In 1927, a young German physicist, Werner Heisenberg (1901–1976), put the idea of limitations on quantum-scale measurements into precise mathematical form. His work, which was one of the first results to come from the new science of quantum mechanics, is called the *Heisenberg uncertainty principle* in his honor. The central concept of the uncertainty principle is simple:

Werner Heisenberg

At the quantum scale, any measurement significantly altering the object being measured.

For example, suppose that you have a particle such as an electron in an atom and want to know where it is *and* how fast it's moving. The uncertainty principle tells us that it is impossible to measure both the position and the velocity with infinite accuracy at the same time so that there is always an uncertainty to our knowledge of some aspects of the subatomic world.

The reason for this state of affairs is that every measurement changes the object being measured. Just as the car in the tunnel could not be the same after the first measurement was made on it, so too does the quantum object change. The result is that as you measure one property such as position more and more exactly, your knowledge of a property such as velocity gets fuzzier and fuzzier.

The uncertainty principle doesn't say that we can't know a particle's location with great precision. It's possible, at least in principle, for the uncertainty in position to be zero, which would mean that we know the exact location of a quantum particle. In this case, however, the uncertainty in the velocity has to be infinite. Thus, at the point in time when we know exactly where the particle is, we have no idea whatsoever how fast it is moving. By the same token, if we know exactly how fast the quantum particle is moving, we cannot know where it is. It could, quite literally, be in the room with us or in China.

In practice, every quantum measurement involves trade-offs. We accept some fuzziness in the location of the particle and some fuzziness in the knowledge of the velocity, playing the two off against one another to get the best solution to whatever problem it is we're working on. We cannot have precise knowledge of both quantities at the same time, but we can know either one as accurately as we like at any time.

Let's look a little more closely at the differences between the world of our intuition and the quantum world. Our intuition, based on experience in the macroscopic world, suggests that a measurement doesn't affect an object being measured. According to that view, we should be able to have exact, simultaneous knowledge of both the position and velocity of an object such as a car or a baseball. In the quantum world, we cannot.

Heisenberg put his notion into a simple mathematical relationship, which is a complete and exact statement of the **uncertainty principle.** (Note that this relationship is given in terms of momentum p rather than velocity v, where $p = mv$, and m is the particle's mass.)

1. In words:

   The error or uncertainty in the measurement of an object's position, multiplied by the error or uncertainty in that object's momentum, must be greater than a constant (called Planck's constant).

2. In an equation with words:

   (Uncertainty in position) × (Uncertainty in momentum) > h

   where h is a number known as *Planck's constant.*

3. In an equation with symbols:

   $$\Delta x \times \Delta p > h$$

   where x represents the position of the particle and p its momentum. The Greek letter Δ (delta) is customarily used to represent the spread of values that a variable can have, and hence the uncertainty in our knowledge of that value.

This equation is a precise, shorthand way of saying that you can never know both the position and momentum (or velocity) of an object with perfect accuracy.

The difference between our everyday world and the world inside the atom hangs on the question of the numerical value of Planck's constant h, the number on the right side of Heisenberg's equation. In SI units (see Appendix A), Planck's constant has a value of 6.63×10^{-34} joule-seconds. This is a very small number, which is why we never notice the uncertainty principle in daily life.

The important point about the Heisenberg relationship is not the exact value of Planck's constant h, but the fact that h is greater than zero. Look at it this way. If you make more and more precise measurements about the location of a particle, you determine its position more and more exactly, and the uncertainty in position, Δx, must get smaller and smaller. In this situation, it follows that the uncertainty in velocity, Δv, has to get bigger and bigger (note that $\Delta p = m\Delta v$). In fact, we can use the uncertainty principle to calculate exactly our uncertainty in velocity for a given uncertainty in position, and vice versa.

Develop Your Intuition: Uncertainties in Time and Energy

Heisenberg developed a second form of the uncertainty principle, which said that the uncertainty in an object's energy multiplied by the uncertainty in the time interval for measuring that energy is always greater than Planck's constant h.

$$\Delta E \times \Delta t > h$$

Use this relation to determine how long you could own a new car that's created out of nothing.

The amazing thing about quantum mechanics is that you can get a real answer to this question. The first step is to estimate how much mass your car probably has. We won't be greedy and ask for a luxury limousine, so let's say 1000 kilograms will do. We know that mass can be converted to energy according to Einstein's equation $E = mc^2$ (Chapter 12). If you work out the numbers with $m = 1000$ kg and $c = 3 \times 10^8$ m/s, you get an energy E of 9×10^{19} joules or about 10^{20} J.

That's a huge amount of energy, but according to the uncertainty principle you could have an uncertainty in energy of this amount for a very short time. That means that you could (in theory) produce a car out of nothing for this short time. How short a time? The time interval would be Planck's constant, 6.63×10^{-34} joule-seconds, divided by the uncertainty in energy, which we just calculated to be about 10^{20} J. So this uncertainty in energy could exist for about 6×10^{-54} seconds. Hardly long enough to wait around for.

This example may seem pretty far-fetched and ridiculous, and in some respects it is. But the idea of something created out of nothing for a very short time is not at all ridiculous; in fact, it happens pretty often. In high-energy experiments, radiation fields can occur with such high energy that tiny particles appear, formed as mass converted from the energy. If the mass is small enough and the energy is high enough, these particles exist long enough to be detected before they disappear back into radiation. Physicists call them "virtual particles" since they are there but not there. Despite their brief existence, physicists have measured their effect on real objects. Virtual particles are just one part of the strange world of quantum mechanics.

Uncertainty in the Macroscopic and Microscopic Worlds

The best way to understand why we do not have to worry about the uncertainty principle in our everyday lives is to calculate the uncertainty of measurements in two separate situations: large objects and very small objects.

Small Uncertainties with Large Objects

EXAMPLE 22-1

A moving automobile with a mass of 1000 kilograms is located in an intersection that is 5 meters across. How precisely can you know how fast the car is traveling?

REASONING AND SOLUTION: We can solve this problem by noting that if the car is somewhere in an intersection 5 m across, then the uncertainty in position of the car is about equal to 5 m. We know the car's mass and uncertainty in position, so we can calculate the uncertainty in velocity:

Δx (Uncertainty in position) $\times \Delta p$ (Uncertainty in momentum) $> h$

Δx (Uncertainty in position) $\times [\Delta v$ (Uncertainty in velocity) $\times m$ (mass)$] > h$

We rearrange this equation to solve for uncertainty in velocity:

$$\Delta v > \frac{h/m}{\Delta x}$$

$$> \frac{(6.63 \times 10^{-34} \text{ J-s})/1000 \text{ kg}}{5 \text{ m}}$$

$$> \frac{(6.63 \times 10^{-37} \text{ J-s})/\text{kg}}{5 \text{m}}$$

$$> 1.33 \times 10^{-37} \text{ m/s}$$

Thus the uncertainty in the velocity of the automobile must be greater than 1.33×10^{-37} m/s (note that the unit J-s/kg-m is equivalent to m/s; see Problem 2 at the end of the chapter). This uncertainty is extremely small. Theoretically, we could know the velocity of the car to an accuracy of 37 decimal places! In practice, however, we have no method of measuring velocities with present or foreseeable human technology to an accuracy remotely approaching this. The uncertainty is for all practical purposes indistinguishable from zero. Therefore, for objects with significant mass such as automobiles, the effects of the uncertainty principle are totally negligible. The equation confirms our experience that Newtonian mechanics works perfectly well in dealing with everyday objects. ●

Large Uncertainty with a Small Object

EXAMPLE 22-2

Contrast Example 22-1 with the uncertainty in velocity of an electron in an atom, located within a volume about 10^{-10} meters on a side. To what accuracy can we measure the velocity of that electron?

REASONING AND SOLUTION: The mass of an electron is 9.11×10^{-31} kg. If we take the uncertainty in position to be 10^{-10} m, then according to the uncertainty principle,

$$\Delta v > \frac{h/m}{\Delta x}$$

$$> \frac{(6.63 \times 10^{-34} \text{ J-s})/(9.11 \times 10^{-31} \text{ kg})}{10^{-10} \text{ m}}$$

$$> 7.3 \times 10^6 \text{ m/s}$$

This uncertainty is very large indeed. The mere fact that we know that an electron is somewhere in an atom means that we cannot know its velocity to within a million meters per second—almost 10% of the speed of light!

For ordinary-sized objects such as cars and bowling balls, whose mass is measured in kilograms, the number on the right side of the uncertainty relation is so small that we can treat it as zero. Only when the masses get very small, as they do for particles such as the electron, does the number on the right get big enough to make a practical difference. ●

 PROBABILITIES

The uncertainty principle has consequences that go far beyond simple statements about measurement. In the quantum world, we must radically change how we describe events. Consider an everyday example in which the uncertainties are much easier to picture than those associated with Heisenberg's equation. Think of a batter hitting a ball during a nighttime baseball game.

Imagine yourself at a big-league ball game under the lights of a great stadium. Cheering fans fill the stands, roving vendors sell their food and drink, and the pitcher and batter play out their classic duel. The pitcher stares the batter down, winds up, and hurls a fastball. But the batter is ready and pounces on the pitch. The ball leaps off the bat with a sharp crack. And then all the lights go out.

Where will the ball be in 5 seconds? If you were an outfielder, this would be more than a philosophical question. You would need to know where to go to make your catch, even in the dark. In a Newtonian world, you would have no problem doing this. If you knew the position and velocity of the ball at the instant the lights went out, a few calculations could tell you exactly where the ball would be at any time in the future.

If you were a quantum outfielder in an atom-sized ball field, on the other hand, you would have a much harder time of it. You couldn't know both the position and velocity of the quantum ball when the lights go out; at best you could put some bounds on them. For example, you might be able to say something like, "The ball is somewhere inside this 3-foot circle around home plate, traveling with a horizontal speed between 30 and 70 feet per second." This result means that when you have to guess where it would be in 5 seconds, you wouldn't be able to do so exactly. In Newtonian terms, if it were traveling 30 feet per second and located at the far end of the 3 foot circle, then it would be 147 feet (147 feet = 30 ft/s × 5 s − 3 ft) from the plate. If, on the other hand, it were traveling 70 feet per second and located at the near end of the circle, then it would be 353 ft (353 ft = 70 ft/s × 5 s + 3 ft) from the plate. Hence, it could be anywhere between 147 and 353 feet from the plate. The best you could do would be to predict the likelihood, or **probability,** that the ball would be anywhere in the outfield. You could present these probabilities on a graph such as the one shown in Figure 22-1. In this graph, the most likely place to find the baseball is near the spot where the probability is highest, but the ball could be some other place instead.

This example shows that the uncertainty principle requires a description of quantum-scale events in terms of probabilities. Just like the baseball in our example of the darkened stadium, there must be uncertainties in the position and velocity for every quantum object when we first start observing it, and hence there are uncertainties at the end—uncertainties that can be dealt with by reporting probabilities.

The graph of probabilities shown in Figure 22-1 can be thought of as a wave where the amplitude of the wave corresponds to the probability of finding a particle at a specific point. For this reason, such a set of probabilities is referred to as a *wave function.* (Technically, the square of the amplitude of the wave at some point gives the probability of finding a particle at that point.) The equations of quantum mechanics, in fact, take this resemblance a step further and actually describe the way that a probability wave changes over time. In the case of our baseball stadium, for example, they would describe how the wave evolved from one describing the probabilities of finding the ball near home plate to the one shown in Figure 22-1. For this reason, quantum mechanics is sometimes called "wave mechanics."

The Austrian physicist Erwin Schrödinger (1887–1961) first wrote down the ground-breaking equation that describes the probability wave, and so it is called the *Schrödinger equation.* While the precise form of this equation is complex, Schrödinger's equation plays the same role in quantum mechanics that Newton's second law does for ordinary mechanics and Maxwell's equations do for electricity and magnetism. It describes quantitatively how a physical system evolves over time in response to outside influences.

This aspect of Schrödinger's equation is extremely important. It tells us that we cannot think of quantum events in the same way that we think of normal events

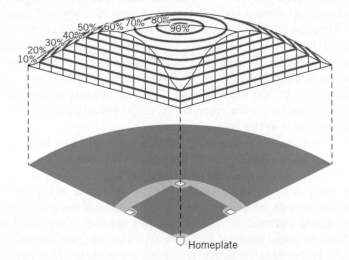

FIGURE 22-1. The wave function that represents the position of the quantum baseball. The height of the surface represents the probability of finding the baseball at a given point. Each contour represents the probability of finding the baseball within the area enclosed. For example, there is a 90% probability that the baseball will be found within the curve labeled 90%, an 80% probability that it will be within the curve labeled 80%, and so on. The most likely place to find the baseball is near the peak of the wave function, where the probability is highest.

in our everyday world. In particular, we have to rethink what it means to talk about concepts such as regularity, predictability, and causality at the quantum level.

 # WAVE-PARTICLE DUALITY

It turns out that quantum objects sometimes act like particles and sometimes act like waves. This dichotomy is known as the problem of *wave-particle duality,* and it has puzzled some of the best minds in science. To understand wave-particle duality, think about how particles and waves behave in our macroscopic world.

The Double-Slit Test

In our everyday world, energy travels either as a wave (see Chapter 14) or as a particle. Particles transfer energy through collisions, while waves transfer energy through collective motion of the media or electromagnetic fields. Every aspect of the everyday world can be neatly divided into particles or waves, and many experiments can be used to determine whether a phenomenon is a particle or a wave. The most famous of these experiments uses a double-slit apparatus, which consists of a barrier that has two slits in it, each of which will allow a particle (or a wave) to pass through the barrier (Figure 22-2). If particles such as baseballs are thrown from the left side, a few will make it through the slits, but most will

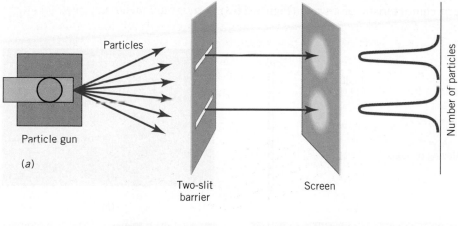

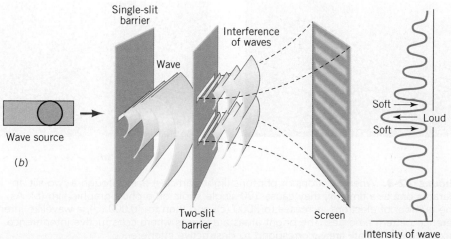

FIGURE 22-2. (*a*) The two-slit experiment may be used to determine whether something is a wave or a particle. A stream of particles striking the barrier will accumulate in the two regions directly behind the slits. (*b*) When waves converge on two narrow slits, however, constructive and destructive interference results in a series of peaks.

bounce off. If you were standing on the other side of the barrier you would expect to see the baseballs coming through more or less in the two places shown, accumulating in two areas behind the barrier. You wouldn't expect to see many particles (baseballs) winding up between the slits.

However, if waves of sound were coming from the left side, you would expect to see the results of constructive and destructive interference (see Chapter 14). Rather than the two areas of baseballs, we would see perhaps half a dozen regions of loud sound beyond the barrier, interspersed with regions of soft sound, a situation illustrated by the wave height shown in Figure 22-2*b*.

Now, let's use the same arrangement to see whether light behaves as a particle or a wave. In Chapter 21 we learned that light is emitted in discrete bundles of energy called *photons*. Photons behave like particles in the sense that they can be localized in space. You can set up experiments in which a photon is emitted at one point and then received somewhere else after an appropriate lapse of time. However, if you shine light—a flood of photons—on the two-slit apparatus, you will definitely get an interference pattern, like the one shown in Figure 22-2*b*. In that experiment, photons act like waves. The big question: How can photons sometimes act like waves and sometimes act like particles?

You can make the problem even more puzzling by setting up the apparatus so that only one photon at a time comes through the slits. If you do this, you find that each photon arrives at a specific point at the screen behind the slits— behavior you would expect of a particle. If you allow photons to accumulate over long periods of time, however, they arrange themselves into an interference pattern characteristic of a wave (Figure 22-3).

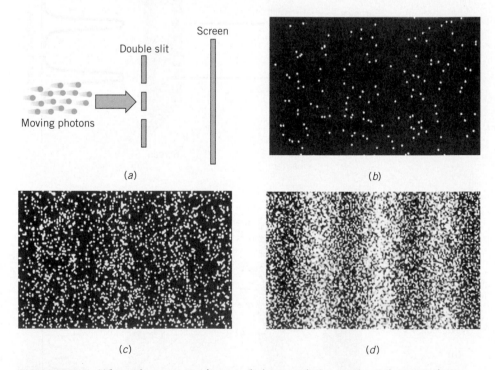

FIGURE 22-3. When electrons or photons (light particles) pass through a two-slit apparatus one at a time (*a*), they cause 100 single spots on a photographic film (*b*). As the number of electrons increases to 3000 (*c*), and then to 70,000 (*d*), a wavelike interference pattern emerges. The bright areas are places where constructive interference occurs, and the dark areas correspond to destructive interference.

The French physicist Louis de Broglie (1892–1987) put this wavelike feature of quantum objects into mathematical form. He asked a simple question: If we are to think about quantum objects as both waves and particles, how are the particlelike properties (such as momentum) related to the wavelike properties (such as wavelength)? The result of his work is known as the *de Broglie relation.*

1. In words:

The higher the momentum an object has (if we think of it as a particle), the shorter its wavelength (if we think of it as a wave).

2. In an equation with words:

The momentum of a quantum object is inversely proportional to its wavelength.

3. In an equation with symbols:

$$p = \frac{h}{\lambda}$$

where p is the momentum, λ is the wavelength, and h is Planck's constant.

You could do a similar series of experiments with any quantum object—electrons, for example, or even atoms. They all exhibit the properties of both particles and waves, depending on what sort of experiment is done. If you perform an experiment that tests the particle properties of these things, they look like particles. If you perform an experiment to test their wave properties, they look like waves. Whether you see quantum objects as particles or waves seems to depend on the experiment that you do.

Some experimenters have gone so far as to try to trick quantum particles such as electrons into revealing their true identity by using modern fast electronics to decide whether a particle- or wave-type experiment is being done *after* the quantum object is already on its way into the apparatus. Scientists who do these experiments find that the quantum object seems to "know" what experiment is being done, because the particle experiments always turn up particle properties and the wave experiments always turn up wave properties.

At the quantum level, the objects that we talk about are neither particles nor waves in the classical sense. In fact, we can't really visualize them at all because we have never encountered anything like them in our everyday experience. They are a third kind of object, neither particle nor wave, but exhibiting the properties of both. If you persist in thinking about them as if they were baseballs or surf coming onto a beach, you will quickly lose yourself in confusion.

It's a little bit like finding someone who has seen only the colors red and green in her entire life. If she has decided that everything in the world has to be either red or green, she will be totally confused by seeing the color blue. What she has to realize is that the problem is not in nature, but in her assumption that everything has to be either red or green.

In the same way, the problem of wave-particle duality arises from our assumption that everything has to be either a wave or a particle. If we allow ourselves the possibility that quantum objects are entities that we have never encountered before and that they therefore might have unencountered properties, then the puzzle vanishes. However, it vanishes only if we agree that we won't try to draw a picture of these objects or pretend that we can actually visualize what they are.

Connection
Interference of Electrons

The classic test for deciding if energy is being transmitted by particles or by waves is to see if you can detect interference effects. If you can, then you're looking at a wave; if you can't, you're looking at particles. Now, we've just learned that particles such as electrons can behave as if they're waves. So can they exhibit constructive and destructive interference?

Yes, they can. The first experiments that detected interference of electron beams were done in 1927 by C. Davisson and L. Germer and in 1928 by G. P. Thomson. (It's a lovely coincidence of history that G. P. Thomson shared a Nobel Prize in physics for demonstrating that electrons could act as waves, and his father, J. J. Thomson, received a Nobel Prize for discovering electrons as particles.) And the wave nature of electrons has been applied in a more practical way than just in laboratory experiments: it is the basis for the electron microscope.

In an electron microscope, electron beams are bent by electric and magnetic fields in the same way that light rays are bent by lenses. Optical microscopes use wavelengths of about 500 nm, which means they can resolve details of a specimen of about a few hundred nanometers—that corresponds to the length of a few thousand atoms side by side. However, by using de Broglie's relation, we can determine that an electron moving at about 10% of the speed of light (3×10^8 m/s) has a wavelength in the neighborhood of 0.1 nm, which is 100 times smaller than the typical size of an atom.

There are two main kinds of electron microscopes (Figure 22-4). Transmission electron microscopes (TEMs) send electron beams through a thin slice of a specimen, such as a biological cell wall. Scanning electron microscopes (SEMs) bounce electron beams off a specimen and collect the beams reflected at various angles, producing more of a three-dimensional effect. Both kinds of instruments are now regular parts of research in biology and materials science. ●

Connection
The Photoelectric Effect

When photons of sufficient energy strike some materials, their energy can be absorbed by electrons, which are shaken loose from their home atoms. If the material in question is in the form of a thin sheet, then when light strikes one side, electrons are observed coming out of the other side. This phenomenon is called the *photoelectric effect* and it finds applications in numerous everyday devices. (See Physics Around Us on page 471.)

One aspect of the photoelectric effect played a major role in the history of quantum mechanics. The time between the arrival of the light and the appearance of the electrons is extremely short—far too short to be explained by the gradual buildup of electromagnetic wave energy that shakes the electron loose. In fact, Albert Einstein pointed out that this rapid response depends on the particlelike nature of the photon. He argued that the interaction between the light and the electron is something like the collision between two billiard balls, with one ball shooting out instantly after the collision. It was this work, which led to our modern concept of the photon, that was the basis for Einstein's Nobel Prize in 1921.

The conversion of light energy into electric current is used in many familiar devices. For example, in a digital camera one photoelectric device measures the

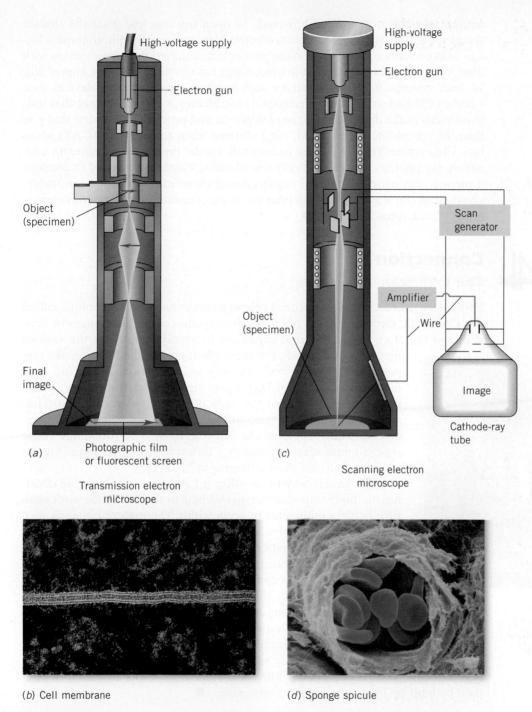

(a) Transmission electron microscope

High-voltage supply

Electron gun

Object (specimen)

Final image

Photographic film or fluorescent screen

(b) Cell membrane

(c) Scanning electron microscope

High-voltage supply

Electron gun

Scan generator

Amplifier

Wire

Object (specimen)

Image

Cathode-ray tube

(d) Sponge spicule

FIGURE 22-4. Two main kinds of electron microscopes. (a) Transmission electron microscopes (TEMs) send electron beams through a thin slice of a specimen such as a biological cell wall (b). In contrast, (c) scanning electron microscopes (SEMs) bounce electron beams off a specimen and collect the beams reflected at various angles, producing more of a three-dimensional effect for microscopic samples such as a sponge spicule (d).

amount of light to determine how wide to open the lens and what the shutter speed should be, and a second photoelectric plate collects the photographic image. Many telephone systems also use photoelectric materials in conjunction with fiber optics, the glass fibers that act like pipes for visible light (see Chapter 20). In such systems, light signals strike sophisticated semiconductor devices (see Chapter 25) and shake loose electrons. These electrons form a current that ultimately drives the diaphragm in your telephone and produces the sound that you hear. In yet another application, computerized axial tomography (CAT) scans (see Connection: The CAT Scan, below) rely on the photoelectric effect to convert X-ray photons into electric currents whose strength can be used to produce a picture of a patient's internal organs. As all these examples show, an understanding of the way that objects interact in the quantum world can have enormous practical consequences. ●

Connection
The CAT Scan

Photoelectric detectors play a crucial role in a modern medical technique called the CAT scan. Ordinary X-ray photographs depend on the differences in density (and therefore in the different capacities to absorb X rays) of the various

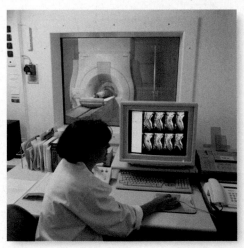

materials in the body. For these photographs, the X rays make one pass through your body in only one direction, producing two-dimensional pictures. They cannot produce a three-dimensional image of the interior of the body, nor can they produce sharp images of the body's organs, whose densities are generally not significantly different from the densities of their surroundings. These shortcomings are overcome by a different X-ray technique known as "computerized axial tomography" (CAT).

The easiest way to visualize a CAT scan is to imagine dividing the body into slices perpendicular to the backbone, with each slice being a millimeter or so in width. The material in each slice is probed by successive short bursts of X rays, lasting only a few milliseconds each, that cross the slice in different directions. Each part of the slice is thus traversed by many different X-ray bursts. Each burst of X rays contains the same number of photons when it starts, and the ones that go all the way through the body (i.e., those not absorbed by material along their path) are measured by a photoelectric device (see Connection: The Photoelectric Effect, above).

A boy having a CAT scan; a video monitor is in the foreground. A CAT scan of a human skull and brain is shown on the right.

Once all the data on a given slice have been obtained, a computer works out the density of each point of the body and produces a detailed cross section along that particular slice. A complete picture of the body (or a specific part of it) can then be built up by combining successive slices. ●

WAVE-PARTICLE DUALITY AND THE BOHR ATOM

Treating electrons as waves helps explain why only certain orbits are allowed in atoms (see Chapter 21). As we have seen, every quantum object displays a simple relationship between its speed (when we think of it as a particle) and its wavelength (when we think of it as a wave). It turns out that for electrons, protons, and other quantum objects, a faster speed always corresponds to a shorter,

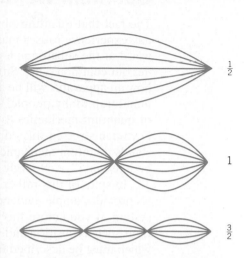

FIGURE 22-5. A vibrating string adopts the regular pattern known as a standing wave. These photos and diagrams illustrate fixed patterns with $\frac{1}{2}$, 1, and $\frac{3}{2}$ wavelengths.

more energetic wavelength (or a higher frequency). This idea is given quantitative form in the de Broglie relationship.

If you think of an electron as a particle, then you can treat its motion around an atom's nucleus in the same Newtonian way that you treat the motion of Earth in orbit around the Sun. That is, for any given distance from the nucleus, the electron must have a precise velocity to stay in a stable orbit. Provided it is moving at such a velocity, it stays in that orbit, just as Earth stays in a stable orbit around the Sun. Any slower and it must adopt a higher orbit; any faster and it moves closer to the nucleus.

However, if we choose to think about the electron as a wave, a different set of criteria can be used to decide how to put the electron into its orbit. A wave on a straight string (on a guitar, for example) vibrates only at certain frequencies that depend on the length of the string (Figure 22-5). These frequencies correspond to fitting $\frac{1}{2}$, 1, and $\frac{3}{2}$ wavelengths on the string in the figure. Now imagine bending the guitar string around into a circle. In this case, you will be able to fit only certain standing waves on the circular string, as shown in Figure 22-6.

You can now ask a simple question: Are there any orbits for which the wave and particle descriptions are consistent? In other words, are there orbits for which the velocity of the electron (when we think of it as a particle) is appropriate to the orbit, while at the same time the electron wave (when we think of it as a wave) fits onto the orbit, given the relation between wavelength and velocity?

When you do the mathematics, you find that the only orbits that satisfy these twin conditions are the Bohr orbits. That is to say, the only orbits allowed in the atom are those for which it makes no difference whether we think of the electron as a particle or a wave. In a sense, then, the wave-particle duality exists in our minds, and not in nature—nature has arranged things so that what we think doesn't matter.

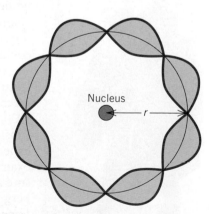

FIGURE 22-6. An electron in orbit about an atom adopts a standing wave like a vibrating string. This illustration shows a standing wave with four wavelengths fitting into the orbit's circumference.

Quantum Weirdness

The fact that quantum objects behave so differently from objects in our everyday experience causes many people to worry that nature has somehow become weird at the subatomic level. The description of particles in terms of waves defies our common sense. Situations in which a photon or electron seems to "know" how an apparatus will be arranged before the arranging is done seem wrong and unnatural. Many people, scientists and nonscientists alike, find the conclusions of quantum mechanics to be quite unsettling. The American physicist Richard Feynman stressed this point when he said, "I can safely say that nobody understands quantum mechanics. . . . Do not keep saying to yourself, 'But how can it be like that?' . . . Nobody knows how it can be like that."

In spite of this rather disturbing situation, the success of quantum mechanics provides ample evidence that it is a correct way of describing an atomic-scale system. If you ignore this fact, you can get into a lot of trouble. Newtonian notions such as position and velocity just aren't appropriate for the quantum world, which must be described from the beginning in terms of waves and probabilities. Quantum mechanics thus becomes a way of predicting how subatomic objects change in time. If you know the state of an electron now, you can use quantum mechanics to predict the state of that electron in the future. This process is identical to the application of Newton's laws of motion in the macroscopic world. The only difference is that in the quantum world, the "state" of the system is described in terms of a probability.

In the view of most working scientists, quantum mechanics is a marvelous tool that allows us to do all sorts of experiments and build all manner of new and important pieces of equipment. The fact that we can't visualize the quantum world in familiar terms seems a small price to pay for all the benefits we receive.

Physics in the Making
A Famous Exchange

Many people are disturbed by the fact that nature at the subatomic level must be described in terms of probabilities. When quantum mechanics was first developed in the early twentieth century, many physicists were also troubled. Even Albert Einstein, who contributed one of the key ideas of quantum mechanics, could not accept what it was telling us about the world.

Einstein and Bohr were lifelong friends as well as colleagues. However, Einstein spent a good part of the last half of his life trying to refute quantum mechanics, while Bohr defended it. At major physics conferences in 1927 and in 1930, the two men exchanged ideas every day. As other physicists described it, Einstein would come down to breakfast every morning with a beautiful thought experiment he had devised for which quantum mechanics would not work. For example, he might come up with a situation in which he thought he could measure a particle's position and velocity with complete accuracy despite the Heisenberg uncertainty relation. Bohr would think about the problem all day, talking to people, trying to find flaws with Einstein's reasoning. Every evening at dinner, he would have the solution worked out and show Einstein how the uncertainty relation held or other ideas of quantum mechanics still worked.

Einstein's most famous statement from this period was, "I cannot believe that God plays dice with the universe." Confronted once too often with this aphorism, Bohr is supposed to have replied, "Albert, stop telling God what to do." ●

Niels Bohr and Albert Einstein discuss quantum mechanics during a physics conference in Brussels, Belgium, in October 1930.

THINKING MORE ABOUT

Quantum Mechanics: Uncertainty and Human Beings

The ultimate Newtonian view of the universe was the concept of the Divine Calculator (see Chapter 5). Given the position and velocity of every particle in the universe, this imaginary being could predict every future state of those particles. The difficulty with this concept is that if the future of the universe is laid out with clockwork precision, it allows no room for human action. No one can make a choice about what he or she will do because that choice is already determined and exists (in the mind of the Divine Calculator) before it is made.

Quantum mechanics gives us one way to get out of this particular bind. Heisenberg tells us that although we might be able to predict the future if we knew the position and velocity of every particle exactly, we can never actually get those two numbers at the same time. The Divine Calculator in a quantum world is doomed to wait forever for the input data with which to start the calculation.

One area where the uncertainty principle is starting to play a somewhat unexpected role is in the old philosophical argument about the connection between the mind and the brain. The brain is a physical object, an incredibly complex organ that processes information in the form of nerve impulses. The problem: What is the connection between the physical reality of the brain—the atoms and structures that compose it—and the consciousness that we all experience?

Many scientists and philosophers have argued that the brain is no more than a physical structure. These thinkers have run into a problem, however, because if the brain is purely a physical object, its future states should be predictable. Recently, scientists (most notably Roger Penrose of Cambridge University) have argued that quantum mechanics can introduce a kind of unpredictability that squares better with our perceptions of our own minds.

Think about how the workings of the brain might be unpredictable at the quantum level. Why might that uncertainty make it difficult (or even impossible) to make precise predictions of the future state of the brain?

Summary

Matter and energy at the atomic scale come in discrete packets called quanta. The rules of **quantum mechanics,** the laws that allow us to describe and predict events in the quantum world, are disturbingly different from Newton's laws of motion.

At the quantum scale, unlike our everyday experience, any measurement of the position or velocity of a particle causes the particle to change in unpredictable ways. The mere act of measurement alters the thing being measured. Werner Heisenberg quantified this situation in the **uncertainty principle,** which states that the uncertainty in the position of a particle multiplied by the uncertainty in its

momentum must be greater than a small positive number. Unlike the Newtonian world, you can never know the exact position and velocity of a quantum particle.

These uncertainties preclude us from describing atomic-scale particles in the classical way. Instead, quantum descriptions are given in terms of **probabilities** that an object is in one state or another. Furthermore, quantum objects are not simply particles or waves, a dichotomy familiar to us in the macroscopic world. They represent something completely different from our experience, incorporating properties of both particles and waves.

Key Terms

probability The likelihood that a certain event or outcome will occur. (p. 478)

quantum mechanics The branch of physics devoted to the study of very small systems in which physical quantities come in discrete bundles called quanta. (p. 472)

uncertainty principle A physical law that places limits on the accuracy to which certain quantities (momentum and position, for example) can be measured simultaneously. (p. 474)

Key Equations

Δx (Uncertainty in position) $\times \Delta p$ (Uncertainty in momentum) $> h$

$$p = \frac{h}{\lambda}$$

Review

1. What does "quantum mechanics" mean?

2. Give three examples of properties that are quantized at the scale of an electron.

3. What are the three essential parts of every physical measurement?

4. In what way is a measurement at the quantum scale of an electron different from a measurement at the large scale of everyday objects?

5. What is the Heisenberg uncertainty principle?

6. Under what circumstances can you know the position of an electron with great accuracy?

7. Under what conditions can you know the velocity of an electron with great accuracy?

8. The equation form of the uncertainty principle is $\Delta x \times \Delta v > h/m$. What does each variable stand for? Restate this equation in your own words.

9. Why is quantum mechanics sometimes called "wave mechanics"?

10. What is a wave function?

11. What role does probability play in describing subatomic events?

12. What is wave-particle duality? Give an everyday example.

13. Which properties of electrons are particlelike? Which are wavelike?

14. Light is emitted in discrete bundles called photons. Does a photon behave like a particle or a wave? Explain.

15. What is a double-slit experiment? What is the difference between a baseball that goes through a slit and a wave that goes through?

16. How is the momentum of a quantum object related to its wavelength?

17. What is the de Broglie relation?

18. Explain how the photoelectric effect works. Does it depend on the wave nature or the particle nature of light?

19. Give several real-life examples of devices that depend on the photoelectric effect.

20. How does a CAT scan work? How does it differ from an ordinary X-ray photograph?

21. What is an allowed orbit? What two conditions must be satisfied for an electron to be in an allowed orbit?

22. How does wave-particle duality explain the allowed orbits of electrons in atoms?

23. What is quantum weirdness?

24. Why did Albert Einstein use playing dice as an analogy for quantum mechanics?

25. What is the Divine Calculator? Are its predictions consistent with quantum mechanics? Explain.

Questions

1. John measures the position of an electron (Δx) to an accuracy of $\pm 10^{-9}$ m, while Jill measures the position of another electron to an accuracy of $\pm 10^{-10}$ m. After these measurements, who is more unsure of the electron's speed? Explain.

2. Your friends, John and Jean, are both driving from Chicago to Des Moines. You know that Jean is on the road, and you know when she left Chicago. On the

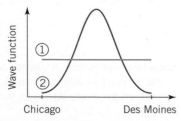

other hand, you know that John is on the road, but you have no idea when he left. The figure shows two wave functions. Which one is for John and which is for Jean? Explain.

3. A freight train leaves Dallas, traveling at about 100 miles per hour in a straight line. It makes a 400-mile trip, stopping at 100, 200, and 300 miles to unload freight. These stops take about 1 hour each. Reproduce the figure and make

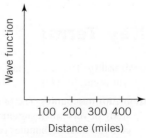

an approximate graph of the wave function for the following three cases:

   a. You know that the train has left Dallas, but you do not know when. You are aware, however, that it has not finished its trip.

   b. You know that the train is about 4 hours into its trip.

   c. You know that the train is about 6 hours into its trip.

4. Little Annie walks to school every day. There is a candy store on the way. Reproduce the figure and graph Annie's wave function for the following three cases:

   a. It's the middle of the day and Annie is supposed to be at school, but it is known that she occasionally sneaks out and goes to the candy store.

   b. It is nighttime and she is sleeping at home.

   c. She is on her way home, but since she has no money we know she will not stop at the candy store.

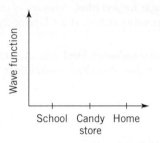

5. Sketch a possible probability diagram for the final resting position of a golf ball on a driving range. Assume that the golf tee is the starting point and that an average drive is 250 feet.

6. Chaotic systems are, for all practical purposes, unpredictable (see Chapter 5). How does this sort of unpredictability differ from that associated with quantum mechanics?

7. If you threw baseballs through a large two-slit apparatus, would you produce a diffraction pattern? Why or why not?

8. There was once a humorous poster showing a picture of a bed with the caption, "Heisenberg may have slept here." In what way is this an inaccurate representation of Heisenberg's uncertainty principle?

9. A hydrogen atom and a uranium atom are moving at the same speed. Which one has the longer wavelength?

10. An electron and a proton are traveling at the same speed. Which one has more momentum? Which one has a longer wavelength?

11. A small amount of water is brought to a boil in a microwave oven and then removed. An accurate mercury thermometer is taken from the refrigerator and used to take the water's temperature. The reading on the thermometer is 98°C even though the temperature of the water was 100°C. Why was this measurement not accurate? How is this measurement similar to what happens in quantum mechanics?

12. Johnny spends most of his time indoors. Other than an occasional trip to the bathroom, he spends his time sleeping and watching TV in his bedroom. He never visits the kitchen or the living room. If Johnny were a quantum object, how would you describe his wave function?

13. Why are ultraviolet photons more effective at inducing the photoelectric effect than visible light photons?

Problems

1. A ball (mass 0.1 kg) is thrown with a speed between 20.0 and 20.1 m/s. How accurately can we determine its position?

2. In Example 22-1, we converted the unit J-s/kg-m to the unit of velocity (m/s) without comment. Demonstrate the equivalence of these two units.

3. An atom of iron (mass 10^{-25} kg) travels at a speed between 20.0 and 20.1 m/s. How accurately can we determine its position? Could we ever actually measure a position to this accuracy? How does the uncertainty in position compare to the size of an atom? Of a nucleus?

Investigations

1. Look up the doctrine of predestination in an encyclopedia. Does it have a logical connection to the notion of the Divine Calculator? Which came first historically?

2. Werner Heisenberg was a central, and ultimately controversial, figure in German science of the 1930s and 1940s. Read a biography of Heisenberg. Discuss how his early work in quantum mechanics influenced his prominent scientific role in Nazi Germany.

3. What changes in artistic movements were taking place during the period around 1900 (just before the discoveries of quantum mechanics) and in the mid-twentieth century? Are there any connections between the artistic and scientific movements of those times?

4. Some people interpret the Heisenberg uncertainty principle to mean that you can never really know anything for certain. Do you agree or disagree?

WWW Resources

See the *Physics Matters* home page at **www.wiley.com/college/trefil** for valuable web links.

1. **http://www.aip.org/history/heisenberg/** Online exhibit on Werner Heisenberg and the Uncertainty Principle by the American Institute of Physics.

2. **http://www.colorado.edu/physics/PhysicsInitiative/Physics2000/atomic_lab.html** A presentation of the classic experiments of quantum physics as animated tutorials and simulations. This section includes quantum interference and the Bose-Einstein condensate.

3. **http://www.colorado.edu/physics/PhysicsInitiative/Physics2000/quantumzone/photoelectric.html** Presentation of the classic experiments of quantum physics as animated tutorials and simulations. This section includes the photoelectric effect.

4. **http://www.aip.org/history/einstein/** An online exhibit of the American Institute of Physics describing the life and legacy of Albert Einstein.

5. **http://www.aip.org/history/electron/** An online exhibit describing the discovery of the electron by J.J. Thompson from the American Institute of Physics.

6. **http://www-groups.dcs.st-and.ac.uk/~history/HistTopics/The_Quantum_age_begins.html** A history of quantum mechanics with biographies of the major scientists from the MacTutor History of Mathematics archive at the University of St Andrews, Scotland. Includes entries on Bohr, Einstein, and other figures.

7. **http://www.colorado.edu/physics/PhysicsInitiative/Physics2000/quantumzone/index.html** The quantum atom: a lavishly illustrated and Java-rich (and often slow) site includes a detailed storyline describing the spectral lines, the Bohr atom, and related topics.

23 | Chemical Bonds and Physical Properties

KEY IDEA

The properties of a material depend on the atoms from which it is made and how they are bonded together.

PHYSICS AROUND US . . . The Mystery of Dropped Objects

It's early morning and you're still a little groggy. As you reach for the cereal box, your hand brushes against a dish on the table, knocking it to the floor. It shatters into pieces, and your morning becomes just a little longer while you sweep them up.

Later that evening you decide to bake some cook-

ies. As you pull one cookie tray out of the drawer, another one falls to the floor with a clatter. The metal tray doesn't break the way the dish did, however, so you just pick it up and put it back in the drawer.

Why do the plate and the tray behave so differently when they're dropped?

MATERIALS AND THE MODERN WORLD

The materials people use, perhaps more than any other facet of a culture, define the technical sophistication of their society. We speak of the most primitive human cultures as Stone Age societies and recognize Bronze Age and Iron Age peoples as progressively more advanced.

College student in a dorm room, surrounded by all sorts of materials.

Take a moment and look around your room. How many different kinds of materials do you see? The lights and windows employ glass—a brittle, transparent material. The walls may be made out of gypsum, a chalk-like mineral that has been compressed in a machine and placed between sheets of heavy paper. Your chair probably incorporates several materials, including metal, wood, woven fabric, and glues.

Many of these materials would not have been familiar to Americans 200 years ago, when almost everything was made from fewer than a dozen common substances: wood, stone, pottery, glass, animal skin, natural fibers, and a few basic metals such as iron and copper. However, thanks to the discoveries of chemists, the number of everyday materials has increased by a thousandfold in the past two centuries. Cheap and abundant steel—much stronger than iron and rust-resistant, as well—transformed the nineteenth-century world with railroads and skyscrapers. Aluminum provided a lightweight metal for thousands of applications. The development of rubber, synthetic fibers, and a vast array of plastics affected every kind of human activity from industry to sports. Brilliant new pigments enlivened art and fashion, while new medications cured many ailments and prolonged lives. And in our electronic age, the application of semiconductor materials has changed life in the United States in ways that our eighteenth-century ancestors could not have imagined.

Chemists take natural elements and compounds from earth, air, and water and devise thousands of useful materials. They succeed, in part, because materials display so many different properties: color, smell, hardness, luster, flexibility, heat capacity, solubility in water, texture, melting point, strength—the list goes on and on. Each new material holds the promise of doing some job more cheaply or more safely or otherwise better than any other.

Based on our understanding of atoms and the ways they bond together, we now realize that the properties of every material depend on three essential features:

1. The kinds of atoms of which it is made.

2. The way those atoms are arranged.

3. The way the atoms are bonded to one another.

In this chapter we look at different properties of materials and see how they relate to their atomic architecture. We examine the strength of materials—how well they resist outside forces. In the next chapter, we look at the ability of materials to conduct electricity and whether they are magnetic. Finally, in Chapter 25, we describe what are perhaps the most important new materials in modern society: the semiconductor and the microchip.

ELECTRON SHELLS AND THE CHEMICAL BOND

Think about how two atoms might interact. You know that the atom is mostly empty space, with a tiny, dense nucleus surrounded by swift electrons. If two atoms approach one another, their outer electrons—the "border guards" if you will—encounter one another first. Whatever holds two atoms together thus involves primarily those outer electrons. In fact, the outer electrons play such an important role in determining how atoms combine that they are given the special name "valence electrons." Chemical bonding often involves an exchange or sharing of valence electrons, and the number of electrons in an atom's outermost shell determines what is called the atom's "valence." Chemists often express the importance of the number of outer electrons by saying that valence represents the combining power of a given atom.

It turns out that by far the most stable arrangement of electrons—the electron configuration of lowest energy—is a completely filled outer shell. It is a fact of chemical life that different electron shells hold different numbers of electrons, which gives rise to the structure of the periodic table of the elements (see Chapter 21 and Figure 23-1). A glance at the periodic table (see Appendix C) tells us that atoms with a total of 2, 10, 18, or 36 electrons (the atoms that appear in the extreme right-hand column) have filled shells and very stable configurations. Atoms with this many electrons in their outermost shells are inert gases (also called noble gases), which do not combine readily with other materials. Indeed, helium, neon, and argon, with atomic numbers 2, 10, and 18 and thus completely filled electron shells, are the only common elements that do not ordinarily react with other elements.

Every object in nature tries to reach a state of lowest energy, and atoms are no exception. Atoms that do not have the magic number of electrons (2, 10, 18, etc.) are more likely to react with other atoms to produce a state of lower combined energy. You are familiar with this process in many other natural systems. For example, if you put a ball on top of a hill, the ball will tend to roll down to the bottom, creating a system of lower gravitational potential energy. Similarly, a compass needle tends to align itself spontaneously with Earth's magnetic field, thereby lowering its magnetic potential energy. In exactly the same way, when two or more atoms come together the electrons tend to rearrange themselves to minimize the chemical potential energy of the entire system. This situation may

1 H Hydrogen Valence, +1							2 He Helium Valence, 0
3 Li Lithium Valence, +1	4 Be Beryllium Valence, +2	5 B Boron Valence, +3	6 C Carbon Valence, +4	7 N Nitrogen Valence, −3	8 O Oxygen Valence, −2	9 F Fluorine Valence, −1	10 Ne Neon Valence, 0
11 Na Sodium Valence, +1	12 Mg Magnesium Valence, +2	13 Al Aluminum Valence, +3	14 Si Silicon Valence, +4	15 P Phosphorus Valence, −3	16 S Sulfur Valence, −2	17 Cl Chlorine Valence, −1	18 Ar Argon Valence, 0

FIGURE 23-1. The first three rows of the periodic table showing element names, atomic numbers, and principal valences. Positive valences indicate the number of electrons an element can donate or share with other elements to form compounds; negative valences indicate the number of electrons an element can accept or share to form compounds.

require that they exchange or share electrons. As often as not, that process involves rearrangements with a total of 2, 10, 18, or 36 electrons.

Chemical bonds result from any redistribution of electrons that leads to a more stable configuration between two or more atoms, especially that of a filled electron shell.

> *Most atoms adopt one of three simple strategies to achieve a filled shell: they give away electrons, accept electrons, or share electrons.*

If the bond formation takes place spontaneously, without outside intervention, energy is released in the reaction. The burning of wood or paper (once their temperature has been raised high enough) is a good example of this sort of process. The heat you feel when you put your hands toward a fire derives ultimately from the chemical potential energy that is given off as electrons and atoms are reshuffled. Alternatively, atoms may be pushed into new configurations by adding energy to systems. Much of industrial chemistry, from the smelting of iron to the synthesis of plastics, operates on this principle.

Types of Chemical Bonds

Atoms link together by three principal kinds of chemical bonds—ionic, metallic, and covalent—all of which involve redistributing electrons between atoms. In addition, three types of attractive forces—polarization, van der Waals interactions, and hydrogen bonding—can result from the shifting of electrons within their atoms or groups of atoms. Each type of bonding or attraction corresponds to a different way of rearranging electrons, and each produces distinctive properties in the materials it forms.

Ionic Bonds We have seen that atoms with 2, 10, 18, or 36 electrons are particularly stable. By the same token, atoms that differ from these magic numbers by only one electron in their outer shells are particularly reactive—in effect, they are "anxious" to fill or empty their outer shells. Such atoms tend to form **ionic bonds,** chemical bonds in which the electric force between two oppositely charged ions holds the atoms together.

Ionic bonds often form as one atom gives up an electron while another receives it. For example, sodium (a soft, silvery white metal) has 11 electrons in an electrically neutral atom—2 in the lowest shell, 8 in the next, and a single electron in its outer shell. Sodium's best bonding strategy, therefore, is to lose one electron. Element 17, chlorine (a yellow-green toxic gas), is one electron shy of a filled shell. Highly corrosive chlorine gas reacts with almost anything that can give it an extra electron. When you place sodium in contact with chlorine gas, the result is predictable: in a fiery reaction, each sodium atom donates its extra electron to a chlorine atom (Figure 23-2).

In the process of this vigorous electron exchange, atoms of sodium and chlorine become electrically charged—they become ions. Neutral sodium has 11 positive protons in its nucleus, balanced by 11 negative electrons. By losing an electron, sodium becomes an ion with one unit of positive charge, shown as Na^+ in Figure 23-2. Similarly, neutral chlorine has 17 protons and 17 electrons. The addition of an extra negative electron creates a chloride ion with one unit of negative charge, shown as Cl^- in the figure. The mutual electrical attraction of positive sodium and negative chloride ions is what forms the ionic bonds between

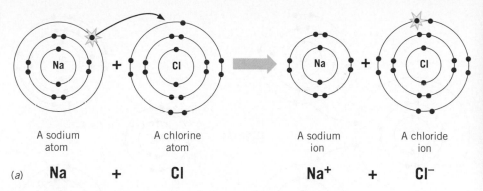

(a) **Na   +   Cl   Na⁺   +   Cl⁻**

FIGURE 23-2. (a) Sodium, a highly reactive element, readily transfers its single valence electron to chlorine, which is one electron shy of the "magic" number 18. In these diagrams, electrons are represented as dots in shells around a nucleus. (b) The result of this fiery reaction is the ionic compound sodium chloride, or ordinary table salt.

(b)

sodium and chlorine. The resulting compound, sodium chloride (common table salt), has properties totally different from either sodium or chlorine.

Under normal circumstances, sodium and chloride ions lock together into a crystal, a regular arrangement of atoms such as the one shown in Figure 23-3. Alternating sodium and chloride ions form an elegant repeating structure in which each Na^+ is surrounded by six Cl^-, and vice versa.

Ionic bonds may involve more than a single electron transfer. For example, element 12, magnesium, donates two electrons to oxygen, which has six valence electrons. In the resulting compound, MgO (magnesium oxide), both atoms have stable filled shells of eight electrons, and the ions, Mg^{2+} and O^{2-}, form a strong ionic bond. Ionic bonds involving the negative oxygen ion O^{2-} and positive ions, such as aluminum (Al^{3+}), magnesium (Mg^{2+}), and iron (Fe^{2+} or Fe^{3+}), are found in many everyday objects: in most rocks and minerals, in china and glass, and in bones and egg shells.

The ionic bonds in these compounds can be very strong, but only in certain ways. You can picture how this works by thinking about Tinkertoys. A Tinkertoy structure can be quite strong: when assembled, it is difficult to break one apart by just pushing in the directions of the sticks. But Tinkertoy bonds break easily if you twist or snap the sticks. In the same way, ionic bonds hold atoms together, but if for some reason the atoms become displaced, the bond can't hold them very well. As a consequence, ionic-bonded materials such as rock, glass, and eggshells are usually quite brittle. These materials are strong in the sense that you can pile a lot of weight on them. But once they shatter and the ionic bonds are broken, they can't be put back together again. This is why the dish described in Physics Around Us (page 491) shattered when it hit the floor.

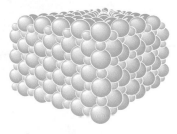

⚪ Sodium ion (Na^+)

⚪ Chloride ion (Cl^-)

FIGURE 23-3. The atomic structure of a sodium chloride crystal consists of a regular pattern of alternating sodium and chloride ions.

Ionic Bonding of Three Atoms

Magnesium chloride, which plays an important role in some types of batteries, is an ionic-bonded compound with one part magnesium to two parts chlorine ($MgCl_2$). How are the electrons arranged in this compound?

REASONING: From the periodic table (see Appendix C), magnesium and chlorine are elements 12 and 17, respectively. Magnesium, therefore, has 10 electrons (2 + 8) in inner shells and 2 valence electrons. Chlorine has 10 electrons (2 + 8)

EXAMPLE
23-1

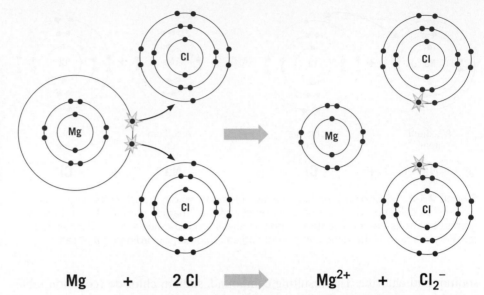

FIGURE 23-4. Magnesium and chlorine neutral-atom electron configurations (*left*), and their configurations after electrons have been transferred from the magnesium to the chlorine atoms (*right*).

$$Mg \quad + \quad 2\,Cl \quad \Longrightarrow \quad Mg^{2+} \quad + \quad Cl_2^-$$

in its inner shells and 7 electrons in the outer one, meaning that it is 1 electron short of a filled outer shell (Figure 23-4).

SOLUTION: Magnesium has two electrons to give and chlorine seeks one electron to achieve stable filled outer shells. Thus magnesium gives one electron to each of two chlorine atoms, and the resulting Mg^{2+} ion attracts two Cl^- ions to form $MgCl_2$. ●

Metallic Bonds Atoms in an ionic bond transfer electrons directly—electrons are on permanent loan from one atom to another. Atoms in a metal also give up electrons, but they use a very different bonding strategy. In a **metallic bond,** electrons are redistributed so that many atoms share them.

Sodium metal, for example, is made up entirely of individual sodium atoms. All of these atoms begin with 11 electrons, but they release one to achieve the more stable 10-electron configuration. The extra electrons move away from their parent atoms to float around the metal, forming a kind of sea of negative charge. In this negative electron sea, the positive sodium ions adopt a regular crystal structure, as shown in Figure 23-5.

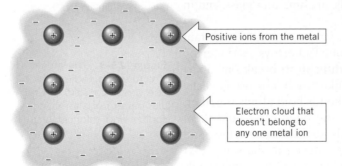

Positive ions from the metal

Electron cloud that doesn't belong to any one metal ion

FIGURE 23-5. Metallic bonding, in which a bond is created by the sharing of electrons among several metal atoms.

You can think of the metallic bond as one in which each atom shares its outer electron with all the other atoms in the system. Picture the free electrons as a kind of loose glue in which the metal ions are placed. In fact, the idea of a metal as a collection of marbles (the ions) in a sea of stiff, gluelike liquid provides a useful analogy.

Metals are formed by almost any element or combination of elements in which large numbers of atoms share electrons to achieve a more stable electron arrangement. Metals are characterized by their shiny luster and ability to conduct electricity. These properties are due to the ability of the loosely held electrons to interact with electromagnetic waves and fields. Some metals, such as aluminum, iron, and copper, are familiar from everyday experience. However, many elements can form a metallic state when the conditions are right, including some that we normally think of as gases, such as hydrogen or oxygen at very high pressure. (The planet Jupiter consists mostly of metallic hydrogen.) In fact, the great majority of chemical elements can occur in a metallic state. In addition, two or more elements can combine to form a metal "alloy," such as brass (a mixture of copper and zinc) or bronze (an alloy of copper and tin). Modern specialty-steel alloys often contain more than half a dozen different elements in carefully controlled proportions.

The special nature of the metallic bond explains many of the distinctive properties we observe in metals. If you attempt to deform a metal by pushing on the marble-and-glue bonding system, atoms will gradually rearrange themselves and come to some new configuration—the metal is malleable. It's hard to break a metallic bond just by pushing or twisting because the atoms are able to rearrange themselves. Thus, when you hammer on a piece of metal you leave indentations but do not break it, in sharp contrast to what happens when you hammer on a ceramic plate.

In the next chapter we examine more closely the electric properties of materials held together by metallic bonds. We see that this particular kind of bond produces materials through which electrons—electric current—can flow.

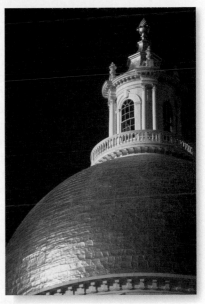

Gold metal is so soft and malleable that it can be hammered paper-thin and applied to surfaces—an art known as gilding. The dome of the State House in Boston, Massachusetts, is gilded in this manner.

Covalent Bonds In an ionic bond, one atom donates electrons to another on more or less permanent loan. In a metallic bond, on the other hand, atoms share some electrons throughout the material. In between these two types of bond is the extremely important **covalent bond,** in which well-defined clusters of neighboring atoms, called *molecules,* share electrons. These strongly bonded groups may consist of anywhere from two atoms to many millions.

The simplest covalently bonded molecules contain two atoms of the same element, such as the diatomic gases hydrogen (H_2), nitrogen (N_2), and oxygen (O_2). In the case of hydrogen, for example, each atom has a relatively unstable single electron. Two hydrogen atoms can pool their electrons, however, to create a more stable two-electron arrangement. The two hydrogen atoms must remain close to one another for this sharing to continue, so a chemical bond is formed, as shown in Figure 23-6. Similarly, two oxygen atoms, each with eight electrons (six in the outer shell), share two pairs of electrons.

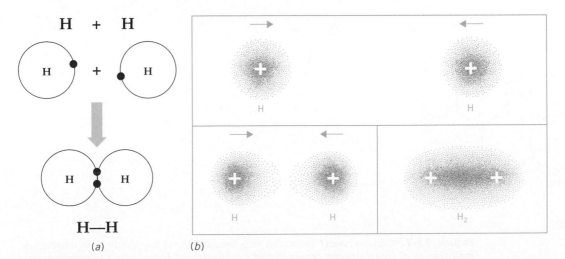

FIGURE 23-6. Two hydrogen atoms become an H_2 molecule by sharing each of their electrons in a covalent bond. This bonding may be represented schematically in a dot diagram (*a*) or by the merging of two atoms with their electron clouds (*b*).

Hydrogen, oxygen, nitrogen, and other covalently bonded molecules have lower chemical potential energy than isolated atoms have because electrons are shared. The negative electrons are attracted to two positive nuclei, not one, which reduces their potential energy. These molecules are less likely to react chemically than are the isolated atoms.

The most fascinating of all elements that form covalent bonds is carbon, which forms the backbone of all life's essential molecules. Carbon, with two electrons in its inner shell and four in its outer shell, presents a classic case of a half-filled shell. When carbon atoms approach one another, therefore, a real question arises as to whether they ought to accept or donate four electrons to achieve a more stable arrangement. You could imagine, for example, a situation where some carbon atoms give four electrons to their neighbors, while other carbon atoms accept four electrons, to create a compound with strong ionic bonds between C^{4+} and C^{4-}. Alternatively, carbon might become a metal in which every atom releases four electrons into an extremely dense electron sea. But neither of these things happens.

In fact, the strategy that lowers the energy of the carbon–carbon system the most is for the carbon atoms to share their outer electrons. Once bonds between carbon atoms have formed, the atoms have to stay close to each other for the sharing to continue. Thus the bonds generated are just like the bond in the hydrogen molecule. The case of carbon is unusual, however, because the shape of its electron shells allows a single carbon atom to form covalent bonds with up to four other atoms by sharing one of its four valence electrons with each. A *single bond* (shown as C—C) forms when one electron from each atom is shared, while a *double bond* (shown as C=C) results when two electrons from each atom are shared.

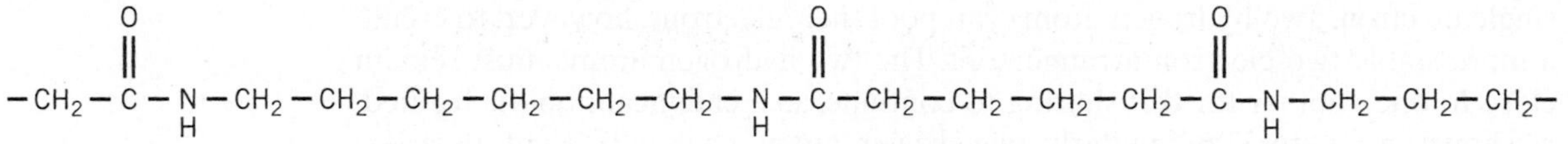

(a)

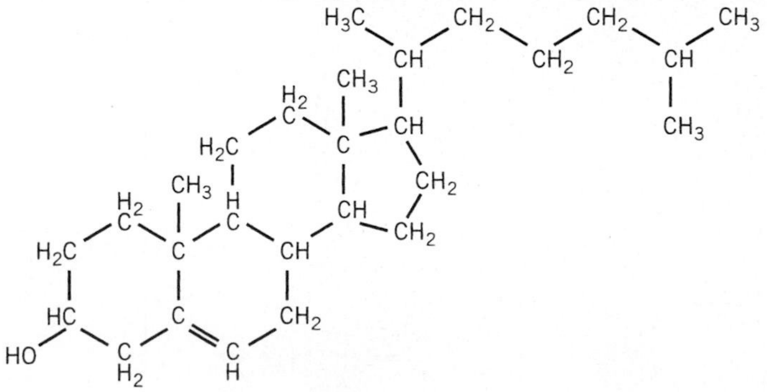

(b)

FIGURE 23-7. Carbon-based molecules may adopt almost any shape. The molecules may consist of long, straight chains of carbon atoms that form fibrous materials such as nylon (*a*), or they may incorporate complex rings and branching arrangements that form lumpy molecules such as cholesterol (*b*).

By forming bonds among several adjacent carbon atoms, you can make rings, long chains, branching structures, planes, and three-dimensional frameworks of carbon in almost any imaginable shape. There is virtually no limit to the complexity of molecules you can build from such carbon–carbon bonding (Figure 23-7). So important is the study of carbon-based molecules that chemists have given it a special name, *organic chemistry*.

Connection
Chemical Bonds in the Human Body

All the molecules in your body and in every other living thing are held together at least in part by covalent bonds in carbon chains. Covalent bonds also drive much of the chemistry in the cells of your body and play a role in holding together the DNA molecules that carry your genetic code. It would not be too much of an exaggeration to say that the covalent bond is the bond of life.

Let's consider one simple example. The food you eat is converted in your body to molecules of a sugar called glucose, with the chemical formula $C_6H_{12}O_6$ (that is, it contains 6 carbon atoms, 6 oxygen atoms, and 12 hydrogen atoms). When cells in your body need energy—for sending nerve impulses, for activating muscles, or for any bodily activity—these glucose molecules are burned as fuel in a combustion reaction known as respiration. In the end, respiration is very similar to the combustion of wood or gasoline, producing the products carbon dioxide and water. This reaction releases energy stored in the covalent bonds of the glucose molecule. You breathe air in order to obtain oxygen for respiration and to get rid of the water vapor and carbon dioxide produced. ●

Develop Your Intuition: The Element of Life

Life on Earth is based on the properties of the element carbon. Looking at the first three rows of the periodic table in Figure 23-1, do you see any other candidate elements that might form the basis of life elsewhere?

In the periodic table, the place to look for similar elements is in the same vertical column. Right underneath carbon in the periodic table is the element silicon, with the same arrangement of four electrons in its outer shell. Silicon forms a wide variety of compounds, as carbon does. Many of these silicon compounds are directly analogous to carbon compounds.

The major difference between carbon and silicon is that silicon has three shells of electrons instead of carbon's two shells. The result is that the electrons in silicon are spread out more in space and thus form longer and weaker bonds than carbon does. These weaker bonds make silicon compounds more reactive than similar carbon compounds. For example, low-weight carbon molecules are used as fuel in gasoline while higher-weight carbon molecules form waxes. However, the analogous low-weight silicon compounds are very dangerous: they can ignite or explode spontaneously in air. In addition, silicon compounds do not have high molecular weights: the silicon bonds are too weak for more than about eight atoms in a chain. Silicon forms a wide variety of compounds with oxygen and is present in most rocks and minerals; the oxygen atoms help form stable bonds between the silicon atoms and atoms of other elements.

Even with these differences, science fiction writers sometimes imagine life forms based on silicon, due to the great variety of compounds it can form. The astronomer and science fiction writer Fred Hoyle wrote a novel called *The Black Cloud,* in which he imagined an intelligent being in the form of a huge interstellar cloud of gas consisting of carbon and silicon compounds. But most biologists with interests in other life forms agree that silicon as the basis for life is likely to remain fiction rather than fact.

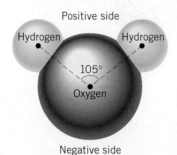

FIGURE 23-8. The water molecule and its polarity.

Polarization and Hydrogen Bonds Ionic, metallic, and covalent bonds form strong links between individual atoms within molecules. However, molecules also experience forces that hold one molecule to another. In many cases, the electric forces are such that, although the molecule by itself is electrically neutral, one part of the molecule has more positive or negative charge than another. For example, in water the electrons tend to spend more time around the oxygen atom than around the hydrogen atoms. This uneven electron distribution has the effect of making the oxygen side of the water molecule more negatively charged and the two Mickey Mouse ears of the hydrogen atoms more positively charged (Figure 23-8). Atomic clusters of this type, with a positive and negative end, are called *polar molecules* (see Chapter 16).

The electrons of an atom or molecule brought near a polar molecule such as water tend to be pushed away from the negative side and shifted toward the positive side. Consequently, the side of an atom facing the negative end of a polar molecule becomes slightly positive. This subtle electron shift, called *polarization,* gives rise to an electrical attraction between the negative end of the polar molecule and the positive side of the neighboring molecule. This electron movement thus creates an attraction between molecules, even though the atoms and molecules in this scheme all may be electrically neutral. One of the most important consequences of forces due to polarization is the ability of water to dissolve many materials. Water, made up of strongly polar H_2O molecules, exerts forces that make it easier for ions such as Na^+ and Cl^- to separate from one another and surround themselves with water molecules instead. The result is that salt dissolves in water.

A process related to the forces of polarization leads to the *hydrogen bond.* Hydrogen bonds are weak bonds that may form after a hydrogen atom links to an atom of certain other elements (oxygen or nitrogen, for example) by a covalent bond. Because of the kind of rearrangement of electric charge just described, hydrogen may become polarized and develop a slight positive charge that attracts another atom to it. You can think of the hydrogen atom as a kind of bridge in this situation, causing a redistribution of electrons that in turn holds the larger atoms or molecules together. Individual hydrogen bonds are weak, but in many molecules they occur repeatedly and therefore play a major role in determining the molecule's shape and function. Note that while all hydrogen bonds require hydrogen atoms, not all hydrogen atoms are involved in hydrogen bonds.

Hydrogen bonds are common in virtually every biological substance, from everyday materials such as wood, plastics, silk, and candle wax to the complex structures of every cell in your body. Hydrogen bonds in every living thing link the two sides of the DNA double helix together, although the sides themselves are held together by covalent bonds. Ordinary egg white is made from molecules whose shape is determined by hydrogen bonds; when you heat the material—when you fry an egg, for example—you break these hydrogen bonds. As a result, the molecules rearrange themselves so that instead of a clear gelatinous liquid you have a white solid.

Van der Waals Forces Hydrogen bonds exist because atoms or molecules can become polarized as their electrons shift to one side or another and thus create local electric charges. In the molecules we've discussed so far, that electric charge is more or less permanently locked into polar molecules in a fixed or static arrangement. Another force between molecules, called the *van der Waals force,* results from the polarization of electrically neutral atoms or molecules that are not themselves polar.

When two atoms or molecules are brought near one another, every part of one atom or molecule feels an electric force exerted on it by all parts of the other. For example, an electron in one atom is repelled by the electrons of an adjacent atom, but is attracted to the adjacent nucleus. The net result of these forces exerted on the electron may be a temporary shift of the electron. The same thing happens to every electron in any nearby atom or molecule, and the net result is that every electron is constantly shifting because of the presence of others.

What is remarkable is that sometimes this mutual, dynamic deformation can give rise to a net attractive force. In compounds where this happens, even if all the molecules are neutral and nonpolar, the sum of attractive forces wins out over repulsive forces and weak bonds are formed. This weak force that binds two atoms or molecules together is the van der Waals force.

If you take a piece of clay and rub it between your fingers, your fingers pick up a slick coating of the material, even though the clay crumbles easily. The reason for this behavior is that the clay is made up of sheets of atoms. Within each sheet, atoms are held together by strong ionic and covalent bonds. However, one sheet is held to another by comparatively weak van der Waals forces. This situation is not unlike the way a stack of photocopying paper will stick together on a dry day. Each sheet of paper is strong, but the stack of paper sticks together because of much weaker electrostatic forces. It is easy to pull the stack apart, but very difficult to rip the stack in two. When you crumble clay in your fingers, therefore, you are breaking weak van der Waals forces between layers but preserving the stronger bonds that hold each layer together. The clay stains on your hands are thin sheets of atoms, held together by ionic and covalent bonds; the crumbling is due to the breaking of the van der Waals bonds.

Many other examples of van der Waals forces can be seen in everyday life. If you rub talcum powder on your body, for example, you use a layered material not unlike the clay just discussed. Similarly, when you write with a lead (really graphite, a form of carbon) pencil, van der Waals–bonded layers of graphite (Figure 23-9) are transferred from the pencil to the paper. As you draw the pencil across the paper, you break the van der Waals forces and leave behind dark graphite sheets on the paper. Van der Waals forces also link molecules in many everyday liquids and soft solids, from candle wax to Vaseline and other petroleum products.

Contrast the behavior of the graphite in your pencil to the behavior of a diamond, which is also made from pure carbon. In a diamond, all the carbon atoms are locked together by covalent bonds in a complex three-dimensional array

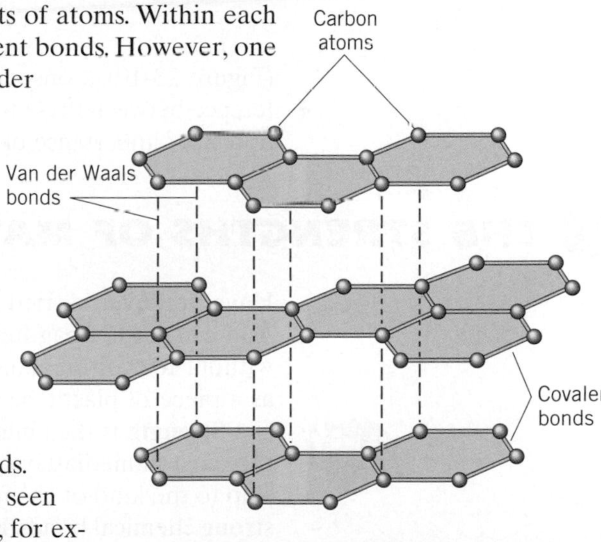

FIGURE 23-9. Graphite, a form of carbon that serves as the lead in your pencil, contains layers of carbon atoms strongly linked to one another by covalent bonds (represented by double solid lines). The separate layers are held together by much weaker van der Waals bonds (represented by dashed lines).

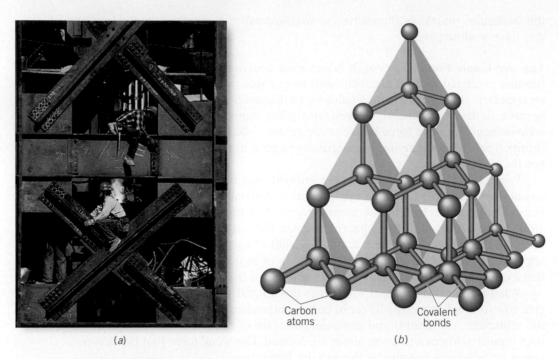

Carbon atoms   Covalent bonds

(a)   (b)

FIGURE 23-10. The girder framework of a skyscraper (a) and the crystal structure of diamond (b) are both strong because of numerous very strong connections. In diamond, these connections are covalent carbon–carbon bonds.

(Figure 23-10). Consequently, diamond is the hardest material known. The difference between these two materials, both made from exactly the same atoms, illustrates the importance of chemical bonds in determining how a substance behaves.

THE STRENGTHS OF MATERIALS

Have you ever carried a heavy load of groceries in a thin plastic grocery bag? You can cram a bag full of heavy bottles and cans and lift it by its thin handles without fear of breakage. How can something as light, flexible, and inexpensive as a piece of plastic be so strong?

Strength is the ability of a solid to resist changes in shape. Strength is one of the most immediately obvious material properties and it bears a direct relationship to the kind of chemical bonds present. A strong material must be made with strong chemical bonds. By the same token, a weak material, like a defective chain, must have weak links between some of its atoms. While no type of bond or attraction is universally stronger than the other kinds, van der Waals forces are generally the weakest. Any material with van der Waals forces will have one or more particularly soft directions of bonding and you will probably be able to pull the material apart with your hands. You experience this softness whenever you use baby powder, graphite lubricant, or soap.

By contrast, many strong materials, such as rocks, glass, and ceramics, are held together primarily by ionic bonds. The next time you see a building under construction, look at the way beams and girders link diagonally to form a rigid framework. Chemical bonds in strong materials do the same thing. A three-dimensional network of ionic bonds in these materials holds them together like a framework of steel girders.

The strongest materials we know, however, incorporate long chains and clusters of carbon atoms held together by covalent bonds. The extraordinary strength of natural spider webs, synthetic Kevlar (used to make bulletproof vests), diamonds, and your plastic shopping bag all stem from the strength of covalent bonds to carbon atoms.

Different Kinds of Strength

Every material is held together by the bonds between its atoms. When an outside force is applied to a material, the atoms must shift their positions in response. The bonds stretch and compress, and an equal and opposite force is generated inside the material to oppose the force that is imposed from the outside, in accordance with Newton's third law of motion. The strength of a material is thus related to the size of the force it can withstand when it is pushed or pulled.

Kevlar is lightweight, flexible, and strong enough to stop a bullet; it is the material from which bulletproof vests are made.

Material strength is not a single property because there are different ways of placing an object under stress. Scientists and engineers recognize three very different kinds of strength when characterizing a material:

1. Its ability to withstand crushing (*compressive strength*).

2. Its ability to withstand pulling apart (*tensile strength*).

3. Its ability to withstand twisting (*shear strength*).

Your everyday experience should convince you that these three properties are often quite independent. For example, a loose stack of bricks can withstand crushing pressures—you can pile tons of weight on it without having the stack collapse because each brick pushes down on the one beneath it but cannot exert a strong force in the sideways direction. But the same stack of bricks has little resistance against twisting; indeed, a child can topple it. A rope, on the other hand, is extremely strong when pulled, but has little strength under twisting or crushing.

The point at which a material is no longer strong enough to resist external forces and begins to bend, break, or tear is called its *elastic limit*. We see examples of this behavior every day. When you break an egg, crush an aluminum can, overstretch a rubber band, or fold a piece of paper, you exceed an elastic limit and permanently change the object. When the materials in your body exceed their elastic limit, the consequences can be catastrophic. Our bones if put under too much stress may break, while our arteries if put under pressure that is too high may rupture in an aneurysm.

A material's strength is a result of the type and arrangements of chemical bonds. Think about how you might design a structure using Tinkertoys that would be strong under crushing, pulling apart, or twisting. The strongest arrangement would have lots of short sticks with triangular patterns. Nature's strongest structure, diamond, adopts this strategy: it is exceptionally strong under all three kinds of stress because of its three-dimensional framework of strong carbon–carbon bonds (see Figure 23-10). Glass, ceramics, and most rocks, which also feature rigid frameworks of chemical bonds, are relatively strong. However, many plastics, such as the one used in your shopping bag, have strong bonds in only one direction and thus are strong when stretched, but have little strength when twisted or crushed. Materials with layered atomic structures, in which planes of atoms are arranged like a stack of paper, are generally strong when squeezed, but quite weak under other stresses. Thus the strength of a material depends on the kinds of atoms in it, the way they are arranged, and the kinds of chemical bonds that hold the atoms together.

Develop Your Intuition: Modern Building Materials

Architects specify a wide variety of different materials in contemporary building design. If you were planning a modern museum or concert hall that needed to be elegant as well as strong, would you be more likely to choose concrete, steel, or glass as building material?

Actually, all three materials are widely used in modern architecture, which is testimony to the progress of materials science over the last few decades. Steel has long been the workhorse of modern building design, but when used with imagination, steel shows elegance as well as strength. The John Hancock Center in Chicago, completed in 1970, shows one aspect of steel as an elegant material; another is the Pompidou Center, in Paris, finished in 1977. In the Hancock Building, many of the steel supports are in plain view on the building's exterior, instead of being hidden inside.

Concrete is usually considered to be a material used for its brute strength in withstanding compression, as in most large dams around the world. However, concrete can also soar in seemingly delicate arches, such as the Gateway

(a)

(b)

(c)

Building materials: (a) steel in the John Hancock Building, Chicago; (b) concrete in the Sydney Opera House, Australia; (c) glass in the Rose Center of the American Museum of Natural History, New York.

Arch in St. Louis, built in 1965 from steel reinforced with concrete. Perhaps even more graceful is the roof of the Opera House in Sydney, Australia, completed in 1973, which consists of concrete arches joined together to form a double series of overlapping shells.

Glass has long been used for windows, but over the years it has been strengthened for use as walls in large skyscrapers. Recent advances have led to glass that retains its transparency as well as gaining strength, as in the Rose Center for Earth and Space in New York City, which opened in 2000.

Modern materials science has advanced to the point that strength and elegance can be combined in steel, concrete, or glass. The only limit is the vision and imagination of the architect.

Composite Materials

As we have seen, some materials resist compression better than tension, while other materials can withstand great tensile forces but crack easily under compression. *Composite materials* combine the properties of two or more materials. The strength of one of the constituents is used to offset the weakness of another, resulting in a material whose strengths are greater than those of any of its components. Plywood, one of the most common composite materials, consists of thin wood layers glued together with alternating grain direction. The weakness of a single thin sheet of wood is compensated by the strength of the neighboring sheets. Not only is plywood much stronger than a solid board of the same dimension, but it can also be produced from much smaller trees by slicing thin layers of wood off a rotating log, like removing paper from a roll.

Reinforced concrete is a common composite material in which steel rods (with great tensile strength) are embedded in a concrete mass (with great compressive strength). A similar strategy is used in fiberglass, formed from a cemented mat of glass fibers. New carbon-fiber composites provide extraordinarily strong and lightweight materials for industry and sports applications.

The modern automobile features a wide variety of composite materials. Windshields of safety glass are layered to resist shattering and reduce sharp edges in a collision. Tires are intricately formed from rubber and steel belts for strength and durability. Car upholstery commonly mingles natural and artificial fibers, and dashboards often employ complex laminated surfaces. The bodies of many cars are formed from fiberglass or other molded lightweight composite. And, as we see later, all of a modern automobile's electronics, from radio to ignition, depend on semiconductor composites of extraordinary complexity.

Reinforced concrete is used in construction for its combined strength in resisting both tension and compression.

Physics in the Making
The Discovery of Nylon

Nature's success in making strong, flexible fibers inspired scientists to try the same thing. American chemist Wallace Carothers (1896–1937) began thinking about polymer formation while a graduate student in the 1920s. At the time, no one was sure how natural fibers formed, or what kinds of chemical bonds were involved. Carothers wanted to find out.

The chemical company DuPont took a gamble by naming Carothers head of its new fundamental research group in 1928. No pressure was placed on him to produce commercial results, but within a few years his team had developed the synthetic rubber neoprene. By the mid-1930s they had devised a variety of extraordinary polymers (see Chapter 9), including nylon, the first human-made fiber. Carothers also demonstrated conclusively that polymers in nylon are covalently bonded chains produced from small molecules, each of which has six carbon atoms, linked together.

DuPont made a fortune from nylon and related synthetic fibers. Nylon is inexpensive to manufacture and has many advantages over natural fibers. It can be melted and squeezed out of spinnerets to form strands of almost any desired size—for example, threads, rope, surgical sutures, tennis racket strings, and paintbrush bristles. These fibers can be made smooth and straight, like fishing line, or rough and wrinkled, like wool, to vary the texture of fabrics. Nylon fibers can also be kinked with heat to provide permanent folds and pleats in clothing. The melted polymer can even be injected into molds to form durable parts such as tubing or zipper teeth.

Sadly, Wallace Carothers did not live to see the impact of his extraordinary discoveries. Suffering from increasingly severe bouts of depression and convinced that he was a failure as a scientist, Carothers took his own life in 1937, just a year before the commercial introduction of nylon.

● CHEMICAL REACTIONS AND THE FORMATION OF CHEMICAL BONDS

Atoms and small molecules can come together to form larger molecules. Large molecules can break up into atoms and/or smaller molecules. We call these processes **chemical reactions.** When we take a bite of food, light a match, wash our hands, or drive a car, we initiate chemical reactions. Earth's chemical reactions include rock formation and rock weathering, atmospheric weathering, soil formation, water erosion, and the cycling of elements around the world. Every moment of every day, countless chemical reactions in every cell of our bodies sustain life.

All chemical reactions involve the rearrangement of the atoms in elements and compounds, as well as the rearrangement of electrons to break and form chemical bonds. Such reactions can be expressed as a simple equation:

$$\text{Reactants} \rightarrow \text{Products}$$

All such reactions must balance, so that the total number and kind of atoms are the same on both sides. For example, oxygen and hydrogen can form water by the reaction

$$2H_2 + O_2 \rightarrow 2H_2O$$

This reaction balances because each side has four hydrogen atoms and two oxygen atoms. In the process of this reaction we can observe both chemical changes (the rearrangement of atoms) and physical changes (the transformation of hydrogen and oxygen gases into liquid water with different properties).

Chemical Reactions and Energy:
Rolling down the Chemical Hill

Why do chemical reactions take place at all? The fundamental reason, as so often happens with natural phenomena, has to do with energy, as described by the laws of thermodynamics (see Chapters 12 and 13).

Consider, for example, one of the electrons in the neutral sodium atom shown in Figure 23-2*a*. This electron is moving around the nucleus, so it has kinetic energy. In addition, the electron possesses potential energy because it is a certain distance from the positively charged nucleus. Finally, the electron has an additional component of potential energy because of the electrical repulsion between it and all the other electrons in the atom. This situation is analogous to the small contribution to Earth's potential energy from the gravitational attractions of the other planets. The sum of these three energies—the kinetic energy associated with the electron's motion, the potential energy associated with its attraction to the nucleus, and the potential energy associated with the other electrons—is the total energy of the single electron in its shell.

The atom's total energy is the sum of the energies of all its electrons. For the isolated sodium atom in Figure 23-2, the total energy is the sum of the energy of the 11 electrons; for the chlorine atom, it is the sum of the energies of the 17 electrons. The total energy of the sodium–chlorine system is the sum of the individual energies of the two atoms.

Now think about what happens to the energy of the sodium–chlorine system after the ionic bond has formed. The force on each electron is now different than it was before. For one thing, the number of electrons in each atom has changed; for another, the atoms are no longer isolated, so electrons and protons in the sodium atom can exert forces on electrons in the chlorine, and vice versa. Consequently, the energies of all the electrons shift a little due to the formation of the bond. This means that each electron finds itself in a slightly different position with regard to the nucleus, moving at a slightly different speed, and experiences a slightly different set of forces than it did before. The total energy of each electron is different after the bond forms, the total energy of each atom is different, and the total energy of the system is different.

Whenever two or more atoms come together to form chemical bonds, the total energy of the system is different after the bonds form than it was before. Two possibilities exist: either (1) the final energy of the two atoms is less than the initial energy or (2) the final energy is greater than the initial energy.

The reaction that produces sodium chloride from sodium and chlorine is an example of the first kind of reaction, in which the total energy of the electrons in the system is lower after the two atoms have come together. According to the first law of thermodynamics, the total energy must be conserved, and that difference in energy is given off during the reaction in the form of heat, light, and sound (there is an explosion). A chemical reaction that gives off energy in some form is said to be *exothermic*.

Many examples of exothermic reactions occur in everyday life. The energy that moves your car is given off by the explosive chemical combination of gasoline and oxygen in the car's engine when it is ignited by a spark. The chemical reactions in the battery that runs your Walkman also produce energy, although in this case some of the energy is in the form of kinetic energy of electrons in a wire. At this moment, cells in your body are breaking down molecules of the sugar glucose to supply the energy you need to live.

If the final energy of the electrons in a reaction is greater than the initial energy, then you have to supply energy to make the chemical reaction proceed. Such reactions are said to be *endothermic*. The chemical reactions that go on when you are cooking (frying an egg, for example, or baking a cake) are of this type. You can put the ingredients of a cake together and let them sit for as long as you like, but nothing happens until you turn on the oven and supply energy in the form of heat. When the energy is available, electrons can move around and rearrange their chemical bonds. The result: a cake, where before there was only a mixture of flour, sugar, and other materials.

You can think of chemical reactions as being analogous to a ball lying on the ground. If the ball happens to be at the top of a hill, it lowers its potential energy by rolling down the hill, giving up the excess energy in the form of frictional heat. If the ball is at the bottom of a hill, you have to do work on it to get it to the top. In the same way, exothermic reactions are systems that "roll down the hill," going to a state of lower energy and giving off excess energy in some form. Endothermic reactions, on the other hand, have to be "pushed up the hill," and hence absorb energy from their surroundings.

Develop Your Intuition: Hot Packs and Cold Packs

Sports trainers and coaches often carry instant cold packs or hot packs for treating minor sprains and injuries. You knead the pack with your fingers to produce the low or high temperatures and then apply the pack to the injury. How do these packs work?

Both packs contain two separate compartments before they are activated. The cold pack contains water in one compartment and crystals of the compound ammonium nitrate in the other. The hot pack has water in one compartment and crystals of magnesium sulfate or calcium chloride in the other. In both cases, kneading the pack breaks the partition between the compartments, allowing the crystals to dissolve in the water.

The interesting phenomena here are the different directions of heat transfer in these two solution processes. Ammonium nitrate absorbs heat as it dissolves; it is one of the few substances that requires energy to dissolve but does so spontaneously anyway. The result is a cold pack. On the other hand, the dissolving of magnesium sulfate or calcium chloride is an exothermic reaction and gives off heat. In both cases, once the crystals have dissolved, the packs slowly return to room temperature and do not renew their effects with further kneading. Once the crystals have dissolved, they do not return to their dry state by themselves.

A chemical cold pack absorbs heat when the inside reactants are allowed to mix together; a hot pack releases heat.

Connection

The Clotting of Blood

Whenever you get a cut that bleeds, your blood begins a remarkable and complex sequence of chemical reactions called "clotting." Normal blood is a liquid crowded with cells and chemicals that distributes nutrients and energy throughout your body. Blood flows freely through the body's circulatory system. However, when that system is breached and blood escapes, the damaged cells cause the release of a molecule called prothrombin.

Prothrombin itself is inactive, but other blood chemicals convert it into the active chemical thrombin. The thrombin reacts to break apart other normally stable chemicals that are always present in blood and thus produces small molecules that immediately begin to bind together into chains. The newly formed material, called fibrin, congeals quickly and forms a tough fiber net that traps blood cells and seals the break in minutes.

Clotting reactions differ, depending on the nature of the injury and the presence of foreign matter in the wound. Biologists have discovered more than a dozen separate chemical reactions that may occur during the process. Several diseases and afflictions may occur if some part of this complex chemical system is not functioning properly. Hemophiliacs lack one of the key clotting chemicals and so may bleed continuously from even small cuts. Some lethal snake venoms, on the other hand, work by inducing clotting in a closed circulatory system. ●

THINKING MORE ABOUT

Atoms in Combination: Life Cycle Costs

Every month, chemists around the world develop thousands of new materials and bring them to market. Some of these materials do a particular job better than those they replace, some do jobs that have never been done before, and some do jobs more cheaply. However, all of them share one property: when the useful life of the product of which they are a part is over, they will have to be disposed of in a way that is not harmful to the environment. Until very recently, engineers and planners had given little thought to this problem.

For example, think about the battery in your car. The purchase price covers the cost of mining and processing the lead in its plates, pumping and refining the oil that was made into its plastic case, assembling the final product, and so on. When that battery reaches the end of its useful life, all these materials have to be dealt with responsibly. For example, if you throw the battery into a ditch somewhere, the lead may wind up in nearby streams and wells.

One way of dealing with this sort of problem is to recycle materials—pull the lead plates out of the battery, process them, and then use them again. But even in the best system, some materials can't be recycled, either because they have become contaminated with other materials during use or because we don't have the technologies to recycle them. These materials have to be disposed of in a way that isolates them from the environment. The question becomes, "Who pays?"

Traditionally in the United States, the person who does the dumping—in effect, the last user—must see to the disposal. In some European countries, however, a new approach is being introduced. Called "life cycle costing," this approach is built around the proposition that once a manufacturer uses a material, he or she owns it forever and is responsible for its disposal. The cost of a product such as a new car, then, has to reflect the fact that someday that car may be abandoned and the manufacturer will have to pay for its disposal.

Life cycle costing increases the price of commodities, contributing to inflation in the process. What do you think the proper trade-off is in this situation? How much extra cost should be imposed up front compared to eventual costs of disposal?

Summary

Atoms link together by **chemical bonds,** which form when a rearrangement of electrons lowers the potential energy of the electron system, particularly by the filling of outer electron shells. **Ionic bonds** lower chemical potential energy by the transfer of one or more electrons to create atoms with filled shells. The positive and negative ions created in the process bond together through electrostatic forces. In metals, on the other hand, isolated electrons in the outermost shell wander freely throughout the material and create **metallic bonds. Covalent bonds** occur when adjacent atoms or molecules share bonding electrons. Hydrogen bonding and van der Waals forces are special cases involving the distortion of electron distributions to create electrical polarity—regions of slightly positive and negative charge that can bind together.

All materials, from building supplies and fabrics to electronic components and food, have properties that arise from the kinds of constituent atoms and the ways those atoms are bonded together. The high **strength** of materials such as stone and synthetic fibers relies on interconnected networks of ionic or covalent bonds, while many soft and pliable materials such as wax and graphite incorporate weak van der Waals forces. Composite materials, such as plywood, fiberglass, and reinforced concrete, merge the special strengths of two or more materials.

Chemical bonds break and form during **chemical reactions,** which may involve the synthesis or decomposition of chemical elements or compounds.

Key Terms

chemical bonds The forces that hold atoms together in stable configurations to form molecules. (p. 494)

chemical reaction The formation or breaking apart of chemical bonds. (p. 506)

covalent bond Chemical bond formed when two or more atoms in a molecule share electrons. (p. 497)

ionic bond Chemical bond formed when one atom gives up one or more electrons to another atom, creating an electrical attraction between the atoms. (p. 494)

metallic bond Chemical bond formed when many atoms share the same electrons. (p. 496)

strength The ability of a solid to resist changes in shape. (p. 502)

Review

1. The basic physical properties of all materials depend on three essential features. What are they?

2. Which is more important in explaining how and why elements combine, the nucleus or the electrons?

3. What is a valence electron? Why are valence electrons so important in understanding how elements combine?

4. What are inert gases? How many valence electrons does an inert gas have, and why does this make them so chemically unreactive?

5. Does an element with an outer shell full of valence electrons have a higher or lower energy state than an element with an unfilled outer shell? How does this affect its stability, and what is the significance of this?

6. What are three simple strategies atoms adopt to achieve a full valence shell of electrons?

7. What are three types of chemical bonds? What do they all share in common?

8. Describe the ionic bond. Give an example of a material that results from such a bond.

9. Is an ionic bond generally strong? What force holds it together?

10. Do you think the compound sodium chloride (table salt) is lower in energy as a compound than as two separate elements? Explain.

11. Can an ionic bond involve more than a single electron transfer? Explain.

12. Why are materials composed of ionic bonds often brittle?

13. Describe the metallic bond. How is it similar to an ionic bond? How does it differ?

14. One description of a metal is a collection of positive ions held together in a crystal structure by the glue of free-floating electrons. What properties of metal follow from the structure of the metallic bond? Why can you hammer or pull a metal without it shattering or breaking easily?

15. Describe the covalent bond. How does this bond differ from an ionic bond?

16. What types of bonds do carbon atoms form?

17. The study of the chemistry of carbon has its own name. What is this name, and why is so much attention given to the study of this one particular element?

18. What is a single bond? A double bond?

19. What is polarization? What is a polar molecule?

20. Describe the hydrogen bond. How does it differ from ionic and covalent bonds?

21. How does the polar nature of water make it an effective dissolving agent?

22. Are all hydrogen atoms involved in hydrogen bonds?

23. Describe the van der Waals force.

24. How are van der Waals forces like hydrogen bonds? How are they different?

25. Graphite and diamond are both made of carbon. How do they differ in structure, and what properties do they exhibit as a result of this difference?

26. What determines the strength of a material? Which types of chemical bonds are strongest?

27. What are the three types of strength of materials recognized by scientists and engineers? Are these types of strength independent of one another?

28. What is the elastic limit of a material?

29. What is a composite material? What are the benefits of such materials?

30. What is the difference between a composite material and a compound? Give an example of each.

31. Identify objects around you that use the three kinds of chemical bonding or the three kinds of attraction discussed in this chapter. Which objects incorporate two or more kinds of these bonds or attractions?

32. Chemical reactions involve the rearrangement of the atoms in compounds and elements. What does it mean to say that a reaction must be balanced?

33. Why do chemical reactions occur? What is the role of energy in making them happen?

34. What is an exothermic reaction? Is heat given off or absorbed in this type of reaction?

35. What is an endothermic reaction? Is heat given off or absorbed in this type of reaction?

36. How is the energy change in a chemical reaction analogous to a ball on the ground in a hilly area?

Questions

1. Is hydrogen or helium a better choice to fill a balloonlike airship, or blimp? Compare the advantages and disadvantages of each gas, keeping in mind the properties that flow from their respective atomic structures.

2. What types of chemical bonds are the strongest and why?

3. Why are covalent bonds so prevalent in biological molecules?

4. If you had your choice, would you build a house with ionic-bonded materials or with materials held together by van der Waals forces?

5. Do the properties of a newly formed chemical compound tend to differ from or be the same as the properties of the individual elements that compose it? Give several examples to support your answer. Is your answer different in the case of an alloy?

6. Diamonds and graphite are both made from carbon atoms. Why is graphite so much weaker?

7. An ionic bond is formed when an electron from one atom is transferred to another atom. Explain why NaF (sodium fluoride) forms an ionic bond.

8. Explain why NaCl (sodium chloride) forms an ionic bond. Why do magnesium and chlorine form $MgCl_2$ more readily than MgCl?

9. A molecule of methane (CH_4) has a carbon atom that is bonded to four hydrogen atoms. The C—H bonds are covalent, meaning that the carbon atom shares an electron with each hydrogen atom. Explain why methane is a good candidate for a covalently bonded molecule.

10. Write the chemical formula for the following covalent compounds: carbon chloride (carbon and chlorine) and hydrogen chloride (hydrogen and chlorine).

11. Write the chemical formula for the ionic compound magnesium oxide (magnesium and oxygen).

12. Potassium iodide can be used as a thyroid-blocking agent in the event of a radiation emergency. Write the chemical formula for the ionic compound potassium iodide (potassium and iodine).

13. Magnesium (Mg) and bromine (Br) form an ionic compound. What is its chemical formula? Which element becomes the positive ion in this compound?

14. Lead (Pb) and sulfur (S) combine to form an ionic compound. What is its chemical formula? Which element is the positive ion in this compound?

15. Aluminum (Al) and chlorine (Cl) combine to form an ionic compound. What is its chemical formula? Which element is the positive ion in this compound?

16. Explain why chlorine is much more reactive than argon, even though they appear in adjacent spots on the periodic table.

Problems

1. Some elements readily form ionic or covalent bonds and some elements do not participate in chemical bonding at all. In the table below, identify each element by its atomic number (see the periodic table in Appendix C), indicate (with a Yes or No) whether or not the element is likely to participate in chemical bonding, and give your reasoning.

Atomic Number	Element Name	Chemical Bonding	Reason
1			
2			
6			
7			
8			
10			
11			
13			

Atomic Number	Element Name	Chemical Bonding	Reason
14			
16			
17			
18			
19			
20			
26			
29			
47			
79			
82			
86			

Investigations

1. What materials were used for the construction of buildings, furniture, and transportation devices in the United States 200 years ago? What modern technologies would be difficult or impossible if we could only use those materials?

2. Gold is often sold in a less than pure form as an alloy. What does it mean to say that you are buying, say, a necklace that is 14-carat gold? What is a carat and what other material(s) is gold alloyed with and why? Similarly, what is sterling silver?

3. Why is fluorine so effective in helping to prevent tooth decay? Investigate how fluorine acts to do this and examine the history of its use, including its controversial introduction into municipal water systems that started in the late 1940s. Over the years, what concerned opponents? Did the concerns differ according to whether the critic was conservative or liberal?

4. Dissect a disposable diaper. How many kinds of materials can you identify? What are the key properties of each? What kind of chemical bonding might contribute to the distinctive properties of these materials? Investigate the arguments for and against using disposable diapers.

5. Investigate how steel is made. Follow the process of production from extraction of the ore from the ground, to the separation of the iron from the ore, to the steel mill, and through the finishing processes. What are some of the different types of steel produced? What are some of the additives used in various alloys? Pay particular attention to the properties of the materials involved at each stage of the process as you explore this broad topic.

6. Visit a sports equipment store. Learn about the new materials that are used in tennis rackets, football helmets, and sports clothing.

7. Write a short story in which a new material with unique properties plays a central role.

8. What kinds of materials do surgeons use to replace broken hip bones? What are the advantages of this material?

 WWW Resources

See the *Physics Matters* home page at **www.wiley.com/college/trefil** for valuable web links.

1. **http://www.sfu.ca/person/lower/TUTORIALS/chembond/index.html** The "All about chemical bonding" tutorial by Stephen Lower of Simon Fraser University.

2. **http://www.chemistry.org/portal/a/c/s/1/acsdisplay.html?DOC=sitetools\periodic_table.html** A sophisticated periodic table of the elements site by the American Chemical Society. See especially the filling of electron shells.

3. **http://educ.queensu.ca/~science/main/concept/chem/c07/C07TPSU3.html** A discussion conceptualizing kinds of bonds as everyday social situations.

4. **http://www.kathysnostalgiabilia.com/lemelson.htm** The history of nylon and its impact on the U.S. chemical industry and culture.

5. **http://physics.uwstout.edu/strength/indexfbt.htm** A complete online course devoted to the strength of materials.

6. **http://micro.magnet.fsu.edu/electromag/java/atomicorbitals/index.html** *Atomic Orbitals*. Short and sweet visualization of electronic orbital shapes from Florida State University.

24 Electric and Magnetic Properties of Materials

KEY IDEA

The electric and magnetic properties of materials depend on the behavior of electrons in the chemical bonds that hold the material together.

PHYSICS AROUND US . . . Electricity on the Go

Portable electronics are the newest big thing. Advances in technology have changed computers from the size of a large room to a small laptop that you can fit inside a briefcase. Television sets used to be good-sized pieces of furniture; now they're so small and light that you can fit them in your car to entertain passengers. Large turntables and high-fidelity systems have been replaced by CD players with headphones so you can listen to music as you walk; telephones fit in your pocket and don't need wires; cameras don't need film but can download pictures right to your computer. All this new technology has come about from advances in materials science.

We have talked about how atomic structure and molecular forces determine the physical properties of materials, including strength and chemical reactivity. In light of this knowledge, you may not be surprised to learn that atomic and molecular bonding also affects the electric and magnetic properties of substances. What makes some materials conduct electricity while others do not? What kinds of materials allow us to pack so many electrical functions into such a small space? The answers to questions such as these all boil down to understanding the behavior of electrons in chemical bonds.

⬤ ELECTRIC PROPERTIES OF MATERIALS

Almost every aspect of our technological civilization depends on electricity, so scientists have devoted a good deal of attention to materials that are useful in electric systems. If the job at hand is to send electric energy from a power plant to a distant city, for example, then we need a material that can carry the electric energy without much loss. On the other hand, if the job is to put a covering over a wall switch so that we are not endangered by electric shock when we turn on a light, then we want a material that does not conduct electricity at all. In other words, several different kinds of materials contribute to any electric device.

Conductors

Any material capable of carrying electric current—that is, any material through which electric charges can flow freely—is called an **electrical conductor,** as discussed in Chapter 18. Metals, such as the copper that carries electricity through the building in which you are now sitting, are the most common conductors, but many other materials also conduct electricity. For example, salt water contains ions of sodium (Na^+) and chlorine (Cl^-), which are free to move if they become part of an electric circuit. We can find out if a material conducts electricity by making it part of an electric circuit and seeing if current flows through it.

The arrangement of a material's electrons determines its ability to conduct electricity. Recall that in the case of metals some electrons are bonded fairly loosely and are shared by many atoms. If you connect a copper wire across the terminals of a battery, these electrons are free to move in response to the battery's voltage. They flow from the negative pole toward the positive pole of the battery,

As we have seen in Chapter 18, the motion of electrons in electric currents is seldom smooth. Under normal circumstances, electrons moving through a metal collide continuously with the much heavier ions in that metal. In each of those collisions, electrons lose some of the energy they have gained from the battery, and that energy is converted to the faster vibration of ions, which we perceive as heat. In Chapter 18, we call this phenomenon **electrical resistance.** Even very good conductors have some electrical resistance.

The electrical conductor with which you are most likely to come into contact is copper. For example, the wires that carry electricity around most homes, schools, and office buildings are made of copper. Copper is an excellent conductor of electricity and is relatively inexpensive, so it is widely used for ordinary household and commercial circuits, as well as in most appliances. For some uses, such as the power lines that carry electricity overland from power plants to cities, aluminum is often used. Aluminum is not as good a conductor as copper, so more electric energy is lost to heat in these wires, but the low cost and light weight of aluminum more than makes up for that loss. In a few instances where cost is no object (in some circuits in satellites, for example), engineers occasionally use gold, which, while much more expensive than copper, is also a better conductor.

Power lines are often made of aluminum to conduct large amounts of electricity cheaply.

Develop Your Intuition: Electrical Attraction of Liquids

Liquid carbon tetrachloride, which in the past was used for dry cleaning clothing, is a symmetrical molecule. It contains four chlorine atoms surrounding a carbon atom. If you hold an electrically charged rod next to a thin stream of carbon tetrachloride, nothing happens. Suppose you try the same thing with chloroform, which used to be used as an anesthetic in hospitals. This molecule has the same structure as carbon tetrachloride, but with one of the chlorine atoms replaced with hydrogen, so it is no longer symmetrical. Now if you hold a charged rod next to a thin stream of chloroform, the liquid bends toward the rod. Why? What is happening? What do you think would happen if you held the charged rod next to a thin stream of water?

Because the chloroform molecule is not symmetrical, it is slightly polar (see Chapter 16). The electrically charged rod attracts charges of the opposite sign and repels charges of the same sign. The molecules turn so that the ends attracted to the rod are closer to it and feel a stronger force than the other ends of the molecules do. As a result, the stream of polar molecules is deflected toward the rod. This does not happen with the symmetrical molecules of carbon tetrachloride, so the stream of liquid passes the rod undeflected.

Water is a polar molecule, as we discussed in Chapter 16. As you might expect, then, a stream of water is deflected toward a charged rod.

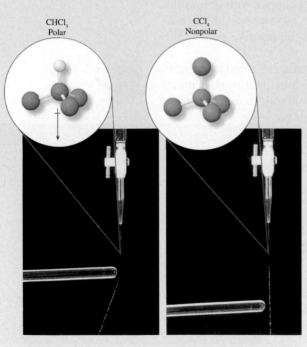

(a) Chloroform   (b) Carbon tetrachloride

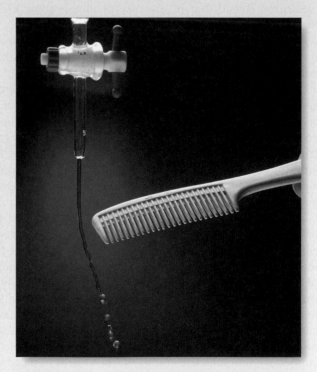

(c) Water

An electrically charged rod can deflect a stream of liquid composed of polar molecules, such as chloroform (a) and water (c), whereas a nonpolar liquid such as carbon tetrachloride (b) is not affected.

Insulators

Many materials incorporate chemical bonds in which few electrons are free to move in response to the push of an electric field. For example, in rocks, ceramics, and many biological materials such as wood and hair, the electrons are bound tightly to one or more atoms by ionic or covalent bonds (see Chapter 23). It takes considerable energy to pry electrons loose from those atoms—energy that is normally much greater than the energy supplied by a battery or an electric outlet. These materials do not conduct electricity unless they are subjected to an extremely high voltage that can pull the electrons loose. If they are made part of an electric circuit, no electricity flows through them. We call these materials **electrical insulators.**

The primary use of insulators in electric circuits is to channel the flow of electrons and to keep people from touching wires that are carrying current (Figure 24-1). For example, the shields on your light switches and household power outlets and the casings for most car batteries are made from plastic, a reasonably good insulating material that has the added advantages of low cost and flexibility. Similarly, electrical workers use protective rubber boots and gloves when working on dangerous power lines. In the case of high-power lines, glass or ceramic components are used to isolate the current because of their superior insulating ability.

Semiconductors

Many materials in nature are neither good conductors nor perfect insulators. We call such materials **semiconductors.** As the name implies, a semiconductor carries electricity but does not carry it very well. Typically, the resistance of silicon—

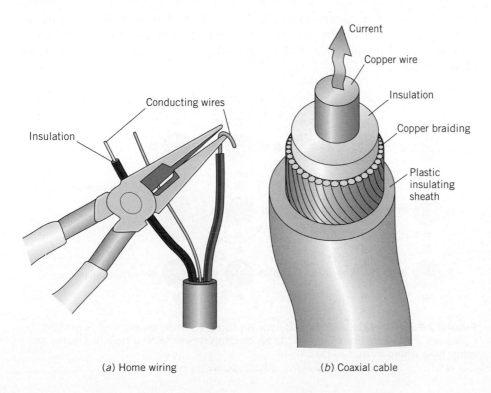

(a) Home wiring   (b) Coaxial cable

FIGURE 24-1. (a) Ordinary household electrical wiring consists of a conducting metal core surrounded by an insulating layer of plastic. The wire without insulation does not normally carry current but is present as a safety feature. (b) A coaxial cable, which is used to carry information in the form of a varying electric current, includes an inner wire and insulation with a surrounding mesh of copper wire and insulation. This outer layer reduces interference from outside sources of electricity.

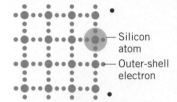

FIGURE 24-2. (a) A normal silicon crystal displays a regular pattern of silicon atoms. Some of its electrons are shaken loose by atomic vibrations; these electrons are free to move around and conduct electricity.

a common semiconductor—is a million times higher than the resistance of a conductor such as copper. Nevertheless, silicon is not an insulator because some of its electrons do flow in response to an applied voltage. Why should this be?

In a silicon crystal (Figure 24-2), all the electrons are taken up in the covalent bonds that hold each silicon atom to its neighbors. At low temperatures, these electrons are locked into bonds and we would expect the material to be an insulator. However, these bonds are relatively weak, so at room temperature the ordinary vibrations of the silicon atoms shake a few of the covalent bonding electrons loose. Think of the electrons as picking up a little of the vibrational energy of the atoms. These *conduction electrons* are free to move around the crystal. If the silicon is made part of an electric circuit, a modest number of conduction electrons are free to move through the solid.

When a conduction electron is shaken loose, it leaves behind a defect in the silicon crystal—the absence of an electron. This missing electron is called a *hole*. Just as electrons move in response to electric charges, so too can holes (see Figure 24-3). An electron can jump from one bond to fill a hole, leaving another hole in its original position. Then another electron can jump to that hole, and so on.

The motion of holes in semiconductors is similar to what you see in a traffic jam on a crowded expressway. A space opens up between two cars, after which one car moves up to fill the space, then another car moves up to fill that space, and so on. You could describe this sequence of events as the successive motion of cars. But you could just as easily (in fact, from a mathematical point of view, more easily) say that the space between cars—the hole—moves backward down the line. In the same way, you can either describe the effects of the successive jumping of electrons from one atom to another or talk about the hole moving through the material.

Although there are relatively few semiconducting materials in nature, they have played an enormous role in the microelectronics industry, as we see in Chapter 25. All of the information products you encounter everyday—computers, cell phones, CD players, etc.—depend on these few materials for their operation.

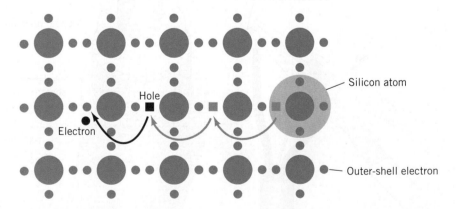

FIGURE 24-3. A hole in a semiconductor is produced when an electron is missing. Holes can move, just like electrons. As an electron moves to fill a hole, it creates another hole where it used to be.

Superconductors

Some materials cooled to extremely low temperatures, sometimes within a few degrees of absolute zero, exhibit a property known as **superconductivity**—the complete absence of any electrical resistance. Below some very cold critical temperature, electrons in these materials are able to move without surrendering any of their energy to the atoms. This phenomenon, discovered in the Netherlands in 1911, was not understood until the 1950s. Today, superconducting technologies provide the basis for a billion-dollar-a-year industry worldwide. The principal reason for this success is that once a material becomes superconducting and is kept cool, current can flow in it forever. This behavior means that if you take a loop of superconducting wire and hook it up to a battery to get the current flowing, the current continues to flow even if you take the battery away.

In Chapter 17, we learned that current flowing in a loop creates a magnetic field. If we make an electromagnet out of superconducting material and keep it cold, the magnetic field is maintained at no energy cost except for the refrigeration. Indeed, superconductors provide strong magnetic fields much more cheaply than any conventional copper-wire electromagnet because they don't heat up from electrical resistance. Superconducting magnets are used extensively in many applications where very high magnetic fields are essential—for example, in particle accelerators (see Chapter 27) and in magnetic resonance imaging systems for medical diagnosis (see Connection: Magnetic Resonance Imaging, page 520). Perhaps they will eventually be used in everyday transportation.

How is it that a superconducting material can allow electrons to pass through without losing energy? The answer in at least some cases has to do with the kind of electron–ion interactions that occur. At very low temperatures, heavy ions in a material don't vibrate very much and can be thought of as being more or less fixed in one place. As a fast-moving electron passes between two positive ions, the ions are attracted to the electron and start to move toward it. By the time the ions respond, however, the electron is long gone. Nevertheless, when the ions move close together, they create a region in the material with a more positive electric charge than normal. This region attracts a second electron and pulls it in. Thus the two electrons can move through the superconducting material somewhat like the way two bike racers move down a track, with the front rider overcoming air resistance so that the second rider can ride in the quieter air behind the leader (a strategy called "drafting").

At the very low temperatures at which a material becomes superconducting, electrons hook up in pairs, and the pairs start to interlock like links of a complex tangled matrix. While individual electrons are very light, the whole collection of interlocked electrons in a superconductor is quite massive. If one electron encounters an ion, the electron can't easily be deflected. In fact, to change the velocity of any electron, which you would have to do to get energy from it, you would have to change the velocity of all the electrons. Because this can't be done, no energy is given up in such collisions and electrons simply move through the material together. If the temperature is raised, however, the ions vibrate more vigorously and are no longer able to perform the delicate minuet required to produce the electron pairs. Thus, above the critical temperature, superconductivity breaks down.

Connection

Searching for New Superconductors

Until the mid-1980s, all superconducting materials had to be cooled in liquid helium, an expensive and cumbersome refrigerant that boils at a few degrees above absolute zero. The reason was that none of these materials was capable of sustaining superconductivity above about 20 kelvins. Acting on a hunch, scientists K. Alexander Müller and J. Georg Bednorz of IBM's Zurich, Switzerland, research laboratory began a search for new superconductors. Traditional superconductors are metallic, but Bednorz and Müller decided instead to focus on oxides—chemical compounds, such as most rocks and ceramics, in which oxygen participates in ionic bonds. It was an odd choice, because oxides make the best electrical insulators, although a few unusual oxides do conduct electricity.

Working with little encouragement from their peers and with no formal authorization from their employers, the scientists spent many months mixing chemicals, baking them in an oven, and testing for superconductivity. The breakthrough came on January 27, 1986, when a small black wafer of baked chemicals was found to become superconducting at greater than 30 degrees above absolute zero—a temperature that shattered the old record and ushered in the era of "high-temperature" (although still extremely cold) superconductors. Their compound of copper, oxygen, and other elements seemed to defy all conventional wisdom, and this discovery began a frantic race to study and improve the novel material.

FIGURE 24-4. A magnet floats magically above a black disk made from a new high-temperature superconductor. The clouds in the background form above the cold liquid-nitrogen refrigerant.

Today, many scientists are attempting to synthesize new oxides closely related to those first described by Bednorz and Müller, while others struggle to devise practical applications for these new materials. Some recently developed compounds superconduct at temperatures as high as 160 degrees above absolute zero (Figure 24-4). It may soon be possible to make commercially useful electric devices out of these materials.

Perhaps equally important, high-temperature superconductors have taken superconductivity from the domain of a few specialists and brought it into classrooms around the world. As a new generation of scientists grows up with these new superconductors, new questions will be asked and exciting new ideas and inventions are sure to be found. As the Connection: Magnetic Resonance Imaging section illustrates, the main commercial use of superconductors to date has been the production of magnets for use in medicine. In the future, the use of superconductors in the next generation of interurban trains (see the Connection: Maglev Trains, page 522) could also become very important. ●

Connection

Magnetic Resonance Imaging

The ability to produce strong magnetic fields has led to an important advance in the ability of physicians to diagnose illness. Called "magnetic resonance imaging" (MRI), this procedure allows the physician to obtain a detailed image of the interior of your body in a noninvasive way. The development of superconducting magnets has helped make MRI more widespread as a diagnostic tool.

To understand how MRI works, you need to recall what we have learned in Chapter 17—that moving electric charges can produce magnetic fields. If you think of a charged particle such as the proton as rotating, then it, too, constitutes

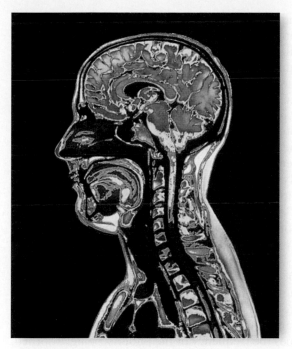

(a)

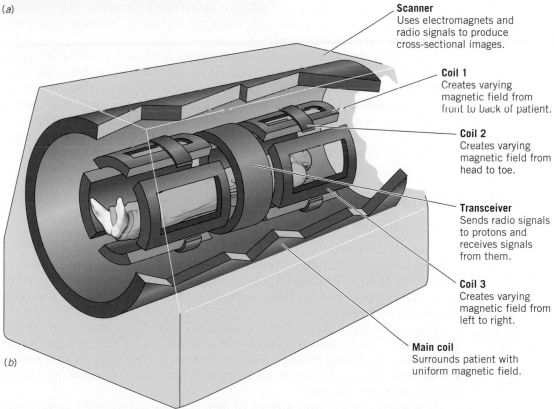

Scanner
Uses electromagnets and
radio signals to produce
cross-sectional images.

Coil 1
Creates varying
magnetic field from
front to back of patient.

Coil 2
Creates varying
magnetic field from
head to toe.

Transceiver
Sends radio signals
to protons and
receives signals
from them.

Coil 3
Creates varying
magnetic field from
left to right.

Main coil
Surrounds patient with
uniform magnetic field.

(b)

(a) A magnetic resonance image (MRI) of the human head and shoulders demonstrates
the ability of this technique to produce pictures of the body's soft tissues. (b) The
heart of the MRI machine consists of powerful magnets that create a varying
magnetic field. This field interacts with molecules in your body.

a moving electric charge. Consequently, the proton has its own dipole magnetic field and we can think of each proton as being like a tiny bar magnet. If the proton is in an external magnetic field, then it turns out that the laws of quantum mechanics (see Chapter 22) require that this magnet be oriented in only one of two ways. Roughly speaking, the north pole of the proton's "bar magnet" can line up with the external field (i.e., its north pole can point toward the north pole of the external magnet) or the proton's "bar magnet" can line up against that field. Physicists say that the proton must be either "spin up" or "spin down."

The orientation in which the proton's magnetic field is aligned with the external field has a slightly lower energy than the orientation in which it is aligned against that field. Under normal circumstances, therefore, there are slightly more protons in the spin up orientation (aligned with the external field) than in the spin down orientation.

Suppose we flood a region of the body with radio-frequency photons. Some protons in the body will absorb photons and flip from spin up to spin down. (For this to happen, the photon has to have precisely the right energy—the energy that corresponds to the difference between the energies of the spin up and spin down orientations.) By noting how much of this particular radiation is absorbed, scientists can tell how many protons there are in that particular region. If the strength of the external field changes, the amount of energy needed to make the proton flip changes as well, as does the frequency of the photons being absorbed. Since protons form the nuclei of hydrogen atoms and hydrogen atoms (in water, for example) are very common in the body, this technique provides a good way of examining the body's tissues.

In an MRI system, superconducting magnets produce a powerful magnetic field that increases in strength from one side of the body to the other. By monitoring the absorption of photons beamed to different spots, computers can put together detailed images of the interior of the body. In particular, MRI allows physicians to see soft tissue, which X rays cannot do. In addition, in MRI the patient is exposed only to magnetic fields and radio waves, rather than to potentially harmful X rays. ●

Connection

Maglev Trains

Another area in which the ability of superconductors to produce powerful electromagnets is used is in transportation. Today, there are many areas in the United States—the Boston–Washington, D.C. corridor, for example, or the region between San Francisco and San Diego—where travel volume is very high and many travelers favor trains.

If trains are to be used for interurban travel, then the faster they go, the better. Currently, the limit on a train's speed is set by energy loss through friction between the wheels and the rails, as well as through the flexing of both the wheels and the rails. The availability of superconducting magnets, however, provides a way to get around this limit.

If a magnet moves over a piece of metal, the electrons in the metal move and create a current. This current, in turn, produces its own magnetic field—one that opposes the change due to the first magnet. You can think of this effect as being due to an induced magnet in the metal, as shown. If the north pole of the original magnet is closest to the metal, then this induced magnet will have its north pole up. The repulsive force between the two magnets—one original and the other due to the movement of electrons in the metal—pushes upward on the original magnet. If the force is strong enough, it can actually balance the force of gravity and keep the magnet floating above the metal. This process is called *magnetic levitation,* or "maglev" for short.

The idea of a maglev train (see Figure 24-5) is that the interaction between the superconducting magnet in the train and in the metal rail results in an upward force through the levitation process, essentially floating the train a few inches above a metal track. Without the need for wheels to touch the track, the train is able to overcome present limits on speed. Speeds in excess of 300 mph are expected for maglev systems. A maglev train would leave the station on ordinary wheels, but as its speed increased the levitation force would eventually get large enough so that the train would take off and literally fly to its destination.

Another aspect of maglev train systems is that by controlling currents in the electromagnets on the train, you can induce opposite magnetic poles in the guide rail in front of each electromagnet. With this system, each electromagnet is pulled forward by the forces between the magnetic poles. By adjusting the timing of the polarity of magnets in the train, you can adjust the speed of the train. Or you

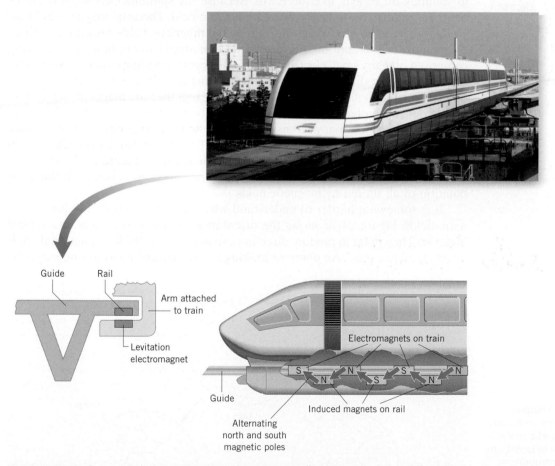

FIGURE 24-5. Superconducting magnets can magnetically levitate a train, while other magnets provide a push-pull effect that accelerates the train to speeds of hundreds of miles per hour. This superconductor technology has been used for commercial trains in the United States.

can reverse the polarity of the induced magnets in the rail so they are the same as the electromagnets, thus slowing the train down to a stop.

In 2002, the first permanent commercial maglev system in the world was installed at Virginia Commonwealth University in Richmond, Virginia. ●

MAGNETIC PROPERTIES OF MATERIALS

The magnets that lie at the heart of most electric motors and generators, although critical to almost everything we do, are not much evident in our everyday lives. Similarly, we are usually unaware of the magnets that drive our stereo speakers, telephones, and other audio systems. Even refrigerator magnets and compass needles are so common that we take them for granted. But why do some common materials, such as iron, display strong magnetism, while other substances seem to be unaffected by magnetic fields?

In Chapter 17, we learned that one of the fundamental laws of nature is that every magnetic field is due, ultimately, to the presence of electric currents. It turns out that every electron has a spin—that is, you can roughly picture an electron as spinning on an axis, like the Earth. Because the spinning electron constitutes a moving charge, the spin produces a magnetic field. The total magnetic field associated with each electron is the sum of the magnetic fields associated with its spin and its orbital motion. Because of this, an atom can be thought of as being composed of many small electromagnets, each corresponding to one orbiting electron, each with a different strength and pointing in a different direction. The total magnetic field of the atom arises by adding together the magnetic fields of all the tiny electron electromagnets.

It turns out that many atoms have magnetic fields that closely approximate the dipole type (originally shown in Figures 16-9 and 16-10). Thus each atom in the material can be thought of as a tiny dipole magnet (Figure 24-6). The magnetic field of a solid material such as a piece of lodestone arises from the combination of all these tiny magnetic fields.

It is somewhat harder to understand why most materials do not have magnetic fields. Figure 24-7a shows the orientation of atomic magnets in a typical material. They point in random directions, so at a place outside the material, their effects tend to cancel. An observer looking at the material measures no magnetic

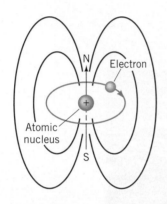

FIGURE 24-6. The dipole magnetic field of atoms takes the same form as that of larger magnets.

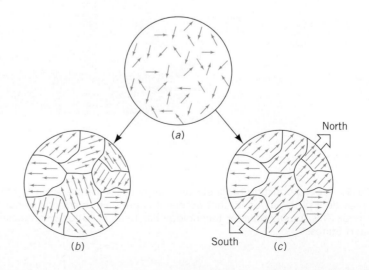

FIGURE 24-7. Different magnetic behavior in materials. (a) Nonmagnetic materials have random orientations of spins. (b) Ferromagnetic materials with randomly oriented domains are not magnetic. (c) A permanent magnet has more uniformly oriented atomic spins.

field, and a compass placed outside the material is not deflected. This explains how materials made up of tiny magnets can, as a whole, be nonmagnetic.

Nevertheless, given the fact that atoms are inherently magnetic, it should come as no surprise that materials often display magnetic properties, either in isolation or when they are immersed in an external magnetic field. There are, in fact, three important classes of material magnetism: **ferromagnetism,** *paramagnetism,* and *diamagnetism.*

Ferromagnetism

In a few materials, including iron, cobalt, and nickel metals, the angular momentum vectors associated with the electrons in the atoms line up with one another. This effect imposes a kind of order on the atoms. As a result, the atomic magnets associated with these atoms line up as well (Figure 24-7b). Typically, all the atoms in a region that measures about a thousand atoms on a side are aligned in this way. Such a region is called a *ferromagnetic domain.*

In a normal piece of iron, atoms within a specific domain all line up pointing in the same direction, but the orientations of the domains are random. You do not measure a magnetic field in this material because the magnetic fields due to different domains cancel one another out. However, in special cases, such as when iron cools from very high temperature in the presence of a strong magnetic field, some of the neighboring domains may line up and thus reinforce one another. Only when most of the magnetic domains line up (as shown in Figure 24-7c) do you get a material that exhibits a strong external magnetic field—the arrangement that occurs in permanent magnets.

(a)

We can understand how ferromagnets form by thinking about a piece of very hot iron. Because of the high temperature, the atoms are moving around vigorously, and there is no chance for the atomic magnets to align and reinforce one another. As the temperature is lowered, however, the random motion slows down and the magnetic force can take over to influence the atoms' orientations. At high temperature, the effects of this force are overwhelmed by thermal motion, but at low temperature it creates a situation where atomic magnetic fields reinforce one another. The temperature where this transition takes place is called the *Curie point* of the metal, after the French physicist Pierre Curie (1859–1906).

This picture of permanent magnets explains many of their features. It explains, for example, how you can turn an ordinary piece of iron into a magnet by stroking it with another magnet. This process aligns some (but usually not all) of the domains in the direction of the stroking, so that they produce an external magnetic field. This description also explains why heating (or sometimes just hammering) a magnet can destroy its properties. Adding energy in this way jostles the domains, randomizing their directions so that their magnetic fields cancel.

Some alloys developed in recent years can produce very powerful permanent magnets after proper magnetization. These alloys contain smaller domains (and more of them than in most ferromagnetic materials) making it easier to align them.

Paramagnetism

In most materials, the atomic magnets are arranged randomly and their magnetic fields cancel one another out. However, in some materials, when an external magnetic field is applied, the atomic magnets line up in such a way as to reinforce that field. These materials, called "paramagnets," do not normally display magnetic properties, but will do so in the presence of a magnetic field.

(b)

(a) Magnets made from new alloys can attract one another from either side of your finger. (b) Using magnetic shoes to climb the side of a steel-hulled ship.

Pierre Curie spent the early part of his career exploring the nature of magnetic materials and, indeed, gave us most of our current understanding of that field. In his later career, he teamed with his wife, Marie Sklodowska Curie, in the study of radioactivity (see Chapter 26), for which they became the only husband and wife team to share a Nobel Prize (in 1903). Marie Curie went on to become the first person to win two Nobel Prizes (she received her second, in chemistry, in 1911).

Pierre Curie found that the extra magnetic field produced when atoms line up in a paramagnet is proportional to the applied magnetic field—that is, the stronger the external field, the more the atoms tend to line up. He also discovered that if the temperature of a paramagnet is raised, the extra magnetic field decreases. This happens because as the temperature goes up, the increased thermal motion of the atoms starts to destroy the alignment. These results are summarized in *Curie's law:*

1. In words:

The magnetization of a material increases as the magnetic field increases and decreases with temperature.

2. In an equation with words:

$$\text{Magnetization} = \text{Curie constant} \times \frac{\text{Magnetic field}}{\text{Temperature}}$$

3. In an equation with symbols:

$$M = C \frac{B}{T}$$

where M is the extra magnetic field, or magnetization, of the material, B is the applied magnetic field, T is the temperature (in kelvins), and C is a number known as Curie's constant. Curie's constant is always the same for a given material, but varies from one material to another.

Develop Your Intuition: Magnetic Attraction of Liquids

Oxygen becomes a liquid at a temperature of 90 kelvins and nitrogen becomes a liquid at 77 kelvins. If you were to pour a stream of liquid oxygen between the poles of a magnet, it would be attracted to the magnet just as if the oxygen were iron filings. However, a stream of liquid nitrogen would pass between the poles as if nothing had happened. How can we explain this different behavior?

(a)　　　　　　(b)

(a) Liquid oxygen poured between the poles of a magnet is attracted to the poles, building up a blockage suspended in the magnetic field. (b) Liquid nitrogen passes right between the poles of a magnet with no attraction.

Nitrogen and oxygen are right next to each other in the periodic table. Nitrogen contains five electrons in its outer shell and oxygen contains six. Careful counting of electrons in molecular nitrogen (N_2) shows that the electrons pair up in this molecule very neatly, with no unpaired electrons. Thus, the magnetic fields of pairs of electrons in nitrogen tend to cancel one another out, and the atom has no net magnetic field. But molecular oxygen (O_2) does not pair up nicely; it is left with two single electrons of the same spin that cannot pair up together. The result is that oxygen is paramagnetic and is attracted to the poles of a magnet.

Diamagnetism

In some materials (bismuth is one of the most common) the magnetic fields associated with the electrons cancel one another so that the atom has no net magnetic field. If such an atom is placed in a magnetic field, however, the motion of the electrons changes to oppose this field. Thus, the magnetic field generated in the material is in a direction opposite to the magnetic field that is imposed from the outside. Such a material is said to be a "diamagnet."

THINKING MORE ABOUT

High-Temperature Superconductors

The discovery of high-temperature superconductors in 1987 created a firestorm of sensationalistic news stories. Optimistic researchers predicted a new age of inexpensive energy, fast transportation, and futuristic applications in medicine, communications, and computer technology. Bold headlines estimated the advance to be worth billions of dollars, while *Time* magazine featured superconductors as their cover story. The government quickly poured tens of millions of dollars into superconductor research while many venture capitalists, hoping to profit from the discovery, invested large sums into new companies.

It didn't take long for the hype to turn into a more realistic assessment of daunting technological hurdles. All the new superconductors are ceramic materials, which are rigid and brittle. They behave very differently from the flexible metal used to conduct electricity in wire. Furthermore, all these materials prove to be rather unstable; for example, they tend to break down when exposed to water. Thus, in spite of the extraordinary ability

Superconductors in the news. (*Time*, May 1987).

of high-temperature superconductors to transmit electricity without loss, it has proven extremely difficult to shape them into reliable wires or useful devices. The dream of a multibillion-dollar industry has failed to materialize.

How should scientists handle the announcement of potentially exciting but unproven new discoveries? How should the press deal with futuristic speculations? Should the government respond quickly to fund research on potentially significant technologies?

Summary

The electric properties of materials depend on the kinds of constituent atoms and the bonds they form. For example, **electrical resistance**—a material's resistance to the flow of an electric current—depends on the mobility of bonding electrons. Metals, which are characterized by loosely bonded outer electrons, make excellent **electrical conductors,** while most materials with tightly held electrons in ionic and covalent bonds are good **electrical insulators.** Materials, such as silicon, that conduct electricity, but not very well, are called **semiconductors.** At very low temperatures, some compounds lose all resistance to electron flow and become **superconductors.**

Magnetic properties also arise from the collective behavior of atoms. While most materials are nonmagnetic, **ferromagnets** have domains in which electron spins are aligned with one another. In paramagnets, the magnetic dipoles associated with atoms line up to reinforce imposed magnetic fields. In diamagnetic materials, atomic electrons produce a magnetic field that opposes imposed fields.

Key Terms

electrical conductor A material in which charges are free to move from place to place. (p. 515)

electrical insulator A material in which charges are not free to move from place to place. (p. 517)

electrical resistance The tendency of a material to resist the flow of electric charges through it. (p. 515)

ferromagnetism The magnetic properties of a material associated with the spontaneous alignment of domains of magnetic fields of its atoms. (p. 525)

semiconductor A material that is neither a good electrical conductor nor a good insulator. (p. 517)

superconductivity The property of having no electrical resistance. (p. 519)

Review

1. What aspects of your life depend on electricity and, hence, on the electric properties of materials?

2. What is an electrical conductor?

3. How does an electrical conductor actually transmit electricity? What particle is moving through the material, and exactly what allows a material that is a good conductor to be one?

4. Give several examples of materials that are good conductors.

5. Describe the origin of electrical resistance in a material. What is it, and how is heat generated due to it?

6. Do materials that are good conductors exhibit any electrical resistance?

7. Aluminum is often used in the main utility power lines that take electricity from power plants to cities, while copper is used to wire a home. Why is this? When might gold be preferred to conduct electricity?

8. What is the hallmark of a material that is an electrical insulator? What types of bonds does such a material generally have, and how does this affect the transmission of electricity?

9. Why is plastic commonly used as an insulator? What other materials are also used?

10. What is a semiconductor? How does it differ from a conductor or insulator?

11. How does the strength of silicon bonds help to account for its semiconducting properties? Where do the conducting electrons in silicon come from?

12. What is a hole in a semiconductor, and how does this facilitate the conduction of electricity?

13. What is so unusual about superconductors? Under what conditions do materials exhibit superconductivity?

14. Why does a superconducting magnet use less electric power than a traditional copper-wire electromagnet?

15. How does a superconducting material allow electrons to pass through it without losing energy? What is the role of low temperature and electron–ion interactions in this process?

16. What is magnetic resonance imaging (MRI)? How does it work?

17. What benefits does MRI offer to patients and physicians in comparison to X-rays photographs?

18. What is magnetic levitation?

19. Why are magnetic levitation trains more efficient than conventional diesel trains?

20. What is the role of electrons and electric current in producing magnetic fields at the atomic level?

21. What is ferromagnetism? What substance exhibits this?

22. What is a ferromagnetic domain? How are the atoms within a single domain oriented, and how do the many separate domains within a material line up when the material as a whole is magnetic? When it is nonmagnetic?

23. What is the Curie point of a metal?

24. What is paramagnetism? What must be present for a material to exhibit this property?

25. In a paramagnetic material, what is the strength of the magnetism in the material proportional to?

26. How does temperature affect the magnetic field of a paramagnetic field? What is Curie's law?

27. What is diamagnetism?

Questions

1. Metals are generally good conductors of electricity. Why is this the case? Can you relate this generality to the type of bonding that occurs in metals? Explain.

2. Take the point of view of an electron moving among other electrons and atoms in a material. Describe your motion in an insulator, a conductor, a superconductor, and a semiconductor.

3. How is it that salt water can conduct electricity? Should absolutely pure water conduct electricity? How about ice?

4. When you go in to have an MRI done, the technician always tells you to remove your watch, pens, and other metal objects from your pockets. Why is this request made?

5. How can a hole moving through a semiconductor be like an electric charge moving through the same material? Explain.

6. If all atoms have electrons that are in motion about an atom, why aren't all materials magnetic?

7. Compare the way that the motion of electrons in a diamagnetic material creates an opposing magnetic field with the way electrons in a copper wire accomplish the same end.

8. A normal piece of iron produces no external magnetic field. Suppose a piece of iron consisted of one very large domain instead of many small ferromagnetic domains. Would this piece of iron produce an external magnetic field? Why?

9. A paramagnetic material will acquire a magnetization if it is placed in an applied magnetic field. What happens to the magnetization of a paramagnetic material if the temperature of the material is doubled? What if the temperature and the applied magnetic field are simultaneously doubled?

10. Is air an electrical conductor or an insulator? Give an example of electrical conduction through air.

11. If the temperature of a semiconductor is increased, does its electrical resistance increase or decrease? Explain.

12. Based on Ohm's law (Chapter 18), how much current would you expect to run though a superconductor if the voltage across it were 100 volts? Is this possible? What do you think would really happen?

13. What is the significance of one material having a larger Curie constant than another?

14. The Pauli exclusion principle (Chapter 21) says that no two electrons can occupy the same energy state unless their spins point in opposite directions. Use the Pauli exclusion principle to argue that atoms with an even number of electrons tend to have small Curie constants.

Investigations

1. Shortly after the discovery of high-temperature superconductivity, many newspapers and TV shows ran features on how these new materials would change society. In what ways might superconductivity change society? Historically, what other new materials have caused significant changes in human societies?

2. Why does a magnet become demagnetized when you repeatedly hit it with a hammer? In what other ways can you destroy a permanent magnet? Why aren't permanent magnets permanent?

3. Compare the absorption of a photon that occurs in the MRI process to the absorption process that occurs in the Bohr atom.

4. Every year, one or two promising new materials capture public attention. Scan recent issues of *Science News* and identify one such material. Who made it? How might it be used?

5. Imagine that you are a science fiction writer. Concoct a description of a new material with unique (but plausible) properties and describe how that material might change a society.

6. Seek out a licensed electrician and examine a new construction project where the wiring is installed yet still visible. Ask about all the precautions taken to avoid electrocution. What types of materials are their clothes, ladders, and tools made of? Trace the flow of electricity from the street, into the construction site, through the various appliances, and back out again. What materials are conductors, what provides resistance in the circuit, and where is insulation important?

7. Research the status of magnetic levitation trains like the one now operating in Japan. How does it operate? How fast might it go? When did such a train begin operating in North America?

8. Silicon is the best-known semiconducting material. What are some other semiconducting materials and compounds in use today? Who uses them, and why are they used instead of silicon?

9. The term magnetic resonance imaging (MRI) is now used instead of the original term nuclear magnetic resonance (NMR). Why do you think that happened? Can you think of other examples where technologies have been renamed or redescribed so that they are more palatable to the public? How about military technologies?

 # WWW Resources

See the *Physics Matters* home page at **www.wiley.com/college/trefil** for valuable web links.

1. **http://www.owlnet.rice.edu/~hkic/superconductors/** A site devoted to history, theory, and applications of superconductivity.

2. **http://www.ornl.gov/reports/m/ornlm3063r1/contents.html** *A Teacher's Guide to Superconductivity for High School Students,* from Oak Ridge National Laboratory.

3. **http://micro.magnet.fsu.edu/electromag/java/filamentresistance/index.html** A short and sweet animated applet demonstrating electron flow through a metal conductor from Florida State University.

4. **http://micro.magnet.fsu.edu/electromag/electricity/resistance.html** Complete tutorial on resistance from Florida State University.

25 Semiconductor Devices and Information Technology

KEY IDEA

Semiconductor devices enable us to transmit and process information quickly, reliably, and conveniently.

PHYSICS AROUND US . . . Keeping in Touch

Your best friend from high school attends college three states away, but you like to stay in touch. You turn on your computer, enter your e-mail system, and send him a message telling him about your week. Later in the day he sends you an e-mail back. It feels good to stay in touch when you're far away from friends.

Later that day you log on to the Web to do research for a term paper about government. Your search takes you to Web pages posted by organizations in many countries and on several continents. All this information, including photos, diagrams, and references to other sources, appears on your computer screen in seconds.

You don't give it a second thought, but all of that information is brought right into your room by a huge network of computers and electronic storage systems that spans the entire globe—a network made possible by the technology of semiconductors.

MICROCHIPS AND THE INFORMATION REVOLUTION

Every material has dozens of different physical properties. We've already seen how strength, electrical conductivity, and magnetism all result from the properties of individual atoms and how those atoms bond together. We could continue in this vein for many more chapters, examining optical properties, elastic properties, thermal properties, and so on. But such a treatment would miss another key idea about materials: understanding how atomic interactions affect the properties of materials can lead to the development of new materials, and new materials can lead to new technologies that change society.

Of all the countless new materials discovered in the twentieth century, none has transformed our lives more than silicon-based semiconductors. In personal computers, auto ignitions, portable radios, sophisticated military weaponry, and countless other devices, microelectronics is a hallmark of our age. Indeed, semiconductors have fundamentally changed the way that we communicate one of society's most precious resources—information. The key to this revolution is our ability to fashion complex crystals atom by atom from silicon, a material that is produced from ordinary beach sand.

Doped Semiconductors

The element silicon by itself is not a very useful substance in electric circuits. What makes silicon useful and has driven our modern microelectronic technology is a process known as doping. **Doping** is the addition of a minor impurity to an element or compound. The idea behind silicon doping is simple. When silicon is melted before being made into circuit elements, a small amount of some other material is added to it. One common additive is phosphorus, an element that has five valence (bonding) electrons, as opposed to the four valence electrons of silicon.

When the silicon crystallizes to form the structure shown in Figure 25-1a, the phosphorus is taken into the crystalline structure. However, of the five valence electrons in each phosphorus atom, only four are needed to make bonds to

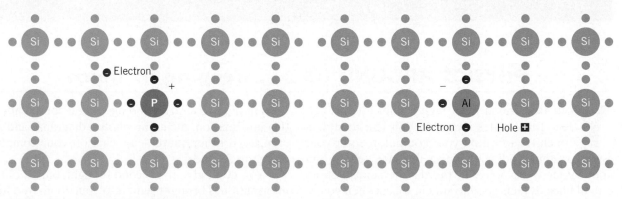

(a) Phosphorus-doped silicon n-type semiconductor   (b) Aluminum-doped silicon p-type semiconductor

FIGURE 25-1. (a) Phosphorus-doped silicon n-type semiconductors and (b) aluminum-doped silicon p-type semiconductors are formed from silicon crystals with a few impurity atoms.

silicon atoms in the crystal. The fifth electron is not locked in at all. In this situation, it does not take long for the extra electron to be shaken loose and to wander off into the body of the crystal. This action has two important consequences: (1) conduction electrons are introduced into the material, and (2) the phosphorus ion that has been left behind has a positive charge. A semiconductor doped with phosphorus is said to be an "n-type semiconductor," because the moving charge is a negative electron.

Alternatively, silicon can be doped with an element such as aluminum, which has only three valence electrons (Figure 25-1b). In this case, when the aluminum is doped into the crystal structure there is one less valence electron compared to the silicon atoms it has replaced in the crystal. This missing electron—a hole—creates a material that can now more easily carry an electric current. The hole need not stay with the aluminum; it is free to move around within the semiconductor, as described in Chapter 24. Once it does so, the aluminum atom, which has now acquired an extra electron, has a negative charge. This type of material is called a "p-type semiconductor," because a positive hole—a missing negative electron—acts as the moving charge.

Diodes

You can understand the basic workings of a microchip by conducting an experiment in your mind. Imagine taking a piece of n-type semiconductor and placing it against a piece of p-type semiconductor. As soon as the two types of material are in contact, how will electrons move?

Near the contact, negatively charged electrons will diffuse from the n-type semiconductor over into the p-type, while positively charged holes will diffuse back the other way. Thus, on one side of the boundary will be a region where negative aluminum ions—ions locked into the crystal structure by the doping process—acquire an extra electron. Conversely, on the other side of the boundary is an array of positive phosphorus ions, each of which has lost an electron but is nonetheless locked into the crystal.

A semiconducting device such as this—formed from one p and one n region—is called a **diode** (see Figure 25-2). Once a diode is constructed, a permanent electric field tends to push electrons across the boundary in only one direction, from the n-type side to the p-type side. As electrons are pushed "with the grain" in the diode, from negative to positive, the current flows through normally. When the current is reversed, however, the electrons are blocked from going through by the presence of the built-in electric field. Thus the diode acts as a one-way gate, allowing the electric current through in only one direction.

Semiconductor diodes have many uses in technology. One use can be found in almost any electronic device that is plugged into a wall outlet. As we have seen in Chapter 18, electricity is sent to homes in the form of alternating current (AC). It turns out, however, that most home electronic devices such as televisions and stereos require direct current (DC). A semiconductor diode can be used to convert the alternating current into direct current by blocking off half of it. In fact, if you examine the inside of almost any electronic gear, the power cord leads directly to a diode and other components that convert pulsing AC into steady DC. A semiconductor diode used in this way is called a *rectifier* (see Chapter 18).

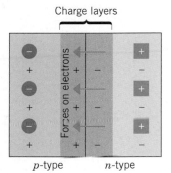

FIGURE 25-2. A semiconductor diode consists of an n-type region and a p-type region. Electrons in this diode can flow easily from the negative to the positive region. The built-in electric field, labeled E, blocks electrons from flowing the opposite way. The result is a one-way valve for electrons.

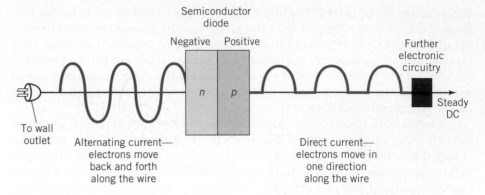

FIGURE 25-3. A diode converts alternating current from the wall into direct current in most electronic devices. Half of the alternating current passes through the diode, but the other half is blocked. Pulses of current from the diode enter a condenser, which stores the electrons and emits a steady DC current.

Figure 25-3 illustrates how a rectifier works. On the left, normal AC power enters into the system; recall that in AC, electrons flow first one way and then the other. If the current is in the half of the cycle with electrons flowing to the right in Figure 25-3, then these electrons pass through the rectifier. During the other half of the AC cycle, with electrons flowing to the left in the figure, electrons cannot move through the rectifier. Consequently, the output of the rectifier is a series of peaks, as shown, with the current in each peak moving in the same direction to the right. Further electronic circuitry then converts these peaks into a smooth DC voltage, as shown on the right. (Note that in most modern rectifiers, the current flowing in the wrong direction is not simply thrown away but converted to current flowing in the correct direction by more complex electric circuits.)

Connection

Photovoltaic Cells and Solar Energy

Recall from Chapter 22 that when some photoelectric materials absorb photons, electrons may be liberated from that material. This photoelectric effect is displayed by some semiconducting diodes, which may play an important role in the energy future of the United States. In such a device, called a *photovoltaic cell* (or solar cell), a thin layer of *p*-type material overlays a thicker layer of *n*-type. How might such a device operate to produce electricity?

Sunlight striking the top *n*-type layer shakes electrons loose from the crystal structure by the photoelectric effect. The electrons are attracted to the locked-in positive charges in the *n*-type material and repelled by the locked-in negative charges in the *p*-type material. They are then accelerated through the *n–p* boundary and pushed out into an external circuit. Thus, while the sun is shining, the photovoltaic cell acts in the same way as a battery. It provides a constant push for electrons and moves them through an external circuit. If large numbers of photovoltaic cells are put together, they can generate enormous amounts of current.

Photovoltaic cells have many uses today. Your hand calculator, for example, may very well contain a photovoltaic cell that recharges the batteries (it's the small dark band just above the buttons). Photovoltaic cells are also used in

(a)

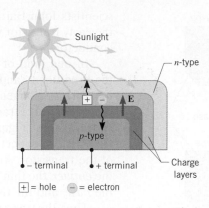

Sunlight

n-type

+ − **E**

p-type

− terminal   + terminal   Charge layers

+ = hole   − = electron

(b)

(a) The Sun's energy is converted to electricity by photovoltaic panels at a southern California generating plant. (b) Photoelectric materials use the energy of photons to liberate electrons, which form an electric current.

regions where it is hard to bring in traditional electricity—for example, to pump water in remote sites or to provide electricity in backcountry areas of national parks. At the moment, it costs several times as much to get a watt of electricity from a solar cell as from the most expensive conventional generating plants, but this cost is falling as the technology improves and markets increase.

To understand how solar electricity might become a reality, you have to know a little about the way electricity is used in the United States. Certain demands exist all the time—people need to run computers, keep their streets and homes lit, run their subways, and perform countless other tasks at all hours of the day and night. Thus, utility companies need to supply a certain amount of electricity round the clock, day in and day out. This demand is called *base-load* electricity. Because the base-load generating plants operate for a large fraction of the time, base-load electricity is relatively cheap.

However, on certain days the demand for electricity soars—think of a hot August afternoon when everyone is running air conditioning, for example. Utility companies need to be able to meet this *peak-load* demand, but to do so they need to build plants that are not used all the time. Thus, peak-load electricity is more expensive to generate than base-load electricity.

Electricity demand peaks on hot summer afternoons, so experts who study energy policy have suggested that this is how solar energy will enter the electricity market. In this case, solar power would be competing with the more expensive peak-load generators, which makes the economic requirements somewhat less stringent. ●

The Transistor

The device that drives the entire information age and perhaps more than any other has been responsible for the transformation of our modern society is the **transistor.** Invented just two days before Christmas 1947 by Bell Laboratory

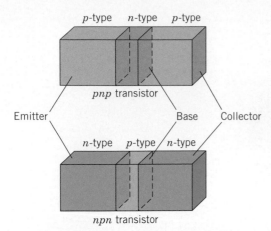

FIGURE 25-4. A *pnp* and an *npn* transistor.

scientists John Bardeen, Walter Brattain, and William Shockley, the early transistor was simply a sandwich of *n*- and *p*-type semiconductors.

In one kind of transistor, two *p*-type semiconductors form the bread of a sandwich, while the *n*-type semiconductor is the meat. This arrangement is abbreviated as the *pnp* configuration. Another kind of transistor uses the *npn* configuration. Both kinds of transistors (Figure 25-4) control the flow of electrons. Electrical leads connect to each of the three semiconductor regions of the transistor. An electric current goes into the region called the *emitter*, the thin slice of semiconductor in the middle is called the *base,* and the third semiconductor section is the *collector*.

Thus, the transistor has two built-in electric fields, one at each *p–n* junction. The idea of the transistor is that a small amount of electric charge running into or out of the base can change these electric fields—in effect, opening and closing the gates of the transistor. The best way to think of the transistor is to use a pipe that carries water as an analogy. The electric current that flows from emitter to collector is like water that flows through the pipe, and the base is like a valve in the pipe. A small amount of energy applied to turning the valve can have an enormous effect on the flow of water. In just the same way, a small amount of charge run onto the base can have an enormous effect on the current that runs through the transistor. These properties of a transistor lead to two of its most important uses—as an amplifier and as a switch.

The Transistor as Amplifier One use for the transistor is to amplify weak electric currents. For example, in your tape deck small electric currents are created when the magnetized tape is run past the tape heads. These small currents can be fed into the base of a transistor (Figure 25-5). As the number of electrons in the base increases, the main current flowing through the transistor decreases—in effect, the valve is closed a little. Similarly, when electrons are removed from the base, the current through the transistor increases—the valve is open. Thus,

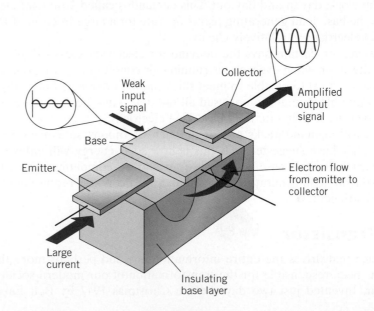

FIGURE 25-5. A transistor acting as an amplifier. A weak input signal sent to the base region modifies the stronger current passing from emitter to collector. The result is that the strong current carries an amplified version of the input signal.

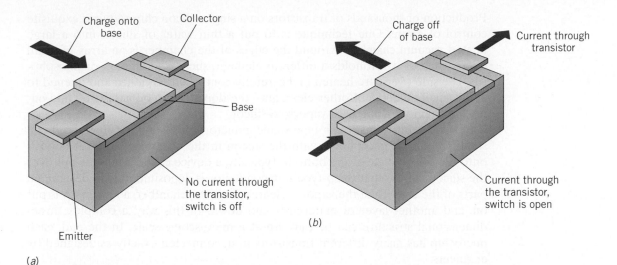

FIGURE 25-6. A transistor acting as a switch. (a) When current is supplied to the base, no current flows through the transistor; the switch is off. (b) When current does not flow to the base, a current can flow through the transistor; the switch is on.

the small signal from the tape can be impressed on the much larger current that is flowing from the emitter to the collector. It is this larger current that runs the speakers that produce the music you hear. A device that takes a small current and converts it into a large one is called an "amplifier."

The Transistor as Switch As important as the transistor's amplifying properties are, probably its most important use has been as a switch. If you run enough negative charge onto the base it can repel any electrons that are trying to get through—in effect, you can close the valve. Thus moving an electric charge onto the base shuts off the flow of current through the transistor, whereas running an electric charge off the base turns the current back on (Figure 25-6). In this manner the transistor acts as an electron switch and it can be used to process information in computers—surely the most important electronic device developed in the twentieth century.

Microchips

Individual diodes and transistors still play a vital role in modern electronics, but these devices have been largely replaced by much more complex arrays of *p*- and *n*-type semiconductors, called **microchips** or integrated circuits (Figure 25-7). Microchips may incorporate hundreds or thousands of transistors in one integrated circuit specially designed to perform a specific function. For example, an integrated circuit microchip lies at the heart of your pocket calculator or microwave oven control. Similarly, arrays of integrated circuits store and manipulate data in your personal computer, and they regulate the ignition in all modern automobiles.

The first transistors were bulky things, about the size of a golf ball. However, today a single microchip the size of a grain of rice can integrate hundreds of thousands of these devices. California's Silicon Valley has become a well-known center for the design and manufacture of these tiny integrated circuits.

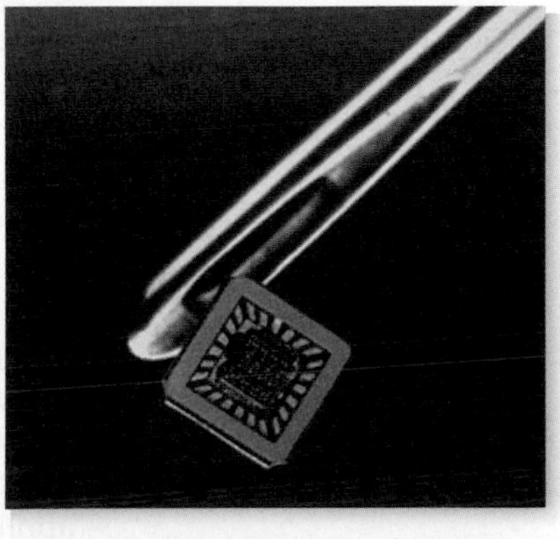

FIGURE 25-7. A microchip incorporates many complex circuits built into a single piece of silicon. For comparison, the eye of an ordinary sewing needle is shown in the background.

Production of thousands of transistors on a single silicon chip requires exquisite control of atoms. One technique is to put a thin wafer of silicon into a large heated vacuum chamber. Around the edges of the chamber is an array of small ovens, each of which holds a different element, such as aluminum or phosphorus. The side ovens are heated in a carefully controlled sequence and opened to allow small amounts of other elements—the dopants—to be vaporized and enter the chamber along with vaporized silicon.

If you want to make a *p*-type semiconductor, for example, you could mix a small amount of phosphorus with the silicon in the chamber and let it deposit onto the silicon plate at the bottom. Typically, a device called a mask is put over the silicon chip so that the *p*-type semiconductor is deposited only in designated parts of the chip. Then the vapor is cleared from the chamber, a new mask is put on, and another layer of material is laid down. In this way, a complex three-dimensional structure can be built up at a microscopic scale. In the end, each microchip has many different transistors in it, connected exactly as designed by engineers.

The reason that electronic devices have gotten so small is that engineers have gotten very good at creating smaller and smaller transistors on their microchips. This ability, in turn, depends on the ability to create finer and finer lines on the masks used in the fabrication process. Given the present rate of miniaturization, scientists have estimated that by the year 2040 the size of transistors will have reached a fundamental limit in which the lines will be one single atom across!

Physics in the Making
From the Transistor to the Integrated Circuit

When the transistor was invented in 1947, engineers quickly saw its advantages over the previous generation of electronic components, which were mostly vacuum tubes. Vacuum tubes in large numbers generated a lot of heat, requiring bulky cooling fans, and they burned out just like common lightbulbs. Transistors were small, reliable, cool, and efficient.

However, transistors presented problems of their own. Connections between transistors were made by the labor-intensive job of soldering wires into place under a microscope. And although the transistors themselves were reliable, it was all too easy for a wire to come loose. With electronics designers calling for more transistors in more complicated circuits, devices were hardly more practical than before.

The breakthrough came in 1958, when Jack Kilby, an electrical engineer at Texas Instruments, realized that you could create other circuit elements such as resistors and capacitors on the same slice of semiconductor material as the transistor. All the wires would be connecting elements on the same piece of material. His original prototype was a thin piece of germanium less than one-half inch long, containing five separate components connected by tiny wires. His patent application claimed that his *integrated circuits* could reduce the space taken up by electronic circuits by a factor of 60.

However, people still had to connect all the little wires—no easy task—and they were so small that they could break easily. In 1959 the second leap forward was made by Robert Noyce of Fairchild Semiconductors. Noyce realized that you could build the connections themselves into the circuit, using thin layers of metal

(a)   (b)

(a) An electronics engineer examines an enlarged map of a complex integrated circuit. (b) Individual microscopic n- and p-type semiconducting regions decorate a silicon wafer containing integrated circuits.

to connect pieces of semiconductor. Noyce and colleagues at Fairchild developed a way of depositing different types of material on the base piece of silicon, covering up, or masking, areas where the next layer was not wanted. This technique became the basis for all modern microchip manufacture. Noyce went on to found the Intel Company, which is one of the largest chip manufacturers in the world.

Today, computers use microprocessor chips containing over 1 million transistors. For example, Intel's i486 microprocessor measures 0.414 by 0.649 inches and contains 1,180,235 transistors.

Kilby shared the Nobel Prize in Physics in 2000, one of the few prizes awarded for applied physics rather than pure research. Most science historians believe that Noyce would have shared the award if he had still been alive. (Noyce died in 1990, and Nobel Prizes are not awarded posthumously.) Nevertheless, few developments in physics have changed the world as much as their work in creating the microchip revolution. ●

● INFORMATION

The single most important use of semiconducting devices is in the storage and manipulation of information. In fact, the modern revolution in information technology—the development of arrays of interconnected computers, global telecommunications networks, vast data banks of personal statistics, digital recording, and the credit card—is a direct consequence of materials science.

While it may appear strange to say so, almost all the media we normally consider as conveying information—the printed or spoken word, pictures, or music, for example—can be analyzed in terms of their information content and manipulated by the microchips we've just discussed. The term "information," like many words, has a precise meaning when it is used in the sciences—a meaning that is somewhat different from colloquial usage. In its scientific context, information is measured in a unit that is called the "binary digit," or **bit.**

You can think of the bit as the two possible answers to any simple question: yes or no, on or off, up or down. A single transistor used as a switch, for example, can convey one bit of information—it is either on or off. Any form of communication contains a certain number of bits of information. As we'll see shortly, the computer is simply a device that stores and manipulates this kind of information.

What is the information content of a single letter of the alphabet? From the point of view of an information theorist, the answer is simply the minimum number of questions with yes-or-no answers that are needed to identify a letter of the alphabet unambiguously. Let's see how you might go about asking such questions. Here are five questions you might ask to specify the letter *E.*

1. Is it in the first half of the alphabet? (yes)

2. Is it one of the first six letters of the alphabet? (yes)

3. Is it one of the first three letters of the alphabet? (no)

4. Is it *D*? (no)

5. Is it *E*? (yes)

From this simple example, you can see that you need the answers to at most five questions—5 bits of information—to specify unambiguously a single letter of the alphabet. (You might be lucky and guess the answer in fewer questions, but five questions are always enough to pinpoint any one of the 26 letters.) We can say, then, that the information content of a single letter of the alphabet is 5 bits.

As a matter of fact, $2 \times 2 \times 2 \times 2 \times 2 = 32$ different objects that can be specified using 5 bits of information. Thirty-two items is not enough, however, to also handle all the numbers, capitalization, punctuation marks, and other symbols used in writing. Six bits of information can specify $2 \times 2 \times 2 \times 2 \times 2 \times 2 = 64$ different items, and you could argue that everything you need to specify on a printed page can be included in those 64. Thus the information content of a single ordinary printed symbol is 6 bits. If you were using switches to store the information on a normal printed page, you would need six of them lined up in a row to specify each symbol.

The average word is six letters long, so the information content of a typical word is 36 bits. The average printed page of a novel contains about 500 words, corresponding to an information content of almost 20,000 bits. In this scheme, a 200-page book thus contains about 4 million bits, or 4 megabits, of information.

Historically, switches in computers were lumped together in groups of eight. Such a group is capable of storing 8 bits of information, or 1 **byte.** In terms of this unit, a 200-page book contains 500,000 bytes, or one-half of a megabyte.

This way of thinking about information—as a string of 0s and 1s—is ideally suited to a machine whose main working part is a transistor, which can be either on or off. Information represented in this way is said to be in **binary** form. It is a special case of information in **digital** form, which refers to a system that can have only a finite number of states.

Connection

Is a Picture Really Worth a Thousand Words?

Pictures and sounds can be analyzed in terms of information content, just as words can. Your television screen, for example, works by splitting the picture into small units called "pixels." In North America, the picture is split up into 525 segments on the horizontal and vertical axes, giving a total of about 275,000 pixels (in rounded numbers) for one picture on the TV screen. Your eye integrates these dots into a smooth picture. Every color can be thought of as a combination of the three colors red, green, and blue, and it is usual to specify the intensity of each of these three colors by a number that requires 10 bits of information to be recorded. (In practice, this procedure means that the intensity of each color is specified on a scale of about 1 to 1000.) Thus each pixel requires 30 bits to define its color, and the total information content of a picture on a TV screen is

$$275,000 \text{ pixels} \times 30 \text{ bits} = \sim 8 \text{ million bits}$$

Thus it requires about 8 megabits, or 1 megabyte, to specify a single frame on a TV picture. We should note that a TV picture typically changes 30 times a second, so the total flow of information on a TV screen may exceed 200 million bits per second.

It thus would appear that a picture is worth much more than a thousand words. In fact, if a word contains 36 bits of information, then the picture will be worth 8 million bits per picture divided by 36 bits per word, which equals about 220,000 words per picture.

The old saying, if anything, underestimates the truth! ●

Connection

Playing Your CD

Much more than writing and pictures can be expressed digitally. For example, every time you play your favorite CD you rely on a digital representation of sound.

As we have seen in Chapter 15, sound is a pressure wave in the air, and the pressure of the air at your eardrum varies rapidly. To convert a variation like this to digital form, we sample the wave at equal time intervals and represent the pressure during that interval as a single number. The procedure, in essence, changes the variation in pressure from a smooth form to a stepped form. If the time intervals are short enough and the measurement of pressure sufficiently accurate, the ear is not able to tell the difference between the two curves.

When the recording for your favorite CD was made, the pressure wave was sampled 44,100 times each second. To convert the pressure to a digital number, the range of possible pressures was split into about 64,000 intervals and the averaged pressure of the wave at a given time was assigned to one of these intervals. As shown in Problem 11 at the end of this chapter, a 16-bit number can represent any one of these 64,000 intervals. Thus, the information content of a second of play on your CD is

$$\text{Information} = (16 \text{ bits per interval}) \times (44,100 \text{ intervals per second})$$
$$= 705,600 \text{ bits per second}$$

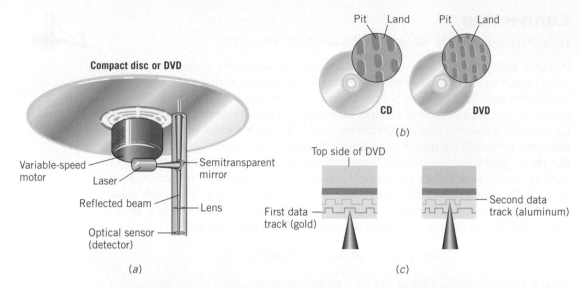

FIGURE 25-8. (*a*) A CD or DVD player bounces a laser beam off the bottom of the disc, where it is reflected to a detector or scattered by a pit. (*b*) A DVD has smaller pits and lands spaced closer together, so it can fit more data on the disc. (*c*) The DVD also has two layers of data tracks. The inner one is of aluminum; the second is of semitransparent gold. The combination gives DVDs their bronzelike color.

Thus, sound of CD quality contains more information than the written word, but less than a television picture.

To convert the digital signal back to sound, the disc that you put into your CD player has a series of pits in its surface (Figure 25-8). A laser beam plays on the disc's surface. If there is no pit on the surface, the beam is reflected into a detector. If there is a pit, the beam is deflected and goes unrecorded. Thus, the sequence of 0s and 1s of the digital signal are converted into a "beam-received" or "beam-not-received" signal by the laser. This signal, then, becomes the electric current that drives the speakers and produces the sounds you enjoy.

Over the past few years, digital video discs (DVDs) have become popular for storing and playing entire movies. A DVD is similar to a CD except that the pits are smaller and closer together. In addition, a DVD made with two layers of data tracks; the laser beam adjusts to a slightly higher power setting to read the inner track. These differences allow a DVD to store about ten times as much data as a CD. ●

In less than a quarter of a century, the computer has evolved from a specialized research aid to an essential tool for business and education.

Computers

A **computer** is a machine that stores and manipulates information. The information is stored in the computer in microchips, each of which incorporates many thousands of interconnected transistors that act as switches and carry information. In principle, a machine with a few million transistors could store the text for this entire book. In practice, however, computers do not normally work in this way. They have a *central processing unit (CPU)* in which transistors store and manipulate relatively small amounts of information at any one time. When the information is ready to be stored—for example, when you have finished working on text in a word processor or writing a program to perform a calculation—it is removed from the CPU and stored elsewhere. For example, it might be stored in the form of magnetically oriented particles on a compact disc or a hard drive. In these cases, 1 bit of information is no longer a switch that is on or off but a bit of magnetic material that has been oriented either north pole up or north pole down.

The ability to store information in this way is extremely important in modern society. As just one example, think about making an airline reservation online. You enter a Web site with access to all commercial flight information and request a specific destination (at the lowest possible price). The site you entered polls the computers of several different airlines. Each of these computers stores strings of bits that represent different flights on different days, the seating assignments, ticket arrangements, and often the address and phone number of every passenger who will be flying on the particular day when you want to fly. When you change your reservation, make a new one, or perform some other manipulation, the information is taken out of storage, brought to the central processing unit, manipulated by changing the exact sequence of bits, and then put back into storage. This process—the storage and manipulation of vast amounts of data—forms the very fabric of our modern society.

You've undoubtedly noticed that the speed and information capacity of computers has increased astonishingly over the past few decades. In the mid-1980s the very best personal computers could store a few hundred thousand bits of information. By the mid-1990s typical personal computers held billions of bits. Today, lightweight laptop computers outperform the most advanced supercomputers of a decade ago with memories that exceed tens of billions of bits. These advances are primarily the result of many improvements in materials and their processing at the atomic scale—a field called *nanotechnology*. New fine-grained magnetic materials have greatly increased the capacity of information storage devices such as hard disks, while improved semiconductor processing techniques have dramatically reduced the size of individual *n*- and *p*-type domains. The result is smaller, more powerful computers. In this way, advances in materials science play a direct role in our lives.

Connection

Jim Trefil Gives His Car a Tune-Up

As a student, I acquired the first of a long string of Volkswagen Beetles. Now let me tell you, my friends, that was a sweet car! There were never any problems with the cooling system, for the simple reason that there wasn't any—the engine was cooled by the air flowing by. And almost any repair could be made by someone with reasonable mechanical ability and a set of tools. While in graduate school, I spent many happy hours under my car, adjusting this or that.

But I never work on my cars any more. When I look under the hood now, all I see is a complex array of computers and microchips—nothing a person can get a wrench around. Yet the car I drive today, provided everything is working, is much more user-friendly than my old Volkswagen. The flow of gasoline to the cylinders, for example, is regulated by a small onboard computer, rather than by a clumsy mechanical carburetor.

A classic VW Beetle.

This personal story about cars turns out to be a pretty good allegory for the way in which the science of materials has developed in the twentieth century. In the beginning, industry turned out big, relatively simple things that were easy to understand and work with—iron wheels for railroads, steel springs for car suspensions, wooden chairs and tables for the home. Today, industry turns out items that perform the same jobs better, but that are made from new kinds of materials such as plastics, composites, and semiconductors. Instead of manipulating large chunks of material, we now control the way atoms fit together. Like modern cars, modern materials do their job well, but they cannot be made (or, usually, repaired) by a simple craftsperson working with simple tools.

So while the materials we use are becoming better at what they do and easier for us to use, it becomes harder and harder for us to understand what those materials are. I was able to fix my Volkswagen myself, but there is no way I can look under the hood of my present-day car and shift the atoms around in its microchip. In a sense, the improved performance of modern materials has been bought at the price of our ability to understand them. ●

Connection

Magnetic Data Storage

Computers rely on the ability to store information on magnetic media such as hard drives and compact discs. Remarkably, for the past several decades this capacity has doubled almost every year. This memory storage capacity (called "areal density") is measured in terms of the number of bits of information that could be stored on 1 square inch of material. (For historical reasons, the semiconductor industry does not always use metric units.) In mid-2001, the best commercially available storage was about 32 billion bits per square inch, and by mid-2002 capacity exceeded 50 billion bits per square inch. Experts predict that magnetic storage capabilities will surpass 1 trillion bits per square inch sometime in 2006.

Here's an amazing fact: according to some calculations, the total amount of information stored in all forms (cuneiform tablets, manuscripts, books, floppy disks, etc.) since the beginning of time is about 2.5 quadrillion bits. Some scholars predict that by 2006 we will be adding this much to the total stored information every 2 months.

The basic principle behind magnetic storage is the same as that behind the cassette tape. Tiny grains of magnetic material are lined up in a gel, and then a pickup head gets the information back out by detecting the directions of these grains' magnetic fields. The recording and reading heads move over the disk in a way analogous to the way an old-fashioned record-player arm moved over an LP record. Today's devices employ heads that float several atomic widths above the disk. The closer together you can get head and storage surface, the better off you are. The disk itself is a high-tech piece of equipment consisting of several layers of different materials. The working part, where the magnetic grains are, is about 200 atoms deep.

To compress more and more data onto a disk, these grains have to get smaller and smaller (in some laboratory systems they approach the size of individual atoms!). This process is inherently limited, however: if the grains are too small, then ordinary movement associated with thermal motion can erase the information minutes after it is written. In any case, current technology is expected to achieve 100 billion bits per square inch in the near future. As we show in the example and problems at the end of this chapter, the implications of this level of data storage are truly staggering. ●

Connection

Computers and the Internet

The worldwide network of computers known as the Internet has changed the way most of the world communicates. You can send personal e-mail to friends or computer files to co-workers that get to their destination within seconds. Doctors can look up a patient's X-ray films or MRI scans from the hospital database and show them on the computer screen to discuss treatment; researchers can find copies of past articles about virtually any topic as background for their own work. With all this information available from your computer, just how does the Internet work?

The heart of the Internet is a network of supercomputers located in various places around the world (Figure 25-9). They store data and are linked by conventional cables, optical cables, and even radio links to communications satellites. To access the data on these computers, you subscribe to an Internet service provider (ISP), which offers you a starting place from which to navigate the Web. When you tell your computer where you want to go, the information goes to a router maintained by the ISP that interprets each Web address as a location on a particular computer in the Internet and connects you to that location.

Of course, you can also download files from the Web. Often you can do so most easily through a server computer operated by your school or company. This server is connected by high-speed lines to the main backbone computers of the Internet, so you can simply connect to the server and access or download the material from there. ●

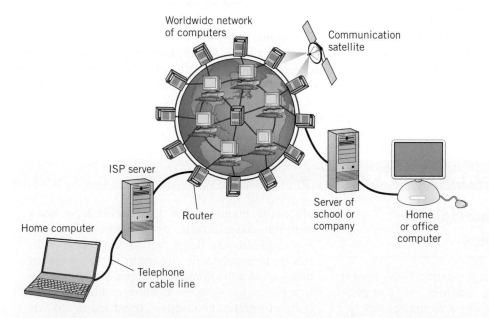

FIGURE 25-9. The Internet links numerous personal computers through a network of supercomputers located at various regions around the world.

Connection
The Computer and the Brain

When computers first came into public awareness, there was a general sense that we were building a machine that would in some way duplicate the human brain. Concepts such as *artificial intelligence* were sold (some would say oversold) on the basis of the idea that computers would soon be able to perform all those functions that we normally think of as being distinctly human. In fact, this scenario has not come to pass. The reason has to do with the difference between the basic unit of the computer, which is the transistor, and the basic unit of the brain, which is the nerve cell, or neuron.

The transmission of electric signals between the brain's neurons is fundamentally different from that between elements in ordinary electric circuits (see Chapter 18). This difference in signal transmission alone, however, does not make a brain different from a computer. A computer normally performs a sequential series of operations—that is, a group of transistors takes two numbers, adds them together, feeds that answer to another group of transistors that performs another manipulation, and so on. Some computers are now being designed and built that have some parallel capacity—machines in which, for example, addition and other manipulations are done at the same time rather than one after the other. Nevertheless, the natural configuration of computers is to have each transistor hooked to, at most, a couple of others.

A nerve cell in the brain, however, operates in quite a different manner. Each of the brain's trillions of nerve cells connects to a thousand or more different neighboring nerve cells. Whether a nerve cell decides to fire—whether or not the signal moves out along the axon—depends in an unknown way on a complex integration of all the signals that come into that cell from thousands of other cells.

Data, from *Star Trek: The Next Generation*, is an android. Computers that think like people have been a staple of science fiction stories, but could they really be built?

This complex arrangement means that the brain is a highly interconnected system, more interconnected than any other system known in nature. In fact, because the brain has trillions of cells and each cell can have thousands of connections, there are on the order of 1,000,000,000,000,000 connections among brain cells. Building a computer of this size and level of connectedness is at present totally beyond the capability of technology. ●

THINKING MORE ABOUT

Properties of Materials: Thinking Machines

One of the most intriguing questions about the ever-increasing abilities of complex computers is whether or not a computer can be built that could in some way mimic, or even replace, the human brain. Computers have been built that can add faster or remember more than any single human being, but these specific abilities by themselves do not seem to be crucial in developing a machine that thinks. The real question is whether or not a machine can be designed that is, by general consensus, regarded as alive or conscious.

World chess champion Gary Kasparov defeated the specially enhanced computer program Deep Blue in 1996. However, in 1997, Deep Blue defeated Kasparov in a rematch.

British mathematician Alan Turing (1912–1954) proposed a test to address this question. Called the "Turing test," it operates this way: A group of human beings sit in a room and interact with something through some kind of computer terminal. They might, for example, type questions into a keyboard and read answers on a screen. Alternatively, they could talk into a microphone and hear answers played back to them by some kind of voice synthesizer. These people are allowed to ask the hidden something any questions they like. At the end of the experiment, they have to decide whether they have been talking to a machine or a human being. If they can't tell the difference and the something is a machine, the machine is said to have passed the Turing test.

As of this date, no machine has passed the test (there have been occasional contests in Silicon Valley in which machines were put through their paces). But what if a machine did actually pass? Would that mean we had invented a truly intelligent machine? John Searle, a philosopher at the University of California at Berkeley, has recently challenged the whole idea of the Turing test as a way of telling if a machine can think by proposing a paradox he calls the "Chinese room."

The Chinese room works like this: An English-speaking person sits in a room and receives typed questions from a Chinese-speaking person in the adjacent room. The English-speaking person does not understand Chinese, but has a large manual of instructions. The manual might say, for example, that if a certain group of Chinese characters are received, then a second group of Chinese characters should be sent out. The English-speaking person could, at least in principle, pass the Turing test if the instructions were sufficiently detailed and complex. Obviously, however, the English speaker has no idea of what he or she is doing with the information that comes in or goes out. Thus, argues Searle, the mere fact that a machine passes the Turing test tells you nothing about whether it is aware of what it is doing.

Do you think a machine that can pass the Turing test must be aware of itself? Do you see any way around Searle's argument about the Chinese room? What moral and ethical problems might arise if human beings could indeed make a machine that everyone agreed has consciousness?

Summary

Doping with small amounts of another element modifies semiconductor material, usually silicon. Phosphorus doping adds a few mobile electrons to produce an n-type semiconductor, while aluminum doping provides positive holes in p-type semiconductors. Devices formed by juxtaposing n- and p-type semiconductors act as switches and valves for electricity. A **diode** joins single pieces of n- and p-type material, for example, to act as a one-way valve for current flow. **Transistors,** which incorporate a pnp or npn semiconductor sandwich, act as amplifiers or switches for current. **Microchips** can combine up to millions of n and p regions in a single integrated circuit.

Semiconductor technology has revolutionized the storage and use of information. Any information can be reduced to a series of simple yes-or-no questions, or **bits.** Eight-bit words, called **bytes,** are the basic unit of **digital** information used by most modern **computers.**

Key Terms

binary Presenting information as a string of 0s and 1s, representing off or on. (p. 540)

bit The smallest unit of information storage. (p. 540)

byte Eight bits of information (p. 540)

computer A machine that stores and manipulates information. (p. 542)

digital Presenting information in a numerical system that can have a finite number of states. Binary is one form of digital information. (p. 540)

diode A semiconductor device formed from one p-type (positive charge carrier) and one n-type (negative charge carrier) region; typically used as a rectifier. (p. 533)

doping The addition of small amounts of an impurity to an element or compound to enhance its conduction properties. (p. 532)

microchip A complex array of p- and n-type semiconductors that constitutes a tiny integrated circuit. (p. 537)

transistor An element in an electric circuit that regulates current or voltage flow. (p. 535)

Review

1. Semiconductors remain the anchor of the information revolution. What are some of the devices they are used in?

2. Identify five objects in your home that use semiconductors. What other kinds of materials with special electric properties are found in all of these five objects?

3. What common chemical element is used to make semiconductors? What aspect of this element's structure allows it to function as a semiconductor?

4. What is doping? What are the consequences of having an extra electron available from a doping agent?

5. Phosphorus is a common doping agent. What is it about the electron structure of phosphorus that makes it so useful as such an agent?

6. What does the term "n-type semiconductor" stand for? Why is a semiconductor doped with phosphorus called an *n-type semiconductor*?

7. Why is a semiconductor doped with aluminum called a "p-type semiconductor?" How does this type of semiconductor work?

8. What happens when an *n*-type and a *p*-type semiconductor are placed back to back? What happens to the charges of each? What is a device that works this way called?

9. In which direction will a permanent electric field push an electron in a diode?

10. How does a diode act as a one-way gate?

11. What is a rectifier, and how does it convert the AC current that enters your home into the DC current used by many household appliances?

12. What is a photovoltaic cell?

13. How does a photovoltaic cell generate electric current? When is the use of such a cell most desirable and efficient?

14. What is the difference between base-load electricity requirements and peak-load requirements? How does this affect the feasibility of solar power?

15. How is a modern transistor designed and what is its function? Describe the importance of this device to the information technology industry.

16. What is a microchip? Why is this object also called an integrated circuit?

17. Compare the size of the original microchips to their size today.

18. Describe a process by which a microchip is made. How can a *p*-type chip be made?

19. How can you amplify a weak electric current with a transistor?

20. How can a transistor act as a switch?

21. What is the unit of information that scientists typically use? What does this represent, and why is this so useful for information devices that rely on electrical circuitry? In other words, how can such a unit of information be expressed using electricity?

22. What is the information content of a single letter of the alphabet in bits?

23. How many bits equal a byte?

24. How can sound wave pressure be converted into digital form?

25. Describe how a digital signal is converted back to sound in a CD player.

26. How does a computer store information? What is the connection between tiny magnetic particles and a bit of information?

27. What is the role of the CPU in a computer?

28. What is nanotechnology, and how has it helped increase the capacity of information systems?

29. Compare a transistor with the nerve cells in your brain. What implications do the differences between these two have for the development of artificial intelligence, if any?

30. What is the Turing test? What does it tell us about the ability of a machine to think?

Questions

1. If water in a pipe is analogous to electricity in a wire, what plumbing equipment is analogous to a diode? To a transistor? A water storage tank is analogous to which electrical device?

2. Would a silicon semiconductor doped with boron be *n*- or *p*-type? How about one doped with arsenic? (*Hint:* Look at the periodic table.)

3. Does the Turing test seem like a reasonable way to judge whether a computer has consciousness? Explain.

4. In order to make an *n*-type semiconductor, silicon can be doped with a small amount of phosphorus. Why is arsenic also a good element to use as a dopant? (*Hint:* see the periodic table of elements.)

5. Compare the electrical conductivity of two phosphorus-doped silicon semiconductors where one has twice the amount of phosphorus as the other.

6. Would a diode result from taking two *n*-type semiconductors and placing them together?

7. The figure represents a graph of voltage versus time for an electric power source. Make a graph of current versus time if this source is connected to a diode rectifier. Make another graph of current versus time if the diode is reversed.

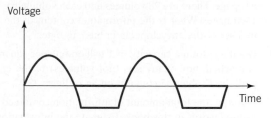

8. Which would be more effective at making a photovoltaic cell work, infrared light or ultraviolet light? Explain.

9. Which method most likely requires more digital storage capacity, storing the words of a song using a word-processing program or storing an actual recording of the song on a CD? Explain.

10. You have five pennies and five nickels in a hat. You draw them out randomly and flip each coin. How many bits of information are required to record the sequence of coins drawn from the hat (e.g., penny, nickel, nickel, penny, ...)? How many bits are required to record the sequence of coins and the sequence of heads/tails?

Problem-Solving Example

EXAMPLE 25-1

Recording Your Life

Suppose that a person has a miniature camera implanted at birth and that every 10 seconds the camera takes a picture of what that person sees. Suppose that the picture is in black and white and is somewhat grainier than that on commercial television, so that its information content is 1 Mbit. If information storage technology reaches 1 trillion bits per square inch, how much of that person's life could be recorded on a disk that is 12 inches across?

SOLUTION: If information is coming in at the rate of 1 Mbit every 10 seconds, then the total information generated in 1 day is

1 Mbit × 6 pictures per minute × 60 minutes per hour × 24 hours per day = 8640 Mbits
= 8.64 Gbits

A 12-inch disk has a radius of 6 inches and an area of

$$A = \pi r^2 = 3.14 \times 36$$
$$= 113 \text{ in}^2$$

At 1 trillion bits per square inch, the disk can hold an amount of information I equal to

$$I = 113 \text{ in}^2 \times 10^{12} \text{ bits/in}^2$$
$$= 1.13 \times 10^{14} \text{ bits}$$

Thus, the total number of days of information that can be stored on the disk is

$$\frac{1.13 \times 10^{14} \text{ bits}}{8.64 \times 10^9 \text{ bits/day}} = 13,078 \text{ days}$$

which is 35 years. Thus, at this rate, a person's entire life could be recorded on two 12-inch disks! ●

Problems

1. There is an effort in the world today to convert television into so-called high-definition TV (HDTV). In HDTV, the picture is split up into as many as 1100 by 1100 (as opposed to 525 by 525) pixels. What is the information content of an HDTV picture? What is the information content that must be transmitted each second in an HDTV broadcast?

2. Construct a set of yes-or-no questions to specify any letter of the alphabet, both upper- and lowercase, and all digits from 0 to 9.

3. Construct a set of yes-or-no questions to specify any state in the United States.

4. How many seconds do you have to listen to a CD to receive as much information as is contained in an average book?

5. The *Encyclopedia Britannica* contains about 1800 words per page. There are 28 volumes and each volume has about 1000 pages. What is the information content of the words in a set of the encyclopedia in bits? In bytes?

6. If a data storage capacity of 1 trillion bits per square inch is reached, how many sets (not volumes) of the encyclopedia in Problem 5 could fit on an ordinary CD?

7. Estimate the total amount of information contained in the printed words in this book. Estimate the information content of the illustrations in this book.

8. Rosa writes a 20-page paper for her extra-credit history grade. Each page has an average of 26 lines, with 12 words per line.

a. How many bits of information has Rosa generated in her paper?

b. Can she store her paper on a normal-mode 3.5-inch floppy disk?

9. The Cyrillic alphabet (used to write Russian and some other Eastern European languages) was devised in the ninth century and had 43 letters, whereas the alphabet used for modern Russian has 30. How many bits would it take to specify a single letter in each of these alphabets?

10. The version of modern written Japanese called "kanji" has 1945 different characters. How many bits would it take to specify a specific kanji character? Compare this to the number required in languages that have alphabets.

11. Show that a 16-bit number can represent any of 64,000 states, as it does in a CD recording.

Investigations

1. Some applications of information technology present real ethical challenges. For example, 24-hour video surveillance in public places is being increasingly used, both for simple security and, for example, to ticket a driver who runs a red light. The ultimate goal of some security agencies is for computerized pattern recognition technology to accurately match an image on a camera to an image stored in a photographic database, allowing police to uniquely identify wanted criminals and apprehend them. How does pattern recognition technology work and how accurate is it currently? What are the potential benefits and what are the potential pitfalls? How do you feel about having such technology in use everywhere? How prevalent is its use now? Considering the technological feasibility, the ethics, and the politics of it, how widespread do you think the use of this technology will be within 10 years?

2. The Internet is one of the most widely recognized applications of information technology. How does it affect your life daily and, especially, how does it affect how you socialize and how you gather information? Do you trust the information found on it? What are your standards of proof with electronic information? How do you yourself verify facts with it? From a social standpoint, would your interactions with other people be at all different without the Internet and its associated e-mail and chat rooms that allow quick online communication? Do you think this technology drastically improves your life or just makes it different?

3. What is DNA computing? How does it work? What are the potential advantages and disadvantages of this type of technology?

4. In this chapter the issue of whether a machine could ever think was raised; the Turing test was mentioned as one test for how we might probe this question and the Chinese room was offered as a criticism of this test. Investigate the history of the search for thinking machines. What are the criteria for a machine that actually thinks? How did human intelligence and the brain evolve over time? Could a machine ever duplicate this process? Summarize the arguments for and against the proposition that it will eventually be possible for machines to think.

5. The ENIAC (Electronic Numerical Integrator and Computer), built in the 1940s, was one of the first all-electronic digital computers. Investigate the ENIAC and compare it to an ordinary desktop PC in use today. How powerful is a PC today compared to the ENIAC? How large is each? In terms of materials that are involved, what was used instead of the typical semiconductors and microchips in use today? What kind of maintenance did the ENIAC require, and how was information stored?

WWW Resources

See the *Physics Matters* home page at **www.wiley.com/college/trefil** for valuable web links.

1. **http://www.pbs.org/transistor/** A historical/tutorial online exhibit accompanying the PBS program, *Transistorized!*, co-produced with the American Institute of Physics.

2. **http://www.101science.com/transistor.htm** A transistor tutorial, containing simulations describing typical circuits, applications, manufacturers, and standards.

3. **http://korea.park.org/Japan/NTT/MUSEUM/html_st/ST_menu_4_e.html** *The Basics of Electronic Communication*, a cartoon-illustrated tutorial by the NTT Digital Museum.

4. **http://www.computer50.org/** The University of Manchester site dedicated to the 50th anniversary of the birth of the modern computer.

5. **http://www.computer-museum.org/exhibits/pccomeshome/index.html** *The Computer Comes Home: A History of Personal Computing.* Online exhibit of the Computer Museum of America.

6. **http://micro.magnet.fsu.edu/electromag/java/cd/index.html** *How a CD Works,* from Florida State University.

7. **http://micro.magnet.fsu.edu/electromag/java/harddrive/index.html** *How A Hard Drive Works,* another short and simple tutorial from Florida State University.

26 | The Nucleus of the Atom

KEY IDEA

The mass of an atom is concentrated in its nucleus, which is held together by powerful forces.

PHYSICS AROUND US . . . Enjoying Empty Spaces

It's great to be lying on the beach, lulled by the sound of the surf, soaking up the Sun. There's a great feeling of peacefulness in the wide-open spaces, with no crowding of people or buildings, just wind and air and water. Away from the pressures of school and work, time seems to stand still.

In such a relaxing setting, it's hard to imagine that your own body is also mostly empty space. Every atom has electrons whizzing around with various amounts of energy, but inside the cloud of electrons is almost nothing but empty space. Yet at the very center of each atom, in a tiny volume only about one-hundred-thousandth of the size of the electron cloud, is the nucleus that holds almost all the mass of the atom.

Most of the time, the nucleus stays in the middle of the atom and nothing happens. Electrons collide and change energy, atoms bond with other atoms, and the nuclei just go along for the ride, never changing. But some nuclei do change, and they emit highly energetic particles. In fact, thousands of energetic particles are passing unnoticed through your body every second. Some of those speeding particles are damaging your cells, breaking apart bonds in the molecules that control critical functions of metabolism and cell division.

But don't lose a moment worrying about this ubiquitous background radioactivity. Since the dawn of life, low levels of radioactivity in rocks, soils, oceans, and air have bathed every living thing. This radioactivity is a natural part of the everyday environment on Earth, like the ocean and beach, and our bodies have evolved mechanisms to repair the damage it causes. But radioactivity also reveals much about the inner structure of the atom.

EMPTY SPACE AND ENORMOUS ENERGY

Imagine that you are holding a basketball, while 25 kilometers (about 15 miles) away, a few grains of sand whiz around. And imagine that all of the vast intervening space—enough to contain a fair-sized city—is absolutely empty. In some respects, that's what the inside of an atom is like, though on a much smaller scale. The basketball is the nucleus and the grains of sand represent the electrons. (Remember, however, that both nuclei and electrons display characteristics of both particles and waves.) The atom, with a diameter 100,000 times that of its nucleus, is almost entirely space.

In previous chapters we explored the properties of atoms in terms of their electrons. Chemical reactions, the way a material responds to electricity, and even the very shape and strength of objects depend on the way that electrons in different atoms interact with one another. In terms of our analogy, all of the properties of the atoms that we have studied so far result from interactions that take place 25 kilometers from where the basketball-sized nucleus is sitting. The incredible emptiness of the atom is a key to understanding two important facts about the relation of the atom to its nucleus:

1. **What goes on in the nucleus of an atom has almost nothing to do with the atom's chemistry, and vice versa.** The chemical bonding of an atom's electrons has virtually no effect on what happens to the nucleus. In most situations you can regard the orbiting electrons and the central nucleus as two separate and independent systems.

2. **The energies available in the nucleus are much greater than those available among electrons.** The particles inside the nucleus are tightly locked in. It takes a great deal more energy to pull them out than it does to remove an electron from an atom. When a particle is released from the nucleus, it carries a lot of energy.

The enormous energy we can get from the nucleus follows from the equivalence of mass and energy (which we described in Chapter 12 and discuss in more detail in Chapter 28). This relationship is defined in Einstein's most famous equation, $E = mc^2$. Remember that the constant c, the speed of light, is a very large number (3×10^8 meters per second) and that this large number is squared in Einstein's equation to give an even larger number. Thus even a very small mass is equivalent to a very large energy.

Einstein's equation tells us that a given amount of mass can be converted into a specific amount of energy in any form, and vice versa. This statement is true for any process involving energy—whenever additional energy E is stored in an object, the mass of the object increases. For example, when hydrogen and oxygen combine to form water, the mass of the water molecule is a tiny bit less than the sum of the masses of the original atoms. This missing mass has been converted to energy holding the atoms together in the molecule. Similarly, when an archer draws a bow, the mass of the bow increases by a tiny amount because of the increased elastic potential energy in the bent material.

The change in mass of objects in everyday events such as these is so small that it is customarily ignored and we speak of the various forms of energy without thinking about their mass equivalents. In nuclear reactions, however, we cannot ignore the mass changes. For example, a nuclear reactor can transform fully

20% of the mass of a proton into energy in each reaction by a process we soon will discuss. Thus nuclear reactions can convert significant amounts of mass into energy, while chemical reactions, which involve only relatively small changes in electric potential energy, involve only infinitesimal changes in mass. This difference explains why an atomic bomb, which derives its destructive force from nuclear reactions, is so much more powerful than conventional explosives, which depend on chemical reactions in materials such as TNT.

THE ORGANIZATION OF THE NUCLEUS

As we have seen in Chapter 21, Ernest Rutherford discovered the atomic nucleus by observing how fast-moving particles scatter off gold atoms. In later experiments with even faster atomic "bullets," physicists found that atomic nuclei sometimes break into smaller fragments. Thus, like the atom itself, the nucleus is made up of smaller pieces. The most important particles in the nucleus are the proton and the neutron. Approximately equal in mass, the proton and neutron can be thought of as the primary building blocks of the nucleus (Figure 26-1).

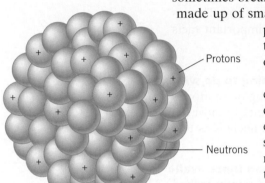

Protons

Neutrons

FIGURE 26-1. The massive atomic nucleus incorporates positively charged protons and electrically neutral neutrons.

The **proton** (from Latin for "the first one") has a positive electric charge of +1 and was the first nuclear constituent to be discovered and identified. The number of protons determines the electric charge of the nucleus. An atom in its electrically neutral state has as many negative electrons in orbit as protons in the nucleus. Thus the number of protons in the nucleus determines the chemical identity of an atom.

However, when physicists began studying nuclei, they quickly found that the mass of a nucleus is significantly greater than the sum of the masses of its protons. In fact, for most atoms, the nucleus is more than twice as massive as its protons. What does this observation imply? Physicists concluded that the additional mass had to be in a then undiscovered particle in the nucleus. We now realize that this extra mass is supplied by a particle with no electric charge called the **neutron** (for "the neutral one"). The neutron has approximately the same mass as the proton. Thus a nucleus with equal numbers of protons and neutrons has twice the mass of the protons alone.

The mass of a proton or a neutron is about 2000 times the mass of an electron. Therefore, almost all the mass of the atom is contained within the protons and neutrons in its nucleus. You can think of things this way: electrons give an atom its size, but the nucleus gives an atom its mass.

Element Names and Atomic Numbers

The most important fact in describing any atom is the number of protons in the nucleus—the **atomic number.** This number defines which element you are dealing with. For example, all atoms of gold (atomic number 79) have exactly 79 protons. In fact, for scientists, the name "gold" is simply a convenient shorthand for "atoms with 79 protons." Every element has its own atomic number: all hydrogen atoms have just 1 proton, carbon atoms must have 6 protons, and so on. The periodic table of the elements that we discussed in Chapter 21 can be thought of as a chart in which the number of protons in the atomic nucleus increases as we read from left to right and from top to bottom.

The fixed number of positively charged protons in an atom dictates the number and arrangement of the atom's electrons and thus its chemical properties. In this way, protons define the atom's chemical behavior.

Isotopes and the Mass Number

Each element has a fixed number of protons, but the number of neutrons may vary from atom to atom. In other words, two atoms with the same number of protons may have different numbers of neutrons. Such atoms are said to be **isotopes** of the element, and they have different masses. The total number of protons and neutrons in an atom is called the **mass number.**

The nucleus of every element exists in several different isotopes, each with a different number of neutrons. For example, the most common isotope of carbon has six neutrons, so it has a mass number of 12 (6 protons + 6 neutrons), usually written ^{12}C or carbon-12 and called "carbon twelve." Other isotopes of carbon, such as carbon-13, with seven neutrons, and carbon-14, with eight neutrons, are heavier than carbon-12. However, they have the same electron arrangements and, therefore, the same chemical behavior. A neutral carbon atom, whether carbon-12, carbon-13, or carbon-14, must have six electrons in orbit to balance its six protons.

The complete set of all the isotopes—every known combination of protons and neutrons—is often illustrated on a graph that plots number of protons versus number of neutrons (Figure 26-2). Several features are evident from this graph. First, every chemical element has many known isotopes—in some cases

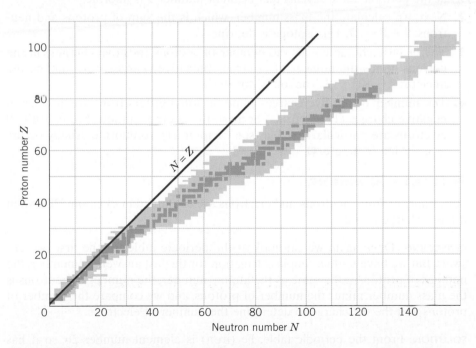

FIGURE 26-2. A chart of the isotopes. Stable isotopes appear in green, and radioactive isotopes are in gold. Each of the approximately 2000 known isotopes has a different combination of protons (Z, on the vertical scale) and neutrons (N, on the horizontal scale). Isotopes of the light elements (toward the bottom left of the chart) have similar numbers of protons and neutrons and thus lie close to the diagonal $N = Z$ line at 45 degrees. Heavier isotopes (on the upper right part of the chart) tend to have more neutrons than protons and thus lie well below this line.

dozens of them. Close to 2000 isotopes have been documented, compared to the hundred or so different chemical elements. This graph also reveals that for any particular isotope, the number of protons is not generally the same as the number of neutrons. While many light elements, up to about calcium (with 20 protons), often have nearly equal numbers of protons and neutrons, heavier elements tend to have more neutrons than protons. This fact plays a key role in the phenomenon of radioactivity, as we shall see.

Inside the Atom

EXAMPLE
26-1

An atom has nine protons and eight neutrons in its nucleus and 10 electrons in orbit.

1. What element is it?

2. What is its mass number?

3. What is its electric charge?

4. How is it possible that the number of protons and electrons is different?

REASONING: We can find the first three answers by looking at the periodic table (Appendix C). For the last answer, we refer back to Chapter 21 and the discussion of stable electron states.

SOLUTION:

1. The element name depends on the number of protons, which is nine. A glance at the periodic table reveals that element number 9 is fluorine.

2. Next, we calculate the mass number, which is the sum of protons and neutrons: $9 + 8 = 17$. This isotope is fluorine-17.

3. The electric charge equals the number of protons (positive charges) in the nucleus minus the number of electrons (negative charges) surrounding the nucleus: $9 - 10 = -1$. The ion is thus F^{-1}.

4. The number of positive charges (nine protons) differs from the number of negative charges (10 electrons) because this atom is an ion. Atoms with 10 electrons are particularly stable (see Chapter 21), so fluorine usually occurs as a -1 ion in nature. ●

A Heavy Element

EXAMPLE
26-2

How many protons, neutrons, and electrons are contained in the atom ^{56}Fe when it has a charge of $+2$?

REASONING: Once again we can look at the periodic table for the first two answers, but we have to do a simple calculation for the last answer. Remember, the number of protons is the same as the atomic number, the number of neutrons is the mass number minus the number of protons, and we compare the number of protons and the $+2$ charge to determine the number of electrons.

SOLUTION: From the periodic table, Fe (iron) is element number 26, so it has 26 protons.

The number of neutrons is the mass number minus the number of protons: $56 - 26 = 30$ neutrons.

The number of electrons surrounding the nucleus is equal to the number of protons minus the charge on the ion, which in this case is $+2$. Thus there are $26 - 2 = 24$ electrons in orbit. ●

The Strong Force

In Chapter 16 we learned that one of the fundamental laws of electricity is that like charges repel one another. If you think about the structure of the nucleus for a moment, you will realize that the nucleus is made up of a large number of positively charged objects (the protons) in close proximity to one another. Why doesn't the electrical repulsion between the protons push them apart and disrupt the nucleus completely?

The nucleus can be stable only if there is an attractive force capable of balancing or overcoming the electrical repulsion at the incredibly small scale of the nucleus. Whatever the force is, it must be vastly stronger than gravity or electromagnetism, the only two forces we've encountered up to this point. For this reason it is called the **strong force.** The strong force must operate over only very short distances (characteristic of the size of the nucleus) because our everyday experience tells us that the strong force doesn't act on large objects. Both with respect to its magnitude and its range, the strong force is somehow confined to the nucleus. In this respect, the strong force is unlike electricity or magnetism.

The strong force has another distinctive feature. If you weigh a dozen apples and a dozen oranges, their total weight is simply the sum of the individual pieces of fruit. But this is not true of protons and neutrons in the nucleus. The mass of the nucleus is always slightly less than the sum of the masses of the protons and neutrons. When protons and neutrons come together, some of their mass is converted into the energy that binds them together. We know this must be true, because it requires energy to pull most nuclei apart. This *binding energy* varies from one nucleus to another. The iron nucleus is the most tightly bound of all nuclei. This fact is important in the life cycle of stars, which can obtain energy by combining lighter nuclei into heavier ones (a process called *fusion,* discussed later in this chapter) until they form iron. Because iron is the most stable of nuclei, stars cannot obtain energy by combining iron nuclei into something else and so fusion in the star stops.

 RADIOACTIVITY

The vast majority of atomic nuclei in objects around you—more than 99.999% of the atoms in our everyday surroundings—are stable. In all probability, the nuclei in those atoms will not change over the lifetime of the universe. But some kinds of atomic nuclei are not stable. Uranium-238, for example, which is the most common isotope of the rather common element uranium, has 92 protons and 146 neutrons in its nucleus. If you put a block of uranium-238 on a table in front of you and watch it for a while, you will find that a few of the uranium nuclei in that block will disintegrate spontaneously. One moment there is a normal uranium atom in the block and the next moment there are fragments of smaller atoms and no uranium atom. At the same time, fast-moving particles speed away from the uranium block into the surrounding environment. This spontaneous release of energetic particles is called **radioactivity** or **radioactive decay.** The emitted particles themselves are referred to as *radiation.* The term *radiation* used in this sense is somewhat different from the electromagnetic radiation that we introduced in Chapter 11. In this case, *radiation* refers to whatever is produced from the spontaneous decay of nuclei, be it electromagnetic waves or actual particles with mass.

What Is Radioactive?

Almost all the atoms around you are stable, but most everyday elements have at least a few isotopes that are radioactive. Carbon, for example, is stable in its most common isotopes carbon-12 and carbon-13; however, carbon-14, which constitutes about one in every 10^{12} carbon atoms in living things, is radioactive. A few elements, such as uranium, radium, and thorium, have no stable isotopes at all. Even though most of our surroundings are composed of stable isotopes, a quick glance at the chart of isotopes (Figure 26-2) reveals that most of the 2000 or so known natural and laboratory-produced isotopes are unstable and undergo radioactive decay of one kind or another.

Physics in the Making

Becquerel and Curie

The nature of radioactivity was discovered in 1896 by Antoine Henri Becquerel (1852–1908), who studied minerals that incorporate uranium and other radioactive elements. He placed some of these samples in a drawer of his desk along with an unexposed photographic plate and a metal coin. When he developed the photographic plate some time later, the silhouette of the coin was clearly visible. From this photograph he concluded that some unknown form of radiation had traveled from the sample to the plate. The coin seemed to have absorbed the radiation and blocked it off, but the radiation that got around the coin and reached the plate delivered enough energy to cause the chemical reactions that normally go into photographic development. Becquerel knew that whatever had exposed the plate must have originated in the minerals and traveled at least as far as the plate.

Becquerel's discovery was followed by an extraordinarily exciting time for chemists, who began an intensive effort to isolate and study the elements from which the radiation originated. The leader in the field we now call radiochemistry was also one of the best-known scientists of the modern era, Marie Sklodowska Curie (1867–1934). Born in Poland and married to Pierre Curie, a distinguished French physicist (see Chapter 24), she conducted her pioneering research in France, often under extremely difficult conditions because of the unwillingness of her colleagues to accept her. She worked with tons of exotic uranium-bearing minerals from mines in Bohemia, and she isolated minute quantities of previously unknown elements such as radium and polonium. One of her crowning achievements was the isolation of 22 milligrams of pure radium chloride, which became an international standard for measuring radiation levels. She also pioneered the use of X rays for medical diagnosis during World War I.

For her work, Madame Curie became the first scientist to be awarded two Nobel prizes, one in physics and one in chemistry. One of the most common units used to measure radioactivity is named the curie, in her honor. She also was one of the first scientists to die from prolonged exposure to radiation, whose harmful effects were not known at that time. Her fate, unfortunately, was shared by many other pioneers in nuclear physics. ●

The Curie family, with Marie Sklodowska, Pierre, and their child, Irene. Both parents received the Nobel Prize in physics in 1903, and Marie was the sole recipient of the 1911 prize in chemistry. Their daughter received the 1935 Nobel Prize with her husband, Frederic Joliot-Curie.

The Kinds of Radioactive Decay

Physicists who studied radioactive rocks and minerals soon discovered three distinct kinds of radioactive decay, each of which changes the nucleus in its own characteristic way and each of which plays an important role in modern science and technology. These three kinds of radioactivity were dubbed *alpha, beta,* and *gamma radiation* (after the first three letters of the Greek alphabet) to indicate that they were unknown and mysterious when first discovered (Figure 26-3).

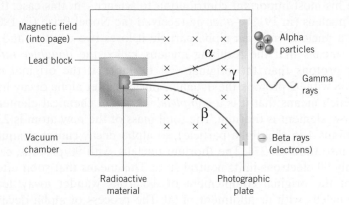

FIGURE 26-3. The three common types of radioactivity are designated alpha, beta, and gamma radiation. Alpha particles have a positive charge and beta rays have a negative charge, so they are deflected in opposite directions by a magnetic field. Gamma rays have no charge and pass through a magnetic field undeflected.

Alpha Decay Some radioactive decays involve the emission of a relatively large and massive particle composed of two protons and two neutrons. Such a particle is exactly the same as the nucleus of a helium-4 atom. It is called an "alpha particle," and the process by which it is emitted is called **alpha decay.** (An alpha particle is usually represented in equations and diagrams by the Greek letter alpha, α.)

Ernest Rutherford, whom we've met as the discoverer of the nucleus, discovered the nature of alpha decay in the first decade of the twentieth century. His simple and clever experiment, sketched in Figure 26-4, began with placing a small amount of radioactive material known to emit alpha particles in a sealed

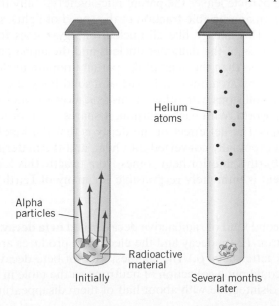

FIGURE 26-4. The Rutherford experiment led to the identification of the alpha particle, which is the same as a helium nucleus.

tube. After several months, careful chemical analysis revealed the presence of a small amount of helium in the tube—helium that hadn't been present when the tube was sealed. From this observation, Rutherford concluded that alpha particles must be associated with the helium atom. Today we say that Rutherford observed the emission of the helium nucleus in radioactive decay, followed by the acquisition of two electrons to form an atom of helium gas.

Rutherford received the Nobel Prize in chemistry for his chemical studies and his work in sorting out radioactivity. He is one of the few people in the world who made his most important contribution to science—in this case, the discovery of the nucleus (in 1911)—*after* he received the Nobel Prize (in 1908).

When a nucleus emits an alpha particle it loses two protons and two neutrons. This means that the resulting nucleus, called the *daughter nucleus,* has two fewer protons than the original. For example, if the original nucleus is uranium-238 with 92 protons, the daughter nucleus after alpha decay has only 90 protons, which means that it is a *completely different* chemical element. In this case, the new element is thorium. The total mass of the new atom is 234 (238 in uranium minus 4 in the alpha particle), so alpha decay causes uranium-238 to transform into thorium-234. The thorium nucleus, with 90 protons, can accommodate only 90 electrons in its neutral state. This means that soon after the decay, two of the original complement of electrons wander away, leaving the daughter nucleus with its allotment of 90. The process of alpha decay reduces the mass and changes the chemical identity of the decaying nucleus.

Radioactivity is nature's philosopher's stone. According to medieval alchemists, the philosopher's stone was supposed to turn lead into gold (among other wonderful properties). The alchemists never found their stone because almost all their work involved what we today call *chemical reactions;* that is, they were trying to change one element into another by manipulating electrons. Given what we now know about the structure of atoms, we realize that they were approaching the problem from the wrong end. If you really want to change one chemical element into another, you have to manipulate the nucleus, precisely what happens in the process of radioactivity.

When the alpha particle leaves the parent nucleus, it typically travels at very high speed (often at an appreciable fraction of the speed of light), so it carries a lot of kinetic energy. This energy, like all nuclear energy, comes from the conversion of mass: the mass of the daughter nucleus and the alpha particle added together is somewhat less than the mass of the parent uranium nucleus. If the alpha particle is emitted by an atom that is part of a solid body, then the particle undergoes a series of collisions with other atoms as it moves from the parent nucleus into the wider world. With each collision, it shares some of its kinetic energy with other atoms. The net effect of the decay is that the kinetic energy of the alpha particle is eventually converted into heat, and the material warms up. About one-half of Earth's interior heat comes from exactly this kind of energy transfer, and this heat is ultimately responsible for many of Earth's major surface features.

Beta Decay The second kind of radioactive decay, called **beta decay,** involves the emission of an electron. (Beta decay and the electron it produces are usually denoted by the Greek letter beta, β.) The simplest kind of beta decay is for a single neutron. If you could put a collection of neutrons on the table in front of you, they would start to disintegrate with about half of them disappearing in the first

10 minutes or so. The most obvious products of this decay are a proton and an electron. Both particles carry an electric charge and are therefore very easy to detect. This production of one positive and one negative particle from a neutral one does not change the total electric charge of the entire system.

In the 1930s, when beta decay of the neutron was first seen in a laboratory, the experimental equipment available at the time easily detected and measured the energies of the electron and proton. Physicists looking at beta decay were troubled to find that the process appeared to violate the law of conservation of energy, as well as conservation of momentum. When they added up the momenta of the proton and electron, the result was less than the momentum of the original neutron. If only the electron and proton were given off, the conservation of momentum would be violated.

Rather than face this possibility, physicists at the time followed the lead of Wolfgang Pauli (see Chapter 6) and postulated that another particle had to be emitted in the decay, a particle that they could not detect at the time, but that carried away the missing momentum. It wasn't until 1956 that physicists were able to detect this missing particle—the *neutrino* ("little neutral one")—in the laboratory. This particle has no electric charge, travels almost as fast as the speed of light, and, if stopped, would have almost no mass. Today, at giant particle accelerators (see Chapter 27), neutrinos are routinely produced in other experiments.

When beta decay takes place inside a nucleus, one of the neutrons in the nucleus is converted into a proton, an electron, and a neutrino. (Note that a neutron does not consist of a proton, electron, and neutrino. These particles are formed only at the moment the decay reaction takes place.) The lightweight electron and the neutrino speed out of the nucleus while the proton remains. The electron that comes off in beta decay is not one of the electrons that originally circled the nucleus in an allowed orbit: the electron emitted from the nucleus comes out so fast that it is long gone from the atom before any of the electrons in orbit have time to react. The new atom has a net positive charge, however, and eventually acquires a stray electron from its environment.

The net effect of beta decay is that the daughter nucleus has approximately the same mass as the parent (it has the same total number of protons plus neutrons), but has one more proton and one less neutron. It is therefore a different element than it was before. For example, carbon-14 (six protons) undergoes beta decay to become an atom of nitrogen-14 (seven protons). If you were to place a small pile of carbon-14 powder—it looks like black soot—in a sealed jar and come back in 20,000 years, most of the powder would have disappeared and the jar would be filled with colorless, odorless nitrogen gas. Beta decay, therefore, is a transformation in which the chemical identity of the atom is changed, but its mass is virtually the same before and after. (Remember, the electron and neutrino that are emitted are extremely lightweight and make almost no difference in the atom's total mass.)

What force in nature could cause an uncharged particle such as the neutron to fly apart? The force is certainly not gravitational attraction between masses. It is not the electromagnetic force that causes oppositely charged particles to fly away from one another. And beta decay seems to be quite different from the strong nuclear force that holds protons together in the nucleus. In fact, beta decay is an example of the operation of the fourth fundamental force in nature, called the *weak force*. (We discuss the weak force again in Chapter 27.)

A safety officer in protective clothing uses a Geiger counter to examine waste for radioactivity.

Gamma Radiation The third kind of radioactivity, called **gamma radiation,** is different in character from alpha and beta decay. (Gamma decay and gamma radiation are usually denoted by the Greek letter gamma, γ.) *Gamma ray* is simply a generic term for a very energetic photon—electromagnetic radiation. In Chapter 19 we have seen that all electromagnetic radiation comes from the acceleration of charged particles, and that is what happens in gamma radioactivity.

When an electron in an atom shifts from a higher energy level to a lower one, we know that a photon is emitted, typically in the range of visible or ultraviolet light. In just the same way, the particles in a nucleus can shift between different energy levels. These shifts, or nuclear quantum leaps, involve energy differences thousands or millions of times greater than those of orbiting electrons. When particles in a nucleus undergo shifts from higher to lower energy levels, some of the emitted gamma radiation is in the range of X rays, while others are even more energetic.

A nucleus emits gamma rays any time its protons and neutrons reshuffle. Neither the protons nor the neutrons change their identity, so the daughter atom has the same mass, the same isotope number, and the same chemical identity as the parent. Nevertheless, this process produces highly energetic radiation.

Moving Down the Chart of the Isotopes

The three kinds of radioactivity, summarized in Table 26-1, affect the nuclei of isotopes. Gamma radiation changes the energy of the nucleus without altering the number of protons and neutrons—the isotope doesn't change into a different element. Alpha and beta radiation result in a new element because they change the numbers of protons and neutrons. Think about how alpha and beta radiation might be represented on the chart of isotopes (Figure 26-2). Alpha decay removes two protons and two neutrons, so we shift diagonally down and to the left by two units on the chart. Alpha decay thus provides a way to move from heavier to lighter elements.

Beta decay, on the other hand, results in one less neutron and one more proton. On the chart of the isotopes, we move diagonally up and to the left by one unit. Beta decay provides a way to reduce the ratio of neutrons to protons, which is generally greater for heavier elements. Combinations of beta and alpha decays, therefore, provide a means to move from heavier radioactive isotopes to lighter stable isotopes.

TABLE 26-1 Types of Radioactive Decay		
Type of Decay	**Particle Emitted**	**Net Change**
Alpha	Alpha particle	New element with two fewer protons and two fewer neutrons
Beta	Beta particle	New element with one more proton and one fewer neutron
Gamma	Photon (gamma ray)	Same element, but with less energy

Radiation and Health

Why is nuclear radiation so dangerous? Alpha, beta, and gamma radiation all carry a great deal of energy—enough energy to damage the molecules that are essential to the workings of the cells that make up your body. The most common type of damage is *ionization,* the stripping away of one or more of an atom's electrons. An ionized atom cannot bond in its normal way, and any structures that depend on that atom—a cell wall or a piece of genetic material, for example—will be damaged. Prolonged exposure to ionizing radiation can so disrupt an organism's cells that it dies. You can think of the damage caused by the different kinds of radiation as being analogous to the damage different sorts of vehicles would inflict in moving through a crowded alley (Figure 26-5). The alpha particle is like a truck that demolishes everything in its path but doesn't get very far. Beta radiation is like a car that does less damage but goes farther, while gamma radiation penetrates the farthest.

It takes a great deal of radiation to cause sickness or death. Only in unusual circumstances, such as the aftermath of nuclear weapons used on the Japanese cities of Hiroshima and Nagasaki at the end of World War II or the nuclear reactor accident at Chernobyl in Ukraine in 1986, do people die shortly after exposure. However, there are possible long-term effects of exposure to lower levels of radiation. While such exposure does not cause immediate death, it may affect

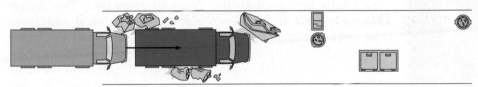

The truck doesn't get far, but destroys whatever it hits,

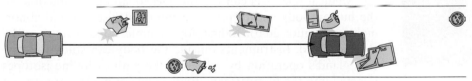

The car travels farther than the truck,
doing less damage per foot traveled than the truck does.

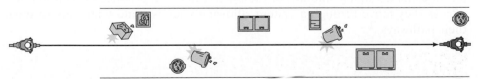

The motorcycle makes it through the alley,
doing less damage per foot traveled.

FIGURE 26-5. The damage to atoms and molecules from different kinds of radiation can be compared to the damage to objects in an alleyway caused by different types of vehicles. The truck is analogous to an alpha particle, the car is analogous to a beta particle, and the motorcycle is analogous to a gamma ray. Although you might conclude that the gamma ray does the least amount of damage, its high energy and ability to penetrate deeply makes it especially dangerous in large quantities.

a body's ability to replace old cells and can increase the risk of cancer. Radiation may also damage the genetic material that carries the coded information required to reproduce. Birth defects may not appear until decades after parental exposure to radiation.

No one can escape high-energy radiation entirely. Radioactive decay of uranium and other elements in rocks and soils, as well as cosmic radiation from space, bombard us all the time. Radioactive elements even occur in our bones and tissues. Certain kinds of radiation—medical X rays, for example, and radioactive isotopic tracers used for diagnosis—have saved countless lives. Human beings have always lived in a radioactive environment. Nevertheless, no matter how small the added risk, it is wise to minimize unnecessary exposure to sources of radioactivity.

APPLICATIONS OF RADIOACTIVITY

The Use of Radioactive Tracers

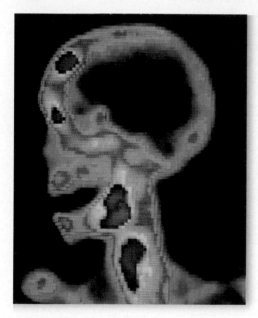

FIGURE 26-6. Radioactive tracers at work. The patient has been given a radioactive tracer that concentrates in the bone and emits radiation that can be measured on a film. The dark spot in the front part of the skull indicates the presence of a bone cancer.

Today, numerous radioactive materials find special uses in medicine and industry because all radioactive isotopes are also chemical elements. The chemistry of atoms is governed by their electrons, while the radioactive properties of a material are totally unrelated to the chemical properties. This means that a radioactive isotope of a particular chemical undergoes the same chemical reactions as a stable isotope of that same element. For example, if a radioactive isotope of iodine or phosphorus is injected into your bloodstream, it collects at the same places in your body as stable iodine or phosphorus do. The only difference is that the radioactive isotope can be detected with ease.

Medical scientists can use this fact to study the functions of the human body and to make diagnoses of diseases and abnormalities (Figure 26-6). Iodine, for example, concentrates in the thyroid gland. Instruments outside the body can study the thyroid gland's operation by following the path of iodine isotopes injected into the bloodstream. Radioactive or nuclear tracers are also used extensively in the earth sciences, in industry, and in other scientific and technological applications to follow the exact chemical progressions of different elements. Small amounts of radioactive material produce measurable signals as they move through a system, allowing scientists and engineers to trace their pathways.

Connection

The Science of Life: Robert Hazen's Broken Wrist

I once had an experience that gave me a whole new perspective on radioactivity. A decade ago, while playing beach volleyball, I dove for a ball and bent back my wrist. It hurt a lot, but it was early in the season, so I taped up the wrist and kept on playing. After a couple of weeks it didn't hurt too much, so I forgot about the injury.

Years later, when the wrist started hurting again, I went to a doctor, who said, "Your wrist has been broken for a long time. When did it happen?" Because the break was so old, my doctor had to find out whether the broken bone surfaces were still able to mend. They sent me to a specialized hospital facility where I was given a shot of a fluid containing a radioactive phosphorus compound—a compound that accumulates on active growth surfaces of bone. After a few minutes, this material circulated through my body and some of the phosphorus compound concentrated on unset regions of my wrist bones. Radioactive molecules constantly emitted particles that moved through my skin to the outside; as I lay on a table, an image of my broken wrist glowed on the overhead monitor. The process produced a clear picture of the fracture, so my doctor was able to reset the bones. My wrist has healed, and I'm back to playing volleyball. ●

Half-Life

Left to itself, a single nucleus of an unstable isotope eventually decays in a spontaneous event. That is, the original nucleus persists up until a specific time, then radioactive decay occurs and from that point on you see only the fragments of the decay.

Watching a single nucleus undergo decay is like watching one kernel in a batch of popcorn. Each kernel pops at a specific time, but not all the kernels pop at the same time. Even though you can't predict when any one kernel will pop, you can predict the average time it takes for a kernel to pop. A collection of radioactive nuclei behaves in a similar way. Some nuclei decay almost as soon as you start watching; others persist for much longer times. The percentage of nuclei that decay in each second after you start watching remains more or less the same.

Physicists use the term **half-life** to describe the average time it takes for one-half of a batch of radioactive isotopes to undergo decay. For example, if you have 100 nuclei at the beginning of your observation and it takes 20 minutes for 50 of them to undergo radioactive decay, then the half-life of that nucleus is 20 minutes. If you were to watch that sample for another 20 minutes, however, not all the nuclei would have decayed. You would find that you had about 25 nuclei at the end. After another 20 minutes you would most likely have 12 or 13, and so on (Figure 26-7).

Saying that a nucleus has a half-life of 1 hour does *not* mean that all the nuclei sit there for 1 hour, at which point they all decay. The nuclei, like the popcorn kernels in our example, decay at different times. The half-life is simply an indication of how long on average it will be before an individual nucleus decays.

The situation is something like giving identical coins to, say, 10,000 people and seeing who can flip heads the most consecutive times. Everyone left in the game flips a coin every minute. After the first minute, about one-half the people have flipped tails and are out of the game, while the other half have flipped heads and continue in the game. (We assume fair coins for which the probability of landing heads or tails is 50%.) After 2 minutes, one-half of the people left after the first minute are still playing, or one-quarter of the original 10,000 people. Radioactive decay is similar to this game because after every half-life, half of the nuclei that have not decayed are still available to decay in the next half-life.

Radioactive nuclei display a wide range of half-lives. Some nuclei, such as uranium-222, are so unstable that they persist only a tiny fraction of a second. Others, such as uranium-238, have half-lives that range into the billions of years,

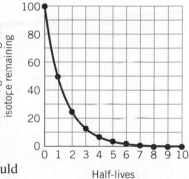

FIGURE 26-7. The graph shows the percentage of radioactive nuclei left in a sample as the number of half-lives increases.

comparable to the age of Earth. Between these two extremes you can find a radioactive isotope that has almost any half-life you wish.

We do not yet understand enough about the nucleus to be able to predict half-lives. On the other hand, the half-life is a fairly easy number to measure and therefore can be determined experimentally for any nucleus. The fine print on most charts of isotopes (expanded versions of Figure 26-2) usually includes the half-life for each radioactive isotope.

Radiometric Dating

Radioactive decay has provided scientists who study Earth and human history with one of their most important methods of determining the age of materials. This remarkable technique, which depends on our knowledge of the half-life of radioactive materials, is called **radiometric dating.**

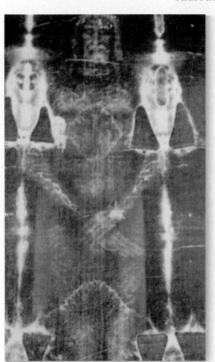

The Shroud of Turin, with its ghostly image of a man, was dated by carbon-14 techniques to centuries after the death of Christ.

The best-known radiometric dating scheme involves the isotope carbon-14. Every living organism takes in carbon during its lifetime. At this moment your body is taking the carbon in your food and converting it to tissue, and the same is true of all other animals. Plants are taking in carbon dioxide from the air and doing the same thing. Most of this carbon, about 99%, is in the form of carbon-12, while perhaps 1% is carbon-13. But a certain small percentage, no more than one carbon atom in every trillion, is in the form of carbon-14, a radioactive isotope of carbon with a half-life of about 5700 years.

As long as an organism is alive, the carbon-14 in its tissues is constantly renewed in the same small proportion that is found in the general environment. All the isotopes of carbon behave the same way chemically, so the proportions of carbon isotopes in the living tissue are the same everywhere, for all living things. When an organism dies, however, it stops taking in carbon of any form. From the time of death, therefore, the carbon-14 in the tissues is no longer replenished. Like a ticking clock, carbon-14 disappears atom by atom to form an ever-smaller percentage of the total carbon. We can determine the approximate age of a bone, piece of wood, cloth, or other carbon-containing object by carefully measuring the fraction of carbon-14 that remains and comparing it to the amount of carbon-14 that we know must have been in that material when it was alive. If the material happens to be a piece of wood taken from an Egyptian tomb, for example, we have a pretty good estimate of how old the artifact is and, probably, when the tomb was built.

Carbon-14 dating often appears in the news when a reputedly ancient artifact is shown to be from more recent times. In one highly publicized experiment, the Shroud of Turin, a fascinating cloth artifact reputed to be involved in the burial of Jesus, was shown by carbon-14 techniques to date from the fourteenth century A.D.

Develop Your Intuition: Dating Stonehenge

In Chapter 3, we described the monument known as Stonehenge and gave an age for it. This age came from the carbon dating technique we've just described. The monument is made of stone; how do you suppose this dating was done?

You might consider measuring radioactive elements within the stones themselves, perhaps some radioactive isotopes of silicon or oxygen. But that would tell you the age of the stones, not the time at which they were used to form Stonehenge. The actual dating of Stonehenge involved measuring amounts of carbon-14 in some of the artifacts found at the site, including tools made out of animal bone and charcoal from buried campfires. Archeologists have found digging tools made from deer antlers and cattle shoulder bones buried at the bottom of ditches used to help raise the stones. Even a human skeleton and fragments of pottery have been found in holes clearly dug to help maneuver the great stones into place. Radiometric dating of all these objects has given us the date of about 2800 B.C. for the building of the monument.

Carbon-14 dating has been instrumental in mapping human history over the last several thousand years. When an object is more than about 50,000 years old, however, the amount of carbon-14 left in it is so small that this dating scheme cannot be used. This means that carbon-14 can only be used to find the age of objects that are relatively young.

In order to date rocks and minerals that are millions of years old, scientists must rely on similar techniques that use radioactive isotopes of much greater half-life. Among the most widely used radiometric clocks in geology are those based on the decay of potassium-40 (half-life of 1.25 billion years), uranium-238 (half-life of 4.5 billion years), and rubidium-87 (half-life of 49 billion years). In these cases, we measure the total number of atoms of a given element, together with the relative percentage of a given isotope, to determine how many radio-active nuclei were present at the beginning.

The oldest human fossils are too ancient to be dated by carbon-14 methods. An alternative technique, potassium-argon dating, is employed for dating the rocks in which these skulls, which are up to 3.7 million years old, were found.

Connection

Dating a Frozen Mammoth

Russian paleontologists occasionally discover beautifully preserved mammoths frozen in Siberian ice. Carbon isotope analyses from these mammoths often show that only about one-quarter of the original carbon-14 is still present in the mammoth tissues and hair. If the half-life of carbon-14 is 5700 years, how old is the mammoth?

To solve this problem it is necessary to determine how many half-lives have passed, with the predictable decay rate of carbon-14 serving as a clock. In this case, only one-quarter of the original carbon-14 isotopes remain ($\frac{1}{4} = \frac{1}{2} \times \frac{1}{2}$), so the carbon-14 isotopes have passed through two half-lives. After 5700 years, about one-half of the original carbon-14 isotopes remain. Similarly, after another 5700 years only one-half of these remaining carbon-14 isotopes (or one-quarter of the original amount) remain. The age of the preserved mammoth is thus two half-lives, or about 11,400 years. ●

● DECAY CHAINS

When a parent nucleus decays, the daughter nucleus is not necessarily stable (it is radioactive). In fact, in the great majority of cases the daughter nucleus is as unstable as the parent. The original parent decays into the daughter, the daughter decays into a second daughter, on and on, perhaps for more than a dozen different radioactive events. Even if you start with a pure collection of atoms of the same isotope of the same chemical element, nuclear decay guarantees that eventually you'll have many different chemical species in the sample. A series of decays of this sort is called a "decay chain." The sequence of decays continues until a stable isotope appears. Given enough time, all the atoms of the original element eventually decay into that stable isotope.

To get a sense of a decay chain, consider the case of uranium-238 with a half-life of approximately 4.5 billion years. Uranium-238 decays by alpha emission into thorium-234, another radioactive isotope. In the process, uranium-238 loses two protons and two neutrons. Thorium-234 undergoes beta decay (half-life of 24.1 days) into protactinium-234 (half-life of about 7 hours), which in turn undergoes beta decay to uranium-234. Each of these beta decays results in the conversion of a neutron into a proton and an electron. After three radioactive decays, we are back to uranium, albeit a lighter isotope with a 247,000-year half-life.

The rest of the uranium decay chain is shown in Figure 26-8. It follows a long path through eight different elements before it winds up as stable lead-206. Given

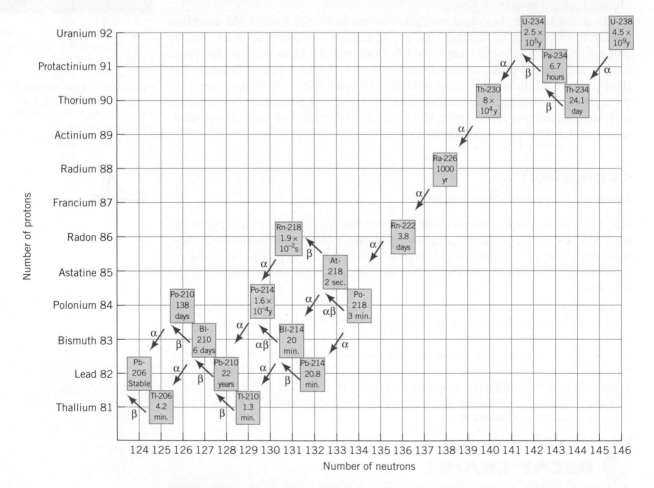

FIGURE 26-8. The uranium-238 decay chain. The nuclei in the chain decay by both alpha and beta emission until they reach lead-206, a stable isotope. Some isotopes may undergo either alpha or beta decay, as indicated by splits in the chain. Nevertheless, all paths eventually arrive at lead-206 after 14 decay events.

enough time, all the uranium-238 now on Earth will eventually decay into lead-206. Earth is only about 4.5 billion years old, however, so there's only been time for about one-half of the original uranium to decay. For the next several billion years we can expect to have all the members of the uranium decay chain in existence on Earth.

Indoor Radon

The uranium-238 decay chain is not an abstract concept, of interest only to theoretical physicists. In fact, the widely publicized health concerns over indoor radon pollution is a direct consequence of the uranium decay chain. Uranium is a fairly common element—about 2 grams out of every ton of rocks at Earth's surface are uranium. The first steps in the uranium-238 decay chain produce thorium, radium, and other elements that remain sealed in ordinary rocks and soils. The principal health concern arises from the production of radon-222, about halfway along the path to stable lead. Radon is a colorless, odorless, inert gas that does not chemically bond to its host rock.

As radon is formed, it seeps out of its mineral host and moves into the atmosphere, where it undergoes alpha decay (half-life of about 4 days) into polonium-218 and a dangerous sequence of short-lived, highly radioactive isotopes. Historically, winds and weather quickly dispersed radon atoms and they posed no serious threat to human health. However, in our modern age of well-insulated, tightly sealed buildings, radon gas can seep in and build up, occasionally to hundreds of times normal levels, in poorly ventilated basements. Exposure to such high radon levels is dangerous because each radon atom undergoes at least five more radioactive decay events in just a few days.

The solution to the radon problem is relatively simple. First, any basement or other sealed-off room should be tested for radon. Simple test kits are available at your local hardware store. If high levels of radon are detected, then the area's ventilation should be improved.

ENERGY FROM THE NUCLEUS

Most scientists who worked on understanding the nucleus and its decays were involved in basic research (see Chapter 1). They were interested in acquiring knowledge for its own sake. But, as frequently happens, knowledge pursued for its own sake is quickly turned to practical use. This certainly happened with the physics of the nucleus.

The atomic nucleus holds vast amounts of energy. One of the defining achievements of the twentieth century was the understanding of and ability to harness that energy. Two very different nuclear processes can be exploited in our search for energy—processes called nuclear fission and nuclear fusion.

Nuclear Fission

Fission means "splitting" and nuclear fission means the "splitting of a nucleus." In most cases, energy is required to tear apart a nucleus. However, some heavy isotopes have nuclei that can be split apart into products that have less mass than the original nucleus. From such nuclei, energy can be obtained from the mass difference.

The most common nucleus from which energy is obtained by fission is uranium-235, an isotope of uranium that constitutes about 7 of every 1000 uranium atoms in the world. If a neutron hits uranium-235, the nucleus splits into two roughly equal-sized large pieces and several smaller fragments. Among these fragments are two or three more neutrons. If these neutrons go on to hit other uranium-235 nuclei, the process is repeated and a *chain reaction* can begin, with each split nucleus producing the neutrons that cause more splittings. By this basic process, uranium can produce large amounts of energy.

The device that allows us to extract energy from controlled nuclear fission is called a **nuclear reactor** (Figure 26-9). The uranium in most reactors contains primarily uranium-238, but it has been processed so it contains much more uranium-235 than it would if it were found in nature. This uranium has been processed into long fuel rods, about the thickness of a lead pencil, surrounded by a metallic protector. Typical reactors incorporate many thousands of fuel rods. Between the fuel rods is a substance called a "moderator," usually water, whose function is to slow down the neutrons that leave the rods.

The nuclear reactor works like this: A neutron strikes a uranium-235 nucleus in one fuel rod, causing that nucleus to split apart. These fragments include several fast-moving neutrons. Fast neutrons are very inefficient at producing fission, but as the neutrons move through the moderator they slow down. In this way, they can initiate other fission events in other uranium-235 nuclei. A chain reaction in a reactor proceeds as neutrons cascade from one fuel rod to another. In the process, the energy released by the conversion of matter goes into heating the fuel rods and the water. The hot water is pumped to another location in the nuclear plant, where it is used to produce steam.

The steam is used to run a generator to produce electricity, as described in Chapter 17 (see Figure 17-6). In fact, the only significant difference between a nuclear reactor and a coal-fired generating plant is the way in which steam is made. In a nuclear reactor, the energy to produce steam comes from the con-

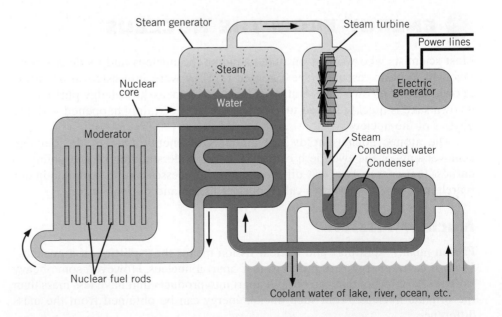

FIGURE 26-9. A nuclear reactor, shown here schematically, produces heat that converts water to steam. The steam powers a turbine, just as in a conventional coal-burning plant.

The nuclear power plant at Three Mile Island, near Harrisburg, Pennsylvania, had to shut down after suffering a partial meltdown. Safety measures ensured that only a very small amount of radioactive material was released into the environment.

version of mass in uranium nuclei; in a conventional generating plant it comes from the burning of coal.

Nuclear reactors must keep a tremendous amount of nuclear potential energy under control while confining dangerously radioactive material. Modern reactors are thus designed with numerous safety features. For example, the water that is in contact with the uranium is sealed in a self-contained system and does not touch the rest of the reactor. Another built-in safety feature is that nuclear reactors cannot function without the presence of the moderator. If there should be an accident in which the water evaporated from the reactor vessel, the chain reaction would shut off. Thus a reactor cannot explode and is *not* analogous to the explosion of an atomic bomb (see Connection: Nuclear Weapons, page 574).

The most serious accident that can occur at a nuclear reactor involves processes in which the flow of water to the fuel rods is interrupted. When this happens, the enormous heat stored in the central part of the reactor can cause the fuel rods to melt. Such an event is called a *meltdown*. In 1979, a nuclear reactor at Three Mile Island in Pennsylvania suffered a partial meltdown, but released only a very small amount of radioactive material to the environment. In 1986, a less carefully designed reactor at Chernobyl, Ukraine, underwent a meltdown accompanied by large releases of radioactivity.

Fusion

Fusion refers to a process in which two nuclei come together (fuse) to form a third, larger nucleus. Under special circumstances it is possible to push two nuclei together and make them fuse. When elements with low atomic numbers fuse, the mass of the final nucleus is less than the mass of its constituent parts. In these cases, it's possible to extract energy from the fusion reaction by conversion of that "missing" mass.

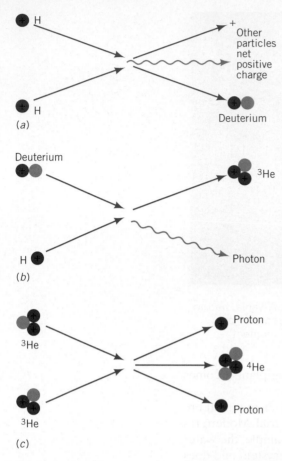

The most common fusion reaction combines four hydrogen nuclei to form a helium nucleus (Figure 26-10). (Remember that an ordinary hydrogen nucleus is a single proton with no neutron. Thus we use the terms "hydrogen nucleus" and "proton" interchangeably.) Note that the reaction requires several steps, starting with two protons fusing to form deuterium, which is the isotope of hydrogen that contains one neutron. This is followed by a second collision in which two deuterium nuclei form helium-3, and a final collision produces helium-4. This nuclear reaction powers the Sun and other stars and thus is ultimately responsible for all life on Earth.

You cannot just put hydrogen in a container and expect it to form helium, however. Two positively charged protons must collide with tremendous speed in order to overcome their electrostatic repulsion and allow the strong force to kick in. (Remember, the strong force operates only over extremely short distances.) In the Sun, high pressures and temperatures in the star's interior trigger the fusion reaction. The sunlight falling outside your window is generated by the conversion of 600 million tons of hydrogen into helium each second. The helium nucleus has a mass about one-half percent less than the original hydrogen nuclei had. The "missing" mass is converted into the energy that eventually radiates out into space.

FIGURE 26-10. A fusion reaction releases energy as nuclei combine. Hydrogen nuclei enter into a multistep process. (*a*) First two protons combine to form a deuterium nucleus. (*b*) Then two deuterium nuclei combine to form helium-3. (*c*) Finally, two helium-3 nuclei combine to form helium-4. The red balls are protons; the blue ones are neutrons.

Connection
Fusion in Stars

Stars are incredibly massive compared to Earth and the other planets that form our solar system. In fact, over 99.9% of all the mass in the solar system is located in the Sun. So even this high rate of converting hydrogen to helium (600 million tons per second) will continue for literally billions of years. However, all the hydrogen in the Sun's core eventually will get used up, leaving only helium in its place. What will happen then?

The high pressure in the Sun's interior is due to the gravitational attraction of its own mass, as with any star. The inward pressure is kept in check by the heat generated by fusion, which acts to expand the Sun's gas outward. When all the hydrogen is used up, the gravitational force will squeeze the Sun's interior further, generating greater pressure and greater temperature to the point that helium nuclei begin to fuse and form carbon atoms.

In a star such as the Sun, fusion never goes beyond the stage of helium fusing to carbon. The carbon core becomes extremely dense, but even the high pressure and temperature (about 300 million kelvins) are not enough for carbon nuclei to fuse. Eventually, the nuclear fusion reactions cease and the star collapses down to a small hot object called a "white dwarf star." Astronomers expect this is how the Sun will eventually end.

However, in more massive stars than the Sun, the greater mass does generate the higher pressure and temperature needed for carbon fusion. In fact, a series of fusion reactions take place in concentric shells within the star, with fusion of carbon, oxygen, neon, and so on until an inner core of iron forms. As we mentioned earlier in this chapter, iron has the most stable nucleus of all the elements. The fusion of iron does not produce energy; instead, less energy is produced than you have to put in. So fusion stops even in the largest stars, even at temperatures of several billion kelvins.

When fusion stops in a star, gravitational collapse takes over, the core becomes unstable, and the star's life ends in an epic explosion called a "supernova." During the actual explosion, the incredible energy released (about 10^{43} joules!) does briefly allow fusion of heavy nuclei to take place. Astronomers think that all the elements beyond lithium are ejected into the galaxy in these explosions, and that all the elements beyond iron in the periodic table are formed by fusion reactions during the supernova explosion itself. Astronomers observe the gaseous debris of such explosions in the form of a planetary nebula. ●

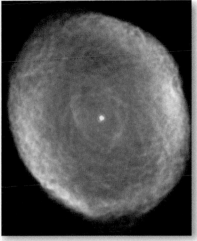

Gaseous debris from exploding stars produce a planetary nebula such as the spectacular Spirograph Nebula.

Since the 1950s, there have been many attempts to harness nuclear fusion reactions to produce energy for human use. The problem has always been that it's very difficult to get protons to collide with enough energy to overcome the electrical repulsion between them and initiate the nuclear reaction.

One promising but technically difficult method is to confine protons in a very strong magnetic field while heating them with high-powered radio waves. Alternatively, powerful lasers have been used to heat pellets of liquid hydrogen. This rapid heating produces shock waves in the liquid, and these shock waves raise the temperature and pressure to the point where fusion is initiated.

Several rather expensive programs are now under way in the United States, Europe, and Japan, where researchers are investigating ways to produce

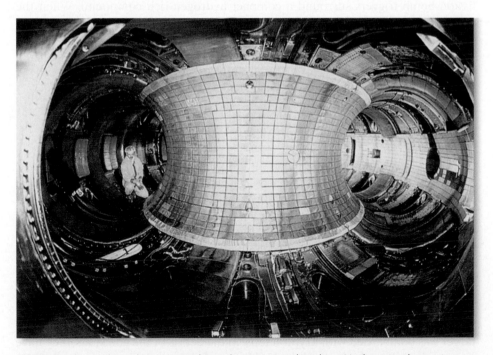

The Princeton Tokamak is a ring-shaped magnetic chamber. Hydrogen plasma is confined in this ring during attempts to achieve nuclear fusion reactions.

commercially feasible nuclear fusion reactors, perhaps sometime in the twenty-first century. The development is slow and costly, but, if researchers are successful, then the energy crisis will be over forever because there is enough hydrogen in the oceans to power fusion reactions virtually forever.

Connection
Nuclear Weapons

Whenever humans have discovered a new source of energy, that source has quickly been turned into weapons. Nuclear energy, the most concentrated energy source known, is no exception. The earliest atomic bombs, developed in the United States as part of World War II's Manhattan Project, relied on fission reactions in concentrated uranium-235 or plutonium-239. The problem faced by the designers of the atomic bomb was that there has to be a certain number of uranium-235 atoms to sustain a chain reaction to the point where large amounts of energy can be released. This quantity of uranium-235, called the *critical mass,* is about 52 kilograms—a chunk of uranium about the size of a cantaloupe.

At the instant an atomic bomb explodes, a critical mass of the radioactive isotope is compressed into a small volume by a conventional chemical explosive. A fission chain reaction then releases a flood of neutrons, which splits much of the nuclear material and releases a destructive wave of heat, light, and other forms of radiation. The first atomic bombs carried the explosive power of many thousands of tons of chemical explosive.

Shortly after World War II, the United States and the Soviet Union developed the hydrogen bomb, which relies on nuclear fusion. In a hydrogen bomb, fission-bomb triggers surround a compact hydrogen-rich compound. When the atomic fission bombs are set off, they trigger much more intense fusion reactions in the hydrogen. Unlike fission bombs, which are limited by the critical mass of nuclear explosive, hydrogen bombs have no intrinsic limit to their size. The largest nuclear warheads, which can fit in a closet, carry payloads equivalent to many millions of tons (megatons) of conventional explosive. ●

Connection
Superheavy Elements

Uranium, with 92 protons, is the heaviest element commonly found in nature. However, ever since the mid-twentieth century, physicists have been able to build heavier ones in the laboratory. If you look at the periodic table of the elements in Appendix C, all the elements past uranium are seen only in specialized experiments. The technique used is to bombard a heavy nucleus such as lead with other nuclei and hope that, occasionally, enough protons and neutrons will stick together to form the nucleus of a still heavier atom. Although these heavier atoms are unstable and decay quickly, they can last long enough to be identified by their spectra.

Groups in California and Germany have run experiments in which a lead target was bombarded with krypton ions. In the debris of the collisions, they have seen evidence for elements up to 116. Some of these superheavy isotopes have half-lives less than 1 thousandth of a second and are the most massive and complex nuclei ever seen. ●

THINKING MORE ABOUT

The Nucleus: Nuclear Waste

When power is generated in a nuclear reactor, many more nuclear changes take place than those associated with the chain reaction itself. Fast-moving debris from the fission of uranium-235 strikes other nuclei in the system—both the ordinary uranium-238 that makes up most of the fuel rods and the nuclei in the concrete and metal that make up the reactor. In these collisions, the original nuclei may undergo fission or absorb neutrons to become isotopes of other elements. Many of these newly produced isotopes are radioactive. The result is that even when all the uranium-235 has been used to generate energy, a lot of radioactive material remains in the reactor. This sort of material is called *high-level nuclear waste*. (The production of nuclear weapons is another source of this kind of waste.) The half-lives of some of the materials in the waste can run to hundreds of thousands of years. How can we dispose of this waste in a way that keeps it away from living things?

The management of nuclear waste begins with storage. Power companies usually store spent fuel rods at a reactor site for tens of years to allow the short-lived isotopes to decay. At the end of this period the long-lived isotopes that are left behind must be isolated from the environment. Scientists have developed techniques for incorporating these nuclei into stable solids, either minerals or glass. The idea is that the electrons in radioactive isotopes form the same kind of bonds as stable isotopes, so that with a judicious choice of materials radioactive nuclei can be locked into a solid mass for long periods of time.

Plans now call for nuclear waste disposal by the incorporation of radioactive atoms into stable glass that is surrounded by successive layers of steel and concrete. These stable containers are to be buried deep under the Earth's surface in stable rock formations. Ultimately, if a long sequence of public hearings, construction permits, and other hurdles are passed, the United States Department of Energy hopes to confine much of the nation's nuclear waste at the Yucca Mountain repository in a remote desert region of Nevada. The hope is that long-lived wastes can be sequestered from the environment until after they are no longer dangerous to human beings.

The Yucca Mountain project continues to be a controversial subject. Supporters of the site argue that a single, remote long-term site is vastly preferable to the present 131 temporary repositories now located in 39 different states. Such scattered sites are difficult to monitor and protect from terrorist threats. Opponents of Yucca Mountain counter that hauling thousands of tons of nuclear waste on interstate highways poses a far greater danger to the public than the present sites. Some geologists, furthermore, fear that Yucca Mountain may be subject to occasional earthquakes and that its location, less than 100 miles from Las Vegas, is not sufficiently remote. In 2002, Congress voted to proceed with the construction of the waste disposal site at this location.

What should we do with our increasing quantities of nuclear waste? What responsibility do we have to future generations to ensure that the waste we bury stays where we put it? Should the existence of nuclear waste restrain us in our development of nuclear energy? Should we, as some scientists argue, keep nuclear waste materials at the surface and use them for applications such as medical tracers and fuel for reactors?

Summary

The nucleus is a tiny collection of massive particles, including positively charged **protons** and electrically neutral **neutrons.** The nucleus plays a role independent of the orbiting electrons that control chemical reactions, and the energies associated with nuclear reactions are much greater. The number of protons—the **atomic number**—determines the nuclear charge and, therefore, the type of element; each element in the periodic table has a different number of protons. The number of neutrons plus protons—the **mass number**—determines the mass of the **isotope.** Nuclear particles are held together by the **strong force,** which operates only over extremely short distances.

While most of the atoms in objects around us have stable, unchanging nuclei, many isotopes are **radioactive**—they spontaneously change through **radioactive decay.** In **alpha decay,** a nucleus loses two protons and two neutrons. In **beta decay,** a neutron spontaneously transforms into a proton, an electron, and a neutrino. A third kind of radioactivity, involving the emission of energetic electromagnetic radiation, is called **gamma radiation.** The rate of radioactive decay is measured by the **half-life,** which is the time it takes for one-half of a collection of isotopes to decay. Radioactive half-lives provide the key for **radiometric dating** techniques based on carbon-14 and other isotopes. Unstable isotopes are also used as radioactive tracers in medicine and other areas of science. Indoor radon pollution and nuclear waste are two problems that arise from the existence of radioactive decay.

We can produce nuclear energy in two ways. **Fission** reactions, as controlled in **nuclear reactors,** produce energy when heavy radioactive nuclei split apart into fragments that together weigh less than the original isotopes. **Fusion** reactions, on the other hand, combine light elements to make heavier ones, as in the conversion of hydrogen into a smaller mass of helium in the Sun. In each case, the lost nuclear mass is converted into energy.

Key Terms

alpha decay The spontaneous release of an alpha particle (two protons and two neutrons) from an atomic nucleus. (p. 559)

atomic number The number of protons in the nucleus of an atom. (p. 554)

beta decay The spontaneous transformation of a neutron into a proton in an atomic nucleus, accompanied by the release of an electron and a neutrino. (p. 560)

fission The splitting of a nucleus into two or more smaller pieces; usually associated with the splitting of uranium to get energy. (p. 569)

fusion The process in which two nuclei join together to form a larger nucleus. (p. 571)

gamma radiation The spontaneous release of a high-energy photon from an atomic nucleus. (p. 562)

half-life The time it takes for one-half of a sample of radioactive material to decay. (p. 565)

isotope Atoms of the same element whose nuclei have the same number of protons but a different number of neutrons. (p. 555)

mass number The number of protons plus the number of neutrons in the nucleus of an atom. (p. 555)

neutron One of the two particles that make up the atomic nucleus; it has no electric charge. (p. 554)

nuclear reactor A device that allows us to extract energy from nuclear fission in a controlled fashion. (p. 570)

proton One of the two particles that make up the atomic nucleus; it has a charge of +1. (p. 554)

radioactivity or **radioactive decay** The spontaneous release of energetic particles from an atomic nucleus. (p. 557)

radiometric dating The determination of the age of materials using the known half-lives of radioactive elements. (p. 566)

strong force One of the four fundamental forces; it is the force that binds the nucleus together. (p. 557)

Review

1. What was the hypothesis behind Rutherford's experiment on alpha decay? What did he prove?

2. What particles make up the nucleus of an atom? Compare the size of a nucleus to the size of an atom.

3. Which are heavier, electrons or protons?

4. Which is easier to remove from an atom, an electron or a proton?

5. Which is most responsible for the chemistry that an atom undergoes, its nucleus or its electrons? Explain.

6. What determines the electric charge of the nucleus? What determines its mass?

7. What fact about atomic nuclei suggests the existence of the neutron?

8. What is an isotope?

9. What does the atomic number of an isotope represent? The mass number?

10. What is the strong force and how do we know it exists?

11. How is the strong force different from gravity and electromagnetism?

12. What happens to atomic nuclei during radioactive decay?

13. What is alpha decay? How does it change the nucleus?

14. What is beta decay?

15. Why does beta decay not change the total electric charge of an atom?

16. What is a neutrino? What led physicists to hypothesize its existence?

17. What is gamma radiation?

18. How does gamma radiation differ from alpha and beta radiation?

19. Why is radiation dangerous? How does it affect biological tissue?

20. Explain the term "half-life."

21. Explain how radioactivity can be used to date (a) a piece of wood and (b) a rock.

22. What is a decay chain? How do decay chains help explain the dangers presented by radon?

23. How is the principle of conservation of energy seen in (a) fission reactions and (b) fusion reactions?

24. What is nuclear fission? How can we obtain energy from it?

25. What is a chain reaction?

26. How does a nuclear reactor work? How do the fuel rods work, and what is the function of the moderator?

27. What are some of the safety features built into nuclear reactors, and how do they function?

28. "Critical mass" is a term that is widely used outside of nuclear science. What is its everyday meaning, and how does that relate to its scientific meaning?

29. How do fusion reactions produce energy? What is the "missing" mass, and what is its significance?

30. How does the Sun generate energy?

31. Do nuclear power plants use fission or fusion?

32. What are superheavy elements and how are they made? Are they stable?

33. What is a radioactive tracer?

34. How are radioactive tracers useful in medicine? Give an example.

35. What is nuclear waste? Why is it a serious problem for society?

Questions

1. Mixing copper and zinc atoms forms the alloy brass. What would form if you fused the nucleus of a copper atom with the nucleus of a zinc atom?

2. Reacting hydrogen and oxygen atoms together produces water, H_2O. What would result from the fusion of two hydrogen nuclei with an oxygen nucleus?

3. An atom has six neutrons, six protons, and six electrons. What element is it? Is it an ion?

4. An atom has 143 neutrons, 92 protons, and 91 electrons. What element is it? Is it an ion?

5. We know that the strong force acts over very short distances. Suppose the range of the strong force were twice what it is now. (In other words, it would attract the same particles with the same force as now, even though they were twice as far away.) Do you think this would have the effect of increasing or decreasing the half-life of uranium-238? Explain.

6. Why does radioactivity seem to be more common with heavier elements?

7. Almost all of the atoms with which we come into daily contact have stable nuclei. Given that most known isotopes are unstable, how could this state of affairs have arisen?

8. Discuss the pros and cons of nuclear power. Refer to specific concepts discussed in this chapter.

9. Can nuclear radiation escape from nuclear power plants? If so, how?

10. Suppose you are a scientist from the future who has discovered the ruins of the Empire State Building. How would you go about estimating the date when it was built?

11. Carbon-14 decays by beta decay with a half-life of 5700 years. What does a carbon-14 nucleus become when it undergoes beta decay? If an ancient campfire were analyzed, and it was found to have only about one-eighth the carbon-14 that is normally found in living things, how long ago was that campfire extinguished?

12. A radioactive isotope is found to decay to one-sixteenth its original amount in 12 years. What is the half-life of this isotope?

13. A radioactive isotope is found to decay to one-eighth its original amount in 30 years. What is the half-life of this isotope?

14. Is it possible for a radioactive nucleus to decay two times and end up as the same element that it started as? Explain. What if the nucleus were restricted to alpha and beta decays only?

15. Hydrogen-3 (also known as tritium) decays by beta decay and has a half-life of about 12 years. What does hydrogen-3 become when it beta decays? If you had a sample of 160 grams of hydrogen-3, how much would still be hydrogen-3 after 36 years?

16. When uranium-238 emits an alpha particle during radioactive decay, what element results?

17. Uranium-235 has a half-life of about 700 million years. What can you say about the likelihood of generating nuclear power from uranium billions of years from now?

18. Suppose you gathered 100,000 of your closest friends to play a game at your local pro-football stadium. Each per-

son has a coin. Every person flips his or her coin once each hour, and everyone that flips tails has to leave. About how many people are left after 1 hour? 2 hours? 3 hours? Can you say exactly how long it will be until everyone is gone?

19. You are given 16 grams of the isotope cobalt-60. Cobalt-60 is radioactive and is used in the process of food irradiation. It has a half-life of 5 years and decays by alpha decay. How long will it take until only 1 gram of cobalt-60 is left? What has the cobalt-60 turned into?

20. If a U-238 nucleus split into two identical pieces and then each piece underwent an alpha decay, what would the two final pieces be?

21. Suppose you are given 10 grams each of two different radioactive isotopes, A and B. You monitor the amount left as a function of time and plot your results (see the figure). What is (approximately) the half-life of each isotope?

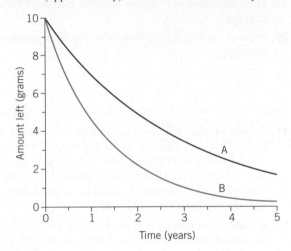

22. A 52-kilogram sphere of pure uranium-235 constitutes a critical mass. If this sphere were flattened out into a thin disk, would it still be a critical mass? Explain.

23. Why can't carbon dating be used to estimate the age of dinosaur bones? Why can't it be used to date the rocks that made Stonehenge?

24. What change takes place within the nucleus during beta radiation? If a hydrogen-3 nucleus decays by beta emission, what is the resulting nucleus?

25. Carbon dating is based on the radioactive element carbon-14, which has a half-life of about 5700 years. If the skeletal remains of a human are found to have only one-eighth the amount of carbon-14 that a living human is expected to have, approximately how old is this skeleton?

26. Cesium-137 is radioactive, with a half-life of 30 years, and is used in the food irradiation process. Cesium-137 decays by beta radiation. What does cesium-137 turn into when it decays? How long would it take for 60 grams of cesium-137 to decay to 30 grams?

27. The heaviest naturally occurring element is uranium, atomic number 92. In terms of the forces within the nucleus, why don't heavier elements exist in appreciable amounts?

Problems

1. Use the periodic table to identify the element, atomic number, mass number, and electric charge of each of the following combinations.

   a. 6 protons, 7 neutrons, 10 electrons
   b. 6 protons, 8 neutrons, 10 electrons
   c. 9 protons, 8 neutrons, 10 electrons
   d. 9 protons, 8 neutrons, 9 electrons

2. Use the periodic table to determine how many protons and neutrons are in each of the following atoms.

   a. C-13   c. Ag-108
   b. Ni-56   d. Rn-222

3. The average atomic weight of cobalt atoms (atomic number 27) is actually slightly greater than the average atomic weight of nickel atoms (atomic number 28). How could this situation arise?

4. One atomic mass unit (amu) equals 1.66×10^{-27} kg. Calculate the equivalent of 1 amu in terms of energy in joules.

5. When 1 gram of hydrogen is converted into helium through nuclear fusion, the difference between the masses of the original hydrogen and the final helium is 0.0072 g. How much energy is released when 1 gram of hydrogen undergoes nuclear fusion?

6. Imagine that a collection of 1,000,000 atoms of uranium-238 was sealed in a box at the formation of Earth 4.5 billion years ago. Use the uranium-238 decay chain to predict some of the things you would find if you opened the box today.

7. An ancient archaeological site was being dated using the carbon-14 dating methods on a fire pit. Only 12.5% of the original carbon-14 was detected. How old is this ancient site?

8. Strontium-90 undergoes beta decay with a half-life of 28.8 years. What is the product of this decay? Include in your answer the number of protons (atomic number), the number of neutrons, and the mass number of both strontium-90 and its daughter decay product.

9. Bismuth-214 undergoes beta decay with a half-life of 19.7 minutes. Repeat Problem 8 for the beta decay of this isotope.

10. Thorium-230 undergoes alpha decay with a half-life of about 80,000 years. What is the product of this decay? Include the atomic number and the mass number of both thorium-230 and the product of the alpha decay. Also, if there are 10 grams of thorium-230 to start with, how much of this will remain after 160,000 years has passed? After 240,000 years?

11. Polonium-210 undergoes alpha decay in the uranium decay chain. What is the product of this decay? Include the atomic number and the mass number in your answer.

12. Thorium-232 undergoes three beta decays and one alpha decay in a decay chain. Does the final daughter element depend on the order of the decay modes?

13. If gamma radiation is released from an isotope, how does this change the atomic number of the isotope? The mass number?

14. Isotope X has a half-life of 100 days. A sample is known to have contained about 1,500,000 atoms of isotope X when it was put together, but is now observed to have only about 100,000 atoms of isotope X. Estimate how long ago the sample was assembled. Explain the relevance of this problem to the technique of radiometric dating.

Investigations

1. Read a historical account of the Manhattan Project. What was the principal technical problem in obtaining the nuclear fuel? Why did chemistry play a major role? What techniques are now used to obtain nuclear fuel?

2. What is the current status of U.S. progress toward developing a depository for nuclear waste? How do your representatives in Congress vote on matters relating to this issue?

3. What sorts of isotopes are used for diagnostics in your local hospital? Where are supplies of those radioisotopes purchased? What are the half-lives of the isotopes, and how often are supplies replaced? What is the hospital's policy regarding the disposal of radioactive waste?

4. How much of the electricity in your area comes from nuclear reactors? What fuel do they use? Where are the used fuel rods taken when they are replaced? If the facility offers public tours, visit the reactor and observe the kinds of safety procedures that are used.

5. Obtain a radon test kit from your local hardware store and use it in the basement of two different buildings. How do the values compare? Is either at a dangerous level? If the values differ, what might be the reason?

6. Only about 90 elements occur naturally on Earth, but scientists are able to produce more elements in the laboratory. Investigate the discovery and characteristics of one of these synthetic elements.

7. Soon the United States government will take over responsibility for the nuclear wastes of the 50 states. What options do we have for waste storage? Do you think all the waste should be stored in one place? Should we try to separate and use the radioactive isotopes? What are the factors—social, political, and economic—that will help determine what happens to this nuclear waste? What is the current status of Yucca Mountain as you read this? Do you think the impact of the events of September 11, 2001, changed the debate on this? If so, how?

WWW Resources

See the *Physics Matters* home page at **www.wiley.com/college/trefil** for valuable web links.

1. **http://www.aip.org/history/curie/** An online exhibit, *Marie Curie and the Science of Radioactivity,* by the American Institute of Physics.

2. **http://www.aip.org/history/sakharov/** An online exhibit dedicated to Andrei Sakharov, "father of the Soviet hydrogen bomb," who later became a dissenter and received the Nobel Peace Prize.

3. **http://www.colorado.edu/physics/PhysicsInitiative/Physics2000/isotopes/index.html** A lavishly illustrated and Java-rich (and often slow) site includes a detailed storyline describing isotopes and radioactive decay.

4. **http://www.lbl.gov/abc/index.html** The *ABCs of Nuclear Science,* a lavish site containing tutorials and simulations from Lawrence Berkeley National Laboratory.

5. **http://FusEdWeb.pppl.gov/CPEP/chart.html** *Plasma Physics and Fusion,* another lavish site containing tutorials and simulations from Lawrence Livermore Berkeley National Laboratory.

6. **http://www.xray.hmc.psu.edu/rci/centennial.html** A site indexing the many sites devoted to the celebration of 100 years of medical radiology since the discovery of the X ray in 1895.

7. **http://ippex.pppl.gov/fusion/default.htm** *About Fusion!* A tutorial from the internet plasma physics education experience.

27 | The Ultimate Structure of Matter

KEY IDEA

All matter is made of quarks and leptons, which are the fundamental building blocks of the universe.

PHYSICS AROUND US . . . Looking at Sand and Stars

Ancient poets did not have the language to express really large numbers. So they would refer to the number of stars in the sky or leaves in a forest or grains of sand on the shore. And if you look at the night sky or go into the woods in summer or let a handful of sand sift through your fingers, these seem to be truly enormous numbers.

But they're not. Next time you lie on a beach, look at a single tiny grain of sand and think about its microscopic structure. Imagine that you could magnify that grain a thousandfold, a millionfold, or even more. What would you see?

At 1000 magnification, the rounded grain would appear rough and irregular, but no hint of its atomic structure could be seen. At 1 million magnification, individual atoms, each about one ten-billionth of a meter across, would begin to be visible. At 1 trillion magnification, the atomic nuclei would appear as tiny points, surrounded by almost nothing. There are more protons, electrons, and neutrons in a grain of sand than there are grains of sand on any beach you can imagine.

But is that it? Or is there some incredibly tiny structure to the particles that make up atoms? What are the ultimate building blocks of matter?

OF WHAT IS THE UNIVERSE MADE?

The Library

The next time you head over to the library, wander through the stacks and think about what constitutes the fundamental building blocks of the library. Your first reaction might be to say that books are the fundamental building blocks—row after row, shelf after shelf, of bound volumes. But a library is not just a collection of books: the volumes are arranged with an order to them. You could describe the set of rules that dictates how books are arranged in libraries—the Dewey decimal system or the Library of Congress classification scheme, for example. Thus a complete description of a library at this most superficial level includes two things: books as the fundamental building blocks and rules about how the books are organized.

Inside a book, the various volumes are not as different from one another as they might seem at first. They are all made of an even more fundamental unit—the word. You could argue that words are the fundamental building blocks of the library. Also, as was the case for the cataloged books, we require a set of rules, called grammar, that tells us how to put words together to make books. Words and grammar, then, take you down to a more basic level in your probe of a library's reality.

You probably wouldn't be content very long with the notion of words as the fundamental building blocks, because all of the thousands of words are different combinations of a small number of more fundamental things—letters. Only 26 letters (at least in the English alphabet) provide the building blocks for all the thousands of words on all the pages in all the books of the library. Furthermore, we need a set of rules (spelling) that tells us how to put letters together into

(a)

(b)

The electrons of an atom or molecule brought near a polar molecule such as water will tend to be pushed away from the negative side and shifted toward the positive side. Consequently, the side of an atom facing the negative end of a polar molecule will become slightly positive. This subtle electron shift, called *polarization*, in turn will give rise to an electrical attraction between the negative end of the polar molecule and the positive side of the other molecule. The electron movement thus creates an attraction between the atom and the molecule, even though the atoms and molecules in this scheme all may be electrically neutral. One of the most important consequences of forces due to polarization is the ability of water to dissolve many materials. Water, made up of strongly polar molecules, exerts forces that make it easier for ions such as Na^+ and Cl^- to dissolve.

A process related to the forces of polarization leads to the **hydrogen bond**, a weak bond that may form after a hydrogen atom links to an atom of certain other elements (oxygen or nitrogen, for example) by a covalent bond. Because of the kind of rearrangement of electrical charge described above, hydrogen may become polarized and develop a slight positive charge, which attracts another atom to it. You can think of the hydrogen atom as a kind of bridge in this situation, causing a redistribution of electrons that, in turn, holds the larger atoms or molecules together. Individual hydrogen bonds are weak, but in many molecules they occur repeatedly and therefore play a major role in determining the molecule's shape and function. Note that while all hydrogen bonds require hydrogen atoms, not all hydrogen atoms are involved in hydrogen bonds.

Hydrogen bonds are common in virtually all biological substances, from everyday materials such as wood, plastics, silk, and candle wax, to the complex structures of every cell in your body. As we shall see in Chapter 23, hydrogen bonds in every living thing link the two sides of the DNA double helix together, although the sides themselves are held together by covalent bonds. Ordinary egg white is made from molecules whose shape is determined by hydrogen bonds, and when you heat the material—when you fry an egg, for example—hydrogen bonds are broken and the molecules rearrange themselves so that instead of a clear liquid you have a white gelatinous solid.

gen or n

arrangem

arized an

(c)

At first glance, the fundamental units of a library might appear to be books (a), but a closer inspection reveals that books are made from words (b), which in turn are made from letters (c).

words. The discovery of letters and spelling would provide perhaps the ultimate description of a library and its organization.

So the library can be described in this way: We use spelling to tell us how to put letters together into words. Then grammar tells us how to put words together into books. Finally, we use organizing rules to tell us how to put books together into a library.

As we shall see, this building-block approach is how scientists attempt to describe the entire physical universe.

Reductionism

How many different kinds of material can you see when you look up from this book? You may see a wall made of cinder blocks, a window made of glass, and a ceiling made of fiberglass panels. Outside the window you may see grass, trees, blue sky, and clouds. We encounter thousands of different kinds of materials every day. They all look different—what possible common ground could there be between a cinder block and a blade of grass? They all look different, but are they really?

For at least two millennia, people who have thought about the physical universe have asked this question. Is the universe just what we see, or is there some underlying structure, some basic stuff, from which it's all made? In some respects, this is one of the most fundamental of scientific questions.

Philosophers refer to the quest for the ultimate building blocks of the universe as *reductionism*. Reductionism is an attempt to reduce the seeming complexity of nature by first looking for underlying simplicity and then trying to understand how that simplicity gives rise to the observed complexity. This pursuit is a continuation of an old intellectual belief that the appearances of the world do not tell us its true nature but that its true nature can be discovered by the application of thought and, in the case of science, experiment and observation.

The Greek philosopher Thales (625?–546 B.C.) suggested that all materials are made of water. This supposition was based on the observation that in everyday experience water appears as a solid (ice), a liquid, and a gas (water vapor). Thus, alone among the common substances, water seemed to exhibit all the states of matter (see Chapter 9). Thales's followers later expanded this notion to include the familiar four elements of the ancient Greeks—earth, fire, air, and water. This was the first attempt to find the basic building blocks of the universe.

The Building Blocks of Matter

To many people, the library analogy presents a profoundly satisfying way of describing complex systems. Some would even argue that everything you could possibly want to know about the physical organization of the library is contained in letters and their organizing principles. In just the same way, physicists want to describe the complex universe by identifying the most fundamental building blocks and deducing the rules by which they are put together.

At first, you might say the most fundamental building block of the universe is the atom. All the myriad solids, liquids, and gases are made of just 100 or so different kinds of chemical elements. The complexity of materials that appears to the senses results from the many combinations of these relatively few kinds of atoms. The rules of chemistry tell us how atoms bind together to make all of the materials we see.

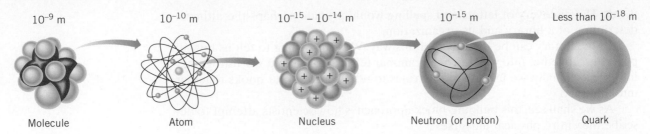

10^{-9} m	10^{-10} m	$10^{-15} - 10^{-14}$ m	10^{-15} m	Less than 10^{-18} m
Molecule	Atom	Nucleus	Neutron (or proton)	Quark

FIGURE 27-1. The modern picture of the fundamental building blocks of the universe. Molecules are made from atoms, which contain nuclei, which are made from elementary particles, which in turn are made from quarks.

However, early in the twentieth century physicists learned that atoms are not really fundamental but are made up of smaller, more fundamental bits—protons, neutrons, and electrons. These particles arrange themselves according to their own set of rules, with massive neutrons and protons in the positively charged nucleus and negatively charged electrons in orbit around the nucleus. A picture of the universe with only these three fundamental building blocks is very simple and appealing. Protons and neutrons together form nuclei, and electrons orbit the nucleus to form atoms. Electrons combine and interact with one another to form all the materials we know about.

 However, just as words and grammar gave us a false level of simplicity in the analogy of the library, this simple picture of the universe based on just three particles didn't stand up to more detailed experiments and observations. As we have hinted, the nucleus contains more than just protons and neutrons, although physicists did not understand the full implications of this statement until the post–World War II era. If we are going to follow the reductionist line in dealing with the universe, we have to start thinking about what makes up the nucleus. By common usage, the particles that make up the nucleus, together with particles such as the electron, were called *elementary particles* to reflect the belief that they are the basic building blocks of the universe (Figure 27-1). The study of these particles and their properties is the domain of a field known as **high-energy physics,** or **elementary-particle physics.** (We see later that the study of elementary particles requires the use of very high energies.)

● DISCOVERING ELEMENTARY PARTICLES

Nowhere in nature is the equivalence of mass and energy more obvious than in the interactions of elementary particles. Imagine that you have a source of protons traveling at very high velocities, approaching the speed of light. This source might be astronomical in nature, or it might be a machine that accelerates particles. Once the proton has been accelerated, it has a very high kinetic energy. If this high-energy proton collides with a nucleus, the nucleus can be split apart. In this process, some kinetic energy of the original proton can be converted into mass according to the equation $E = mc^2$ (see Chapter 12). When this happens, new kinds of particles that are neither protons nor neutrons can be created.

Cosmic Rays

During the 1930s and 1940s, physicists used a natural source of high-energy particles, called **cosmic rays,** to study the structure of matter. Cosmic rays are particles (mostly protons) that rain down continuously on the atmosphere of Earth after they are emitted by stars in our galaxy and in other galaxies.

Space is full of cosmic rays. When they hit the atmosphere, they collide with molecules of oxygen or nitrogen and produce sprays of very fast-moving secondary particles. These secondary particles, in turn, can make further collisions and produce even more particles, building up a cascade in the atmosphere. It is not uncommon for a single incoming particle to produce billions of secondary particles by the time the cascade reaches the surface of Earth. Indeed, on average, several of these rays pass through your body every minute of your life.

Physicists in the 1930s and 1940s set up their apparatuses on high mountaintops and observed what happened when fast-moving primary cosmic rays or slightly slower-moving secondary particles collided with nuclei. A typical apparatus incorporated a gas-filled chamber several centimeters across (Figure 27-2). A thin sheet of target material such as lead was located midway in the chamber. Cosmic rays occasionally collided with one of the nuclei in the piece of lead, producing a spray of secondary particles. By studying particles in that spray, physicists hoped to understand what was going on inside the target nucleus.

By the early 1940s, when the international effort in physics research shut down temporarily because of World War II, physicists working with these cosmic ray experiments had discovered particles in addition to the proton, neutron, and electron. And when the research effort started up again after the war, these discoveries multiplied as more and more particles were found in the debris of nuclear collisions, both by cosmic ray physicists and by those working at the new particle accelerators (which we discuss shortly).

The net result of these discoveries was that the nucleus can no longer be considered to be a simple bag of protons and neutrons. Instead, we have to think of the nucleus as a very dynamic place. All kinds of newly discovered elementary particles in addition to protons and neutrons are found there. These exotic particles are created in the interactions inside the nucleus, and they give up their energy (and, indeed, their very existence) in subsequent interactions to make other kinds of particles. This constant dance of the elementary particles inside the nucleus has been well documented since these early explorations.

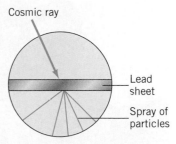

FIGURE 27-2. In a typical cosmic ray experiment, cosmic rays hit a lead nucleus, producing a spray of particles.

Connection
Detecting Elementary Particles

If elementary particles are even smaller than an individual nucleus, how do we know they're there? Experimental physicists have raised detection of elementary particles to a fine art over the years. Nevertheless, the basic technique used in any detection process is the same: the particle in question interacts with matter in some way, and we measure the changes in matter that result from that interaction.

If an elementary particle has an electric charge, it may tear electrons loose as it goes by an atom. Thus, a charged elementary particle moving through material such as a photographic emulsion leaves a string of ions in its wake, much as a speedboat going across a lake leaves a trail of troubled water. The earliest kinds of detectors were simple photographic plates; the plates were angled so the particle's path would lie in the plane of the emulsion.

A more modern detection method is to allow the particles to pass though a grid of thin conducting wires (usually made of gold). As a particle passes a wire, it exerts a force on the electrons in the metal, creating a small pulse of current. By measuring the time when this pulse arrives at the end of the wire, and by putting together such information from many wires, a computer can reconstruct the particle's path with high precision.

Uncharged particles such as neutrons are much more difficult to detect because they do not leave a string of ions in their path. Typically, the passage of an uncharged particle cannot be detected directly; instead, we wait until it collides with something. If that collision produces charged particles, then we can detect them by the techniques just outlined and can work backward and deduce the property of the uncharged particle. ●

Particle Accelerators: The Essential Tool

For a time, physicists had to sit around and wait for nature to supply high-energy particles (in the form of cosmic rays) so that they could study the fundamental structure of matter. The arrival of cosmic rays could not be controlled and it could be very time-consuming waiting for one to hit. Physicists realized that they had to build machines that could produce streams of artificial cosmic rays—**particle accelerators** that scientists could turn on and off at will and that would take the place of the sporadic cosmic rays in experiments. At the beginning of the 1930s at the University of California at Berkeley, Ernest O. Lawrence began producing a new kind of accelerator called a *cyclotron,* an invention for which he won the 1939 Nobel Prize in physics.

The cyclotron works by applying an electric force to groups of charged particles, usually electrons or protons, that are forced to move in circles by large magnets (Figure 27-3). In this system, the magnets supply a centripetal force that keeps the particles moving in circles. In one region of the circular path, an electric field accelerates the particles in the direction of their motion, increasing their speed. As the particles pass through this region over and over again, they are eventually accelerated almost to the speed of light. Once they have acquired this much kinetic energy, they are allowed to collide with other particles. These collisions provide the interactions that physicists study.

Lawrence's first cyclotron was no more than a dozen centimeters across and produced energies that were pretty puny by today's standards. Modern particle accelerators are huge high-tech structures, capable of producing particle energies as high as all but the most energetic cosmic rays.

(a)

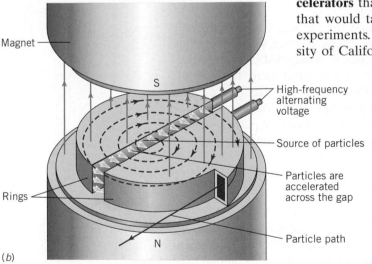

(b)

FIGURE 27-3. (a) Ernest O. Lawrence posed in the 1930s with his invention, the cyclotron, which was the first particle accelerator. (b) The cyclotron accelerates charged particles in a spiral path by generating an intense magnetic field.

Labels in figure (b): Magnet; Rings; S; N; High-frequency alternating voltage; Source of particles; Particles are accelerated across the gap; Particle path

In the type of accelerator called a *synchrotron,* the main working part is a large ring of magnets that keep the accelerated particles moving in a circular track. As we have seen in Chapter 16, magnetic fields exert a force on moving charged particles. That force tends to make charged particles move in a circular track. As a particle in a synchrotron moves around the circle, the large electromagnets are adjusted to keep its track within a small chamber (typically several centimeters on a side) in which a near-perfect vacuum has been produced. This chamber, in turn, is bent into the large circle that marks the particle's orbit. Each time the particles come around to a certain point, an electric field boosts their energy. As the velocity increases, the field strength in the magnets is also increased to compensate, so that the particles continue around the same circular track. Eventually, the particles reach the desired speed and they are brought out into an experimental area where they undergo collisions. Note that in a synchrotron the particles move in a circle of constant radius, while in a cyclotron the particles spiral outward.

The four-story-high Fermilab CDF particle detector. Particle detectors have gone from being small, desktop pieces of apparatus to huge high-technology instruments such as the one shown.

As the energy required to stay at the frontier of particle physics increases, so too does the size of accelerators. The highest-energy accelerator in the world today is at the Fermi National Accelerator Laboratory (Fermilab) outside Chicago, Illinois. There, protons move around a ring almost 2 kilometers (about 1 mile) in diameter and achieve energies of 1 trillion electron volts. Even higher energies are expected to be produced at the Large Hadron Collider in Geneva, Switzerland. Located in an underground tunnel about 8 km (5 miles) across, it will accelerate protons to over 7 trillion electron volts when it becomes operational in 2004.

The *linear accelerator* provides an alternative strategy for making high-velocity particles. This device relies on a long, straight vacuum tube into which electrons are injected. The electronics are arranged so that an electromagnetic wave travels down the tube, and electrons ride this wave more or less the way a surfer

(a)   (b)

(a) Fermilab in Illinois and (b) Stanford Linear Accelerator Center (SLAC) in California are two of the world's most powerful accelerators.

rides a wave on the ocean. The largest linear accelerator in the world, at the Stanford Linear Accelerator Center in California, is about 3 kilometers (almost 2 miles) long.

Connection

Accelerators in Medicine

The ability to build machines that accelerate charged particles has had an important effect in many areas of medicine, most notably in the treatment of cancer. Often the goal of this treatment is to destroy malignant cells in tumors, and subjecting those cells to high-energy X rays or gamma rays is a particularly effective way of doing this for some cancers.

To produce a beam of gamma rays for cancer therapy, a small accelerator produces an intense beam of high-speed electrons. These electrons are then directed into a block of heavy metal such as copper, which stops them abruptly. As we learned in Chapter 19, electrically charged objects that are accelerated (or, in this case, decelerated) emit electromagnetic waves. In the case of electrons accelerated to an appreciable fraction of the speed of light and suddenly stopped, those waves are in the form of gamma rays. The direction of the electron beam is arranged so that the gamma rays pass through the tumor, killing cells as they pass through. In a treatment known as gamma knife surgery, the gamma rays are focused through holes in a shield worn by the patient, so that their energy is concentrated on the tumor cells and damage to healthy cells is minimized.

A form of cancer therapy still under development involves using beams of accelerated protons, which are directed to hit a metal target and produce another kind of elementary particle, called a "pi-meson." When beams of pi-mesons enter tissue, the distance they travel depends on their energy. Thus the accelerator can be set so that the particle beam stops in the tumor itself, delivering large amounts of energy to a relatively small region. ●

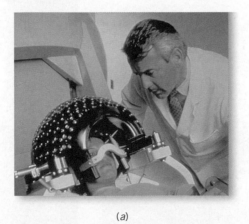

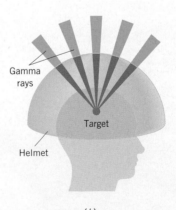

(a)　　　　　　　　　　　　　　(b)

Gamma knife surgery employs a beam of gamma rays, which is focused through holes in a shield worn by the patient (a). In this manner the gamma-ray energy is concentrated on the tumor cells and damage to healthy cells is minimized (b).

THE ELEMENTARY PARTICLE ZOO

At the beginning of the 1960s, the first generation of modern particle accelerators began to produce copious results, and the list of elementary particles began to grow rapidly. The list now numbers in the hundreds. A few important groups of particles are summarized in the following sections and in Table 27-1.

Leptons

Leptons are elementary particles that do not participate in the strong force that holds the nucleus together, and they are not part of the nuclear maelstrom. Instead, they interact by the weak nuclear force, as we describe later in this chapter. Some leptons have electric charge, so they also react by the electromagnetic force. We have encountered two leptons so far—the *electron,* which is normally found in orbit around the nucleus rather than in the nucleus itself, and the *neutrino* (see Chapter 6), a light neutral particle that hardly interacts with matter at all. Since the 1940s, physicists have discovered four additional kinds of leptons, for a total of six. If you keep in mind that the electron and the neutrino are typical leptons, you will have a pretty good idea of what the others are like. The six leptons seem to be arranged in pairs—in each pair there is a particle like the electron, which has a mass, and a specific kind of neutrino, which has a very small mass.

Hadrons

All the different kinds of particles that exist inside the nucleus are referred to collectively as **hadrons** ("strongly interacting ones"). The array of these particles is truly spectacular. Hadrons include particles that are stable, such as the proton, particles that undergo radioactive decay in a matter of minutes, such as the neutron (which undergoes beta decay), and still other particles that undergo radioactive decay in 10^{-24} seconds. This third kind do not live long enough to travel across even a single nucleus! Some hadrons carry an electric charge, while

TABLE 27-1	Summary of Elementary Particles	
Type	**Definition[a]**	**Examples**
Leptons	Interact by weak and electromagnetic forces	Electron, neutrino
Hadrons	Interact by strong, weak, and electromagnetic forces	Proton, neutron, roughly 200 others
Antiparticles	Particles with the same mass as the corresponding particles but with opposite charge and other properties	Positron

[a]All elementary particles that have mass also interact through the gravitational force, but this is so weak compared to the other forces that it is usually ignored.

Matter–antimatter annihilation reactions are often employed by science fiction writers to power futuristic spaceships, such as the starship *Enterprise* in the *Star Trek* series.

others are neutral. But all these particles are subject to the strong force and all participate in holding the nucleus together; thus, they help make the physical universe possible.

Antimatter

For every particle that we see in the universe, it is possible to produce an antiparticle. Every particle of **antimatter** has the same mass as its matter twin, but the particles have opposite charge and opposite magnetic characteristics. For example, the antiparticle of the electron is a positively charged particle known as the *positron*. It has the same mass as the electron, but a positive electric charge. Antinuclei, composed of antiprotons and antineutrons and orbited by positrons, can form antiatoms. (Such antiatoms have been produced in the laboratory.)

When a particle collides with its antiparticle, both masses are converted completely to energy in a process called *annihilation*, the most efficient and violent process that we know in the universe. The original particles disappear, and this means that energy appears as a spray of rapidly moving particles and electromagnetic radiation. Science fiction writers have long adopted this fact in their descriptions of futuristic weapons and power sources. (The starship *Enterprise* on *Star Trek*, for example, has matter and antimatter pods as its power source.

Although antimatter is fairly rare in the universe, it is routinely produced in particle accelerators. High-energy protons or electrons strike nuclear targets, and the energy of the particles is converted to equal numbers of other particles and antiparticles. Thus the existence of antimatter is verified daily in laboratories.

Physics in the Making

The Discovery of Antimatter

In 1932, Carl Anderson, a young physicist at the California Institute of Technology, performed a rather straightforward cosmic ray experiment of the type described in this chapter. Cosmic rays entered a type of detector called a "cloud chamber." In Anderson's cloud chamber, a cosmic ray particle would move through a moisture-laden gas, leaving behind a string of ions. Pulling out a piston at the bottom of the chamber lowered the gas pressure, causing the liquid (usually alcohol) that had been in gaseous form to condense out into droplets. The ions acted as nuclei for the condensation of these droplets, so that the path of the particle was marked by a string of droplets in the chamber.

The key innovation in Anderson's experiment was the positioning of the cloud chamber between the poles of powerful magnets. These magnets caused electrically charged cosmic rays to move in curved tracks, with the amount of curving dependent on the particle's mass, speed, and charge. Furthermore, the tracks of positively and negatively charged particles curved in opposite directions under the influence of the magnetic field.

Soon after he switched on his apparatus, Anderson saw tracks of particles whose mass seemed to be identical to that of the electron, but whose tracks curved in the opposite direction from those of electrons being detected (Figure 27-4). This feature, he concluded, had to be the result of a "positive electron," a phrase he contracted to *positron*. Although no one realized it at the time, Anderson was the first human being to see antimatter. ●

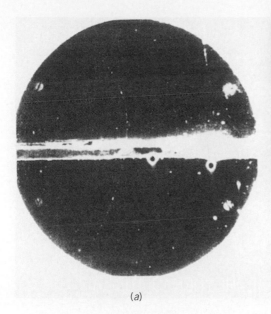

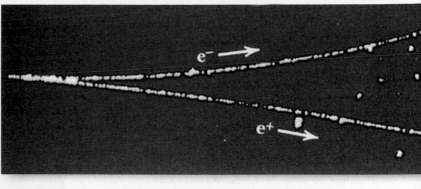

(b)

FIGURE 27-4. Carl Anderson identified the positron (the antiparticle of the electron) from the distinctively curved path left in a bubble chamber. In Anderson's original photograph (a) the positron path curves upward and to the left. In a more recent photograph (b) an electron (e⁻) and a positron (e⁺) curve in opposite directions in a magnetic field.

(a)

Develop Your Intuition: Curved Particle Tracks

How might Anderson have interpreted his results if he had seen tracks of particles curving in the same direction as the electrons' tracks, but curving a different amount? (*Hint:* Remember Newton's second law of motion.)

The direction of curvature in a magnetic field depends on the sign of the particle's electric charge, as we have seen in Chapter 16. If the track curves in the same direction as electrons' track, it must have the same sign of charge; that is, negative charge. However, the amount of curvature—that is, the change in velocity, or acceleration—depends on the mass and velocity of the particle. If we assume the particle has the same speed as the electron, and if we assume the particle track curves more than that of an electron, the particle must have a larger acceleration for the same amount of magnetic force; that is, the particle must have less mass than an electron has. If the track curves less than that of an electron, the particle must have more mass than the electron has. Anderson would have discovered a new particle, but not the positron.

Connection

How Does the Brain Work?

The study of elementary particles often seems quite abstract, but situations do arise where elementary particles play a very important role in understanding the real world. The relatively recent technology of *positron emission tomography (PET),* for example, is helping scientists probe the mysterious workings of the brain.

In this medical technique, molecules such as glucose (a sugar that the body uses to provide energy for its cells) are created using an unstable isotope of an element such as oxygen and then are injected into a patient's bloodstream. Organs in the body, including the brain, take up these molecules. They go to the

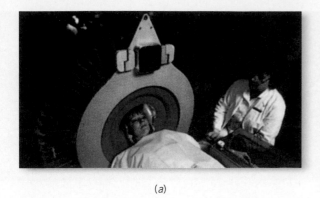

(a)

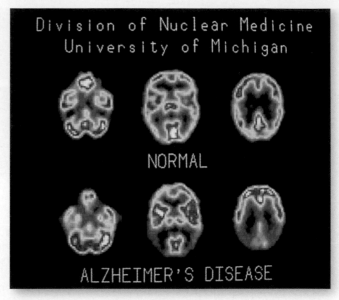

(b)

FIGURE 27-5. Positron emission tomography, commonly called the PET scan, reveals activity in the human brain. (a) A patient undergoing a PET scan. (b) Scans of a normal brain reveal bright spots where large amounts of glucose are being used by the brain. By contrast, a scan of a person suffering from Alzheimer's disease shows decreased brain activity.

parts of the brain that need glucose; that is, the parts that require extra energy at the time (see Figure 27-5).

The isotopes chosen for this technique are ones that emit a positron, the antiparticle of an electron, when they decay. These positrons quickly "annihilate" with ordinary electrons in the surrounding tissue, emitting two gamma rays that are relatively easy to detect from outside the body. A PET scan works like this: After the radioactive glucose is injected into the bloodstream, the patient is asked to do something—talk, read, do mathematical problems, or just relax. Each of these activities uses a different region of the brain. Scientists watching the emission of positrons can see those regions of the brain light up as they are used. In this way, scientists use antimatter to study the normal working of the human brain without disturbing the patient, as well as to detect and study abnormalities that can perhaps be treated. ●

QUARKS

When chemists understood that the chemical elements could be arranged in the periodic table, it wasn't long before they realized what caused this regularity. Atoms of different chemical elements were not elementary, as Dalton had suggested, but were structures made up of things more elementary still. The same thing is true of the hundreds of elementary hadrons, or nuclear particles. They are not themselves elementary, but are made up of units more elementary still—units that are given the name **quark** (pronounced "quork"). First suggested in the late 1960s by American physicist Murray Gell-Mann, quarks have come to be accepted by physicists as *the* fundamental building blocks of hadrons. Even though they never have been (and probably cannot be) isolated in the laboratory, the concept of quarks has brought order and predictability to the complex zoo of elementary particles. (It is important to remember that only hadrons, not leptons, are made from quarks.)

Quarks are different from other elementary particles in several ways. Unlike any other known particle, they have fractional electric charge, equal to $\pm\frac{1}{3}$ or $\pm\frac{2}{3}$ the charge on an electron or proton. In this model of matter, quarks in pairs or triplets make up all the hadrons, but once they are locked into these particles, no amount of experimental machination can ever pry them loose. Quarks existed as free particles only briefly in the very first stages of the universe (see Chapter 29).

In spite of these strange properties, the quark picture of matter is a very appealing one. Why? Because instead of our having to deal with dozens of different hadrons, all atomic nuclei are reduced to only six kinds of quarks (and six antiquarks). The quarks, like many things in elementary-particle physics, have been given fanciful names: up, down, strange, charm, top, and bottom (see Table 27-2). Physicists have seen evidence for elementary particles that contain all these six quarks, although only as pairs or triplets in any single particle.

From these six simple particles, all of the hadrons that we know about—all those hundreds of particles that whiz around inside the nucleus—can be made. The proton, for example, is the combination of two up quarks and one down quark, while the neutron is the combination of two down quarks and one up quark. In this scheme, the charge on the proton, equal to the sum of the charges on its three quarks, is

$$\frac{2}{3} + \frac{2}{3} + \left(-\frac{1}{3}\right) = +1$$

while the charge on the neutron is

$$\frac{2}{3} + \left(-\frac{1}{3}\right) + \left(-\frac{1}{3}\right) = 0$$

In the more exotic particles, pairs of quarks circle one another in orbit, like some impossibly tiny star system.

TABLE 27-2 **Quark Properties**

Name of Quark	Symbol	Electric Charge[a]
Down	d	$-\frac{1}{3}$
Up	u	$+\frac{2}{3}$
Strange	s	$-\frac{1}{3}$
Charm	c	$+\frac{2}{3}$
Bottom	b	$-\frac{1}{3}$
Top	t	$+\frac{2}{3}$

[a]Quarks with the same charge differ from one another in mass and other properties.

Develop Your Intuition: The Quark Model

Hadrons can be categorized into two families of particles: baryons, which consist of three quarks, and mesons which consist of a quark and an antiquark. The xi baryon (*xi* is a Greek letter) is made of two strange quarks and one up quark. The K meson is made of an up quark and a strange antiquark. What are the charges of these particles?

For the xi particle, we can look up the charges of the three quarks in Table 27-2 and add them together. We get

$$\frac{2}{3} + \left(-\frac{1}{3}\right) + \left(-\frac{1}{3}\right) = 0$$

The xi baryon is neutral; it has no electric charge. For the K meson, we have to remember that an antiquark has the opposite charge of its corresponding quark. The strange quark has a charge of $-\frac{1}{3}$, so the strange antiquark must have a charge of $+\frac{1}{3}$. Adding the two charges, we get

$$\frac{2}{3} + \frac{1}{3} = 1$$

This particular K meson has a charge of $+1$, the same as the proton.

Quarks and Leptons

The quark model gives us a picture of the universe that restores the kind of simplicity that was brought by both Dalton's atoms and Rutherford's nucleus. All the elementary particles in the nucleus are made from various combinations of six kinds of quarks and their antiquarks. These elementary particles are then put together to make the nuclei of atoms. The six different leptons—primarily the electrons—are located outside the nucleus. Different atoms interact with one another to produce what we see in the universe. In this scheme, the quarks and leptons are the letters of the universe; they are the basic stuff from which everything else is made. The fact that there are six leptons and six quarks has not escaped the notice of physicists. This phenomenon is built into almost all theories of elementary particles. The question of *why* nature should be arranged this way remains unanswered.

Quark Confinement

It would be nice to be able to study individual quarks in the laboratory, and physicists have conducted extensive searches for them. Yet there has been no generally accepted experimental isolation of a quark, and many particle theorists suspect that quarks can never be pried loose from confinement within a hadron. In these theories, once a quark has been taken up into a particle, it can move from one particle to another during a nuclear reaction but it can never escape confinement in a hadron and be isolated from other quarks.

You can hit elementary particles as hard as you like in an attempt to shake the quarks loose, but every time you start to pull out a quark, you've also supplied enough energy to the system to make more quarks and antiquarks, and those new quarks are immediately taken up into ordinary hadrons. If you hit one particle hard enough, you wind up with lots of other elementary particles.

THE FOUR FUNDAMENTAL FORCES

In our excursion into the library, finding the letters of the alphabet wasn't enough to explain what we saw. We had to know the rules of spelling and grammar by which letters are converted into words and words made into books. In the same way, if we are going to understand the fundamental nature of the universe, we have to understand not only the quarks and leptons, but also the forces that influence them and make them behave the way they do.

One useful analogy is to think of the quarks and leptons as the bricks of the universe. The universe appears to be built of these two different kinds of bricks that are arranged in different ways to make everything we see. But you cannot build a house using bricks alone. There has to be something like mortar to hold the bricks together. The mortar of the universe—the things that hold the elementary particles together and organize the physical universe into the structures we know—are the forces. At the moment, we know of only four fundamental forces in nature. Two of these, *gravity* (Chapter 5) and *electromagnetism* (Chapter 17), were known to nineteenth-century physicists and are part of our everyday experience. They are forces with infinite range—that is, objects such as stars and planets can exert these forces on one another even though they are very far apart.

The other two forces are less familiar to us because they operate in the realm of the nucleus and the elementary particles. They have a range comparable to the size of the nucleus (or smaller) and hence play no role in our everyday experience. The *strong force* holds the nucleus together, while the *weak force* is responsible for processes such as beta decay (see Chapter 26) that tear nuclei and elementary particles apart.

Each of the four fundamental forces is different from the others in strength and range (see Table 27-3). The important point about the four forces is that whenever anything happens in the universe, whenever an object changes its motion, it happens because one or more of these forces is acting.

Force as an Exchange

We know that forces cause matter to accelerate—nothing happens without a force. We know of four: the gravitational force, the electromagnetic force, the strong force, and the weak force. Each has its own distinctive effects on nature. We have not, however, asked how these forces work.

TABLE 27-3	The Four Forces		
Force	**Relative Strength**[a]	**Range**	**Gauge Particles**
Gravity	10^{-39}	Infinite	Graviton
Electromagnetic force	$\frac{1}{137}$	Infinite	Photon
Strong force	1	10^{-13} cm	Gluon
Weak force	10^{-5}	10^{-15} cm	W and Z

[a]Relative to the strong force.

FIGURE 27-6 The exchange of a ball between two skaters provides an analogy for the exchange of a gauge particle. Skater A, who throws the ball, recoils, and skater B recoils when the ball reaches her. Thus both skaters change velocity, and, by Newton's first law, we say that a force acts between them.

The modern understanding of forces may be thought of schematically as illustrated in Figure 27-6. Every force between two particles corresponds to the exchange of a third kind of particle, called a *gauge particle* for historical reasons. That is, a first particle (an electron, for example) interacts with a second particle (say, another electron) by the exchange of a gauge particle. The gauge particles produce the fundamental forces, such as electricity, that hold everything together.

In Chapter 4 we used the analogy of someone standing on skates throwing baseballs to explain Newton's third law of motion. Suppose a person on skates throws a baseball, and another person standing on skates catches the baseball some distance away. The person who threw the baseball would recoil, as we have discussed. The person who subsequently caught the baseball would also recoil. We can describe the situation this way: Two people stand still before anything happens. After some time, the two people are moving away from each other. From Newton's first law, we conclude that a repulsive force has acted between those two people. Yet it's very clear in this analogy that the repulsive force is intimately connected with (a physicist would say "mediated by") the exchange of the baseball.

In just the same way, we believe that every fundamental force is mediated by the exchange of some kind of gauge particle (Figure 27-7). For example, the electromagnetic force is mediated by the exchange of photons. That is, for example, the magnet holding notes onto your refrigerator is exchanging huge numbers of photons with atoms inside the refrigerator metal to generate the magnetic force.

In the same way, the gravitational force is mediated by particles called "gravitons." Right now you are exchanging large numbers of gravitons with Earth, an exchange that prevents you from floating up into space. The four fundamental forces and the gauge particles that are exchanged to generate each of them are listed in Table 27-3.

The two familiar forces of gravity and electromagnetism act over long distances because they are mediated by massless, uncharged particles (of which the

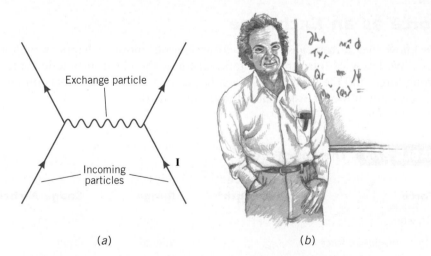

(a) (b)

FIGURE 27-7 (a) Exchange diagrams, introduced by physicist Richard Feynman, provide a model for particle interactions and the fundamental forces. Two incoming particles (such as two electrons) exchange a gauge particle (a photon) and thus are deflected by the force. (b) Richard Feynman (1918–1988) was one of the greatest physicists and physics teachers of the twentieth century.

familiar photon is one). The weak interaction, on the other hand, has a short range because it is mediated by the exchange of very massive particles—the W and Z *particles*—that have masses about 80 times that of the proton. Like the photon, the W and Z are particles that can be seen in the laboratory—they were first discovered in 1983 and are now routinely produced at accelerators around the world.

The situation with the strong force is a bit more complicated. The force that holds quarks together is mediated by particles called *gluons* (they glue the hadrons together). These particles are supposed to be massless, like the photon; but, like the quarks, they are confined to the interior of particles.

Unified Field Theories

Although a universe with six kinds of quarks, six kinds of leptons, and four kinds of forces may seem to be a relatively simple one, physicists have discovered an even greater underlying simplicity. The four fundamental forces turn out not to be as different from one another as their properties might at first suggest. The current thinking is that all four of these fundamental forces may simply be different aspects of a single underlying force.

Some physicists suggest that the four forces appear to be different because we are observing them at a time when the universe has been around for a long time and is at a relatively low temperature. The situation is somewhat analogous to freezing water. When water freezes, it can adopt many apparently different forms—powdered white snow, solid ice blocks, delicate hoarfrost on tree branches, and a smooth, slippery layer on the sidewalk. You might interpret these forms of frozen water as very different things, and in some respects they are distinct. But heat them up and they are all simply water.

Similarly, the four forces look different at the relatively low temperatures of our present existence. However, if you could heat matter up to trillions of degrees, then the different forces would not appear different at all. Theories in which fundamental forces are seen as different aspects of one force are called **unified field theories.**

The first unified field theory in history was Isaac Newton's synthesis of Earthly gravity and the circular motions observed in the heavens. To medieval philosophers, Earthly and heavenly motions seemed as different as the strong and electromagnetic forces do to us. Nevertheless, they were unified in Newton's theory of universal gravitation. In the same way, physicists today are working to unify the four fundamental forces.

The general idea of these theories is that if the temperature can be raised high enough—that is, if enough energy can be pumped into an elementary particle—the underlying unity of the forces will become clear. At a few laboratories around the world, it is possible to take protons and antiprotons (or electrons and positrons), accelerate them to extremely high energies, and let them collide. (As we have noted, proton–antiproton collisions involve the process of annihilation between particle and antiparticle as well as an ordinary collision.) When these collisions occur, for a brief moment the temperature in the volume of space about the size of a proton is raised to temperatures that have not been seen in the universe since it was less than 1 second old. In the resulting maelstrom, particles are produced that can be accounted for only if the electromagnetic and weak forces become unified.

In 1983, experiments at the European Center for Nuclear Research (CERN) and the Stanford Linear Accelerator Center (SLAC) demonstrated that this kind

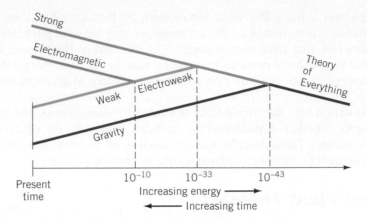

FIGURE 27-8. The four forces become unified at extremely high temperatures, equivalent to those at the beginning of the universe. At 10^{-43} second after the moment of creation, the universe had already cooled sufficiently for gravity to separate from the other three forces. The strong force separated at 10^{-33} second, while the weak and electromagnetic forces separated at 10^{-10} second.

of unification does occur. When protons and antiprotons (at CERN) or electrons and positrons (at SLAC) were accelerated and allowed to collide head-on, W and Z particles were seen in the debris of the collisions. Not only were the reactions seen, but the properties of the resulting particles and their rates of production were exactly those predicted by the first unified field theories.

The expectations of today's physicists regarding the unification of forces can be illustrated in a simple flow diagram (Figure 27-8). Scientists have already seen the unification of the electromagnetic and weak force in their laboratories. The resulting force, which physicists call the *electroweak force,* will be studied in great detail by the next generation of particle accelerators. At much higher energies, energies that will probably never be attained in Earth-based laboratories, we expect the strong force to unify with the electroweak. The theories that make up this prediction constitute the *standard model*—the best model we have today of elementary particles and their interactions. Physicists have accumulated a fair amount of experimental evidence supporting the standard model.

Finally, at still higher energies, we hope to see the force of gravity unify with the strong electroweak force. No theory yet describes this unification successfully, and attempts to develop a successful model are very much a frontier issue. Scientists have, only half in jest, started to refer to theories that combine all four forces as TOEs—*theories of everything.*

THINKING MORE ABOUT

The Theory of Everything

As often happens at a research frontier, many scientists are devoting great effort into finding a theory of everything. Some of them joke about finding the ultimate equation that explains everything, that can be written on the back of an envelope (or, better yet, on a T-shirt). Of several intriguing candidates for this ultimate theory, perhaps the best known is called *string theory.* String theory pictures quarks and leptons as being vibrations on tiny stringlike structures. Experts in this field regard their search for such a theory as the culmination of a 2000-year-old quest for a basic understanding of the universe.

However, other physicists argue that finding a theory of everything, while it would answer questions in particle physics, would not be very useful in other areas. Knowing that a silicon microchip or the Florida Everglades are ultimately made of quarks and leptons doesn't help you much in dealing with practical questions about electronics or the environment. They argue that too much attention is paid to glamorous endeavors such as the search for the ultimate theory and not enough to things that might actually pay off in human terms.

This isn't a conflict between basic and applied research, but between different areas of basic research. How much attention do you think ought to be paid to searching for the solutions to fundamental problems such as the ultimate nature of the universe and how much to basic research into fields such as the study of complex behavior systems? Do you think that there ought to be a limit on how much government support is directed toward theories of everything? Why?

Summary

High-energy physics, or **elementary-particle physics,** deals with bits of matter that we cannot see and with forces and energies totally outside our everyday experience. Nevertheless, the study of the subatomic world holds the key to understanding the structure and organization of the universe.

All matter is made up of atoms, which are made up of even smaller particles—electrons, protons and neutrons—but these are not the most fundamental building blocks of the universe. Physicists originally examined collisions between energetic **cosmic rays** and nuclei to study elementary particles. They now employ **particle accelerators,** including synchrotrons and linear accelerators, to collide charged particles at near-light speeds. These scientists have discovered hundreds of subatomic particles.

One class of particles, the **leptons** (including the electron and neutrino), are not subject to the strong force and thus do not participate in holding the nucleus together. Nuclear particles called **hadrons** (including the proton and neutron), according to present theories, are made from **quarks,** which are particles that have fractional electric charge and cannot exist alone in nature. Together, leptons and quarks are the most fundamental building blocks of matter that we know. Each of these particles has an **antimatter** particle, such as the positron, the positively charged antiparticle of the electron.

The four known forces—gravity, electromagnetism, the strong force, and the weak force—cause particle interactions that lead to all the organized structures we see in the universe. Particle interactions are mediated by the exchange of gauge particles, with a different gauge particle for each of the different forces.

While the four known forces appear to us to be quite different from one another, we believe that early in the universe, when temperatures were extremely high, the four forces were unified into a single force. At the forefront of modern physics research is the search for a **unified field theory** that describes this single force.

Key Terms

antimatter Any substance that annihilates with an equal amount of ordinary matter, resulting in a complete conversion to electromagnetic energy. (p. 590)

cosmic rays High-speed particles (mostly protons) that originate in space and travel throughout the universe. (p. 585)

hadron Any particle that exists within the nucleus of the atom; hadrons interact with the strong force. (p. 589)

high-energy physics or **elementary-particle physics** The study of elementary particles and their properties. (p. 584)

lepton Elementary particle that does not participate in the strong nuclear force; the electron and the neutrino are examples of leptons. (p. 589)

particle accelerator A scientific instrument that increases the speed of charged particles. (p. 586)

quark The fundamental particle that is the building block of all hadrons. (p. 592)

unified field theory A theory that sees the fundamental forces as different aspects of the same force. (p. 597)

Review

1. What is reductionism?

2. How is the search for elementary particles an example of reductionism? Why is reductionism appealing?

3. What are the fundamental building blocks of a library? Why is there more than one correct answer?

4. Why is "high-energy physics" an appropriate alternative name for "elementary-particle physics"?

5. What are cosmic rays? Where do they come from and how do they interact with matter here on Earth?

6. How were scientists in the 1930s and 1940s able to make use of cosmic rays to study elementary particles? What types of devices did they use to accomplish this?

7. How do scientists detect the presence of subatomic particles? How can charged particles be detected? Uncharged particles?

8. The first particle accelerator was the cyclotron used in the 1930s. What did this device do? Why was it invented and how did it work?

9. What is a synchrotron?

10. How does a linear accelerator differ from a synchrotron?

11. How might you detect the presence of a charged elementary particle?

12. Small particle accelerators can be used to kill cancer cells. How is this done?

13. What are leptons? How do we know they exist?

14. Why are leptons said to be weakly interacting particles?

15. What are hadrons and where are they located?

16. What force are all hadrons subject to?

17. Why are there so many different kinds of hadrons, but only a few kinds of leptons? Are hadrons or leptons more elementary?

18. What observations led Carl Anderson to conclude that he had discovered a particle with the same mass as the electron, but with a positive electric charge?

19. What is antimatter and how do we know it exists?

20. What is a positron? What is its charge and mass?

21. When matter meets antimatter, what happens? What is this called?

22. What is a PET scan? How does it employ elementary particles to enable the activity of a brain to be viewed?

23. Why is it appealing for physicists to think that there are only six kinds of quarks, as opposed to more than 100 hadrons?

24. How do quarks differ from other elementary particles? Is there any way to prove that quarks exist? Explain.

25. What are the similarities in the modern argument over the reality of quarks and the nineteenth-century argument over the reality of atoms? What are the differences?

26. Describe how quarks and leptons are put together to make all the matter we see.

27. What does it mean to say that a quark is "confined"?

28. Name the four fundamental forces. How is it that these forces are often considered the mortar of the universe?

29. What is the strong force? The weak force? Where do these act?

30. Each of the fundamental forces can be considered an exchange of particles. Explain how this could be.

31. What is a gauge particle?

32. What specific particle is exchanged to generate each of the four fundamental forces?

33. How do we know that gravity and electromagnetic force act over enormous distances while the strong and weak forces act only at very short range?

34. What does it mean to say that all four fundamental forces were unified?

35. What is a unified field theory? Give an example.

36. Identify what might be considered the fundamental units and rules of organization of (a) a grocery store and (b) a parking garage. How many levels of organization can you identify? (Remember, not all questions have only one correct answer.)

Questions

1. Which particle–antiparticle interaction releases more energy: an electron–positron annihilation or a proton-antiproton annihilation? How does the law of conservation of energy come into play?

2. Some particle accelerators accelerate particles around in circular paths up to speeds very close to the speed of light. It is advantageous to make the diameters of these circles as large as possible. Why?

3. When an electron and a positron annihilate, they form a pair of photons. Suppose a friend told you that sometimes, instead of two photons, the positron and electron annihilate into a proton and a neutron. Why is this impossible?

4. How many quarks are there in a helium-4 nucleus? How many quarks are there in a carbon-12 nucleus?

5. A neutron is made of three quarks. Is it possible that two up quarks and a down quark could make a neutron? Explain.

6. As a particle passes through a bubble chamber, it leaves tracks. The figure represents the tracks of a particle (moving left to right) that decayed into two particles as it passed through the chamber. One of the particles curved upward in the magnetic field and the other kept going straight. What is the charge of the original particle before it decayed? Explain why these tracks are impossible.

7. A particle moving left to right decays into two other particles. The paths of the particles are shown in the figure. Why could this figure not possibly represent a real situation? (*Hint:* What conservation law is violated?)

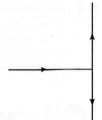

Problems

1. Arrange the four fundamental forces according to strength from strongest to weakest. What is the range of each of these forces?

2. What is the electric charge of an antiproton? An antineutron? Why?

3. A hadron called the sigma particle is made from two down quarks and one strange quark. What is the charge of the sigma particle?

4. Are any leptons made up of quarks? Explain.

5. Mesons are made from a quark and an antiquark. A particle called the pi-meson is made from an up quark and an anti-down quark. What is the charge of this particle?

6. A proton and an antiproton, both at rest with respect to one another, mutually annihilate into two gamma rays. How much energy is produced in this annihilation? (*Hint:* How are mass and energy related to one another?)

7. One naturally occurring reaction in a radioactive decay process is nitrogen-12 (atomic number 7) decaying into carbon-12 (atomic number 6) plus an unknown particle. What are the properties (atomic mass, atomic number, and charge) of this unknown particle, and how does it compare with the properties of an electron?

8. Some astronomers have theorized that galaxies (matter) and antigalaxies (antimatter) have existed. Each galaxy contains about 10^{11} masses of the Sun (the mass of the Sun is 1.9×10^{30} kg). Calculate the total energy released if a matter–antimatter galaxy pair annihilated.

9. Temperature can be related to energy by the equation $E = \frac{1}{2}kT$, where $k = 1.38 \times 10^{-23}$ J/K and the temperature T is in kelvins. What is the difference in the temperatures generated between an electron–positron annihilation and a proton–neutron annihilation?

Investigations

1. The superconducting supercollider (SSC) was planned to be perhaps the last of the great particle accelerators. The ultimate goal was to see the details of the unification of the weak and electromagnetic forces and, perhaps, to learn why particles have mass. The project was started and then in 1994 was killed when Congress failed to allocate the money for the project after much debate. Discuss the SSC controversy in terms of differences between basic and applied research.

2. How did your congressional representative and senators vote on SSC funding in Congress? Why did they vote that way? Why do you agree or disagree with their vote?

3. Locate the nearest PET-scan facility and arrange a visit. Where do the physicians obtain the special form of glucose used in the procedure? What kind of educational training would you need to operate such a facility?

4. Watch an episode of *Star Trek* and discuss the use of matter and antimatter in the propulsion system of the *Enterprise.*

Can you find any other uses of antimatter in science fiction stories?

5. What does it mean that the fundamental building blocks of the universe are things we can never isolate and study? Does that mean they aren't real? You might want to think about the question of the reality of atoms for a historical precedent to this situation.

6. There is a significant amount of information on particle physics and elementary particles available on the Internet. Go to your favorite search engine and type in the keywords "elementary particles" or "high-energy physics." Explore the sites devoted to these topics. What are the major research accelerators throughout the United States and the world? What types of equipment do they have, and what are some of the specific projects they are currently working on? Can you find any good tutorials that help further explain and illustrate the concepts in this chapter?

7. Particle physics is often described with such words as charm, strangeness, up, down, the eightfold way, and many other seemingly peculiar terms for such a serious scientific subject. What are the advantages of using such language to describe the elementary particle zoo? What are, perhaps, some of the disadvantages? Does language make a difference in, say, funding or understandability? Using examples from the world of high-energy physics, comment on the use of this descriptive language.

 # WWW Resources

See the *Physics Matters* home page at **www.wiley.com/college/trefil** for valuable web links.

1. http://www.aip.org/history/lawrence/ The history of the original accelerator and its inventor from the American Institute of Physics.

2. http://particleadventure.org/particleadventure/ The *CPEP Particle Adventure*—the best, most lavish site for particle physics education, by Lawrence Berkeley National Laboratory.

28 | Albert Einstein and the Theory of Relativity

KEY IDEA

All observers, no matter what their frame of reference, see the same laws of nature.

PHYSICS AROUND US . . . The Airport

You walk down the ramp into the plane and get into your seat. It's been a long day—classes and maybe an exam or two. You doze off for a moment and then wake with a start. For a moment, it appears to you that the plane has started to move backward, away from the gate. Then, as you become more alert, you realize that your plane isn't moving at all but another plane is pulling into the gate next to yours.

During that brief moment between sleep and waking, you were seeing the world through eyes unaf-

fected by years of experience. You were realizing that there is always more than one way to view any kind of uniform motion. One way is to say that you are stationary and the other plane is moving with respect to you. But you could also say that the other plane is stationary and you are moving with respect to it.

Which point of view is right?

One of the great scientific discoveries of the early twentieth century, the theory of relativity, grew out of thinking about this sort of question.

FRAMES OF REFERENCE

A **frame of reference** is the physical surroundings from which you observe and measure the world around you. If you read this book at your desk or in an easy chair, you experience the world from the frame of reference of your room, which seems firmly rooted to solid Earth. If you read on a train or in a plane, your frame of reference is the vehicle, which moves with respect to Earth's surface. If you could imagine yourself in an accelerating spaceship in deep space, your frame of reference would be different still. In each of these reference frames you are what scientists call an "observer." An observer looks at the world from a particular frame of reference, with anything from casual interest to a full-fledged laboratory investigation of phenomena that leads to a determination of natural laws.

For human beings who grow up on Earth's surface, it is natural to think of the ground as a fixed, immovable frame of reference and to refer all motion to it. After all, train or plane passengers don't think of themselves as stationary while the countryside zooms by. However, as we have seen in the opening Physics Around Us section, there are indeed times when we lose this prejudice and see that the question of who is moving and who is standing still is largely one of definition.

From the point of view of an observer in a spaceship above the solar system, there is nothing solid about the ground you're standing on. Earth is rotating on its axis and moving in an orbit around the Sun, while the Sun itself is performing a stately rotation around the galaxy. Thus, even though a reference frame fixed in Earth may seem "right" to us, there is nothing special about it.

Descriptions in Different Reference Frames

Different observers in different reference frames may provide very different accounts of the same event. To convince yourself of this idea, think about a simple experiment. While riding on a train, take a coin out of your pocket and flip it. You know what will happen—the coin goes up in the air and falls straight back into your hand, just as it would if you flipped it while sitting in a chair in your

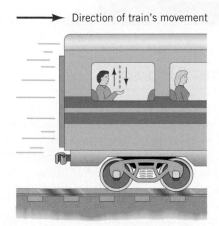

Apparent direction of coin's fall
(a) Frame of reference: inside the train

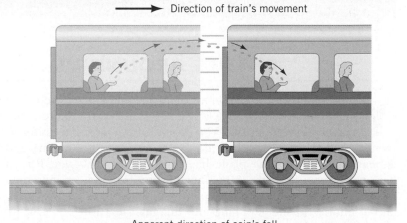

Apparent direction of coin's fall
(b) Frame of reference: outside the train

FIGURE 28-1. The path of a coin flipped in the air depends on the observer's frame of reference. (a) A rider in the car sees the coin go up and fall straight down. (b) An observer on the street sees the coin follow an arching path.

room (Figure 28-1*a*). But now ask yourself this question: how would a friend standing near the tracks, watching your train go by, describe the flip of the coin? To your friend, it appears that the coin goes up into the air, but by the time it comes down the car has traveled some distance down the tracks. As far as your friend on the ground is concerned, the coin has traveled in an arc (Figure 28-1*b*).

So you, sitting in the train, say the coin has gone straight up and down, while someone on the ground says it has traveled in an arc. You and the ground-based observer would describe the path of the coin quite differently, and you would both be correct in your respective frames of reference. The universe we live in possesses this general feature—different observers describe the same event in different terms, depending on their frames of reference.

Does this mean that we are doomed to live in a world where nothing is fixed, where everything depends on the frame of reference of the observer? Not necessarily. The possibility exists that even though different observers give different descriptions of the same event, they agree on the underlying laws that govern it. Even though the observers disagree on the path followed by the flipped coin, they may very well agree that motion in their frame is governed by Newton's laws of motion and the law of universal gravitation.

 ## THE PRINCIPLE OF RELATIVITY

Albert Einstein came to his theories of relativity by thinking about a fundamental contradiction between Newton's laws and Maxwell's equations. You can see the problem by thinking about a simple example. Imagine you're on a moving railroad car and you throw a baseball. What speed does the baseball have according to an observer on the ground?

If you throw the ball forward at 40 kilometers per hour while on a train traveling 100 kilometers per hour, the ball appears to a ground-based observer to travel 140 km/h—that is, 40 km/h from the ball plus 100 km/h from the train. On the other hand, if you throw the ball backward, the ground-based observer sees the ball moving at only 60 km/h—the train's 100 km/h minus the ball's 40 km/h. In our everyday world, we just add the two speeds to get the answer, and this notion is reflected in Newton's laws.

Albert Einstein (1879–1955).

Now suppose that instead of throwing a ball you turn on a flashlight and measure the speed of the light coming from it. In Chapter 19 we noted that the speed of light is built into Maxwell's equations. If every observer is to see the same laws of nature, they all have to see the same speed of light. In other words, the ground observer would have to see light from the flashlight moving at 300,000 km/s and not 300,000 km/s plus 100 km/h. In this case, velocities wouldn't add, as our intuition tells us they must.

Albert Einstein thought long and hard about this paradox, and he realized that it could be resolved in only three possible ways:

1. The laws of nature are not the same in all frames of reference (an idea Einstein was reluctant to accept on philosophical grounds).

2. Maxwell's equations are wrong and the speed of light depends on the speed of the source emitting the light (in spite of abundant experimental support for the equations).

3. Our intuitions about the addition of velocities are wrong, in which case the universe might be a very strange place indeed.

Einstein focused on the third of these possibilities.

The idea that the laws of nature are the same in all frames of reference is called the *principle of relativity,* and it can be stated as follows:

Every observer must experience the same natural laws.

This statement is the central assumption of Einstein's **theory of relativity.** Hidden beneath this seemingly simple statement lies a view of the universe that is both strange and wonderful. The extraordinary effort required to understand the consequences of this one simple assumption occupied Einstein during much of the first decades of the twentieth century.

We can begin to understand Einstein's work by recalling what Isaac Newton had demonstrated three centuries earlier—that all motions fall into one of two categories, uniform motion or accelerated motion (Chapter 3). Einstein therefore divided his theory of relativity into two parts—one for each of these kinds of motion. The first part, published by Einstein in 1905, is called **special relativity** and deals with all frames of reference in uniform motion relative to one another— reference frames that do not accelerate. It took Einstein another decade to complete his treatment of **general relativity,** mathematically a much more complex theory, which applies to any reference frame, whether or not it is accelerating relative to another.

At first glance, the underlying principle of relativity seems obvious, perhaps almost too simple. Of course the laws of nature are the same everywhere—that's the only way that scientists can explain how the universe behaves in an ordered way. But once you accept that central assumption of relativity, be prepared for some surprises. Relativity forces us to accept the fact that nature doesn't always behave as our intuition says it must. You may find it disturbing that nature sometimes violates our sense of the way things should be. But you'll have little problem with relativity if you just accept the idea that the universe is what it is and not necessarily what we think it should be.

Another way of saying this is to note that our intuitions about how the world works are built up from experience with things that are moving at modest speeds—a few hundred, or at most a few thousand, miles per hour. None of us has any experience with things moving near the speed of light, so when we start examining phenomena in that range, our intuitions won't necessarily apply. As with our examination of the concepts of quantum mechanics at the scale of atoms and molecules, we shouldn't be surprised by anything we find.

Relativity and the Speed of Light

As the example of the train and the flashlight shows, one of the most disturbing aspects of the principle of relativity has to do with our everyday notions of speed. According to the principle, any observer, no matter what his or her reference frame, should be able to confirm Maxwell's description of electricity and magnetism. Because the speed of light is built into these equations, it follows that

The speed of light, c, is the same in all reference frames.

Strictly speaking, this statement is only one of many consequences of the principle of relativity. However, so many of the surprising results of relativity follow from this statement that it is often accorded special status and given special attention in discussions of relativity.

Physics in the Making

Einstein and the Streetcar

Newton and his apple have entered modern folklore as a paradigm of unexpected discovery. A less well known incident led Albert Einstein, then an obscure patent clerk in Berne, Switzerland, to relativity.

One day, while riding home in a streetcar, he happened to glance up at a clock on a church steeple (Figure 28-2). In his mind he imagined the streetcar speeding up, moving faster and faster, until it was going at almost the speed of light. Einstein realized that if the streetcar were traveling at the speed of light, it would appear to someone on the streetcar that the clock had stopped. A passenger looking at the clock would always see the same thing—for him or her, the clock would be "frozen." On the other hand, a clock moving with Einstein— his pocket watch, for example—would still tick away the seconds in its usual way. Perhaps, Einstein thought, time as measured on a clock, just like motion, is relative to one's frame of reference.

In later years, Einstein liked to talk to other physicists about his ideas as he continued to develop the theory of relativity. Eve Curie, in her biography of her mother Marie Curie, tells of Einstein walking with Madame Curie and her two young daughters during a visit. Einstein delighted the children because he talked about wanting to know what it would be like to ride on a beam of light. But Einstein was quite serious about these ideas and they led him to revolutionize our basic ideas of time and space. ●

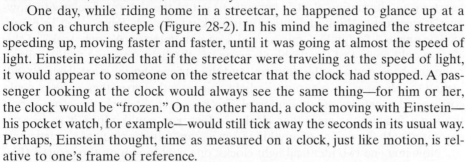

FIGURE 28-2. Albert Einstein, moving away from a clock tower, imagined how different observers might view the passage of time. If Einstein were traveling at the speed of light, for example, the clock would appear to him to have stopped, even though his own pocket watch would still be ticking.

SPECIAL RELATIVITY

Time Dilation

Think about how you measure time. The passage of time can be measured by any kind of regularly repeating phenomenon—a swinging pendulum, a beating heart, or an alternating electric current. To get at the theory of relativity, however, it's easiest to think of a rather unusual kind of clock. Suppose, as in Figure 28-3, we have a flashbulb, a mirror, and a photon detector. A "tick–tock" of this clock would consist of the flashbulb going off, the light traveling to the mirror, bouncing back down to the detector, and then triggering the next flash. By adjusting the distance, *d,* between the light source and mirror, these pulses could correspond to any desired time interval. This unusual light clock, therefore, serves the same function as any other clock—in fact, you could adjust it to be synchronized with the ticking of anything from a grandfather's clock to a wristwatch.

Now imagine two identical light clocks: one next to you on the ground (Figure 28-3*a*) and the other whizzing by in a spaceship (Figure 28-3*b*). Imagine, further, that the mirrors are adjusted so that both clocks would be ticking at the same rate if they were standing next to one another. How would the moving clock look to you?

Standing on the ground, you would see the ground-based clock ticking along as the light pulses bounce back and forth between the mirror and detector. When you looked at the moving clock, however, you would see the light following a longer, zigzag path. If the speed of light is indeed the same in both frames of reference, it should appear to you that the light in the moving frame takes longer to travel the zigzag path from flashbulb to detector than the light on the ground-based clock. Consequently, from your point of view on the ground, the moving clock must tick more slowly. The two clocks are identical, but the moving clock runs slower. This surprising phenomenon, known as **time dilation,** is a direct consequence of relativity. Once we accept the idea that the laws of physics (including the speed of light) must be the same in all reference frames, we have to accept the fact that time is not the same for observers moving at different speeds.

Remember that each observer regards the clock in his or her own reference frame as completely normal, while all other clocks appear to be running slower. Thus, paradoxically, while we observe the spaceship clock as slow because it is

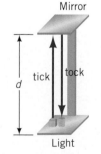

(*a*) Stationary light clock

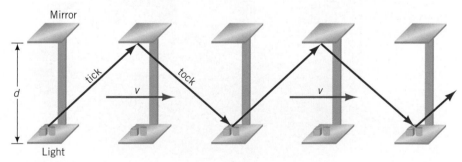

(*b*) Moving light clock

FIGURE 28-3. A light clock incorporates a flashing light and a mirror. A light pulse from a flashbulb bounces off the mirror and returns to trigger the next pulse. Two light clocks, one stationary (*a*) and one moving (*b*), illustrate the phenomenon of time dilation. Light from the moving clock must travel farther, and so the moving clock appears to the stationary observer to tick more slowly.

moving, observers in the speeding spaceship see the Earth-based clock moving and believe that the Earth-based clock is running more slowly than theirs.

Relativity's prediction of time dilation can be tested in a number of ways. Physicists have actually documented relativistic time dilation by comparing two extremely accurate atomic clocks, one on the ground and one strapped into a jet aircraft. Even though jets travel at a paltry hundred-thousandth of the speed of light, the difference in the time recorded by the two clocks can be measured and has verified the predictions of relativity.

Time dilation can also be observed with high-energy particle accelerators that routinely produce unstable subatomic particles (see Chapter 27). The normal half-life of these particles is well known. However, when accelerated to near the speed of light, these particles last much longer because of the relativistic slow-down in their decay rates.

Thus, although the notion that moving clocks run slower than stationary ones violates our intuition, it seems to be well documented by experiment. Why, then, aren't we aware of this effect in everyday life? To answer that question, we have to ask how big an effect time dilation is. How much do moving clocks slow down?

LOOKING DEEPER

The Size of Time Dilation

We have tried, in general, to talk about science in everyday terms and stay away from formulas in this book. But we have now run into a rather fundamental question that requires some simple mathematics to answer. In this section, you'll be able to follow the kind of thought process used by Einstein when he first formulated his revolutionary theory.

Consider the two identical light clocks in Figure 28-3, one moving at a velocity v relative to the ground and one stationary on the ground. Each clock has a flashbulb-to-mirror separation distance of d. (The various symbols we are using are summarized in Table 28-1.) Note that we use the terms "moving" and "stationary" for convenience; this same derivation holds for any two clocks moving relative to one another.

The notation for the time it takes for light to travel the distance d from the flashbulb to its opposite mirror—that is, one tick of the stationary clock—is a little trickier because we have to keep track of which clock we're looking at and from which reference frame we're looking. We will use two subscripts—the first subscript to tell us whether the clock is on the ground (G) or moving (M), and the second subscript to indicate whether the observer is on the ground or moving. Thus, t_{GG} is the time for one tick of the ground-based clock as measured by an observer on the ground. On the other hand, t_{MG} is the

TABLE 28-1 **Symbols for Deriving Time Dilation**

Symbol	Description
v	Velocity of the moving light clock relative to the ground
d	Distance between the clock's flashbulb and mirror
t_{GG}	Time for one tick (ground clock, ground observer)
t_{MG}	Time for one tick (moving clock, ground observer)
t_{GM}	Time for one tick (ground clock, moving observer)
t_{MM}	Time for one tick (moving clock, moving observer)
c	Speed of light, a constant

time for one tick of the moving clock from the point of view of this ground-based observer. According to the principle of relativity, all observers see clocks in their own reference frames as normal. Or, in equation form,

$$t_{GG} = t_{MM}$$

As ground-based observers, we are interested in determining the relative values of t_{GG} and t_{MG}—what we see as ticks of the stationary versus the moving clocks. In the stationary ground-based frame of reference, one tick is simply the time it takes light to travel the distance d:

$$\text{Time} = \frac{\text{Distance}}{\text{Speed}}$$

Substituting values for the light clock into this equation,

$$\text{Time for one tick} = \frac{\text{Flashbulb-to-mirror distance}}{\text{Speed of light}}$$

or

$$t_{GG} = \frac{d}{c}$$

where c is the standard symbol for the speed of light.

We have argued that to the observer on the ground it appears that the light beam in the moving clock travels on a zigzag path, as shown in Figure 28-3b, and that this makes the moving clock appear to run more slowly. In what follows, we show how to take an intuitive statement such as this and convert it into a precise mathematical equation. We begin by labeling the dimensions of our two clocks.

The moving clock travels a horizontal distance of $v \times t_{MG}$ during each of its ticks. In order to determine the value of t_{MG}, we must first determine how far light must travel in the moving clock as seen by the observer on the ground. As illustrated in Figure 28-4, we know the lengths of the two shortest sides of a right triangle. One side has length d, representing the vertical distance between flashbulb and mirror (a distance, remember, that is the same in both frames of reference). The other side is $v \times t_{MG}$, which corresponds to the distance traveled by the moving clock as observed in the stationary frame of reference. The distance traveled by the moving light beam in one tick is represented by the hypotenuse of this right triangle and is given by the Pythagorean theorem.

1. In words:

The square of the length of a right triangle's long side equals the sum of the squares of the lengths of the other two sides.

2. In an equation with words (applied to our light clock):

The square of the distance light travels during one tick equals the sum of the squares of the flashbulb-to-mirror distance and the horizontal distance the clock moves during one tick.

3. In an equation with symbols:

$$(\text{Distance light travels})^2 = d^2 + (v \times t_{MG})^2$$

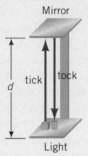

(a) Stationary light clock

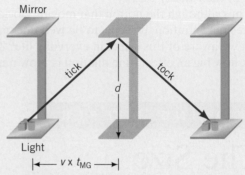

(b) Moving light clock

FIGURE 28-4. Light clocks with dimensions labeled. Both the stationary clock (a) and the moving clock (b) have flashbulb-to-mirror distance d. During one tick the moving clock must travel a horizontal distance $v \times t_{MG}$.

We can begin to simplify this equation by taking the square roots of both sides:

$$\text{Distance light travels} = \sqrt{d^2 + (v \times t_{MG})^2}$$

Remember, time equals distance divided by velocity. So the time it takes light to travel this distance t_{MG}, is given by the distance $\sqrt{d^2 + (v \times t_{MG})^2}$ divided by the velocity of light c:

$$t_{MG} = \frac{\sqrt{d^2 + (v \times t_{MG})^2}}{c}$$

We now must engage in a bit of algebraic manipulation. First, square both sides of this equation.

$$t_{MG}^2 = \frac{d^2}{c^2} + \frac{v^2 t_{MG}^2}{c^2}$$

But we saw previously that $t_{GG} = d/c$, so, substituting gives us

$$t_{MG}^2 = t_{GG}^2 + \frac{v^2 t_{MG}^2}{c^2}$$

Dividing both sides by t_{MG}^2 gives

$$\frac{t_{MG}^2}{t_{MG}^2} = \frac{t_{GG}^2}{t_{MG}^2} + \frac{v^2 t_{MG}^2/c^2}{t_{MG}^2}$$

or

$$1 = \left(\frac{t_{GG}}{t_{MG}}\right)^2 + \left(\frac{v}{c}\right)^2$$

Finally, regrouping yields

$$t_{MG} = \frac{t_{GG}}{\sqrt{1 - \left(\frac{v}{c}\right)^2}}$$

This equation expresses in mathematical form what we said earlier in words—that moving clocks appear to run slower. It tells us that t_{MG}, the time it takes for one tick of the moving clock as seen by an observer on the ground, is equal to the time it takes for one tick of an identical clock on the ground divided by a number less than 1. Thus, the time required for a tick of the moving clock is always greater than that for a stationary clock.

The expression $\sqrt{1 - (v/c)^2}$ is called the *Lorentz factor* and this number appears over and over again in relativistic calculations. In the case of time dilation, the Lorentz factor arises from an application of the Pythagorean theorem.

An important point to notice is that if the velocity of the moving clock is very small compared to the speed of light, the quantity $(v/c)^2$ becomes very small and the Lorentz factor is almost equal to 1. In this case, the time on the moving clock is almost equal to the time on the stationary one, as our intuition demands that it should be. Only when speeds get very high do the effects of relativity become important; close to the speed of light, the clock appears to stop ticking altogether.

LOOKING DEEPER

How Important Is Relativity?

To understand why we aren't aware of relativity in everyday life, let's calculate the size of the time dilation for a clock in a car moving relative to the ground at 70 km/h (about 50 miles per hour).

The first problem is to convert the familiar speed in kilometers per hour to a speed in meters per second so we can compare it to the speed of light. There are 60 minutes/hour × 60 seconds/minute = 3600 seconds in an hour, so a car traveling 70 km/h is moving at a speed of

$$70 \text{ km/h} = \frac{70,000 \text{ m}}{3600 \text{ s}}$$

$$= 19.4 \text{ m/s}$$

For this speed, the Lorentz factor is

$$\sqrt{1 - \left(\frac{19.4}{300,000,000}\right)^2} = 0.9999999999999999$$

Thus the passage of time for a stationary car and a speeding car differs by only one part in the sixteenth decimal place.

To get an idea of how small the difference is between the ground clock and the moving one in this case, we can note that if you watched the moving car for a time equal to the age of the universe, you would observe it running 10 seconds slow compared to your ground clock.

However, for an object traveling at 99% of the speed of light, the Lorentz factor is

$$\sqrt{1 - \left(\frac{v}{c}\right)^2} = \sqrt{1 - (0.99)^2}$$

$$= \sqrt{0.0199}$$

$$= 0.1411$$

In this case, you would observe the stationary clock to be ticking about seven times as fast as the moving one—that is, the ground clock would tick about seven times while the moving clock ticked just once.

This numerical example illustrates a very important point about relativity. Our intuition and experience tell us that the exterior clock on our local bank doesn't suddenly slow down when we view it from a moving car. Consequently, we find the prediction of time dilation to be strange and paradoxical. But all of our intuition is built up from experiences at very low velocities—none of us has ever moved at an appreciable fraction of the speed of light. For the everyday world, the predictions of relativity coincide precisely with our experience. It is only when we get into regions near the speed of light, where that experience isn't relevant, that the "paradoxes" arise.

Connection
Space Travel and Aging

While humans presently do not experience the direct effects of time dilation in their day-to-day lives, at some future time they might. If we ever develop interstellar space travel at near-light speed, then time dilation may wreak havoc with family lives (and genealogists' records).

Imagine a spaceship that accelerates to 99% of the speed of light and goes on a long journey. While 15 years seem to pass for the crew of the ship, more than a century goes by on Earth. The space explorers return almost 15 years older than when they left, but biologically younger than their great-grandchildren! Friends and family would all be long-since dead.

If we ever enter an era of extensive high-speed interstellar travel, people may drift in and out of other people's lives in ways we can't easily imagine. Parents and children could repeatedly leapfrog one another in age, and the notion of relatedness could take on complex twists in a society with widespread relativistic travel. ●

Length Contraction

Many results from relativity run counter to our intuition. They can be derived by procedures similar to (but more complicated than) the one we just gave for working out time dilation. In fact, using arguments like those we have presented, Einstein showed that moving objects must appear to be shorter than stationary ones (see Figure 28-5).

The equation that relates the ground-based observer's measurement of a stationary object's length, L_{GG}, to that observer's measurement of the length of an identical moving object, L_{MG}, is

$$L_{MG} = L_{GG} \times \sqrt{1 - (v/c)^2}$$

The term on the right side of this equation is the familiar Lorentz factor that we derived from our study of light clocks. The equation tells us that the length of the moving ruler can be obtained by multiplying the length of the stationary ruler by a number less than 1, and thus must appear shorter. This phenomenon is known as **length contraction.**

Note that the height and width of the moving object do not appear to change—only the length along the direction of motion. Thus, a basketball moving at speeds near the speed of light would take on the appearance of a pancake. Computer simulations have shown that because length contracts in the direction of motion, but not in other directions, the appearance of objects seen by an observer moving at speeds near the speed of light can become warped in unexpected ways.

Length contraction is not just an optical illusion. While relativistic shortening doesn't affect most of our daily lives, the effect is real. Physicists who work at particle accelerators inject bunches of particles into their machines. As these particles approach light speed, the bunches are observed to contract according to the Lorentz factor, an effect that must be compensated for.

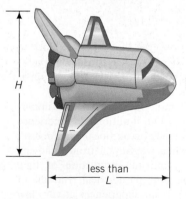

(a) Spaceship at rest

(b) Spaceship at high speed

FIGURE 28-5. A spaceship in motion appears to contract in length, L, along the direction of motion. However, the height, H, and width of the ship do not appear to change.

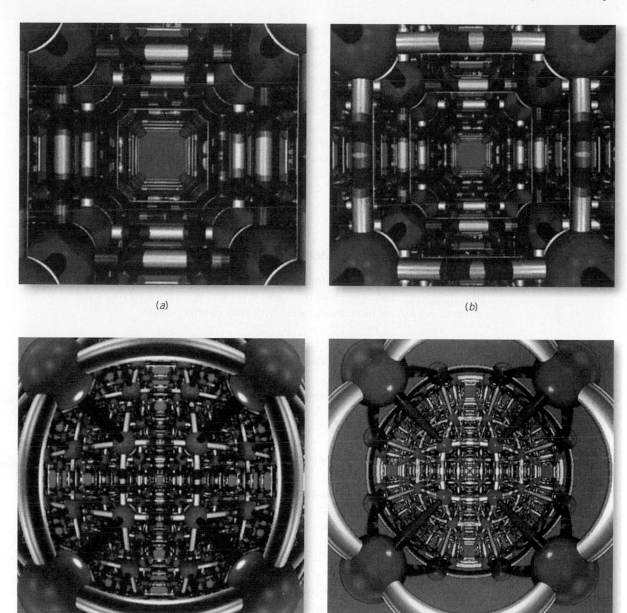

(a)   (b)

(c)   (d)

A series of four computer-generated images shows the changing appearance of a network of balls and rods as it moves toward you at different speeds. (a) At rest, the normal view. (b) At 50% of light speed, the array appears to contract. (c) At 95% of light speed, the lattice has curved rods. (d) At 99% of light speed, the network is severely distorted.

So What About the Train and the Flashlight?

Now that we understand a little about how relativity works, we can go back and unravel the paradox we discussed earlier in this chapter—the problem of how both an observer on the ground and an observer on the train could see light from a flashlight moving at the same speed.

Velocity is defined as distance traveled divided by the time required for the travel to take place. Since both length and time appear to be different for different observers, it should come as no surprise that the rule that tells us how to add velocities (such as the velocity of the light and of the train) might be more complicated than we would expect. The simple intuition that tells us that we should add the velocity of the train to the velocity of the ball, like our notions of time and space, is valid at small velocities but breaks down for objects moving near the speed of light. For those objects, a more complex addition has to be done, and, when it is, we find that both observers see the light moving at a velocity of c.

RELATIVISTIC DYNAMICS

Mass and Relativity

Perhaps the most far-reaching consequence of Einstein's theory of relativity was the discovery that mass, like time and distance, is relative to one's frame of reference. So far we have been faced with two strange ideas:

1. Clocks run fastest for stationary objects, whereas moving clocks slow down. As we observe a clock that approaches the speed of light, the clock appears to slow down and stop.

2. Distances are greatest for stationary objects; moving objects shrink in the direction of motion. As we observe an object that approaches the speed of light, the length of that object appears to shrink and approach zero.

Einstein showed that a third consequence followed from his principle:

3. Mass is lowest for stationary objects; moving objects become more massive. As we observe an object that approaches the speed of light, its mass appears to increase and approach infinity.

Einstein showed that if the speed of light is a constant in all reference frames—which must follow from the central assumption of the theory of relativity—then an object's mass depends on its velocity. The faster an object travels, the greater its mass and the harder it is to deflect from its course. If a ground-based observer measures an object's stationary or rest mass, m_{GG}, then the apparent mass, m_{MG}, of that object moving at velocity v is

$$m_{MG} = \frac{m_{GG}}{\sqrt{1 - \left(\dfrac{v}{c}\right)^2}}$$

Once again the Lorentz factor comes into play. As we observe an object approach the speed of light, its mass appears to us to approach infinity.

Mass and Energy

Time, distance, and mass—all quantities that we can easily measure in our homes or laboratories—actually depend on our frame of reference. But not everything in nature is so variable. The central tenet of relativity is that natural laws must apply in every frame of reference. Light speed is constant in all reference frames in accord with Maxwell's equations. Similarly, the first law of thermodynamics—the idea that the total amount of energy in any closed system is constant—must hold, no matter what the frame of reference. Yet here, Einstein's description of the universe seems to run into a problem. He claims that the observed mass depends on your frame of reference. But, in that case, kinetic energy—defined as mass times velocity squared—could not follow the conservation of energy law. In Einstein's treatment, faster frames of reference seem to possess more energy than slower ones. Where does the extra energy come from?

Conservation of energy appears to be violated because we have missed one key form of energy in our equations: mass itself. In fact, Einstein was able to show that the amount of energy contained in any mass turns out to be the mass times a constant.

1. In words:

All objects contain a rest energy (in addition to any kinetic or potential energy), which is equal to the object's rest mass times the speed of light squared.

2. In an equation with words:

$$\text{Rest energy} = \text{Rest mass} \times (\text{Speed of light})^2$$

3. In an equation with symbols

$$E = mc^2$$

This familiar equation has become an icon of our modern age because it defines a new form of energy. It says that mass can be converted to energy, and vice versa. Furthermore, the amounts of energy involved are prodigious (because the constant, the speed of light squared, is so large). A handful of nuclear fuel can power a city; a fist-sized chunk of nuclear explosive can destroy it.

Until Einstein traced the implications of special relativity, the nature of mass and its vast potential for producing energy was hidden from us. Now more than 10% of all electric power in the United States is produced in nuclear reactors that confirm the predictions of Einstein's theory every day of our lives (see Chapter 26).

GENERAL RELATIVITY

Special relativity is a fascinating and fairly accessible intellectual exercise, requiring little more than an open mind and a lot of basic algebra. General relativity, which deals with all reference frames, including accelerating ones, is much more challenging in its full rigor. While the details are tricky, you can get a pretty good feeling for Einstein's general theory by thinking about the nature of forces.

The Nature of Forces

Begin by imagining yourself in a completely sealed spaceship far from the reaches of gravity that is accelerating at exactly 1 *g*—Earth's gravitational acceleration. Could you devise any experiment that would reveal one way or the other if you were on Earth or accelerating in deep space?

If you dropped this book on Earth, the force of gravity would cause it to fall to your feet (Figure 28-6a). However, if you dropped the book in the accelerating spaceship, then Newton's first law tells us that it will keep moving with whatever speed it had when it was released. The floor of the ship, still accelerating, will therefore come up to meet it (Figure 28-6b). To you, standing in the ship, it appears that the book falls, just as it does if you are standing on Earth.

From an external frame of reference these two situations would involve very different descriptions. In the first case, the book falls due to the force of gravity; in the second, the spaceship accelerates up to meet the free-floating book. But no experiment you could devise in your reference frame could distinguish between acceleration in deep space and the force of Earth's gravitational field.

In some deep and profound way, therefore, gravitational forces and acceleration are equivalent. Einstein went a step further by recognizing that calling something "gravity" versus calling it "acceleration" is a purely arbitrary decision, based on our choice of reference frame. Whether we think of ourselves as stationary on a planet with gravity or accelerating on spaceship Earth makes no difference in the passage of events. This idea is often called the "principle of equivalence:" observations made in a gravitational field can be duplicated exactly in an appropriately accelerated reference frame, far from the reaches of gravity.

Although this connection between gravity and acceleration may seem a bit abstract, you already have had experiences that should tell you it is true. Have you ever been in an elevator and felt momentarily heavier when it starts up or momentarily lighter when it starts down? If so, you know that the feeling we call "weight" can indeed be affected by acceleration.

The actual working out of the consequences of the equivalence of acceleration and gravity is complicated, but a simple analogy can help you visualize the difference between Einstein's and Newton's views of the universe. In the Newtonian universe, forces and motions can be described by a ball rolling on a perfectly flat surface with neatly inscribed grid lines. The ball rolls on and on, following a line exactly, unless an external force is applied. If, for example, a large

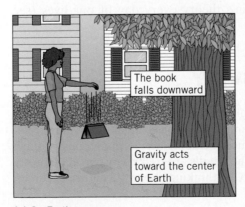

(a) On Earth

(b) In an accelerating spaceship

FIGURE 28-6. (a) If you drop a book at Earth's surface, the force of gravity causes it to fall. (b) However, if you drop the same book in an accelerating spaceship, it will keep moving with whatever speed it had when it was released, as the floor of the accelerating ship comes up to meet it. Standing in the ship, it appears that the book falls, just as it does on Earth.

mass rests on the surface, the rolling ball changes its direction and speed—it accelerates in response to the force of gravity. Thus, for Newton, motion occurs along curving paths in a flat universe. This is shown by the example of Earth revolving around the Sun in Figure 28-7a.

The description of that same event in general relativity is very different. In this case, we say that the heavy object distorts the surface. Peaks and depressions on the surface influence the ball's path, deflecting it as it rolls across the surface. For Einstein, the ball moves in a straight line across a curved universe. This is shown by the example of the Earth–Sun system in Figure 28-7b.

Given these differing views, Newton and Einstein would give very different descriptions of physical events. For example, Newton would say that the Moon orbits Earth because of an attractive gravitational force between the two bodies (Figure 28-7a). Einstein, on the other hand, would say that space has been warped

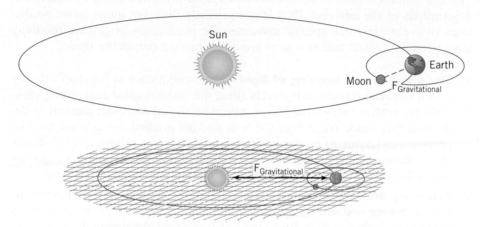

(a) Newtonian universe: gravitational forces in a flat universe

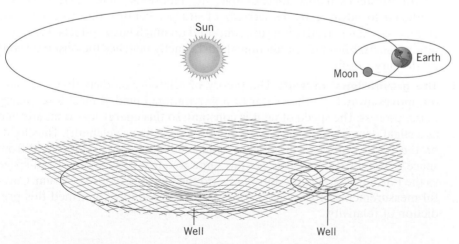

(b) Einstein's universe: motion in a curved universe.

FIGURE 28-7. Newtonian and Einsteinian universes treat the motion of planets in orbit in different ways. In the Newtonian scheme (a), a planet travels in uniform motion along curved paths in a flat universe. In the Einsteinian universe (b), a planet's mass distorts the universe; it moves in a straight line across a curved surface.

A computer-generated image of a gravity field. A mass distorts an otherwise flat grid to form depressions or "gravity wells."

in the vicinity of the Earth–Moon system and this warping of space governs the Moon's motion (Figure 28-7*b*). In the relativistic view, space deforms around the Sun, and planets follow the curvature of space like marbles rolling around in the bottom of a curved bowl.

We now have two very different ways of thinking about the universe. In the Newtonian universe, forces cause objects to accelerate. Space and time are separate dimensions that are experienced in very different ways. This view more closely matches our everyday experience of how the world seems to be. In Einstein's universe, objects move according to distortions in space, and the distinction between space and time depends on your frame of reference.

Predictions of General Relativity

The mathematical models of Newton and Einstein are not just two equivalent descriptions of the universe. They lead to slightly different quantitative predictions of events. In three specific instances, the predictions of general relativity have been confirmed, and new tests are being carried out all the time.

1. **The gravitational bending of light** One consequence of Einstein's theory is that light can be bent as it travels along the warped space near strong gravitational centers such as the Sun. Einstein predicted the exact amount of deflection that would occur near the Sun, and his prediction was confirmed by precise measurements of star positions during a solar eclipse in 1919. Today these measurements are made with much more precision by measuring the deflection of radio waves emitted by distant galaxies called *quasars*.

2. **Planetary orbits** In Newton's solar system, the planets adopt elliptical orbits, with long and short axes that rotate slightly because of the perturbing influence of other planets. Einstein's calculations make nearly the same prediction, but his axes advance slightly more than Newton's from orbit to orbit. In Einstein's theory, for example, the innermost planet, Mercury, was predicted to advance by 43 seconds of arc per century due to relativistic effects—a small perturbation superimposed on much larger effects due to the other planets. Einstein's prediction almost exactly matches the observed shift in Mercury's orbit.

3. **The gravitational redshift** The theory of relativity predicts that as a photon moves away from the center of a gravitational field, it must lose energy in the process. The speed of light is constant, so this energy loss is manifested as a slight decrease in frequency (a slight increase in wavelength). Thus lights on the Earth's surface will appear slightly redder if they are observed from space than they do on Earth. By the same token, a light shining from space to the Earth will be slightly shifted toward the blue end of the spectrum. Careful measurements of laser light frequencies have amply confirmed this prediction of relativity.

Connection
Black Holes

According to general relativity, light can be bent toward a large mass by its gravitational attraction. Suppose light approaches an extremely large mass, with a very powerful gravitational field. Would it be possible for the light to be trapped within the gravitational field so it can't escape? What would such an object look like?

One of the most surprising predictions of general relativity is that objects of extremely strong gravitational attraction can, in fact, occur. Astronomers call such objects "black holes" because they are so dense, so concentrated, that nothing, not even light, can escape from them.

Theorists have talked about three different kinds of black holes—galactic, stellar, and quantum. Observations of the motion of stars and gases near the centers of galaxies have confirmed that most galaxies (including our own Milky Way) have huge black holes at their centers—objects with masses a million times or more the mass of the sun. These are galactic black holes. Recent observations have supplied strong circumstantial evidence for such an object (Figure 28-8).

According to present-day theories of astronomy, a large star—some 30 times as massive as the Sun—can collapse into a black hole after all its nuclear fuel has been used up. These are stellar black holes. The problem with confirming this theory is that since light can't escape from a black hole, you can't see it. The only way to detect a black hole is to find one that is attracting a large amount of material into itself. This process would release such a tremendous amount of energy that it would be visible in the X-ray and gamma-ray parts of the electromagnetic spectrum. There are several candidates for stellar black holes whose X-ray emissions seem to match those predicted.

Some of the unified field theories discussed in Chapter 27 predict the existence of black holes whose dimensions are much smaller than those of an elementary particle. At the moment, these quantum black holes remain in the realm of theoretical speculation. ●

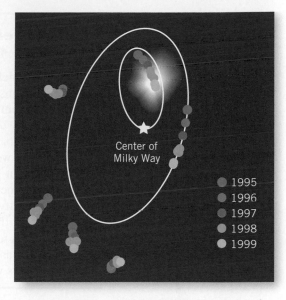

FIGURE 28-8. Stars near the center of our galaxy are observed to orbit rapidly around an exceptionally massive, yet unseen, object. Astronomers suspect that this object is a black hole 3 million times the mass of the Sun.

Recent advances in highly sensitive electronics are now providing more opportunities for researchers to measure the predictions of general relativity. One of the most intriguing tests will be conducted as part of a satellite mission in the near future. Meticulously machined quartz spheres will be set into rotation and carefully measured. According to general relativity, these spheres should develop a small wobble as they rotate in Earth's gravitational field. Sensitive electronics will detect any perturbations of this sort.

As scientists get better and better at making precision measurements, more and more tests of the extremely small differences between Newtonian and relativistic predictions of physical events can be made. A few years ago, for example, a group of scientists at institutions in the Washington, D.C. area proposed an experiment in which light from a laser would be sent out over the city from the University of Maryland, reflected from a mirror on top of the National Cathedral, and received by detectors at the Naval Research Laboratory. Because of the rotation of Earth, there should be a tiny difference in travel time between light traveling east and light traveling west. The theory of relativity predicts an additional, tinier difference as well. With good enough clocks and short enough laser pulses, these sorts of differences can be measured, and they provide just as good a check on general relativity as instruments in a satellite. This experiment, and many others like it, all deliver the same verdict: whenever our experimental techniques are good enough to test the predictions of general relativity, the theory is confirmed.

Develop Your Intuition: The Global Positioning System

We briefly described the Global Positioning System (GPS) in Chapter 5. The system consists of 24 satellites in orbit around Earth at distances of about 27,000 km from Earth's center, whizzing along at speeds of about 3.9 km/s. The system is designed to provide locations on Earth that are accurate to within 2 meters. Is the theory of relativity significant for the GPS?

The GPS works by sending radio signals from two or more satellites to the receiver on Earth. The accuracy we're looking for in distance requires an accuracy in time of about 6 nanoseconds (ns). This makes the time corrections due to relativity very important. We won't work out the details here, but it turns out that the correction factor due to time dilation is about 0.08 ns and the correction factor due to the gravitational redshift is about 0.16 ns. If corrections such as these were not taken into account by the system, the accumulated errors would exceed the required accuracy in less than 1 minute.

Physics in the Making

Who Can Understand Relativity?

Einstein's theory of relativity is extraordinary. When first introduced, the theory was difficult to grasp, in part because it relied on some complex mathematics that were unfamiliar to many physicists at the time. Furthermore, while the theory made specific predictions about the physical world, most of those predictions were exceedingly difficult to test. Soon after the theory's publication, it became conventional wisdom that only a handful of geniuses in the world could understand it.

One of those who did understand both the theory and its significance was Arthur Eddington, one of the leading physicists and astronomers of the time. The story is told that after a meeting of the Royal Society, Eddington was approached by a member, who said, "Well! Professor Eddington, you must be one of the three people in the world who understand relativity." Eddington looked puzzled and said, "Oh, I don't know . . ." The member continued, "Come, Professor Eddington, don't be modest." And Eddington replied, "On the contrary! I am wondering who the third person might be."

Einstein did make one very specific prediction, however, that could be tested. His proposal that the strong gravitational field of the Sun would bend the light coming from a distant star was different from other theories. The total eclipse of the Sun in 1919, just after the end of the First World War, gave scientists the chance to test Einstein's prediction. Eddington himself led a British expedition to observe the eclipse and, sure enough, the apparent position of stars near the Sun's disk was shifted by exactly the predicted amount.

Around the world, front-page newspaper headlines trumpeted Einstein's success. He became an instant international celebrity and his theory of relativity became a part of scientific folklore. Attempts to explain the revolutionary theory to a wide audience began almost immediately.

Few scientists may have grasped the main ideas of general relativity in 1915, when the full theory was first unveiled, but that certainly is not true today. The basics of special relativity are taught to tens of thousands of college freshmen every year, while hundreds of students in astronomy and physics explore general relativity in its full mathematical splendor.

If this subject intrigues you, you might want to read some more, watch TV specials or videos about relativity, or even sign up for one of those courses! ●

THINKING MORE ABOUT

Relativity: Was Newton Wrong?

The theory of relativity describes a universe about which Isaac Newton never dreamed. Time dilation, contraction of moving objects, and mass as energy play no role in his laws of motion. Curved space-time is alien to the Newtonian view. Does that mean that Newton was wrong? Not at all.

In fact, all of Einstein's equations reduce exactly to Newton's laws of motion, at speeds significantly less than the speed of light. This feature was shown specifically for time dilation in Looking Deeper: The Size of Time Dilation on page 609. Newton's laws, which have worked so well in describing our everyday world, fail only when dealing with extremely high velocities or extremely large masses. Astronomers must work with relativity theory routinely to explain their observations of stars and galaxies, which have large mass and move at extremely high speed. Such conditions do not apply to most activities on Earth. Thus Newton's laws

represent an extremely important special case of Einstein's more general theory.

Science often progresses in this way, with one theory encompassing previous valid ideas. For example, Newton merged Galileo's discoveries about Earth-based motions and Kepler's laws of planetary motion into his unified theory of gravity. Someday, Einstein's theory of relativity may be incorporated into an even grander view of the universe.

Some modern philosophers have argued that because scientific ideas change over time, they are nothing more than social conventions with no grounding in any external reality. In an extreme version of this view, what we have called *laws of nature* have little more meaning than the agreement that a red traffic light means "stop." How would you answer this argument? (*Hint:* You might want to compare the change in science that occurred when the Copernican system replaced its predecessors to the relation between Einstein and Newton.)

Summary

Every observer sees the world from a different **frame of reference.** Descriptions of actual physical events are different for different observers, but the **theory of relativity** states that all observers must see the universe operating according to the same laws. Because the speed of light is built into Maxwell's equations, this principle requires that all observers must measure the same speed of light in their frames of reference.

Special relativity deals with observers who are not accelerating with respect to one another, while **general relativity** deals with observers in any frame of reference whatsoever. In special relativity, simple arguments lead to the conclusion that moving clocks appear to tick more slowly than stationary ones—a phenomenon known as **time dila-**

tion. Furthermore, moving objects appear to get shorter in the direction of motion—the phenomenon of **length contraction.** Finally, moving objects become more massive than stationary ones, and an equivalence exists between mass and energy, as expressed by the famous equation $E = mc^2$.

General relativity begins with the observation that the force of gravity is equivalent to acceleration and describes a universe in which heavy masses warp the fabric of space-time and affect the motion of other objects. There are three classic tests of general relativity—the bending of light rays passing near the Sun, the changing orientation of the orbit of Mercury, and the redshift of light passing through a gravitational field.

Key Terms

frame of reference A set of physical surroundings from which events are observed and measured. (p. 604)

general relativity The part of relativity theory that deals with accelerated reference frames and gravity. (p. 606)

length contraction The observed shortening of an object that is moving with respect to the observer. (p. 612)

principle of relativity The principle that states that the laws of physics are the same in all frames of reference. (p. 605)

special relativity The part of relativity theory that deals with events in reference frames moving uniformly. (p. 606)

theory of relativity The physical laws that govern the measurement of time and space as observed in differing reference frames. (p. 606)

time dilation The slowing of time relative to an observer in a different reference frame. (p. 608)

Key Equations

Time dilation: $t_{MG} = \dfrac{t_{GG}}{\sqrt{1 - \left(\dfrac{v}{c}\right)^2}}$

Length contraction: $L_{MG} = L_{GG} \times \sqrt{1 - \left(\dfrac{v}{c}\right)^2}$

Mass effect: $m_{MG} = \dfrac{m_{GG}}{\sqrt{1 - \left(\dfrac{v}{c}\right)^2}}$

Rest mass: $E = mc^2$

Review

1. What is a frame of reference? What are some of the frames of reference that you have been in today?

2. What is the central idea of Einstein's theory of relativity?

3. Imagine arriving by spaceship at the solar system for the first time. Identify three different frames of reference that you might choose to describe Earth.

4. What is the difference between special and general relativity?

5. Does the speed of light depend at all on your frame of reference or on how fast the source of light is moving when it is emitted? Explain.

6. What is time dilation?

7. What are two examples of time dilation in the real world?

8. How might it be possible for a child to be older than its parent?

9. How fast does something have to be moving for time dilation to be appreciable? Why don't we normally notice this effect?

10. What is the Lorentz factor? When is it most commonly used?

11. According to an observer on the ground, how does the length of a moving object along the line of motion appear compared to the length of an identical object on the ground? How about the height and width of the same object?

12. Is the length of contraction simply an optical illusion? How do we know this?

13. An observer on the ground sees a fast-moving object above him. How does the mass of this object compare to the mass of an identical object at rest on the ground?

14. Does relativity allow anything to travel faster than the speed of light? Explain.

15. What is the relation between the mass of an object and its energy?

16. Which has greater mass, a flexed bow pulled back as if ready to shoot an arrow, or the same bow unflexed with the same string loosely attached?

17. How can we say that gravitational forces and acceleration are equivalent?

18. How is the warping of space around a massive object an equivalent but different description of the gravitational force generated by that object?

19. What is the difference between a curved and a flat universe?

20. Did Einstein disprove Newton's laws of motion? Explain.

21. What are the three specific predictions of general relativity that have been confirmed? In each case, what would Newton have predicted that differs from the general relativity prediction?

Questions

1. For each of the following situations, specify whether you are in a uniformly moving reference frame or an accelerated reference frame.
   a. You are standing still on the surface of Earth.
   b. You are floating deep in space, far from the effects of gravity.
   c. You are in your car, slowing down to make a stop.

2. Imagine taking a ride on a perfectly quiet train that rides on perfectly smooth and straight tracks. If the train is moving at a constant speed and you throw a ball straight up, will it appear to you that it falls straight back down? What if the train is accelerating forward?

3. You are riding on a flatbed truck moving at 50 kilometers per hour. You have two identical guns, one aimed forward and one aimed backward, and you fire them at the same time. According to an observer on the ground, which bullet is moving faster? According to an observer on the truck, which bullet is moving faster?

4. You are riding on a flatbed truck moving at 100 kilometers per hour. You have two identical lasers, one aimed forward and one aimed backward. According to an observer on the ground, which laser light moves faster? According to an observer on the truck, which laser light moves faster?

5. A meterstick moves by at a very high rate of speed. As it flies by, you measure it with an identical meterstick and find that its length is only 70 cm, as shown in the figure. According to you, are the masses of the metersticks the same or different? If different, which one is more massive?

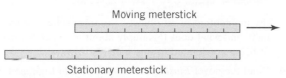

Moving meterstick

Stationary meterstick

6. You and a friend buy two identical watches. Some days later you see your friend traveling relative to you at 25% of the speed of light. Is your friend's watch running faster or slower than your watch? Does your friend agree with you? Explain.

7. You take your pulse while sitting in your room and you measure 60 beats per second. All else being equal, what would your pulse measure if you took it while on a fast-moving train? Explain.

8. Does relativity theoretically allow you to go backward in time into the past?

9. In your own words, explain why no object traveling at less than the speed of light can be accelerated all the way to, or faster than, the speed of light.

10. As you ride in an elevator, when is the apparent acceleration greater than the acceleration due to gravity outside the elevator, on the surface of Earth? When is the apparent acceleration smaller? What does this experience tell you about acceleration and gravity?

11. If you are in a spaceship far from the reaches of gravity, under what conditions will it feel to you as if the spaceship were sitting stationary on Earth's surface?

12. As you look up into the sky, you see a planet and a star. The star appears to be at location B in the figure. Assume that the planet creates very strong gravity. Where is the star actually located: toward A, at B, or toward C?

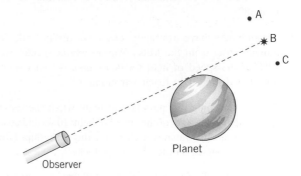

Planet

Observer

13. Due to length contraction, you notice that a train passing by appears to be shorter than when it is stationary. What do the people on the train observe about you?

14. Why is it that we don't ordinarily notice the bending of light?

15. To an outside observer, would you appear to age faster on top of a mountain or at sea level? Explain.

16. You've decided to let your sister, a NASA astronaut, cook the Thanksgiving turkey this year. Normally the turkey takes 6 hours to cook, but your sister decides to cook it on her spaceship while traveling at close to the speed of light. According to your watch, she was gone for 6 hours. Is the turkey overcooked, undercooked, or just right? Explain.

17. Because of relative motion, you notice a friend's clock running slowly. How does your friend view your clock?

18. Someone shines a light while moving toward you at 1000 m/s. With what speed will the light strike you? (The speed of light is 300,000,000 m/s.)

19. If you can do 20 pushups on the surface of Earth, how many can you do in a spaceship, far from gravity, accelerating at g?

20. You are jealous of your younger brother, who looks very young for his age. You are interested in reducing the rate at which you age relative to him. Having heard of general relativity and the effect of gravity on time, you decide that you need to spend more time at an altitude that makes you age more slowly relative to your brother. Given the choice, would you work as a park ranger high in the mountains, or as a taxi cab driver at sea level? (Take into account the effects of gravity only.) Explain your reasoning.

Problems

1. While running at 10 mph directly toward Patrick, Lisa passes a basketball to him. Patrick is stationary when he receives the ball, which is moving at a speed of 35 mph. How fast did Lisa throw the ball?

2. Giselle is traveling on her bicycle at a speed of 10 mph when a car passes her. From her frame of reference, she estimates that the car is going 45 mph toward her and 45 mph away from her. What is the speed of the car from the frame of reference of someone standing on the ground?

3. Astronomers have routinely observed distant galaxies moving away from the Milky Way galaxy at speeds over 10% of the speed of light (30,000 km/s). At what speed does this distant light reach astronomers?

4. Anna is watching the stars late at night when she sees a spaceship pass at 80% of the speed of light; 10 seconds pass on Earth as she watches a clock on the spaceship. How much time passes on the spaceship clock?

5. Elliott is traveling by a building at 150,000 km/s, moving along the width of the building. Elliott measures the building to be 50 m wide and 100 m tall. What is the height and width of the building as measured by a person standing at rest next to the building?

6. An interplanetary spaceship has windows that are 3 meters wide. How fast must it pass by a planet so that an observer on that planet measures the width of the windows to be 1.5 meters?

7. You are traveling 80 km/h and you throw a ball 40 km/h with respect to yourself. What is the ball's apparent speed to a person standing by the road when the ball is thrown in each of the following ways?

    a. Straight ahead
    b. Sideways
    c. Backward

8. Calculate the Lorentz factor for objects traveling at 1%, 50%, and 99.9% of the speed of light.

9. What is the apparent mass of a 1-kg object that has been accelerated to 99% of light speed?

10. If a moving clock appears to be ticking one-half as fast as normal, at what percentage of light speed is it traveling?

11. Draw a picture illustrating how a spaceship passing Earth might look at 1%, 90%, and 99.9% of light speed.

12. If you were able to extract 100% of the energy available in 1 kilogram of hydrogen, how much energy would you have? How much energy would be available from 1 kilogram of uranium if the same 100% efficiency were attained in this extraction?

Investigations

1. Read a biography of Albert Einstein. What were his major scientific contributions? For what work did he receive the Nobel Prize?

2. Take a bathroom scale into an elevator in a tall building, stand on it, and record your weight under acceleration and deceleration. Why does the scale reading change?

3. In Chapter 6 we discussed the concept of momentum, where $p = mv$. How did Einstein redefine momentum with yet another application of the Lorentz factor? What happens to the momentum of an object as it approaches the speed of light, and what does this imply about an object's capability of being accelerated to this speed?

4. Read the novel *Einstein's Dreams* by Alan Lightman. Each of the chapters explores different time–space relationships. Which chapters teach you something about Einstein's theory of relativity?

5. Investigate the influence of Einstein's theory of relativity on twentieth-century art and philosophy.

WWW Resources

See the *Physics Matters* home page at **www.wiley.com/college/trefil** for valuable web links.

1. **http://www.aip.org/history/einstein/** The online exhibit *Albert Einstein: Image and Impact* by the American Institute of Physics.

2. **http://www.learner.org/vod/index.html?sid=42&pid=611&po=42** *The Lorentz Transformations*, a thirty-minute streamed video program from the CPB-Annenberg series *The Mechanical Universe*. Cable modem or faster connection and free registration required.

3. **http://www.learner.org/vod/index.html?sid=42&pid=613&po=43** *Velocity and Time*, a thirty-minute streamed video program from the CPB-Annenberg series *The Mechanical Universe*. Cable modem or faster connection and free registration required.

4. **http://www.walter-fendt.de/ph11e/timedilation.htm** A time dilation Java applet simulation by Walter Fendt.

5. **http://www.mira.org/fts0/s_system/161/text/txt001z.htm** A site on gravitational lenses from the Monterey Institute of research in Astronomy.

6. **http://www.curtin.edu.au/curtin/dept/phys-sci/gravity/index2.htm** The Exploring Gravity tutorial site from Australia's Curtin University of Technology.

29

Cosmology

KEY IDEA

The universe began billions of years ago in the big bang and has been expanding ever since.

PHYSICS AROUND US . . . A Glowing Charcoal Fire

Think about the last time you grilled hamburgers on a charcoal fire. If you observed closely, you may have noticed that the color of the coals in the fire changed depending on how hot the fire was. They are ordinarily red, but in a roaring blaze they can glow blue-white. Then, as the fire starts to go out, the coals glow a dull orange and, eventually, stop glowing altogether.

However, even when the coals aren't glowing, they are giving off energy in the form of infrared radiation, which you can feel if you put your hand out to the fire. Even the next day, you might still be able to feel the radiation given off by the cooling embers.

Believe it or not, a phenomenon like this charcoal fire led twentieth-century scientists to a completely new understanding of the structure and history of the universe in which we live.

GALAXIES

On any given night, as we look into the sky with modest-sized telescopes, we can see that the hazy band of the **Milky Way** is composed of countless millions of stars. Those stars appear as tiny pinpricks of light. But lots of other less distinct objects appear as fuzzy masses, too distant to resolve. Those cloudlike objects, called "nebulae," were the subject of intense debate in the early twentieth century.

The Nebula Debate

Some astronomers thought nebulae are nearby dust clouds that are illuminated by other stars. In that case, they would be fairly close by and have no resolvable structures, even using the most powerful telescope. Other astronomers suggested that nebulae are much more distant clusters of stars. They are composed of lots of individual stars, but are much too far away for those stars to be resolved.

This controversy came into sharp focus on April 26, 1920, when rival American astronomers Harlow Shapley and Heber D. Curtis engaged in a public debate at the National Academy of Sciences. The younger Shapley had come into prominence by discovering that the Milky Way is much larger than previously thought—more than 100,000 light-years across. Given that immense size, he couldn't imagine that much more distant objects existed. It's ironic that Shapley, who had dramatically increased the accepted size of the known universe, just couldn't accept how much larger the universe really is.

Curtis, on the other hand, argued that nebulae are much more distant collections of stars like our own. He saw the distinctive shapes of nebulae with spiral arms and central bulges, and concluded that a cloud wasn't likely to adopt that shape. However, Curtis believed Shapley was in error about the size of the Milky Way and actually thought the universe to be smaller than Shapley thought it is, so each man got something right and something wrong.

The more senior Curtis is said to have been the more eloquent of the two, and some attendees found him to have been more persuasive. But in science, debaters don't settle controversies. The scientific method demands that independently verifiable measurements and observations be the ultimate arbiter of any scientific question. The fact of the matter is that no one could make such observations with the tools available before 1920.

Distance and the Standard Candle

The basic problem was that astronomers were unable to tell how far away the nebulae are. This is actually a general problem in astronomy. We see the sky as a two-dimensional array, and there is no way of telling whether a star or nebula appears dim because it doesn't give off much light or because it is far away. An important tool that astronomers use to estimate distances—to supply that third dimension—is called the *standard candle*.

A standard candle is any object whose energy output is known. A 100-watt lightbulb, for example, is a splendid standard candle because we know exactly how much light it puts out each second. By comparing the known output of the standard candle to the amount of light we actually receive from it, we can tell how far away it is. We show an example of such a calculation at the end of this chapter.

Henrietta Swan Leavitt (1868–1921) headed the department of photographic stellar photometry at the Harvard College Observatory.

In astronomy, a fascinating type of star called a *Cepheid variable* is often used as a standard candle. These stars, the first of which was discovered in the constellation Cepheus, show a regular behavior of steady brightening and dimming over a period of weeks or months. Henrietta Leavitt (1868–1921) of Harvard College Observatory showed that the absolute magnitude (that is, the stars' luminosity) of these stars is related to the time it takes for them to go through the dimming-brightening-dimming sequence. Thus we can watch a Cepheid variable for a while and deduce how much energy it is pouring into space. This measurement, together with knowledge of how much energy we actually receive, tells us how far away it is.

Physics in the Making
The Women of Harvard Observatory

Edward Pickering became the director of the Harvard College Observatory in 1877. His mission, as he saw it, was to collect and categorize as many astronomical facts as possible. In particular, the observatory had piles of photographic plates of stellar spectra in its rooms, pictures of light taken from thousands upon thousands of stars. Pickering was chosen to administer a fund for creating a catalog of these spectra, which meant hiring people to carefully look over photographic plate after photographic plate and devise a system for categorizing them—a tedious task, indeed.

Pickering proceeded to hire a staff of all women assistants. Other astronomers joked about "Pickering's harem," but Pickering said he believed

Annie Jump Cannon (*foreground with magnifying glass*) and Henrietta Swan Leavitt (*sitting to the right of Cannon*) contributed important studies of the spectroscopy of stars at the Harvard College Observatory.

women were more suitable than men for this careful and repetitive work. What he did not say was that women worked for far less money than did men. They were paid 25–35 cents an hour for their work, which was actually a decent wage for women at that time. In addition, some women, including students, started to work at the observatory as unpaid volunteers so they could learn more about astronomy.

The *Henry Draper Catalogue* was published by the observatory in nine volumes, starting in 1918, and contained the spectra of 225,300 stars. But what was even more significant was the development of several women assistants into some of the most important contributors of their time to the progress of astronomy.

- Antonia Maury, a student of the first woman astronomer in the United States (Maria Mitchell), devised the first classification scheme for stars. Her work enabled later astronomers to develop the basic categories of stars, from white dwarf to red giant.

- Annie Jump Cannon developed the standard classification scheme for stellar spectra, from bright new stars to dying old ones. This sequence also turned out to be a guide to the temperatures and luminosities of stars.

- Henrietta Swan Leavitt discovered the relationship between period and luminosity of Cepheid variable stars, which became the standard candle that enabled later astronomers to measure distances to other galaxies.

It is no exaggeration to say that the work of these three women formed the foundation of modern astronomy. ●

Connection
Astronomical Distances

Miles and kilometers are not much use when trying to describe distances between stars or galaxies. It would be like trying to count the number of centimeters between New York and Boston. Instead, astronomers use distance units called the "light-year" and the *parsec*. What are these units? Is a light-year a unit of time or a unit of distance?

A light-year is not a unit of time but a unit of distance—the distance light travels in 1 year. You know that light travels very fast, at 300,000 km/s. If you do the math, you'll find out that a light-year turns out to be 9.46×10^{12} kilometers. (That's 10 trillion kilometers in round numbers.)

This may seem to you like a really enormous distance, and it is in terms of distances on Earth. But in terms of stars, even a light-year isn't that much. For example, the closest star to the Sun, called Proxima Centauri, is more than 4.2 light-years away. That means if a beam of light was emitted from Proxima Centauri when you graduated from high school, it wouldn't arrive on Earth until after you had graduated from a typical college program. And that's just the closest star, not the more distant ones.

When you start to consider galaxies, even light-years don't do the job. A typical galaxy such as the Milky Way is about 100,000 light-years across. Our nearest neighbor galaxy, called the Andromeda galaxy (located in the constellation Andromeda), is about 2.5 million light-years away. Astronomers use a base unit, the parsec, for these distances; a parsec equals 3.26 light-years. Then they build on that unit and use kiloparsecs (kpc) and megaparsecs (Mpc), which equal 1000 and 1 million parsecs, respectively. Let's face it, the universe is a pretty big place! ●

Edwin Hubble and the Discovery of Galaxies

As so often happens in science, improved instruments were the key to discovery. In this case a larger telescope, quite literally, resolved the issue. In 1900 the world's largest telescopes were reflectors with mirrors in the range of 50 to 60 inches in diameter—not large enough to reveal nebular structure. However, the Carnegie Institution of Washington decided to build a mammoth new telescope with a mirror that was an unprecedented 100 inches in diameter on Mount Wilson, near Los Angeles, California. At the time, Mount Wilson was a lonely outpost on the outskirts of a small city. Today the Los Angeles metropolitan area has almost engulfed it, but in those days it afforded astronomers a chance to look at the sky through clear, unpolluted air.

In 1919, the young American astronomer Edwin Hubble (1889–1953), fresh from distinguished service in the First World War, went to work at Mount Wilson and used this magnificent instrument to tackle the mystery of the nebula. The new Hooker telescope allowed him to see individual stars in some nebulae, which no one had been able to do before. Hubble was able to measure the distance to the nebulae by identifying some of those stars as Cepheid variables, which, as we have explained, astronomers use as a standard candle.

Edwin Hubble (1889–1953) at the 100-inch Hooker telescope of California's Mount Wilson Observatory.

It turned out that the Cepheid variable stars were extremely faint, so the distance to the nearest one, located in the Andromeda nebula, was some 2 million light years, far outside the bounds of the Milky Way. Thus, with a single observation, Hubble established one of the most important facts about the universe we live in: it is made up of billions of galaxies, of which the Milky Way is but one.

We now know that each of these countless **galaxies** is an immense collection of millions to hundreds of billions of stars, together with gas, dust, and other materials, that is held together by the forces of mutual gravitational attraction. In making these discoveries, Hubble set the tone for a century of progress in the new branch of science called **cosmology,** which is devoted to the study of the structure and history of the entire universe.

The Milky Way is a rather typical galaxy. As shown in Figure 29-1, it is a flattened disk about 100,000 light-years across. A central bulge known as the nucleus holds most of our galaxy's hundreds of billions of stars. Bright regions in the disk, known as spiral arms, mark areas where new stars are being formed. About 75% of the brighter galaxies in the sky are of this spiral type (Figure 29-2).

Galaxies such as the Milky Way can be thought of as quiet, homey places, where the process of star formation and death goes on in a stately, orderly way. But a small number of galaxies—perhaps 10,000 among the billions known—are quite different, containing unusual objects, and are referred to collectively as "active" galaxies. The most spectacular of these unusual objects are the *quasars* (for quasi-stellar radio sources). Quasars are wild, explosive, violent objects, where as yet unknown processes pour vast

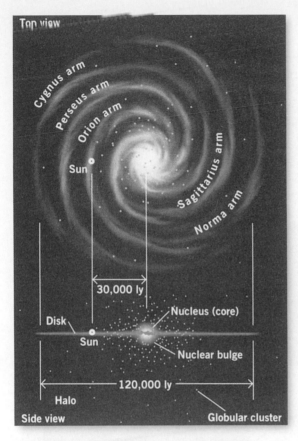

FIGURE 29-1. A map of the Milky Way galaxy, showing the nucleus and spiral arms.

FIGURE 29-2. The Whirlpool galaxy is a typical spiral galaxy, with a bright core and spiral arms where new stars are forming.

amounts of energy into space each second from an active center no larger than our solar system. Astronomers suggest that the only way to generate this kind of energy is for the center of a quasar to be occupied by an enormous black hole (see Chapter 28), with a mass millions of times greater than that of the Sun. The energy of quasars is generated by huge amounts of mass falling into this center. Because they are so bright, quasars are the most distant objects we can see in the universe. (See Looking at Astronomical Energies on page 632.)

● THE REDSHIFT AND HUBBLE'S LAW

Hubble's recognition of galaxies other than our own Milky Way wasn't the end of his discoveries. When he looked at the light from nearby galaxies, he noticed that the distinctive colors emitted by different elements seem to be shifted toward the red (long-wavelength) end of the spectrum, compared to light emitted by atoms on Earth. Hubble interpreted this **redshift** as an example of the Doppler effect (see Chapter 14), the same phenomenon that causes the sound of a car whizzing past to change its pitch. Hubble's observation meant that distant

Looking at Astronomical Energies

There is nothing in human experience to compare with some of the events astronomers see in deep space. An exploding star (supernova) can radiate more energy than the Sun has emitted in its entire 5-billion-year lifetime. The centers of active galaxies outshine entire normal galaxies, with their billions of stars. Quasars are some of the most energetic objects known and are thought to result from black holes colliding when two galaxies come together. But the ultimate energy is the Big Bang itself, the beginning of our universe.

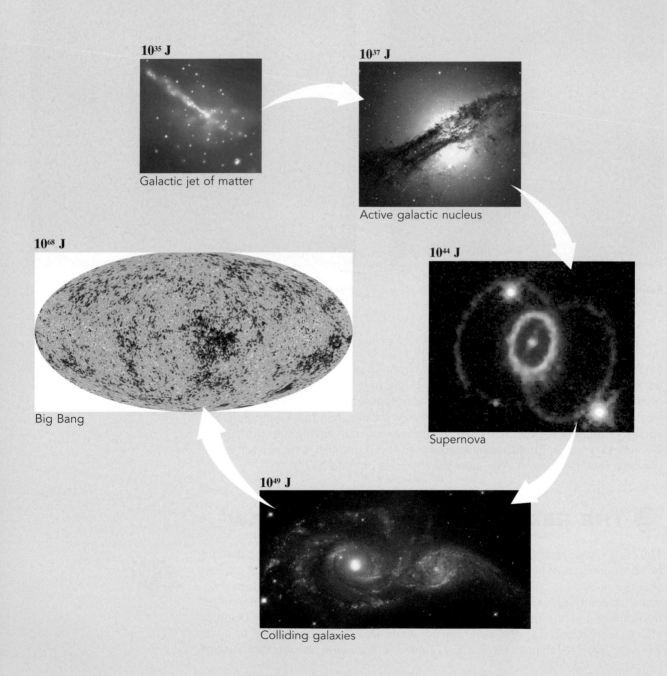

10^{35} J
Galactic jet of matter

10^{37} J
Active galactic nucleus

10^{44} J
Supernova

10^{49} J
Colliding galaxies

10^{68} J
Big Bang

galaxies are moving away from Earth. Furthermore, Hubble noticed that the more distant a galaxy, the faster it is moving away from us (Figure 29-3).

On the basis of measurements of a few dozen nearby galaxies, Hubble suggested that a simple relationship exists between the distance of an object from Earth and that object's speed away from Earth. The farther away a galaxy is, the

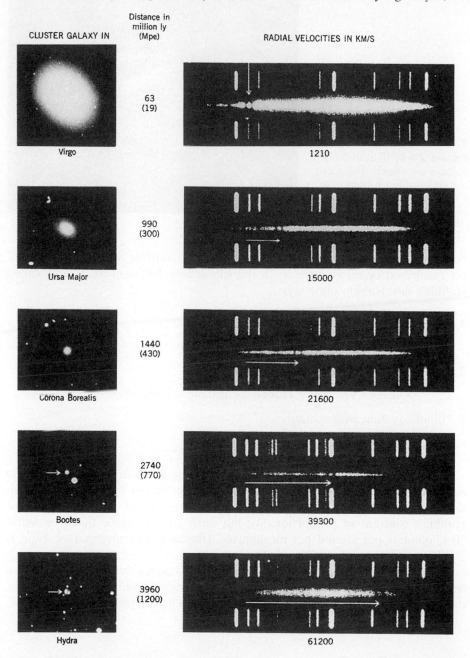

CLUSTER GALAXY IN	Distance in million ly (Mpe)	RADIAL VELOCITIES IN KM/S
Virgo	63 (19)	1210
Ursa Major	990 (300)	15000
Corona Borealis	1440 (430)	21600
Bootes	2740 (770)	39300
Hydra	3960 (1200)	61200

FIGURE 29-3. Photographs of galaxies as seen through a telescope (*left*), with spectra of those galaxies (*right*). The distance to each galaxy in megaparsecs is also given. Double dark lines in the spectra, characteristic of the calcium atom, are shifted farther to the right (toward the red) the farther away the galaxy is. Thus more-distant galaxies are traveling away from us at higher velocities. This phenomenon was used by Edwin Hubble to derive his law.

Milky Way galaxy

FIGURE 29-4. Illustration of Hubble expansion. The more distant a galaxy is from Earth, the faster it moves away from us.

faster it moves away from us (Figure 29-4). This statement, which has been amply confirmed by measurements in the subsequent half-century, is now called **Hubble's law.** Hubble's law says:

1. In words:

The farther away a galaxy is, the faster it recedes.

2. In an equation with words:

Galaxy's velocity = Hubble's constant × Distance to galaxy

3. In an equation with symbols:

$$v = H \times d$$

Hubble's law tells us that we can determine the distance to galaxies by measuring the redshift of the light we receive, whether or not we can make out individual stars in them. Astronomers continue to debate the exact value of Hubble's constant of proportionality, but most experts agree that it is about 70 kilometers per second per megaparsec. (Recall that a megaparsec, Mpc, is 1 million parsecs, or 3.3 million light-years.)

One way of interpreting Hubble's constant is to notice that if a galaxy were to travel from the location of the Milky Way to its present position with a velocity v, then the time it would take to make the trip would be distance divided by speed:

$$t = \frac{d}{v}$$

Substituting for v from Hubble's law,

$$t = \frac{d}{H \times d}$$

$$= \frac{1}{H}$$

Thus the Hubble constant provides a rough estimate of the time that the expansion has been going on and, hence, of the age of the universe. The current best estimate of the age of the universe is 13.7 billion years.

Connection
Analyzing Hubble's Data

In his original sample, Hubble observed 46 galaxies, but was able to determine distances to only 24. Some of his data are given in Table 29-1.

How does one go about analyzing data such as these? One common way is to make a graph. In this case (Figure 29-5), the vertical axis is the velocity of recession of the galaxy and the horizontal axis is the distance to the galaxy. Figure 29-5a shows the data as originally plotted by Hubble, while Figure 29-5b presents a more recent compilation of many galaxies.

Looking at the original data, we see that the general trend of Hubble's law is obvious—the farther you go to the right (i.e., the farther away the galaxies are), the higher the points (i.e., the faster the galaxies are moving away). You also notice, however, that the points do not fall on a straight line but are scattered. Confronted with this sort of situation, you can do one of two things. You can assume that the scattering is due to experimental error and that more accurate experiments will verify that the points fall on a straight line; or you can assume that the scatter is a real phenomenon and try to explain it. Hubble took the first alternative, so the only problem left was to find the line about which experimental error was scattering his data.

The way this step is usually done is to find the line that comes closest to all the data points. The slope of this line, which measures how fast the velocity increases for a given change in distance, is the best estimate of Hubble's constant. ●

TABLE 29-1	Some of Hubble's Data
Distance to Galaxy (megaparsecs)	**Velocity (km/s)**
1.0	620
1.4	500
1.7	960
2.0	850
2.0	1090

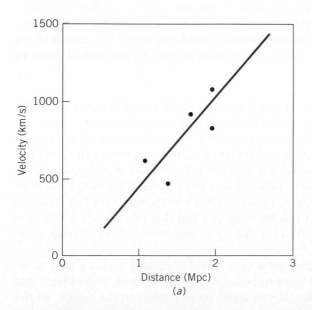
(a)

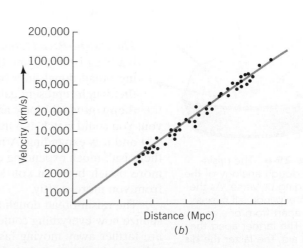
(b)

FIGURE 29-5. (a) Hubble's original distance-versus-velocity relationship. (b) A modern version, using many more galaxies.

THE BIG BANG

Hubble's law reveals an extraordinary aspect of our universe: it is expanding. Nearby galaxies are moving away from us and faraway galaxies are moving away even faster. The whole thing is blowing up like a balloon. This startling fact leads us, in turn, to perhaps the most amazing discovery of all. If you look at our universe expanding today and imagine moving backward in time (think of running a videotape in reverse), you can see that at some point in the past the universe must have started out as a very small object. In other words:

The universe began at a specific time in the past
and has been expanding ever since.

This picture of the universe—that it began from a small, dense collection of matter and has been expanding ever since—is called the **big bang theory.** This theory constitutes our best idea of what the early universe was like.

Think how different the big bang theory of the universe is from the theories of the Greeks or the medieval scholars or even the great scientists of the nineteenth century whose work we have studied. To them, Earth went in stately orbit around the Sun, and the Sun moved among the stars, but the collection of stars you can see at night with your naked eye or with a telescope was all that there was. Suddenly, with Hubble's work, the universe grew immeasurably. Our own collection of stars, our own galaxy, is just one of perhaps 100 billion known galaxies in a universe in which galaxies are flying away from one another at incredible speeds. It is a vision of a universe that began at some time in the distant past and will, presumably, end at some time in the future.

Some Useful Analogies

The big bang picture of the universe is so important that we should spend some time thinking about it. Many analogies can be used to help us picture what the expanding universe is like, and we'll look at two. Be forewarned, however: none of these analogies is perfect. If you pursue any of them far enough they fail, because none of them captures the entirety and complexity of the universe in which we live. And yet each of the analogies can help us understand some aspects of that universe.

FIGURE 29-6. The raisin-bread dough analogy of the expanding universe. As the dough expands, all raisins move apart from one another. The farther apart the raisins are, the faster the distance increases.

The Raisin-Bread Dough Analogy One standard way of thinking about the big bang is to imagine the universe as being analogous to a huge vat of rising bread dough in a bakery (Figure 29-6). If raisins scattered throughout the dough represent galaxies, and if you're standing on one of those raisins, then you would look around you and see other raisins moving away from you. You could watch as a nearby raisin moves away because the dough between you and it is expanding. A nearby raisin wouldn't be moving very fast, because there isn't much expanding dough between you and that raisin. Because there is more dough between you and more distant raisins, they will be moving away from you more swiftly.

The raisin-bread dough analogy is very useful because it makes it easy to visualize how everything could seem to be moving away from us, with objects that are farther away moving faster. If you stand on any raisin in the dough, all the other raisins look as though they're moving away from you. This analogy thus

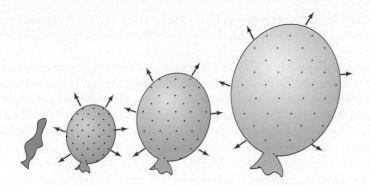

FIGURE 29-7. The expanding-balloon analogy of the universe. All points on the surface of the expanding balloon move away from one other. The farther apart the points, the faster they move apart.

explains why Earth seems to be the center of the universe. It also explains why this fact isn't significant—*every* point appears to be at the center of the universe.

But the expanding dough analogy fails to address one of the most commonly asked questions about the Hubble expansion: what is outside the expansion? A mass of bread dough, after all, has a middle and an outer surface; some raisins are nearer the center than others. But we believe that the universe has no surface, no outside and inside, and no unique central position. In this regard, the surface of an expanding balloon provides a better analogy.

The Expanding-Balloon Analogy Imagine that you live on the surface of a balloon in a two-dimensional universe. You would be absolutely flat, living on a flat-surface universe (similar to the way we are three-dimensional, living in a three-dimensional universe). Evenly spaced points cover the balloon's surface, and one of these points is your home. As the inflating balloon expands, you observe that every other point moves away from you— the farther away the point, the faster away it moves (Figure 29-7).

Where is the edge of the balloon? What are the inside and outside of the balloon in two dimensions? The answers, at least from the perspective of a two-dimensional being on the balloon's surface, are that every point appears to be at the center, and the universe has no edges, no inside, and no outside. The two-dimensional being experiences one continuous, never-ending surface. We live in a universe of higher dimensionality, but the principle is the same: our universe has no center and no inside versus outside.

The balloon analogy is also useful because it can help us visualize another question that is often asked about the expanding universe: what is it expanding *into*? If you think about being on the balloon, you realize that you could start out in any direction and keep traveling. You might come back to where you started, but you would never come to an end. There would never be an "into." The surface of a balloon is an example of a system that is bounded, but that has no "outside" (as seen in two dimensions). Similarly, the universe includes all space and is not expanding into anything.

Evidence for the Big Bang

In Chapter 1 we pointed out that every scientific theory must be tested and have experimental or observational evidence backing it up. The big bang theory provides a comprehensive picture of what our universe might be like, but are there sufficient observational data to support it? In fact, three pieces of evidence make the big bang idea extremely compelling to scientists: universal expansion, the cosmic microwave background, and the relative abundances of light elements.

The Universal Expansion Edwin Hubble's observation of universal expansion provided the first strong evidence for the big bang theory. If the universe began from a compact source and has been expanding, then you would expect to see the expansion going on today. The fact that we do see such an expansion is taken as evidence for a big bang event in the past. It is not, however, conclusive evidence. Many other theories of the universe have incorporated an expansion, but not a specific beginning in time. During the 1940s, for example, scientists proposed a steady-state universe. Galaxies in this model move away from one another, but new galaxies are constantly being formed in the spaces that are being vacated. Thus the steady-state model describes a universe that is constantly expanding and forming new galaxies, but with no trace of a beginning.

Because of the possibility of this kind of theory, the universal expansion, in and of itself, does not compel us to accept the big bang theory.

The Cosmic Microwave Background In 1964, Arno Penzias and Robert W. Wilson, two scientists working at Bell Laboratories in New Jersey, used a primitive radio receiver to scan the skies for radio signals. Their motivation was simple. They worked during the early days of satellite broadcasting, and they were measuring microwave radiation to document the kinds of background signals that might interfere with radio transmission. They found that whichever way they pointed their receiver, they heard a faint hiss in their apparatus. There seemed to be microwave radiation falling on Earth from all directions. We now call this radiation the **cosmic microwave background radiation.**

At first, Penzias and Wilson suspected that this background noise might be a fault in their electronics or even interference caused by droppings from a pair of pigeons that had nested inside their funnel-shaped microwave antenna. However, a thorough testing and cleaning made no difference in the odd results. A constant influx of microwave radiation of wavelength 7.35 centimeters flooded Earth from every direction in space. And so the scientists asked where is this radiation coming from?

In order to understand the answer to their question, you need to remember that every object in the universe that is above the temperature of absolute zero emits some sort of radiation (see Chapter 11). As we saw in the Physics Around Us section that opens this chapter, a coal in a fire may glow white-hot and emit the complete spectrum of visible electromagnetic radiation. As the fire cools, it gives out light that is first concentrated in the yellow, then orange, and eventually dull red range. Even after it no longer glows with visible light, you can tell that the coal is giving off radiation by holding out your hand to it and sensing the infrared or heat radiation that still pours from the dying embers. As the coal cools still more, it gives off wavelengths of longer and longer radiation.

One way to think about the cosmic microwave background, then, is to imagine that you are inside a cooling coal on a fire. No matter which way you look, you'll see radiation coming toward you, and that radiation shifts from white to orange to red light and, eventually, all the way down to microwaves as the coal cools.

In 1964, a group of theorists at Princeton University (not far from Bell Laboratories) pointed out that if the universe had indeed begun at some time in the past, then today it would still be giving off electromagnetic radiation in the microwave range. In fact, the best calculations at the time indicated that the radiation would be characteristic of an object at a few degrees above absolute zero. When Penzias and Wilson got in contact with these theorists, the reason they couldn't get rid of the microwave signal became obvious. Not only was it a real signal, it was evidence for the big bang itself. For their discovery, Penzias and Wilson shared the Nobel Prize in physics in 1978—not a bad outcome for a measurement designed to do something else entirely!

We have said before that it is possible to imagine theories, such as the steady-state theory, in which the universe is expanding but has no beginning. However, it is impossible to imagine a universe that does not have a beginning but that produces the kind of microwave background we're talking about. Thus Penzias and Wilson's discovery put an end to the steady-state theory.

In 1989, the Cosmic Background Explorer satellite measured the microwave background to extreme levels of accuracy. The purpose of this measurement was to see, in great detail, whether the predictions of the big bang theory about the nature of the cosmic microwave background radiation were correct. These data established beyond any doubt that we live in a universe where the average temperature is 2.7 kelvins. This finding reaffirmed the validity of the big bang theory in the minds of scientists.

In 2003, new data from a satellite called the Wilkinson Microwave Anisotropy Probe (WMAP) was released. Launched in June of 2001, this satellite has been collecting data of unprecedented accuracy by making extremely detailed measurements of small differences in the microwaves reaching Earth from different directions in space. Many of the detailed results about the age and composition of the universe used in this chapter come from an analysis of the WMAP data.

The Abundance of Light Elements The third important piece of evidence for the big bang theory comes from studies of the abundances of light nuclei in the universe. For a short period in the early history of the universe, as we see at the end of this chapter, atomic nuclei could form from elementary particles. Cosmologists believe that the only nuclei that could have formed in the big bang are isotopes of hydrogen, helium, and lithium (the first three elements, with one, two, and three protons in their nuclei, respectively). All elements heavier than lithium were formed later in stars.

The conditions necessary for the formation of light elements were twofold. First, matter had to be packed together densely enough to allow collisions that would produce a fusion reaction (see Chapter 26). Second, the temperature had to be high enough for those reactions to happen, but not so high that nuclei created by fusion would be broken up in subsequent collisions. In an expanding universe, the density of matter decreases rapidly because of the expansion, and each type of nucleus can form only in a very narrow range of conditions. Calculations based on density and collision frequency, together with known nuclear reaction rates, make rather specific predictions about how much of each isotope could have been made before matter spread too thinly. Thus the cosmic abundances of elements such as deuterium (the hydrogen isotope with one proton and one neutron in its nucleus), helium-3 (the helium isotope with two protons and one neutron), and helium-4 (with two protons and two neutrons) comprise another test of our theories about the origins of the universe.

In fact, studies of the abundances of these isotopes find that they agree quite well with the predictions made in this way. The prediction for the primordial abundance of helium-4 in the universe, for example, is that it cannot have exceeded 25% of all atoms. Observations of helium abundance are quite close to this prediction. If the abundance of helium differed by more than a few percent from this value, the theory would be in serious trouble.

THE EVOLUTION OF THE UNIVERSE

Our vision of an expanding universe leads us to peer back in time, to the early history of matter and energy. What can we say about the changes that must have taken place during the past 15 billion years?

Some General Characteristics of an Expanding Universe

Have you ever pumped up a bicycle tire with a hand pump? If you have, you may have noticed that after you've run the pump for a while, the barrel gets very hot. All matter heats up when it is compressed.

The universe is no exception to this rule. A universe that is more compressed and denser than the one in which we live would also be hotter on average. In such a universe, the cosmic background radiation would correspond to a temperature much higher than 2.7 K (which is what it is today), and the wavelength of the background radiation would be shorter than 7.35 cm.

When the universe was younger, it must have been much hotter and denser than it is today. This cardinal principle guides our understanding of how the universe evolved. In fact, the big bang theory we have been discussing is often called the "hot big bang" to emphasize the fact that the universe began in a very hot, dense state and has been expanding and cooling ever since.

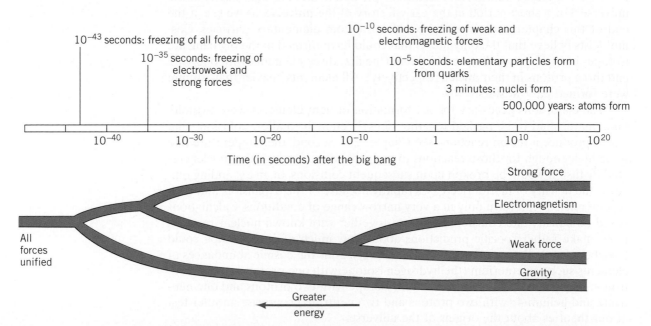

FIGURE 29-8. The sequence of "freezings" in the universe since the big bang. The earliest freezings involve the splitting of forces, while later freezings involve forms of matter.

In Chapter 9 we saw that changes of temperature may correspond to changes of state in matter. For example, if you cool water it eventually turns into ice at the freezing point. In just the same way, modern theories claim, as the universe cooled from its hot origins, it went through changes of state very much like the freezing of water. We refer to these dramatic changes in the fabric of the universe in technical language as "phase transitions," or, more colloquially, as *freezings,* even though they are not actually changes from a liquid to a solid state.

A succession of freezings dominates the history of the universe. Six distinct episodes occurred, and each had its own unique effect on the universe that we live in. Between each pair of freezings was a relatively long period of steady and rather uneventful expansion. Once we understand these crucial transitions in the history of the universe, we will have come a long way toward understanding why the universe is the way it is.

Let's look at the transitions in order, from the earliest to the most recent, as summarized in Figure 29-8. The first set of freezings involves the unification of forces we discussed in Chapter 27, and the last set involves the coming together of particles to form more complex structures, such as nuclei and atoms.

10^{-43} Second: The Freezing of All Forces

Cosmologists calculate that the first freezing after the beginning of the universe took place at about 10^{-43} second. (This is a really small number: 0.0000000000000000000000000000000000000000001 second!) Before this time, there was only a single unified force. At 10^{-43} second, the gravitational force split off from the strong–electroweak force, so there were two fundamental forces acting in nature.

We can't reproduce in our laboratories the unimaginably high temperatures that existed at this freezing, and we do not have successful theories that describe the unification of gravity with the other forces. Thus, this earliest freezing remains both the theoretical frontier and the limit of our knowledge about the universe at the present time.

10^{-35} Second: The Freezing of the Electroweak and Strong Forces

Unified field theories that describe the behavior of all matter and forces (see Chapter 27) tell us that before the universe was 10^{-35} second old, the strong force was unified with the electroweak force. At 10^{-35} second, the strong force split off from the electroweak force. That is, before this time there were only two fundamental forces acting in the universe (the strong–electroweak force and the gravitational force), but after this time there were three.

Two important events are associated with the freezing at 10^{-35} seconds: the elimination of antimatter and inflation.

The Elimination of Antimatter Antimatter (see Chapter 27) is fairly rare in the universe we live in. It wasn't until the twentieth century that scientists were able to identify antimatter, and we presently have compelling evidence that no large collections of antimatter exist anywhere in the universe. For example, spaceships have landed on the Moon, Mars, and Venus, and if any of those bodies had been made of antimatter, those spaceships would have been annihilated in a massive burst of gamma rays. Because they were not, we conclude that none of those planets is made of antimatter.

By the same token, the solar wind, composed of ordinary matter, is constantly streaming outward from the Sun to the farthest reaches of the solar system. If any objects in the solar system were made of antimatter, the protons in the solar wind would be annihilating with materials in that body and we would see evidence of it. The entire solar system, therefore, is made of ordinary matter. By the same type of argument, scientists have been able to show that our entire galaxy is made of ordinary matter and that no clusters of galaxies anywhere in the observable universe are made of antimatter.

The question, then, is this: if antimatter is indeed simply a mirror image of ordinary matter and if antimatter appears in our theories on an equal footing with ordinary matter, as it does, then why is there so little antimatter in the universe?

Unified field theories give us an explanation of this striking feature of the cosmos. In experiments, we find one instance of a particle that decays preferentially into matter over antimatter—a particle whose decay products more often contain more matter than antimatter. This particle is called the K^0_L ("K-zero-long"), one of the many heavy particles (such as protons and neutrons) that were discovered in the second half of the twentieth century.

If you take the theories that are successful in explaining this laboratory phenomenon and extrapolate them to the very early universe, you find that there were about 100,000,001 protons made for every 100,000,000 antiprotons. In the maelstrom that followed the big bang, the 100,000,000 antiprotons annihilated with 100,000,000 protons, leaving only a sea of intense radiation to mark their presence. From the collection of leftover protons, all the matter in the universe (including Earth and its environs) was made. This discovery allowed physicists to explain the puzzling absence of antimatter in the universe, and in the 1980s led to a burst of interest in the evolution of the early universe.

Inflation According to the most widely accepted versions of the unified field theories, the freezing at 10^{-35} second was accompanied by an incredibly rapid (but short-lived) increase in the rate of expansion of the universe. This short period of rapid expansion is called *inflation,* and theories that incorporate this phenomenon are called "inflationary theories."

One way to think about inflation is to remember that changes in volume are often associated with changes of state. Water, for example, expands when it freezes, which explains why water pipes may burst open when the water in them freezes in very cold weather. In the same way, scientists argue, the universe underwent a period of very rapid expansion during the time when the strong force froze out from the electroweak. Roughly speaking, at this time the universe went from being much smaller than a single proton to being about the size of a grapefruit, an incredible increase in size of about 10^{50} times (Figure 29-9).

Inflation explains another puzzling feature of the universe. We have repeatedly observed that the cosmic microwave background (which is an index of the temperature of the universe) is remarkably uniform. The temperatures associated with microwaves coming from one region of the sky differ from those coming from another region by no more than 1 part in 1000. But calculations based on a uniform rate of expansion say that different parts of the universe would not have been close enough together to establish a common temperature.

In the inflationary theory, the resolution of this problem is simple. Before 10^{-35} second, all parts of the universe were in contact with one another because the universe was much smaller than you would have guessed based on a uniform

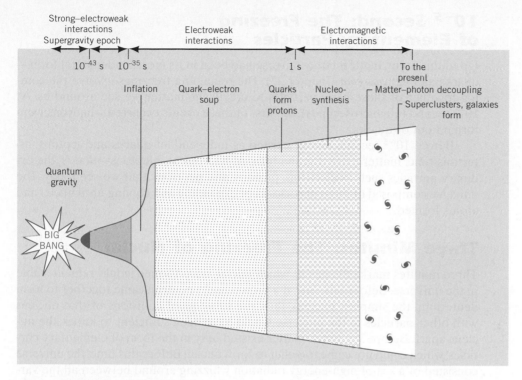

FIGURE 29-9. The evolution of the universe through the succession of freezings. Note the rapid expansion associated with the inflationary period.

rate of expansion. There was time to establish equilibrium before inflation took over and increased the size of the universe. The temperature equilibrium, established early, was preserved through the inflationary era and is seen today in the uniformity of the microwave background.

Thus the coming together of the theories of elementary particle physics and the study of cosmology has produced solutions to long-standing problems and questions about the universe.

10^{-10} Second: The Freezing of the Weak and Electromagnetic Forces

Before 10^{-10} second (that's 1 ten-billionth of a second), the weak and the electromagnetic forces were unified. In other words, before 10^{-10} second, there were only three fundamental forces operating in the universe. These were the strong, gravitational, and electroweak forces. After 10^{-10} second, the full complement of four fundamental forces was present.

The time of 10^{-10} second also marks another milestone in our discussion of the evolution of the universe. The modern particle accelerators of high-energy physics can just barely reproduce the incredible concentration of energy associated with that event. This means that from this point forward it is possible to have direct experimental checks of the theories that describe the evolution of the universe.

10^{-5} Second: The Freezing of Elementary Particles

Up to this point, matter in the universe had been in its most fundamental form— quarks and leptons (see Chapter 27). The remaining freezings involve the coming together of those basic particles to create the matter we see around us. At 10^{-5} second (10 microseconds), the first of these events occurred—hadrons were formed out of quarks.

Before 10^{-5} second, matter existed as independent quarks and leptons; after this time, matter existed in the form of hadrons and leptons—that is, the ordinary particles such as electrons, protons, and neutrons that we see today. The universe composed of these particles kept expanding and cooling until nuclei and atoms formed.

Three Minutes: The Freezing of Nuclei

Three minutes marks the age at which nuclei, once formed, could remain stable in the universe. Before this time, if a proton and a neutron came together to form deuterium, the simplest nucleus, then the subsequent collisions of that nucleus with other particles in the universe would have been sufficient to knock the nucleus apart. Before 3 minutes, matter existed only in the form of elementary particles, which could not come together to form nuclei. Before this time, the universe consisted of a sea of high-energy radiation whizzing around between all the various species of elementary particles we discussed in Chapter 27.

At 3 minutes, a short burst of nucleus formation occurred, as we have discussed earlier. Thus from 3 minutes on, the universe was littered with nuclei, which formed part of the plasma, the hot fluid mixture of electrons and simple nuclei that was the material of the early universe. Remember, however, that only nuclei of the light elements—hydrogen, helium, and lithium—were made at this time.

Before One Million Years: The Freezing of Atoms

The most recent transition occurred gradually between the time that the universe was a few hundred thousand and 1 million years old. At this time, the background temperature of the hot, dense universe was so great that electrons could not settle within atomic orbits to form atoms. Even if an atom formed by chance, its subsequent collisions were sufficiently violent that the atom could not stay together. Thus all of the universe's matter was in the form of plasma.

This freezing of atoms marks an extremely important point in the history of the universe because it is a point at which radiation such as light was no longer locked into the material of the universe. You know from experience that light can travel long distances through the atmosphere (which is made of atoms). But light cannot travel freely through plasma, which quickly absorbs light and other forms of radiation. Thus when atoms formed, the universe became transparent and radiation was released. It is this radiation, cooled and stretched out, that we now see as the cosmic microwave background. Thus, the cosmic microwave background is a picture of the universe at this time.

The formation of atoms marks an important milestone for another reason. Before this event, if clumps of matter happened to begin forming (under the influence of gravity, for example), they would absorb radiation and be blown apart. This means that there must have been a window of opportunity for the formation of galaxies. If galaxies are made of ordinary matter, they couldn't have started to come together out of the primordial gas cloud until atoms had formed, about 500,000 years after the big bang. By that time, however, the Hubble expansion had spread matter out so thinly that the ordinary workings of the force of gravity would not have been able to make a universe of galaxies, clusters, and superclusters. Known as the "galaxy problem," this puzzle remains the great riddle that must be answered by cosmologists.

Another way of stating this problem is to compare the clumpiness of matter in the universe to the smoothness of the cosmic background radiation. The background radiation seems to be pretty much the same no matter which way you look in the universe. This uniformity argues that the universe had a smooth, regular beginning. How can this statement be reconciled with the lumpy structure we see when we look at the distribution of matter?

 DARK MATTER

As complete as this history of the early universe may seem, significant gaps in our understanding of the evolution of the universe remain. Some of these gaps were closed with the development of unified field theories and the inflationary scheme of the universe. However, the problem of explaining the existence of galaxies, clusters, and superclusters remains.

It now appears that all the impressive luminous objects in the sky constitute less than 10% of the matter in the universe, perhaps a good deal less than 10%. The rest of the matter exists in forms that we cannot see, but whose effects we can measure. This mysterious new kind of material is called **dark matter.**

The easiest place to see evidence for dark matter is in galaxies such as our own Milky Way. Far out from the stars and spiral arms that we normally associate with galaxies, we can still see a diffuse cloud of hydrogen gas. This gas gives off radio waves, so we can detect its presence and its motion. In particular, we can tell how fast it is rotating. When we do these sorts of measurements, a rather startling fact emerges. In Chapter 3 we saw that Kepler's laws implied that any object orbiting around a central body under the influence of gravity will travel slower the farther out it is. The distant planet Jupiter, for example, moves more slowly in orbit around the Sun than does Earth. Similarly, you would expect that when hydrogen molecules are far enough away from the center of a galaxy, these more distant atoms would move more slowly than those closer in. Even though we can see these hydrogen atoms out to distances three times and more the distance from the center of the Milky Way to the end of the spiral arms, no one has ever seen the predicted slowing down.

The only way to explain this phenomenon is to say that those hydrogen atoms are still in the middle of the gravitational influence of the galaxy. This means that luminous matter—the bright stars and spiral arms—is not the only thing that is exerting a gravitational force. Something else, something that makes up at least 90% of the mass of the galaxy and that extends far beyond the stars, exerts a gravitational force and affects the motion of the hydrogen we observe. Studies of other galaxies show the same effect and scientists have found evidence that dark matter exists in the voids between galaxies as well. The most recent estimates tell us that about 4% of the matter in the universe is ordinary stuff such as that found in stars, and that about 23% is in the form of dark matter. The rest of the matter in the universe is in a newly discovered form called "dark energy," which we discuss in the next section.

Dark matter is strange, indeed. It does not interact through the electromagnetic force. If it did, it would absorb or emit photons and it wouldn't be "dark" in the sense we're using the term here. Yet because we know that it exerts a gravitational attraction, we can conclude that this unseen stuff must be a form of matter—matter that interacts with ordinary matter only through the gravitational force. Detecting dark matter, and finding out what it is, remains a very active research field today.

The existence of dark matter might help us understand a key event in the early history of the universe, because dark matter could have formed into clumps before atoms formed. In the first several hundred thousand years after the big bang, photons blew apart collections of luminous matter that were trying to form galaxies, but light would not have affected the clumping of dark matter. Therefore, when atoms formed and luminous matter could clump together, that matter found itself in a universe in which large clusterings of dark matter already existed. The luminous matter would simply have fallen into these clusters and would not have had to form under the influence of its own gravitational attraction. Thus, if dark matter exists and if dark matter formed clumps early in the history of the universe, the problem of structure is solved.

DARK ENERGY AND THE END OF THE UNIVERSE

When most people think about the Hubble expansion they wonder whether the universe will continue to expand forever or will someday fall back in on itself. In the case of eternal expansion, astronomers say that the universe is open; in the case of eventual collapse, they say that the universe is closed. Finally, in the intermediate case between the two, in which universal expansion slows down, but never quite stops, the universe is said to be flat. Can we tell which kind of universe we live in?

The way that scientists have approached this question is to note that the force pulling back on distant galaxies is the gravitational attraction associated with the mass of the universe. If there is enough mass, the outward motion of the galaxy will be slowed down and reversed. If there isn't enough mass, the expansion will be slowed but never stopped. The luminous matter of the universe—stars, nebulae, and dust clouds—is only about 0.1% of the mass needed to close the universe. Even counting dark matter brings us only up to about 20% to 30% of the needed value. Just from counting mass, then, we would conclude that we live in an open universe.

In 1999, astronomers announced a surprising result that confirms this conclusion. Using a new standard candle called a Type Ia supernova, they have measured the distances to the most distant galaxies. Using the fact that light from those galaxies has been traveling toward us for billions of years, they can compare the rate of expansion of the universe as it was long ago to what it is today. The surprising result is that the expansion of the universe isn't slowing down at all—in fact, it's speeding up! This observation suggests that there may be a new kind of force acting over the vast distances between the galaxies. This force seems to behave like a kind of antigravity that pushes galaxies apart. Cosmologists call this new discovery **dark energy.** As is the case with dark matter, we don't yet know what it is, and because of its recent discovery dark energy is, at the moment, even more mysterious than dark matter. Whatever it is, however, the results from the WMAP program previously discussed indicate that it makes up approximately 73% of the mass of the universe. (Remember that mass and energy are related through Einstein's equation $E = mc^2$).

Although it may seem that cosmologists are piling mystery upon mystery with these new discoveries, in fact each new mystery gets us closer to an understanding of our universe. The process is a little like baking a cake—you have to know what all the ingredients are before you can succeed. With the discovery of dark energy, cosmologists believe that they have completed the list of ingredients and can get down to the job of figuring out how they all fit together.

THINKING MORE ABOUT

Cosmology: The History of the Universe

The story of the big bang that we have just recounted has one clear feature: there were no human beings around to observe any of the events we've just described. In 1999, creationists on the Kansas Board of Education used this fact as a reason to ban questions about the big bang from statewide high school scientific achievement tests. Let's think for a moment about the kind of evidence we require to establish the existence of events in the past.

How do you know there was an event called the American Civil War? No one alive today actually took part in the Civil War, yet no one suggests that we should doubt its existence. The reason is that there is all sorts of evidence in the form of texts, artifacts, documents, and even recorded stories told by survivors before they died. The weight of this evidence is so overwhelming that the existence of the war is universally accepted.

But what about events farther back in time—the Crusades, for example, or the Thirty Years War? The evidence here is weaker than that for the Civil War. What about events that occurred before the invention of writing—the arrival of the first humans in North America, for example? Here the evidence is exclusively in the form of archaeological data. And what about geological events where the evidence is in the rocks themselves?

The evidence for the big bang has been outlined in this chapter. How does it compare to the evidence for other events in the past? How much evidence is required to establish the existence of such events? Why do you suppose so few scientists agree with the decision of the Kansas school board? (By the way, this decision was reversed by a new school board elected in 2002.)

Summary

Early in the twentieth century, Edwin Hubble made two extraordinary discoveries about the structure and behavior of the universe, the science we call **cosmology.** First, he demonstrated that our home, the collection of stars known as the **Milky Way,** is just one of countless **galaxies** in the universe, each containing billions of stars. By measuring the **redshift** of galaxies, he also discovered that these distant objects are moving away from one another. According to **Hubble's law,** the farther the galaxy, the faster it is moving away. This relative motion implies that the universe is expanding.

One theory that accounts for universal expansion is the **big bang theory**—the idea that the universe began at a specific moment in time and has been expanding ever since. Evidence from the **cosmic microwave background radiation** and the relative abundances of light elements, in addition to expansion, support the big bang theory.

At the moment of creation, all forces and matter were unified in one unimaginably hot and dense volume. As the universe expanded, however, a series of six "freezings" led to the universe we see today. Freezings at 10^{-43} second, 10^{-35} second, and 10^{-10} second caused a single unified force to split progressively into the four forces we observe today: the gravitational, strong, electromagnetic, and weak forces. At that early stage of the universe, when all matter and energy were contained in a volume no larger than a grapefruit, matter was in its most elementary form of quarks and leptons.

At 10^{-5} second, the quarks bonded together to form heavy nuclear particles such as protons and neutrons. Subsequent freezings saw these particles first fuse into nuclei at 3 minutes and ultimately join with electrons to form atoms at 500,000 years. Stars, which formed from those atoms, then could begin the processes that provided all the other chemical elements.

Understanding **dark energy** that pushes galaxies apart and the search for **dark matter**—mass that we cannot see with our telescopes—is a research frontier that may help us determine whether or not the universe will continue expanding forever.

Key Terms

big bang theory The theory that the universe was, at one time, a very small, dense collection of matter and energy that has been expanding ever since. (p. 636)

cosmic microwave background radiation The electromagnetic radiation from space that is thought to be a remnant of the big bang. (p. 638)

cosmology The branch of science devoted to the study of the history and structure of the universe. (p. 630)

dark energy Newly discovered energy that has the effect of pushing galaxies apart. (p. 647)

dark matter The unseen matter in the universe that has been postulated to account for the observed structure of the universe. (p. 645)

galaxy a collection of gas, dust, and millions or billions of stars all held together by gravity; there are billions of galaxies in the universe. (p. 630)

Hubble's law The law that relates the distance between Earth and a galaxy to the galaxy's recession speed. (p. 634)

Milky Way The galaxy that is home to Earth and our solar system. (p. 627)

redshift The shift toward lower frequencies of the spectra of galaxies moving away from us. (p. 633)

Key Equation

Hubble's Law: Galaxy's velocity = Hubble's constant × Distance to galaxy

Review

1. What is cosmology? How does it differ from astronomy?

2. What is a standard candle? How can it be used to measure distances between stars?

3. Why are Cepheid variables frequently used as a standard candle?

4. What is a galaxy? How does a galaxy differ from a star?

5. What galaxy do we live in? How large is it?

6. How did Edwin Hubble discover that there are galaxies in the universe other than the Milky Way?

7. What does the term "quasar" stand for? What is thought to be at the center of these types of galaxies?

8. How does the light we see from distant stars become redshifted? What does this redshift say about the universe?

9. Describe Hubble's law. How did Hubble discover it?

10. What is the best current estimate of the age of the universe? How was this estimated?

11. What is the big bang theory? Why is it sometimes called the "hot big bang"?

12. Two analogies were presented to explain the expansion of the universe. One presents the expanding universe as expanding dough filled with raisins, while the other is an analogy of the universe as an expanding balloon. What are the strengths and weaknesses of each of these analogies?

13. What kinds of evidence support the big bang theory?

14. How did the now-discarded steady-state theory explain the expansion of the universe?

15. Why was the steady-state theory of the universe abandoned? How does this episode fit into the discussion of the scientific method in Chapter 1?

16. What is the significance of the discovery by Penzias and Wilson of cosmic microwave background radiation?

17. What is the average temperature of the universe we live in? How do we know this?

18. If the universe were hotter than the temperature in Review Question 17, would the universe be older or younger? Similarly, what would the relative age be if the universe were colder than this?

19. Which elements do cosmologists think were formed in the big bang? How abundant are these elements today, and what does this imply?

20. Make a table with ages of the universe in the left column (10^{-43} seconds, 10^{-35} seconds, 10^{-10} seconds, 10^{-5} seconds, 3 minutes, 500,000 years) and the major events in the history of the universe in the right column.

21. Why is the universe composed of matter and not antimatter?

22. What is inflation? When did it occur, and what puzzling feature of the universe does it help explain?

23. What do we mean when we say that light was locked into the material of the universe before atoms formed? When was this light released?

24. What is the galaxy problem?

25. Dark matter probably makes up over 90% of the matter in the universe. What is it and what evidence is there that it exists?

26. How can dark matter help provide an explanation for the clumpiness of matter in the universe?

27. What will be the fate of the universe? Is the universe open, closed, or flat?

28. What measurements or observations contribute to the question of whether the universe is closed or open?

29. Some advances in our knowledge have been made possible through better equipment, such as Hubble's discoveries using the 100-inch Hooker telescope at Mount Wilson. What other major discoveries in cosmology have relied on improvements in existing apparatus?

30. Describe how the universe is moving at present. Is it accelerating or decelerating?

31. What is dark energy?

Questions

1. Why does Earth seem to be at the center of the Hubble expansion?

2. If the universe is closed, describe the results that some future Hubble will get when he looks through a telescope during the period of contraction. Will he still see other galaxies? Will he still see a redshift?

3. Suppose that scientists were able to travel to a different universe. The figure shows a distance-versus-velocity graph for galaxies measured in the two universes. The solid line represents the data for our universe; the dotted line is the data for the new universe. Which universe is older? Explain.

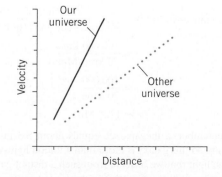

4. If a life form on a planet in a distant galaxy measured the Hubble constant from its location, would you expect that it would get the same value that we measure here from Earth? (We'll assume that we use the same units of measurement.) Why would a different Hubble constant cause scientists to question our current understanding of the universe?

5. Suppose that a new experiment showed that the wavelength of the cosmic background radiation were slightly shorter than its previously measured value. How would that change our estimate of the average temperature of the universe?

6. What is meant by the term "freezings" in the explanations of the current cosmological model? How can enormously high temperatures prevent the formation of particles such as nuclei or atoms?

7. If the universe were expanding more rapidly than it presently is, would the Hubble constant be affected? If so, how?

8. If all that you knew was the energy per square meter that the Sun radiated on the surface of Earth, could you determine the distance from the Sun to Earth?

9. We say that galaxies moving away from us are redshifted. What would we say about galaxies if they were all moving toward us?

Problem-Solving Examples

EXAMPLE 29-1

Measuring Distance with a Standard Candle

A 100-watt lightbulb shines in the center of a large darkened room. We have a light meter that measures 20 cm by 5 cm (0.01 m²). The light meter detects a power of 0.007 watt falling on it when the light is on. What is the distance between the light and the meter?

REASONING AND SOLUTION: We solve this problem by assuming that the 100 watts of power that the lightbulb emits radiates equally in all directions (Figure 29-10). Imagine a sphere of radius R centered on the bulb and including the detector. (Note that R, the radius of the sphere, is also the unknown distance between the light and the light meter.) The total surface area of the sphere is:

$$\text{Area} = 4\pi R^2$$

The fraction of that surface covered by the detector is:

$$\text{Fraction of area} = \frac{\text{Area of detector}}{\text{Area of sphere}}$$

$$= \frac{0.01 \text{ m}^2}{4\pi R^2}$$

Similarly, the fraction of the total light from the bulb that falls on the meter is:

$$\text{Fraction of light} = \frac{\text{Light falling on meter}}{\text{Total light emitted}}$$

$$= \frac{0.007 \text{ watt}}{100 \text{ watts}}$$

$$= 7.0 \times 10^{-5}$$

To find R, we note that these two fractions must be equal to one another (that is, the fraction of light falling on the meter must be the same as the fraction of the imaginary sphere that the meter covers), or

$$\text{Fraction of area} = \text{Fraction of light}$$

$$= \frac{0.01 \text{ m}^2}{4\pi R^2} = 7.0 \times 10^{-5}$$

So the distance we seek is

$$R = 3.37 \text{ m} \bullet$$

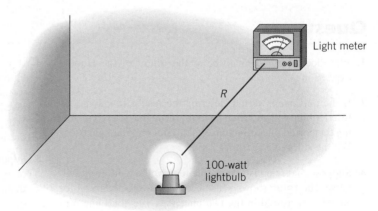

FIGURE 29-10. If you know how much light a standard candle emits, then you can determine its distance.

Light meter

R

100-watt lightbulb

EXAMPLE 29-2

The Distance to a Receding Galaxy

Astronomers discover a new galaxy and determine from its redshift that it is moving away from us at approximately 100,000 km/s (about one-third the speed of light). Approximately how far away is this galaxy? Assume a value of 70 km/s/Mpc for the Hubble constant.

REASONING: According to Hubble's law, a galaxy's distance equals its velocity divided by the Hubble constant.

SOLUTION:

$$\text{Distance (in Mpc)} = \frac{\text{Velocity (in km/s)}}{\text{Hubble's constant (in km/s/Mpc)}}$$

$$= \frac{100,000 \text{ km/s}}{70 \text{ km/s/Mpc}}$$

$$= \frac{100,000}{70} \text{ Mpc}$$

$$= 1428 \text{ Mpc}$$

Remember, a megaparsec equals about 3.3 million light-years, so this galaxy is more than 4 billion light-years away. The light that we observe from such a distant galaxy began its trip about the time that our solar system was born. ●

Problems

1. In February 1987, a supernova was seen to explode in the Large Magellanic Cloud, a small galaxylike structure near the Milky Way galaxy. The supernova was about 170,000 light-years from Earth. (A light year is the distance light travels in 1 year. Take the speed of light to be 3×10^8 m/s.)

   a. How far away was this explosion in kilometers? In miles?

   b. When did this explosion actually occur?

   c. If it were somehow possible for you to drive your car along some stellar highway in the sky all the way to the location of this explosion, how long would it take you to get there, assuming you drove at 100 km/h (about 62 miles/h)?

2. Using the Sun as a standard candle, calculate the distance from the Sun to Venus if the energy detected per square meter on Venus is 2.89 kW/m^2. (Take the energy emitted by the Sun to be 4.24×10^{23} kW.)

3. Suppose that you observe a Cepheid variable to have a period of about 100 days and, hence, a luminosity of 6.4×10^{28} W. Suppose also that the amount of light you get from that star corresponds to an energy flow of about 2×10^{-14} W/m^2 at the location of your telescope. How far is that star from Earth?

4. What is the number of kilometers in 1 parsec? The number of miles?

5. Assuming a Hubble constant of 70 km/s/Mpc, what is the approximate velocity of a galaxy 10 Mpc away? 250 Mpc away? 5000 Mpc away?

6. If a galaxy is 700 Mpc away, how fast is it receding from us?

7. An observer on one of the raisins in our bread-dough analogy measures distances and velocities of neighboring raisins. The data are listed next:

Distance (cm)	Velocity (cm/h)
0.5	1.02
0.9	2.00
1.4	2.90
2.1	4.05
3.0	5.90
3.4	7.10

   Plot these data on a graph and use the plot to estimate a Hubble constant for the raisins.

8. From the data in Problem 7, estimate the time that has elapsed since the dough started rising. Estimate the largest and smallest values of this number consistent with the data.

9. Some theories say that during the inflationary period, the scale of the universe increased by a factor of 10^{50}. Suppose your height were to increase by a factor of 10^{50}. How tall would you be? Express your answer in light-years and compare it to the size of the observable universe, which is roughly 30 billion light-years across.

10. Suppose a proton (diameter about 10^{-13} cm) were to inflate by a factor of 10^{50}. How big would it be? Convert the answer to light-years and compare it to the size of the observable universe, which is roughly 30 billion light-years across.

11. How fast is a galaxy 5 billion light-years from Earth moving away from us? What fraction of the speed of light is this?

Investigations

1. The Milky Way is a band of stars that, as seen from Earth in the summer months, stretches all the way across the sky. Given what you know about galaxies, why do you suppose that our own galaxy appears this way to us? Who was the first natural philosopher to figure this out?

2. Will the constellation Andromeda be above the horizon tonight? If so, go out and try to spot the Andromeda galaxy.

3. Look up the "Great Attractor." How does the existence of such an object fit in with the concept of the Hubble expansion? How would you modify the raisin-bread dough analogy to put in the Great Attractor?

4. Investigate the cosmologies of other societies. How do they think the universe began? Do they predict how it will end?

5. What agencies or organizations fund cosmological research? What was the role of the Carnegie Institution of Washington in Edwin Hubble's research?

 # WWW Resources

See the *Physics Matters* home page at **www.wiley.com/college/trefil** for valuable web links.

1. **http://universeadventure.org/** The Contemporary Physics Education Project (CPEP) *Universe Adventure*, by Lawrence Berkeley National Laboratory.

2. **http://map.gsfc.nasa.gov/m_uni.html** *Cosmology: The Study of The Universe,* a NASA site tutorial and collection of NASA educational resources and sites on cosmology.

3. **http://www.time.com/time/time100/scientist/profile/hubble.html** The Edwin Hubble biography from *Time* magazine.

4. **http://www.damtp.cam.ac.uk/user/gr/public/bb_cosmo.html** *A Brief History of Observational Cosmology,* from Cambridge University.

5. **http://www.pbs.org/wnet/hawking/html/home.html** A rich site to accompany *Stephen Hawking's Universe,* a PBS special.

6. **http://www.astro.ubc.ca/people/scott/cmb.html** *The Cosmic Microwave Background,* from the University of British Columbia.

7. **http://imagine.gsfc.nasa.gov/docs/science/mysteries_l1/origin_destiny.html** Origin and Destiny of the Universe, from NASA's *Imagine the Universe.*

Appendix A

UNITS AND NUMBERS

 ## THE INTERNATIONAL SYSTEM OF UNITS

Within SI, units are based on multiples of 10. Thus the centimeter is one-hundredth the length of a meter, the millimeter one-thousandth, and so on. In the same way, a kilometer is 1000 meters, a kilogram is 1000 grams, and so on. This organization differs from that of the English system, in which a foot equals 12 inches and a yard is 3 feet. A list of metric prefixes follows.

Metric Prefixes

If the prefix is:	Multiply the basic unit by:	If the prefix is:	Divide the basic unit by:
giga- (G)	1 billion (thousand million)	deci- (d)	10
mega- (M)	1 million	centi- (c)	100
kilo- (k)	1000	milli- (m)	1000
hecto- (h)	100	micro- (μ)	1 million
deka- (da)	10	nano- (n)	1 billion

UNITS OF LENGTH, MASS, AND TEMPERATURE

Next we give the conversion factors between SI and English units of length and mass.

Length and Mass Conversion from SI to English Units

To get:	Multiply:	By:
inches	meters	39.4
feet	meters	3.281
miles	kilometers	0.621

Length and Mass Conversion from English to SI Units

To get:	Multiply:	By:
meters	inches	0.0254
meters	feet	0.3048
kilometers	miles	1.609

For example, a distance of 5 miles can be converted to kilometers by multiplying by the factor 1.609:

$$5 \text{ miles} \times 1.609 = 8.05 \text{ kilometers}$$

To convert between Celsius and Fahrenheit degrees, use the following formulas:

$$°F = 1.8(°C) + 32 \qquad °C = 0.55(°F - 32)$$

where °F and °C stand for degrees Fahrenheit and Celsius, respectively. To find temperatures in the Kelvin scale, simply add 273.15 to the temperature on the Celsius scale.

UNITS OF FORCE, ENERGY, AND POWER

Once the basic units of mass, length, time, and temperature have been defined, the units of other quantities such as force and energy follow. Recall the energy units that we have defined in the text:

joule	a force of 1 newton acting through 1 meter
foot-pound	a force of 1 pound acting through 1 foot
calorie	energy required to raise the temperature of 1 gram of water by 1°C
British Thermal Unit (BTU)	energy required to raise the temperature of 1 pound of water by 1°F
kilowatt-hour	1000 joules per second for 1 hour
newton	a force required to accelerate a mass of 1 kilogram at a rate of 1 meter per second per second

Power units are:

watt	1 joule per second
horsepower	550 foot-pounds per second

Conversion factors between SI and English units for energy and power follow.

Energy and Power Conversion from SI to English Units

To get:	Multiply:	By:
BTUs	joules	0.00095
calories	joules	0.2390
kilowatt-hours	joules	2.78×10^{-7}
foot-pounds	joules	0.7375
horsepower	watts	0.00134
pounds	newtons	0.2248[a]

[a]Recall that the weight of a 1-kilogram mass is 9.806 newtons.

Energy and Power Conversion from English to SI Units

To get:	Multiply:	By:
joules	BTUs	1055
joules	calories	4.184
joules	kilowatt-hours	3.6×10^{6}
joules	foot-pounds	1.356
watts	horsepower	745.7
newtons	pounds	4.448

 # POWERS OF 10

Powers of ten notation allows us to write very large or very small numbers conveniently, in a compact way. Any number can be written by following three rules:

1. Every number is written as a number between 1 and 10 followed by 10 raised to a power, or an exponent.

2. If the power of 10 is positive, it means "move the decimal point this many places to the right."

3. If the power of 10 is negative, it means "move the decimal point this many places to the left."

Thus, using this notation, five trillion is written 5×10^{12}, instead of 5,000,000,000,000. Similarly, five-trillionths is written 5×10^{-12}, instead of 0.000000000005.

Multiplying or dividing numbers in powers of 10 notation requires special care. If you are multiplying two numbers, such as 2.5×10^3 and 4.3×10^5, you multiply 2.5 and 4.3, but you add the two exponents:

$$(2.5 \times 10^3) \times (4.3 \times 10^5) = (2.5 \times 4.3) \times 10^{3+5}$$
$$= 10.75 \times 10^8$$
$$= 1.075 \times 10^9$$

When dividing two numbers, such as 4.3×10^5 divided by 2.5×10^3, you divide 4.3 by 2.5, but you subtract the denominator exponent from the numerator exponent:

$$\frac{4.3 \times 10^5}{2.5 \times 10^3} = \frac{4.3}{2.5} \times 10^{5-3}$$
$$= 1.72 \times 10^2$$
$$= 172$$

 SELECTED MATHEMATICAL FORMULAS

Circumference of a circle, radius R:

$$C = 2\pi R$$

Area of a circle, radius R:

$$A = \pi R^2$$

Volume of a sphere, radius R:

$$V = \frac{4}{3}\pi R^3$$

Surface area of a sphere, radius R:

$$A = 4\pi R^2$$

Volume of a cylinder, radius R, height h:

$$V = \pi R^2 h$$

Appendix B

SELECTED PHYSICAL CONSTANTS

Average acceleration due to gravity at Earth's surface:

$$g = 9.8 \text{ m/s}^2$$

Gravitational constant:

$$G = 6.67 \times 10^{-11} \text{ N-m}^2/\text{kg}^2$$

Coulomb's law constant:

$$k = 9 \times 10^9 \text{ N-m}^2/\text{C}^2$$

Charge on electron:

$$e = 1.6 \times 10^{-19} \text{ C}$$

Proton mass:

$$m_p = 1.66 \times 10^{-27} \text{ kg}$$

Electron mass:

$$m_e = 9.11 \times 10^{-31} \text{ kg}$$

Speed of light in a vacuum:

$$c = 3.00 \times 10^8 \text{ m/s}$$

Planck's constant:

$$h = 6.63 \times 10^{-34} \text{ J-s}$$

Boltzmann's constant:

$$k = 1.38 \times 10^{-23} \text{ J/K}$$

Astronomical unit (mean distance from Earth to the Sun):

$$\text{AU} = 1.46 \times 10^{11} \text{ m}$$

Light-year:

$$1 \text{ ly} = 9.46 \times 10^{12} \text{ km}$$
$$= 6.3 \times 10^4 \text{ AU}$$

Parsec:

$$1 \text{ pc} = 3.3 \text{ light-years}$$

Power output of the Sun:

$$P = 4.24 \times 10^{23} \text{ kW}$$

Mass of the Sun:

$$M_{\text{sun}} = 1.989 \times 10^{30} \text{ kg}$$

Mass of Earth:

$$M_E = 5.974 \times 10^{24} \text{ kg}$$

Mass of the Moon:

$$M_{\text{moon}} = 7.348 \times 10^{22} \text{ kg}$$

Radius of the Sun:

$$R_{\text{sun}} = 6.96 \times 10^5 \text{ km}$$

Radius of Earth:

$$R_E = 6.378 \times 10^3 \text{ km}$$

Radius of the Moon:

$$R_{\text{moon}} = 1.738 \times 10^3 \text{ km}$$

Appendix C

PERIODIC TABLE AND ATOMIC WEIGHTS

PERIODIC TABLE OF THE ELEMENTS

Atomic number

metal   metalloid   nonmetal

	IA (1)																		VIIIA (18)
1	1 H 1.00794	IIA (2)											IIIA (13)	IVA (14)	VA (15)	VIA (16)	VIIA (17)		2 He 4.00260
2	3 Li 6.941	4 Be 9.01218											5 B 10.811	6 C 12.011	7 N 14.00674	8 O 15.9994	9 F 18.99840		10 Ne 20.1797
3	11 Na 22.98977	12 Mg 24.3050	IIIB (3)	IVB (4)	VB (5)	VIB (6)	VIIB (7)	VIIIB (8)	(9)	(10)	IB (11)	IIB (12)	13 Al 26.98154	14 Si 28.0855	15 P 30.97376	16 S 32.066	17 Cl 35.4527		18 Ar 39.948
4	19 K 39.0983	20 Ca 40.078	21 Sc 44.95591	22 Ti 47.88	23 V 50.9415	24 Cr 51.9961	25 Mn 54.9380	26 Fe 55.847	27 Co 58.93320	28 Ni 58.69	29 Cu 63.546	30 Zn 65.39	31 Ga 69.723	32 Ge 72.61	33 As 74.92159	34 Se 78.96	35 Br 79.904		36 Kr 83.80
5	37 Rb 85.4678	38 Sr 87.62	39 Y 88.90585	40 Zr 91.224	41 Nb 92.90638	42 Mo 95.94	43 Tc 98.9072	44 Ru 101.07	45 Rh 102.90550	46 Pd 106.42	47 Ag 107.8682	48 Cd 112.411	49 In 114.82	50 Sn 118.710	51 Sb 121.75	52 Te 127.60	53 I 126.90447		54 Xe 131.29
6	55 Cs 132.90543	56 Ba 137.327	*La 57 138.9055	72 Hf 178.49	73 Ta 180.9479	74 W 183.85	75 Re 186.207	76 Os 190.2	77 Ir 192.22	78 Pt 195.08	79 Au 196.96654	80 Hg 200.59	81 Tl 204.3833	82 Pb 207.2	83 Bi 208.98037	84 Po 208.9824	85 At 209.9871		86 Rn 222.0176
7	87 Fr 223.0197	88 Ra 226.0254	†Ac 89 227.0278	104 Rf (261)	105 Db (262)	106 Sg (266)	107 Bh (264)	108 Hs (269)	109 Mt (268)	110 Ds (271)	111 Uuu (272)	112 Uub (285)		114 Uuq (289)					

Periods

Alkali metals   Alkaline earth metals

Halogens   Noble or inert gases

*	58 Ce 140.115	59 Pr 140.90765	60 Nd 144.24	61 Pm 144.9127	62 Sm 150.36	63 Eu 151.965	64 Gd 157.25	65 Tb 158.92534	66 Dy 162.50	67 Ho 164.93032	68 Er 167.26	69 Tm 168.93421	70 Yb 173.04	71 Lu 174.967
†	90 Th 232.0381	91 Pa 231.0359	92 U 238.0289	93 Np 237.0482	94 Pu 244.0642	95 Am 243.0614	96 Cm 247.0703	97 Bk 247.0703	98 Cf 242.0587	99 Es 252.083	100 Fm 257.0951	101 Md 258.10	102 No 259.1009	103 Lr 260.105

Names of elements 110–114 are temporary. Official names and symbols must be approved by the International Union of Pure and Applied Chemistry.

TABLE OF ATOMIC WEIGHTS

Name	Symbol	Atomic Number	Atomic Weight	Name	Symbol	Atomic Number	Atomic Weight
Actinium	Ac	89	227.0	Indium	In	49	114.8
Aluminum	Al	13	26.98	Iodine	I	53	126.9
Americium	Am	95	243.1	Iridium	Ir	77	192.2
Antimony	Sb	51	121.8	Iron	Fe	26	55.85
Argon	Ar	18	39.95	Krypton	Kr	36	83.80
Arsenic	As	33	74.92	Lanthanum	La	57	138.9
Astatine	At	85	210.0	Lawrencium	Lr	103	260.1
Barium	Ba	56	137.3	Lead	Pb	82	207.2
Berkelium	Bk	97	247.1	Lithium	Li	3	6.941
Beryllium	Be	4	9.012	Lutetium	Lu	71	175.0
Bismuth	Bi	83	209.0	Magnesium	Mg	12	24.31
Bohrium	Bh	107	(264)	Manganese	Mn	25	54.94
Boron	B	5	10.81	Meitnerium	Mt	109	(268)
Bromine	Br	35	79.90	Mendelevium	Md	101	258.1
Cadmium	Cd	48	112.4	Mercury	Hg	80	200.6
Calcium	Ca	20	40.08	Molybdenum	Mo	42	95.94
Californium	Cf	98	242.06	Neodymium	Nd	60	144.2
Carbon	C	6	12.01	Neon	Ne	10	20.18
Cerium	Ce	58	140.1	Neptunium	Np	93	237.05
Cesium	Cs	55	132.9	Nickel	Ni	28	58.69
Chlorine	Cl	17	35.45	Niobium	Nb	41	92.91
Chromium	Cr	24	52.00	Nitrogen	N	7	14.01
Cobalt	Co	27	58.93	Nobelium	No	102	259.1
Copper	Cu	29	63.55	Osmium	Os	76	190.2
Curium	Cm	96	247.1	Oxygen	O	8	16.00
Darmstadtium	Ds	110	(271)	Palladium	Pd	46	106.4
Dubnium	Db	105	(262)	Phosphorus	P	15	30.97
Dysprosium	Dy	66	162.5	Platinum	Pt	78	195.1
Einsteinium	Es	99	252.1	Plutonium	Pu	94	244.1
Erbium	Er	68	167.3	Polonium	Po	84	209.0
Europium	Eu	63	152.0	Potassium	K	19	39.10
Fermium	Fm	100	257.1	Praseodymium	Pr	59	140.9
Fluorine	F	9	19.00	Promethium	Pm	61	144.9
Francium	Fr	87	223.0	Protactinium	Pa	91	231.0
Gadolinium	Gd	64	157.25	Radium	Ra	88	226.0
Gallium	Ga	31	69.72	Radon	Rn	86	222.0
Germanium	Ge	32	72.61	Rhenium	Re	75	186.2
Gold	Au	79	197.0	Rhodium	Rh	45	102.9
Hafnium	Hf	72	178.5	Rubidium	Rb	37	85.47
Hassium	Hs	108	(269)	Ruthenium	Ru	44	101.1
Helium	He	2	4.003	Rutherfordium	Rf	104	(261)
Holmium	Ho	67	164.9	Samarium	Sm	62	150.4
Hydrogen	H	1	1.008	Scandium	Sc	21	44.96

TABLE OF ATOMIC WEIGHTS (*continued*)

Name	Symbol	Atomic Number	Atomic Weight	Name	Symbol	Atomic Number	Atomic Weight
Seaborgium	Sg	106	(266)	Tin	Sn	50	118.7
Selenium	Se	34	78.96	Titanium	Ti	22	47.88
Silicon	Si	14	28.09	Tungsten	W	74	183.85
Silver	Ag	47	107.9	Ununbium	Uub	112	(285)
Sodium	Na	11	22.99	Ununquadium	Uuq	114	(289)
Strontium	Sr	38	87.62	Unununium	Uuu	111	(272)
Sulfur	S	16	32.07	Uranium	U	92	238.0
Tantalum	Ta	73	180.95	Vanadium	V	23	50.94
Technetium	Tc	43	98.91	Xenon	Xe	54	131.3
Tellurium	Te	52	127.6	Ytterbium	Yb	70	173.0
Terbium	Tb	65	158.9	Yttrium	Y	39	88.91
Thallium	Tl	81	204.4	Zinc	Zn	30	65.39
Thorium	Th	90	232.0	Zirconium	Zr	40	91.22
Thulium	Tm	69	168.9				

Elements by Name, Symbol, Atomic Number, and Atomic Weight.

(Atomic weights are given to four significant figures for elements below atomic number 104.)

Names of elements 110–114 are temporary. Official names and symbols must be approved by the International Union of Pure and Applied Chemistry.

Glossary

Note: Numbers in parentheses refer to the page on which the term is found. **Bolded** page numbers refer to key terms.

absolute zero The lowest possible temperature, at which no energy can be extracted from atoms. **(p. 228)**

absorption The conversion of electromagnetic wave energy into heat, resulting in a reduction (partial or complete) of the wave strength. **(p. 426)**

absorption lines The frequencies missing from white light after it passes through a material. (p. 458)

acceleration The change in velocity divided by the time it takes for that change to occur. Acceleration can involve changes of speed, changes in direction, or both. **(p. 60,** 81)

acceleration due to gravity *(g)* The velocity change of a freely falling body at the Earth's surface. **(p. 66)**

active optics Optical devices (such as the mirrors in a reflecting telescope) that are able to move in order to focus and make corrections. (p. 424)

additive primary colors *See* **primary additive colors.**

air resistance The resultant force that occurs in the opposite direction of an object's movement, when the object moves through air; this force increases in proportion to the object's speed; also known as *drag.* (p. 82)

alkali metals The highly reactive elements in the left-hand column of the periodic table. (p. 463)

alkaline earths Elements in the second column of the periodic table; they combine with oxygen to form colorless compounds with high melting temperatures. (p. 463)

alpha decay The spontaneous release of an alpha particle (two protons and two neutrons) from an atomic nucleus. **(p. 559)**

alpha particle A helium-4 nucleus; frequently emitted during radioactive decay. (p. 450)

alternating current (AC) Electric current that changes direction periodically in a circuit. **(p. 366)**

ampere (amp) The physical unit used to define a quantity of electric current. **(p. 375)**

amplitude The maximum displacement from equilibrium of a wave medium or a vibrating body. **(p. 293)**

amplitude modulation (AM) The process of encoding a radio wave with information by varying its strength or amplitude. (p. 404)

angle of incidence The angle that the direction of incoming radiation makes with a line drawn perpendicular to the surface. (p. 421)

angle of reflection The angle that the direction of reflected radiation makes with a line drawn perpendicular to the surface. (p. 421)

angular frequency The number of radians traversed in 1 second. (p. 143)

angular momentum The moment of inertia of a body, times its angular velocity. **(p. 147)**

angular speed The angle through which an object has moved about the axis of rotation, divided by the time it takes it to go through that angle. **(p. 141)**

annihilation Process by which matter and antimatter come together and have all their mass converted into energy. (p. 590)

antimatter Any substance that annihilates with an equal amount of ordinary matter, resulting in a complete conversion to electromagnetic energy. **(p. 590)**

antinodes The points of a standing wave where the amplitude of the oscillations of the medium is a maximum. (p. 299)

applied research The type of research performed by scientists with specific and practical goals in mind. This research is often translated into practical systems by large-scale research and development projects. **(p. 14)**

Archimedes' principle The statement that the upward force exerted on an object immersed in a fluid (the buoyant force) is equal to the weight of the fluid that the object displaces. (p. 213)

artificial intelligence The concept of creating a machine or computer that performs functions normally associated with human intelligence. (p. 546)

astronomy The study of stars, planets, and other objects in space. (p. 13)

atom The tiniest particle of matter that retains the chemical properties of an element. **(p. 187)**

atomic number The number of protons in the nucleus of an atom. (p. 463, **554)**

aurora borealis and **aurora australis** The light displays (northern lights and southern lights) produced when the charged particles in the Van Allen belt enter the Earth's atmosphere. (p. 347)

average velocity The total distance traveled divided by the total time it takes to travel that distance. (p. 61)

Avogadro's principle The statement that equal volumes of any gas at the same temperature and pressure contain the same number of gas molecules. **(p. 188)**

axis of rotation The line through the center of an object, around which everything else rotates. **(p. 139)**

base One of the three semiconductor regions of a transistor. (p. 536)

base load The electric power that needs to be supplied around the clock. (p. 535)

basic research The type of research performed by scientists who are interested simply in finding out how the world works, in knowledge for its own sake. **(p. 14)**

battery A device that uses stored chemical energy as an electric power source. **(p. 354)**

beats The slow, pulsing sound heard when two sound waves of slightly different frequencies interfere. (p. 322)

Bernoulli effect The effect by which the pressure exerted by a fluid decreases as the fluid velocity increases. (**p. 215**)

beta decay The spontaneous transformation of a neutron into a proton in an atomic nucleus, accompanied by the release of an electron and a neutrino. (**p. 560**)

big bang theory The theory that the universe was, at one time, a very small, dense collection of matter and energy that has been expanding ever since. (**p. 636**)

binary Presenting information as a string of 0s and 1s, representing off or on. (**p. 540**)

binary digit *See* **bit**.

binding energy The energy required to separate a nucleus into its constituent parts. (p. 557)

biology The study of living systems. (p. 13)

biosphere The part of the Earth's surface and atmosphere that is capable of supporting life. (p. 282)

bit The smallest unit of information storage. (**p. 540**)

Bohr atom A model of the atom proposed in 1913 by Danish physicist Niels Bohr; the Bohr atom revolutionized our understanding of physics. (**p. 451**)

buoyant force The upward force on an object due to the pressure of a fluid. (**p. 213**)

byte Eight bits of information. (**p. 540**)

calorie The amount of heat required to raise the temperature of 1 gram of water by 1 degree Celsius. (p. 235)

capacitance The ratio of stored charge to voltage across a capacitor. (p. 381)

capacitor A charge configuration that consists of equal amounts of opposite charge separated by some distance; usually used as a device in an electric circuit. (p. 380)

center of mass (center of gravity) The point of support on which an object can be balanced. (**p. 152**)

central processing unit (CPU) The main part of a computer that stores and manipulates relatively small amounts of information. (p. 542)

centripetal acceleration Acceleration directed inward toward the center of a circle, perpendicular to the velocity. (p. 69)

centripetal force The force in circular motion, directed toward the center of the circle, that keeps an object following a curved or circular path. (**p. 103**)

Cepheid variable A star that shows a regular pattern of brightening and dimming over a period of weeks or months; used in astronomy as a standard candle. (p. 628)

chain reaction A self-sustaining sequence of nuclear fission in which each fissioning nucleus produces the neutrons that cause more splitting. (p. 570)

change of phase A process by which a material, without changing its constituent atoms or molecules, changes their arrangement. (**p. 196**)

chaos theory The field of study devoted to systems in nature that can be described in simple Newtonian terms but whose futures are extremely sensitive to initial conditions, and which are, for all intents and purposes, unpredictable. (p. 111)

charge *See* **electric charge**.

chemical bonds The forces that hold atoms together in stable configurations to form molecules. (p. 187, **494**)

chemical reaction The formation or breaking apart of chemical bonds. (**p. 506**)

chemistry The study of atoms in combination. (p. 13)

chromatic aberration The effect that results when different colors are focused at different locations by a lens. (p. 434)

circuit breaker A device that prevents too much current from flowing through it by creating an open circuit. (**p. 382**)

classical mechanics The field of study that focuses on how any object is affected by any force. (p. 80)

classical physics The study of classical mechanics, thermodynamics, electricity, and magnetism. (p. 367)

coefficient of linear expansion A quantity that relates the temperature change with the corresponding length change of a material. (**p. 232**)

collector One of the three semiconductor regions of a transistor. (p. 536)

composite materials Materials that combine the properties of several materials in order to improve their working characteristics. (p. 505)

compound A material that is made up of two or more elements. (**p. 187**)

compression The condition in which the atoms of a material are squeezed closer together, due to an external force. (**p. 219**)

compressive strength The ability of a solid to withstand crushing. (p. 503)

computer A machine that stores and manipulates information. (**p. 542**)

concave mirror A curved mirror that bows inward, away from the incoming light. (p. 423)

conduction The transfer of heat due to atomic or molecular collisions. (**p. 240**)

conduction electrons The electrons in a conductor or semiconductor that are free to move from place to place. (p. 518)

conservation law Statement that a quantity is constant in nature. (**p. 124**)

conservation of angular momentum The physical law that states that in the absence of external torques, the angular momentum of any system must stay constant over time. (**p. 149**)

conservation of electric charge The physical law that states that the total amount of charge in the universe does not change. (**p. 338**)

constant An unchanging value, often determined by measurements, that defines a quantitative mathematical relationship. (p. 4)

constellations Closely spaced groups of stars. (p. 51)

convection Heat transfer due to the motion of a liquid or a gas. (**p. 242**)

convection cell A region of a fluid that is either rising or sinking due to the heat convection process. (**p. 242**)

converging lens A lens that bends light towards its axis. (p. 432)

conversion factor Established mathematical quantity used to shift from one system of units to another. (**p. 37**)

convex mirror A curved mirror that bows outward, toward the incoming light. (p. 423)

cosmic microwave background radiation The electromagnetic radiation from space that is thought to be a remnant of the big bang. (**p. 638**)

cosmic rays High-speed particles (mostly protons) that originate in space and travel throughout the universe. (p. 346, **585**)

cosmology The branch of science devoted to the study of the history and structure of the universe. (**p. 630**)

coulomb The physical unit used for quantifying amounts of electric charge. (p. 338)

Coulomb's law The law of physics that defines the force between electrically charged objects. (**p. 339**)

covalent bond Chemical bond formed when two or more atoms in a molecule share electrons. (**p. 497**)

crest The highest point of a wave; commonly associated with water waves. (p. 293)

critical mass The minimal amount of fissionable material that makes a chain reaction possible. (p. 574)

crystal A material whose atoms are arranged in an ordered, repeating pattern. (p. 191)

cubed A term describing a number multiplied twice by itself. (p. 33)

Curie point The temperature below which neighboring atoms in a material begin to align themselves into ferromagnetic domains. (p. 525)

Curie's law The law that states that the magnetic field produced by a paramagnet is proportional to the applied magnetic field. (p. 526)

cycle One complete back-and-forth vibration or oscillation. (p. 290)

cyclotron A type of particle accelerator in which the particles spiral outward between two magnets, gaining energy through the repeated application of a high-frequency voltage. (p. 586)

dark energy Newly discovered energy that pushes galaxies away from one another, causing the universe to expand more rapidly as time goes by. (**p. 647**)

dark matter The unseen matter in the universe that has been postulated to account for the observed structure of the universe. (**p. 645**)

daughter nucleus The nucleus that is a product of radioactive decay. (p. 560)

de Broglie relation The mathematical formula that relates the wavelength to the momentum of a quantum object. (p. 481)

deceleration The process of slowing down; because the process involves a change in velocity, it is actually a form of acceleration. (p. 61)

decibel (dB) A unit of the intensity of sound. (**p. 319**)

density The mass per unit volume of a substance; it is a measure of how much material is packed into a given volume. (**p. 204**)

diamagnetism The magnetic properties of a material associated with the alignment of atoms in the presence of an applied magnetic field; in diamagnetism the induced field opposes the applied field. (p. 525)

diffuse scattering The scattering of light in many different directions. (p. 421)

digital Presenting information in a numerical system that can have a finite number of states. Binary is one form of digital information. (**p. 540**)

dimensional analysis The process of ensuring that the units in a problem come out correctly. (p. 60)

diode A semiconductor device formed from one p-type (positive) and one n-type (negative) region; typically used as a rectifier. (**p. 533**)

dipole field The field created by an electric or magnetic dipole. (p. 344)

direct current (DC) Electric current that is always in the same direction in a circuit. (**p. 366**)

direct relationship The simplest relationship between two variables, in which the two variables change together. (p. 31)

dispersion The phenomenon that different wavelengths of light refract different amounts when entering a medium. (**p. 434**)

diverging lens A lens that bends light away from its axis. (p. 432)

Divine Calculator Proposed by French mathematician Pierre Simon Laplace (1749–1827), the idea that with a known position and velocity of every atom in the universe, and with infinite computational power, the position and velocity of every atom in the universe for all times could be predicted. (p. 111)

doping The addition of small amounts of an impurity to an element or compound to enhance its conduction properties. (**p. 532**)

Doppler effect A shift in the observed frequency of a wave due to the motion of the source of the wave, the observer, or both. (**p. 299**)

double bond A covalent bond in which two electrons from each atom are shared. (p. 498)

drag *See* **air resistance**.

drift velocity The speed with which the electrons in a circuit move from one place in the circuit to another. (p. 375)

echolocation The process of determining the distance to objects by bouncing sound waves off those objects. (p. 317)

efficiency A measure of how much useful work you can get from an engine compared to the amount of energy put into it; it is equal to the work done by an engine divided by the heat input to the engine. (**p. 273**)

elastic limit The point at which Hooke's law no longer describes the elastic properties of a material because too much force is applied; the maximum force or torque that can be applied to a solid without bending or breaking it. (p. 218, 503)

elasticity A term used to describe the way materials respond to forces. (p. 218)

electric charge The property of an object or particle that quantifies its response to electric and magnetic phenomena; charges are either *positive* or *negative*. (**p. 334**)

electric circuit An unbroken path of material that carries electricity. (**p. 374**)

electric current A flow of charged particles. (**p. 354**)

electric field A measure of how much electric force a positive electric charge would experience at a particular location in space; it is defined as the force per unit charge. (**p. 342**)

electric force The force exerted on charged objects by other charged objects. (p. 339)

electric generator A device that uses electromagnetic induction to produce electricity. (**p. 365**)

electric motor A device that does work by running electric current through coils of wire in the presence of permanent magnets. (**p. 358**)

electric potential The quantity in electric circuits that is analogous to pressure in water flowing through pipes; it is the potential energy per unit charge across a region in a circuit; also called voltage. (**p. 375**)

electrical conductor A material in which charges are free to move from place to place. (**p. 515**)

electrical insulator A material in which charges are not free to move from place to place. (**p. 517**)

electrical resistance (measured in **ohms**) The quantity that defines how hard it is to run electric current through an object; the tendency of a material to resist the flow of electric charges through it. (**p. 375, 515**)

electricity A general term used to define the presence and or motion of electric charges. (**p. 334**)

electromagnet A device that creates a magnetic field by running electric current through coils of wire. (**p. 357**)

electromagnetic force *See* **electromagnetism**.

electromagnetic induction The effect in which a changing magnetic field causes electric current in a loop of wire. (**p. 361**)

electromagnetic radiation *See* **electromagnetic wave**.

electromagnetic spectrum The entire collection of electromagnetic waves, from the shortest wavelength (gamma rays) to the longest wavelength (radio waves) and everything in between. (**p. 403**)

electromagnetic wave or **electromagnetic radiation** A wave that incorporates electric and magnetic fields that oscillate together. (**p. 399**)

electromagnetism One of the four fundamental forces in nature; associated with electric charges and electromagnetic radiation. (p. 354, 595)

electron A negatively charged fundamental particle; it is one of the primary building blocks of the atom. (**p. 449**)

electron shells *See* **shells**.

electroweak force The force that results from the unification of the weak force and the electromagnetic force. (p. 598)

element A substance that cannot be broken down into other substances by chemical means. (**p. 187**)

elementary particles The particles that constitute the basic building blocks of all matter. (p. 584)

elementary-particle physics *See* **high-energy physics**.

ellipse A curve drawn so that the sum of the distances from any point on the curve to two fixed points is always the same. (p. 53)

emitter One of the three semiconductor regions of a transistor. (p. 536)

endothermic reaction A chemical reaction that takes place only if energy is supplied. (p. 508)

energy The ability to do work. (**p. 166**)

English system The traditional system of units that has roots going back into the Middle Ages, still predominantly used today in the United States. (p. 35)

entropy A measure of the disorder in a system. (**p. 278**)

equation The definition of a precise mathematical relationship between two or more measurements. (**p. 29**)

excited state In a system with quantized energy levels, any energy level with more energy than the ground state. (p. 453)

exothermic reaction A chemical reaction that gives off energy. (p. 507)

experiment The manipulation of some aspect of nature to observe the outcome. (**p. 3**)

experimentalist A scientist who manipulates nature with controlled experiments. (p. 11)

ferromagnetic domain A region of a material, typically about a thousand atoms on a side, in which the angular momentum vectors of the atoms are lined up. (p. 525)

ferromagnetism The magnetic properties of a material associated with the spontaneous alignment of domains of magnetic fields of its atoms. (**p. 525**)

fiber optics A technology that uses long and thin glass fibers to carry light great distances using the principle of total internal reflection. (**p. 425**)

field researcher A scientist who goes into natural settings to observe nature at work. (p. 11)

first law of thermodynamics The law of physics that states the relationship between heat, work, and changes in the internal energy of a closed system. (**p. 261**)

fission The splitting of a nucleus into two or more smaller pieces; usually associated with the splitting of uranium to get energy. (**p. 569**)

fluid A term used to refer collectively to both liquids and gases. (p. 194)

focal length The distance from a mirror or a lens to the focal point. (p. 423)

focal point The point at which parallel rays will converge in a mirror or lens system. (p. 423)

focus A fixed point used as the basis for drawing an ellipse. (p. 54)

foot The basic unit of length in the English system. (p. 35)

foot-pound The unit of work in the English system, corresponding to the work done in lifting a weight of 1 pound 1 foot upward against the force of gravity. (p. 165)

force (measured in newtons) A physical effect that can produce a change in an object's state of motion. (**p. 81**)

Fourier analysis or **Fourier synthesis** The technique of taking a complex wave and breaking it down into a sum of simple, single-frequency waves. (p. 326)

frame of reference A set of physical surroundings from which events are observed and measured. (**p. 604**)

freezings or **phase transitions** Moments in the evolution of the universe that are marked by a dramatic change in the nature of the matter and forces that constitute it. (p. 641)

frequency The number of vibrations per second in oscillations and waves. (**p. 290, 293**)

frequency modulation (FM) The process of encoding a radio wave with information by varying its frequency. (p. 404)

frequency of rotation The number of times an object completes a rotation in a given amount of time. (**p. 140**)

fundamental or **first harmonic** The lowest frequency mode of a vibrating body. (p. 325)

fuse A device that prevents too much current from flowing through it by melting a piece of metal that is part of the circuit. (p. 382)

fusion The process in which two nuclei join together to form a larger nucleus. (**p. 571**)

g See **acceleration due to gravity**.

G See **gravitational constant**.

galaxy A collection of gas, dust, and millions or billions of stars all held together by gravity; there are billions of galaxies in the universe. (**p. 630**)

gamma radiation The spontaneous release of a high-energy photon from an atomic nucleus. (**p. 562**)

gamma rays Electromagnetic waves with the shortest wavelength; they are usually emitted in nuclear or particle reactions. (**p. 413**)

gas A material that retains neither its shape nor its volume, but expands to fill any container in which it is placed. (**p. 193**)

gauge particles Particles that regulate the forces between other particles by being exchanged. (p. 596)

general relativity The part of relativity theory that deals with accelerated reference frames and gravity. (**p. 606**)

geology The study of the history, evolution, and present state of our home, planet Earth. (p. 13)

geosynchronous orbit An orbit in which an object completes one revolution around the Earth in 24 hours; the object appears to hover over the same spot on the Earth's surface, because that spot also completes a revolution in 24 hours. (p. 108)

glasses Solids in which atoms that are near each other are arranged regularly, but in which there is no long range order. (p. 191)

gluons Gauge particles that hold quarks together. (p. 597)

gravitational constant (*G*) The exact numerical relation between the masses of two objects and the distance between them, on the one hand, and the gravitational force between them, on the other. (**p. 98**)

gravitational potential energy The energy a body has by virtue of its position in a gravitational field. (**p. 169**)

gravity The attractive force that acts between any two objects in the universe. (**p. 97, 595**)

ground state The lowest energy level in a system with quantized energy states. (p. 453)

grounding The process of connecting an element of a circuit to the ground to provide a safe path for the current in the case of an overload. (**p. 385**)

hadron Any particle that exists within the nucleus of the atom; hadrons interact with the strong force. (**p. 589**)

half-life The time it takes for one half of a sample of radioactive material to decay. (**p. 565**)

harmonics *See* **overtones**.

heat The energy transferred from one body to another due to a difference in temperature between the two bodies. (**p. 234**)

heat capacity A measure of the change in temperature of an object on adding or removing heat; the amount of heat required to raise the temperature of the object by 1° C. (**p. 235**)

heat insulator A material that does not conduct heat well; it has a low thermal conductivity. (p. 241)

heat transfer The process by which thermal energy moves from one place to another. (**p. 240**)

Heisenberg uncertainty principle *See* **uncertainty principle**.

hertz The unit of measure of rotational frequency, corresponding to one complete period every second; the physical unit equal to one cycle per second. (**p. 140, 293**)

high-energy physics or **elementary-particle physics** The study of elementary particles and their properties. (**p. 584**)

high-level nuclear waste Radioactive materials that are a direct byproduct of the fission process in nuclear reactors and in nuclear weapons production. (p. 575)

high-temperature reservoir *See* **temperature reservoir**.

hole A defect in a crystal formed by the absence of an electron. (p. 518)

Hooke's law The law that states that the harder you pull on a material, the more it stretches. (p. 218)

horsepower The unit of power in the English system, defined as 550 foot-pounds per second. (p. 167)

Hubble's law The law that relates the distance between Earth and a galaxy to the galaxy's recession speed. (**p. 634**)

hydrogen bond A weak chemical bond due to polarization forces; developed when hydrogen forms a covalent bond with another atom. (p. 500)

hypothesis A tentative, educated guess, after summarizing experimental and observational results, about how the world works for the behavior under study. (**p. 4**)

ideal gas law The law that relates the pressure, volume, and temperature of a gas. (**p. 217**)

impulse The product of a force multiplied by the time over which it acts. (**p. 121**)

impulse–momentum relationship Restatement of Newton's second law, in which impulse equals the change in momentum. (**p. 121**)

incompressible Not capable of being compressed; as in an *incompressible* fluid. (p. 209)

index of refraction Defined for a specific material, the ratio of the speed of light in a vacuum to the speed of light in that material; it is a measure of how much light slows down as it enters the material. (**p. 428**)

inertia The tendency of an object to remain in uniform motion–to resist changes in its state of motion. (**p. 82**)

inflation A short period of rapid expansion early in the history of the universe. (p. 642)

infrared radiation Electromagnetic waves with wavelengths in the range of about 1 micron to 1 millimeter; they are what we feel as radiant heat. (**p. 407**)

instantaneous velocity The velocity at a specific time. (p. 61)

insulator *See* **electrical insulator; heat insulator**.

integrated circuit Electronic circuit containing many transistors. (p. 538)

intensity A measure of the energy of a wave; it is measured in watts per square meter. (p. 319)

interference The result of having two or more waves in the same place; constructive interference results in a wave with a larger amplitude; destructive interference results in a wave with a smaller amplitude. (**p. 296**)

internal energy *See* **thermal energy**.

internal reflection *See* **total internal reflection.**

International System or **SI** (Système International) A widely used, internally consistent system of units within the metric system; also known as the *metric system*. (**p. 35**)

inverse relationship The relationship between two variables in which one variable increases as another decreases, and vice versa. (p. 31)

inverse square relationship The relationship between two variables in which one variable increases to a squared amount as another variable decreases, and vice versa. (p. 34)

inversely proportional The relationship between two variables in which one variable increases in the same measure as another variable decreases, and vice versa. (p. 32)

ion An atom that has lost or gained electrons, thus acquiring an electric charge. (**p. 337**)

ionic bond Chemical bond formed when one atom gives up one or more electrons to another atom, creating an electrical attraction between the atoms. (**p. 494**)

ionization The stripping away of one or more of an atom's electrons. (p. 563)

isotopes Atoms of the same element whose nuclei have the same number of protons but a different number of neutrons. (**p. 555**)

joule The SI unit of work, corresponding to a force of 1 newton acting through 1 meter. (**p. 164**)

Kepler's laws of planetary motion Three basic mathematical statements about the solar system: *Kepler's first law of planetary motion* states that the planets have elliptical orbits with one focus at the Sun; *Kepler's second law* says that for a given time interval, the swept-out area is the same, no matter where the planet is in its orbit; *Kepler's third law* expresses the relationship between a planet's distance from the Sun and its period as a simple equation that allows scientists to predict the behavior of orbiting objects. (**p. 53**, 54)

kilogram The SI unit of mass. (p. 35)

kilowatt The unit of 1000 watts (corresponding to an expenditure of 1000 joules per second). (**p. 167**)

kinetic energy The energy a body has by virtue of its motion. (**p. 169**)

laser Acronym for *l*ight *a*mplification by *s*timulated *e*mission of *r*adiation; a device that creates a very intense, coherent, and collimated beam of light. (p. 460)

latent heat of fusion The amount of heat required to change 1 gram of a solid material to a liquid when the solid is at its melting temperature; equivalently, the amount of heat that must be removed from 1 gram of liquid material to turn it into a solid when the liquid is at its freezing temperature. (**p. 238**)

latent heat of vaporization The amount of heat required to change 1 gram of a liquid material to a gas when the liquid is at its boiling temperature; equivalently, the amount of heat that must be removed from 1 gram of a gaseous material to turn it into a liquid when the gas is at its condensation temperature. (**p. 238**)

law of compound motion Galileo's proposition that motion in one dimension has no effect on motion in another dimension. (**p. 67**)

law of conservation of energy The law that states that in a closed system, the total amount of all forms of energy remains the same. (**p. 174**)

law of conservation of momentum The law that states that if no external forces act on a system, then the total momentum of that system remains the same. (**p. 124**)

law of nature An overarching statement of how the universe works, following repeated and rigorous observation and testing of a hypothesis or group of related hypotheses. (**p. 6**)

laws of thermodynamics *See* **first law; second law.**

length contraction The observed shortening of an object that is moving with respect to the observer. (**p. 612**)

lens A piece of transparent material designed to bend and focus light. (**p. 432**)

Lenz's law The physical law that allows one to determine the direction of the induced current during electromagnetic induction. (p. 362)

lepton Elementary particle that does not participate in the strong nuclear force; the electron and the neutrino are examples of leptons. (**p. 589**)

lift The net upward force on a wing due to the pressure difference between the top and the bottom of that wing. (**p. 215**)

light Electromagnetic radiation detectable by the human eye, with wavelengths from about 400 to 700 nanometers in length. (**p. 399**)

linear accelerator A type of particle accelerator in which the particles are accelerated in a straight line. (p. 587)

linear momentum Another term used for *momentum* (the product of mass times velocity) when the object is understood to move in a straight line. (**p. 118**)

liquid A material that maintains a constant volume, but assumes the shape of its container. (**p. 192**)

liter The SI unit of volume. (p. 35)

load The place where useful work gets done in a circuit. (p. 379)

longitudinal wave A wave in which the motion of the medium is in the same direction as the wave propagation. (p. 295)

Lorentz factor The factor $\sqrt{1 - (v/c)^2}$ that often occurs in relativity when quantities in different frames of reference are related to one another. (p. 611)

loudness A measure of how loud a sound is perceived by humans. (**p. 319**)

low-temperature reservoir *See* **temperature reservoir.**

machine Device that changes the direction or magnitude (or both) of an applied force. (p. 177)

maglev *See* **magnetic levitation**.

magnet Material or object that possesses a magnetic property such that it is attracted to or repelled from another magnet. (**p. 343**)

magnetic field A measure of the effect that a magnet has on the space surrounding it, or the effect that would be felt by a magnet if it were at a particular location in space. (**p. 344**)

magnetic field lines Imaginary lines in space that give the direction of a compass needle. (**p. 344**)

magnetic force The force that a magnet exerts on other magnets or on moving electric charges. (**p. 343**)

magnetic levitation (maglev) The magnetic repulsion between a permanent magnet and a metal due to the motion of the permanent magnet. (p. 523)

magnetism The group of phenomena related to magnets and magnetic fields. (p. 343)

mass The amount of matter contained in an object, independent of where that object is found. (**p. 84**)

mass number The number of protons plus the number of neutrons in the nucleus of an atom. (**p. 555**)

mechanics The branch of physics that deals with motions of material objects. (**p. 56**)

meltdown The melting of the fuel rods in a nuclear reactor during an uncontrolled heating of the core. (p. 571)

metallic bond Chemical bond formed when many atoms share the same electrons. (**p. 496**)

meter The SI unit of length. (p. 35)

metric system *See* **International System**.

microchip A complex array of *p*-type (positive) and *n*-type (negative) semiconductors that constitutes a tiny integrated circuit. (**p. 537**)

microscope A device designed to magnify images, usually with a series of lenses. (p. 433)

microwaves Electromagnetic waves with wavelengths between about 1 millimeter and 1 meter. (**p. 405**)

Milky Way The galaxy that is home to Earth and our solar system. (**p. 627**)

mirage An optical illusion created by the refraction of light. (p. 430)

mirror A device that scatters light by reflection. (**p. 421**)

mixture A combination of two or more substances in which each substance retains its own chemical identity. (**p. 188**)

modern physics The study of relativity, quantum mechanics, and nuclear and particle physics. (p. 367)

molecule Two or more atoms bound together by electric forces (chemical bonds). (**p. 187**)

moment of inertia The quantity that describes the distribution of mass around an axis of rotation. (**p. 146**)

momentum The product of an object's mass and velocity. (**p. 118**)

monsoon The seasonal wind of the Indian Ocean and southern Asia. (p. 236)

nanotechnology The scientific field devoted to developing technologies based on devices that have atomic-scale sizes. (p. 543)

National Science Foundation A federal government agency, with an annual budget of almost $5 billion, that supports research and education in all areas of science. (p. 17)

net force The unbalanced force on an object. (p. 82)

neurotransmitters Molecules that carry signals between cells in the nervous system. (p. 386)

neutrino A very light, neutral particle that interacts very weakly with matter. (p. 589)

neutron One of the two particles that make up the atomic nucleus; it has no electric charge. (p. 451, **554**)

newton (N) The SI unit of force that accelerates a 1-kilogram mass at the rate of 1 meter per second per second. (p. 35, **85**)

Newton's law of universal gravitation Newton's law that states that between any two objects in the universe there is an attractive force (gravity) that is proportional to the masses of the objects and inversely proportional to the square of the distance between them. (**p. 98**)

Newton's laws of motion Three laws that describe how any object in the universe behaves when acted on by any force; the *first law* states that a moving object will continue moving in a straight line at a constant speed, and a stationary object will remain at rest, unless acted on by an unbalanced force; the *second law* states that the acceleration produced on a body by a force is proportional to the magnitude of the force and inversely proportional to the mass of the object; the *third law* states that for every action (force) there is an equal and opposite reaction (force). (**p. 80**)

Newtonian mechanics The area of physics concerned with Newton's discoveries and their development. (p. 80)

Newtonian worldview The philosophical ideas that grew from Newton's contributions to the development of physics. (p. 80)

noble gases Elements in the far right-hand column of the periodic table; they have filled outer shells, so they do not readily react with other elements. (p. 463)

nodes The points of a standing wave where the displacement of the medium is always zero. (p. 299)

north pole *See* **poles.**

northern lights *See* **aurora borealis.**

nuclear fission *See* **fission.**

nuclear fusion *See* **fusion.**

nuclear reactor A device that allows us to extract energy from nuclear fission in a controlled fashion. (**p. 570**)

nucleus A very small, dense, positively charged object at the center of every atom; nuclei are made up of protons and neutrons. (**p. 450**)

observation The act of noting nature without manipulating it. (**p. 3**)

Ockham's Razor The idea that the simplest solution to a problem is most likely to be right. (p. 50)

octave Two frequencies are an octave apart when the higher frequency is twice the lower frequency. (p. 322)

ohm The physical unit that defines the amount of electrical resistance. (**p. 375**)

Ohm's law The relationship among voltage, current, and resistance in a circuit. (**p. 378**)

opaque materials Materials that absorb or scatter electromagnetic radiation. (**p. 426**)

open circuit An incomplete electric circuit; it is incapable of carrying electric current. (p. 382)

optical devices or **optics** Devices used to alter and control electromagnetic radiation. (p. 419)

optics The branch of physics dealing with the manipulation and analysis of electromagnetic waves. (**p. 419**)

orbitals *See* **shells.**

organic chemistry The study of carbon-based molecules. (p. 499)

overtones (harmonics) A series of oscillations in which the frequency of each oscillation is an integral multiple of the fundamental frequency. (**p. 325**)

parabola The shape of the curve followed by an object thrown in a gravitational field. (p. 67)

parallel circuit A circuit, or part of a circuit, that consists of two or more loads linked together along different loops of wire. (**p. 391**)

paramagnetism The magnetic properties of a material associated with the alignment of atoms in the presence of an applied magnetic field; in paramagnetism the induced field reinforces the applied field. (p. 525)

parsec A unit of length equal to 3.3 light years. (p. 629)

particle accelerator A scientific instrument that increases the speed of charged particles. (**p. 586**)

Pascal's principle The statement that an increase of pressure of a static fluid in one place is transmitted immediately to every part of the fluid. (**p. 211**)

Pauli exclusion principle The statement that no two electrons in an atom can occupy the same state at the same time. (p. 463)

peak load The maximum demand on a power-generating source. (p. 535)

period The time it takes a planet to complete one full orbit around the Sun, or a vibrating object to complete one full oscillation. (p. 54, **289**)

period of rotation The time it takes for an object to make one complete rotation. (**p. 140**)

periodic A term that describes phenomena that occur at regular intervals. (p. 8)

periodic table of the elements The systematic organization of the known chemical elements in terms of their electron configurations. (**p. 462**)

PET *See* **positron emission tomography**.

phase transitions *See* **freezings**.

phases of matter The different forms that matter can take; solid, liquid, and gas are the most common. (**p. 190**)

photoelectric effect The ejection of electrons from a material due to electromagnetic radiation (photons) striking that material. (p. 482)

photon The quantum of electromagnetic radiation; a particlelike bundle of light. (**p. 453**)

photosynthesis The process of using sunlight as a form of energy to form complex organic molecules. (p. 251)

photovoltaic cell (solar cell) A semiconductor device that converts sunlight into electric energy. (p. 534)

physics The branch of science devoted to the search for laws that describe the most fundamental aspects of nature: matter, energy, forces, motion, heat, light, and other phenomena. (**p. 12**)

pitch A measure of how the frequency of sound waves is perceived by humans; high pitch means high frequency. (**p. 320**)

Planck's constant A fundamental physical constant that specifies the extent to which quantum behavior affects nature. (p. 474).

plasma An energetic gaslike state of matter made up of ions, electrons, and neutral particles. (p. 195)

plastics Synthetic polymer materials formed primarily from petroleum. (p. 192)

polar molecules Clusters of atoms that have a positive end and a negative end. (p. 500)

polarization The occurrence of equal but opposite amounts of charge separated from each other; the shift of electrons within atoms or molecules giving them a positive end and a negative end. (p. 341, 500)

poles (north and south) The two ends of a magnet; one is called the *north pole* and one is called the *south pole*. (**p. 343**)

polymers Very long molecules commonly found in biological materials and plastics. (p. 192)

positron (from *positive electron*) The antiparticle of the electron. (p. 590)

positron emission tomography (PET) A medical technique used to study the brain by injection of positron-emitting isotopes into the bloodstream. (p. 591)

pound The basic unit of weight in the English system. (p. 35)

power The amount of work done divided by the time it takes to do it, or the energy expended divided by the time it takes to expend it. (**p. 166**)

power stroke The downward motion of a piston when the spark plug fires during the operation of an internal combustion engine. (p. 274)

prediction The use of hypotheses to test how a particular system will behave. (**p. 5**)

pressure A force divided by the area over which the force acts. (**p. 207**)

primary additive colors The three colors (red, blue, and green) which, if added together in the right proportions, can create any color. (p. 436)

principle of relativitiy The statement that the laws of nature are the same in all frames of reference. (p. 606)

probability The likelihood that a certain event or outcome will occur. (**p. 477**)

proportional In a direct relationship, when both variables increase or decrease together. (p. 31)

proton One of the two particles that make up the atomic nucleus; it has a charge of +1. (p. 451, **554**)

pseudoscience The types of inquiry, such as extrasensory perception (ESP), unidentified flying objects (UFOs), astrology, crystal power, reincarnation, and the myriad claims of psychic phenomena, that fail the elementary test that defines science. (**p. 9**)

quadrant An observational instrument shaped like a large sloping device, something like a gunsight, that determines each star's or planet's position. (p. 53)

quanta Discrete bundles of energy; fundamental amounts of discrete physical quantities. (p. 452)

quantize To restrict a variable to discrete values. (p. 472)

quantum leap or **quantum jump** The process of changing location without having to traverse any of the positions in between; usually in reference to electrons changing energy levels in an atom. (**p. 453**)

quantum mechanics The branch of physics devoted to the study of very small systems in which physical quantities come in discrete bundles called quanta. (**p. 472**)

quark The fundamental particle that is the building block of all hadrons. (**p. 592**)

quasar An unusually active galaxy that pours vast amounts of energy into space. (p. 630)

radian A unit of measure of an angle, equal to the length of a circular arc subtended by the angle divided by the radius of the circle. (p. 142)

radiation Heat transfer due to the emission and absorption of electromagnetic waves between two bodies at different temperatures; also, particles emitted in radioactive decay. (**p. 244**, 557)

radio waves Electromagnetic waves with the longest wavelength, used to broadcast radio signals. (**p. 403**)

radioactivity or **radioactive decay** The spontaneous release of energetic particles from an atomic nucleus. (**p. 557**)

radiometric dating The determination of the age of materials using the known half-lives of radioactive elements. (**p. 566**)

ray The path taken by a beam of electromagnetic radiation; a line drawn perpendicular to the wavefront. (**p. 419**)

rectifier An electric device that converts alternating current to direct current; an element in an electric circuit that allows electric current to pass in only one direction. (**p. 386**, 533)

redshift The shift toward lower frequencies of the spectra of galaxies moving away from us. (**p. 631**)

reductionism An attempt to reduce the seeming complexity of nature by first looking for underlying simplicity and then trying to understand how that simplicity gives rise to the observed complexity. (p. 583)

reference frame *See* **frame of reference**.

reflecting telescope A telescope that uses a mirror arrangement to focus the incoming light. (p. 433)

reflection The return of light from a surface on which it falls. (**p. 421**)

refracting telescope A telescope that uses a lens arrangement to focus the incoming light. (p. 433)

refraction The change in direction and speed of a wave as it enters a different medium. (p. 313, **427**)

relativity *See* **theory of relativity; general relativity; special relativity**.

reproducible A term referring to the method of conducting and reporting of observations and experiments so that anyone with the proper equipment can verify the results. (p. 7)

research and development (R&D) The process of bringing new discoveries to practical use, often in industrial or governmental laboratories. (**p. 14**)

resonance The buildup of large vibrations in an oscillating system, due to the application of a force with a frequency that matches the natural or resonant frequency of the oscillator. (**p. 302**)

resonant frequency The frequency at which a system will vibrate if left to itself. (p. 302)

restoring force A force that always pushes an object towards its equilibrium position. (p. 290)

resultant force The net force in those situations in which two or more forces act in different directions. (p. 82)

right-hand rule Any rule that gives the direction of a vector quantity by using the fingers of the right hand. (p. 345)

rotational motion The spinning motion that occurs when an object rotates about an axis located within it, such as an axis through its center of mass. (**p. 139**)

scalar Any quantity that can be expressed as a single number and without a direction. (**p. 26**)

scale In music, a specific sequence of frequencies used for playing music. (p. 322)

scattering The process of changing the direction (and sometimes the properties) of a wave as it encounters an obstacle. (**p. 421**)

Schrödinger equation The equation that is used to calculate the wave function of a quantum system. (p. 478)

science A method for answering questions about the working of the physical world. (**p. 2**)

scientific method A cycle of collecting observations (data), identifying patterns and regularities in the data (synthesis), forming hypotheses, and making predictions, which lead to more observations. (**p. 2**)

scientist A person who studies questions about our world. (**p. 2**)

second The basic unit of time in both the SI and English systems. (p. 35)

second law of thermodynamics The law of physics that places restrictions on the ways heat and other forms of energy can be transformed and used to do work. (**p. 271**)

seismic waves Waves in the Earth's crust generated by earthquakes. (p. 294)

semiconductor A material that is neither a good electrical conductor nor a good insulator. (**p. 517**)

series circuit A circuit, or part of a circuit, that consists of two or more loads linked together along a single loop of wire. (**p. 391**)

shear strength The ability of a material to withstand twisting. (p. 503)

shells The pattern of arrangements into which electrons organize themselves in an atom; also known as *orbitals*. (p. 463)

shielding The effect that causes electric fields to be significantly less strong inside a conductor. (p. 342)

short circuit A low-resistance path that causes potentially dangerous amounts of current in the circuit. (**p. 385**)

SI *See* **International System**.

simple harmonic motion The oscillatory motion caused by the action of a restoring force; the restoring force must be proportional to the displacement from equilibrium in order to produce simple harmonic motion. (**p. 290**)

simple harmonic oscillator An oscillator that exhibits simple harmonic motion. (p. 290)

single bond A covalent bond in which one electron from each atom is shared. (p. 498)

solar cell, solar panel *See* **photovoltaic cell**.

solid A rigid material that has a definite shape and volume. (**p. 190**)

south pole *See* **poles**.

special relativity The part of relativity theory that deals with events in reference frames moving at constant velocities. (**p. 606**)

specific heat The quantity of heat required to raise the temperature of 1 gram of a material by 1° C. (**p. 235**)

spectroscopy The analysis of emission and absorption spectra of materials to determine their chemical properties. (**p. 458**)

spectrum The total collection of photons emitted by an atom; the distribution of the frequencies of photons emitted by a radiating system. (**p. 457**)

speed The distance an object travels divided by the time that it takes to travel that distance. (**p. 58**)

speed of light, *c* The speed at which all electromagnetic waves travel when in a vacuum, 3.00×10^8 m/s. (**p. 401**)

spontaneously Resulting from a natural impulse or tendency. (p. 272)

squared A term describing a number multiplied by itself. (p. 32)

standard candle Any object whose energy output is known; used in astronomy to estimate distances to extraterrestrial objects. (p. 627)

standard model The theory that predicts the unification of the strong and the electroweak forces. (p. 598)

standing wave A wave pattern that can be created by the interference of two waves moving in opposite directions; as the name implies, the wave pattern does not move. (p. 299)

static electricity Electric charges that are at rest; also used to describe the force due to electric charges that are at rest. (**p. 334**)

step-down transformer, step-up transformer *See* **transformer**.

strength The ability of a solid to resist changes in shape. (**p. 502**)

string theory A theory that pictures the quarks and leptons as vibrations on tiny stringlike structures. (p. 598)

strong force One of the four fundamental forces; it is the force that binds the nucleus together. (**p. 557**, 595)

superconductivity The property of having no electrical resistance. (**p. 519**)

synchrotron A type of particle accelerator in which strong magnets force the particles to follow a circular track while it is accelerated. (p. 587)

system A collection of matter and energy that is controlled in such a way that its physical properties can be studied; systems can be *open, closed,* or *isolated.* (**p. 260**)

system of units Units assigned to fundamental quantities such as mass (or weight), length, time, and temperature. (**p. 35**)

Système International (SI) *See* **International System**.

technology The application of science to specific commercial or industrial goals. (**p. 14**)

temperature A quantity that reflects how vigorously atoms or molecules are moving and colliding in a material. (**p. 228**)

temperature reservoir A body so large that large quantities of heat can be added to it (**high-temperature reservoir),** or removed from it (**low-temperature reservoir),** without changing its temperature; the Atlantic ocean is a good approximation of a temperature reservoir. (p. 275)

temperature scale A standard of measurement for estimating temperature; familiar examples are the Fahrenheit and the Celsius scales. (**p. 228**)

tensile strength The ability of a solid to withstand pulling apart. (p. 503)

tension The condition in which the atoms of a material are pulled further apart, due to an external force. (**p. 219**)

terminal velocity The speed attained by an object falling under the influence of gravity, when the downward pull of gravity is balanced by the upward force of the resistance of the medium through which the body moves. (p. 83)

theorist A scientist who spends time imagining how the universe might work in areas where no detailed explanations exist. (p. 11)

theory A description of the world that covers a relatively large number of phenomena and has met and explained many observational and experimental tests. (**p. 4**)

theory of everything A theory in which all of the fundamental forces are seen as different aspects of a single force. (p. 598)

theory of relativity The physical laws that govern the measurement of time and space as observed in differing reference frames. (**p. 606**)

thermal conductivity The ability of a material to transfer heat. (**p. 241**)

thermal energy or **internal energy** The energy of an object that results from the vibrations of individual atoms and molecules. (**p. 234**, 252)

thermocouple A temperature sensor based on the electric properties of metals. (p. 231)

thermodynamics The study of heat and energy. (**p. 185**)

thermometer A device used to measure temperature. (**p. 229**)

timbre The characteristic quality of sound that depends on the relative strength of the different frequencies that make up that sound. (p. 326)

time dilation The slowing of time relative to an observer in a different reference frame. (**p. 608**)

torque The force applied perpendicular to a line from the axis of rotation, multiplied by the distance from the axis of rotation; the application of torque changes angular momentum. (**p. 143**)

total internal reflection The process in which light, if it is at a great enough incident angle, will totally reflect from a surface with a lower density than the one in which it is traveling. (p. 425)

total momentum The sum of the momenta of all the objects in a system. (**p. 122**)

transformer A device that converts a high voltage to a low voltage (**step-down transformer),** or a low voltage to a high voltage (**step-up transformer). (p. 387**)

transistor A semiconducting element in an electric circuit that regulates current flow. (**p. 535**)

transmission The process by which a wave passes through a material (even though it may lose some of its energy). (**p. 427**)

transparent materials Materials that transmit electromagnetic waves. (**p. 427**)

transverse wave A wave in which the motion of the medium is perpendicular to the direction of the wave propagation. (p. 295)

trophic level A level in the food chain hierarchy. All organisms that get their food from the same source belong to the same trophic level. (**p. 258**)

trough The lowest point of a wave; commonly associated with water waves. (p. 293)

ultraviolet radiation Electromagnetic waves with wavelengths from about 100 nanometers to 400 nanometers; they are known to cause sunburn. (**p. 411**)

unbalanced force A force, not cancelled by another force, acting on an object and causing an acceleration. (p. 82)

uncertainty principle A physical law that places limits on the accuracy to which certain quantities (momentum and position, for example) can be measured simultaneously. (**p. 474**)

unified field theory A theory that sees the fundamental forces as different aspects of the same force. (**p. 597**)

uniform motion Motion at a constant speed in a single direction. (**p. 61**)

Van Allen belts The path of cosmic rays curved by the Earth's magnetic field; named after American physicist James Van Allen. (p. 346)

van der Waals force The force due to the polarization of electrically neutral atoms or molecules that are not themselves polar. (p. 501)

vector A quantity that requires two numbers in its definition—a magnitude and a direction. (**p. 26**)

velocity A vector quantity that has the same numerical value as speed but also includes information about the direction of travel. (**p. 60**)

vibrations The oscillating, periodic motions of a medium or body forced from a position of stable equilibrium. (**p. 289**)

virtual image An image created by a diverging lens where light rays do not actually originate on the object itself. (p. 432)

visible light The specific frequencies of electromagnetic waves that can be detected by the human eye; they include all colors of the rainbow. (**p. 409**)

volt The physical unit used to quantify the amount of electric potential. (**p. 375**)

voltage (measured in **volts**) Synonymous with *electric potential*. (**p. 375**)

volume The quantity of space an object occupies. (p. 33)

W and Z particles Very massive gauge particles that mediate the weak force. (p. 597)

watt The SI unit of power, defined as the expenditure of 1 joule of energy in 1 second. (**p. 167**)

wave A disturbance of a medium that travels without a net displacement of the medium itself, as in a sound wave. (**p. 291**)

wave function A mathematical formula that is related to the probability of finding a particle in a particular location. (p. 478)

wave mechanics *See* **quantum mechanics**.

wave-particle duality The aspect of subatomic entities by which they sometimes exhibit the properties of particles and sometimes exhibit the properties of waves. (p. 479)

wavelength The distance between two adjacent corresponding points (crests, for example) of a wave. (**p. 293**)

weak force One of the four fundamental forces in nature; it is responsible for some kinds of radioactive decay. (p. 595)

weight The force of gravity on an object. (**p. 100**)

work The product of the force exerted on an object times the distance over which it is exerted (**p. 163**)

work-energy theorem The statement that the total potential and kinetic energy of an object in a given state is equal to the work that was done to bring the object to that state. (**p. 175**)

X rays Electromagnetic waves with wavelengths from about 0.1 nanometer to 100 nanometers. (**p. 411**)

Z particles *See* **W and Z particles**.

Photo Credits

168 (center): Romilly Lockyer/The Image Bank/Getty Images. Page 168 (right): David Joel/Stone/Getty Images. Page 169: Runk/Schoenberger/Grant Heilman Photography. Page 171 (top left): PhotoDisc, Inc./Getty Images. Page 171 (top rght): Digital Vision/Getty Images. Page 171 (center left): Corbis Images. Page 171 (center right): Martyn Goddard/Corbis Images. Page 171 (bottom): Public Domain. Page 172: Les David Manevitz/SUPERSTOCK. Page 178: Neil Beer/Corbis Images.

Chapter 9
Opener: Jan Halaska/Photo Researchers. Page 188: Susumu Nishinga/Science Photo Library. Page 189: Charles D. Winters/Photo Researchers. Page 190: Galen Rowell/Corbis Images. Page 191 (top): Geoff Tompkinson/Science Photo Library/Photo Researchers. Page 191 (bottom): Fabrik Studios/Index Stock. Page 194 (top left): David Houser/Corbis Images. Page 194 (top right): Vince Streano/Corbis Images. Page 194 (bottom): Courtesy JSC/NASA. Page 196: Peter Arnold, Inc. Page 198 (top): Courtesy of Digiray Corporation. Page 198 (bottom): Michael Hochella.

Chapter 10
Opener: Photofest. Page 205 (top left): Topham/The Image Works. Page 205 (top center): Custom Medical Stock Photo. Page 205 (top right): Myrleen Ferguson Cate/PhotoEdit. Page 205 (bottom left): Tony Freeman/PhotoEdit. Page 205 (bottom center): Andy Washnik. Page 205 (bottom right): Rachel Epstein/The Image Works. Page 207: Corbis Images. Page 209: Richard Megna/Fundamental Photographs. Page 210 (top): Bettman/Corbis Images. Page 210 (bottom): Robert Cameron/Stone/Getty Images. Page 211: Phil Prosen/The Image Bank/Getty Images. Page 212: Courtesy NASA. Page 214 (top): Rick Rickman/Duomo Photography, Inc. Page 214 (bottom): Ralph A. Clevenger/Corbis Images. Page 215: Gary Gladstone/The Image Bank/Getty Images. Page 221: Dr. Patricia J. Shulz/Peter Arnold, Inc.

Chapter 11
Opener: Royalty-Free/Corbis Images. Page 230 (top left): Courtesy National Institute of Standards and Technology. Page 230 (top right): Courtesy J.F. Allen, St. Andrews University. Page 230 (center left): NASA Media Services. Page 230 (center right): Digital Vision/Getty Images. Page 230 (bottom): Photo Researchers. Page 231 (top): Tony Freeman/PhotoEdit. Page 231 (bottom): PhotoDisc, Inc./Getty Images. Page 233: Breck Kent/Animals Animals/Earth Scenes. Page 235 (top): Steve Taylor/Stone/Getty Images. Page 235 (bottom): Andy Washnik. Page 236: ThinkStock/SUPERSTOCK. Page 241: Charles Thatcher/Stone/Getty Images. Page 242: Douglas Stone/Corbis Images. Page 243: Astrid & Hanns-Frieder Michler/Photo Researchers. Page 244 (top): Art Wolfe/Stone/Getty Images. Page 244 (bot-

tom): Galen Rowell// Mountain Light Photography, Inc. Page 245: F. Stuart Westmorland/Stone/Getty Images.

Chapter 12
Opener: Paul A. Souders/Corbis Images. Page 253 (left): The Image Bank/Getty Images. Page 253 (center): W.J. Scott/H. Armstrong Roberts, Inc./ NYC. Page 253 (right): Ken Straiton/Corbis Stock Market. Page 254: Greg Stott/Masterfile. Page 255: ©Estate of Harold Edgerton, courtesy of Palm Press, Inc. Page 256: Shahn Kermani/Liaison Agency, Inc./Getty Images. Page 262: Jim Cummins/Taxi/Getty Images. Page 265: Ken Graham/Bruce Coleman, Inc.

Chapter 13
Opener: Charles Gupton/Corbis Images. Page 271 (top left): David Leah/Stone/Getty Images. Page 271 (top right): Bruce Forster/Stone/Getty Images. Page 271 (bottom): Damir Frkovic/Masterfile. Page 272: Dale Durfee/Stone/Getty Images Page 278: D. Boone/Corbis Images. Page 279: Paul Silverman/Fundamental Photographs.

Chapter 14
Opener: Tom McCarthy/PhotoEdit. Page 292: NOAA. Page 294: Michael Townsend/Photographers Choice/Getty Images. Page 297 (top): Manfred Cage/Peter Arnold, Inc. Page 297 (bottom): Aaron Haupt/Photo Researchers. Page 298 (top): Royalty-Free/Corbis Images. Page 298 (bottom): Education Development Center. Page 300: Warren Faidley/International Stock Photo. Page 301: Andrew David Hazy/Rochester Institute of Technology. Page 302: AFP/Corbis Images. Page 303 (left): ©AP/Wide World Photos. Page 303 (right): ©AP/Wide World Photos. Page 304 (left): Iwasa/Sipa Press. Page 304 (right): Andrew Rafkind/Stone/Getty Images.

Chapter 15
Opener: Marshall John/Corbis Images. Page 312: ©AP/Wide World Photos. Page 316 (left): Don Fawcett, K. Saito, K. Hama/Photo Researchers. Page 316 (right): Bonnier Alba. Page 317: Stephen Dalton/Animals Animals NYC. Page 319 (left): Yoav Levy/Phototake. Page 319 (right): ISM/Phototake. Page 321: James Baldocchi. Page 322: Bob Krist/Stone/Getty Images. Page 323: PhotoDisc, Inc./Getty Images. Page 330: Kelly-Mooney Photography/Corbis Images.

Chapter 16
Opener: John Lund/The Image Bank/Getty Images. Page 335: Peter Menzel/Stone/Getty Images. Page 336: ©Antman/The Image Works. Page 343: Paul Silverman/Fundamental Photographs. Page 345: Andy Washnik. Page 346: Pal Hermansen/The Image Bank/Getty Images. Page 348: D. Blackwill & D. Maratez/Visuals Unlimited.

Chapter 17

Opener: Digital Vision/Getty Images. Page 354: The Royal Institute. Page 355: Corbis-Bettmann. Page 356: Spencer Grant/PhotoEdit. Page 358: ©Arthur Hill/Visuals Unlimited. Page 360: Transition Region and Coronal Explorer, Trace, is a Mission of The Stanford-Lockheed Institute For Space Research, Part of the NASA Small Explorer Program. Page 366: AIP Emilio Segre Visual Archives.

Chapter 18

Opener: Roger Ressmeyer/Corbis Images. Page 375: Richard Nowitz/Corbis Images. Page 376 (top left): Photo-Disc, Inc./Getty Images. Page 376 (top right): Tony Freeman/PhotoEdit. Page 376 (center left): Lester Lefkowitz/Corbis Images. Page 376 (center right): PhotoDisc, Inc./Getty Images. Page 376 (bottom): Lester Lefkowitz/Corbis Images. Page 378: SUPERSTOCK. Page 381: Courtesy NTE Electrics, Inc. Page 383: Tony Freeman/PhotoEdit. Page 389 (left): Roger Ressmeyer/Corbis Images. Page 389 (center): ThinkStock/SUPERSTOCK. Page 389 (right): Courtesy Linksys.com.

Chapter 19

Opener: Photonica. Page 405: Aero Graphics Inc./Corbis Images. Page 407: Courtesy Daedalus Enterprises, Inc. Page 409 (top): David Parker/Photo Researchers. Page 409 (center): ©Brokaw Photography/Visuals Unlimited. Page 409 (bottom left): ©Gregory K. Scott/Photo Researchers. Page 409 (bottom right): Photo Disc, Inc/Getty Images. Page 412: ©Pete Saloutos/Corbis Stock Market. Page 413 (left): NASA Media Services. Page 413 (right): NASA/Goddard Space Flight Center/Photo Researchers.

Chapter 20

Opener: Spencer Grant/PhotoEdit. Page 420 (left): Michael Keller/Corbis Images. Page 420 (center): Phil Degginger/Stone/Getty Images. Page 420 (right): V.C.L./Taxi/Getty Images. Page 421: © Ron Levy /Global Image Group. Page 422 (top): Michael S. Yamashita/Corbis Images. Page 422 (bottom): Ariel Skelley/Corbis Images. Page 423 (left): Roger Ressmeyer/Corbis Images. Page 423 (center): Kevin Schafer/Corbis Images. Page 423 (right): Paul Edmondson/Corbis Images. Page 424: Roger Ressmeyer/Corbis Images. Page 425 (top): Exploratorium, www.exploratorium.edu. Page 425 (bottom): Lester Lefkowitz/Corbis Images. Page 429: Fundamental Photographs. Page 430: Peter Christopher/Masterfile. Page 431: Pete Turner/The Image Bank/Getty Images. Page 432: NASA/Corbis Images. Page 433: Voller Ernst/eStock Photo. Page 434 (left): Corbis-Bettmann. Page 434 (right): Pat O'Hara/Corbis Images. Page 435 (left): "Correction of Chromatic Abberations by the Multi-Layer Diffractive Optical Element" Drawing. ©Canon Inc. All Rights Reserved. Used by Permission. Page 435 (right):

©Canon Inc. All Rights Reserved. Used by Permission. Pages 438 & 439: PhotoDisc, Inc./Getty Images.

Chapter 21

Opener: Duomo/Corbis Images. Page 448 (top left): David Young-Wolff/PhotoEdit. Page 448 (top center): George B. Diebold/Corbis Images. Page 448 (top right): Dana White/PhotoEdit. Page 448 (bottom left): Mark Harwood/Stone/Getty Images. Page 448 (bottom right): Brand X Pictures/Getty Images. Page 451: AIP Emilio Segre Visuals Archives/Margarethe Bohr Collection. Page 454: Yoav/Phototake. Page 455: D. Boone/Corbis Images. Page 456: Stuart Schneider, Fluorescent Mineral Mueseum Online, ©2002. Page 458: Courtesy Bausch & Lomb. Page 462 (left): Alexander Tsiaras/Photo Researchers. Page 462 (right): ©Ray Nelson/Phototake. Page 466: Courtesy IBM Almaden Research Center.

Chapter 22

Opener: AFP/Corbis Images. Page 473: ©Mark E. Gibson/Visuals Unlimited. Page 480: Courtesy A. Tonomura, J. Endo, T. Matsuda, and T. Kawasaki, Am. J. Phys. 57(2): 117, February 1989. Page 483 (left): Don W. Fawcett/Photo Researchers. Page 483 (right): P. Motta & S. Correr/Photo Researchers. Page 484: Marx/Taxi/Getty Images. Page 485: Courtesy Education Development Center. Page 485: Niels Bohr Archive.

Chapter 23

Opener: PhotoDisc, Inc./Getty Images. Page 492: International Stock Photo. Page 495: Michael Watson. Page 497: Joseph Sohm/Corbis Images. Page 502: Bruce Hands/Stone/Getty Images. Page 503: Duane Newton/PhotoEdit. Page 504 (left): Peter Pearson/Stone/Getty Images. Page 504 (top right): Catherine Karnow/Corbis Images. Page 504 (bottom right): Alan Schein/Corbis Images. Page 505: Billy E. Barnes/PhotoEdit. Page 508: Richard Megna/Fundamental Photographs.

Chapter 24

Opener & Page 515: PhotoDisc, Inc./Getty Images. Page 516 (left): Stephen Frisch. Page 516 (right): Charles D. Winters/Photo Researchers. Page 520: University of Birmingham/Science Photo Library/Photo Researchers. Page 521: David Job/Stone/Getty Images. Page 523: ©AP/Wide World Photos. Page 525 (top): ©Sergio Piumatti. Page 525 (bottom): ©Trident Technologies. Reproduced with permission. Page 526: Courtesy Bassam Z. Shakhashiri. Page 527: Illustration by Philip Castle, Time Inc./Time Life Pictures/Getty Images.

Chapter 25

Opener: David Young-Wolff/PhotoEdit. Page 535: Volker Steger/Photo Researchers. Page 537: Uniphoto, Inc. Page

539 (left): Time Life Pictures/Getty Images. Page 539 (right): Adam Hart-David/Photo Researchers. Page 542: Stewart Cohen/Stone/Getty Images. Page 543: David Reed/ Corbis Images. Page 546: David Carson/Corbis Sygma. Page 547: ©AP/Wide World Photos.

Chapter 26

Opener: Robert Y. Ono/Corbis Images. Page 558: Courtesy College Physicians of America. Page 562: Tom Raymond/ Stone/Getty Images. Page 564: Science Photo Library/ Photo Researchers. Page 566: Patrick Mesner/Liaison Agency, Inc./Getty Images. Page 567: John Reader/Science Photo Library/Photo Researchers. Page 571: ©Breck P. Kent/Animals Animals/Earth Scenes. Page 573 (top): NASA and the Hubble Heritage Team. Page 573 (bottom): Courtesy Princeton University Plasma Physics Lab.

Chapter 27

Opener: PhotoDisc, Inc./Getty Images. Page 582 (left): Stone/Getty Images. Page 582 (center & right): Courtesy John Wiley & Sons, Inc. Page 586: Courtesy Lawrence Berkeley Laboratory. Page 587 (top): CERN/Science Photo Library/Photo Researchers. Page 587 (bottom left): Courtesy Fermi National Accelerator Laboratory. Page 587 (bot-tom right): Bill W. Marsh/Photo Researchers. Page 588: Courtesy Elekta Instruments, Inc. Page 590: Photofest. Page 591 (left): Courtesy Carl Anderson, California Institute of Technology. Page 591 (right): Courtesy Lawrence Berkeley Laboratory, University of California. Page 592 (left): Hank Morgan/Rainbow. Page 592 (right): Visuals Unlimited.

Chapter 28

Opener: Chet Gordon/The Image Works. Page 613: ©1989 Ping-Kang Hsiung. Page 618: Imtek Imagineering/ Masterfile.

Chapter 29

Opener: FoodPix/Getty Images. Page 628: Harvard College Observatory. Page 630: Courtesy AIP Emilio Segre Visual Archives and Hale Observatories. Page 631: NASA/ AURA/STScI/Hubble Heritage Team. Page 632 (top left): NASA /SAO/R. Kraft et al. Page 632 (top right): European Southern Observatory. Page 632 (center left): NASA/ GSFC. Page 632 (center right): Dr. Christopher Burrows, ESA/STScI and NASA. Page 632 (bottom): NASA and The Hubble Heritage Team (STScI). Page 633: Courtesy Palo-mar Observatory, California Institute of Technology

Index

CONVERSION FACTORS

Units of Length, Mass, and Temperature

Length and Mass Conversion from SI to English Units

To get:	Multiply:	By:
inches	meters	39.4
feet	meters	3.281
miles	kilometers	0.621

Length and Mass Conversion from English to SI Units

To get:	Multiply:	By:
meters	inches	0.0254
meters	feet	0.3048
kilometers	miles	1.609

Conversion between Celsius and Fahrenheit Degrees

$°F = 1.8(°C) + 32$ $°C = 0.55(°F - 32)$

Units of Force, Energy, and Power

Energy and Power Conversion from SI to English Units

To get:	Multiply:	By:
BTUs	joules	0.00095
calories	joules	0.2390
kilowatt-hours	joules	2.78×10^{-7}
foot-pounds	joules	0.7375
horsepower	watts	0.00134
pounds	newtons	0.2248

Energy and Power Conversion from English to SI Units

To get:	Multiply:	By:
joules	BTUs	1055
joules	calories	4.184
joules	kilowatt-hours	3.6×10^6
joules	foot-pounds	1.356
watts	horsepower	745.7
newtons	pounds	4.448

SELECTED PHYSICAL CONSTANTS

Acceleration due to gravity at Earth's surface:
$g = 9.8$ m/s^2

Gravitational constant:
$G = 6.67 \times 10^{-11}$ N-m^2/kg^2

Coulomb's law constant:
$k = 9 \times 10^9$ N-m^2/C^2

Charge on electron:
$e = 1.6 \times 10^{-19}$ C

Proton mass:
$m_p = 1.66 \times 10^{-27}$ kg

Electron mass:
$m_e = 9.11 \times 10^{-31}$ kg

Speed of light in a vacuum:
$c = 3.00 \times 10^8$ m/s

Planck's constant:
$h = 6.63 \times 10^{-34}$ J-s

Boltzmann's constant:
$k = 1.38 \times 10^{-23}$ J/K

Astronomical unit (mean distance from Earth to Sun):
$AU = 1.46 \times 10^{11}$ m

Light-year:
1 ly $= 9.46 \times 10^{12}$ km
$\quad\quad = 6.3 \times 10^4$ AU

Parsec:
1 pc $= 3.3$ light-years

Power output of Sun:
$P = 4.24 \times 10^{23}$ kW

Mass of Sun:
$M_{sun} = 1.989 \times 10^{30}$ kg

Mass of Earth:
$M_E = 5.974 \times 10^{24}$ kg

Mass of Moon:
$M_{moon} = 7.348 \times 10^{22}$ kg

Radius of Sun:
$R_{sun} = 6.96 \times 10^5$ km

Radius of Earth:
$R_E = 6.378 \times 10^3$ km

Radius of Moon:
$R_{moon} = 1.738 \times 10^3$ km